Methods in Microbiology
Volume 31

Recent titles in the series

Volume 23 *Techniques for the Study of Mycorrhiza*
JR Norris, DJ Reed and AK Varma

Volume 24 *Techniques for the Study of Mycorrhiza*
JR Norris, DJ Reed and AK Varma

Volume 25 *Immunology of Infection*
SHE Kaufmann and D Kabelitz

Volume 26 *Yeast Gene Analysis*
AJP Brown and MF Tuite

Volume 27 *Bacterial Pathogenesis*
P Williams, J Ketley and GPC Salmond

Volume 28 *Automation*
AG Craig and JD Hoheisel

Volume 29 *Genetic Methods for Diverse Prokaryotes*
MCM Smith and RE Sockett

Volume 30 *Marine Microbiology*
JH Paul

Methods in Microbiology

Volume 31
Molecular Cellular Microbiology

Edited by

Philippe Sansonetti

Pathogen and Molecular Microbiology Unit
INSERM U389, Pasteur Institute,
25–28 rue du Docteur Roux
75724 Paris, Cedex 15
France

and

Arturo Zychlinsky

Max Planck Institute for Infection Biology,
Schummanstrasse 21/22, Campus Charité Mitte
Berlin D-10117, Germany

ACADEMIC PRESS

An Elsevier Science Imprint

San Diego San Francisco New York
Boston London Sydney Tokyo

This book is printed on acid-free paper.

Academic Press
An Elsevier Science Imprint
Harcourt Place, 32 Jamestown Road, London NW1 7BY, UK
http://www.academicpress.com

Academic Press
An Elsevier Science Imprint
525 B Street, Suite 1900, San Diego, California 92101-4495, USA
http://www.academicpress.com

ISBN 0-12-521531-2 (Hardback)
ISBN 0-12-619071-2 (Comb bound)

A catalogue record for this book is available from the British Library

Cover photograph: Taken from Chapter 27, figure 1(f), courtesy of B. Lemaitre.

Typeset by Newgen Imaging Systems (P) Ltd., Chennai, India
Printed and bound in Great Britain by Bookcraft, Midsomer Norton

02 03 04 05 06 07 BC 9 8 7 6 5 4 3 2 1

Contents

Series Advisors

Contributors

Abrami Laurence Department of Biochemistry, University of Geneva, 30 quai E. Ansermet, 1211 Geneva 4, Switzerland

R Athman Laboratoire de Morphogenèse et Signalisation Cellulaires, UMR 144 CNRS/Institut Curie, Paris, France

M Ausubel Frederick Department of Genetics, Harvard Medical School and Department of Molecular Biology, Massachusetts General Hospital, Boston, MA 02114

Bäckhed Fredrik Microbiology and Tumorbiology Center, Karolinska Institute, Stockholm, Sweden

Badizadegan Kamran Department of Pathology, Harvard Medical School, Boston, MA 02115, USA

M Barnhart Michelle Department of Molecular Microbiology, Washington University School of Medicine, St Louis, MO, USA

B Bliska James State University of New York at Stony Brook, New York, USA

Boquet Patrice INSERM U452, Faculté de Médecine, 28 Avenue de Valombrose, Nice, France. E-mail: boquet@unice.fr

Bougnères Laurence Unité de Pathogénie Microbienne Moléculaire, Institut Pasteur, 28 rue du Dr Roux, 75724 Paris, Cedex 15, France

Boujemaa-Paterski Rajaa Dynamique du Cytosquelette, Laboratoire d'Enzymologie et Biochimie Structurale, CNRS, Gif-sur-Yvette, 91198 France

Bourdoulous Sandrine CNRS UPR 415, Institut Cochin de Génétique Olivier Moléculaire, 22 rue méchain, 75014 Paris, France

H D Brightbill Department of Microbiology, Immunology and Molecular Genetics, Division of Dermatology, UCLA School of Medicine, University of California, LA

Carlier Marie-France Dynamique du Cytosquelette, Laboratoire d'Enzymologie et Biochimie Structurale, CNRS, Gif-sur-Yvette, 91198 France
E-mail: carlier@lebs.cnrs-gif.fr

R Collier John Department of Microbiology and Molecular Genetics, Harvard Medical School, Boston, MA 02115, USA

R Cornelis Guy Christian de Duve Institute of Cell Pathology, Université catholique de Louvain, Av Hippocrate, 74, B1200 Brussels, Belgium

P Cossart Unite des Interactions Bacteries-Cellules, Institut Pasteur, Paris, France

Couraud Pierre Olivier CNRS UPR 415, Institut Cochin de Génétique Olivier Moléculaire, 22 rue méchain, 75014 Paris, France

Dahlfors Agneta Richter Microbiology and Tumorbiology Center, Karolinska Institute, Stockholm, Sweden

de Bernard Marina CNR, Centro CNR Biomembrane and Dipartimento di Scienze Biomediche Sperimentali, Via G. Colombo 3, 35121, Padova, Italy

DeLeo Frank Laboratory of Human Bacterial Pathogenesis, Rocky Mountain Laboratories, NIAID, NIH, Hamilton, MT 59840

Dermine Jean-François Départment de pathologie et biologie cellulaire, Universite de Montréal, CP 6128, Succ. Centre ville, Montréal, QC, Canada, H3C 3J7

Desjardins Michel Départment de pathologie et biologie cellulaire, Université de Montréal, CP 6128, Succ. Centre ville, Montréal, QC, Canada, H3C 3J7
E-mail: michel.desjardins@umontreal.ca

S Detweiler Corrella Department of Microbiology and Immunology, Stanford University Medical Center, Stanford, CA 94305, USA

E Dixon Jack University of Michigan, Ann Arbor, MI, USA

Duclos Sophie Départment de pathologie et biologie cellulaire, Université de Montréal, CP 6128, Succ. Centre ville, Montréal, QC, Canada, H3C 3J7

Duménil Guillaume Department of Molecular Biology and Microbiology, Tufts School of Medicine, 136 Harrison Avenue, Boston, MA 02111, USA

Falkow Stanley Department of Microbiology and Immunology, Stanford University Medical Center, Stanford, CA 94305, USA

B Finlay Brett Centre d'Immunology INSERM-CNRS-Univ. Med. de Marseille-Luminy, Marseille, France

Fivaz Marc Department of Biochemistry, University of Geneva, 30 quai E. Ansermet, 1211 Geneva 4, Switzerland

T Gewirtz Andrew Epithelial Pathobiology Division, Department of Pathology and Laboratory Medicine, Emory University, Atlanta, GA 30322

F Gisou van der Goot Department of Biochemistry, University of Geneva, 30 quai E. Ansermet, 1211 Geneva 4, Switzerland

I Gordon Jeffrey Department of Molecular Biology, Pharmacology, Pathology and Immunology, Washington University School of Medicine, St Louis, MO 63110

Gorvel Jean-Pierre Biotechnology Laboratory and Departments of Biochemistry and Microbiology, University of British Columbia, Canada

P Gounon and Hautefort Isabelle Institute of Food Research, Norwich Research park, Colney Lane, Norwich, NR4 7UA. E-mail: isabell.hautefort@bbsrc.ac.uk

Heffron Fred and C D Hinton JayE-mail: jay.hinton@bbsrc.ac.uk Institute of Food Research, Norwich Research park, Colney Lane, Norwich, NR4 7UA

Hobert Michael Epithelial Pathobiology Division, Department of Pathology and Laboratory Medicine, Emory University, Atlanta, GA 30322

V Hooper Lora Department of Molecular Biology, Pharmacology, Pathology and Immunology, Washington University School of Medicine, St Louis, MO 63110

J Hultgren Scott Department of Molecular Microbiology, Washington University School of Medicine, St Louis. MO USA

R Isberg Ralph Department of Molecular Biology and Microbiology Tufts University School of Medicine, 136 Harrison Avenue, M & V 409 Boston, MA 02111, USA

S H E Kaufmann Max-Planck-Institute for Infection Biology, Schumannstr. 21-22, D-10117 Berlin, Germany

M Lecuit Unité des Interactions Bactéries-Cellules, Institut Pasteur, Paris, France

Lemaitre Bruno Centre de Génétique Moléculaire du CNRS 91198 Gif-sur-Yvette, France

I Lencer Wayne Department of Pediatrics, Harvard Medical School, Boston, MA 02115, USA

Levin Petra Anne Department of Biology, Washington University, St Louis, MO 63130, USA

D Louvard Laboratoire de Morphogenèse et Signalisation Cellulaires, UMR 144 CNRS/Institut Curie, Paris, France

L Madara James Epithelial Pathobiology Division, Department of Pathology and Laboratory Medicine, Emory University, Atlanta, GA 30322

Marlin Didier Epithelial Pathobiology Division, Department of Pathology and Laboratory Medicine, Emory University, Atlanta, GA 30322

Meister Marie UPR9022 du CNRS, Institut de Biologie Moleculaire et Cellulaire, 15 rue Descartes, 67084 Strasbourg Cedex, France

Méresse Stéphane Biotechnology Laboratory and Departments of Biochemistry and Microbiology, University of British Columbia, Canada

C Mills Jason Department of Molecular Biology, Pharmacology, Pathology and Immunology, Washington University School of Medicine, St Louis, MO 63110

R L Modlin Department of Microbiology, Immunology and Molecular Genetics, Division of Dermatology, UCLA School of Medicine, University of California, LA

Montecucco Cesare CNR, Centro CNR Biomembrane and Dipartimento di Scienze Biomediche Sperimentali, Via G. Colombo 3,35121, Padova, Italy

Nassif Xavier INSERM U411, Laboratoire de Microbiologie, Faculté de Médecine Necker-Enfants Malades, 156 Rue de Vaugirard, 75730 Paris, France

M Nauseef William Department of Internal Medicine and the Inflammation Program, University of Iowa, Iowa City, IA 52242

S Neish Andrew Epithelial Pathobiology Division, Department of Pathology and Laboratory Medicine, Emory University, Atlanta, GA 30322

J Niewöhner Xerion Pharmaceutical Gmbh, Martinsried, Germany

Normark Staffan Microbiology and Tumorbiology Center, Karolinska Institute, Stockholm, Sweden

Orth Kim University of Michigan, Ann Arbor, MI, USA

J Pizarro-Cerdá Unité des Interactions Bactéries-Cellules, Institut Pasteur, Paris, France

A Reed Katherine Epithelial Pathobiology Division, Department of Pathology and Laboratory Medicine, Emory University, Atlanta, GA 30322

S Robine Laboratoire de Morphogenèse et Signalisation Cellulaires, UMR 144 CNRS/ Institut Curie, Paris, France

Rossetto Ornella CNR, Centro CNR Biomembrane and Dipartimento di Scienze Biomediche Sperimentali, Via G. Colombo 3,35121, Padova, Italy

A Roth Kevin Department of Molecular Biology, Pharmacology, Pathology and Immunology, Washington University School of Medicine, St Louis, MO 63110

Sansonetti Philippe Unité de Pathogénie Microbienne Moléculaire, Institut Pasteur, 28 rue du Dr Roux, 75724 Paris, Cedex 15, France

U E Schaible Max-Planck-Institute for Infection Biology, Schumannstr. 21-22, D-10117 Berlin, Germany

D Schilling Joel Department of Molecular Microbiology, Washington University School of Medicine, St Louis, MO USA

Srivastava Amit Department of Molecular Biology and Microbiology Tufts University School of Medicine, 136 Harrison Avenue, M & V 409 Boston, MA 02111, USA

S Stappenbeck Thaddeus Department of Molecular Biology, Pharmacology, Pathology and Immunology, Washington University School of Medicine, St Louis, MO 63110

A Swanson Joel Department of Microbiology and Immunology, University of Michigan Medical School, Ann Arbor, MI 48109-0620. E-mail: Joeljswan@umich.edu

Tan Man-Wah Department of Genetics and Department of Microbiology and Immunology, Stanford University School of Medicine, Stanford, CA 94305

S Stappenbeck Thaddeus Department of Molecular Biology and Immunology, Stanford University, Washington University School of Medicine, Stanford, CA 94305

Tötemeyer Sabine Christian de Duve Institute of Cell Pathology, Université catholique de Louvain, Av Hippocrate, 74, B1200 Brussels, Belgium

Tran Van Nhieu Guy Unité de Pathogénie Microbienne Moléculaire, Institut Pasteur, 28 rue du Dr Roux, 75724 Paris, Cedex 15, France. E-mail: gtranvan@pasteur.fr

Tzou Phoebe Centre de Génétique Moléculaire du CNRS 91198 Gif-sur-Yvette, France

S Weiss David Skirball Institute and Department of Microbiology, New York University School of Medicine, 540 First Avenue, New York, NY 10016

Weiss Jerrold Department of Internal Medicine and the Inflammation Program, University of Iowa, Iowa City, IA 52242

Wiesner Sebastian Dynamique du Cytosquelette, Laboratoire d'Enzymologie et Biochimie Structurale, CNRS, Gif-sur-Yvette, 91198 France

H Wong Melissa Department of Molecular Biology, Pharmacology, Pathology and Immunology, Washington University School of Medicine, St Louis, MO 63110

J Worley Micah and Zychlinsky Artura Skirball Institute and Department of Microbiology, New York University School of Medicine, 540 First Avenue, New York, NY 10016

Introduction

This volume is intended as a travel guide for both microbiologists and cell biologists in their incursions into the study of pathogenic microbes and their hosts. It attempts to organize current affairs in cellular microbiology through a survey of current methods.

In the last 20 years, the analysis of infectious diseases has made two giant leaps forward. The first one was brought about by the introduction of molecular genetics in the early 1980s enabling the development of *in vitro* cell assays to screen for mutants and transformants. These efficient *in vitro* systems quickly supplanted the use of animal models for genetic screenings. Most importantly, the infection of eukaryotic cells *in vitro* set the stage for the adoption of concepts and methods of cell biology for the analysis of host–microbe interactions.

The second major step occurred when scientists made microbes meet their hosts. Studying the molecular 'cross talk' between microbes and eukaryotic cells forced researchers to manipulate the actin cytoskeleton, microfilaments, vesicular compartments and cellular organelles. These forays into new territory yielded insights into adherence, cell-entry, intracellular trafficking, the pathways of the cell's life-cycle, differentiation and death. The resulting symbiosis between microbiology and cell biology became known as cellular microbiology.

In spite of its fast moving character, we thought this discipline required its first attempt to collect and organize the major elements of its methodology. The idea was to provide a reference that would be useful to newcomers in the field, especially young scientists. The resulting collection focuses on current technical needs such as imaging technologies, cellular biochemistry, establishment and exploitation of cell assay systems, as well as promising new areas like global analysis of genome expression and proteomic analysis of cellular compartments.

We asked contributors to this volume to share their methodological expertise anchored in their own research. The large variety of experimental systems is mirrored by a large variety of currently used methods. Some disorder and heterogeneity are the inevitable side-effect, but aren't those in fact intrinsic to the study of cellular microbiology? This book therefore represents a platform to which further developments in structural biology, immunology, genomics and proteomics, as well as integrative *in vivo* approaches will be easily added.

We hope that '*Molecular Cellular Microbiology*' will serve our community and we wish to express our sincere thanks to all contributors for adapting to the spirit of this volume by contributing the best possible update in their own field.

1 Ratiometric Fluorescence Microscopy

Joel A Swanson
Department of Microbiology and Immunology, University of Michigan Medical School, Ann Arbor, MI 48109-0620, USA

◆◆

CONTENTS

◆◆◆◆◆◆ INTRODUCTION

For many experimental questions about host–microbe interactions, it is valuable to quantify chemistries inside living host cells. Ratiometric fluorescence microscopy is a spectroscopic method to measure and localize intracellular analytes. In essence, cellular compartments of interest are labeled with fluorophores and located in a fluorescence microsocope, then two digital fluorescence images are collected, of either two spectrally distinct fluorophores or two parts of the spectrum of a single fluorophore. These digital images are then combined into a third image that represents, at each digital picture element (pixel), the ratio of analyte-sensitive fluorescence to analyte-insensitive fluorescence. By converting each pixel into a concentration index, ratiometric imaging corrects for differences in cell or organelle thickness, also called pathlength, and for loss of fluorescence intensity due to photobleaching, two confounding factors that can make single intensity measurements uninformative about analyte concentrations (Bright *et al.*, 1989; Silver, 1998).

Most applications of ratiometric imaging have been microscopic measurements of pH or intracellular free calcium concentrations, using fluorescent dyes whose excitation or emission spectra change as a function of analyte concentration. Ratiometric imaging can also be used to measure relative concentrations of proteins inside cytoplasm. For

METHODS IN MICROBIOLOGY, VOLUME 31
ISBN 0-12-521531-2

example, the local concentration of actin inside cells has been inferred by ratiometric imaging of macrophages injected with rhodamine-labeled actin and fluorescein dextran (Swanson *et al.*, 1999). The rhodamine-actin is incorporated into cytoskeletal structures, such as phagosomes, and its concentration relative to other regions of cytoplasm is determined by imaging the fluorescein dextran, which distributes uniformly throughout the cytoplasm and serves to normalize for differences in cell thickness. The ratio image (rhodamine/fluorescein) is a two-dimensional projection of the cell which shows the relative concentrations of rhodamine-actin in cytoplasm. Ratiometric imaging microscopy has also been used to detect resonance energy transfer between molecules (Herman, 1989; Adams *et al.*, 1991; Sorkin *et al.*, 2000), and to measure fluorescence anisotropy of fluorescently labeled proteins inside cells (Gough and Taylor, 1993).

This chapter describes applications of ratiometric fluorescence microscopy to the study of host–microbe interactions. It provides a complete description of methods for measuring intracellular pH, and more general descriptions of related ratiometric methods.

◆◆◆◆◆◆ EQUIPMENT

Ratiometric fluorescence microscopy requires a research grade fluorescence microscope, with filter sets for fluorescence excitation and emission at different wavelengths of light, an attached camera capable of recording low light images, and software to control image acquisition and processing. Within these constraints is a wide range of options for equipment and expense. A schematic diagram of such a system is shown in Figure 1.1.

Fluorescent probes have characteristic excitation (absorption) and emission spectra. The excitation spectrum is of shorter wavelength (higher energy) than the emission spectrum; hence, fluorescence is detected in the microscope using filters and mirrors that selectively expose samples to the shorter wavelengths of light and detect the longer wavelengths of fluorescence. This is achieved using a broad spectrum light source (e.g. mercury arc lamp), excitation filters that transmit a portion of the excitation spectrum, dichroic mirrors that reflect excitation light and transmit the longer wavelengths of fluorescence, and emission filters that transmit a portion of the fluorescence emission spectrum. For example, a filter set for fluorescein consists of an excitation filter that transmits 480 to 490 nm light (exc. 485 nm), a dichroic mirror that reflects wavelengths less than 500 nm and transmits wavelengths greater than 500 nm (500 nm longpass), and an emission filter that transmits 505 to 525 nm light (em. 510 nm). Fluorophores with different spectra, such as Texas Red, require different filter combinations and dichroic mirrors with different selectivity (e.g. exc. 580 nm; dichroic 590 nm; em. 645 nm). The filters of standard fluorescence microscopes are commonly arranged as groups of filters in cassettes (a.k.a. filter cubes), which can be moved into the light path manually or electronically.

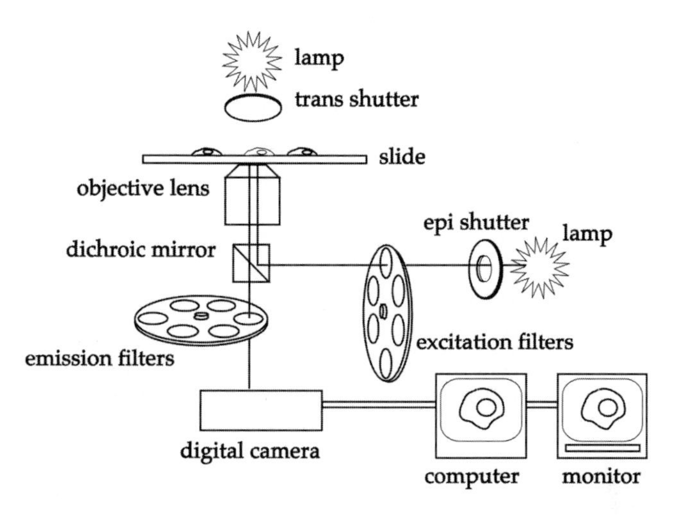

Figure 1.1. Schematic diagram of a ratiometric fluorescence microscope, showing cells on the upper surface of a coverslip (slide), mounted in an inverted fluorescence microscope. Illumination is controlled by shutters: transmitted light by the trans shutter, epifluorescence by epi shutter. Excitation and emission filters in filter wheels can be positioned independently. The dichroic mirror assembly contains the dichroic mirror plus an additional emission filter between the mirror and the filter wheel. The light path for epifluorescence is indicated.

For ratiometric imaging, one must be able to compare different fluorescent signals in a single sample, quickly. For measuring pH with fluorescein, different regions of its excitation spectrum are sampled by measuring the fluorescence excited at two wavelengths. This is done either using two separate fluorescein filter cubes, which differ only in their excitation filters (exc. 440 nm vs. exc. 485 nm), or with one filter cube that lacks an excitation filter, and two excitation filters in a filter wheel between the light source and the dichroic mirror. In the latter arrangement, the fluorescein fluorescence at each of the two excitation wavelengths is obtained simply by moving the excitation filter wheel.

Ratiometric imaging of two fluorophores with distinct excitation and emission spectra uses a dichroic mirror that reflects and transmits light at multiple wavelengths. For example, for fluorescein and Texas Red, the dichroic reflects the excitation wavelengths 485 nm and 575 nm, and transmits the emission wavelengths 528 nm and 633 nm. The emission filter in the filter cube transmits 528 nm and 633 nm light. Thus, during image collection, the dichroic mirror remains stationary, and the excitation filter wheel alternates between the excitation filters for fluorescein and Texas Red. Some fluorophore combinations, such as cyan fluorescent protein (CFP) and yellow fluorescent protein (YFP), require additional filters in an emission filter wheel to achieve full spectral discrimination of the fluorophores.

Ratiometric imaging is usually applied to living cells, which are most easily studied using inverted microscopes. A temperature-controlled stage incubator can contain cells in dishes or, preferably, on coverslips that are mounted in chambers (e.g. Leiden chambers). Many studies of interactions between bacteria and eukaryotic cells require manipulations or addition of solutions during an experiment, a constraint favoring incubation chambers that can be accessed without compromising safety. Leiden chambers can be viewed using a second coverslip positioned as a lid; this slows evaporation of buffers but allows easy access for changing solutions.

Any ratiometric imaging microscope must be optimized to minimize light-induced damage to cells. This can be achieved by inserting neutral density filters and shutters in the excitation light path, to limit the intensity and duration of cell exposure to light, and by using sensitive cameras, such as cooled CCD cameras (see Inoue and Spring, 1997 and Berland *et al.*, 1998 for descriptions of cameras). Light detection can be further enhanced by positioning an image intensifier in front of the camera (e.g. VideoScope Intl., Dulles, VA).

Proper ratiometric imaging requires computer-controlled systems for collecting, storing and processing digital images of fluorescence. Some of this can be done with NIH image software (available free at http://rsb.info.nih.gov/nih-image/). In addition, many image collection and image processing programs are offered by commercial vendors.

Although many and various configurations are feasible for ratiometric imaging microscopy, the methods described in this chapter employ a complete ratiometric fluorescence microscope system. In our lab, this consists of an inverted fluorescence microscope (Nikon TE-300), equipped with a tungsten lamp for transmitted light and a 100 W mercury arc lamp for epiillumination. It also includes a temperature-controlled stage (Harvard Apparatus, Inc., Holliston, MA), sliders containing two neutral density filters, shutters for trans- and epifluorescence illumination (Uniblitz, Rochester, NY), filter wheels for both excitation and emission filters (Lambda 10-2, Sutter Instruments, Novato, CA), and a cooled, digital CCD camera (Quantix, Photometrics, Tuscon, AZ). Filters and dichroic mirrors are from Omega Optical (Brattleboro, VT). Image acquisition, storage and processing are computer-controlled through Metamorph software (Universal Imaging, West Chester, PA), and networked computers are available for data analysis, using MatLab (The Mathworks, Inc., Natick, MA) and Excel (Microsoft, Inc., Seattle, WA), and for further image processing, using Adobe Photoshop and Premier (Adobe Systems, Inc., San Jose, CA). More extensive descriptions of equipment for ratiometric imaging microscopy can be found in related texts (Inoue and Spring, 1997; Dunn and Maxfield, 1998).

◆◆◆◆◆◆ RATIOMETRIC MEASUREMENT OF pH

For many questions relevant to microbial pathogenesis, pH is measured in vacuolar compartments, which consist of pinosomes, phagosomes, endosomes, lysosomes and the replication vacuoles of intracellular pathogens. The prototypical fluorescent probe for vacuolar pH is fluorescein, which emits green light (510 nm), and which can be excited at the pH-sensitive wavelength of 485 nm or at the pH-insensitive wavelength of 440 nm. Fluorescein can be delivered into the vacuolar compartment by endocytosis of fluorescein-labeled dextran (see Protocol 1), or by first cross-linking it chemically to the surface of a particle or microbe, using reactive molecules such as fluorescein isothiocyanate (FITC) or 5-(4,6-dichlorotriazynl) aminofluorescein (5-DTAF), then allowing phagocytosis of that particle or microbe (Steele-Mortimer *et al.*, 1999). Fluorescently labeled compartments of interest are identified in the microscope, then three digital images are collected: a transmitted light image (phase-contrast or Nomarski), a 440 nm image (excite 440 nm, emit 510 nm), and a 485 nm image (excite 485 nm, emit 510 nm; Figure 1.2a). In subsequent image processing, the 485 nm image is

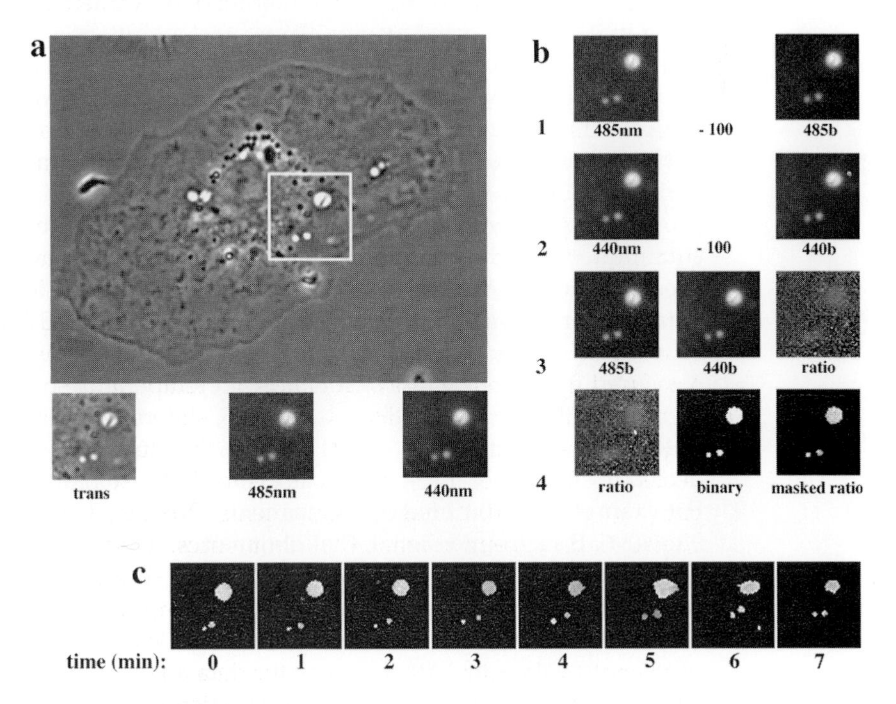

Figure 1.2. Image processing for the measurement of pH in phagosomes containing *Listeria monocytogenes*. (a) A macrophage infected with *L. monocytogenes* in the presence of fluorescein dextran is shown by phase-contrast microscopy. The region of the image showing a bacterium in a phagosome is indicated by the square, and the three component images used for ratiometric processing are shown below the larger image. (b) The processing steps for obtaining a ratio image: (1, 2) background signals are subtracted from the 485 nm image (1) and the 440 nm image (2); (3) the background-subtracted 485 nm image is divided by the background-subtracted 440 nm image, then multiplied by 1000 to obtain the ratio image; (4) the ratio image is masked by a binary image to exclude regions of the image with little or no fluorescent signal, where ratio values simply reflect noise and consequently interfere with image scaling. (c) A ratio time-series generated from image sets collected at one-minute intervals and processed as in (b), showing changes in pH over time. The upper vacuole was a *L. monocytogenes*-containing phagosome whose pH decreased (indicated by a decrease in ratio brightness from mins 0 to 4) then increased as the vacuole was perforated by the bacterium (mins 5 and 6).

divided by the 440 nm image at each corresponding pixel to obtain an image showing the fluorescence ratios in a two-dimensional projection of the cell (Figure 1.2b and Protocol 2). The fluorescence ratios are later converted to pH by experimental calibration with ionophores and buffers. Controlled repetition of the collection protocol can be used to obtain time-lapse movies of cells and corresponding measurements of compartment pH.

The best method for calibrating ratiometric images of pH probes is to measure fluorescently labeled cells in which the pH in all compartments has been equilibrated to defined values. For example, at the end of an experiment, cells with fluorescein dextran-labeled vacuoles are incubated in a pH-clamping buffer (115 mM KCl, 1 mM $MgCl_2$, 15 mM MES, 15 mM HEPES, pH 4.5) containing the ionophores nigericin and valinomycin (each at 10 μM; prepared using 10 mM stock solutions in DMSO; Note: ionophores should be handled carefully). After allowing 10 min for pH in intracellular compartments to equilibrate to that of the clamping buffer, several sets of 440 nm and 485 nm images are collected and stored. The buffer is then changed to a new clamping buffer at pH 5.0, an equilibration period of 10 min follows, then another set of images is collected. This procedure is repeated until images have been obtained from cells equilibrated to pH 4.5, 5.0, 5.5, 6.0, 6.5 and 7.0. Subsequent image processing of fluorescently labeled compartments in pH-clamped cells obtains the ratio values that correspond to each pH; and these values can be used to make a standard curve for calibrating the experimentally measured fluorescence ratios. A handsome example of such a curve is in Figure 4 of Bright *et al.* (1987).

Reliable measurements of pH require that the fluorescent signal be detectably bright yet not so bright as to exceed the digital range of values. For example, a 12-bit image contains pixel intensity values ranging from 0 to 4096. Background signal, that obtained when the camera collects an image with all shutters closed, can give average pixel values around 100 (background must be determined empirically for each system). If too much light from a cell reaches the camera, the images contain some pixels with values of 4096 (i.e. saturated pixels). Such images should not be used for ratiometric measurements because the saturated pixels are most likely under-reporting the actual fluorescence in those regions. Pixel saturation can be a problem when calibrating at higher pH values (bright 485 nm images) after experimentally measuring acidic compartments (dim 485 nm images). This can be avoided by using neutral density filters and adjusting image collection times to obtain low average pixel values, between 300 and 1000 for 12-bit images, in the experimental measurements.

Excitation of fluorophores inside cells leads to photo-oxidative damage, which can substantially alter measured values. Macrophages are especially susceptible to light-induced toxicity. For example, when we measure pH at 10- or 15-s intervals in a pinosome pulse-labeled with fluorescein dextran, we observe an initial drop in pH during the first 5 min after pinosome formation, then a slow rise in pH over the next 15 min. If instead of measuring one organelle we sample different organelles during the time course, measuring no organelle more than once, then combine these measurements into a time-course derived from population averages, we find that macropinosomes acidify rapidly, but do not realkalinize. This indicates that the extended exposure of the individual labeled pinosome to excitation light inhibited its acidification machinery. Limiting light exposure with shutters, neutral density filters and low-light cameras can reduce this phototoxicity, but will not eliminate it completely. Therefore, population sampling is recommended as a check for light-induced measurement artefacts.

Different probes measure different ranges of pH. The pK_a of fluorescein is 6.4. Consequently, its most effective range for ratiometric measurements is between pH 5.3 and 7.5. Outside of that range, changes in pH do not produce substantial changes in fluorescence ratio. Several probes with fluorescein-like excitation and emission spectra are available for measuring different pH ranges. BCECF has a pK_a of 7.0, and is consequently suitable for measuring pH in neutral compartments (Bright *et al.*, 1987; Silver, 1998). Oregon Green has a pK_a of 5.0, which makes it an effective probe for lysosomal pH (pH 4.0 to 6.0). Extended ranges for pH measurements have been obtained by combining fluorophores, such as fluorescein and Oregon Green, which are used as mixtures to label particles, then measured and calibrated *in situ* (Downey *et al.*, 1999).

Because these fluorescent probes bind protons, they can also buffer pH. Thus, the presence of too much probe in a cellular compartment can shift the pH of that compartment toward the pK_a of the probe. One should label compartments with no more probe than is necessary to obtain a measurable signal.

A variety of other fluorescent probes for pH are available and worth consideration. Factors to consider include the effective pH range of the probe relative to the biological pH range in question, the filters and dichroic mirrors necessary for detection of that probe, the mechanism of probe delivery to and retention in target compartments, and the relative sensitivity of the probe to other chemical species. Additionally, the efficacy of a fluorophore can be affected by its quantum yield, which is the ratio of the number of photons emitted per number of photons absorbed. A fluorophore with a low quantum yield may only be detectable at high concentrations, and high concentrations of probe can alter concentrations of the parameter being measured.

Many probes have excitation spectra different from that of fluorescein, necessitating different, dedicated filter sets. Some, such as SNARF, have pH-dependent fluorescence emission spectra. Measuring pH with emission ratio probes requires one excitation filter and two emission filters. If such measurements are performed using an emission filter wheel, one should be aware that changing emission filters can cause the projected images to shift relative to each other. Software can correct this by re-aligning corresponding pixels.

Each combination of micro-organism and host cell will require its own protocol for compartment labeling. Rates of internalization vary greatly between different eukaryotic cell types, and any given host cell will internalize different microbes at different rates. Therefore, before attempting to measure vacuolar chemistries, one should determine empirically how long cells should be incubated with microbes and pulse-labeling medium to obtain fluorescently labeled phagosomes. In addition, some kinds of intracellular vacuoles form as tight-fitting phagosomes with very little fluid volume between the surface of the microbe and the vacuolar membrane. Tight-fitting phagosomes may require labeling with higher concentrations of probe than is needed to label pinosomes or spacious phagosomes (Alpuche-Aranda *et al.*, 1992).

◆◆◆◆◆◆ RATIOMETRIC MEASUREMENT OF CALCIUM

Calcium is most commonly measured in the cytosolic space, where regulation of its concentration is critical for many signaling mechanisms. Many measurements of calcium relevant to microbial pathogenesis do not require precise determination of subcellular free calcium concentrations in different regions of cytoplasm, and simpler measurements of relative changes in intracellular calcium are often sufficient for determining how a microbe is affecting a host cell. These can be obtained using probes whose fluorescence intensity changes upon binding calcium, but whose excitation or emission spectra remain the same. Dyes like Calcium Green or Fluo-3 can be quantified digitally, and these can reliably report relative changes in total intracellular free calcium concentrations in response to a stimulus.

Some questions, however, require quantitative measurement of localized changes in free calcium. Calcium can be measured ratiometrically using fluorophores such as fura-2, whose fluorescence excitation spectrum varies between calcium-bound and calcium-free states (Silver, 1998; Tsien, 1999). In practice, fura-2 is introduced into cytoplasm as an acetoxymethyl ester (fura-2/AM), which readily diffuses across cellular membranes. Upon hydrolysis by cytosolic esterases it is converted to fura-2, which is membrane impermeant and consequently remains trapped intracellularly. Pairs of digital fluorescence images from probe-loaded cells, collected at 510 nm emission and excited alternately at 340 nm and 380 nm, report calcium-dependent changes in the fura-2 excitation spectrum. Processed images representing 340/380 nm fluorescence ratios can later be calibrated by incubating labeled cells in calcium ionophore (ionomycin), recording images after adjusting calcium to maximum and minimum values, then calibrating the experimental measurements from these determinations (Silver, 1998). Calibrated ratiometric images display a two-dimensional map of intracellular free calcium concentration. An extended description of fluorescent calcium indicators and their calibration is available in Tsien (1999).

Microscopy for imaging fura-2 requires UV excitation light, which imposes constraints on the experimental system. One must have appropriate filter sets and dichroic mirrors, as well as objective lenses that can transmit UV light. Because it also excites many endogenous fluorescent molecules, UV excitation can produce autofluorescence and cause considerable photodamage to cells. Concerns about phototoxicity resulting from overexposure to light, noted above for pH probes, apply to calcium probes as well.

As with probes for pH, it is possible to load cells with buffering levels of calcium probe. Dyes like fura-2 are essentially calcium buffers; and as such they can dampen intracellular calcium signals or inhibit cellular activities that require increases in calcium. Moreover, because fura-2 has a high affinity for calcium, as well as a very high diffusion coefficient, it can redistribute calcium throughout cells rapidly and cause otherwise localized calcium responses to appear diffuse (Tsien, 1999). Other analogues of fura-2 have been developed that have lower affinities for calcium. They allow detection of faster calcium responses and can correct for the buffering effects produced by higher affinity calcium probes. Information about many of these probes is available from Molecular Probes, Inc. (Eugene, OR; http://www.probes.com/).

Although fura-2 is an excellent probe for cytosolic calcium in many kinds of cells, it can give misleading results in others. Macrophages, for example, contain organic anion transporters that actively export many cytosolic fluorescent dyes across membranes out of the cell or into endocytic organelles. Cytosolic fura-2 is quickly redistributed by these mechanisms. A modified version of fura-2 has been developed, furaPE-3, which is not redistributed by organic anion transporters. It can be delivered into cells as the acetoxymethyl ester (FuraPE-3AM). Alternatively, dextran conjugates of probes are not redistributed by the organic anion transporter; therefore, probes like fura-dextran that can be loaded into cytoplasm, by microinjection or scrape-loading (see Protocol 3), should remain cytosolic and reliably report calcium.

Calcium has been difficult to measure inside vacuolar compartments because the calcium affinities of all fluorescent probes vary with pH and ionic strength. Accurate measurement of vacuolar calcium requires knowledge of the pH of the compartment, the fluorescence ratio of the calcium probe, plus that probe's binding affinities for calcium at different pH. This has been achieved recently using mixtures of pH probes (fluorescein dextran and Oregon Green dextran), and pH-calibrated ratiometric calcium probes (fura-2-conjugated dextran), together with filter sets and custom dichroic mirrors for sequential measurements of 340 nm, 380 nm, 450 nm and 488 nm images (em. 535 nm) (Christensen *et al.*, 2001).

Fluorescent probes for ratiometric measurement of other analytes, such as magnesium, sodium and potassium have been developed and have seen limited application (Molecular Probes, Inc., Eugene, OR). Fully interpretable ratiometric measurements using these probes require methods for calibrating them intracellularly, as well as methods for identifying and excluding confounding artifacts, such as sensitivity to other ion species.

◆◆◆◆◆◆ RATIOMETRIC MEASUREMENT OF PROTEIN DISTRIBUTIONS

Ratiometric imaging has also been used to estimate relative concentrations of proteins inside cells. This approach was originally applied to studies of fluorescent analogues of actin inside amoebae (Taylor and Wang, 1978), and it has been used recently to localize actin and YFP-labeled chimeric proteins inside macrophages. An example is shown in Figure 1.3, which shows digital images of RAW 264.7 macrophages expressing CFP and a chimera of YFP and the pleckstrin homology domain of Bruton's tyrosine kinase (YFP-Btk-PH), which binds specifically to phosphatidylinositol 3,4,5-trisphosphate (PI 3,4,5-P_3) inside cells (Varnai *et al.*, 1999). In unstimulated macrophages, and during the first stages of macropinosome formation in response to macrophage colony-stimulating factor, both CFP and YFP-Btk-PH are distributed uniformly through the cytoplasm

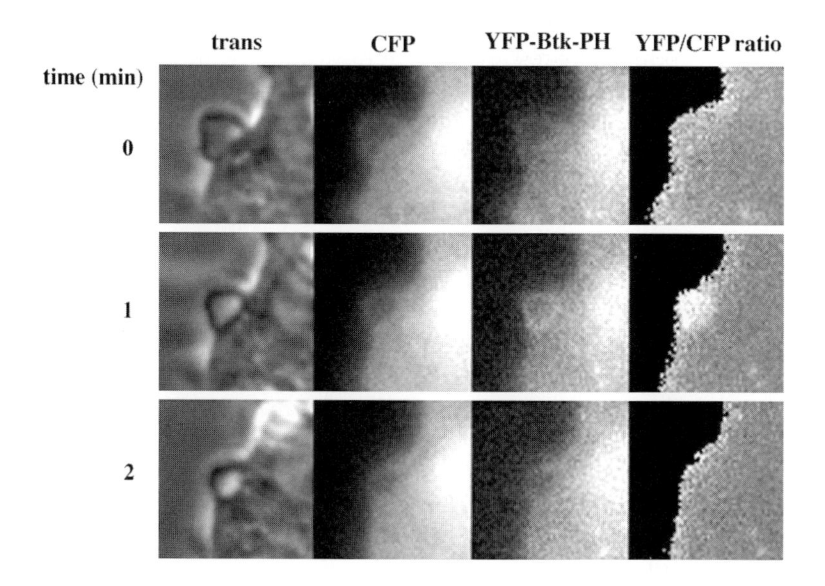

Figure 1.3. Ratiometric imaging of PI 3,4,5-P_3 distributions using YFP-Btk-PH chimeras. A RAW 264.7 macrophage expressing CFP and YFP-Btk-PH was imaged by phase-contrast (trans) and by CFP and YFP fluorescence at 15-s intervals during the formation of a macropinosome. The variable brightness of the CFP and YFP-Btk-PH images indicate differences in pathlength, with the brighter areas on the right side of each panel corresponding to the thicker regions of the cells. The ratio images show how differences in pathlength are corrected by processing. Before and during the early stages of pinosome formation, the YFP/CFP ratios in the cell were all similar, indicating uniform intracellular distribution of both CFP and YFP-Btk-PH. At the one-minute timepoint (1 : 00), and at timepoints of 0 : 30, 0 : 45 and 1 : 15 (data not shown), the ratio around the macropinosome increased by 50% above the rest of the cytoplasm. This indicates increased localization of YFP-Btk-PH to the pinosome and, by inference, locally increased concentrations of PI 3,4,5-P_3. The ratio returned to basal levels by the two-minute time point.

(Figure 1.3). Although the component fluorescent images are uneven, due to regional differences in cell thickness, the ratiometric image (YFP-Btk-PH/CFP) is flat. Later, YFP-Btk-PH concentrates on the forming macropinosome, due to localized generation of PI 3,4,5-P_3, then returns to a uniform distribution shortly after the pinosome forms. Ratiometric imaging of cells expressing CFP plus either non-chimeric YFP or YFP-Btk-PH containing a point mutation in the PI 3,4,5-P_3-binding domain are uniformly flat and do not change during stimulation. These important controls indicate that the increases in fluorescence ratios in cells containing CFP and YFP-Btk-PH reflect changes in PI 3,4,5-P_3 levels. Ratiometric imaging methods can be superior to confocal microscopic methods in that they can localize changes more rapidly, more sensitively, and more precisely.

Collection protocols for imaging protein distributions are essentially similar to those used for measuring pH or calcium, except that both excitation and emission filters must be changed for each component image. The images shown in Figure 3 were obtained as described in Protocol 2, collecting and storing at 15-s intervals a phase-contrast image, a CFP fluorescence image and a YFP fluorescence image. The ratio images were prepared afterward from the CFP and YFP images, using the processing steps outlined in Protocol 2. CFP images were taken using the 436 nm excitation (Omega Optical #440AF21) and the 474 nm emission filters (#480AF30), and the YFP images were taken using the 513 nm excitation (#500AF25) and the 530 nm emission filters (#535AF26). The dichroic mirror (#465-543DBDR) selectively transmitted the two fluorescence wavelengths and reflected the two excitation wavelengths, and an additional emission filter in the cassette allowed transmission of both YFP and CFP fluorescence (#465-543DBEM). The cassette emission filter allows viewing of the CFP and YFP fluorescence through the eyepiece (which does not have the emission filter wheel in its light path). Although that filter does not provide full ocular discrimination of the CFP and YFP signals, it provides an important protection for the eyes. One must none the less take care never to look through the eyepiece when changing filters; to avoid direct exposure of the eyes to excitation light that could be caused by accidental selection of the wrong excitation filter or filter cassette.

◆◆◆◆◆◆ RATIOMETRIC MEASUREMENT OF RESONANCE ENERGY TRANSFER

The close proximity of two fluorophores, or of two fluorescently labeled macromolecules, can be detected and measured using resonance energy transfer (RET) (Herman, 1989; Tsien *et al.*, 1993; Lakowicz, 1999). If one fluorophore, the donor, has an emission spectrum overlapping the excitation spectrum of a second, acceptor fluorophore, and if the two fluorophores are within about 2–6 nm of each other, then the donor can transfer the resonance energy of its excited state to the acceptor and cause it to fluoresce. Adams *et al.* (1991), developed a RET-based sensor for cyclic AMP that measured the energy transfer of a fluorescein-labeled catalytic subunit of protein kinase A (FlC) to a rhodamine-labeled regulatory subunit of the same enzyme (RhR). In low cyclic AMP, protein kinase A existed as a holoenzyme of two catalytic (FlC) and two regulatory subunits (RhR); excitation of the fluorescein in FlC allowed energy transfer to the rhodamine of the RhR. When cyclic AMP concentrations increased, the regulatory subunits of protein kinase A (PKA) bound cyclic AMP and dissociated from the catalytic subunits, increasing the distance between the fluorescein and rhodamine such that energy transfer was lost. Hence, increases in intracellular cyclic AMP could be measured as the decreased energy transfer in FlCRhR-loaded cells.

Recently, the spectrally distinct variants of green fluorescent protein (GFP) have been used for RET-based detection systems. PKA was labeled in an analogous manner to that described above, using enhanced blue fluorescent protein (EBFP)-labeled regulatory domains (donor) and GFP-labeled catalytic domains (acceptor); radiometric imaging microscopy was then used to measure intracellular cyclic AMP (Zaccolo *et al.*, 2000). GFP-based donor–acceptor pairs have been expressed inside cells to detect other intracellular chemistries, such as intracellular free calcium concentrations (Miyawaki *et al.*, 1999), activation of protein kinase A (Nagai *et al.*, 2000), protein dimerization (Sako *et al.*, 2000; Schmid *et al.*, 2001), and the activation of caspases (Mahajan *et al.*, 1999). In addition, RET between expressed fluorescent proteins and chemically labeled proteins have been used to measure localized activation of Rho-family GTPases inside motile fibroblasts (Kraynov *et al.*, 2000).

All of the above systems used ratiometric imaging microscopy to detect RET inside cells. Different imaging systems and image processing algorithms were used to obtain these measurements (Gordon *et al.*, 1998; Schmid *et al.*, 2001), and there is presently no conventional method for quantifying RET. Two general strategies are outlined below.

When RET occurs between two fluorophores, the fluorescence of the donor decreases and that of the acceptor increases. In cells containing both donor and acceptor fluorophores (subscript DA), RET can be detected by measuring the ratio of fluorescence at two filter combinations: donor excitation, acceptor emission (DA_{DA}) and donor excitation, donor emission (DD_{DA}). RET increases the ratio DA_{DA}/DD_{DA}. This method requires independent control experiments measuring cells containing both fluorophores but no RET. In addition, ratios that indicate RET can be confirmed by experimentally photobleaching the acceptor fluorophore, using prolonged exposure to the acceptor excitation light, then determining that the ratio values have decreased (with acceptor molecules bleached, RET cannot occur).

In this lab, RET is quantified by collecting three component images, using three fluorescence filter combinations: donor excitation, donor emission (DD_{DA}), acceptor excitation, acceptor emission (AA_{DA}), and donor excitation, acceptor emission (DA_{DA}). For studies of RET between CFP and YFP, we use the same filter wheels and dichroic mirrors used for ratiometric imaging of these fluorophores. Quantifying RET requires independent measurements of the three images from cells expressing only the donor (subscript D) or the acceptor (subscript A). Thus, in separate experiments, one obtains values for DD_D, DA_D, AA_D, DD_A, DA_A, and AA_A. Assuming that the filter sets distinguish the two fluorophores sufficiently, then the measured values for AA_D and DD_A should be close to zero. Correction factors for spectral overlap, independent of RET, can then be defined as $C_D = DA_D/DD_D$, for the donor, and $C_A = DA_A/DD_A$, for the acceptor. These independently determined correction factors can be used to measure RET ratiometrically by processing the data images according to the formula: $R = DA_{DA}/((C_D \times DD_{DA}) + (C_A \times AA_{DA}))$. If the donor and acceptor molecules are both in the cell but not close enough for RET, then R should be approximately 1.0. If RET is occurring, then R should be greater than 1 (up to 3.0, in our measurements). Photobleaching the acceptor, by prolonged exposure to filter combination AA_{DA}, should decrease R values to 1.0.

◆◆◆◆◆◆ THE RELATIVE MERITS OF OTHER TECHNOLOGIES

The interaction between a microbe and a host cell can be visualized in many ways, and each different method provides a distinct kind of information about that interaction. Electron microscopy provides high-resolution images, sometimes of host–pathogen interactions in natural settings. Light microscopy offers lower resolution, but can allow continuous observation of biology as it occurs. There are many circumstances in which other imaging techniques are preferable to, or substantially augment, ratiometric fluorescence microscopy of living cells.

Electron microscopy and immunoelectron microscopy. The resolution obtainable by electron microscopy greatly exceeds that attainable by conventional fluorescence microscopy. It is therefore superior to light microscopy for many questions requiring information about fine structure. Immunoelectron microscopy provides high-resolution localization of protein antigens inside cells; as such it can be used to confirm localization of proteins by ratiometric methods (Maunsbach, 1998). The principal limitations of electron microscopy are that only fixed cells can be viewed and that sample sizes are generally smaller than those for light microscopic studies, which makes quantification by stereological or morphometric methods arduous.

Immunofluorescence. Immunofluorescence can be used to localize protein antigens in fixed cells. Its advantage over electron microscopic and ratiometric fluorescence microscopic methods is that it allows large samples to be analyzed relatively quickly (Araki and Swanson, 1998). Moreover, several distinct kinds of molecule can be viewed independently in a single cell, using different fluorophores to mark different antibodies or probes (Mies *et al.*, 1998). Immunofluorescence localizations have been quantified using ratiometric methods (Araki *et al.*, 2000). Because cells must be fixed and extracted for most immunofluorescence applications, temporal aspects of some interactions between microbes and host cells may be more difficult to define than with methods that study living cells.

Confocal fluorescence microscopy. Ratiometric imaging produces a two-dimensional projection of fluorescence in and near a focal plane inside a cell; consequently, it compresses somewhat the third dimension (z-axis). For very thick cells, contributions from out-of-focus fluorescence can obscure information about subcellular chemistries along that z-axis. Confocal fluorescence microscopy corrects for pathlength variation by excluding all light originating outside a narrow plane of focus; consequently, it provides high z-axis resolution. By correcting for pathlength variation in this way, the confocal microscope reduces the need for ratiometric measurements; single intensity measurements can often suffice. For thick cells, such as some kinds of epithelial monolayers, confocal fluorescence microscopy is preferable to ratiometric imaging for quantitative measurements of fluorophore distribution (Boyde *et al.*, 1998). For very thin cells such as macrophages or endothelial cells, however, the fluorescence intensities in a confocal image can be misleading, as the z-axis resolution of the confocal microscope is 0.5 μm or greater, and cells are often as thin as 0.15 μm. In general, ratiometric imaging is more sensitive than confocal microscopy; consequently, it requires less excitation light and allows more rapid image acquisition.

Fluorescence lifetime imaging microscopy. Many fluorescent molecules do not change their absorbance or emission spectra upon binding an analyte, but do change the average time they remain in the excited state after absorbing photons (the fluorescence lifetime, τ, is the average time required for the fluorescence intensity from a population of fluorophores to decrease to 1/e) (Lakowicz, 1999). For example, Calcium Green does not change its excitation or emission spectra upon binding calcium, as does Fura-2, but its fluorescence lifetime changes from 1 ns to 4 ns, and this change can be measured in subcellular regions using a fluorescence lifetime imaging microscope. A time-domain fluorescence lifetime microscope collects images at defined delay intervals after brief excitation pulses (1 picosecond). Subsequent image processing allows computation of the rates of fluorescence decay for all corresponding pixels in a digital image. These processing steps can be simplified into a ratiometric measurement that compares the fluorescence obtained at two different delay times after an excitation pulse. Methods for measuring fluorescence lifetime of intracellular fluorophores can be used to measure pH, intracellular free calcium concentrations (Koester *et al.*, 1999), resonance energy transfer (donor lifetimes shorten during energy transfer (Bastiaens and Squire, 1999)), and fluorescence anisotropy (Lakowicz, 1999).

Acknowledgments

I thank Adam Hoppe for his suggestions and his critique of the text. Microscopic studies in this lab are supported by NIH grants AI-35950 and RR-14039.

References

Adams, S. R., Harootunian, A. T., Buechler, Y. J., Taylor, S. S. and Tsien, R. Y. (1991). *Nature* **349**, 694–697.

Alpuche-Aranda, C. M., Swanson, J. A., Loomis, W. P. and Miller, S. I. (1992). *Proc. Nat. Acad. Sci. USA* **89**, 10079–10083.

Araki, N., Hatae, T., Yamada, T. and Hirohashi, S. (2000). *J. Cell Sci.* **113**, 3329–3340.

Araki, N. and Swanson, J. A. (1998). In *Cell Biology. A Laboratory Handbook* (J. E. Celis, ed.), second edn. Vol. 2, pp. 495–500. Academic Press, San Diego.

Bastiaens, P. I. H. and Squire, A. (1999). *Trends Cell Biol.* **9**, 48–52.

Berland, K., Jacobson, K. and French, T. (1998). In *Video Microscopy* (G. Sluder and D. E. Wolf, eds), Vol. 56, pp. 20–44. Academic Press, New York.

Boyde, A., Gray, C. and Jones, S. J. (1998). In *Cell Biology. A Laboratory Handbook* (J. E. Celis, ed.), second edn, Vol. 3, pp. 179–186. Academic Press, San Diego.

Bright, G. R., Fisher, G. W., Rogowska, J. and Taylor, D. L. (1987). *J. Cell Biol.* **104**, 1019–1033.

Bright, G. R., Fisher, G. W., Rogowska, J. and Taylor, D. L. (1989). In *Fluorescence Microscopy of Living Cells in Culture* (Y.-L. Wang and D. L. Taylor, eds), Vol. 29B, pp. 157–192. Academic Press, New York.

Christensen, K. A., Myers, J. T. and Swanson, J. A. (2001). submitted.

Downey, G. P., Botelho, R. J., Butler, J. R., Moltyaner, Y., Chien, P., Schreiber, A. D. and Grinstein, S. (1999). *J. Biol. Chem.* **274**, 28436–28444.

Dunn, K. and Maxfield, F. R. (1998). In *Video Microscopy* (G. Sluder and D. E. Wolf, eds), Vol. 56, pp. 217–236. Academic Press, New York,

Gordon, G. W., Berry, G., Liang, X. H., Levine, B. and Herman, B. (1998). *Biophys. J.* **74**, 2702–2713.

Gough, A. H. and Taylor, D. L. (1993). *J. Cell Biol.* **121**, 1095–1107.

Herman, B. (1989). In *Fluorescence Microscopy of Living Cells in Culture. Part B. Quantitative Fluorescence Microscopy – Imaging and Spectroscopy* (D. L. Taylor and Y.-L. Wang, eds), Vol. 30, pp. 219–243, Academic Press, New York.

Inoue, S. and Spring, K. R. (1997). *Video Microscopy, The Fundamentals.* Plenum Press, New York.

Koester, H. J., Baur, D., Uhl, R. and Hell, S. W. (1999). *Biophys. J.* **77**, 2226–2236.

Kraynov, V. S., Chamberlain, C., Bokoch, G. M., Schwartz, M. A., Slabaugh, S. and Hahn, K. M. (2000). *Science* **290**, 333–337.

Lakowicz, J. R. (1999). *Principles of Fluorescence Spectroscopy.* Kluwer Academic/ Plenum, New York.

Mahajan, N. P., Harrison-Shostak, D. C., Michaux, J. and Herman, B. (1999). *Chem. Biol.* **6**, 401–409.

Maunsbach, A. B. (1998). In *Cell Biology. A Laboratory Handbook* (J. E. Celis, ed.), second edn., Vol. 3, pp. 268–276. Academic Press, San Diego.

Mies, B., Rottner, K. and Small, J. V. (1998). In *Cell Biology. A Laboratory Handbook* (J. E. Celis, ed.), second edn, Vol. 2, pp. 469–476. Academic Press, San Diego.

Miyawaki, A., Griesbeck, O., Heim, R. and Tsien, R. Y. (1999). *Proc. Nat. Acad. Sci. USA* **96**, 2135–2140.

Nagai, Y., Miyazaki, M., Aoki, R., Zama, T., Inouye, J., Hirose, K., Iino, M. and Hagiwara, M. (2000). *Nature Biotech.* **18**, 313–316.
Sako, Y., Minohgchi, S. and Yanagida, T. (2000). *Nature Cell Biol.* **2**, 168–172.
Schmid, J. A., Scholze, P., Kudlacek, O., Freissmuth, M., Singer, E. A. and Sitte, H. H. (2001). *J. Biol. Chem.* **276**, 3805–3810.
Silver, R. B. (1998). In *Video Microscopy* (G. Sluder and D. E. Wolf, eds), Vol. 56, pp. 237–251. Academic Press, New York.
Sorkin, A. M. M., Huang, F. and Carter, R. (2000). *Curr. Biol.* **10**, 1395–1398.
Steele-Mortimer, O., Meresse, S., Gorvel, J.-P., Toh, B.-H. and Finlay, B. B. (1999). *Cell Microbiol.* **1**, 33–49.
Swanson, J. A., Johnson, M. T., Beningo, K., Post, P., Mooseker, M. and Araki, N. (1999). *J. Cell Sci.* **112**, 307–316.
Taylor, D. L. and Wang, Y.-L. (1978). *Proc. Natl. Acad. Sci. USA* **75**, 857–861.
Tsien, R. Y. (1999). In *Calcium as a Cellular Regulator* (E. Carafoli and C. Klee, eds), pp. 28–54. Oxford University Press, New York.
Tsien, R. Y., Bacskai, B. J. and Adams, S. R. (1993). *Trends Cell Biol.* **3**, 242–245.
Varnai, P., Rother, K. I. and Balla, T. (1999). *J. Biol. Chem.* **274**, 10983–10989.
Zaccolo, M., DeGiorgi, F., Cho, C. Y., Feng, L., Knapp, T., Negulescu, P. A., Taylor, S. S., Tsien, R. Y. and Pozzan, T. (2000). *Nature Cell Biol.* **2**, 25–29.

Protocol 1. Labeling endocytic compartments with fluorescent probes

1. Plate cells on coverslips for mounting in Leiden chambers.
2. Prepare solution of drink: 1–2 mg/ml fluorescein dextran, avg. mol. wt. 3000 (Molecular Probes, Eugene, OR) in Ringers buffer (RB: 155 mM NaCl, 5 mM KCl, 2 mM $CaCl_2$, 1 mM $MgCl_2$, 2 mM NaH_2PO_4, 10 mM HEPES and 10 mM glucose, pH 7.2–7.4) or medium.
3. To label pinosomes, replace buffer in chamber with drink; incubate 1–5 min. To label phagosomes containing bacteria, include an infectious dose of bacteria in the fluorescent drink and incubate long enough to obtain phagosomes (generally 10–20 min).
4. Wash cells with buffer or medium.
5. Incubate in unlabeled medium for 0–30 min (chase period).
6. View in microscope, taking care not to overexpose cells before collecting digital images (Protocol 2).
7. For labeling later compartments, such as late endosomes or lysosomes, pulses may be longer (15 min to 12 h) and chase periods longer (45 min to 12 h).

NB 1: Different cell types exhibit different rates of endocytosis; and this will affect the pulse periods needed to label pinosomes. Macrophage pinosomes can be labeled with a 1–2 min pulse, whereas pinosomes of fibroblasts or epithelial cells may require longer pulses to be labeled.

NB 2: Buffers or media buffered with carbon dioxide will acidify upon removal from CO_2 incubation chambers; this rapid acidification can inhibit endocytosis dramatically. A period of equilibration may be required to obtain measurable rates of endocytosis in the macrophage.

Protocol 2. Ratiometric measurement of pH in fluorescein-dextran-labeled vacuoles

1. Plate cells onto 25 mm diameter circular coverslips (10^4 cells each) and assemble into Leiden chamber.
2. Label phagosomes or pinosomes with fluorophore by pulse-chase method (Protocol 1).
3. Locate fluorescently labeled cell using 480 nm exc. (510 nm emission). Cut light to sample.
4. Run programmed collection protocol:
   a. Open transmitted light shutter.
   b. Collect transmitted light image (e.g. binned 2 × 2 image, 100 ms exposure).
   c. Close transmitted light shutter.
   d. Store in folder on hard disk (assign file name automatically, e.g. image = T-001, stored as C: Data/Trans/T-001).
   e. Position 450 nm excitation filter by moving excitation filter wheel (510 nm emission filter should already be in position).
   f. Open epi-illumination shutter.
   g. Collect 450 nm image (same size image as T-001, but different exposure time as necessary, e.g. 400 ms).
   h. Store in folder on hard disk (assign file name automatically, e.g. image = B-001 stored as C: Data/Blue/B-001).
   i. Position 485 nm excitation filter by moving excitation filter wheel (510 nm emission filter remains in place).
   j. Collect 485 nm image (same size image as T-001 and B-001, but different exposure time as necessary, e.g. 300 ms).
   k. Close epi-illumination shutter.
   l. Store in folder on hard disk (assign file name automatically, e.g., image = G-001 stored as C: Data/Green/G-001).
5. Repeat three-image series at regular intervals (e.g. creating files T-002, B-002 and G-002, etc.) to make time-lapse series.
6. After all data are collected, move image stacks from data folders to record folders (e.g. C: Records/(*data notebook address*) /expt1).
7. Process images or image stacks (e.g. groups of time-series images consisting of files B-001 to B-n, G-001 to G-n):
   a. Subtract background values from fluorescence images (e.g. B-001 minus 100 = B; G-001 minus 100 = G).
   b. Prepare ratio image by dividing background-subtracted 480 nm image by background-subtracted 450 nm image. Multiply ratio times 1000 (e.g. R = (G divided by B) ×1000).
   c. Convert non-cellular pixels of the ratio image to zero, using a binary mask. This allows optimal display of cell-associated ratio values.
   d. Calibrate experimental ratio measurements by preparing ratio images of labeled cells equilibrated with ionophores (nigericin and valinomycin) in pH-clamping buffer, pH 4.0, followed by images taken from the same cells after changing to pH-clamping buffers at pH 4.5, 5.0, 5.5, etc.
   e. Compare ratio images with transmitted light images.

Protocol 3. Scrape-loading proteins into cytoplasm

1. Plate 10^6cells/dish on 35 mm, bacteriological grade Nunc dishes (not tissue culture plastic).
2. Prepare solutions:
   a. FDx/protein solution: 1 mg/ml FDx10 (fluorescein dextran, avg. mol. wt. 10 000; from Sigma or Molecular Probes), plus protein to be loaded into cells, in PD; 25–150 µl final volume.
   b. Coverslips (25 mm circular, No. 1) in six-well tissue culture dish with medium (37°C).
   c. Warm, divalent cation-free PBS, ~25 ml.
3. Decant medium from cells (about 2 ml) into the cover of the Nunc dish.
4. Rinse dish two times with 2 ml warm PBS—remove.
5. Add 25–75 µl FDx/protein solution.
6. Stir with pointed end of rubber policeman to penetrate unstirred layer of fluid near cells; then scrape with flat end for ~20–30 s.
7. Return medium from lid to dish. Dispense suspended cells to coverslips in six-well dish (50–100 µl per well) .
8. Incubate 10 min in 37°C incubator.
9. Rinse with fresh medium.
10. Incubate 60–120 min in 37°C incubator.
11. Observe FDx10 fluorescence to identify loaded cells. Look for fluorescence in the nucleus.

2 Dissecting Host–Pathogen Molecular Interactions with Microarrays

Corrella S Detweiler and Stanley Falkow
Department of Microbiology and Immunology, Stanford University Medical Center, Stanford, CA 94305, USA

◆◆

CONTENTS

◆◆◆◆◆◆ INTRODUCTION

Microarrays are a powerful tool for measuring DNA or RNA content and will become increasingly available and cost-effective. Here we describe methods for using bacterial cDNA microarrays to explore new territory in pathogen–host interactions. We have included basic protocols for isolating bacterial RNA and for labeling and hybridizing DNA and RNA to microarrays. In addition, the references provide examples of experimental rationale and design. Since array protocols are still being optimized, at the end of this chapter we have included web sites that will serve as sources for protocol updates.

cDNA microarrays, which consist of DNA fragments that have been robotically spotted to glass slides, can be generated by different means. If the genome of the organism has been sequenced, gene-specific primers for each open-reading frame can be used to amplify the DNA for each gene (Chu *et al.*, 1998; Spellman *et al.*, 1998). Alternatively, the spotted DNA can consist of long (~70 nucleotides) oligos (Chambers *et al.*, 1999). If the organism's genome sequence is not known, the DNA fragments can be derived from cDNA library inserts that have been amplified by PCR with

METHODS IN MICROBIOLOGY, VOLUME 31
ISBN 0-12-521531-2

a single set of primers (Iyer *et al.*, 1999). Ideally, the cDNA clones have been sequence verified and the library is non-redundant. Shotgun microarrays, consisting of amplified genomic DNA library inserts, also can be produced if representative cDNA libraries are difficult to generate (Hayward *et al.*, 2000).

DNA or RNA to be hybridized to microarrays is typically labeled with a Cy3 or Cy5 dye (Eisen and Brown, 1999). The consistent use of a reference sample allows investigators to make comparisons between microarrays after normalizing the sample data to the reference data for each microarray (Eisen and Brown, 1999). For bacteria, randomly labeled genomic DNA (derived from the bacterial strain(s) from which the microarray was created) serves as a reliable reference since it will hybridize to every successfully spotted cDNA on the microarray (Salama *et al.*, 2000). Genomic DNA can be used as a reference for both DNA and RNA experiments, as described below.

◆◆◆◆◆◆ GENOMIC PROFILING

Microarrays can be used to detect genomic differences between strains or closely related pathogens. For example, pathogen arrays can be used to profile DNA isolated from a variety of environmental or clinical sources with the aim of identifying genomic regions that correlate or anti-correlate with virulence (Behr *et al.*, 1999; Salama *et al.*, 2000). Pathogen arrays also can be used to monitor changes within a population over time during an infection, and selection against particular genes or blocks of genes can be monitored (Behr *et al.*, 1999). An important caveat is that the presence or absence of only the sequences on the microarray can be assessed; divergent sequences below the detection threshold, which is a function of the hybridization conditions, will not be detectable (Salama *et al.*, 2000). Below is a protocol for labeling genomic DNA and hybridizing it to a microarray. Figure 2.1 shows DNA labeled with this protocol hybridized to a *Helicobacter pylori* whole genome spotted array.

Amino-allyl dUTP labeling of bacterial genomic DNA (Salama *et al.*, 2000)

The DNA used in the following protocol should be of high quality, such as that obtained with cesium chloride centrifugation or commercial genomic DNA preparation kits. Before beginning, prepare Cy3 and Cy5 dyes by resuspending each monofunctional reactive dye (Amersham Pharmacia Biotech, PA 23001 and 25001) in 72 μl dH_2O. Aliquot 4.5 μl of each into separate microfuge tubes and dry in speed vacuum. Store at −20°C.

Klenow incorporation of aadUTP

1. Denature 2 μg DNA in 38 μl dH_2O for 5 min at 99°C. Cool for 5 min on ice.

2. Add
   5 μl 10× Buffer (400 μg ml^{-1} random octamers, 0.5 M Tris-HCL, 100 mM $MgSO_4$, 10 mM DTT)
   5 μl dNTP/dUTP mix (0.5 mM dGTP, dATP, dCTP; 0.2 mM aadUTP (Sigma A-0410), 0.3 mM dTTP)
   2 μl Klenow 5 U μl^{-1}
   Incubate for at least 1 h at 37°C.
3. Purify probe by adding each reaction to a microcon YM-30 (Millipore 42410) containing 450 μl dH_2O. Spin 8 min 12 000 rpm in a microcentrifuge.
4. Discard eluate. Wash twice with 450 μl dH_2O and spin as above.
5. After last wash, spin sample until only a 'smile' of liquid (5–10 μl, which will collect in a crescent around the filter's edge) remains.
6. To collect probe, invert microcon into a fresh microfuge tube and spin 1 min 12 000 rpm. Dry in a speed vacuum.

Label DNA with Cy3 or Cy5

7. Resuspend probe in 4.5 μl dH_2O. Resuspend Cy3 or Cy5 monofunctional reactive dye in 4.5 μl 0.1 M Na Bicarbonate pH 9.0.

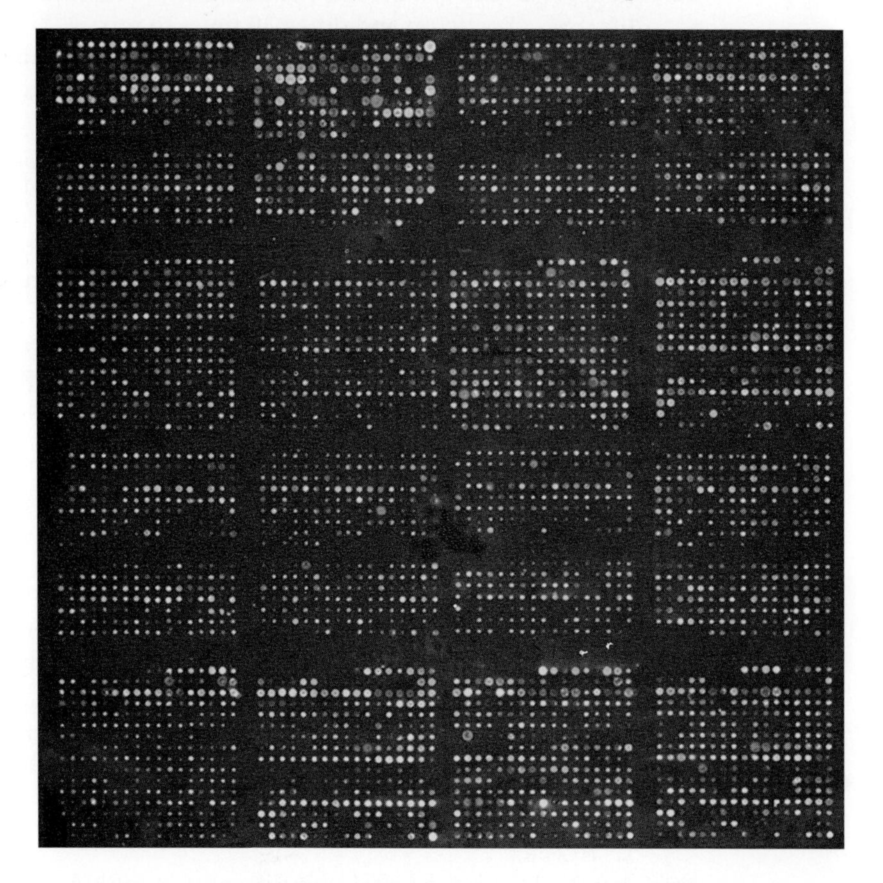

Figure 2.1. DNA labeled with the specified protocol hybridized to a *Helicobacter pylori* whole genome spotted array. (This figure is also reproduced in colour between pages 276 and 277.)

8. Label probe by mixing a dye with the probe. Incubate 1 h at room temperature in the dark. Typically, the reference DNA for each microarray is labeled with Cy3 and the sample DNA with Cy5.
9. Quench the labeling reaction by adding 4.5 μl of 4 M hydroxylamine. Incubate for 15 min at room temperature in the dark.
10. Remove unincorporated dye with Quia-quick PCR purification (Quiagen) as follows. Combine Cy3 and Cy5 reactions. Add 70 μl dH_2O and 500 μl Buffer PB. Apply to Qia-quick column and spin at 13 000 rpm for 30–60 s. Discard flow-through. Wash with 750 μl Buffer PE and spin 30–60 s. Discard flow-through and spin column dry 30–60 s. Transfer column to a fresh microfuge tube. Elute by adding 30 μl Buffer EB to center of filter and let it stand for 1 min at room temperature. Spin at 13 000 rpm for 1 min. Repeat elution step with an additional 30 μl EB.

Prepare probe for hybridization (the following applies for a 22 × 22 mm coverslip; adjust volumes for other coverslip sizes)

11. Dry down eluate in speed vacuum.
12. Resuspend in 11 μl TE. Add 1 μl 25 mg ml^{-1} yeast tRNA. Add 2.55 μl 20 × SSC. Pipette up and down 10 times to mix.
13. Add 0.45 μl 10% SDS, pipette up and down 10 times to mix.
14. Denature for 2 min at 99°C. Cool briefly, centrifuge for 2 min at 13 000 rpm.
15. Add 15 μl, avoiding any precipitate, to the microarray. Spot a total of 15 μl of 3 × SSC near the edges of the slide (away from the coverslip) to keep the chamber humidified.
16. Hybridize overnight at appropriate temperature (60–65°C) in a hybridization chamber (Corning). An appropriate temperature can be determined by labeling the reference DNA with Cy3 and Cy5, combining the reactions and hybridizing them to the same microarray. At the ideal temperature, the signal will be strong and the expected red to green ratio for all spots is one. An optimal SSC concentration can be determined similarly.
17. Wash slides as follows. Place in slide rack. Rinse in 200 ml 0.03% SDS, 2 × SSC in a slide chamber. Wait until the coverslips fall off.
18. Remove slides and blot to paper towel. Transfer slides to fresh slide rack. Rinse in 200 ml 1 × SSC, 5 min, rocking, in a fresh slide chamber.
19. Move rack to a fresh slide chamber containing 200 ml 0.2 × SSC. Let rock for 5 min. Spin slides dry in a rack on a paper towel, 5 min, 500 rpm. Scan slides on a commercial scanner, such as the Axon GenePix 4000a Microarray Scanner.

◆◆◆◆◆◆ MONITORING BACTERIAL mRNA EXPRESSION

Microarrays also are used to examine changes in mRNA levels in bacterial grown in broth under conditions that are relevant to infection (Wilson *et al.*, 1999). To obtain high quality data, it is important to conduct a time course experiment or to titrate the condition of interest. It is also important to compare RNA from bacteria under a variety of related conditions in order to identify specific expression profiles (Krodursky *et al.*, 2000). For example, in a study examining the response of *Sacchromyces cerevisiae* to stress, researchers identified genes with expression levels that change under multiple kinds of stress as well as those that change in response to a particular kind of stress (Gasch *et al.*, 2000).

We have included a protocol for isolating total bacterial RNA and two protocols for converting RNA into labeled cDNA. One uses random hexamers to prime the RNA and one uses gene-specific primers. In the random hexamer protocol, aadUTP is incorporated into the cDNA, which is then labeled with a monofunctional Cy3/C5 dye. The gene-specific primer protocol features direct labeling of the cDNA with Cy3/5 conjugated dUTP. Since the monofunctional Cy3/Cy5 dyes are approximately ten times cheaper than the Cy3/5 dUTP, they likely will become the labeling system of choice in the future, and investigators should check the web sites listed at the end of this chapter for more current protocols. The two labeling protocols also include somewhat different hybridization procedures since they were developed in different laboratories.

Isolation of total RNA from *E. coli*

A protocol for purifying total RNA from bacteria is described below. It uses 4 M guanidinium thiocyanate and acidic phenol to isolate RNA quickly in a single step (Chomczynski, 1993). The RNA remains soluble in the guanidinium thiocyanate, while most proteins and <10 kb DNA fragments remain in the organic phase and large DNA fragments and some proteins remain at the interface (Krieg, 1996). A guanidinium thiocyanate/phenol solution is available commercially (TRIZOL, GibcoBRL 15596) and users can follow the manufactor's instructions. We have also had success with total RNA isolated with commercial kits, such as the Quiagen Rneasy mini bacterial RNA isolation kit, followed by a DNase step.

Perform all steps quickly to minimize RNA degradation. Use Rnase-free or DEPC-treated solutions and Rnase-free microfuge tubes and filter pipette tips. When possible, use Rnase-free plastic instead of glassware. Tris buffers cannot be treated with DEPC. Note that DEPC is a toxin and follow handling instructions on the bottle.

Before beginning, make the denaturing solution as follows. Dissolve 250 g of guanidinium thiocyanate in 293 ml of distilled water, 17.6 ml of 0.75 M sodium citrate, pH 7.0 and 26.4 ml of 10% sarcosyl at 65°C by stirring. Add 4.1 ml of 2-mercaptoethanol. Store at room temperature for up to one month (Krieg, 1996).

Isolation of Bacterial RNA (Chomczynski, 1993; Krieg, 1996)

1. Grow bacteria under appropriate conditions. Harvest the cells as quickly as possible, either by filtration through a 0.2 μm filter or centrifugation in a 15 ml conical tube. Freeze the filter (in 15 ml conical tube) or the pellet in liquid nitrogen immediately. Store at −80°C. The

following protocol is sufficient for approximately 2×10^9 bacterial cells and should yield approximately 50 μg of RNA.

2. Upon removal from −80°C, thaw on ice and add 1 ml of denaturing solution as soon as possible (when it will not freeze). Add sequentially 0.1 ml of 2 M sodium acetate pH4, 1 ml of water-saturated phenol and 0.2 ml of chloroform : isoamly alcohol (49 : 1). Homogenize with a power homogenizer such as the Polytron 1200 to lyse bacteria for approximately 30 s. There is no need to remove the filter.
3. Incubate at 4°C for 15 min. Centrifuge at 10 000*g* for 20 min at 4°C.
4. Transfer the aqueous phase to a new 15 ml conical tube. Be careful not to take material from the interface.
5. Add an equal volume (about 1 ml) of isopropanol. Store at −20°C for 30 min and centrifuge at 10 000*g* for 10 min at 4°C. Aspirate off supernatant.
6. Dissolve the pellet in 0.3 ml of denaturing solution and transfer to a 1.5 ml microfuge tube.
7. Precipitate the RNA with 1 volume (0.3 ml) of isopropanol at −20°C for 30 min. Centrifuge at 10 000*g* for 10 min at 4°C. Aspirate off supernatant.
8. Wash pellet with 1 ml 75% ethanol. Be sure to resuspend the pellet as much as possible to dissolve the salt. Let sit at room temperature for 15 min. Centrifuge at 10 000*g* for 5 min and aspirate off supernatant. Dry the pellet in a speed vacuum.
9. Dissolve pellet in Rnase-free water. Store at −80°C.
10. Calculate the concentration of RNA. Take 2 μl RNA sample and add to 800 μl of DEPC-treated water (do not use a Tris buffer). Use a clean quartz cuvette to avoid Rnase contamination. Read absorbance at 260 nm. Calculate concentration using the following formula:

    $$\text{Concentration} = (\text{A } 260)(400)(40\,000\ \mu\text{g}\ \mu\text{l}^{-1})$$

11. Determine the purity of the RNA. Take 2 μl RNA sample and add 800 μl 10 mM Tris-HCl pH 7.5. Read absorbance at 260 and 280 nm to determine A 260/A 280 ratio. Values between 1.8 and 2.1 indicate that the RNA is clean.
12. The integrity of the RNA should be analyzed on an Rnase-free agarose gel containing ethidium bromide. Load approximately 200 ng–1 ug of each sample and run the gel quickly in a gel-box dedicated to Rnase-free work to minimize degradation. Upon visualization under a UV lamp, both the 23S and 16S ribosomal bands should be distinct. Smearing towards smaller RNAs indicate significant sample degradation during preparation. Such samples should not be used on the microarrays.

Bacterial cDNA synthesis with Random Hexamers and aadUTP, followed by labeling with monofunctional Cy3/Cy5 and hybridization to microarrays

(Adapted from Virgil Rhodius, Carol Gross, Joe De Risi and Holly Baxter, and developed by Rosetta Inpharmatics, Kirkland, WA.)

Primer annealing and cDNA synthesis

Note—use Rnase-free tubes, pipette tips and solutions

1. Annealing step. In 0.5 ml microfuge tubes, mix 16 μg RNA with 10 μg random hexamer (5 μl of 2 μg μl^{-1}) in H_2O (DEPC treated, RNase free) to give a final volume of 20.5 μl.
2. Incubate mixture of RNA and hexamer at 70°C for 10 min.
3. Chill on ice for 10 min.
4. cDNA synthesis reaction: Make up following master mix:

Vol. per reaction (μl)	Reagent	
3	10× StrataScript RT Buffer	Stratagene
0.6	50× aa-dUTP/dNTP mix	1× = 500 μM each dA/dC/dG, 200 μM amino-allyl dUTP, 300 μM dTTP
2	StrataScript RNase H-RT	Stratagene; catalog #600085-51
0.3	RNase Inhibitor (40 U μl^{-1})	Boehringer Mannheim; catalog #799 017
3.6	H_2O	DEPC treated, RNase free
9.5	Total vol/tube	

5. Add 9.5 μl of the Master Mix to each RNA/hexamer mixture (20.5 μl) to give 30 μl final volume.
6. Incubate at room temperature for 10 min.
7. Incubate at 42°C for 1 h 50 min.

RNA hydrolysis

1. Add 10 μl 1 N NaOH (freshly prepared) and 10 μl 0.5 M EDTA (pH 8.0) to the 30 μl RNA/cDNA reaction.
2. Incubate at 65°C for 15 min.
3. Add 25 μl of 1 M Tris pH 7.5 to neutralize the reaction.

Cleanup using microcon-30 filters

Note—Tris inhibits the coupling reaction of the monofunctional NHS-ester Cy-dyes to the free amine groups of the aa-dUTP. Therefore it is essential to remove all the tris before proceeding with the coupling reaction. When using the Microcon filters, do not let sample spin dry: this results in loss of cDNA by binding to the filters.

1. Fill microcon-30 tube with 350 μl H_2O, add sample (~75 μl), and rinse reaction tube with 100 μl H_2O. (Total amount of H_2O added is 450 μl.)

2. Spin at 12 000 rpm for 7–8 min in a microfuge.
3. Check volume in upper chamber. It should be between 50 and 100 μl; if not, spin for additional 1–2 min. Recheck volume. Discard flow-through.
4. Wash two times by adding 450 μl H_2O to upper chamber and recentrifuging at 12 000 rpm for 7–8 min. Each time ensure the volume has reduced to 50–100 μl before proceeding.
5. Elute sample by placing the microcon inverted into a fresh microfuge tube. Centrifuge at 9000 rpm for 2 min.
6. Dry the sample in a speed vacuum (approx. 30 min). Do not over dry. Store dried samples at −20°C. The dried samples can be stored at −20°C for at least 1 month.

Coupling reaction of monofunctional dye (Cy3/Cy5) to aadUTP-cDNA

The Cy3/Cy5 dyes are light sensitive. Therefore, minimize light exposure where possible during the following procedures. In addition, the Cy3/Cy5-cDNA degrades over a few days. Only perform the coupling reaction if it is possible to directly proceed to the hybridization step and then on to scan the microarrays.

The Cy3 dye appears pink, but scans as 'green,' and the Cy5 dye appears blue, but scans as 'red.' By convention, the 'wild type' or control sample is labeled with Cy3 and the experimental sample is labeled with Cy5.

Before beginning, prepare Cy3 and Cy5 dyes by re-suspending each monofunctional reactive dye (Amersham Pharmacia Biotech, PA 23001 and 25001) in 10 μl DMSO. Aliquot 8 × 1.25 μl of each into separate microfuge tubes and dry in speed vacuum in the dark. Store at −20°C.

1. Resuspend cDNA pellet in 9 μl 0.05 M Na Bicarbonate pH 9.0. Let sit for 10–15 min at room temperature to ensure full resuspension.
2. Resuspend aliquot of dye in 1.25 μl DMSO.
3. Mix cDNA and dye together and incubate at room temperature for 1 h in the dark.
4. Add 4.5 μl 4 M hydroxylamine to sample to quench the reaction by preventing further cross-linking of the Cy3 or Cy5 to the cDNA.
5. Incubate at room temperature for 15 min in the dark.

Cleanup with QIA-quick PCR kit

1. Combine desired cDNA samples—one Cy3 sample (14 μl) and one Cy5 sample (14 μl).
2. Add 70 μl H_2O.
3. Add 500 μl PB buffer from kit. (Be sure to rinse original tubes containing the cDNA.)
4. Apply to QIA-quick column. Spin at 13 000 rpm for 30–60 s.
5. Dump flow-through. Add 750 μl PE buffer and spin at 13 000 rpm for 30–60 s.

6. Dump flow-through. Repeat PE buffer wash step.
7. Dump flow-through. Spin for 1 min at 14 000 rpm. (Filter may look a little pink, but it is not always possible to see color.)
8. Transfer to a fresh microfuge tube. Add 30 µl H_2O. Let sit 1 min. Spin at 13 000 rpm for 1 min and save the flow-through.
9. Add an additional 30 µl to column. Let it sit for 1 min. Spin at 13 000 rpm for 1 min and save the flow-through.
10. The solution should be purple at this point. If it is not, the labeling reaction did not work.
11. Dry sample in speed vacuum in the dark. (Approximately 30–40 min to dry.) Store the dried sample in the dark at 4°C. They are stable for 1–2 days.

Hybridization

(Hybridization conditions: cDNA from 16 µg total RNA, 15 µg poly(dI-dC), 3× SSC, 25 mM Hepes (pH 7.0), 0.225% SDS)

1. Set up the following hybridization mix.

   14.1 µl resuspended cDNA in H_2O (dissolve at 65°C, 1–2 min)
   3.3 µl 20× SSC
   2.5 µl poly (dI-dC) (6 µg/µl) (Sigma, P4929)
   1.1 µl 0.5 M Hepes pH 7.0
   21.0 µl Total volume

2. Pre-wet Millipore 0.45 µm membrane with 10 µl H_2O, place in 1.5 ml microcentrifuge tube and spin at 9000 rpm for 2 min in microcentrifuge.
3. Remove flow-through and deposit hybridization mix. Add two droplets to the inside of the Millipore filter case. Spin at 13 000 rpm for 2 min to elute and transfer sample to 0.5 ml microfuge tube.
4. Accurately, add 1 µl 5% SDS to each of the samples. Do not chill samples after addition of SDS; this will cause the SDS to precipitate.
5. Incubate samples at 95°C for 2 min. Allow samples to cool for 5–10 min at room temperature and spin down briefly.

Slide preparation

1. While samples are cooling, place slides in hybridization chamber and remove any dust using compressed air briefly.
2. Clean a Lifterslip (Erie Scientific, 22 × 25-21-4635) with EtOH soaked Kimwipes. Place slip on array using forceps (the dull white strips face downwards). Slowly apply the probe to one corner of the coverslip and let capillary action draw the probe over the entire array.
3. Add a total of 6–10 2 µl drops of 3× SSC at the two ends of the slides removed from the coverslip.

Sample application

1. After the samples have cooled, apply to the array by placing a pipette tip at one end of the coverslip and allow the sample to move up

underneath the coverslip by capillary action. Move the pipette tip repeatedly along the length of the coverslip to avoid any bubbles. Add sample to the other end of the coverslip once completely full underneath, to 'top up' both ends.

2. Place cover on the hybridization chamber and tighten the lid screws carefully to make watertight. Keep the chamber horizontal at all times so as not to disturb the 3× SSC droplets.
3. Carefully lower the hybridization chamber onto a plastic holder in a water bath.
4. Hybridize at 63–65°C for at least 5–6 h, or overnight (12 h max.).

Rinse step

1. Prepare wash solutions in glass slide dishes, with each dish having its own rack.

Wash solution 1:	340 ml Milli-Q water
	10 ml 20× SSC
	1 ml 10% SDS
Wash solution 2:	350 ml Milli-Q water
	1 ml 20× SSC

2. Remove array carefully from the water bath, keeping the chamber level. Dry the chambers with paper towels and 'wick' any water from the chamber seams. Unscrew the chamber and remove array slide.
3. First rinse: Rinse slide in Wash solution 1. Use forceps to move slide gently up and down in the solution until the coverslip is dislodged. Avoid allowing coverslip to scratch the surface of the array. Once coverslip is off and all the slides are in place, shake in solution by plunging rack up and down 10–20 times.
4. Second rinse: Individually transfer slides to Wash solution 2, blotting the base of the slide on a paper towel to avoid carrying over too much SDS. Shake gently in solution a few times.
5. Remove excess liquid by blotting the rack on a paper towel, and then dry array at room temperature by centrifuging at 600 rpm for 5 min.
6. Scan array as soon as the dyes are unstable and degrade differentially.

Bacterial cDNA synthesis with gene-specific primers

When gene-specific primers are available, we use them instead of random hexamers, as they yield a higher signal-to-noise ratio. Gene-specific primers can be generated by pooling the 3-prime primers used to make the microarray. A less expensive way of generating gene-specific primers uses a computer algorithm to predict the minimum number and sequence of short (7–8 nucleotides) primers needed to specifically anneal to all open-reading frames in a given genome. This program found that 37 oligonucleotides, called genome-directed primers, were sufficient to prime all

Mycobacterium tuberculosis genes and generated higher signal on microarrays than random primers (Talaat *et al.*, 2000).

Bacterial cDNA synthesis and direct labeling with gene-specific primers and Cy3/5 dUTP followed by hybridization to microarrays (Peterson *et al.*, 2000)

1. Mix 2 μg total RNA, 1 μg gene-specific primer mix, and water to 10 μl. Incubate at 65°C 10 min, place on ice for 2 min.
2. Add:
   4 μl 5× 1st strand buffer (comes with Superscript II)
   2 μl 0.1 M DTT
   1 μl 0.5 mM dA,C,G, 0.25 mM dUTP
   1 μl 1 mM cy3 or cy5-dUTP (Amersham PA53022 and PA55032)
   2 μl Superscript II (GibcoBRL, 18064-014, 200 U/μl)
   Incubate for 10 min at room temperature followed by 110 min at 42°C.
3. Hydrolyze RNA with 1 μl 1 M NaOH; incubate at 65°C for 10 min. Neutralize with 1 μl 1 N HCl.
4. Combine Cy5/Cy3 reactions. Add 450 μl TE and 1 μl 25 mg ml^{-1} yeast tRNA.
   Add mixture to Microcon YM30 membrane filtration unit.
5. Spin for 8 min at 10 000 rpm in microfuge (NOT highest speed; keep under 14 000 rpm). Discard flow-through.
6. Add 450 ml fresh TE. Spin for 8 min at 10 000 rpm in microfuge (NOT highest speed). Discard Flow-through.
7. Add 450 ml fresh TE. Spin for 8 min at 10 000 rpm in microfuge (NOT highest speed). Spin sample down until volume <10 μl.
8. Invert Microcon unit in fresh collection tube. Collect retained sample by spinning at high speed for 1 min. Some color is usually evident in sample, although it may be faint.
9. Bring up volume to 12 μl with fresh TE. Add 2.55 μl 20× SSC and 0.45 μl 10% SDS. Incubate at 100°C for 2 min. Note, these volumes are for a 22 × 22 mm coverslip.
10. Spin for 2 min at high speed at room temperature to pellet any particulate matter.
11. Place 15 μl (for a 22 × 22 mm coverslip) sample on slide, avoiding particulate matter at bottom of the tube. Cover with coverslip. Place four dots of 3× SSC (total volume 15 ml) on four corners of slide. Hybridize in sealed chamber O/N at 65°C.
12. Wash slides as follows. Place in slide rack. Rinse in 200 ml 0.03% SDS, 2× SSC in a slide chamber. Wait until the coverslips fall off.
13. Remove slides and touch edges to a paper towel. Transfer slides to fresh slide rack. Rinse in 200 ml 1× SSC, 5 min, rocking, in a fresh slide chamber.
14. Move rack to a fresh slide chamber containing 200 ml 0.2× SSC. Let rock for 5 min. Spin slides dry in a rack on a paper towel, for 5 min at 500 rpm. Scan slides on a commercial scanner, such as the Axon GenePix 4000a Microarray Scanner.

Pathogen gene expression in tissue culture or animal models

Microarrays also can be used to study pathogen gene expression in tissue culture cells. For example, researchers studying CMV used microarrays to examine viral gene expression during a time course of tissue culture cell infection. They were able to construct a temporal map of gene expression that included genes that had not been known previously to be temporarily regulated (Chambers *et al.*, 1999).

Either gene-specific primers or genome-directed primers, both of which are described in the previous section, can be used to specifically label bacterial RNA in the presence of excess host RNA. Hybridization temperature influences the amount of non-specific signal obtained, as demonstrated by experiments using different ratios of bacterial RNA to mammalian RNA (Talaat *et al.*, 2000). We have not yet worked out a protocol for labeling bacterial RNA in the presence of eukaryotic RNA, and we think that conditions will vary depending on the ratio of bacterial to eukaryotic RNA and on the primer set used. We recommend that interested workers check http://bugarrays.stanford.edu/or other web sites for updates.

◆◆◆◆◆◆ MONITORING HOST mRNA EXPRESSION

Host microarrays consisting of either the entire host genome or a relevant subset of genes (such as lymphocyte specific) will reveal much about the course of bacterial infection. Early forays into the global examination of host expression necessarily relied on partial genome arrays but yielded important information (Eckmann *et al.*, 2000; Ichikawa *et al.*, 2000; Rosenberger *et al.*, 2000). For example, a comparison of a mouse macrophage cell-line response to wild-type *Salmonella typhimurium* and to purified LPS indicated that most of the observed transcriptional response to the bacteria is LPS mediated (Rosenberger *et al.*, 2000).

Either total or poly-A RNA can be isolated from tissue culture cells or tissue and processed for hybridization to microarrays. An advantage of isolating total RNA from infected cells is that the bacterial RNAs will also be in the preparation and may be processed for hybridization to a pathogen array. Infected tissue culture cells can be harvested by pouring off the media and directly adding Trizol (GibcoBRL, #15596) to the tissue culture flask. Total RNA can be isolated according to the manufacturer's instructions. If the cells (or bacteria) are in suspension, they can be harvested by filtration through a 2 μm pore filter. We recommend using a power homogenizer (such as the Polytron 1200) to lyse cells grown in culture as well as tissues to increase yield. After precipitating the RNA, the amount and quality should be determined as described in the bacterial RNA protocol above (steps 15 and 16). The 28S and 18S ribosomal bands will be prominent upon UV-visualization of the agarose gel if the RNA is intact. If the agarose gel reveals genomic DNA contamination, use a higher Trizol to sample ratio.

A reference RNA can be generated from RNA pooled from multiple cell lines under a variety of conditions. Alternatively, a small amount of RNA from each sample in an experiment can be pooled and used as the reference. The important factors are simply that the same reference be used for all of the samples within an experiment and that this reference yield adequate signal from the majority of the spots on the microarray. If data are to be compared between experiments hybridized with different reference RNA pools, it will be important to calibrate the two reference pools to each other by hybridizing them both to the same microarray. Also, we perform one large reference RNA labeling reaction instead of multiple small ones to reduce variation between arrays.

Direct labeling of total eukaryotic RNA

Adapted from http://genome-www4.stanford.edu/MicroArray7/SMD/

1. Mix 20 μg of total RNA, 4 μg of oligo dT primer (TTTTTTTTTTTTTTTTTTTTVN, HPLC purified) in a total of 10 μl of RNase-free dH2O. Heat to 65°C for 10 min. Cool on ice.
2. Add
   3 μl 5× first strand buffer (comes with Superscript II)
   1.5 μl 0.1 M DTT
   0.3 μl dNTPs (25 mM dATP, dGTP, dCTP; 15 mM dTTP)
   1 μl Superscript II (Gibco-BRL 18064-014, 200 U/uL)
   1.5 μl Cy3 or Cy5 (Amersham, PA53022, PA55022)
   2.7 μl dH_2O
   Incubate at 42°C for 1 h. Add an additional microliter of Superscript II and incubate at 42°C for another hour.
3. Add 15 μl 0.1 M NaOH. Incubate for 10 min at 70°C. Add 15 μl 0.1 M HCl.
4. Combine Cy5/Cy3 reactions.
   Add
   450 μl TE
   1 μl 25 mg ml^{-1} yeast tRNA
   20 μg of Human cot-1 DNA (Gibco-BRL, 15297-011)
   20 μg of poly dA18 (Pharmacia, 27-4110-01)

   Add mixture to Microcon YM30 membrane filtration unit.
5. Follow steps 5–14 of the Bacterial cDNA synthesis and direct labeling with gene-specific primers protocol above.

◆◆◆◆◆◆ ANALYSIS OF MICROARRAY DATA

There are several programs available for identifying and measuring the signal on microarrays. We have used the Axon GenePix and Stanford Scanalyze programs. Scanalyze is available on the web at http://genome-www4.stanford.edu/MicroArray/SMD/.

Once the red (Cy5) and green (Cy3) measurements have been collected, they will need to be stored in a database that allows access to and manipulation of data. This may be best accomplished by a large consortium of laboratories, preferably with parallel interests, so that data generated by different investigators can be compared easily. However, individual laboratories also can use AMAD, a flat file database designed for microarray data that can be downloaded from http://microarrays.org. In the near future, this web site will make available a similar relational microarray database called NOMAD.

Before analyzing data, it is important to normalize the red signal on each microarray with the green signal so that data from different arrays can be compared. One way to do this is to set the medians of the green and red signals to one. Investigators will then need to consider which parameters will be used to assess the quality of the spots and to process their data accordingly. This is discussed in the Scanalyze manual. We also recommend collaborating with scientists who have prior microarray experience and with statisticians.

A common way of analyzing microarray data is to filter data based on, for example, mRNA level changes, and then to cluster the remaining genes. Two ways to cluster the genes are hierarchically and with self-organizing maps, both of which can be done with the Cluster program (Eisen *et al.*, 1998; Tamayo *et al.*, 1999). TreeView, the companion to cluster, graphically displays Cluster results and allows image generation for publication. Both programs are available at http://genome-www4.stanford.edu/MicroArray/SMD/.

◆◆◆◆◆◆ VERIFICATION OF MICROARRAY RESULTS

Microarray experiments yield both false-positives and false-negative signals (Lee *et al.*, 2000). It is therefore critical to verify results. First, experiments must be completely repeated, preferably at least three times, on different occasions. Second, an independent method for checking DNA or mRNA level differences is needed, such as a Southern, Northern, or real-time reverse-transcription PCR. Most importantly, researchers will want to determine the biological relevance of their results with genetic or biochemical experiments.

◆◆◆◆◆◆ *HELICOBACTER PYLORI* GENOME ANALYSIS

An example of the generation and use of bacterial array data is the recent study for which a *Helicobacter pylori* whole genome spotted array was constructed (Salama *et al.*, 2000). Each spot on the array corresponds to the unique segments of individual open reading frames and was generated by PCR using gene-specific primers. The entire genome was spotted twice onto the array to allow for comparison between duplicate spots.

To characterize the genomic diversity of fifteen *H. pylori* strains isolated from multiple sources, DNA was from each strain was isolated, labeled, and hybridized to the arrays as described above (see Figure 2.1). DNA from the strains used to make the microarray was used as a reference. Each DNA sample was hybridized to two or three arrays and the geometric mean of the normalized red/green ratio was calculated from the resulting two to six measurements of each gene. The normalized red/green ratio for each gene had to be above 0.5 in order for that gene to be considered present in a strain. The data were simplified into a binary score where a value of one indicated the gene is present and zero indicated the gene is absent or divergent. The programs Cluster and Treeview were used to hierarchically cluster and display the data (Salama *et al.*, 2000).

The authors found a set of genes (22% of the genome) that varied between *H. pylori* strains, and included restriction modification genes, transposases, and cell surface genes. Many of these genes were located in the two previously described plasticity zones, which are regions that have a different G-C content and contain multiple transposases and endonucleases, suggesting that they serve as insertion sites. However, two-thirds of the strain-specific genes were not in the plasticity zones, and of these, the outer membrane proteins and lipopolysaccharide biosynthesis genes were single-gene insertions. In contrast, restriction modification genes and transposases were most often in cassettes of 2–8 genes, suggesting that both of these gene classes can help additional genes insert with them into the genome. Putative virulence determinants were also identified, as seven genes that are not located near the *H. pylori* pathogenicity island (PAI) were found to be co-inherited with the PAI. One of these genes is babA, a known virulence determinant. Thus, microarray genomic analysis has not only yielded interesting and novel information about *H. pylori*, but has also identified putative virulence determinants (Salama *et al.*, 2000).

◆◆◆◆◆◆ MICROARRAY WEB SITES

Below is a short list of microarray web sites that include information on microarrays as well as links to other microarray web pages.

http://genome-www4.stanford.edu/MicroArray/SMD/
http://microarrays.org/
http://bugarrays.stanford.edu/
http://linkage.rockefeller.edu/wli/microarray/
http://www.genome.wi.mit.edu/MPR/

Acknowledgments

We thank Nina Salama, Karen Guillemin, and Sara Fisher for critiquing this manuscript.

References

Behr, M. A., Wilson, M. A., Gill, W. P., Salamon, H., Schoolnik, G. K., Rane, S. and Small, P. M. (1999) Comparative genomics of BCG vaccines by whole-genome DNA microarray [see comments] *Science* **284**, 1520–1523.

Chambers, J., Angulo, A., Amaratunga, D., Guo, H., Jiang, Y., Wan, J. S., Bittner, A., Frueh, K., Jackson, M. R., Peterson, P. A., Erlander, M. G. and Ghazal, P. (1999) DNA microarrays of the complex human cytomegalovirus genome: profiling kinetic class with drug sensitivity of viral gene expression *J. Virol.* **73**, 5757–5766.

Chomczynski, P. (1993) A reagent for the single-step simultaneous isolation of RNA, DNA and proteins from cell and tissue samples *Biotechniques* **15**, 532–534, 536–537.

Chu, S., DeRisi, J., Eisen, M., Mulholland, J., Botstein, D., Brown, P. O., Herskowitz, I. (1998) The transcriptional program of sporulation in budding yeast [published erratum appears in *Science* 1998 Nov 20; 282(5393): 1421] *Science* **282**, 699–705.

Eckmann, L., Smith, J. R., Housley, M. P., Dwinell, M. B. and Kagnoff, M. F. (2000) Analysis by high density cDNA arrays of altered gene expression in human intestinal epithelial cells in response to infection with the invasive enteric bacteria Salmonella *J. Biol. Chem.* **275**, 14084–14094.

Eisen, M. B. and Brown, P. O. (1999) DNA arrays for analysis of gene expression *Meth. Enzymol.* **303**, 179–205.

Eisen, M. B., Spellman, P. T., Brown, P. O. and Botstein, D. (1998) Cluster analysis and display of genome-wide expression patterns *Proc. Nat. Acad. Sci. USA* **95**, 14863–14868.

Gasch, A. P., Spellman, P. T., Kao, C. M., Carmel-Harel, O., Eisen, M. B., Storz, G. Botstein, D. and Brown, P. O. (2000) Genomic expression programs in the response of yeast cells to environmental changes *Molec. Biol. Cell* **11**, 4241–4257.

Hayward, R. E., Derisi, J. L., Alfadhli, S., Kaslow, D. C., Brown, P. O. and Rathod, P. K. (2000) Shotgun DNA microarrays and stage-specific gene expression in Plasmodium falciparum malaria *Molec. Microbiol.* **35**, 6–14.

Ichikawa, J. K., Norris, A., Bangera, M. G., Geiss, G. K., van't Wout, A. B., Bumgarner, R. E. and Lory, S. (2000) Interaction of pseudomonas aeruginosa with epithelial cells: identification of differentially regulated genes by expression microarray analysis of human cDNAs *Proc. Nat. Acad. Sci. USA* **97**, 9659–9664.

Iyer, V. R., Eisen, M. B., Ross, D. T., Schuler, G., Moore, T., Lee, J. C. F., Trent, J. M., Staudt, L. M., Hudson, J., Jr.Boguski, M. S., Lashkari, D., Shalon, D., Botstein, D. and Brown, P. O. (1999) The transcriptional program in the response of human fibroblasts to serum [see comments] *Science* **283**, 83–87.

Khodursky, A. B., Peter, B. J., Cozzarelli, N. R., Bolstein, D., Brown, P. O. and Yanofsky, C. (2000) DNA microarray analysis of gene expression in response to physiological and genetic changes that affect tryptophan metabolism in *Escherichia coli*. *Proc. Natl. Acad. Sci. USA* **97**(22), 12170–12175.

Krieg, P. A. (1996) A Laboratory Guide to RNA Wiley-Liss, New York.

Lee, M. L., Kuo, F. C., Whitmore, G. A. and Sklar, J. (2000) Importance of replication in microarray gene expression studies: statistical methods and evidence from repetitive cDNA hybridizations *Proc. Nat. Acad. Sci. USA* **97**, 9834–9839.

Peterson, S., Cline, R. T., Tettelin, H., Sharov, V. and Morrison, D. A. (2000) Gene expression analysis of the *Streptococcus pneumoniae* competence regulons by use of DNA microarrays *J. Bacteriol.* **182**, 6192–6202.

Rosenberger, C. M., Scott, M. G., Gold, M. R., Hancock, R. E. and Finlay, B. B. (2000) *Salmonella typhimurium* infection and lipopolysaccharide stimulation induce similar changes in macrophage gene expression *J. Immunol.* **164**, 5894–5904.
Salama, N., Guillemin, K., McDaniel, T. K., Sherlock, G., Tompkins, L. and Falkow, S. (2000) A whole-genome microarray reveals genetic diversity among *Helicobacter pylori* strains *Proc. Nat. Acad. Sci. USA* **97**, 14668–14673.
Spellman, P. T., Sherlock, G., Zhang, M. Q., Iyer, V. R., Anders, K., Eisen, M. B., Brown, P. O., Botstein, D. and Futcher, B. (1998) Comprehensive identification of cell cycle-regulated genes of the yeast *Saccharomyces cerevisiae* by microarray hybridization *Molec. Biol. Cell* **9**, 3273–3297.
Talaat, A. M., Hunter, P. and Johnston, S. A. (2000) Genome-directed primers for selective labeling of bacterial transcripts for DNA microarray analysis *Nature Biotechnol.* **18**, 679–682.
Tamayo, P., Slonim, D., Mesirov, J., Zhu, Q., Kitareewan, S., Dmitrovsky, E., Lander, E. S. and Golub, T. R. (1999) Interpreting patterns of gene expression with self-organizing maps: methods and application to hematopoietic differentiation *Proc. Nat. Acad. Sci. USA* **96**, 2907–2912.
Wilson, M., DeRisi, J., Kristensen, H. H., Imboden, P., Rane, S., Brown, P. O. and Schoolnik, G. K. (1999) Exploring drug-induced alterations in gene expression in Mycobacterium tuberculosis by microarray hybridization *Proc. Nat. Acad. Sci. USA* **96**, 12833–12838.

3 Promoter Traps and Related Methods of Identifying Virulence Factors

Micah J Worley and Fred Heffron
Department of Microbiology, Oregon Health Sciences University, Portland, Oregon 97201

◆◆

CONTENTS

◆◆◆◆◆◆ INTRODUCTION

Often there is little known about bacterial virulence factors except that they are expressed in and required for growth in the host. The focus of this chapter is to identify virulence factors via their expression with emphasis on *Salmonella*. Two new techniques—signature tagged mutagenesis (stm) and *in vivo* expression technology (IVET)—are only briefly described as they are the subject of excellent recent reviews (Chiang *et al.*, 1999; Slauch and Camilli, 2000).

Regulation of expression of a given virulence factor may take place at several levels—transcription, post-transcription, translation, or post-translation. Most often we think of transcriptional regulation of virulence factors. There are few examples of regulation at later steps probably because they have not been studied. In order to determine how an unknown gene is regulated it is necessary to construct transcriptional and translational fusions with a reporter gene. By definition a reporter must provide an advantage in determining expression of specific genes, usually a convenient assay as well as convenient visualization of the expressed phenotype on bacteriological plates. Important reporters include *lacZ* (β-galactosidase), the green fluorescent protein (gfp) from *Aequorea victoria*, the luciferase gene (Lux) from *Vibrio harvei*, the *E. coli*

METHODS IN MICROBIOLOGY, VOLUME 31
ISBN 0-12-521531-2

alkaline phosphatase gene (*phoA*; to identify membrane proteins), calmodulin-dependent adenyl cyclase (to identify expression within the host cell cytoplasm) and specific epitope tags such as 6× his. Selectable reporters such as certain biosynthetic genes and antibiotic resistance genes can be considered as still another class. A promoterless copy of a purine biosynthetic gene is used in one of the IVET selections as this is missing in the host and required for growth of many pathogens. We have developed a selectable reporter to identify genes that contain cleavage signal sequences via fusion to a truncated copy of the *Yersinia* invasin gene (Worley *et al.*, 1998).

There are two general strategies to construct and identify transcriptional and translational fusions to virulence factors. One strategy is to insert a PCR fragment or restriction fragment from a pathogen into a plasmid vector that contains a promoterless copy of a reporter (or a copy also missing its translational start site). The clone bank is then transformed back into the pathogen and examined for expression of the reporter. This has the advantage that the virulence properties of the strain are unlikely to be changed because the wild type gene remains intact. It has the disadvantage that construction of a mutation in a specific gene requires additional steps and, second, the regulation of the gene may not be faithfully duplicated because of the higher number of copies. An example is the IVET method. Alternatively, the reporter can be part of a transposon. Transcriptional or translational fusions are constructed by transposition and identified by screening or selecting for the appropriate expression. The advantage is that a mutation is simultaneously constructed allowing direct determination of the virulence properties of the strain. However, the mutation may affect virulence properties so much that measurement in the host is difficult. An example of this approach is Tn*phoA*. This transposon contains a truncated copy of alkaline phosphatase missing its signal sequence. It is used to identify membrane and periplasmic proteins via formation of active alkaline phosphatase as the active protein is only formed outside the cytoplasm of the cell.

The change that will most effect the way in which virulence factors are identified is the availability of complete DNA sequences of many pathogenic species. Sequence homology searches may identify probable virulence factors. In addition, the sequence allows the construction of DNA microarrays containing sequences that correspond to every gene in the bacterial genome. The array can be probed with RNA isolated from the pathogen to identify expression of various genes during infection. The disadvantage of microarrays for the study of signal transduction is that they only allow identification of genes that are transcriptionally regulated. The analysis of post-transcriptional signal transduction is still being developed (proteomics). DNA microarrays also allow analysis of genetic mutations at the level of the entire genome. Transposon insertions can be constructed throughout a genome and their location can be determined using microarrays. Regions in which no insertions are recovered define genes that are essential for cell viability. Regions, in which no insertions are recovered following infection of the host, define genes that are required for infection.

◆◆◆◆◆◆ IVET (*IN VIVO* EXPRESSION TECHNOLOGY), STM (SIGNATURE TAGGED MUTAGENESIS), AND GFP FUSIONS

It is necessary to describe these approaches to be able to compare and contrast with the methods described in the sections below.

Signature tagged mutagenesis

A type III protein secretion apparatus, the second, was identified in *Salmonella* by signature tagged mutagenesis (Hensel *et al.*, 1995; Shea *et al.*, 1996). In this procedure a collection of about 100 different transposons are constructed that can be distinguished by the presence of a DNA sequence tag. Each transposon is used to construct mutations and then pools of distinguishable transposons are used to infect the host. Following infection the strains are re-isolated and probed for the loss of specific transposons by polymerase chain reaction of their DNA sequence tag. Strains that are lost are likely to be attenuated in a specific step in infection. The single biggest advantage of stm is that it requires only an animal model and the ability to make transposon insertion for it to work. Two disadvantages of stm have become apparent from application to *Salmonella*. First, some steps in infection cause a severe restriction in the number of bacteria that pass to the next step. Only a few orally inoculated *Salmonella* bacteria go on to give systemic infection probably less than 10 (A. Baumler, pers. commun). Consequently when banks of 100 different mutants are fed to mice most cannot be identified in the spleen and liver and the mutants recovered will vary each time. Second, this approach identifies only non-redundant virulence factors. *Salmonella* is filled with redundant copies of genes (see below for examples).

In vivo expression technology (IVET, RIVET)

IVET identifies any gene that is induced within host tissue via fusion of its promoter to a promoterless copy of a gene required for growth *in vivo* (Mahan *et al.*, 1993). A newer version requires only transient expression of a recombination enzyme. The advantage of this technique is that it identifies even redundant virulence factors. However, many bacterial genes induced during infection encode metabolic functions or are merely responding to general stress. In fact, IVET identifies so many genes of dubious significance in understanding pathogenesis, the identification of interesting ones may be difficult.

In vivo expression of green fluorescent protein (gfp)

This was the subject of a recent review by Valdivia and Ramakrishnan (2000). Bacterial genes that are specifically expressed within host cells will include virulence factors as well as biosynthetic genes for nutrients that are in short supply within the eukaryotic cell cytoplasm or vacuole.

Nearly all screens for intracellular expression have been carried out with the green fluorescent protein (gfp) because it can be expressed within living cells without harming them, it is extremely bright, localization even within the cell is possible, and finally, derivatives have been constructed that fluoresce at different wavelengths. Clontech, Quantum Biotechnologies, and other companies are now carrying out most of this research. The one disadvantage of gfp is that its fluorescence is non-enzymatic and thus not as sensitive as a few other reported methods such as fluorescent substrates of β-lactamase (Zlokarnik *et al.*, 1998).

Construction of variants of gfp that are more fluorescent when excited by the Argon laser of the FACS or via the sources available for microscopy was essential because the minimum amount of protein necessary for detection was too high to be practical for many applications. The protein has been altered in a myriad ways. These alterations include: (1) absorbtion and emission spectrum; (2) stability and turnover; (3) folding to native structure; and (4) level of expression via alteration of codons to match those in the target organism (Tsien, 1998; much has also become proprietary). Cormack *et al.* (1996) generated a derivative of the green fluorescent protein (gfp) from *Aequorea victoria* that was 100 times more fluorescent than the native protein as determined in a fluorescent-activated cell sorter (FACS) of bacteria expressing the protein (Tsien, 1998; Valdivia and Ramakrishnan, 2000). The 20 amino acids surrounding the chromophore were mutagenized by a scheme that results in equal probability of change to any amino acid (Cormack and Struhl, 1993). Following multiple rounds of selection by FACS, mutants were isolated that had a shifted absorption maximum to 488 nm and that did not form inclusion bodies. These derivatives were better suited for FACS because the argon laser that is commonly used excites at 488 nm. They have used two of these mutants to construct both promoter trap vectors and a transposon containing a promoterless copy of gfp to identify bacterial proteins expressed following internalization (Valdivia and Falkow, 1997). A mini Tn5 Km cassette containing gfpmut3 (Cormack *et al.*, 1996) was inserted into the suicide vector pGP704. This plasmid is self-transmissible but cannot replicate outside the host SM10lambda pir. Mixing the strain containing this plasmid with a recipient strain and subsequently selecting for kanamycin resistance selects transposition events (see also Suarez *et al.*, 1997 for mTn3-based constructs). The disadvantages of gfp depend largely on the gfp version that is used.

◆◆◆◆◆◆ IDENTIFICATION OF VIRULENCE DETERMINANTS USING CONDITIONS SIMILAR OR IDENTICAL TO THOSE IN THE HOST

This approach has long been used successfully, however, it often depends on luck or ones ability to think like a pathogen. As long as the environment of a particular nitch within the host is known it is possible to design a screen that will mimic it, however the precise environment is usually not known.

Environmental conditions previously identified

Earlier screens in the Mekalanos laboratory for *Vibrio cholera* genes expressed during infection used the pH and salt concentration characteristic of the small bowel that resulted in expression of cholera toxin. The screens identified adherence factors required for colonization of the intestine (Peterson and Mekalanos, 1987; Taylor *et al.*, 1987). In the process, they constructed several broad host range suicide plasmids for delivery of Tn*phoA* that are useful in Gram negative bacteria (Taylor *et al.*, 1989). In *Yersinia* the conditions that appear to mimic the host are temperature and low calcium concentration. These conditions were identified many years ago as they allow secretion of proteins into the media later shown to be secreted by the type 3 apparatus. However, the environmental cues for *Yersinia* that are sensed on cell contact are not known nor are the sensors and how the signal is transduced. Low iron concentration within the host is also commonly used as an environmental cue. For example *Cornyebacterium diphtheria* toxin and shiga toxin encoded by *Shigella dysenteriae* are expressed only in low iron media. Surprisingly, high iron concentration induces the *Salmonella* genes that are critical for resistance to defensins. It will be interesting to see at which step(s) in infection these *Salmonella* genes are induced.

The signal transduction network leading to virulence gene expression is best understood in *Salmonella*. This knowledge will ultimately help us to identify many new virulence factors, to understand the entire network of signal transduction, and to anticipate new microenvironments in the host in which *Salmonella* may reside. Two major enviromental conditions that induce different virulence factors have been identified. One set of conditions are those present within the small bowel that stimulate expression of the genes for cell invasion and destruction of M cells. These conditions include high salt concentration (regulated by *ompR/envZ*), high divalent ion concentration (*phoP/phoQ*), low oxygen tension, growth phase, certain fatty acids (*fadD*) and the presence of other intestinal bacteria (*sirA/barA*; Ahmer *et al.*, 1999; Behlau and Miller, 1993; Lee and Falkow, 1990; Schiemann and Shope, 1991; Tartera and Metcalf, 1993; Rakeman and Miller, 1999; Altier *et al.*, 2000). Another set of conditions are characteristic of the phagosome of cells and stimulate transcription of a different set of genes (described below).

By first identifying mutations that increased the rate of cell invasion and then identifying genes that acted upstream of the first gene identified, Lee and her collegues identified four transcription regulators that integrate different environmental signals (Lee *et al.*, 1992). These *Salmonella* pathogenicity island 1 (spi 1) encoded transcriptional regulators include *hilA*, *hilC*, *hilD*, and *invF*. They act in the order *hilC/hilD* activate *hilA* which in turn activates *invF*. *HilA* was identified first and has been studied in most detail. It is related to regulators of the two component regulator family such as *toxR* in *Vibrio cholera*. Lee's laboratory and others have identified genes that effect expression of *hilA* by constructing *lacZ* fusions to *hilA* itself and to *hilA* regulated genes and then making transposon Tn5 mutations that repress or induce expression (Lucas *et al.*, 2000; Fahlen *et al.* 2000). Tn mutagenesis is carried out via P22 transduction of an F′::Tn5. There is no homology between F and the chromosome, consequently kanamycin resistant transductants represent transposition events. Mutations that also showed a strong effect on cell invasion were further characterized. New genes identified include *phoB*, *fadD*, *fliZ*. In other studies *sirA*, *envZ*, and *phoP* were identified. Epistasis studies demonstrated that the various signals act independently to regulate *hilA*.

A change from acid to slightly alkaline pH was the most important inducing condition for spi I genes identified by Daefler (1999), however, he found that the three type III effectors studied were induced under slightly different conditions. In fact, one of the effectors, *sipC*, is secreted during stationary phase and secretion is independent of expression of the type III secretion apparatus. Certain *Shigella* type III effectors can also leak out even without the type III secretion apparatus being expressed (C. Parsot, pers. commun). A transient acidification may mimic what happens to *Salmonella* during passage through the bowel but is not convenient for identification of the induced genes by a simple genetic screen. DNA microarrays that contain *Salmonella* DNA and are probed with RNA prepared shortly after acidification would be ideal for this.

What happens if the environmental conditions do not faithfully mimic the conditions within the host? Results of the previous studies demonstrate that a given virulence factor is most strongly induced only when several environmental conditions have been met. Specific environmental conditions may have a small or large effect on expression and, furthermore, there are often differences in how various environmental regulators effect expression of genes in the same virulence regulon. An interesting suggestion from the study by Daefler (1999) is that the heterogeneity of response may indicate adaptation of *Salmonella* to many different nitches within the host. This also illustrates why, in spite of not having ideal induction conditions, mutant hunts have usually succeeded.

General strategy

The best way to identify virulence factors is to have identified several already, the scientific equivalent of Catch 22. Transcriptional and translational fusions to these genes are constructed first to identify environmental clues that can be used and subsequently to identify other genes that are similarly regulated. This is the approach that Hensel and his collaborators have used as described next (Deiwick *et al.*, 1999). Genetic fusion between *lacZ* and several spi 2 encoded genes were monitored to determine induction conditions. They found that low magnesium concentration and low phosphate were the cues that most closely mimicked the host. Stationary phase also increased expression but acid pH, such as would be encountered within the phagosome, stimulated export but not transcription of the genes studied. Again, like Daefler (1999) they found that the factors that effect expression differ slightly between different effectors even though they are exported via the same secretion apparatus. Presumably this is important in the pathogenic lifecycle of *Salmonella* perhaps in providing the timing that takes place during infection (e.g. acidification of the phagosome) or the adjustment to different locations within the host during systemic infection. The availability of several independent tests to determine the validity of various identified virulence factors was helpful. For example, spi 2 is not present in other closely related but nonpathogenic enteric bacteria suggesting an important role in systemic infection (*S. typhimurium* vs *S. bangorii*; Hensel *et al.*, 1997).

Similar approaches in *Listeria monocytogenes*, *Shigella* spp., *Staphylococcus aureus* and *Mycobacterium tuberculosis*

The first strategy that Portnoy and his colleagues used to isolate *Listeria* virulence factors was penicillin selection for mutants that did not grow inside the cell. They infected a macrophage cell line with transposon Tn916 mutagenized *Listeria* and killed dividing bacteria with methicillin (Camilli *et al.*, 1989). Most of the mutants lacked expression of listeriolysin O. This gene is absolutely essential for intracellular growth and there are several genes that effect its expression making this a good transposon target. In fact, Gahan and Hill (2000) used expression of listeriolysin as the basis for IVET selection in *Listeria*.

Identification of virulence factors in both *Shigella* spp. and *Listeria monocytogenes* is made easier by a simple plaque assay similar to the plaque assay used to identify viral mutants. When either organism infects a monolayer of cells, cell-to-cell spread takes place with destruction of cells in a contiguous region of the monolayer. The plaques are visualized by staining with a vital stain such as methylene blue. Following mutagenesis, the size and appearance of the plaques formed on the monolayer is used to detect mutations in essential virulence factors. The results of these studies have provided some spectacular insight into intracellular pathogenesis including the ability of *Listeria* to polymerize actin as a means of cell motility, to lyse the phagosome, and to grow in the cytoplasm (Sun *et al.*, 1990).

For *Shigella* the genetic techniques are identical to those used in *E. coli* because of the close similarity of the two organisms. The virulence determinants present on the large *Shigella* plasmid are sufficient to confer very similar virulence properties on *E. coli*. As a consequence most genetic screens in *Shigella* have been limited to the plasmid (Small and Falkow, 1988; Sasakawa *et al.*, 1988; but also see Okada *et al.*, 1991 and Sasakawa *et al.*, 1991 for screens of the chromosome).

Some of the same methods developed in *Salmonella* have been used to identify virulence factors in *Staphylococcus aureus*. Genetic screens for virulence factors have used modified versions of Tn917 containing a promoterless copy of *lacZ* (for useful derivatives of Tn917 see Camilli *et al.*, 1990).

Tn917 is widely used to make mutations in Gram positive organisms. *S. aureus* is one of the most important pathogens and has acquired high levels of antibiotic resistance making therapy difficult. Use of the promoter trap vectors such as IVET and a murine renal abscess model of infection has allowed identification of new virulence factors (Lowe *et al.*, 1998). Two independent studies using signature tagged transposon Tn917 but different animal models have identified different genes and come to different conclusions (Coulter *et al.*, 1998; Mei *et al.*, 1997). The first used a murine bacteremia model and identified many genes that had no sequence homologue in the data base. The second study used three different animal models and identified 237 avirulent mutants of which 10% were avirulent in all models tested. The genes that effected virulence in multiple models were mostly amino acid and peptide transporters thus identifying new targets for antibacterial therapies. The stm studies were carried out by introducing a plasmid that is temperature sensitive for replication and contained signature tagged Tn917 into the strain to be mutagenized. Each strain, containing a different signature tagged transposon, is grown first at the permissive temperature followed by growth at the nonpermissive temperature to select transposition events.

Mycobacterium tuberculosis (MTB) grows slowly, does not transform well, and is a class 3 or 4 pathogen according to the biosafety rules. Furthermore, until recently, allelic replacement was difficult (Kalpana *et al.*, 1991). Given these inherent disadvantages the progress in the last few years in developing new genetic approaches to the organism has been striking. Two methods have been developed to introduce transposons—one uses a defective mycobacterial bacteriophage (Bardarov *et al.*, 1997) and the other uses a temperature-sensitive mycobacterial plasmid (Guilhot *et al.*, 1992, 1994). A new derivative of the plasmid also contains the sacB or levan sucrase gene for negative selection (Pelicic *et al.*, 1997). This later plasmid is the best method of allelic replacement found in mycobacterium. With the development of genetic systems the Guilot laboratory has identified MTB virulence factors using STM (Camacho *et al.*, 1999).

◆◆◆◆◆◆ IDENTIFICATION OF GENES THAT ARE LOCATED UPSTREAM AND DOWNSTREAM OF AN IDENTIFIED VIRULENCE REGULATOR

Once a regulator has been identified the most obvious question is 'what does it regulate?' There are several strategies that can be used to answer this question all of which depend on conditional expression of the regulator.

Transcriptional fusions missing the regulator

The first method is to construct transcriptional fusions in a background missing the regulator and to compare expression once the regulator is returned. Following construction of a random library of 50 000 *lacZ* transcriptional fusions in *Salmonella*, the expression of β-galactosidase was compared between a wild type strain and a strain missing the regulator, *phoP* (Soncini *et al.*, 1996). *PhoP* was supplied from a plasmid that was transduced via bacteriophage P22. They identified 47 *phoP*-induced genes and seven repressed genes. In the Groisman screen they used minimal bacteriological media with relatively low magnesium concentration. Because it was later shown that phoQ senses magnesium concentration it is not clear if this experiment would have worked using rich media. The choice of media for these experiments will be discussed below.

Our laboratory has recently conducted two very successful screens for genes activated by regulators involved in virulence, which provide nice examples of this approach. By looking for genes activated by SirA, we identified new genes important in the intestinal phase of *Salmonella* disease (Ahmer *et al.*, 1999). We then identified new genes important in intracellular, systemic *Salmonella* disease by assaying for genes regulated by the transcription factor SsrB (Worley *et al.*, 2000).

SirA was identified as a response regulator important for optimal expression of SPI-1 genes (Johnston *et al.*, 1996). We hypothesized that SirA may play a broader role in enteropathogenesis than just activating genes within SPI-1. To this end, we conducted a global screen for SirA regulated genes, utilizing the strategy depicted in Figure 3.1. Briefly, we first inactivated the chromosomal copy of SirA through allelic exchange. We then introduced an ectopic SirA copy under the control of pBAD, a tightly regulated, inducible promoter. When grown with glucose as the carbon source, this strain does not produce SirA. However, when grown in the presence of arabinose, SirA is over-expressed from the plasmid. Thus, the presence of SirA is being controlled by carbon source. We then generated *lacZY* transcriptional fusions throughout the genome of this strain with MudJ mutagenesis. Next we patched the mutants in grids

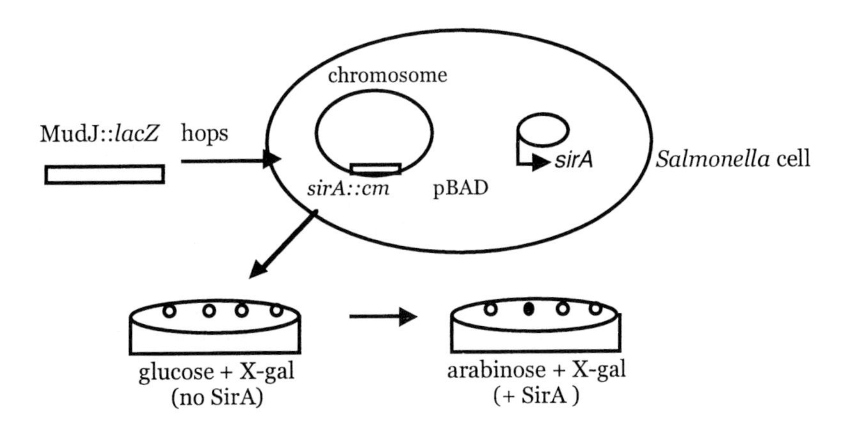

Figure 3.1. Strategy for identifying genes downstream of the *Salmonella* response regulator SirA. A strain was created in which SirA production is controlled by arabinose. *lacZY* transcriptional fusions were generated throughout the chromosome, and the mutants patched onto M_9 glucose versus M_9 arabinose. The colorimetric substrate X-gal allowed the observation of differences in gene expression in response to SirA.

onto M9 plates supplemented with glucose and X-gal and M9 plates supplemented with arabinose and X-gal, and observed color changes.

This screen revealed that SirA, in addition to regulating SPI-1 genes, also regulates virulence genes within SPI-4 and SPI-5. One of the most interesting results of this study was the discovery that SirA regulates these genes through HilA, which is encoded within SPI-1. Until recently, regulators encoded within pathogenicity islands were thought to exclusively activate genes within the islands in which they reside (Morschhauser *et al.*, 1994; Gomez-Duarte and Kaper, 1995).

This result led us to speculate that SsrB, a regulator encoded within SPI-2, may activate genes outside of SPI-2. To address this possibility, we performed a global screen for SsrB regulated genes, in a similar fashion to the one described above. This study identified a global regulon of SsrB regulated genes, within previously undescribed, horizontally acquired regions of the *S. typhimurium* chromosome (Figure 3.2). The extent of genetic cross-talk between horizontal acquisitions revealed in this study was unprecedented. Several new intracellular virulence genes were identified that we are currently characterizing.

There are several important factors to consider before undertaking a study like the two described above. Perhaps most important, is whether or not the organism being studied is amenable to genetic analysis. It must be possible to knock out genes, to introduce plasmids and also to generate reporter fusions. With a genetically intractable organism, this approach cannot be implemented. The choice of regulator is also important. Many regulators, such as PhoP, evolved as housekeeping regulators, and subsequently began activating virulence genes as well. With such a regulator, many genes will be identified that do not contribute to an organism's pathogenic character. The success of the SsrB screen was largely due to its eclectic nature—because it was never a housekeeping regulator, only putative virulence genes were identified. A third consideration is whether or not a regulator can be easily activated. Generally, a sensor kinase autophosphorylates in response to an environmental cue and then phosphorylates (and thus activates) its cognate response regulator, which then binds promoters to regulate transcription. Unphosphorylated response regulators presumably retain some residual affinity for their binding sites, and thus can be driven onto them by mass action when over expressed. However, this can lead to spurious results. For instance, in the SirA screen many genes identified responded to plasmid-encoded SirA, but not chromsomally encoded SirA. It is difficult to interpret the biological significance of such fusions.

While there are limitations to this approach, there are also advantages over other methods. Many *Salmonella* screens have assayed directly for virulence phenotypes—such as defects in the ability to enter normally non-phagocytic epithelial cells, defects in the ability to survive within normally microbicidal macrophages, and defects in the ability to kill in the murine model of Salmonellosis. One major shortcoming of these approaches is that the phenotypes of many virulence genes is masked by other genes with overlapping or even completely redundant functions. Perhaps the best example of such a gene is SopE. SopE is a type III effector

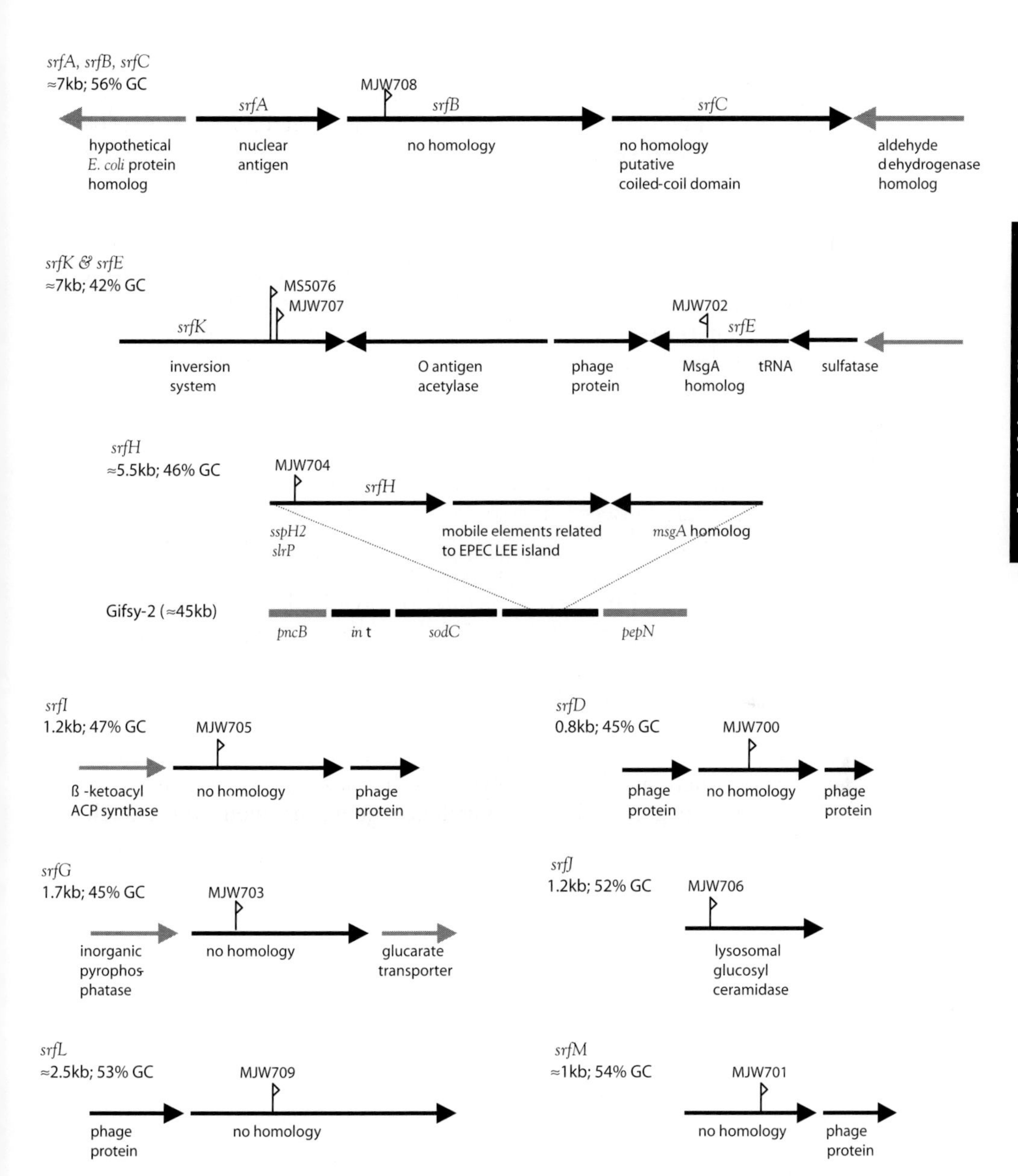

Figure 3.2. Loci identified in the SsrB screen. By globally screening for genes regulated by SsrB, several previously undescribed areas of the genome that are involved in pathogenesis were identified. Maps show as blocks the regions of the various horizontal acquisitions identified. The arrows indicate the orientation of open reading frames. Transposon insertion points are designated with flags that also indicate the transposon orientation. Gene identities or homologies are shown under the blocks. Allele designations, approximate length of the horizontal acquisitions and GC content are shown on the left. The borders of these horizontal acquisitions (if known) are in gray.

that directly engages the host cell signaling machinery, acting as a potent exchange factor for Cdc42 and Rac. This triggers a rearrangement of the actin cytoskeleton (ruffling) that results in bacterial internalization (Hardt *et al.*, 1998). However, SopE mutants are proficient invaders, due in part to complete functional redundancy with a homolog termed SopE2 (Stender *et al.*, 2000). The phenotype of SopE is further masked by a partial functional redundancy with SopB, whose inositol phosphatase activity also promotes bacterial entry. *Salmonella* strains lacking any one of these three effectors can efficiently enter epithelial cells, whereas a strain missing all three is completely defective for entry (Zhou *et al.*, 2001). Important virulence genes such as these cannot be identified by assaying for virulence defects, but can be identified with regulatory screens.

Constitutive mutants

If expression of *phoP*-regulated genes depends not only on the expression level of PhoP but also on its phosphorylation state then it may be preferable to use a mutant of *phoP* that acts as if it is phosphorylated all of the time. A *phoP* constitutive mutant was reported by Kier *et al.* (1979) and was used in a screen similar to that described above (Miller and Mekalanos, 1990; Behlau and Miller, 1993; Belden and Miller, 1994). These authors identified 13 pag genes induced by *phoP*^{c} and 5 repressed prg genes. Many of the genes identified by Miller and Mekalanos were different than those reported by Soncini *et al.* (1996). The success of this approach suggests that isolating a constitutive mutant of the response regulator may be helpful later on. Constitutive mutants of regulators are not difficult to isolate although it requires two unlinked regulated genes as a first step.

Other approaches

The authors' laboratory has tried unsuccessfully to identify the genes regulated by *ompR/envZ* because mutations in either of these two genes strongly effect systemic infection in the mouse. Few *ompR*-regulated genes have already been reported. Gibson *et al.* (1987) identified only three regulated genes. In the authors' laboratory, constructions were made in which *ompR* was conditionally regulated by an arabinose promoter. Induction of expression of *ompR* with arabinose was lethal for the bacteria and partial induction did not yield interesting mutants. Constructions using a hybrid sensor kinase that responds to aspartate (thus controlled by the amount of aspartate in the medium) also failed to identify *ompR* regulated genes. In fact, we could find no condition in which we recovered *ompR* regulated genes although, by this time, we knew that many virulence genes such as *ssrB* are regulated by *ompR* in *Salmonella*. A biochemical approach, although old fashioned, was successful (Oakley *et al.*, 1975; Worley, Heffron and Kenney, unpubl. obs.). Purified ompR or phospho-ompR was first attached to a nitrocellulose filter followed by *Salmonella* DNA restriction fragments to identify those to which *ompR* binds. This approach has identified many new ompR regulated genes and detailed characterization of a few of these has confirmed the regulation. A biochemical approach was also used in *Staphylococcus* to identify new virulence genes.

Tegmark *et al.* (2000) coupled DNA containing specific promoter sequences for known virulence factors to magnetic beads and used increasing salt concentration to identify regulatory proteins that bound tightly to the promoter. The genes corresponding to the bound proteins were identified by N terminal analysis of the purified protein. These authors identified a new virulence regulator (sarA) that had not been found by a genetic approach by any of the genetic approaches described above.

Characterization of regulatory networks

Precisely, what is the signal transduction network between environmental sensors and individual virulence factors? New approaches have been used to study the promoters for virulence genes and these studies have led not only to a better understanding of regulation but also to the identification of new virulence factors.

Salmonella virulence factors that are brought in via horizontal spread invariably make use of one of the several highly conserved regulatory pathways already present in the host bacterium. Those newly acquired genes may include regulators that integrate signals in new ways allowing identification of specific environments within a host (hilA, hilC hilD and invF). Among the global regulators that have been used by pathogens are phoP/phoQ, ompR/envZ, fadD, fruR but there are likely to be many more. Consider one example for how this has evolved from *Salmonella*. Systemic *Salmonella* infection requires the virulence regulator *ssrA/B* encoded within *Salmonella* pathogenicity island 2. Lee and co-workers (2000) showed that *ssrA/B* is regulated by *ompR* after Lundgren *et al.*, (1997) first showed that *ompR* had a dramatic effect on *Salmonella* interaction with cells. Spi 2 encoded effectors are negatively regulated by high salt concentration. At high salt concentration *ompR* is phosphorylated and binds to specific sites in the promoter region of *ssrB* to inhibit expression of this major virulence activator. At the same time it positively regulates ssrA expression (L. Kenney, pers. comm.)! At low osmolarity, as found within host cell phagosomes, the unphosporylated form of *ompR* bind to the ssrB promoter to activate transcription of *ssrB* (L. Kenney, pers. comm.). It appears that *ompR* both activates and represses virulence genes. The fact that *ssrA* and *ssrB* are differently regulated by *ompR* has interesting implications for how this sensor kinase may interact with other regulatory networks in the cell. This study demonstrates how detailed understanding of regulation requires biochemistry and may lead to the discovery of new genes.

◆◆◆◆◆◆ CONCLUSIONS

Almost all analysis of virulence rests on the availability of a good animal model without which virulence factors are identified solely by analogy with genes from other pathogens with no method of verification. An organism that is highly refractory to transformation, such as *Chlamydia*, is almost as perverse as not having an animal model. Manipulations must be carried out in *E. coli* or other organism that are easy to manipulate and the genes re-introduced only as a last resort.

Several new developments will impact the way that virulence factors are identified. *In vitro* transposition reactions for Tn5 and mariner allow transposition events to be selected in most hosts. The complete transposition reaction may be carried out *in vitro* or alternatively a partial reaction is carried out and the active transposition complex returned to the cell by electroporation. In this case, enzymes present within the cell complete the transposition reaction. The availability of DNA microarrays containing sequences for each gene in a pathogen and its host will revolutionize our understanding of host–parasite interaction and lead to the discovery many new genes and pathways.

Identification of virulence factors via promoter traps or by looking upstream or downstream of identified virulence regulators is ideal for pathogens such as *Salmonella* where there are alternative genes for almost every aspect of infection and many bone fide virulence factors may never be found in any other way. But the identification of new virulence factors is always only the starting point for research.

References

Ahmer, B. M., van Reeuwijk, J., Watson, P. R., Wallis, T. S. and Heffron, F. (1999). Salmonella SirA is a global regulator of genes mediating enteropathogenesis. *Mol. Microbiol.* **31**(3), 971–82.

Behlau, I. and Miller, S. I. (1993). A PhoP-repressed gene promotes Salmonella typhimurium invasion of epithelial cells. *J. Bacteriol.* **175**, 4475–84.

Belden, W. J. and Miller, S. I. (1994). Further characterization of the PhoP regulon: identification of new PhoP-activated virulence loci. *Infect. Immun.* **62**(11), 5095–101.

Camacho, L. R., Ensergueix, D., Perez, E., Gicquel, B. and Guilhot, C. (1999). Identification of a virulence gene cluster of Mycobacterium tuberculosis by signature-tagged transposon mutagenesis. *Mol. Microbiol.* **34**(2), 257–67.

Camilli, A., Portnoy, A. and Youngman, P. (1990). Insertional mutagenesis of Listeria monocytogenes with a novel Tn917 derivative that allows direct cloning of DNA flanking transposon insertions. *J. Bacteriol.* **172**(7), 3738–44.

Camilli, A., Paynton, C. R. and Portnoy, D. A. (1989). Intracellular methicillin selection of Listeria monocytogenes mutants unable to replicate in a macrophage cell line. *Proc. Natl. Acad. Sci. USA* **86**(14), 5522–6.

Chiang, S. L., Mekalanos, J. J. and Holden, D. W. (1999). In vivo genetic analysis of bacterial virulence. *Annu. Rev. Microbiol.* **53**, 129–54.

Cormack, B. P. and Struhl, K. (1993). Regional codon randomization: defining a TATA-binding protein surface required for RNA polymerase III transcription. *Science* **262**(5131), 244–8.

Cormack, B. P., Valdivia, R. H. and Falkow, S. (1996). FACS–optimized mutants of the green fluorescent protein (GFP). *Gene* **173**(1 Spec No), 33–8.

Coulter, S. N., Schwan, W. R., Ng, E. Y., Langhorne, M. H., Ritchie, H. D., Westbrock-Wadman, S., Hufnagle, W. O., Folger, K. R., Bayer, A. S. and Stover, C. K. (1998). Staphylococcus aureus genetic loci impacting growth and survival in multiple infection environments. *Mol. Microbiol.* **30**(2), 393–404.

Daefler, S. (1999). Type III secretion by Salmonella typhimurium does not require contact with a eukaryotic host. *Mol. Microbiol.* **31**, 45–51.

Deiwick, J., Nikolaus, T., Erdogan, S. and Hensel, M. (1999). Environmental regulation of Salmonella pathogenicity island 2 gene expression. *Mol. Microbiol.* **31**(6), 1759–73.

Fahlen, T. F., Mathur, N. and Jones, B. D. (2000). Identification and characterization of mutants with increased expression of hilA, the invasion gene transcriptional activator of Salmonella typhimurium. *FEMS Immunol. Med. Microbiol.* **28**(1), 25–35.

Gahan, C. G. and Hill, C. (2000). The use of listeriolysin to identify in vivo induced genes in the gram-positive intracellular pathogen Listeria monocytogenes. *Mol. Microbiol.* **36**(2), 498–507.

Gibson, M. M., Ellis, E. M., Graeme-Cook, K. A. and Higgins, C. F. (1987). OmpR and EnvZ are pleiotropic regulatory proteins: positive regulation of the tripeptide permease (tppB) of Salmonella typhimurium. *Mol. Gen. Genet.* **207**(1), 120–9.

Gomez-Duarte, O. G. and Kaper, J. B. (1995). A plasmid-encoded regulatory region activates chromosomal *eaeA* expression in enteropathogenic. *Escherichia coli. Infect. Immun.* **63**, 1767–1776.

Guilhot, C., Otal, I., Van Rompaey, I., Martin, C. and Gicquel, B. (1994). Efficient transposition in mycobacteria: construction of Mycobacterium smegmatis insertional mutant libraries. *J. Bacteriol.* **176**(2), 535–9.

Guilhot, C., Gicquel, B. and Martin, C. (1992). Temperature-sensitive mutants of the Mycobacterium plasmid pAL5000. *FEMS Microbiol. Lett.* **77**(1–3), 181–6.

Hardt, W. D., Chen, L. M., Schuebel, K. E., Bustelo, X. R. and Galan, J. E. (1998). *S. typhimurium* encodes an activator of Rho GTPases that induces membrane ruffling and nuclear responses in host cells. *Cell* **93**, 815–26.

Hensel, M., Shea, J. E., Gleeson, C., Jones, M. D., Dalton, E. and Holden, D. W. (1995). Simultaneous identification of bacterial virulence genes by negative selection. *Science* **269**, 400–3.

Johnston, C., Pegues, D. A., Hueck, C. J., Lee, A. and Miller, S. I. (1996). Transcriptional activation of Salmonella typhimurium invasion genes by a member of the phosphorylated response-regulator superfamily. *Mol. Microbiol.* **22**(4), 715–27.

Kalpana, G. V., Bloom, B. R and Jacobs, W. R. Jr, . (1991). Insertional mutagenesis and illegitimate recombination in mycobacteria. *Proc. Natl. Acad. Sci. USA* **88**(12), 5433–7.

Kier, L. D., Weppelman, R. M. and Ames, B. N. (1979). Regulation of nonspecific acid phosphatase in Salmonella, phoN and phoP genes. *J. Bacteriol.* **138**(1), 155–61.

Lee, A. K., Detweiler, C. S. and Falkow, S. (2000). OmpR regulates the two-component system SsrA-ssrB in Salmonella pathogenicity island 2. *J. Bacteriol.* **182**(3), 771–81.

Lee, C. A., Jones, B. D. and Falkow, S. (1992). Identification of a Salmonella typhimurium invasion locus by selection for hyperinvasive mutants. *Proc. Natl. Acad. Sci. USA* **89**(5), 1847–51.

Lindgren, S. W., Stojiljkovic, I. and Heffron, F. (1996) Macrophage killing is an essential virulence mechanism of Salmonella typhimurium. *Proc. Natl. Acad. Sci. USA* **93**, 4197–201.

Lowe, A. M., Beattie, D. T. and Deresiewicz, R. L. (1998). Identification of novel staphylococcal virulence genes by in vivo expression technology. *Mol. Microbiol.* **27**(5), 967–76.

Lucas, R. L., Lostroh, C. P., DiRusso, C. C., Spector, M. P., Wanner, B. L. and Lee, C. A. (2000). Multiple factors independently regulate hilA and invasion gene expression in Salmonella enterica serovar typhimurium. *J. Bacteriol.* **182**(7), 1872–82.

Mahan, M. J., Slauch, J. M. and Mekalanos, J. J. (1993). Selection of bacterial virulence genes that are specifically induced in host tissues. *Science* **29**, 686–698.

Mei, J. M., Nourbakhsh, F., Ford, C. W. and Holden, D. W. (1997). Identification of Staphylococcus aureus virulence genes in a murine model of bacteraemia using signature-tagged mutagenesis. *Mol. Microbiol.* **26**(2), 399–407.

Miller, S. I. and Mekalanos, J. J. (1990). Constitutive expression of the phoP regulon attenuates Salmonella virulence and survival within macrophages. *J. Bacteriol.* **172**(5), 2485–90.

Morschhauser, J., Vetter, V., Emody, L. and Hacker, J. (1994). Adhesin regulatory genes within large, unstable DNA regions of pathogenic *Escherichia coli*, cross-talk between different adhesin gene clusters. *Mol. Microbiol.* **11**, 555–566.

Oakley, J. L., Pascale, J. A. and Coleman, J. E. (1975). T7 RNA polymerase: conformation, functional groups, and promotor binding. *Biochemistry* **14**(21), 4684–91.

Okada, N., Sasakawa, C., Tobe, T., Talukder, K. A., Komatsu, K. and Yoshikawa, M. (1991). Construction of a physical map of the chromosome of Shigella flexneri 2a and the direct assignment of nine virulence-associated loci identified by Tn5 insertions. *Mol. Microbiol.* **5**(9), 2171–80.

Pelicic, V., Jackson, M., Reyrat, J. M., Jacobs, W. R. Jr, Gicquel, B. and Guilhot, C. (1997). Efficient allelic exchange and transposon mutagenesis in Mycobacterium tuberculosis. *Proc. Natl. Acad. Sci. USA* **94**(20), 10955–60.

Peterson, K. M. and Mekalanos, J. J. (1988) Characterization of the Vibrio cholerae ToxR regulon: identification of novel genes invoved in intestinal colonization. *Infect Immun.* **56**, 2822–9.

Rakeman, J. L. and Miller, S. I. (1999). Salmonella typhimurium recognition of intestinal environments. *Trends Microbiol.* **7**(6), 221–3.

Sasakawa, C., Kamata, K., Sakai, T., Makino, S., Yamada, M., Okada, N. and Yoshikawa, M. (1988). Virulence-associated genetic regions comprising 31 kilobases of the 230-kilobase plasmid in Shigella flexneri 2a. *J. Bacteriol.* **170**(6), 2480–4.

Schiemann, D. A. and Shope, S. R. (1991). Anaerobic growth of Salmonella typhimurium results in increased uptake by Henle 407 epithelial and mouse peritoneal cells in vitro and repression of a major outer membrane protein. *Infect. Immun.* **59**, 437–40.

Shea, J. E., Hensel, M., Gleeson, C. and Holden, D. W. (1996). Identification of a virulence locus encoding a second type III secretion system in Salmonella typhimurium. *Proc. Natl. Acad. Sci. USA* **93**(6), 2593–7.

Slauch, J. M. and Camilli, A. (2000). IVET and RIVET: use of gene fusions to identify bacterial virulence factors specifically induced in host tissues. *Meth. Enzymol.* **326**, 73–96.

Small, P. L. and Falkow, S. (1988). Identification of regions on a 230-kilobase plasmid from enteroinvasive Escherichia coli that are required for entry into HEp-2 cells. *Infect. Immun.* **56**(1), 225–9.
Soncini, F. C., Garcia Vescovi, E., Solomon, F. and Groisman, E. A. (1996). Molecular basis of the magnesium deprivation response in Salmonella typhimurium: identification of PhoP-regulated genes. *J. Bacteriol.* **178**(17), 5092–9.
Stender, S., Friebel, A., Linder, S., Rohde, M., Mirold, S. and Hardt, W. D. (2000). Identification of SopE2 from *Salmonella typhimurium*, a conserved guanine nucleotide exchange factor for Cdc42 of the host cell. *Mol. Microbiol.* **36**, 1206–21.
Suarez, A., Guttler, A., Stratz, M., Staendner, L. H., Timmis, K. N. and Guzman, C. A. (1997). Green fluorescent protein-based reporter systems for genetic analysis of bacteria including monocopy applications. *Gene* **196**(1–2), 69–74.
Sun, A. N., Camilli, A. and Portnoy, D. A. (1990). Isolation of Listeria monocytogenes small-plaque mutants defective for intracellular growth and cell-to-cell spread. *Infect. Immun.* **58**(11), 3770–8.
Tartera, C. and Metcalf, E. S. (1993) Osmolarity and growth phase overlap in regulation of Salmonella typhi adherence to and invasion of human intestinal cells. *Infect. Immun.* **61**, 3084–9.
Taylor, R. K., Manoil, C. and Mekalanos, J. J. (1989) Broad-host-range vectors for delivery of TnphoA: use in genetic analysis of secreted virulence determinants of Vibrio cholerae. *Journal of Bacteriology* **171,** 1870–8.
Taylor, R. K., Miller, V. L., Furlong, D. B. and Mekalanos, J. J. (1987) Use of phoA gene fusions to identify a pilus colonization factor coordinately regulated with cholera toxin. *Proc. Natl. Acad. Sci. USA* **84,** 2833–7.
Tegmark, K., Karlsson, A. and Arvidson, S. (2000). Identification and characterization of SarH1, a new global regulator of virulence gene expression in Staphylococcus aureus. *Mol. Microbiol.* **37**(2), 398–409.
Tsien, R. Y. (1998). The green fluorescent protein. *Annu. Rev. Biochem.* **67**, 509–44.
Valdivia, R. H., Hromockyj, A. E., Monack, D., Ramakrishnan, L. and Falkow, S. (1996). Applications for green fluorescent protein (GFP) in the study of host–pathogen interactions. *Gene* **173**(1 Spec No), 47–52.
Valdivia, R. H. and Ramakrishnan, L. (2000). Applications of gene fusions to green fluorescent protein and flow cytometry to the study of bacterial gene expression in host cells. *Meth. Enzymol.* **326**, 47–73.
Valdivia, R. H. and Falkow, S. (1997). Fluorescence–based isolation of bacterial genes expressed within host cells. *Science* **277**(5334), 2007–11.
Worley, M. J., Stojiljkovic, I. and Heffron, F. (1998). The identification of exported proteins with gene fusions to invasin. *Mol. Microbiol.* **29**(6), 1471–80.
Worley, M. J., Ching, K. H. and Heffron, F. (2000). Salmonella SsrB activates a global regulon of horizontally acquired genes. *Mol. Microbiol.* **36**(3), 749–61.
Zhou, D., Chen, L. M., Hernandez, L., Shears, S. B. and Galan, J. E. (2001). A *Salmonella* inositol polyphosphatase acts in conjunction with other bacterial effectors to promote host cell actin cytoskeleton rearrangements and bacterial internalization. *Mol. Microbiol.* **39**, 248–260.
Zlokarnik, G., Negulescu, P. A., Knapp, T. E., Mere, L., Burres, N., Feng, L., Whitney, M., Roemer, K. and Tsien, R. Y. (1998). Quantitation of transcription and clonal selection of single living cells with beta-lactamase as reporter. *Science* **279**(5347), 84–8.

4 Molecular Methods for Monitoring Bacterial Gene Expression During Infection

Isabelle Hautefort and Jay CD Hinton
Molecular Microbiology Group, Institute of Food Research, Norwich Research Park, Colney Lane, Norwich NR4 7UA, UK

◆◆◆

CONTENTS

◆◆◆◆◆◆ INTRODUCTION

Bacterial infections involve a myriad of bacterial-encoded virulence determinants, which are produced in response to the host cellular environment. The expression cascade of virulence factors is likely to occur in the right place, at the right stage of infection and at the right level to achieve successful infection by the pathogen. A number of bacterial virulence factors have been shown to alter normal host gene expression, leading to production of new host environmental signals which in turn can cause further bacterial adaptation. Despite our increasing knowledge relating to bacterial virulence and host response determinants, this cross-talk mechanism remains unclear. It has become necessary to develop novel techniques for (i) identifying determinants involved in bacterial virulence, as well as their inducing signals; (ii) measuring their expression levels during infection; and (iii) understanding the interactions of host and bacterial factors. *In Vivo* Expression Technology, Differential Fluorescence Induction and transposon-based approaches have given new insights into bacterial virulence. The increasing availability of genome sequences, of improved reporter systems, the development of new technologies such as *in situ* RT-PCR, DNA microarray technology, as well as the combination of several physical techniques (for example *in situ* hybridization and flow cytometry) promise to provide a clearer understanding of bacterial pathogenicity and to contribute to the development of new anti-microbial agents and vaccines.

METHODS IN MICROBIOLOGY, VOLUME 31
ISBN 0–12–521531–2

In this chapter we aim to give a brief description of currently available techniques used to identify *in vivo* induced (*ivi*) genes and to monitor bacterial gene expression during infection. We also comment on new approaches currently being developed to clarify important aspects of the host/pathogen interaction. For the purposes of this review '*in vivo*' refers to studies involving an animal model, and '*in vitro*' refers to experiments in cultured cells or in laboratory media.

◆◆◆◆◆◆ THE REVOLUTIONARY TALE OF *IVI* GENE IDENTIFICATION

The fact that much virulence gene expression is usually environmentally induced has now been clearly shown for a large number of microbial pathogens. However, researchers have had to rely on *in vitro* systems as a crude representation of the dynamic milieu of the host cell or the complexity of an entire animal. The environment of the mammalian cell is constantly reacting to different signals, and many levels of cellular response contribute to the global homeostatic status of the host. Therefore, the few bacterial genes identified as responding to environmental signals experienced in *in vitro* systems do not reflect the complexities of a living host. These limitations have been overcome in the last decade by the application of new technologies such as *In Vivo* Expression Technology (IVET), Differential Fluorescence Induction (DFI) and Signature Tagged Mutagenesis (STM).

Positive selection *in vivo*

IVET

In Vivo **E**xpression **T**echnology is based on a stringent selection system for identifying genes which are switched on *in vivo*. It involves the random insertion of chromosomal DNA fragments upstream of a promoter-less reporter gene (such as *purA::lacZY*, *cat*, *tnpR*) whose expression allows the survival of the bacteria in an animal model. Only the bacteria that express the reporter gene under the control of an *ivi* gene promoter will survive this procedure (Heithoff *et al.*, 1997; Mahan *et al.*, 1995; Merrell and Camilli, 2000).

IVET has allowed the identification of hundreds of *ivi* genes in a diverse range of bacterial pathogens including *Salmonella enterica* sv. Typhimurium (Mahan *et al.*, 1993), *Pseudomonas aeruginosa* (Wang *et al.*, 1996a,b), *Staphylococcus aureus* (Lowe *et al.*, 1998), *Vibrio cholerae* (Camilli *et al.*, 1994) and *Candida albicans* (Staib *et al.*, 1999) (Table 4.1). IVET applications have not been limited to animal models, but have also involved cultivated cells for pathogens such as *S. typhi* (Staendner *et al.*, 1995) and *S. Typhimurium* (Janakiraman and Slauch, 2000). Because IVET can be used in different models for the same pathogen, *ivi* genes which

Table 4.1 Approaches to identifying bacterial genes induced *in vivo*

Organism	Technique used	Model	Reference
Actinobacillus pleuropneumoniae	STM	Porcine infection	(Fuller *et al.*, 2000b)
A. pleuropneumoniae	IVET	Porcine infection	(Fuller *et al.*, 1999)
Aspergillus fumigatus	STM	Invasive pulmonary infection	(Brown *et al.*, 2000)
Bartonella henselae	DFI	HEp-2 cell invasion	(Lee and Falkow, 1998)
Bordetella bronchiseptica	Differential Display	*In vitro* induced Bvg+ infectious phase	(Yuk *et al.*, 1998)
Brucella suis	STM	Human macrophage	(Foulongne *et al.*, 2000)
Clostridium perfringens	Differential Display	Wild-type v/s virR mutant	(Banu *et al.*, 2000)
Escherichia coli	Differential Display	Heat-shocked cultures	(Gill *et al.*, 1999)
E. coli K1	STM	Human brain vascular endothelial cells	(Badger *et al.*, 2000)
		Gastrointestinal tract of infant rats	(Martindale *et al.*, 2000)
Histoplasma capsulatum	IVET	Mouse and cultivated cells	(Retallack *et al.*, 2000)
Legionella pneumophila	Differential Display	Macrophage infection	(Abu Kwaik and Pederson, 1996)
Listeria monocytogenes	IVET	Mouse	(Gahan and Hill, 2000) (Dubail and Barche, 2000)
Mycobacterium marinum	DFI	Macrophage infection	(Barker *et al.*, 1998) (Ramakrishnan *et al.*, 2000)
M. tuberculosis	Differential Display	Murine macrophage infection	(Ragno *et al.*, 1998)
M. tuberculosis	DFI	Macrophage infection	(Triccas *et al.*, 1999)
M. tuberculosis	SCOTS	Human macrophage	(Graham and Clark-Curtiss, 1999)

Table 4.1 Continued

Organism	Technique used	Model	Reference
M. tuberculosis	Subtractive Hybridization	Macrophage infection	(Plum and Clark-Curtiss, 1994)
Pasteurella multocida	STM	Septicaemia in mouse	(Fuller *et al.*, 2000a)
Proteus mirabilis	STM	CBA mouse	(Zhao *et al.*, 1999)
Pseudomonas aeruginosa	IVET	BALB/c mouse	(Handfield *et al.*, 2000)
		Respiratory mucus derived	(Wang *et al.*, 1996a)
		from Cystic Fibrosis patients	(Handfield *et al.*, 2000)
		Rat lung infection	(Wang *et al.*, 1996b)
		Murine burn wound infection	(Ha and Jin, 1999)
Pseudomonas fluorescens	IVET	Rhizosphere	(Rainey, 1999)
Salmonella enterica sv. Typhi	IVET	Henlé cells	(Staendner *et al.*, 1995)
S. Typhimurium	DFI	Acid shock	(Valdivia and Falkow, 1996)
		Macrophage infection	(Valdivia and Falkow, 1997)
S. Typhi	SCOTS	Macrophage infection	(Daigle *et al.*, 2001)
S. Typhimurium	STM	BALB/c mouse	(Hensel *et al.*, 1995)
S. Typhimurium	IVET	Peyer's patches	(Stanley *et al.*, 2000)
S. Typhimurium	IVET	Murine hepatocyte	(Janakiraman and Slauch, 2000)
S. Typhimurium	SCOTS	Macrophage infection	(Morrow *et al.*, 1999)

S. Typhimurium	Proteomics	Macrophage infection	(Deiwick and Hensel, 1999)
Small intestine and colonic epithelial cells	Differential Display	Association with gut microflora	(Ogawa *et al.*, 2000)
Staphylococcus aureus	STM	Mouse abscess	(Coulter *et al.*, 1998)
		Bacteremia in mouse	(Mei *et al.*, 1997)
		Infected wounds in mouse and endocarditis in rabbit	(Schwan *et al.*, 1998)
S. aureus	IVET	Murine renal abscess	(Lowe *et al.*, 1998)
Staphylococcus aureus	Subtractive Hybridisation	Identification of DNA sequences specific to virulent clinical isolates	(El-Adhami, 1999)
Streptococcus agalactiae	STM	Neonatal rat sepsis	(Jones *et al.*, 2000)
Streptococcus gordonii	IVET	Rabbit endocarditis model	(Kilic *et al.*, 1999)
S. gordonii	IVET	Acidic pH	(Vriesema *et al.*, 2000)
S. gordonii	Differential Display	Saliva v/s brain heart Infusion	(Du and Kolenbrander, 2000)
Streptococcus pneumoniae	STM	BALB/c mouse	(Polissi *et al.*, 1998)
Vibrio cholerae	STM	Suckling mouse	(Chiang and Mekalanos, 1998)
Yersinia enterocolitica	STM	BALB/c mouse	(Darwin and Miller, 1999)

respond to distinct environmental conditions have been identified (e.g. *P. aeruginosa* (see Table 4.1)).

Many *ivi* genes have been characterized in great detail, and include a mixture of genes of known function (for example *pvdI, soxR, ppkA* in *P. aeruginosa* (Ha and Jin, 1999; Handfield *et al.*, 2000b; Motley and Lory, 1999) with genes of unknown function [FUN genes (Hinton, 1997)]. A clear and well illustrated description of IVET and the various classes of genes identified by IVET has recently been presented (Merrell and Camilli, 2000).

The IVET approach has evolved a great deal from the original *purA* and *cat*-based selection systems. A selection, based on transcriptional fusions of *ivi* promoters with the *tnpR* recombinase gene, has allowed identification of *ivi* genes expressed transiently or at a low-level (Camilli *et al.*, 1994; Camilli and Mekalanos, 1995; Merrell and Camilli, 2000). This adaptation of IVET has been named RIVET, **R**ecombinase-based **IVET** (Lee and Camilli, 2000; Slauch and Camilli, 2000).

However, the IVET/RIVET technique possesses a number of limitations. First, it requires the development of genetic tools, restricting the number of applicable pathogens. Second, the selective system itself presents inherent limitations such as the need for an auxotrophic strain (e.g. *purA* mutant), the administration of antibiotic solutions at the appropriate concentrations (e.g. *cat*) or the over-sensitivity of the *tnpR*-based system in some model systems. A tunable RIVET approach has recently been developed to allow the identification of *ivi* genes which show a significant basal expression level *in vitro* that would have been missed with the original technique (Merrell and Camilli, 2000). Based on various mutations introduced to the ribosome binding site of the recombinase, tunable RIVET offers a large choice of reporter systems with varying sensitivity and different thresholds for *ivi* gene selection.

Differential Fluorescence Induction (DFI)

As an alternative to the IVET-based promoter-trap systems, a complementary approach has been developed for *ivi* gene identification. **D**ifferential **F**luorescence **I**nduction (DFI) relies on the expression of the enhanced Green Fluorescent Protein (GFP) as a reporter of promoter activity. GFP derives from the jellyfish, *Aequorea victoria*, and will be described later in more detail. It does not require any substrate or co-factor to fluoresce and thus can be used to monitor gene expression in living samples. DFI involves the cloning of random fusions of genomic DNA fragments upstream of a promoter-less *gfp* gene. Bacteria harbouring *gfp* fusions are pooled and either used to infect cultured mammalian cells (Valdivia and Falkow, 1997). Fluorescence Activated Cell Sorting (FACS) is then used to enrich for green fluorescent mammalian cells allowing isolation of GFP-expressing bacteria. Alternatively, bacteria can be submitted to various stimuli, such as low pH (Valdivia and Falkow, 1996). This technique can be applied to more complex environments including host tissue (Valdivia and Ramakrishnan, 2000 and the IVIF technique described later). However, DFI, like IVET, will not identify genes which are expressed *in vitro* and are also important for virulence *in vivo*. Further experiments are always needed to determine whether *ivi* genes are crucial for bacterial virulence and survival in the host.

Negative selection *in vivo*; Signature Tagged Mutagenesis

Gene inactivation by transposon insertion has been extensively used to assess the role of genes of interest on particular bacterial functions (Kleckner *et al.*, 1977). However, screening of individual mutants to identify genes implicated in host survival or virulence is time consuming when mutants are screened individually. A new technique based on transposon mutagenesis has been developed to overcome this limitation, **S**ignature **T**agged **M**utagenesis (STM).

STM allows the identification of genes required for *in vivo* virulence and survival of the pathogen, based on a negative selection system (Lehoux and Levesque, 2000; Lehoux *et al.*, 1999; Shea *et al.*, 2000; Unsworth and Holden, 2000). The technique involves comparative hybridization following efficient random transposition or use of other insertional genetic tools. Different oligonucleotide sequences are used to mark individual transposons so that each insertion harbours a specific signature tag. This allows the testing of a large number of different mutants at the same time, each mutant being distinguishable by its unique tag sequence (Hensel, 1998).

STM was used for the first time in *S.* Typhimurium (Hensel *et al.*, 1995). A mutant library was generated and tested in mice after intraperitoneal injection of pools of 96 mutants. PCR reactions were performed on DNA extracted from bacteria recovered from spleen and products were used as new probes to be tested on the initial library. The absence of hybridisation revealed mutations that were lethal to the bacteria. This technique allowed identification of the second pathogenicity island (SPI2) of *S.* Typhimurium, which is required for systemic spread and intracellular survival of the pathogen (Cirillo *et al.*, 1998; Shea *et al.*, 1996, 1999). Since then, many virulence genes have been identified by STM in various pathogens (Table 4.1). The related GAMBIT (**G**enomic **A**nalysis and **M**apping **b**y ***I**n vitro* **T**ransposition) allows identification of essential genes *in vitro* and is now being used to screen for genes required for survival *in vivo* in animal models (Chiang *et al.*, 1999).

In an analogous fashion to IVET, the same STM mutant library can also be tested simultaneously in different selective systems (animal model versus cultured cells, or in different animal models). Recently, host-specific virulence determinants were identified by STM in *S.* Dublin after challenge in calves and in mice (Bispham *et al.*, 2001). Although STM does not allow the direct monitoring of the expression of virulence genes during infection, it is ideal for identifying candidate *ivi* genes.

The powerful STM technology is subjected to several limitations that include: (i) the requirement of a random transposition system; (ii) the need to optimize a large number of parameters (the tagging strategy; the complexity and dose of the inocula depending on the administration route; the duration of infection); (iii) the problem of polar effects preventing discrimination between single gene or operon insertions. *In vitro* transposition systems could improve targeting the entire genome of the bacterial pathogen. STM has the significant advantage over IVET and DFI of immediately showing the importance of *ivi* genes for virulence and survival rather than just identifying genes that are switched on *in vivo*.

Direct detection of *IVI* gene expression

The IVET, DFI and STM approaches are very powerful, but are not relevant to genetically-intractable pathogens. Furthermore, these techniques tend to give a 'yes or no' answer, rather than precisely measuring levels of gene expression. Therefore alternative techniques have been developed to detect *ivi* messenger RNA directly. Subtractive Hybridization and Differential Display detect transcripts that are induced *in vivo* and circumvent the need for genetic tools.

Subtractive Hybridization

Initially developed for eukaryotic research (Kurvari *et al.*, 1995), Subtractive Hybridization has subsequently been used to study bacterial pathogens. This approach is based on the isolation of mRNA from bacteria grown under conditions of interest (infected mammalian cells or animal models) and the use of reverse transcriptase to generate cDNA. The cDNA representing the same bacteria grown in broth is used to subtract constitutively-expressed genes by hybridization. The remaining non-hybridized cDNA is used to identify *ivi* genes for the pathogen under study. Subtractive Hybridization analysis has proved to be useful in *Mycobacterium avium* during macrophage infection (Plum and Clark-Curtiss, 1994) as well as in several other organisms (Table 4.1), and is now commonly used to compare genomes between sequenced strains and other strains of interest, allowing identification of differences between pathogens of the same genus that reflect variations in tissue tropism (Dasgupta *et al.*, 2000; Jaufeerally-Fakim *et al.*, 2000; Tinsley and Nassif, 1996; Zhang *et al.*, 2000).

However, several limitations have been found with this approach (Hautefort and Hinton, 2000): first, only two organisms or two gene expression patterns can be compared at once. Second, this technique is hampered by the instability of bacterial mRNA as well as the requirement for high-quality mRNA isolated from small populations of bacteria grown *in vivo*. Finally, transiently-expressed genes may not be well represented; RT-PCR-based amplification can improve the sensitivity of this approach.

Differential Display

Like Subtractive Hybridization, this technique was originally applied to eukaryotic studies (Liang and Pardee, 1992). It is based on the electrophoretic comparison of mRNA profiles of the same organism grown under various environmental and control conditions. Microbial mRNA is extracted from the organism, reverse-transcribed, PCR-amplified and radio-labelled for auto-radiographic analysis. Differential Display has identified host genes induced by infection of a pathogen such as the macrophage response caused by *M. tuberculosis* infection, the gastric cellular response to *Helicobacter pylori* infection or the mucosal defences associated with the gut microflora (Ogawa *et al.*, 2000; Ragno *et al.*, 1997, 1998; Wong *et al.*, 1996).

The absence of polyadenylation of prokaryotic mRNA necessitated modifications of Differential Display for bacterial applications. These involved the use of random primers for reverse transcription and cDNA amplification, and were used for the first time to identify genes of *Legionella pneumophila* that were induced during macrophage infection (Abu Kwaik and Pederson, 1996). Since then Differential Display has been considerably improved (Fleming *et al.*, 1998), leading to the identification of important *ivi* genes in different bacteria (Table 4.1). Northern blotting and total RNA dot blotting are often combined with Differential Display to confirm variations in transcript levels (Gill *et al.*, 1999; Ragno *et al.*, 1998).

Differential Display has several advantages compared with Subtractive Hybridization. (i) It allows the comparison of more than two different conditions at the same time. (ii) Because of the cDNA PCR amplification step, high levels of gene expression are not required, allowing identification of less abundant microbial mRNA with a detection level of at least 10^3 transcript molecules (Fleming *et al.*, 1998). (iii) Unlike IVET and STM, Differential Display can be used to study both up- and down-regulation of gene expression.

Limitations of Subtractive Hybridization and Differential Display and possible alternatives

Subtractive Hybridization and Differential Display do not show whether expression of particular genes is required for a particular aspect of virulence. The exact role of identified genes must be determined subsequently by genetic mutation and virulence studies in appropriate model systems. The analysis of prokaryotic mRNA expression with Differential Display is known to generate false positives, making it very labour-intensive. The efficiency of Differential Display is impeded by the short half-life of prokaryotic mRNA, the absence of RNA polyadenylation and still lacks the sensitivity to detect rare mRNA species.

Alternative approaches for the analysis of bacterial mRNA induced *in vivo* have been described. First, Subtractive Cloning is an updated Subtractive Hybridization method for comparing bacterial gene expression (Carulli *et al.*, 1998; Diatchenko *et al.*, 1999; Sturtevant, 2000). Messenger RNA is reverse-transcribed into cDNA for each condition tested, using different linkers each time. A series of Subtractive Hybridization experiments allow enrichment of *ivi* cDNA that possess a linker sequence at both ends, allowing direct cloning and sequencing of identified *ivi* genes.

Serial **A**nalysis of **G**ene **E**xpression (SAGE) is another useful alternative to Differential Display that can be used to analyse and to simultaneously quantify a large number of transcripts (Velculescu *et al.*, 1995). To demonstrate this strategy, the authors chose a eukaryotic example, pancreatic gene expression. SAGE involves the extraction of tissue mRNA and reverse transcription to cDNA with a biotinylated oligo(dT) primer. Action of appropriate restriction enzymes generates small sequence tags specific to each cDNA molecule, and these are ligated to linker sequences. Compatible linkers allow the concatenation of the different tags and their direct cloning and sequencing. The abundance of each tag reflects the initial proportion of the transcript in the initial mRNA pool. SAGE has been successfully applied to eukaryotes but in principle could also be used in prokaryotes.

Another new hybridization-based method has been developed to compare gene induction during infection of cell culture or animal tissue models (Graham and Clark-Curtiss, 1999). This technique is called **S**elective **C**apture **O**f **T**ranscribed **S**equences (SCOTS), and derives from the initial Subtractive Hybridization approach used in *M. avium* (Plum and Clark-Curtiss, 1994). SCOTS involves two important steps: the first one aims to normalize levels of microbial cDNA obtained from bacteria released from different experimental conditions such as infected host cells or animal tissue. This procedure involves an enrichment step for low abundance transcripts. The second step is based on the comparison of gene expression patterns between two environments. Subtractive Hybridization of normalized cDNA obtained from bacteria grown in the different environments under study reveals preferential expression of genes under one or the other conditions tested. SCOTS has identified numerous macrophage-induced genes in *S. Typhi* (Daigle *et al.*, 2001). In addition, it has been used in other pathogens like *S.* Typhimurium allowing identification of a novel fimbrial operon and putative transcriptional regulator that are absent from the *S. typhi* genome (Morrow *et al.*, 1999). Although SCOTS has the advantage of identifying low abundance mRNA, it cannot be used to quantify *in vivo* gene expression levels because of the cDNA normalization step included in the protocol.

Proteomics

Direct analysis of bacterial gene expression during infection can only reveal events, which have occurred during transcription. Because many levels of gene regulation are post-transcriptional, and contribute directly to translational efficiency it is important to consider expression at the protein level. The obvious goal is to characterize the bacterial proteome during infection, as it genuinely represents the whole bacterial response to the host. Unfortunately, it has not yet been possible to analyse the proteomes of bacteria removed directly from host tissue, due to a combination of factors including the relatively low numbers of bacteria that can be isolated after infection and the lack of a system for protein amplification. Nevertheless, proteomic technology has been used to detect bacterial proteins induced during infection of cultured cells. This approach relies on the specific labelling of bacterial-encoded proteins expressed within mammalian cells, and has been reviewed elsewhere (Cash, 2000; Hautefort and Hinton, 2000). One particular report on the regulation of the *S.* Typhimurium SPI2 locus demonstrates the direction that *in vivo* expression research is likely to take in the near future (Deiwick and Hensel, 1999). They used a combination of reporter genes, regulatory mutants and two-dimensional gel electrophoresis to simultaneously characterize SPI2 expression at the level of the transcriptome and the proteome. Once this approach has been combined with identification of the *in vivo*-induced proteome, it will reveal the role of particular regulators in controlling the expression of IVI proteins.

In Vivo induced antigen technology

An innovative approach has been developed to identify IVI proteins expressed during human infection, that does not rely on *in vitro* or animal model systems. The IVIAT approach (***In Vivo*** **I**nduced **A**ntigen **T**echnology) is based on the immunological screening of an expression library of the pathogen under study with various sera obtained from human patients infected by this organism (Handfield *et al.*, 2000b). Each batch of sera is previously depleted of antibodies raised against constitutively-expressed proteins using whole bacterial cells and total protein extracts. The straightforward techniques used in IVIAT make it applicable to all type of eukaryotic or prokaryotic pathogens (Hautefort and Hinton, 2000). The most attractive aspect of IVIAT is that it uses genuine human infection samples. IVIAT has successfully identified IVI proteins of *Actinobacillus actinomyetemcomitans* in localized juvenite periodontitis (Handfield *et al.*, 2000a) and is currently being used to study other human pathogens such as *P. aeruginosa* and *Candida* spp. However, IVIAT requires identification of patients suffering from infection by the pathogen of interest as well as knowing at which stage of infection the sample has been obtained, which necessitates close medical involvement.

◆◆◆◆◆◆ THE CHALLENGE OF MONITORING GENE EXPRESSION IN INDIVIDUAL CELLS

The ability of bacteria to control and to optimize their growth rate and gene activity by carefully adjusting the cellular composition of DNA, RNA and proteins in response to the prevailing environmental conditions has led to the development of various technical approaches to study gene expression in individual bacterial cells.

In situ PCR

This technology has been more widely used to monitor gene expression in eukaryotic than bacterial systems. The use of *in situ* PCR to study *lac* gene expression in *Salmonella* established the sensitivity of the technique in bacteria (Tolker-Nielsen *et al.*, 1997). More recently, the same group have used *in situ* RT-PCR to visualize expression of *dnaK* in *Methanosarcina mazei* (Lange *et al.*, 2000). The combination of *in situ* PCR with flow cytometry offers the potential of selecting individuals from a bacterial population that are expressing particular genes to certain levels (Chen *et al.*, 2000). A wide range of molecular tools can be used to monitor bacterial growth activity *in situ*, and the subject has been well reviewed (Molin and Givskov, 1999).

Transcriptional reporter systems

β-Galactosidase

The importance of reporter gene systems for measuring bacterial gene expression *in vivo* has already been described (Hautefort and Hinton, 2000). β-Galactosidase has been used for decades, and has proved to be a reliable screening and reporter system (Jacob and Monod, 1961). It remains very useful for studying up- and down-regulation of gene expression, as well as translational regulation. A practical guide to the construction and use of *lac* fusions in *E. coli* has recently been published (Hand and Silhavy, 2000). β-Galactosidase is occasionally used for the analysis of gene expression *in vivo* (Hautefort and Hinton, 2000).

The development of new substrates, including fluorescence-generating substrates has considerably improved the sensitivity of β-galactosidase detection, and has facilitated studies with intact individual bacterial cells (Zhang *et al.*, 1991). The use of β-galactosidase with the Fluorescein di-β-D-galactopyranoside (FDG) substrate was compared to the use of the **G**reen **F**luorescent **P**rotein (GFP) for monitoring *gyrB* expression in *Mycobacterium* species (Rowland *et al.*, 1999). Detection of β-galactosidase with FDG was 70-fold more sensitive than GFP in *M. bovis* BCG when measured by fluorimetry, and comparable to GFP when measured by flow cytometry in individual bacterial cells. However, this study did not use the latest brightly fluorescent derivatives of GFP which are likely to prove more sensitive than β-galactosidase, even with FDG as a substrate. Furthermore, permeabilization of bacterial cells is required to deliver β-galactosidase substrate; due to variations in permeability, the amount of substrate that enters bacterial cells varies from one cell to another, preventing the accurate measurement of changes in gene expression levels in individual bacteria (Nwoguh *et al.*, 1995).

Luciferases

Luciferase catalyses production of photons, and can be used as a reporter of gene expression in appropriate situations. The short half-life of luciferase guarantees real-time observation of gene expression. In *Yersinia* pseudotuberculosis, *luxAB* fusions showed that *yopE* and *yopH* expression was highest during early stages of colonization of either Peyer's patches or the spleen of infected mice (Forsberg and Rosqvist,

1993). Expression of the Lux system derived from *Photorhabdus* has been used to follow the real-time trafficking of pathogens, such as *S.* Typhimurium or *Staphylococcus aureus* in a live mouse (Contag *et al.*, 1995; Francis *et al.*, 2000). However, the use of luciferase for monitoring gene expression *in vivo* has not yet been reported, due to the following problems. First, luciferase is sensitive to the concentration of molecular oxygen. An example of the unreliability of luciferase and *lux* gene fusions to monitor gene expression was shown by the discovery that a p*lac*::*luxAB* fusion in *Salmonella* exhibited a 10-fold reduction in luciferase expression immediately following invasion of epithelial cells. Similar results were observed following contact with tissue culture media, and probably reflect decreased oxygen tension (Maurer *et al.*, 2000). Second, detection of luciferase bioluminescence has not been improved sufficiently to allow measurement of gene expression in single bacterial cells. Finally, fusion of firefly-derived luciferase to eukaryotic gene promoters for monitoring induction upon bacterial invasion produced data that was not consistent with results obtained with β-galactosidase or *cat* reporter systems (Savkovic *et al.*, 2000). The authors observed that bacteria such as EPEC and *S.* Typhimurium that express intact type-III secretion machinery are responsible for the decrease of *luc* activity. Consequently, the use of luciferase to monitor bacterial virulence gene expression has not been widespread.

Green Fluorescent Protein

The Green Fluorescent Protein (GFP) offers the advantage of being naturally fluorescent without the addition of exogenous substrates. Because the fluorescence of GFP is linearly proportional to the amount of GFP protein, it is an attractive reporter for monitoring *in situ* gene expression in single cells. GFP was used as a reporter system for the first time in eukaryotes in 1994 (Chalfie *et al.*, 1994). Its adaptation for microbial studies led to the development of optimized, more soluble, brighter, blue- or red-shifted mutants, extending potential applications (Cormack *et al.*, 1996; Crameri *et al.*, 1996; Heim *et al.*, 1994; Siemering *et al.*, 1996). GFP has now been successfully used to monitor pathogen trafficking within host tissue (Errampalli *et al.*, 1999; Kohler *et al.*, 2000).

However, the major limitation of many GFP mutant proteins is their inability to fold correctly. For example, only 20% of EGFP is folded properly and is able to fluoresce (Scholz *et al.*, 2000). In 1997 an improved GFP cloning cassette was developed for the generation of prokaryotic transcriptional fusions. It contains a *gfp* gene expressing the S65T 'red shift' and the F64L 'protein solubility' amino acid substitutions (Heim *et al.*, 1995; Cormack *et al.*, 1996) flanked by convenient restriction sites, a translational enhancer and a consensus ribosome-binding site with an optimized spacer region (Miller and Lindow, 1997). The authors observed a 40- to 80-fold brighter fluorescence with these *gfp* fusions than with fusions to wild-type GFP. Recently an even brighter GFP protein (GFP^+) has been developed (Scholz *et al.*, 2000). GFP^+ combines the mutations from GFPuv, which confer enhanced folding, and the mutation from

GFPmut1 to give a protein with 130-fold brighter signal than wild type GFP. Various sets of GFP-based vectors have been constructed for Gram-negative bacteria other than *E. coli*. For example, one of these vectors facilitates the generation of GFP transcription fusions in *E. coli* for subsequent introduction into the final host strain, whereas another allows the direct construction of insertions in the host strain (Matthysse *et al.*, 1996). Recent 'promoter-trap' vectors have been developed using *gfp* or *inaZ* as reporter genes (Miller *et al.*, 2000). These vectors are thought to be highly stable in a broad range of bacterial species and also to encode GFP stability variants with different half-lifes, allowing the study of gene repression as well as gene induction.

The use of *gfp* as a reporter of promoter activity has had one significant limitation. A recent comparison was made between the *luxAB* reporter genes from *Vibrio* spp. and the *gfp* reporter system developed by Crameri *et al.* (1996). Both were used in promoter-trap vectors in the cyanobacterium *Synechocystis* sp. PCC6803 (Kunert *et al.*, 2000). For both *isiAB*::*gfp* and *isiAB*::*luxAB* fusions, the induction rate of mRNA synthesis was shown to be identical. However, the *gfp* fusion showed a slow increase and decrease of fluorescence whereas the *luxAB* fusion caused a rapid increase and decrease in luminescence, probably due to the time taken for formation of active fluorophore from GFP protein. Therefore, the use of GFP is not ideal for real-time experiments that involve rapid changes in gene expression. Enhanced versions of GFP are known to become fluorescent much faster than wild-type GFP, which should improve applicability of GFP (Cormack *et al.*, 1996).

◆◆◆◆◆◆ FUTURE PROSPECTS: UNDERSTANDING THE COMPLEX DIALOG BETWEEN THE HOST AND THE PATHOGEN

Microscopy

The development of more sensitive microscopic tools to complement flow cytometric techniques has been essential for the determination of whether particular genes have a specific impact on the host and host cellular response. Recent research has shown that pathogens can interfere with normal host cell-mediated immune response by modifying the inflammatory MAP-Kinase-based cascade that normally leads to the production of cytokines and the elimination of the pathogen (Galan, 1999; Hobbie *et al.*, 1997). More and more examples show specific interactions of bacterial virulence factors ('effector' proteins) and host signalling determinants. This reinforces the need to investigate each step of bacterial pathogenesis from the viewpoint of both the bacteria and the host. Microscopic approaches to visualizing infection have evolved considerably, and the increasing availability of antibodies specific to bacterial- or host-derived determinants has improved our understanding of natural infection processes. Light microscopy has been used for comparing the histopathology of mice infected with wild type *S.* Typhimurium or with a

S. Typhimurium mutant strain not capable of repressing the production of host γ-IFN (Valdivia *et al.*, 2000). Conventional microscopy has also been used to demonstrate that mycobacterial surface proteins are released from bacterial phagosome in infected macrophages (Beatty and Russell, 2000). Recent advances in optical imaging systems have now been successfully applied to answer important microbiological questions (see below), and involve cooled charged-couple-device (CCD) cameras capable of generating colour image data sets containing more information than available from a 35 mm colour reversal film. CCD cameras allow improved spatial resolution, combined with geometric and photometric linearity (Cinelli, 1998; Fung and Theriot, 1998; Entwistle, 1998). Acquisition and data storage capabilities have also been improved; the replacement of film with digital technology has improved image quality by facilitating background subtraction and contrast enhancement on both 2D and 3D images. Dedicated software is now available to perform signal intensity comparison from microscopic images, but this does not yet provide a reliable tool for accurate quantification of gene expression from reporter fusions.

Microscopes now offer much more sensitive detection of weak signals than a few years ago. Confocal microscopes have been developed to register light only from the focal plane, removing the problem of light contamination from out-of-focus regions. This approach has allowed the co-localization of bacteria with host-specific markers. Examples include the proof that *S.* Typhimurium cells are found within phagocytic cells in liver sections (Richter-Dahlfors *et al.*, 1997), and are co-localized with the LAMP-1 vacuolar membrane protein of infected cultured cells (Beuzon *et al.*, 2000; Rathman *et al.*, 1997). Confocal microscopy has also been used to observe interference of *Salmonella* with normal host cell structure, such as the tight junctions in MDCK epithelial cells (Jepson *et al.*, 2000).

Multiphoton microscopy promises to revolutionize the field by overcoming the problem of photobleaching. The technique involves treating the specimen with a stream of pulsed infrared light, at a pulse frequency that allows one dye molecule to absorb two photons at once. This happens only at the focal plane. Combining the energy of the two long-photons brings the dye to its excitation state, generating fluorescence (Denk *et al.*, 1990). The two-photon approach allows photobleaching to be confined to the vicinity of the focal plane, and allows sharp signal localization with greatly improved image quality. In summary, microscopy is generally regarded as a qualitative rather than a quantitative technique.

Flow cytometry

It is clear that bacterial gene expression needs to be accurately measured *in vivo*, and flow cytometry is proving to be an ideal tool for the quantification. The process of flow cytometry involves the carriage of sample cell suspensions in a fluidic system and passage through a laser beam. Forward- and side-scattered light is collected to give information about size and density of each cell. If the cells harbour any fluorescent molecules, the fluorescence intensity of each individual particle is amplified, collected and saved. Flow cytometric approaches are ideal for measuring expression in individual eukaryotic cells initially and remain a powerful tool to monitor a wide range of cellular functions including immune cell maturation or cytokine production. The use of flow cytometry has been developing in the microbiological field, and is becoming increasingly important for measuring population heterogeneity, cell cycle and membrane integrity, etc. (Davey and Kell, 1996; Nebe-von-Caron *et al.*, 2000). Flow cytometry has also been used in studies of pathogenesis, either to follow intracellular spread of pathogens, such as *Shigella flexneri* (Rathman *et al.*, 2000) or to detect bacterial gene expression in individual macrophages, as described for DFI (Valdivia and Falkow, 1997). FACS is ideal for the enrichment of cellular subpopulations. However, flow cytometric analysis and sorting of bacteria is complicated by the small size of bacterial cells, which is close to the current limit of detection (Davey and Kell, 1996; Shapiro, 1995). Nevertheless, analysing and sorting has been reported for many bacterial species (Hewitt *et al.*, 2000; Vives-Rego *et al.*, 2000). Several examples of the use of flow cytometry to monitor bacterial gene expression have been reported in pathogens including several Mycobacterial species (Kremer *et al.*, 1995; Valdivia and Falkow, 1996, 1997; Valdivia and Ramakrishnan, 2000).

The first combination of *in situ* hybridization with flow cytometry has recently been achieved (Chen *et al.*, 2000). RT-PCR was used to reverse transcribe mRNA and to amplify cDNA generated from *P. putida* grown in inducing conditions for expression of the gene of interest, a toluene dioxygenase (*todC1*). PCR products were labelled with fluorescence and used as probes for *in situ* hybridization experiments inside *Pseudomonas* cells. Analysis in flow cytometry of hybridized cells allowed visualization of *todC1* induction as well as indirect estimation of its expression levels in each bacterial cell. Flow cytometry is likely to be the key technique for measuring gene expression in individual bacterial cells.

Global expression profiling

The availability of the Human UniGene database (http://www.ncbi.nlm.nih.gov/UniGene/) and the recent completion of a number of bacterial pathogen genome sequences allow us to monitor mammalian and bacterial transcription at the genomic scale using DNA microarrays, and to produce a 'gene expression profile' for a particular organism under certain environmental conditions (DeRisi *et al.*, 1996; Hughes *et al.*, 2000; Lockhart *et al.*, 1996; Schena *et al.*, 1995).

Mammalian gene microarrays have been used to study host–pathogen interactions from the host's viewpoint, by identifying gene expression patterns induced by the presence of a pathogen (Manger and Relman, 2000). Several *in vitro* studies have explored the effects of infection by *L. monocytogenes* and *S.* Typhimurium on the human cell response (Cohen *et al.*, 2000; Eckmann *et al.*, 2000; Rosenberger *et al.*, 2000). These authors observed an up-regulation of certain genes including several cytokines, kinases, HLA-Class I and transcriptional factors. Other groups have used a more complex approach to compare host response induced by pathogenic strains of bacteria carrying well-defined mutations (Manger and Relman, 2000).

We require improved technology for the *in vivo* study of infection with microarrays, both from the host's and the pathogen's point of view (Lucchini *et al.*, 2001). Advances have been recently made in applying the linear RNA amplification method to extract sufficient eukaryotic mRNA for microarray analyses from small amounts of mammalian tissue (Gonzalez *et al.*, 1999; Wang *et al.*, 2000). However, the use of microarrays for the study of bacterial infection at the level of gene expression is still in its infancy (Cummings and Relman, 2000; Hautefort and Hinton, 2000). The lack of polyadenylated mRNA complicates the isolation of sufficient, good quality bacterial mRNA from complex environments such as mammalian tissue. The successful application of a linear RNA amplification approach to bacterial mRNA on a genomic scale has yet to be reported.

Consequently, the alternative technology of 'real-time' RT-PCR is likely to become the current method of choice for direct monitoring bacterial mRNA during infection. Bubert *et al.* (1999) described the use of RT-PCR to monitor *Listeria* gene expression in cultivated mammalian cells. The combination of this approach with 'real time' PCR (Wei *et al.*, 2001) should result in a sensitive and widely-applicable technique.

Tissue culture models

Although it is ideal to work with genuine animal models of infection, these do not exist for many pathogens. Therefore, cultivated cell-lines still have an important role for understanding microbial pathogenesis. More human or animal cell-lines are becoming available, and are generating intriguing results (see Table 4.2). *S.* Typhimurium benefits from being studied both in cultivated mammalian cells and in the mouse typhoid animal model. Until now, *Salmonella* gene expression has only been studied in cultivated cell lines particularly in the epithelial cells, dendritic cells or macrophages that are encountered by the bacteria during actual host infection. Animal models have been used to study the trafficking of *Salmonella* in host tissue, and to identify new bacterial *ivi* genes. Research on specific cultivated cell lines has proved invaluable for developing our understanding of *Salmonella* pathogenesis, as summarized below.

Host invasion by *S.* Typhimurium not only occurs in epithelial cells (Finlay *et al.*, 1989) but also in phagocytic cells which have been shown to be able to capture *Salmonella* in the gut lumen (Vazquez-Torres *et al.*, 1999). Cultivated epithelial cell-lines have been used to demonstrate that actin rearrangements in the host cell cytoskeleton were induced by *Salmonella*, resulting in characteristic membrane ruffling (Galan and Zhou, 2000; Ginocchio *et al.*, 1994). Actin rearrangement is caused by the *Salmonella* effector, SopE, which is an exchange factor for RhoGTPases (Hardt *et al.*, 1998). Epithelial cells have also been used to show that *Salmonella* participates actively in the restoration of the normal structure of the

Table 4.2 Bacterial determinants involved in host/pathogen interaction

Organism	Host/pathogen interaction	Host and pathogen factors involved	Reference
Campylobacter jejuni	Release of IL-8 by human intestinal epithelial cells	Adherence and/or invasion and Cytolethal Distending toxin (CDT)	(Hickey *et al.*, 2000)
Group A *Streptococcus*	Adhesion to human epithelial cells	Sc1	(Lukomski *et al.*, 2000)
Haemophilus ducreyi	Arrest of infected cells in mitosis only after internalization of HdCDT	Hd Cytolethal distending toxin	(Cortes-Bratti *et al.*, 2000)
HIV type 1	Chemokine coreceptor CCR-5 phenotypic knockout leads to infection resistance to HIV-1	Unknown	(Yang *et al.*, 1997)
Klebsiella pneumoniae	Adhesion to and invasion of epithelial cells	Capsule	(Sahly *et al.*, 2000)
Legionella pneumophila	Phagosome-related stress such as Thymine limitation	Rep Helicase	(Harb and Abu Kwaik, 2000)
Listeria monocytogenes	Intra-macrophage survival and bacterial growth in infected murine organs	OppA	(Borezee *et al.*, 2000)
Mycobacterium bovis BCG	Dendritic cell maturation	Peptidoglycan, arabinogalactan and mycolic acids	(Tsuji *et al.*, 2000)

Salmonella typhi	IL6 production by epithelial cells	Unknown	(Weinstein *et al.*, 1997)
Salmonella Typhimurium	Dendritic cell maturation	LPS	(Svensson *et al.*, 2000)
S. Typhimurium	Actin cytoskeleton rearrangement via Cdc42 and Rac1 RhoGTPAses in rat Ref52 cells	SptP	(Fu and Galan, 1999)
S. Typhimurium	Membrane ruffling via action rearrangement	SopE SopE2	(Hardt *et al.*, 1998) (Baskshi *et al.*, 2000)
S. Typhimurium	LPS-mediated expression of E-selectin and TNFα by endothelial cells and adherent monocytes respectively	LPS-lipidA	(Guo *et al.*, 1997)
S. Typhimurium	Invasion repression by bile	BarA/SirA	(Prouty and Gunn, 2000)
S. Typhimurium	Fusion of bacterial and host cell outer membranes	SipB	(Hayward *et al.*, 2000)
S. Typhimurium	Rapid and delayed induction of apoptosis in macrophages	SipB, InvA and OmpR, SsrB, SpiB respectively	(van Der Velden *et al.*, 2000)
S. Typhimurium	Apoptosis of macrophages	Caspase 1 & SipB	
S. Typhimurium & Dublin	Macrophage lysis	Caspase 1 & SipB	(Hersh *et al.*, 1999)
S. Typhimurium	Aggregation of host endosomal compartments into *lgp-tubules*	SpiC, SseFG, SsaJ,L,M,V,P, SpvR, SifA, *carA*$^{-}$ associated auxotrophy	(Watson *et al.*, 2000) (Guy *et al.*, 2000)
S. Typhimurium	Induction of inducible nitric oxide synthase (iNOS) expression	SipB,C,D & SopE2	(Cherayil *et al.*, 2000)

Table 4.2 Continued

Organism	Host/pathogen interaction	Host and pathogen factors involved	Reference
S. Typhimurium	Prevention of co-localization of iNOS and *Salmonella* in phagocytic cells in mouse organs	SPI-2 proteins like SseB	(Vazquez-Torres *et al.*, 2000)
S. Typhimurium	Bacterial interference with normal cytokine production such as γIFN	Mig14	(Valdivia *et al.*, 2000)
S. Typhimurium	Induction of TNFα expression in human promonocytic cells	FliC	(Ciacci-Woolwine *et al.*, 1998)
S. Typhimurium	HEp-2 epithelial cells	CsrA	(Altier *et al.*, 2000)
S. Typhimurium	Mouse intestinal colonization	LPS	(Licht *et al.*, 1996)
Shigella flexneri	IL-1β production by macrophages	IpaB invasin	(Thirumalai *et al.*, 1997)
Ureaplasma urealyticum	Activation of NF-κB and induction of iNOS in macrophages	Unknown	(Li *et al.*, 2000)
Uropathogenic *E. coli*	CD55 and CD66e clustering around the bacteria and apical actin rearrangement in polarized epithelial Caco-2/TC7 cells	Dr-II adhesin	(Guignot *et al.*, 2000)
Vibrio vulnificus	Adherence to HEp-2 epithelial cells and virulence in iron-overloaded mice	PilD	(Paranjpye *et al.*, 1998)

host cell cytoskeleton after its entry. This involves a second effector protein, SptP, that acts at the level of two RhoGTPases Dc42 and Rac1 (Fu and Galan, 1999). Recently, polarized intestinal epithelial cells have been used to observe the induction of the pro-inflammatory response by *S.* Typhimurium (Gewirtz *et al.*, 1999). The authors have shown that translocation of flagellin across the intestinal epithelial cells was responsible for this induction at the basolateral surface of the cell layer.

Following invasion, *Salmonella* encounters various other immune cell types in the murine model. Macrophages have long been regarded as the main target encountered by *S.* Typhimurium. Recently, Niedergang and collaborators have demonstrated that the uptake of *Salmonella* by dendritic cells does not depend on the same virulence factors required for uptake by macrophages (Niedergang *et al.*, 2000). Furthermore, we now know that *S.* Typhimurium is capable of inducing the maturation of bone-marrow-derived dendritic cells, causing reduction in their capability to present antigens from subsequently encountered pathogens (Svensson *et al.*, 2000). Similar observations were also made in *Mycobacterium bovis* BCG-infected dendritic cells (Tsuji *et al.*, 2000).

Most pathogens will encounter macrophage-mediated phagocytosis at some stage of the infection process. Cultivated macrophages have been used to study several other pathogens, including *Listeria monocytogenes*, and led to the surprising discovery that OppA, encoded by a gene belonging to the oligopeptide permease operon, was important for intracellular survival (Borezee *et al.*, 2000). Cultured cell models allow us to investigate how the pathogen can escape the normal host immune response. One strategy relies on interference with cytokine production by factors such as *Salmonella* LPS lipid A. PhoPQ-regulation of the structural modification of Lipid A is directly responsible for inducing expression of the E-selectin adhesion molecule in endothelial cells, and TNF-α in adherent monocytes (Guo *et al.*, 1997). In macrophages, SifA has recently been shown to be the bacterial determinant responsible for maintaining membrane integrity of the *Salmonella*-containing vacuole (Beuzon *et al.*, 2000).

However, the use of cultured cells does present limitations for the analysis of bacterial and host responses. (i) Most phagocytic and invasion assays require recently isolated primary cells, such as macrophages or bone-marrow-derived dendritic cells, to obtain biologically-relevant data. It is clear that differentiated macrophages rapidly lose important functions associated with host immune response during passage, preventing the acquisition of accurate and reliable data. (ii) Results can differ significantly between individual cell lines, even though they are all derived from the same cell type. A good example of this is the comparison of the importance of vacuole acidification on *S.* Typhimurium survival in different macrophage cell lines. Acidification is unnecessary for pathogen survival in RAW264.7 macrophages, but is crucial in J774 macrophages (Rathman *et al.*, 1996; Steele-Mortimer *et al.*, 2000). This highlights the danger of extrapolating from observations made on a simplified model to understand what happens during infection. (iii) Pure cultures of mammalian cells are generally grown in stabilized media. This situation is completely different to host tissue, where different cell types co-exist and share cross-talk. In summary, cultivated cell lines are convenient for preliminary studies but the results obtained must be confirmed in more complex systems such as animal models, when possible.

Animal models

The limitations of research based on cell lines described in the previous section have fuelled interest in relevant animal models. We know that the host represents a complex and dynamic environment, which is modified during the infection process, presenting a variety of stimuli to which the pathogen must respond if it is to be successful. Techniques like IVET, STM, Subtractive Hybridization or Differential Display have been extensively used with animal models to identify *ivi* genes and the use of these models to focus on the host/pathogen interaction is developing rapidly.

Techniques previously used for *ivi* gene identification are now commonly combined with other approaches to study the host/pathogen dialogue. Differential Display has been used in combination with *in situ* hybridization to characterize differential expression of nervous tissue genes upon *Herpes simplex* virus infection (Wilkinson *et al.*, 2000). Differential Display was used to identify cDNA derived from nervous-tissue-induced genes, and was subsequently used as probe for *in situ* hybridization experiments on nervous tissue sections. This approach allowed co-localization of transcript expression with specific cell-types of the host tissue, and should prove applicable to the identification of pathogen genes exhibiting cell-type specific expression patterns in the infected host.

The role of identified bacterial virulence determinants in animal infection has often been determined by studying the virulence attenuation of particular mutants, as reported for many *S.* Typhimurium studies (Deiwick *et al.*, 1998; Shea *et al.*, 1999). This strategy identifies the stage of pathogenesis at which a particular gene is required, and has been used effectively in combination with competitive index (Unsworth and Holden, 2000). Recently, inactivation of a DFI-identified gene of *S.* Typhimurium, *mig14*, was shown to have no effect on bacterial replication in host tissue early after infection, but to cause rapid clearance from spleen and liver at a later stage (Valdivia *et al.*, 2000). This suggests that *mig14* might be involved in survival of the bacteria in deep tissue.

However, the use of bacterial mutant studies to obtain indirect evidence is not sufficient to establish the role of specific genes in infection. It is crucial to determine where microbial virulence genes are expressed during infection, whether virulence genes exhibit organ- or cell type-specific expression patterns and if their expression varies from one stage of infection to another. This approach requires an optimized reporter system to visualize bacterial gene expression directly in host tissue. The system we have developed is being used to assess the transcriptional response of *Salmonella* SPI1, SPI2 and other virulence genes during mouse infection. This IVIF approach (***In Vivo*** **I**nduction of **F**luorescence)

(Hautefort and Hinton, 2000) uses the bright and stable GFP$^+$ as a transcriptional reporter gene that is fused to various virulence gene promoters and integrated into the chromosome at single copy. In both cultivated cells and in mouse infected liver and spleen, GFP fluorescence can be detected and quantified by fluorescent microscopic and flow cytometric techniques respectively. This GFP-based reporter system allows the study of single bacteria directly isolated from an infected host. Furthermore, IVIF facilitates the co-localization of bacterial gene expression with specific host immune markers. Figure 4.1 shows expression of *ssaG::gfp* within CD-18 labelled phagocytic cells in a section of infected mouse spleen.

Although animal models present many advantages over cultivated cell systems, a major limitation exists for human pathogens: most animal models only partially reproduce the symptoms observed in human infections. Indeed, in many instances, the animal model used does not closely resemble the condition found in human host and erroneous conclusions exist in the literature resulting from extrapolation from animal model to human (Smith, 1998). *Salmonella* has been extensively studied in mouse, although this model does not resemble human gastroenteritis. Very little information is available on the interaction of *S.* Typhimurium with the epithelium of the gastrointestinal (GI) tract in humans, as almost all available data come from animal studies and cell

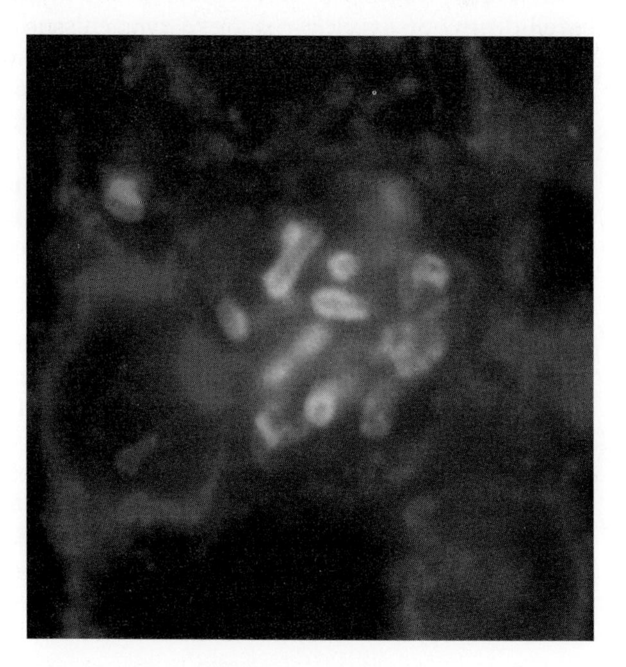

Figure 4.1. Spleen section of BALB/c mouse infected with *S.* Typhimurium. *Salmonella* cells that carry a single copy *ssaG::gfp* transcriptional fusion in the chromosomal *put* locus are detected with an anti-LPS antibody (red fluorescence). An anti-CD18 monoclonal antibody (blue fluorescence) identifies all phagocytic cells. Green fluorescence reflects *ssaG::gfp* expression within phagocytic cells. (This figure is also reproduced in colour between pages 276 and 277.)

culture models (Wallis and Galyov, 2000); current knowledge of the pathogenesis of *Salmonella* in humans is limited to observations from clinical or typhoid fever and experimental infection of volunteers (Hornick *et al.*, 1970a,b; Santos *et al.*, 2001; Tsolis *et al.*, 1999). Access to healthy human tissue samples is limited, and development of improved models must be a priority. We should remember that animal models, which mimic natural pathogen reservoirs (chicken, pigs, calves), are very useful for monitoring bacterial behaviour prior to human infection (Bispham *et al.*, 2001; Tsolis *et al.*, 1999). New animal models of increased relevance are being developed for pathogens initially studied in a different model, such as the use of chicken jejunal loops to study intestinal epithelial invasion by *S. enterica* (Aabo *et al.*, 2000). Intravenous infection of calves and mice by *S.* Dublin has recently allowed comparison of the behaviour of the same SPI2 mutants in different animal hosts (Bispham *et al.*, 2001). In both animal models, the *sseD* and *ssaT* mutants tested were attenuated, although still capable of invading bovine ileal loops.

The knock-out mouse is becoming the animal model of choice for dissecting pathogen interaction with the host immune system. The inactivation of certain modulators of the immune response, such as cytokines in knockout mice, allows their effects on bacterial pathogenesis to be established. Macrophage apoptosis has been described as a common mechanism for evading the immune system and facilitating intracellular survival by *Salmonella* (Monack *et al.*, 1996). Induction of this programmed cell-death by *Salmonella* requires a mechanism that is under investigation. Macrophages have been purified from Caspase-1 knock-out mice, and used for cytotoxicity and binding assays with *Salmonella*, in comparison with macrophages derived from a wild-type mouse. This study showed that the *Salmonella* SipB effector protein induces macrophage apoptosis by binding to the pro-apoptotic enzyme caspase-1 (Hersh *et al.*, 1999; Monack *et al.*, 2000). Knock-out mouse studies have been used to identify two other factors involved in the inhibition of intracellular proliferation by *Salmonella*. Phagocyte oxidase (Phox) and inducible nitric oxide synthase (iNOS) are two enzymes involved in the synthesis of reactive bactericidal oxidants. Phox and iNOS knock-out mouse experiments confirmed a role of these enzymes in microbial proliferation and host survival of *S.* Typhimurium (Mastroeni *et al.*, 2000). Subsequently, the authors showed that SPI2 effector proteins, such as SseB, are involved in targetting iNOS away from the *Salmonella*-containing vacuole (Vazquez-Torres *et al.*, 2000). The host response cascade is also being dissected by introducing the use of appropriate antibodies into the animal model (Mastroeni *et al.*, 1998).

◆◆◆◆◆◆ CONCLUSION

Studies at the level of molecular pathogenesis have considerably accelerated the pace of understanding bacterial infection processes. A wide range of powerful techniques is now available in this field of cellular microbiology, and interest is now extending to pathogens for which very few genetic tools have been available.

IVET, DFI and STM approaches have revolutionized studies of bacterial infection, allowing indirect selection of genes whose expression occurs during infection. Subtractive Hybridization, Differential Display and alternative techniques like SAGE or SCOTS are appealing because they facilitate direct monitoring of mRNA induced during infection. Proteomics allows the correlation of changes in levels of transcription of *ivi* genes with direct analysis of IVI proteins. Immunological-based techniques allow us to look directly at human infections, and to identify IVI proteins that are definitely involved in human infectious disease.

The rapid development of relevant tissue culture cell lines, new animal models, highly-sensitive microscopic and flow cytometric equipment combined with the increased availability of specific antibodies, fluorescent probes and reporter genes has had considerable impact on our ability to understand infection processes *in situ*. Analysis of gene expression in individual micro-organisms is now possible with IVIF for *S.* Typhimurium, and should be applicable to a wide range of pathogens.

The revolutionary development of DNA microarray technology provides a complementary approach for monitoring global levels of gene expression during infection. Microbial infection studies are moving towards a new post-genomic era that will certainly provide new insights into host/pathogen interactions.

Acknowledgement

We are grateful to Roy Bongaerts, Romina Emilianus, Duncan Maskell, Pietro Mastroeni, Francis Mulholland, and to Claudio Nicoletti and the rest of Hinton Laboratory for useful discussions. I.H. is supported by a Training and Mobility of Researchers fellowship from the European Union (contract number ERBFMRXCT9). J.C.D. Hinton is supported by the BBSRC.

References

Aabo, S., Christensen, J. P., Chadfield, M. S., Carstensen, B., Jensen, T. K., Bisgaard, M. and Olsen, J. E. (2000). Development of an *in vivo* model for study of intestinal invasion by *Salmonella enterica* in chickens. *Infect. Immun.* **68**, 7122–7125.

Abu Kwaik, Y. and Pederson, L. L. (1996). The use of differential display-PCR to isolate and characterize a *Legionella pneumophila* locus induced during the intracellular infection of macrophages. *Mol. Microbiol.* **21**, 543–556.

Altier, C., Suyemoto, M. and Lawhon, S. D. (2000). Regulation of *Salmonella enterica* Serovar Typhimurium invasion genes by *csrA*. *Infect. Immun.* **68**, 6790–6797.

Badger, J. L., Wass, C. A., Weissman, S. J. and Kim, K. S. (2000). Application of signature-tagged mutagenesis for identification of *Escherichia coli* K1 genes that contribute to invasion of human brain microvascular endothelial cells. *Infect. Immun.* **68**, 5056–5061.

Bakshi, C. S., Singh, V. P., Wood, M. W., Jones, P. W., Wallisand T. S. and Galyov, E. E. (2000). Identification of SopE2, a *Salmonella* secreted protein which is highly homologous to SopE and involved in bacterial invasion of epithelial cells. *J. Bacteriol.* **182**, 2341–2344.

Banu, S., Ohtani, K., Yaguchi, H., Swe, T., Cole, S. T., Hayashi, H. and Shimizu, T. (2000). Identification of novel VirR/VirS-regulated genes in *Clostridium perfringens*. *Mol. Microbiol.* **35**, 854–864.

Barker, L. P., Brooks, D. M. and Small, P. L. (1998). The identification of *Mycobacterium marinum* genes differentially expressed in macrophage phagosomes using promoter fusions to green fluorescent protein. *Mol. Microbiol.* **29**, 1167–1177.

Beatty, W. L. and Russell, D. G. (2000). Identification of mycobacterial surface proteins released into subcellular compartments of infected macrophages. *Infect. Immun.* **68**, 6997–6541.

Beuzon, C. R., Meresse, S., Unsworth, K. E., Ruiz-Albert, J., Garvis, S., Waterman, S. R., Ryder, T. A., Boucrot, E. and Holden, D. W. (2000). *Salmonella* maintains the integrity of its intracellular vacuole through the action of SifA. *Embo J.* **19**, 3235–3249.

Bispham, J., Tripathi, B. N., Watson, P. R. and Wallis, T. S. (2001). *Salmonella* pathogenicity island 2 influences both systemic salmonellosis and *Salmonella*-induced enteritis in calves. *Infect. Immun.* **69**, 367–377.

Borezee, E., Pellegrini, E. and Berche, P. (2000). OppA of *Listeria monocytogenes*, an oligopeptide-binding protein required for bacterial growth at low temperature and involved in intracellular survival. *Infect. Immun.* **68**, 7069–7077.

Brown, J. S., Aufauvre-Brown, A., Brown, J., Jennings, J. M., Arst, H., Jr. and Holden, D. W. (2000). Signature-tagged and directed mutagenesis identify PABA synthetase as essential for *Aspergillus fumigatus* pathogenicity. *Mol. Microbiol.* **36**, 1371–1380.

Bubert, A., Sokolovic, Z., Chun, S. K., Papatheodorou, L., Simm, A. and Goebel, W. (1999). Differential expression of *Listeria monocytogenes* virulence genes in mammalian host cells. *Mol. Gen. Genet.* **261**, 323–336.

Camilli, A., Beattie, D. T. and Mekalanos, J. J. (1994). Use of genetic recombination as a reporter of gene expression. *Proc. Natl Acad. Sci. USA* **91**, 2634–2638.

Camilli, A. and Mekalanos, J. J. (1995). Use of recombinase gene fusions to identify *Vibrio cholerae* genes induced during infection. *Mol. Microbiol.* **18**, 671–683.

Carulli, J. P., Artinger, M., Swain, P. M., Root, C. D., Chee, L., Tulig, C., Guerin, J., Osborne, M., Stein, G., Lian, J. and Lomedico, P. T. (1998). High throughput analysis of differential gene expression. *J. Cell Biochem. Suppl.* **30–31**, 286–296.

Cash, P. (2000). Proteomics in medical microbiology. *Electrophoresis.* **21**, 1187–1201.

Chalfie, M., Tu, Y., Euskirchen, G., Ward, W. W. and Prasher, D. C. (1994). Green fluorescent protein as a marker for gene expression. *Science* **263**, 802–805.

Chen, F., Binder, B. and Hodson, R. E. (2000). Flow cytometric detection of specific gene expression in prokaryotic cells using *in situ* RT-PCR. *FEMS Microbiol. Lett.* **184**, 291–216.

Cherayil, B. J., McCormick, B. A. and Bosley, J. (2000). *Salmonella enterica* serovar typhimurium-dependent regulation of inducible nitric oxide synthase expression in macrophages by invasins SipB, SipC, and SipD and effector SopE2. *Infect. Immun.* **68**, 5567–5574.

Chiang, S. L. and Mekalanos, J. J. (1998). Use of signature-tagged transposon mutagenesis to identify *Vibrio cholerae* genes critical for colonization. *Mol. Microbiol.* **27**, 797–805.

Chiang, S. L., Mekalanos, J. J. and Holden, D. W. (1999). *In vivo* genetic analysis of bacterial virulence. *Annu. Rev. Microbiol.* **53**, 129–154.

Ciacci-Woolwine, F., Blomfield, I. C., Richardson, S. H. and Mizel, S. B. (1998). *Salmonella* Flagellin Induces Tumor Necrosis Factor Alpha in a Human Promonocytic Cell Line. *Infect. Immun.* **66**, 1127–1134.

Cinelli, A. R. (1998). Flexible method to obtain high sensitivity, low-cost CCD cameras for video microscopy. *J. Neurosci. Methods* **85**, 33–43.

Cirillo, D. M., Valdivia, R. H., Monack, D. M. and Falkow, S. (1998). Macrophage-dependent induction of the *Salmonella* pathogenicity island 2 type III secretion system and its role in intracellular survival. *Mol. Microbiol.* **30**, 175–188.

Cohen, B. A., Mitra, R. D., Hughes, J. D. and Church, G. M. (2000). A computational analysis of whole-genome expression data reveals chromosomal domains of gene expression. *Nat. Genet.* **26**, 183–186.

Contag, C. H., Contag, P. R., Mullins, J. I., Spilman, S. D., Stevenson, D. K. and Benaron, D. A. (1995). Photonic detection of bacterial pathogens in living hosts. *Mol. Microbiol.* **18**, 593–603.

Cormack, B. P., Valdivia, R. H. and Falkow, S. (1996). FACS-optimized mutants of the green fluorescent protein (GFP). *Gene* **173**, 33–38.

Cortes-Bratti, X., Chaves-Olarte, E., Lagergard, T. and Thelestam, M. (2000). Cellular internalization of cytolethal distending toxin from *Haemophilus ducreyi*. *Infect. Immun.* **68**, 6903–6541.

Coulter, S. N., Schwan, W. R., Ng, E. Y., Langhorne, M. H., Ritchie, H. D., Westbrock-Wadman, S., Hufnagle, W. O., Folger, K. R., Bayer, A. S. and Stover, C. K. (1998). *Staphylococcus aureus* genetic loci impacting growth and survival in multiple infection environments. *Mol. Microbiol.* **30**, 393–404.

Crameri, A., Whitehorn, E. A., Tate, E. and Stemmer, W. P. (1996). Improved green fluorescent protein by molecular evolution using DNA shuffling. *Nat. Biotechnol.* **14**, 315–319.

Cummings, C. A. and Relman, D. A. (2000). Using DNA microarrays to study host–microbe interactions. *Emerg. Infect. Dis.* **6**, 513–525.

Daigle, F. , Graham, J. E. and Curtis, R. (2001). Identification of *Salmonella* Typhi genes expressed within macrophages by selective capture of transcribed sequences (SCOTS). *Mol. Microbiol.* **41**, 1211–1222.

Darwin, A. J. and Miller, V. L. (1999). Identification of *Yersinia enterocolitica* genes affecting survival in an animal host using signature-tagged transposon mutagenesis. *Mol. Microbiol.* **32**, 51–62.

Dasgupta, N., Kapur, V., Singh, K. K., Das, T. K., Sachdeva, S., Jyothisri, K. and Tyagi, J. S. (2000). Characterization of a two-component system, *devR-devS*, of *mycobacterium tuberculosis*. *Tuber. Lung Dis.* **80**, 141–159.

Davey, H. M. and Kell, D. B. (1996). Flow cytometry and cell sorting of heterogeneous microbial populations: the importance of single-cell analyses. *Microbiol. Rev.* **60**, 641–696.

Deiwick, J. and Hensel, M. (1999). Regulation of virulence genes by environmental signals in *Salmonella typhimurium*. *Electrophoresis* **20**, 813–817.

Deiwick, J., Nikolaus, T., Shea, J. E., Gleeson, C., Holden, D. W. and Hensel, M., (1998). Mutations in *Salmonella* pathogenicity island 2 (SPI2) genes affecting transcription of SPI1 genes and resistance to antimicrobial agents. *J. Bacteriol.* **180**, 4775–4780.

Denk, W., Strickler, J. H. and Webb, W. W. (1990). Two-photon laser scanning fluorescence microscopy. *Science* **248**, 73–76.

DeRisi, J., Penland, L., Brown, P. O., Bittner, M. L., Meltzer, P. S., Ray, M., Chen, Y., Su, Y. A. and Trent, J. M. (1996). Use of a cDNA microarray to analyse gene expression patterns in human cancer. *Nature Genetics* **14**, 457–460.

Diatchenko, L., Lukyanov, S., Lau, Y. F. and Siebert, P. D. (1999). Suppression subtractive hybridization: a versatile method for identifying differentially expressed genes. *Methods Enzymol.* **303**, 349–380.

Du, L. D. and Kolenbrander, P. E. (2000). Identification of saliva-regulated genes of *Streptococcus gordonii* DL1 by differential display using random arbitrarily primed PCR. *Infect. Immun.* **68**, 4834–4837.

Dubail, I. and Berche, P. The European *Listeria* Genome Consortium and Charbit, A. (2000). Listeriolysin O as a reporter to identify constitutive and *in vivo*-inducible promoters in the pathogen *Listeria monocytogenes. Infect. Immun.* **68**, 3242–3073.

Daigle, F., Graham, J. E. and Curtis, R. The European *Listeria* Genome Consortium and Charbit, A. (2000). Listeriolysin O as a reporter to identify constitutive and *in vivo*-inducible promoters in the pathogen *Listeria monocytogenes. Infect. Immun.* **68**, 3242–3073.

Eckmann, L., Smith, J. R., Housley, M. P., Dwinell, M. B. and Kagnoff, M. F. (2000). Analysis by high density cDNA arrays of altered gene expression in human intestinal epithelial cells in response to infection with the invasive enteric bacteria *Salmonella. J. Biol. Chem.* **275**, 14084–14094.

El-Adhami, W. (1999). Expression of a clone specific DNA sequence from *Staphylococcus aureus* in *Escherichia coli. J. Biotechnol.* **73**, 181–184.

Entwistle, A. (1998). A comparison between the use of a high-resolution CCD camera and 35 mm film for obtaining coloured micrographs. *J. Microsc.* **192**, 81–89.

Errampalli, D., Leung, K., Cassidy, M. B., Kostrzynska, M., Blears, M., Lee, H. and Trevors, J. T. (1999). Applications of the green fluorescent protein as a molecular marker in environmental microorganisms. *J. Microbiol. Methods.* **35**, 187–199.

Finlay, B. B., Fry, J., Rock, E. P. and Falkow, S. (1989). Passage of *Salmonella* through polarized epithelial cells: role of the host and bacterium. *J. Cell Sci. Suppl.* **11**, 99–107.

Fleming, J. T., Yao, W. H. and Sayler, G. S. (1998). Optimization of differential display of prokaryotic mRNA: application to pure culture and soil microcosms. *Appl. Environ. Microbiol.* **64**, 3698–3706.

Forsberg, A. and Rosqvist, R. (1993). *In vivo* expression of virulence genes of *Yersinia pseudotuberculosis. Infect. Agents Dis.* **2**, 275–278.

Foulongne, V., Bourg, G., Cazevieille, C., Michaux-Charachon, S. and O'Callaghan, D. (2000). Identification of *Brucella suis* genes affecting intracellular survival in an *in vitro* human macrophage infection model by signature-tagged transposon mutagenesis. *Infect. Immun.* **68**, 1297–1303.

Francis, K. P., Joh, D., Bellinger-Kawahara, C., Hawkinson, M. J., Purchio, T. F. and Contag, P. R. (2000). Monitoring bioluminescent *staphylococcus aureus* infections in living mice using a novel *luxABCDE* construct. *Infect. Immun.* **68**, 3594–3600.

Fu, Y. and Galan, J. E. (1999). A *Salmonella* protein antagonizes Rac-1 and Cdc42 to mediate host-cell recovery after bacterial invasion. *Nature* **401**, 293–297.

Fuller, T. E., Shea, R. J., Thacker, B. J. and Mulks, M. H. (1999). Identification of *in vivo* induced genes in *Actinobacillus pleuropneumoniae. Microb. Pathog.* **27**, 311–327.

Fuller, T. E., Kennedy, M. J. and Lowery, D. E. (2000a). Identification of *Pasteurella multocida* virulence genes in a septicemic mouse model using signature-tagged mutagenesis. *Microb. Pathog.* **29**, 25–38.

Fuller, T. E., Martin, S., Teel, J. F., Alaniz, G. R., Kennedy, M. J. and Lowery, D. E. (2000b). Identification of *Actinobacillus pleuropneumoniae* virulence genes using signature-tagged mutagenesis in a swine infection model. *Microb. Pathog.* **29**, 39–51.

Fung, D. C. and Theriot, J. A. (1998). Imaging techniques in microbiology. *Curr. Opin. Microbiol.* **1**, 346–351.

Gahan, C. G. and Hill, C. (2000). The use of listeriolysin to identify *in vivo* induced genes in the gram-positive intracellular pathogen *Listeria monocytogenes*. *Mol. Microbiol.* **36**, 498–507.

Galan, J. E. (1999). Interaction of *Salmonella* with host cells through the centisome 63 type III secretion system. *Curr. Opin. Microbiol.* **2**, 46–50.

Galan, J. E. and Zhou, D. (2000). Striking a balance: modulation of the actin cytoskeleton by *Salmonella*. *Proc. Natl Acad. Sci. USA* **97**, 8754–8761.

Gewirtz, A. T., Siber, A. M., Madara, J. L. and McCormick, B. A. (1999). Orchestration of neutrophil movement by intestinal epithelial cells in response to *Salmonella typhimurium* can be uncoupled from bacterial internalization. *Infect. Immun.* **67**, 608–617.

Gill, R. T., Valdes, J. J. and Bentley, W. E. (1999). Reverse transcription-PCR differential display analysis of *Escherichia coli* global gene regulation in response to heat shock. *Appl. Environ. Microbiol.* **65**, 5386–5393.

Ginocchio, C. C., Olmsted, S. B., Wells, C. L. and Galan, J. E. (1994). Contact with epithelial cells induces the formation of surface appendages on *Salmonella typhimurium*. *Cell.* **76**, 717–724.

Gonzalez, P., Zigler, J. S., Jr., Epstein, D. L. and Borras, T. (1999). Identification and isolation of differentially expressed genes from very small tissue samples. *Biotechniques* **26**, 884–892.

Graham, J. E. and Clark-Curtiss, J. E. (1999). Identification of *Mycobacterium tuberculosis* RNAs synthesized in response to phagocytosis by human macrophages by selective capture of transcribed sequences (SCOTS). *Proc. Natl Acad. Sci. USA* **96**, 11554–11559.

Guignot, J., Breard, J., Bernet-Camard, M.-F., Peiffer, I., Nowicki, B. J., Servin, A. L. and Blanc-Potard, A.-B. (2000). Pyelonephritogenic diffusely adhering *Escherichia coli* EC7372 harboring Dr-II adhesin carries classical uropathogenic virulence genes and promotes cell lysis and apoptosis in polarized epithelial Caco-2/TC7 cells. *Infect. Immun.* **68**, 7018–7027.

Guo, L., Lim, K. B., Gunn, J. S., Bainbridge, B., Darveau, R. P., Hackett, M. and Miller, S. I. (1997). Regulation of lipid A modifications by *Salmonella typhimurium* virulence genes *phoP-phoQ*. *Science* **276**, 250–253.

Guy, R. L., Gonias, L. A. and Stein, M. A. (2000). A fluorescence microscopy based genetic screen to identify mutants altered for interactions with host cells. *J. Microbiol. Methods* **42**, 129–138.

Ha, U. W. and Jin, S. G. (1999). Expression of the *soxR* gene of *Pseudomonas aeruginosa* is inducible during infection of burn wounds in mice and is required to cause efficient bacteremia. *Infect. Immun.* **67**, 5324–5331.

Hand, N. J. and Silhavy, T. J. (2000). A practical guide to the construction and use of *lac* fusions in *Escherichia coli*. *Methods Enzymol.* **326**, 11–35.

Handfield, M., Lehoux, D. E., Sanschagrin, F., Mahan, M. J., Woods, D. E. and Levesque, R. C. *In vivo*-induced genes in *Pseudomonas aeruginosa*. *Infect. Immun.* **68**, 2359–2362.

Handfield, M., Brady, L. J., Progulske-Fox, A. and Hillman, J. D. (2000a). Direct probing for *Actinobacillus actinomyetemcomitans in vivo* induced genes in localized juvenite periodontitis. *J. Dent. Res.* **79**, 1557.

Handfield, M., Brady, L. J., Progulske-Fox, A. and Hillman, J. D. (2000b). IVIAT: a novel method to identify microbial genes expressed specifically during human infections. *Trends Microbiol.* **8**, 336–339.

Harb, O. S. and Abu Kwaik, Y. (2000). Essential Role for the *Legionella pneumophila* Rep Helicase Homologue in Intracellular Infection of Mammalian Cells. *Infect. Immun.* **68**, 6970–6541.

Hardt, W. D., Chen, L. M., Schuebel, K. E., Bustelo, X. R. and Galan, J. E. (1998). *S. typhimurium* encodes an activator of Rho GTPases that induces membrane ruffling and nuclear responses in host cells. *Cell* **93**, 815–826.

Hautefort, I. and Hinton, J. C. (2000). Measurement of bacterial gene expression *in vivo*. *Philos. Trans. R. Soc. Lond. B. Biol. Sci.* **355**, 601–611.

Hayward, R. D., McGhie, E. J. and Koronakis, V. (2000). Membrane fusion activity of purified SipB, a *Salmonella* surface protein essential for mammalian cell invasion. *Mol. Microbiol.* **37**, 727–739.

Heim, R., Cubitt, A. B. and Tsien, R. Y. (1995). Improved green fluorescence. *Nature* **373**, 663–664.

Heim, R., Prasher, D. C. and Tsien, R. Y. (1994). Wavelength mutations and post-translational autoxidation of green fluorescent protein. *Proc. Natl. Acad. Sci. USA* **91**, 12501–12504.

Heithoff, D. M., Conner, C. P., Hanna, P. C., Julio, S. M., Hentschel, U. and Mahan, M. J. (1997). Bacterial infection as assessed by *in vivo* gene expression. *Proc. Natl Acad. Sci. USA* **94**, 934–939.

Hensel, M. (1998). Whole genome scan for habitat-specific genes by signature-tagged mutagenesis. *Electrophoresis*. **19**, 608–612.

Hensel, M., Shea, J. E., Gleeson, C., Jones, M. D., Dalton, E. and Holden, D. W. (1995). Simultaneous identification of bacterial virulence genes by negative selection. *Science* **269**, 400–403.

Hersh, D., Monack, D. M., Smith, M. R., Ghori, N., Falkow, S. and Zychlinsky, A. (1999). The *Salmonella* invasin SipB induces macrophage apoptosis by binding to caspase-1. *Proc. Natl Acad. Sci. USA* **96**, 2396–2401.

Hewitt, C. J., Nebe-Von Caron, G., Axelsson, B., McFarlane, C. M. and Nienow, A. W. (2000). Studies related to the scale-up of high-cell-density *E. coli* fed-batch fermentations using multiparameter flow cytometry: Effect of a changing microenvironment with respect to glucose and dissolved oxygen concentration. *Biotechnol. Bioeng.* **70**, 381–390.

Hickey, T. E., McVeigh, A. L., Scott, D. A., Michielutti, R. E., Bixby, A., Carroll, S. A., Bourgeois, A. L. and Guerry, P. (2000). *Campylobacter jejuni* Cytolethal Distending Toxin Mediates Release of Interleukin-8 from Intestinal Epithelial Cells. *Infect. Immun.* **68**, 6535–6541.

Hinton, J. C. (1997). The *Escherichia coli* genome sequence: the end of an era or the start of the FUN? *Mol. Microbiol.* **26**, 417–422.

Hobbie, S., Chen, L. M., Davis, R. J. and Galan, J. E. (1997). Involvement of mitogen-activated protein kinase pathways in the nuclear responses and cytokine production induced by *Salmonella typhimurium* in cultured intestinal epithelial cells. *J. Immunol.* **159**, 5550–5559.

Hornick, R. B., Greisman, S. E., Woodward, T. E., DuPont, H. L., Dawkins, A. T. and Snyder, M. J. (1970a). Typhoid fever: pathogenesis and immunologic control. *N. Engl. J. Med.* **283**, 686–691.

Hornick, R. B., Greisman, S. E., Woodward, T. E., DuPont, H. L., Dawkins, A. T. and Snyder, M. J. (1970b). Typhoid fever: pathogenesis and immunologic control. 2. *N. Engl. J. Med.* **283**, 739–746.

Hughes, T. R., Marton, M. J., Jones, A. R., Roberts, C. J., Stoughton, R., Armour, C. D., Bennett, H. A., Coffey, E., Dai, H. Y., He, Y. D.D., Kidd, M. J., King, A. M., Meyer, M. R., Slade, D., Lum, P. Y., Stepaniants, S. B., Shoemaker, D. D., Gachotte, D., Chakraburtty, K., Simon, J., Bard, M. and Friend, S. H.

(2000). Functional discovery via a compendium of expression profiles. *Cell* **102**, 109–126.

Jacob, F. and Monod, J. (1961). Genetic regulatory mechanisms in the synthesis of proteins. *J. Mol. Biol.* **3**, 318–356.

Janakiraman, A. and Slauch, J. M. (2000). The putative iron transport system SitABCD encoded on SPI1 is required for full virulence of *Salmonella typhimurium*. *Mol. Microbiol.* **35**, 1146–1155.

Jaufeerally-Fakim, Y., Autrey, J. C., Daniels, M. J. and Dookun, A. (2000). Genetic polymorphism in *Xanthomonas albilineans* strains originating from 11 geographical locations, revealed by two DNA probes. *Lett. Appl. Microbiol.* **30**, 287–293.

Jepson, M. A., Schlecht, H. B. and Collares-Buzato, C. B. (2000). Localization of dysfunctional tight junctions in *Salmonella enterica* serovar typhimurium-infected epithelial layers. *Infect. Immun.* **68**, 7202–7208.

Jones, A. L., Knoll, K. M. and Rubens, C. E. (2000). Identification of *Streptococcus agalactiae* virulence genes in the neonatal rat sepsis model using signature-tagged mutagenesis. *Mol. Microbiol.* **37**, 1444–1455.

Kilic, A. O., Herzberg, M. C., Meyer, M. W., Zhao, X. M. and Tao, L. (1999). Streptococcal reporter gene-fusion vector for identification of in vivo expressed genes. *Plasmid.* **42**, 67–72.

Kleckner, N., Roth, J. and Botstein, D. (1977). Genetic engineering *in vivo* using translocatable drug-resistance elements. New methods in bacterial genetics. *J. Mol. Biol.* **116**, 125–159.

Kohler, R., Bubert, A., Goebel, W., Steinert, M., Hacker, J. and Bubert, B. (2000). Expression and use of the green fluorescent protein as a reporter system in *Legionella pneumophila*. *Mol. Gen. Genet.* **262**, 1060–1069.

Kremer, L., Baulard, A., Estaquier, J., Poulain-Godefroy, O. and Locht, C. (1995). Green fluorescent protein as a new expression marker in mycobacteria. *Mol. Microbiol.* **17**, 913–922.

Kunert, A., Hagemann, M. and Erdmann, N. (2000). Construction of promoter probe vectors for *Synechocystis* sp. PCC 6803 using the light-emitting reporter systems GFP and LuxAB. *J. Microbiol. Methods* **41**, 185–194.

Lange, M., Tolker-Nielsen, T., Molin, S. and Ahring, B. K. (2000). *In situ* reverse transcription-PCR for monitoring gene expression in individual *Methanosarcina mazei* S-6 cells. *Appl. Environ. Microbiol.* **66**, 1796–1800.

Lee, A. K. and Falkow, S. (1998). Constitutive and inducible green fluorescent protein expression in *Bartonella henselae*. *Infect. Immun.* **66**, 3964–3967.

Lee, S. H. and Camilli, A. (2000). Novel approaches to monitor bacterial gene expression in infected tissue and host. *Curr. Opin. Microbiol.* **3**, 97–101.

Lehoux, D. E. and Levesque, R. C. (2000). Detection of genes essential in specific niches by signature-tagged mutagenesis. *Curr. Opin. Biotechnol.* **11**, 434–439.

Lehoux, D. E., Sanschagrin, F. and Levesque, R. C. (1999). Defined oligonucleotide tag pools and PCR screening in signature-tagged mutagenesis of essential genes from bacteria. *Biotechniques* **26**, 473–480.

Li, Y.-H., Yan, Z.-Q., Jensen, J. S., Tullus, K. and Brauner, A. (2000). Activation of nuclear factor kappa B and induction of inducible nitric oxide synthase by *Ureaplasma urealyticum* in macrophages. *Infect. Immun.* **68**, 7087–7093.

Liang, P. and Pardee, A. B. (1992). Differential display of eukaryotic messenger RNA by means of the polymerase chain reaction. *Science* **257**, 967–971.

Licht, T. R., Krogfelt, K. A., Cohen, P. S., Poulsen, L. K., Urbance, J. and Molin, S. (1996). Role of lipopolysaccharide in colonization of the mouse intestine by *Salmonella typhimurium* studied by *in situ* hybridization. *Infect. Immun.* **64**, 3811–3817.

Lockhart, D. J., Dong, H., Byrne, M. C., Follettie, M. T., Gallo, M. V., Chee, M. S., Mittmann, M., Wang, C., Kobayashi, M., Horton, H. and Brown, E. L. (1996). Expression monitoring by hybridization to high-density oligonucleotide arrays. *Nat. Biotech.* **14**, 1675–1680.

Lowe, A. M., Beattie, D. T. and Deresiewicz, R. L. (1998). Identification of novel staphylococcal virulence genes by *in vivo* expression technology. *Mol. Microbiol.* **27**, 967–976.

Lucchini, S., Thompson, A. and Hinton, J. C. D. (2001). Microassays for microbiologists. *Microbiol.* **147**, 1403–1414.

Lukomski, S., Nakashima, K., Abdi, I., Cipriano, V. J., Ireland, R. M., Reid, S. D., Adams, G. G. and Musser, J. M. (2000). Identification and characterization of the *scl* gene encoding a group A *streptococcus* extracellular protein virulence factor with similarity to human collagen. *Infect. Immun.* **68**, 6542–6553.

Mahan, M. J., Slauch, J. M. and Mekalanos, J. J. (1993). Selection of bacterial virulence genes that are specifically induced in host tissues. *Science* **259**, 686–688.

Mahan, M. J., Tobias, J. W., Slauch, J. M., Hanna, P. C., Collier, R. J. and Mekalanos, J. J. (1995). Antibiotic-based selection for bacterial genes that are specifically induced during infection of a host. *Proc. Natl Acad. Sci. USA* **92**, 669–673.

Manger, I. D. and Relman, D. A. (2000). How the host 'sees' pathogens: global gene expression responses to infection. *Curr. Opin. Immunol.* **12**, 215–218.

Martindale, J., Stroud, D., Moxon, E. R. and Tang, C. M. (2000). Genetic analysis of *Escherichia coli* K1 gastrointestinal colonization. *Mol. Microbiol.* **37**, 1293–1305.

Mastroeni, P., Harrison, J. A., Robinson, J. H., Clare, S., Khan, S., Maskell, D. J., Dougan, G. and Hormaeche, C. E. (1998). Interleukin-12 is required for control of the growth of attenuated aromatic-compound-dependent salmonellae in BALB/c mice: role of gamma interferon and macrophage activation. *Infect. Immun.* **66**, 4767–4776.

Mastroeni, P., Vazquez-Torres, A., Fang, F. C., Xu, Y., Khan, S., Hormaeche, C. E. and Dougan, G. (2000). Antimicrobial actions of the NADPH phagocyte oxidase and inducible nitric oxide synthase in experimental Salmonellosis. II. Effects on microbial proliferation and host survival *in vivo*. *J. Exp. Med.* **192**, 237–248.

Matthysse, A. G., Stretton, S., Dandie, C., McClure, N. C. and Goodman, A. E. (1996). Construction of GFP vectors for use in gram-negative bacteria other than *Escherichia coli*. *FEMS Microbiol. Lett.* **145**, 87–94.

Maurer, J. J., Doggett, T. A., Burns-Keliher, L. and Curtiss, R. 3rd. (2000). Expression of the *rfa*, LPS biosynthesis promoter in *Salmonella typhimurium* during invasion of intestinal epithelial cells. *Curr. Microbiol.* **41**, 172–176.

Mei, J. M., Nourbakhsh, F., Ford, C. W. and Holden, D. W. (1997). Identification of *Staphylococcus aureus* virulence genes in a murine model of bacteraemia using signature-tagged mutagenesis. *Mol. Microbiol.* **26**, 399–407.

Merrell, D. S. and Camilli, A. (2000). Detection and analysis of gene expression during infection by *in vivo* expression technology. Philos. *Trans. R. Soc. Lond. B. Biol. Sci.* **355**, 587–599.

Miller, W. G., Leveau, J. H. J. and Lindow, S. E. (2000). Improved gfp and inaZ broad-host-range promoter-probe vectors. *Molecular Plant-Microbe Interactions* **13**, 1243–1250.

Miller, W. G. and Lindow, S. E. (1997). An improved GFP cloning cassette designed for prokaryotic transcriptional fusions. *Gene* **191**, 149–153.

Molin, S. and Givskov, M. (1999). Application of molecular tools for *in situ* monitoring of bacterial growth activity. *Environ. Microbiol.* **1**, 383–391.

Monack, D. M., Raupach, B., Hromockyj, A. E. and Falkow, S. (1996). *Salmonella typhimurium* invasion induces apoptosis in infected macrophages. *Proc. Natl Acad. Sci. USA* **93**, 9833–9838.
Monack, D. M., Hersh, D., Ghori, N., Bouley, D., Zychlinsky, A. and Falkow, S. (2000). Salmonella exploits Caspase-1 to colonize Peyer's patches in a murine typhoid model. *J. Exp. Med.* **192**, 249–258.
Morrow, B. J., Graham, J. E. and Curtiss, R., 3rd. (1999). Genomic subtractive hybridization and selective capture of transcribed sequences identify a novel *Salmonella typhimurium* fimbrial operon and putative transcriptional regulator that are absent from the *Salmonella typhi* genome. *Infect. Immun.* **67**, 5106–5116.
Motley, S. T. and Lory, S. (1999). Functional characterization of a serine/threonine protein kinase of *Pseudomonas aeruginosa*. *Infect. Immun.* **67**, 5386–5394.
Nebe-von-Caron, G., Stephens, P. J., Hewitt, C. J., Powell, J. R. and Badley, R. A. (2000). Analysis of bacterial function by multi-colour fluorescence flow cytometry and single cell sorting. *J. Microbiol. Methods* **42**, 97–114.
Niedergang, F., Sirard, J. C., Blanc, C. T. and Kraehenbuhl, J. P. (2000). Entry and survival of *Salmonella typhimurium* in dendritic cells and presentation of recombinant antigens do not require macrophage-specific virulence factors. *Proc. Natl Acad. Sci. USA* **97**, 14650–14655.
Nwoguh, C. E., Harwood, C. R. and Barer, M. R. (1995). Detection of induced beta-galactosidase activity in individual non-culturable cells of pathogenic bacteria by quantitative cytological assay. *Mol. Microbiol.* **17**, 545–554.
Ogawa, H., Fukushima, K., Sasaki, I. and Matsuno, S. (2000). Identification of genes involved in mucosal defense and inflammation associated with normal enteric bacteria. *Am. J. Physiol. Gastrointest. Liver Physiol.* **279**, 492–499.
Paranjpye, R. N., Lara, J. C., Pepe, J. C., Pepe, C. M. and Strom, M. S. (1998). The type IV leader peptidase/N-methyltransferase of *Vibrio vulnificus* controls factors required for adherence to HEp-2 cells and virulence in iron-overloaded mice. *Infect. Immun.* **66**, 5659–5668.
Plum, G. and Clark-Curtiss, J. E. (1994). Induction of *Mycobacterium avium* gene expression following phagocytosis by human macrophages. *Infect. Immun.* **62**, 476–483.
Polissi, A., Pontiggia, A., Feger, G., Altieri, M., Mottl, H., Ferrari, L. and Simon, D. (1998). Large-scale identification of virulence genes from *Streptococcus pneumoniae*. *Infect. Immun.* **66**, 5620–5629.
Prouty, A. M. and Gunn, J. S. (2000). *Salmonella enterica* serovar typhimurium invasion is repressed in the presence of bile. *Infect. Immun.* **68**, 6763–6769.
Ragno, S., Estrada, I., Butler, R. and Colston, M. J. (1997). Regulation of macrophage gene expression following invasion by *Mycobacterium tuberculosis*. *Immunol. Lett.* **57**, 143–146.
Ragno, S., Estrada-Garcia, I., Butler, R. and Colston, M. J. (1998). Regulation of macrophage gene expression by *Mycobacterium tuberculosis*: down-regulation of mitochondrial cytochrome c oxidase. *Infect. Immun.* **66**, 3952–3958.
Rainey, P. B. (1999). Adaptation of *Pseudomonas fluorescens* to the plant rhizosphere. *Environ. Microbiol.* **1**, 243–257.
Ramakrishnan, L., Federspiel, N. A. and Falkow, S. (2000). Granuloma-specific expression of *Mycobacterium* virulence proteins from the glycine-rich PE-PGRS family. *Science* **288**, 1436–1439.
Rathman, M., Barker, L. P. and Falkow, S. (1997). The unique trafficking pattern of *Salmonella typhimurium*-containing phagosomes in murine macrophages is independent of the mechanism of bacterial entry. *Infect. Immun.* **65**, 1475–1485.

Rathman, M., Jouirhi, N., Allaoui, A., Sansonetti, P., Parsot, C. and Tran Van Nhieu, G. (2000). The development of a FACS-based strategy for the isolation of *Shigella flexneri* mutants that are deficient in intercellular spread. *Mol. Microbiol.* **35**, 974–990.

Rathman, M., Sjaastad, M. D. and Falkow, S. (1996). Acidification of phagosomes containing *Salmonella typhimurium* in murine macrophages. *Infect. Immun.* **64**, 2765–2773.

Retallack, D. M., Deepe, G. S., Jr. and Woods, J. P. (2000). Applying *in vivo* expression technology (IVET) to the fungal pathogen *Histoplasma capsulatum*. *Microb. Pathog.* **28**, 169–182.

Richter-Dahlfors, A., Buchan, A. M. J. and Finlay, B. B. (1997). Murine salmonellosis studied by confocal microscopy: *Salmonella typhimurium* resides intracellularly inside macrophages and exerts a cytotoxic effect on phagocytes *in vivo*. *J. Exp. Med.* **186**, 569–580.

Rosenberger, C. M., Scott, M. G., Gold, M. R., Hancock, R. E. and Finlay, B. B. (2000). *Salmonella typhimurium* infection and lipopolysaccharide stimulation induce similar changes in macrophage gene expression. *J. Immunol.* **164**, 5894–5904.

Rowland, B., Purkayastha, A., Monserrat, C., Casart, Y., Takiff, H. and McDonough, K. A. (1999). Fluorescence-based detection of *lacZ* reporter gene expression in intact and viable bacteria including *Mycobacterium* species. *FEMS Microbiol. Lett.* **179**, 317–325.

Sahly, H., Podschun, R., Oelschlaeger, T. A., Greiwe, M., Parolis, H., Hasty, D., Kekow, J., Ullmann, U., Ofek, I. and Sela, S. (2000). Capsule impedes adhesion to and invasion of epithelial cells by *Klebsiella pneumoniae*. *Infect. Immun.* **68**, 6744–6749.

Santos, R. L., Zhang, S., Tsolis, R. M., Kingsley, R. A., Adams, L. G. and Baumler, A. J. (2001). Animal models of *Salmonella* infections: enteritis vs. typhoid fever. *Microbes and Infection*. in press.

Savkovic, S. D., Koutsouris, A., Wu, G. and Hecht, G. (2000). Infection by bacterial pathogens expressing type III secretion decreases luciferase activity: ramifications for reporter gene studies. *Biotechniques* **29**, 514–522.

Schena, M., Shalon, D., Davis, R. W. and Brown, P. O. (1995). Quantitative monitoring of gene expression patterns with a complementary DNA microarray. *Science* **270**, 467–470.

Scholz, O., Thiel, A., Hillen, W. and Niederweis, M. (2000). Quantitative analysis of gene expression with an improved green fluorescent protein. *Eur. J. Biochem.* **267**, 1565–1570.

Schwan, W. R., Coulter, S. N., Ng, E. Y., Langhorne, M. H., Ritchie, H. D., Brody, L. L., Westbrock-Wadman, S., Bayer, A. S., Folger, K. R. and Stover, C. K. (1998). Identification and characterization of the PutP proline permease that contributes to *in vivo* survival of *Staphylococcus aureus* in animal models. *Infect. Immun.* **66**, 567–572.

Shapiro, H. M. (1995). *Practical Flow Cytometry.* Wiley-Liss, New York.

Shea, J. E., Beuzon, C. R., Gleeson, C., Mundy, R. and Holden, D. W. (1999). Influence of the *Salmonella typhimurium* pathogenicity island 2 type III secretion system on bacterial growth in the mouse. *Infect. Immun.* **67**, 213–219.

Shea, J. E., Hensel, M., Gleeson, C. and Holden, D. W. (1996). Identification of a virulence locus encoding a second type III secretion system in *Salmonella typhimurium*. *Proc. Natl Acad. Sci. USA* **93**, 2593–2597.

Shea, J. E., Santangelo, J. D. and Feldman, R. G. (2000). Signature-tagged mutagenesis in the identification of virulence genes in pathogens. *Curr. Opin. Microbiol.* **3**, 451–458.

Siemering, K. R., Golbik, R., Sever, R. and Haseloff, J. (1996). Mutations that suppress the thermosensitivity of green fluorescent protein. *Curr. Biol.* **6**, 1653–1663.

Slauch, J. M. and Camilli, A. (2000). IVET and RIVET: use of gene fusions to identify bacterial virulence factors specifically induced in host tissues. *Methods Enzymol.* **326**, 73–96.

Smith, H. (1998). What happens to bacterial pathogens *in vivo*? *Trends Microbiol.* **6**, 239–243.

Staendner, L. H., Rohde, M., Timmis, K. N. and Guzman, C. A. (1995). Identification of *Salmonella typhi* promoters activated by invasion of eukaryotic cells. *Mol. Microbiol.* **18**, 891–902.

Staib, P., Kretschmar, M., Nichterlein, T., Kohler, G., Michel, S., Hof, H., Hacker, J. and Morschhauser, J. (1999). Host-induced, stage-specific virulence gene activation in *Candida albicans* during infection. *Mol. Microbiol.* **32**, 533–546.

Stanley, T. L., Ellermeier, C. D. and Slauch, J. M. (2000). Tissue-specific gene expression identifies a gene in the lysogenic phage Gifsy-1 that affects *Salmonella enterica* serovar typhimurium survival in Peyer's patches. *J. Bacteriol.* **182**, 4406–4413.

Steele-Mortimer, O., St-Louis, M., Olivier, M. and Finlay, B. B. (2000). Vacuole acidification is not required for survival of *Salmonella enterica* serovar typhimurium within cultured macrophages and epithelial cells. *Infect. Immun.* **68**, 5401–5404.

Sturtevant, J. (2000). Applications of differential-display reverse transcription-PCR to molecular pathogenesis and medical mycology. *Clin. Microbiol. Rev.* **13**, 408–427.

Svensson, M., Johansson, C. and Wick, M. J. (2000). *Salmonella enterica* Serovar typhimurium-induced maturation of bone marrow-derived dendritic cells. *Infect. Immun.* **68**, 6311–6320.

Thirumalai, K., Kim, K. and Zychlinsky, A. (1997). IpaB, a *Shigella flexneri* invasin, colocalizes with interleukin-1 beta-converting enzyme in the cytoplasm of macrophages. *Infect. Immun.* **65**, 787–793.

Tinsley, C. R. and Nassif, X. (1996). Analysis of the genetic differences between *Neisseria meningitidis* and *Neisseria gonorrhoeae*: two closely related bacteria expressing two different pathogenicities. *Proc. Natl. Acad. Sci. USA.* **93**, 11109–11114.

Tolker-Nielsen, T., Holmstrom, K. and Molin, S. (1997). Visualization of specific gene expression in individual *Salmonella typhimurium* cells by *in situ* PCR. *Appl. Environ. Microbiol.* **63**, 4196–4203.

Triccas, J. A., Berthet, F. X., Pelicic, V. and Gicquel, B. (1999). Use of fluorescence induction and sucrose counterselection to identify *Mycobacterium tuberculosis* genes expressed within host cells. *Microbiology* **145**, 2923–2930.

Tsien, R. Y. (1998). The green fluorescent protein. *Annu. Rev. Biochem.* **67**, 509–544.

Tsolis, R. M., Kingsley, R. A., Townsend, S. M., Ficht, T. A., Adams, L. G. and Baumler, A. J. (1999). Of mice, calves, and men. Comparison of the mouse typhoid model with other *Salmonella* infections. *Adv. Exp. Med. Biol.* **473**, 261–274.

Tsuji, S., Matsumoto, M., Takeuchi, O., Akira, S., Azuma, I., Hayashi, A., Toyoshima, K. and Seya, T. (2000). Maturation of human dendritic cells by cell wall skeleton of *Mycobacterium bovis* bacillus Calmette-Guerin: involvement of toll-like receptors. *Infect. Immun.* **68**, 6883–6541.

Unsworth, K. E. and Holden, D. W. (2000). Identification and analysis of bacterial virulence genes *in vivo*. *Philos. Trans. R. Soc. Lond. B. Biol. Sci.* **355**, 613–622.

Valdivia, R. H., Cirillo, D. M., Lee, A. K., Bouley, D. M. and Falkow, S. (2000). *mig-14* is a horizontally acquired, host-induced gene required for *Salmonella enterica* lethal infection in the murine model of typhoid fever. *Infect. Immun.* **68**, 7126–7131.

Valdivia, R. H. and Falkow, S. (1996). Bacterial genetics by flow cytometry: rapid isolation of *Salmonella typhimurium* acid-inducible promoters by differential fluorescence induction. *Mol. Microbiol.* **22**, 367–378.

Valdivia, R. H. and Falkow, S. (1997). Fluorescence-based isolation of bacterial genes expressed within host cells. *Science* **277**, 2007–2011.

Valdivia, R. H. and Ramakrishnan, L. (2000). Applications of gene fusions to green fluorescent protein and flow cytometry to the study of bacterial gene expression in host cells. *Methods Enzymol.* **326**, 47–73.

Van Der Veldem, A. W., Lindgren, S. W., Worley, M. J. and Heffron, F. (2000). *Salmonella* pathogenicity island 1-independent induction of apoptosis in infected macrophages by *Salmonella enterica* serotype typhimurium. *Infect Immun.* **68**, 5702–5709.

Vazquez-Torres, A., Jones-Carson, J., Baumler, A. J., Falkow, S., Valdivia, R., Brown, W., Le, M., Berggren, R., Parks, W. T. and Fang, F. C. (1999). Extraintestinal dissemination of *Salmonella* by CD18-expressing phagocytes. *Nature* **401**, 804–808.

Vazquez-Torres, A., Xu, Y., Jones-Carson, J., Holden, D. W., Lucia, S. M., Dinauer, M. C., Mastroeni, P. and Fang, F. C. (2000). *Salmonella* pathogenicity island 2-dependent evasion of the phagocyte NADPH oxidase. *Science* **287**, 1655–1658.

Velculescu, V. E., Zhang, L., Vogelstein, B. and Kinzler, K. W. (1995). Serial analysis of gene expression. *Science* **270**, 484–487.

Vives-Rego, J., Lebaron, P. and Nebe-von Caron, G. (2000). Current and future applications of flow cytometry in aquatic microbiology. *FEMS Microbiol. Rev.* **24**, 429–448.

Vriesema, A. J.M., Brinkman, R., Kok, J., Dankert, J. and Zaat, S. A.J. (2000). Broad-host-range shuttle vectors for screening of regulated promoter activity in *viridans* group *Streptococci*: Isolation of a pH-regulated promoter. *Appl. Environ. Microbiol.* **66**, 535–542.

Wallis, T. S. and Galyov, E. E. (2000). Molecular basis of *Salmonella*-induced enteritis. *Mol. Microbiol.* **36**, 997–1005.

Wang, E., Miller, L. D., Ohnmacht, G. A., Liu, E. T. and Marincola, F. M. (2000). High-fidelity mRNA amplification for gene profiling. *Nat. Biotechnol.* **18**, 457–459.

Wang, J., Lory, S., Ramphal, R. and Jin, S. (1996a). Isolation and characterization of *Pseudomonas aeruginosa* genes inducible by respiratory mucus derived from cystic fibrosis patients. *Mol. Microbiol.* **22**, 1005–1012.

Wang, J., Mushegian, A., Lory, S. and Jin, S. (1996b). Large-scale isolation of candidate virulence genes of *Pseudomonas aeruginosa* by *in vivo* selection. *Proc. Natl Acad. Sci. USA* **93**, 10434–10439.

Watson, P. R., Gautier, A. V., Paulin, S. M., Bland, A. P., Jones, P. W. and Wallis, T. S. (2000). *Salmonella enterica* serovars Typhimurium and Dublin can lyse macrophages by a mechanism distinct from apoptosis. *Infect. Immun.* **68**, 3744–3747.

Wei, Y., Vollmer, A. C. and LaRossa, R. A. (2001). *In vivo* titration of mitomycin C action by four *Escherichia coli* genomic regions on multicopy plasmids. *J. Bacteriol.* **183**, 2259–2264.

Weinstein, D., O'Neill, B. and Metcalf, E. (1997). *Salmonella typhi* stimulation of human intestinal epithelial cells induces secretion of epithelial cell-derived interleukin-6. *Infect. Immun.* **65**, 395–404.

Wilkinson, R., Tscharke, D. and Simmons, A. (2000). Molecular localisation of a G-protein mRNA using differential display and *in situ* hybridization. *Brain Res. Brain Res. Protoc.* **5**, 290–297.

Wong, L., Lue, M. Y., Chang, C. A., Lin, Y. L. and Chan, E. C. (1996). *Helicobacter pylori* induces gene expression in human gastric cells identified by mRNA differential display. *Biochem. Biophys. Res. Commun.* **228**, 484–488.

Yang, A. G., Bai, X., Huang, X. F., Yao, C. and Chen, S. (1997). Phenotypic knockout of HIV type 1 chemokine coreceptor CCR-5 by intrakines as potential therapeutic approach for HIV-1 infection. *Proc. Natl Acad. Sci. USA* **94**, 11567–11572.

Yuk, M. H., Harvill, E. T. and Miller, J. F. (1998). The BvgAS virulence control system regulates type III secretion in *Bordetella bronchiseptica*. *Mol. Microbiol.* **28**, 945–959.

Zhang, Y. L., Ong, C. T. and Leung, K. Y. (2000). Molecular analysis of genetic differences between virulent and avirulent strains of *Aeromonas hydrophila* isolated from diseased fish. *Microbiology* **146**, 999–1009.

Zhang, Y. Z., Naleway, J. J., Larison, K. D., Huang, Z. J. and Haugland, R. P. (1991). Detecting *lacZ* gene expression in living cells with new lipophilic, fluorogenic beta-galactosidase substrates. *FASEB J.* **5**, 3108–3113.

Zhao, H., Li, X., Johnson, D. E. and Mobley, H. L. (1999). Identification of protease and *rpoN*-associated genes of uropathogenic *Proteus mirabilis* by negative selection in a mouse model of ascending urinary tract infection. *Microbiology* **145**, 185–195.

5 Epithelial Cells: Establishment of Primary Cultures and Immortalization

R Athman, J Niewöhner*, D Louvard and S Robine
Laboratoire de Morphogenèse et Signalisation Cellulaires, UMR 144 CNRS/Institut Curie, Paris, France

Cell Isolation/Immortal

◆◆

CONTENTS

List of abbreviations

cAMP	Adenosine 3′,5′-cyclic monophosphate
BSA	Bovine serum albumin
DNAse	Deoxyribonuclease
DMEM	Dulbecco's modified Eagle's medium
EDTA	Ethylenediamine tetraacetic acid
EGF	Epidermal growth factor
FBS	Fetal bovine serum
HBSS	Hank's buffered saline solution
Hepes	N-[2-hydroxyethyl]piperazine-N′-[2-ethanesulfonic acid]
MDCK	Madin-darby canine kidney
PBS	Phosphate-buffered saline
pfu	Plaque-forming unit
PTH	Parathormone
SCID	Severe combined immunodeficiency
SV40	Simian virus 40
TER	Transepithelial resistance

*JN present address, Xerion Pharmaceutical Gmbh, Martinsried Germany

METHODS IN MICROBIOLOGY, VOLUME 31
ISBN 0–12–521531–2

◆◆◆◆◆◆ INTRODUCTION

Recently, a large number of studies have been focused upon interactions between virulent pathogens and host cells. Microbiologists together with cellular biologists defined a new field as 'cellular microbiology'.

Many cells that are the initial target of pathogens are the epithelial cells that form boundaries between different tissue compartments. They are either epidermal cells or internal epithelial cells such as airway or intestinal/colonic epithelial cells. Mucosal surfaces are a major route of entry of microbial pathogens into the host. One role of this mucosal epithelium is to function as an active mechanical barrier against the external environment. The intestinal epithelium carries out conflicting tasks: (1) absorption of nutrients ions and water; (2) maintenance of a barrier against entry of noxious material, such as antigens, microbes or microbial products. It is now recognized that cultured epithelial cells expressing functions of native epithelia can be used sucessfully to probe fundamental questions in epithelial biology at the cell and molecular levels. Ideally, epithelial cell culture should allow the process of normal cell differentiation and function to be recapitulated, manipulated and observed under controlled conditions.

Unfortunately, a large number of studies have been made using inappropriate epithelial cell cultures in terms of pathogen tissue specific target and phenotypic differentiation of the cell lines used for these studies.

In this chapter, we shall focus on the use of epithelial cell cultures derived from organs relevant as host tissue for pathogen infections, namely, the airway tract, the digestive tract and the kidney. Methods for the isolation, characterization and use of primary cell cultures in reasonable homogeneity and with acceptable short-term viability will be presented. Another approach utilizes monolayer cultures of long-term established epithelial cell lines. It should be pointed out that these cell lines are derived from carcinomas and are transformed cell lines, and thus may not reproduce the full profile of the differentiated state of their *in vivo* counterpart. Long-term cultures can be obtained using immortalizing antigens such as the large T antigen from the SV40 virus and will also be discussed. Finally, primary cell cultures from transgenic or 'knock out' mice as an example of the usefulness of this experimental strategy will be discussed.

◆◆◆◆◆◆ PRIMARY CULTURES OF EPITHELIAL CELLS

Airway epithelial cells

Acute viral infections of the respiratory system are among the most common causes of human disease. The infectious agents are called the acute respiratory viruses and collectively produce a variety of clinical manifestations including rhinitis, tonsillitis, laryngitis and bronchitis. Primary cultures of airway epithelial cells enable one to study the infectious process of some viruses and bacteria implicated in the infection of the respiratory system. Studies concerning infection by *H. influenzae*, an exclusive human pathogen which infects the respiratory epithelium, have shown that the bacteria can initiate cytoskeletal rearrangement within human airway epithelium. This results in internalization of the bacteria within nonciliated human airway epithelial cells by the process of macropinocytosis (Ketterer *et al.*, 1999).

Primary human airway epithelial cells

Cell isolation and culture

According to Ketterer *et al.* (1999), samples are obtained from nasal polyps, normal bronchial tissue or from bronchial brushings taken by bronchoscopy. Tissue samples are treated at 4°C in Ca-free/Mg-free HBSS freshly supplemented with pronase at $1.5\,mg\,ml^{-1}$ and DNAse at $0.1\,mg\,ml^{-1}$. Digestion of the tissue is halted after 48 h by the addition of FBS to 10% (v/v). The resulting cell suspension is washed twice with Airway Medium (AM)*. Samples obtained by bronchial brushing are collected into Ham's F12 medium containing penicillin ($100\,mU\,ml^{-1}$), streptomycin ($100\,\mu g\,ml^{-1}$) and gentamycin ($50\,\mu g\,ml^{-1}$) and stored cold until they could be distributed onto the desired growth surfaces, usually within 24 h of the time of collection. From this point, the brushings and the suspensions of airway epithelial cells are handled similarly. The cell suspensions are pelleted and resuspended in AM at an approximate concentration of 5×10^5 cells ml^{-1}. For submerged cultures, cell suspensions in AM are seeded at 50 to 60 μl onto a sterile 12-mm-diameter glass coverslips previously coated with a solution of $0.5\,mg\,ml^{-1}$ bovine collagen in distilled water. Airway cells are also grown submerged on tissue culture well insert units (BioCoat, Becton Dickinson). In both types of submerged cultures, the medium is replaced with defined Bronchial Epithelial Cell Growth Medium (BGEM, Clonetics Corp., San Diego, CA) following a 24 h initial incubation at 37°C and 5% CO_2. The cells are allowed to grow for 5 to 7 days before the initial medium is replaced with fresh BGEM, replaced twice weekly. Airway cells are grown at an air–fluid interface according to the method of Smith *et al.* (1996). The cell suspensions are seeded at 50 to 60 μl onto the surface of the microporous-in-tissue culture well insert units and incubated as described above. Following the initial 24 h incubation, the medium in the tissue culture wells below the insert units is replaced with Widdicombe's medium (WM) containing 2% Ultroser-G (BioSepra SA, France) and 1% penicillin-streptomycin. The residual AM is aspirated from the upper surface of the insert units, and the incubation is continued. Medium is aspirated daily from the apical surface of the insert units until cell growth prevents leakage of medium from the well. Cell growth is maintained by replacing the WM in the well below the microporous membrane as needed.

Purity To confirm the integrity and the polarization of the epithelial cell monolayer, transepithelial resistance is measured. Based on cell morphology, three cell types can be distinguished: ciliated cells represent about 10 to 15% of the epithelium. The remaining two cell types have differences in the density of microvilli on their surfaces. One cell type has a small number of small microvillus-like structures, while the second is covered with these structures. The surface of human bronchial epithelial cells grown on bovine collagen-covered glass coverslips in submerged cultures appeared morphologically similar to that of cells grown at the air–fluid interface, with the exception of the absence of ciliated cells in the former.

*AM: High-glucose DMEM and Ham's F12 medium combined in equal volumes and supplemented with 5% heat-inactivated FBS, 1% non-essential amino acids, 1% penicillin-streptomycin, and human insulin (5 $\mu g\,ml^{-1}$).

Primary mouse pulmonary epithelial cells

Tracheostomy is performed on 6-week-old mice under ether anesthesia. 1 ml of protease type X (2 $mg\,ml^{-1}$, Sigma) is infused into the lung with a syringe. After incubation for 10 min at room temperature, the lung is withdrawn and minced in PBS. Blocks of the tissues are removed by filtration through four layers of sterilized gauze. Single cells are collected by centrifugation in DMEM supplemented with 10% FBS, and cultivated at 38°C (Itoh *et al.*, 1998).

The digestive tract

The liver is both an endocrine gland, located between two venal segments, releasing metabolites directly in the right auricle, and an exocrine gland that releases bile into the duodenal lumen. The hepatic gland contains at least five cellular types that cooperate: hepatocytes, endothelial cells, Kupffer's cells, lipocytes and pits cells. Hepatocytes represent approximately 70% of the cells and are involved in intrahepatic metabolism, biliary secretion and the metabolism of exogene and endogene molecules. The liver is the target of various diseases such as hepatic colitis, jaundice, cirrhosis and infectious pathologies including viral hepatitis. These cultures may help an understanding of the role of hepatocytes in the intrahepatic metabolism and could be a useful tool for the analysis of viral infection, notably in the preparation and test of vaccines.

The stomach is the target of many diseases including cancer and infections. *H. pylori* is the most common pathogenous agent affecting the stomach. This microaerophilic gram-negative bacterium has been identified as the cause of chronic gastritis and peptic ulcer disease in humans. These diseases are characterized by severe infiltration of neutrophils, lymphocytes, monocytes and plasma cells in the gastric mucosa. Primary cultures of stomachal cells allow the study of the infectious processes of such pathogens.

Most studies on the organization and function of the different cell types that constitute the intestinal mucosa have been carried out with intact organs *in situ* or after isolation for short-term experiments. Further development of such studies requires the use of dynamic systems of homogeneous cell populations that can be monitored over long periods. Different cell lines are now available and can thus be used to analyse the establishment of cell polarity, intestinal transport properties and some metabolic functions. Furthermore, the establishment of primary cultures derived from normal or pathological intestinal tissue can be another useful tool for studying the action of pharmacological agents and to dissect the interaction of intestinal cells with bacteria, viruses and parasites such as salmonella, staphylococcus and botulism.

Hepatocytes

Procedures to obtain primary cultures of hepatocytes have been described by Seglen (1976) and are applicable to any animal, including humans.

Cell isolation and culture

A two-step procedure can be performed with a very simple perfusion apparatus and results in the complete dissociation of the liver within 10–15 min, usually with 90–95% of the hepatocytes remaining structurally intact.

Perfusion The perfusion apparatus contains a perfusion pump, a heating unit necessary to maintain the perfusion buffer's temperature at 37°C, a filter/canula unit and a liver support dish. The perfusate flow in the portal vein is initiated at a moderate rate (20 ml min^{-1}) and is gradually increased to 50 ml min^{-1} to wash out the blood from the liver. The perfusion is continued with recirculating collagenase buffer at 50 ml min^{-1} for 5–10 min or until rupture of the portal vein occurs. During this period, the liver should swell uniformly to approximately twice its original size.

Solutions Perfusion buffer: NaCl 140 mM, KCl 7 mM, Hepes 10 mM, and NaOH 6 mM, pH 7.4 at 37°C.
Collagenase buffer: 0.05% collagenase in $CaCl_2$ $2H_2O$ 5 mM, NaCl 68 mM, KCl 6.6 mM, Hepes 100 mM, and NaOH 66 mM, pH 7.6 at 37°C.

Liver dissociation and hepatocyte purification The liver is transferred to a wide Petri dish containing 80 ml of ice-cold suspension buffer. The liver is held firmly in the portal connective tissue with forceps and the cells are gently scraped. The suspension is filtered through a 250 μm nylon filter to remove connective tissue debris and cell clumps. The cells are incubated in a 20 cm glass Petri dish at 37°C on a tilting platform (10 tilts min^{-1}) for 30 min, to allow damaged cells to repair, dissolve or aggregate. After this preincubation, the cells are cooled to 0°C and the suspension is filtered through a double-layered nylon filter. The hepatocytes are purified by four centrifugations at 0°C in flat-bottomed 150 ml beakers (400 rpm for 2 min), with gentle resuspension in 40 ml ice-cold wash buffer each time.

Finally the resuspended cells are sedimented in ice-cold suspension buffer at the cell concentration desired. The viability of the final cell suspension is checked by trypan blue exclusion.

Purity Fully differentiated hepatocytes can be characterized by their ability to produce specific markers such as albumin, transferrin, cytokeratin 8 and 18. Moreover, Connexin 26 and glucagon are expressed only by hepatocytes.

Solutions Suspension buffer: NaCl 68 mM, KCl 5.4 mM, KH_2PO_4 1 mM, Na_2SO_4 0.7 mM, Hepes 30 mM, Tes 30 mM, Tricine 36 mM, NaOH 52.5 mM, $MgCl_2$ $6H_2O$ 0.65 mM and $CaCl_2$ $2H_2O$ 1.25 mM, pH 7.6 at 37°C.

Wash buffer: perfusion buffer added with $CaCl_2$ $2H_2O$ 1.2 mM, pH 7.4 at 37°C.

Human biliary epithelial cells (BECs)

The epithelial biliary cells line the intrahepatic biliary ducts and form with the hepatocytes the Hering's duct in which the bile passes in transit. Intrahepatic cholangiocarcinoma and other cancers of the biliary tree, including the gallbladder, are often associated with non-neoplastic inflammatory diseases of the biliary tree, such as sclerosing, clonorchiasis and hepatolithiasis. These diseases also involve reparative, non-neoplastic BECs proliferation. Primary cultures of these cells may facilitate the molecular analysis and are necessary for the establishment of a diagnosis before the tumours reach a non-curable stage.

Cell isolation and culture

BECs are isolated from intact donor human livers insuitable for transplantation because of steatosis (Yokomuro *et al.*, 2000). and sliced into 3-cm-thick coronal sections. Septal bile ducts are identified by their yellow color. Several ducts are dissected from the surrounding connective tissue, opened longitudinally, washed in Ca and Mg-free HBSS. They are then placed in HBSS, supplemented with 0.25% collagenase, for 20 min at 37°C. The detached BECs are transferred to a 50 ml conical centrifuge tube and pelleted at 1200 rpm for 10 min and then resuspended in complete DMEM/F12 medium*. 0.5×10^5 to 1×10^5 BECs are seeded on 24 well

*Supplemented with: D-glucose (5.4 g l^{-1}), gentamycin (50 μg ml^{-1}), penicillin G sodium (100 U ml^{-1}), streptomycin sulfate (100 μg ml^{-1}), amphotericin B (250 ng ml^{-1}), Hepes (10 mmol l^{-1}), BSA (2.5 mg ml^{-1}), insulin (10 mg ml^{-1}), transferrin (5.5 mg ml^{-1}), sodium selenite (0.67 mg ml^{-1}), non-essential amino acids (0.1 mmol l^{-1}), L-glutamine (2 mmol l^{-1}), thyroxin (32 ng ml^{-1}), prostaglandin E1 (10 ng ml^{-1}), hydrocortisone (40 ng ml^{-1}), forskolin (10 μmol l^{-1}), bovine pituitary extract (50 μg ml^{-1}), trypsin inhibitor (50 μg ml^{-1}), EGF (10 ng ml^{-1}).

plates precoated with a collagen gel (0.3 ml of a collagen solution [4 mg ml^{-1}] 10 × DMEM/F12 medium and 0.34 N NaOH).

Purity Gamma glutamyl transpeptidase activity (GGT), histologically demonstrated, and immunoreactivity for cytoplasmic CK19 are both constitutive phenotypic properties of bile epithelium.

Human stomachal epithelial cells

Tissues from fetuses of 17 to 20 weeks gestation are obtained from pregnancy terminations (Basque *et al.*, 1999). The stomach is immersed in Leibovitz L-15 dissection medium containing gentamycin and nystatin (40 μg ml^{-1} each), and prepared within 30 min at room temperature.

Cell isolation and culture

Cardiac and pyloric segments are removed from the stomach, leaving the body and fundic regions. Tissue specimens are cut into explants (3 × 3 mm^2) and rinsed with dissection medium. The gastric epithelium is dissociated using a new non-enzymatic technique based on a procedure for recovering cells grown on Matrigel. Explants are immersed in ice-cold Matrisperse (Collaborative Biomedicals, Bedford, MA) for 16 to 20 h and gently agitated for approximately 1 h (4°C), allowing dissociation of gastric cells as intact epithelial sheets or large aggregates. The resulting material is centrifuged and resuspended in culture medium (CM)*. Epithelial aggregates are fragmented mechanically into multicellular clumps, seeded in plastic 6-well multiwell plates (5×10^4 cells in 3 ml) or 24-well plates (1.5×10^4 cells in 1 ml) and left undisturbed for at least 24 h to allow attachment. Culture medium is renewed every 48 h.

Purity Mucous-secreting cells are identified using periodic acid Schiff staining, and pepsinogen-containing cells and parietal cells using Bowie staining. Immunofluorescence staining of various markers such as keratin-18 is used to confirm the purity of epithelial cultures.

*CM: DMEM/Ham's F12 (1 : 1) supplemented with penicillin (50 U ml^{-1}), streptomycin (50 μg ml^{-1}) and 10% FBS.

Intestinal human small and large intestinal epithelial cells

Cell isolation and culture

The mucosa is stripped from the submucosa within 30 min after bowel resection or biopsy and rinsed with PBS several times (Rogler *et al.*, 1998). The mucus is removed by treatment with 1 mM Dithiothreitol for 15 min. After washing with PBS, the mucosa is placed in 1.5 mM EDTA in Ca and Mg-free HBSS and trumbled for 10 min at 37°C. This supernatant, containing debris and mainly villus cells, is discarded. The mucosa is incubated again with EDTA for 10 min at 37°C. The supernatant is collected into a 15 ml tube and the remaining mucosa is vortexed in PBS. This supernatant is also collected since it contains complete crypts, some single cells, and a small amount of debris. To separate intestinal epithelial cells (IEC) from contaminating non-epithelial cells, the solution is allowed to sediment for 15 min. The sedimented cells, mainly complete crypts, are collected and washed twice with PBS. The cells are resuspended in 400 μl minimal essential medium*. The IEC are seeded into collagen A (collagens I and III [9 : 1] from bovine placenta)-coated Millicel-CM cell culture plate inserts suitable for 24-well culture plates with a translucent and permeable membrane at the bottom. The medium containing the isolated cells is placed inside the filter inserts. As the medium is passed through the filter membranes, the cells are forced into rapid contact with the collagen coating of the membranes. Just enough medium is added to cover the cells with a thin film of liquid so that the cells are at the medium/air interface. The cells are incubated at 37°C in air with 10% CO_2. This method enables the structure and function of active IEC in culture to be maintained for at least 1 week.

*Supplemented with Earle's salts, 20% FBS, 10 ng ml^{-1} EGF, ITS (5 μg ml^{-1} insulin, 5 μg ml^{-1} transferrin, 5 ng ml^{-1} selenious acid), 2 mM glutamine, 100 U ml^{-1} penicillin, 100 μg ml^{-1} streptomycin, 100 μg ml^{-1} gentamycin, 2.5 μg ml^{-1} fungizone.

Purity Under the conditions used, the IEC maintained their characteristic morphologic features, tested by light and electron microscopy, for at least 2 weeks, showing viable, cuboid or round-shaped cells with characteristic microvilli and a well-defined terminal web. Their functional capacity can be demonstrated by the activity of alkaline phospahatase and their ability to secrete IL-1. Brush border enzyme activity and protein secretion studies indicate that the first week of culture under these conditions can yield reliable results. This can permit better representation of the *in vivo* situation than very short-term studies on transformed cells.

Rat epithelial intestinal cells

Cell isolation and culture

All the small intestine is withdrawn from six-day-old rats and put in HBSS (Evans *et al.*, 1992). The intestine is slit open and cut into 2–3 mm small fragments. The opened fragments are washed five to eight times, in a T25 flask containing HBSS, by agitating at 150 rpm for 5 min. The fragments are spliced into small pieces < 1 mm^3, using a sharp scalpel blade. They are then transferred into a T25 ml flask with 20 ml of enzyme solution (ES*) and shaken vigorously (150 rpm) for 30 min at 25°C. The solution is pipetted vigorously (with 2 mm bore pipette), approximately 150 times,

*ES: 0.1 mg ml^{-1} dispase and 300 U ml^{-1} collagenase in HBSS pH 7.4.

until it is cloudy and then transferred into a sterile tube. The contents are left to sediment under gravity for 1 min and the solution is carefully removed without the bottom few milliliters. This operation is repeated twice. The last solution is then resuspended in half volume of DMEM-S† and centrifuged at 300 rpm for 3 min. The centrifugation procedure is repeated until the supernatant is clear and the pellet well defined (five–six times). The pellet is finally resuspended in culture medium containing 2 mM L-glutamine, 2.5% FBS, penicillin (100 U ml^{-1}), streptomycin (60 μg ml^{-1}), non-essential amino acids, Insulin (0.25 μg ml^{-1}), EGF (10 ng ml^{-1}).

† DMEM-S: DMEM, 2.5% FBS and 2% sorbitol.

Purity The epithelial cells can be identified using a combination of antisera against cytokeratins 8, 18 and 19.

This method has been applied to transgenic mice lacking the villin gene in our laboratory. Villin is an actin-binding protein located in the brush border of enterocytes and kidney proximal tubule cells. This protein presents two activities *in vitro*: at low calcium concentration, villin initiates the bundling of actin filaments, whereas at high calcium concentration, it induces the severing of these filaments. To test the role of this protein *in vivo*, knock-out mice (vil −/−) have been generated by introducing a null mutation in the villin gene. We have shown that in vil −/− mice the dynamic of the actin cytoskeleton may be impaired after different physiopathological conditions, such as fasting/refeeding or dextran sodium sulfate induced injuries (Ferrary *et al.*, 1999). To complete the studies performed *in vivo*, primary cultures of enterocytes from vil+/+ and vil −/− mice have been established, that allow motility assay as well as infection experiments.

Primary cultures from rat and human colonic tumors

Colonic tumors are taken off and rapidly rinsed with ethanol 70% (Pujuguet *et al.*, 1996). They are then washed in Ham's F10 medium and minced into pieces of approximately 2 mm^3 with a scalpel in PBS. The fragments are then dissociated three times in an enzymatic solution prepared in Ham's F10 medium containing type I collagenase (300 mg ml^{-1}) and DNase I (125 U ml^{-1}), under slight agitation at 37°C for 45 min. After each digestion, the supernatants are pooled, filtered through a stainless steel mesh to remove the undissociated tumor fragments and then centrifuged at 300 rpm for 10 min. The disaggregated cells are washed twice, suspended in Ham's F10 medium supplemented with 10% FBS and 40 mg ml^{-1} gentamycin, and then plated in tissue culture flasks. Although in most of the human tumor cell suspensions, fibroblasts grow rapidly as the prevalent population, it is possible to isolate the tumor cells by detaching the fibroblasts from the plates using an EDTA treatment. In the case of experimental colon cancer models, where the tumor cells come from already characterized and established cell lines, it is possible to grow out mesenchymal cells from the freshly dissociated tumors. This is an important way to study mechanisms involving epithelium/mesenchyme interactions, such as invasion and metastasis.

The kidney

The urinary tract can be affected by different diseases concerning the kidney in particular, electrolyte imbalance, renal insufficiency, glomerulopathy or more extensively reno-ureteral lithiasis and urinary infections, often responsible for chronic pyelonephritis. More than 80% of urinary tract infections in adults are caused by *E. coli*. For these bacteria, different factors of virulence, such as lipopolysaccharides (LPS), hemolysins or various types of fimbriae have been characterized. Multiple lines of evidence have emerged concerning the involvement of proximal tubular epithelial cells in the renal immune response (Wuthrich *et al.*, 1990).

Proximal tubule culture

The kidney is removed from a freshly killed rabbit and rinsed three times in ice-cold Hepes. The kidney is decapsulated and thin 250 μm sections are performed. The sections are then placed in 4.5 ml Hepes and 4.5 ml of RKI medium* to which 0.25 ml of 10% BSA (in Hepes) are added. The sections are disrupted and an aliquot of 1 ml collagenase solution is added. The suspension is incubated with stirring for 45 min at 37°C. The reaction is stopped by adding 10 ml cold Hepes. The solution is strained into a 250 ml beaker and the filtrate is transferred into 50 ml tubes. The tubes are centrifugated at 1000*g* for 3 min at 4°C and the supernatant is discarded. The pellet is resuspended in 30 ml cold Hepes in a single 50 ml tube. This step is repeated three times. Prior to the third speed, the pellet is resuspended in 5% BSA (in Hepes). The tubes are stored at 4°C for 5 min and spun. The last supernatant is discarded and the pellet is resuspended in 80 ml Percoll mixture (40 ml Percoll and 40 ml Hepes 2×). The suspension is pipetted into polycarbonate tubes and spun at 20 000*g* for 30 min at 4°C. The bottom layer of the Percoll gradient, which consists of 98% pure proximal tubules, is removed. The fragments appear to be a mixture of all three proximal segments (S1–S3). The preparation is washed twice with PBS and once in proximal tubule RKI medium, resuspended in an appropriate volume for 5×10^4 fragments per well. The cultures are fed after 72 h. Subsequently, feeding is typically carried out every 48 h for the lifetime of the cultures. The cultures display differentiated transepithelial function for up to 2 weeks.

*RKI medium: a 1:1 mix of Ham's/F12 and DMEM high glucose with no $HCO3^-$, supplemented with human transferrin (5 mg l^{-1}), insulin (5 mg l^{-1}), sodium selenite (50 nM), hydrocortisone (50 nM), penicillin (10^5 U l^{-1}), streptomycin (100 mg l^{-1}).

Proximal tubule RKI medium: identical to RKI but DMEM with low glucose and no $HCO3^-$.

The Percoll gradient method of isolation has the advantage of yielding large quantities of material but the disadvantage of uncertainty regarding the absolute specificity of cell type.

Purity The proximal tubule origin of the cells can be assessed by the expression of villin, a tissue-specific actin-binding protein mainly expressed in epithelial cells developing brush borders.

The proximal tubule monolayers form a low-resistance ($<10\,\Omega\,cm^{-2}$) uniform epithelium and possess a well-developed brush border. They undergo apical membrane depolarization on addition of glucose (indicative of Na^+-coupled co-transport).

Solutions Collagenase solution: 7.5 mg type IV collagenase in 1 ml Hepes.

◆◆◆◆◆◆ PERMANENT EPITHELIAL CELL LINES

From the liver

In vitro polarized cell systems, such as MDCK, Caco-2 and HT-29 lines have been successfully exploited to study how simple epithelial cells establish and maintain their polarity. However, for hepatocytes that display a complex polarity, *in vitro* studies have been hampered for a long time by lack of well-polarized hepatic lines. WIF 12-1 was the first hepatic cell line to exhibit a stable and well characterized polarized phenotype (Cassio and Leao, 1991; Cassio *et al.*, 1991). Although two human hepatoma cell lines HuH7 and the well-studied HepG2 were reported to be polarized (Chiu *et al.*, 1990), neither the frequency with which cells of either line formed bile canaliculi nor the functional properties of these canaliculi was provided. From the WIF 12-1 clone which forms bile canaliculi-like structures, the derivative WIF-B, with improved hepatic characteristics, was generated (Shanks *et al.*, 1994). It has been shown that WIF-B constitutes a good *in vitro* model, not only for structural studies of hepatocyte polarity (Ihrke *et al.*, 1993; Decaens *et al.*, 1996), but also for functional studies of hepatocyte-specific properties, such as vectorial bile acid transport (Bravo *et al.*, 1998), and intercellular communication via connexin 32-constituted gap junctions (Chaumontet *et al.*, 1998).

From the intestine

From normal tisssue

Cell lines have been successfully established from the rat small intestine, including the group of the IEC cell lines IEC-6, IEC-17 and IEC-18, and the cell lines IRD-98 and RIE-1. Recent experiments using *in vitro* association of IEC cells with fetal rat or mouse gut mesenchyme followed by *in vivo* grafting, have shown that IEC cells were able to differentiate into the four main intestinal cell types: absorptive, goblet, endocrine and Paneth's granular cells, indicating that these cells are physiologically relevant (Kedinger *et al.*, 1986). IEC cells have also been used for studying the metabolic functions of intestinal cells and to measure the transport of nutrients such as vitamins, amino acids and hexoses (Adams, 1984; Inui *et al.*, 1980).

From chemically induced intestinal tumors

A limited number of cell lines have been derived from intestine tumors induced in various mammalian species with trophic chemical carcinogens such as 1,2-dimethylhydrazine or N-methyl-N-nitrosourea. These include malignant cell lines from rat small intestine (Martin *et al.*, 1975) and colon (Borman *et al.*, 1982) and from mouse (Brattain *et al.*, 1981) or guinea pig (O'Donnell and Cockerell, 1981) colons. Some of these cell lines are of high interest because of their ability to recapitulate the different stages of human intestinal carcinogenesis (Martin *et al.*, 1987; Pujuguet *et al.*, 1996).

Human colon carcinoma cell lines

Multiple human adenocarcinoma cell lines are available because adenocarcinoma of the colon is a common human malignancy (Fogh and Trempe, 1975; Zweibaum *et al.*, 1991; Neutra and Louvard, 1989).

HT 29

HT-29 was the first human colon carcinoma cell line to be established in culture. Even though these cells are mainly used as in-culture models for studies on human colon cancers, they retain some tissue-specific functions. HT-29 cells display functional receptors for intestinal peptides and hormones, synthesize the receptor of dimeric immunoglobin A and are able to transport dimeric IgA across the cell and to release it on the apical side. These cells have also been shown to be an interesting *in vitro* model for studying the formation of tight junctions. Inducers of differentiation such as sodium butyrate, dimethyl sulfoxide, and N,N-dimethylformamide have been shown to initiate an important increase of the activity of alkaline phosphatase and less frequently an increased activity of other brush border associated hydrolases such as sucrase isomaltase, aminopeptidase N and dipeptidylpeptidase IV. HT-29 cells are able to undergo different patterns of intestinal differentiation depending on modifications of the culture medium. When grown in the absence of glucose, HT-29 cells express a typical enterocyte differentiation. HT29-18 is a subclone isolated from HT-29 cells grown in glucose (Huet *et al.*, 1987). This clone is able to differentiate into several cell types similar to those found in the intestinal mucosa.

Caco-2

Caco-2 is a highly enterocyte-like differentiated cell line. Until recently these cells have been essentialy investigated for their tumorigenic properties in nude or SCID mice and for the characteristics of the growth-related variations of glycogen metabolism. When seeded either on permeable filters or impermeable substrates at high density, they consistently form polarized monolayers joined by tight junctions, with well-developed apical microvilli. Such monolayers survive for up to

30 days in culture before spontaneously detaching from the substrate. Although derived from adult human colon carcinoma, where microvillar hydrolases are not expressed, Caco-2 cells express two disaccharridases and two peptidases typical of the digestive enzymatic repertoire of normal small intestinal villus cells. They also transport ions and water toward the basolateral surface. Despite this resemblance to normal small intestinal absorptive enterocytes, Caco-2 cells are more closely analagous to enterocytes of the normal 15-week human fetal colon, in which microvillar hydrolases are transiently expressed. Through their electrical parameters, ion conductance and permeability properties, Caco-2 monolayers resemble colonic crypt cells. Whereas crypts contain multiple cell types, however, Caco-2 monolayers are homogeneous, suggesting that these cells represent the neoplastic equivalent of the crypt enterocytes normally committed to an absorptive function but arrested in differentiation at the fetal stage. Dedifferentiation is frequently observed during the carcinogenic process and is accompagnied by re-expression of a fetal-like gene pattern.

T84

This cell line is originally derived from a lung metastasis of a human colon carcinoma. T84 cells resemble Caco-2 cells in their ability to form spontaneously polarized monolayers of high electrical resistance. Unlike Caco-2 cells, T84 cells do not form well-developed brush borders and fail to express microvillar membrane hydrolases. In their morphology, electrical parameters, and ion transport activities, T84 monolayers resemble adult colonocytes. T84 monolayers are homogeneous and appear to represent a line derived from committed crypt cells. In both Caco-2 and T84 cells, monolayer formation and differentiation are accelerated on extracellular matrices that enhance attachment. This differentiation is always preceded by the synthesis, the polarized secretion and the assembly of an endogenous basement membrane.

From the kidney

Renal cell physiologists started out to use renal tubule cell lines in the late 1960s when Leighton *et al.*, recognized that MDCK cells, an established cell line derived from normal dog kidney, retained structural and functional features of transporting epithelia. A number of cell lines derived from the whole kidney or bladder cell suspensions of various species (LLCPK, OK, JTC-12, A6, TB-M) are now available, but initially no attempts were made to characterize their origin. Thus, only assumptions based on their morphological and functional properties are made, years after their emergence. Moreover, because the cells were not initally cloned, they have undergone substantial genetic drifts.

MDCK cell line

An unusual feature of these cells is that, while in culture, they retain many of the differentiated properties of normal kidney epithelial cells. Among these are an asymmetric distribution of enzymes and vectorial transport of sodium and water from the apical to the basolateral faces. The latter property gives rise to domes or blisters in confluent cultures. These are transient areas where collected fluid has forced the monolayer to separate from the substratum. Morphologically the cells resemble a typical cuboidal epithelium with spikes on the apical side of the cells. Two different types of the MDCK cell line are known. Type I cells form a tight epithelium with a TER above 2000 $\Omega\,cm^{-2}$. Because of their high electrical resistance transcytosis is more conveniently studied in type I cells. Type II cells form a monolayer of lower resistance (100–200 $\Omega\,cm^{-2}$).

♦♦♦♦♦♦ IMMORTALIZATION

Tools for immortalizing

Treatment of primary cultures with mutagens

A few cell lines have been generated by treating primary cultures with mutagens, such as radiation, in the form of UV light or gamma rays, or chemicals. This approach is designed to induce mutations in the genome in order to alter the pattern of gene expression. However this technique works at very low efficiency (O'Donnell and Cockerell, 1981).

Introduction of oncogenes

The most commonly used viral oncogenes are adenovirus E1A and E1B and SV40 large T antigen.

Adenovirus are DNA viruses, double strand and linear (3 kb). Humans are the natural host of adenoviruses. The gene organisation is complex since there are late (L) and early (E) genes. In contrast to SV40, these two regions are not separated. One of the most important genes is the E1 gene. The product of it is a trans-activator protein of other viral genes but also of some host genes. After infection, the virus is integrated randomly. E1A has an immortalizing power and E1B has a transforming effect on already immortalized cells.

SV40 is a papovavirus. This is a small double stranded circular DNA virus without envelope. The natural host of this virus is the rhesus monkey. SV40 induces tumors when injected into new-born rodents and leads to the transformation of human fibroblasts in culture. The genome of this virus encodes six proteins. The genes are localized on two separated transcription units: a region which contains early genes encoding large T and small t proteins, and a second one containing late genes which encode the capside proteins, VP1, 2 and 3. The transforming power is localized in the early region and is the result of protein T. This transforming capacity has been demonstrated by deletion experiences in the viral genome and by the use of thermosensitive mutants. The large T protein is a 90 kD protein localized in the host nucleus which is associated with p53, thus stabilizing it. The formation of a p53/T complex induces activation of all the genes implicated in viral or genomic DNA replication, even in quiescent cells (Go state). This leads to enhanced cell division. This effect can be mimicked by infection of purified large T protein in non-permissive cells.

Transfection of primary cultures

Utilization of SV40 and recombinant SV40 adenovirus

Oncogene-transformed renal cell lines

Utilization of wild-type SV40 Well-characterized subcultures of rabbit tubule cells were infected with wild-type SV40 strain LP1 (100 pfu/cell) (Ronco *et al.*, 1994). Approximately 3 weeks later, foci of transformed cells were observed in the SV40-infected cultures. The classical criteria of cell transformation had been tested: expression of large T antigen, ability to grow in soft agar, capacity of tumor induction when cells were injected subcutaneously into athymic nude mice and prolonged life span over 90 passages. In this case, the major properties of renal tubule could be preserved after cell transformation including cell morphology, expression of specific markers, pattern of adenylate cyclase stimulation, mineralocorticoid receptors and transport functions. These transformed cells produce cAMP in response to PTH, exhibit Na^+-dependent glucose transport, but the cAMP content on PTH stimulation and the rate of α-methyl-glucose uptake were respectively four and eight times lower than in parental cells. This was partly due to the detrimental effect of SV40 genome on the cell differentiation program, which led the authors to produce a second generation of immortalized cell lines using an inducible transforming vector, namely a temperature-sensitive strain of SV40.

Utilization of temperature-sensitive strain of SV40 The SV40 tsA58 strain has been used to generate a rabbit tubule cell line exhibiting the major characteristics of principal cells of the collecting duct, the main function of which is arginine vasopressin-regulated water reabsorption. In this model, when infected cells are grown at the permissive temperature of 33°C, the nuclear large T antigen is activated and cells show a transformed phenotype. When transferred to the non-permissive temperature of 39.5°C, the viral genome is repressed within 6 h and cells are expected to recover characteristics of normal parental cells.

Targeted oncogenesis in transgenic mice

Targeted oncogenesis in transgenic mice has been proposed as a useful method for producing established lines of differentiated cells (Hanahan, 1988). The expression of the oncogene is driven either by a ubiquitary promoter such as the $H2K^b$ promoter (Jat *et al.*, 1991, Noble *et al.*, 1995) or by a tissue-specific promoter, only active in differentiated cells or their precursors. According to the tissue-specific promoter used, these models of transgenic can provide new types of cultures: differentiated β-pancreatic cell lines derived from pancreatic tumors in which the production of insulin remains controlled by glucose (Efrat *et al.*, 1988, 1991), neuronal cell lines (Mellon *et al.*, 1990) and hepatoma cell lines (Antoine *et al.*, 1992) have been established using this methodology.

Immortomouse

The immortomouse possesses a stably incorporated, conditionally expressed immortalizing gene (Jat *et al.*, 1991; Noble *et al.*, 1995). The gene is a variant of the tumor antigen (Tag) expressed by the A gene from SV40. A temperature-sensitive mutant of this gene (*ts*A58) encodes a thermolabile variant of Tag that is functional at 33°C but rapidly inactivated at body temperature. This mutant allows conditional expression of Tag in transfected cells *in vitro* and has been used successfully on a very wide range of cells. Coupling *ts*A58 to a suitable promoter can limit expression to specific tissues (De Leon *et al.*, 1994, Cairns *et al.*, 1994). Concerning the immortomouse, the aim was to select a promoter to ensure expression in as many tissues as possible. This was achieved by using the major histocompatibility complex $H2K^b$ class I promoter which can be induced to high levels of expression with γ-interferon. The immortomouse is thus a potential source of continuous cell lines (see Table 5.1) for many cell types that express many features of the differentiated phenotypes and that cannot easily be transformed by gene transfection *in vitro*.

Renal proximal tubule cell lines by targeted oncogenesis

Establishment of two immortalized renal cell lines from transgenic mice has been described in Cartier *et al.* (1993). The mice are obtained using a vector containing a 2.7 kb *Bam*H1-*Bcl*I fragment of SV40 early region,

Table 5.1. Establishment of cell lines using the immortomouse

Cell type	Reference
Thymic epithelial cells	Jat *et al.*, 1991
Renal mesangial cells	Barber and Henderson, 1996
Testis Sertoli cells	Walther *et al.*, 1996
Bone marrow mesenchymal cells	Dennis and Caplan, 1996
Vascular smooth and skeletal	Ehler *et al.*, 1995
muscle myo-epithelial cells	Morgan *et al.*, 1994
Liver bile duct epithelial cells	Paradis *et al.*, 1995
Intestinal epithelial cells	Whitehead *et al.*, 1993
	Whitehead and Joseph, 1994
Hippocampus neuro-epithelial cells	Kershaw *et al.*, 1994
Osteoclasts	Chambers *et al.*, 1993
Glial cells	Groves *et al.*, 1993

including the coding sequences of the transforming large tumor (T) and small tumor (t) antigens. This vector was placed under the control of a 3.2 kb fragment of the rat L-pyruvate kinase (L-PK) gene regulatory region in the 5′ flanking region. After purification, the fragment of the L-PK transgene was microinjected into fertilized mouse oocytes. Primary cultures from proximal renal tubules have been established from 3-month-old transgenic mice. The renal capsule is removed and thin sections from the renal cortex are incubated for 1 h at 37°C in DMEM/F12 medium.* The cortical slices are rinsed in this medium, without collagenase, and tubule segments are microdissected out. The convoluted (PCT) and terminal (Pars Recta, PR) segments of superfical proximal tubules are isolated. Pools of two to five segments (0.1 to 0.5 mm) of PCT and PR tubules are rinsed twice in 20 ml of medium, transferred to 24-well trays precoated with rat tail collagen. To ensure cell growth, the tubules are cultivated in a modified culture medium (CM)†. After 3 weeks and two changes of medium, primary cultures reach confluency.

* DMEM/Ham's F12 (1 : 1, v/v) completed with 30 nM sodium selenate, 5 μg ml^{-1} transferin, 2 mM glutamine, 20 mM Hepes, pH 7.4, 20 mM D-glucose, 2% decomplemented FBS and 0.1% (w/v) collagenase type I.
† CM: DMEM/Ham's F12 (1 : 1, v/v), 30 nM sodium selenate, 5 μg ml^{-1} transferin, 2 mM glutamine, 50 nM dexamethasone, 1 nM triiodothyronine, 10 nM EGF, 2% FBS, 20 mM Hepes, pH 7.4.

Transfection of primary cultures versus targeted oncogenesis in transgenic animals

The usefulness of primary cultures for experimentation has been limited by low recoveries, fibroblast contamination and by the fact that the cells survive for only a very limited time. Introduction of SV40 into primary cultured cells is a useful tool to obtain immortalized cells that retain their ability to further differentiate *in vitro*. Immortal cell lines theoretically represent ideal cell culture models, provided that they derive from a single well-defined cell type and they maintain many characteristics of the parental cells. However, differentiated functions that may not be needed for survival in culture are likely to be lost in culture before transformation. Targeted oncogenesis in mice has been proposed as a powerful method of producing established lines of differentiated cells. The principal advantage is the avoidance of viral infection or transfection of primary or subcultured cells for the establishment of the cell lines. The expression of the oncogene is driven by a tissue-specific promoter, only active in differentiated cells or their precursors.

◆◆◆◆◆◆ CONCLUSION

Modern research in biology is fostered by new possibilities to analyse biological processes at different levels of complexity, from molecules to organism. The advent of genomics together with the power of genetics allows investigation of the physiological states and diseases, taking into account multiple parameters. Such approaches should permit improvements in both innovative diagnostic and therapeutic solutions.

In vivo studies using experimental animal models can easily be undertaken but investigations on human beings is constrained. In both cases, the use of suitable and well-defined cellular models derived from living organisms appears essential to perform studies aiming to unravel *in vitro* the molecular mechanisms of pathogen cell interactions.

Moreover, high throughput screening of new drugs can take advantage of reproducible and well-established cell culture models. Since epithelial cells are often the target of infectious agents, the use of the various cell cultures strategies and systems reported in this chapter should allow investigators in numerous fields to undertake new challenges.

References

Adams, J. S. (1984). Specific internalization of 1,25-dihydroxyvitamin D3 by cultured intestinal epithelial cells. *J. Steroid. Biochem.* **20**, 857–862.

Antoine, B., Levrat, F., Vallet, V., Berbar, T., Cartier, N., Dubois, N., Briand, P. and Kahn, A. (1992). Gene expression in hepatocyte-like lines established by targeted carcinogenesis in transgenic mice. *Exp. Cell Res.* **200**, 175–185.

Barber, R. D. and Henderson, R. M. (1996). Inhibition by P1075 and pinacidil of a calcium-independent chloride conductance in conditionally-immortal renal glomerular mesangial cells. *Br. J. Pharmacol.* **119**, 772–778.

Basque, J.-R., Chailler, P., Perreault, N., Beaulieu, J.-F. and Menard, D. (1999). A new primary culture system representative of the human gastric epithelium. *Exp. Cell Res.* **253**, 493–502.

Borman, L. S., Swartzendruber, D. C. and Littlefield, L. G. (1982). Establishment of two parental cell lines and three clonal cell strains from rat colonic carcinoma. *Cancer Res.* **42**, 5074–5083.

Brattain, M. G., Brattain, D. E., Fine, W. D., Khaled, F. M., Marks, M. E., Kimball, P. M., Arcolano, L. A. and Danbury, B. H. (1981). Initiation and characterization of cultures of human colonic carcinoma with different biological characteristics utilizing feeder layers of confluent fibroblasts. *Oncodev. Biol. Med.* **2**, 355–366.

Bravo, P. and Bender, V. and Cassio, D. (1998). Efficient in vitro vectorial transport of a fluorescent conjugated bile acid analogue by polarized hepatic hybrid WIF-B and WIF-B9 cells. *Hepatology* **27**, 576–583.

Cairns, L. A., Crotta, S., Minuzzo, M., Moroni, E., Granucci, F., Nicolis, S., Schiro, R., Pozzi, L., Giglioni, B. and Ricciardi-Castagnoli, P. (1994). Immortalization of multipotent growth-factor dependent hemopoietic progenitors from mice transgenic for GATA-1 driven SV40 tsA58 gene. *EMBO J.* **13**, 4577–4586.

Cartier, N., Lacave, R., Vallet, V., Hagege, J., Hello, R., Robine, S., Pringault, E., Cluzeaud, F., Briand, P., Kahn, A. and Vandewalle, A. (1993). Establishment of renal proximal tubule cell lines by targeted oncogenesis in transgenic mice using the L-pyruvate kinase-SV40 (T) antigen hybrid gene. *J. Cell Sci.* **104**, 695–704.

Cassio, D., Hamon-Benais, C., Guerin, M. and Lecoq, O. (1991). Hybrid cell lines constitute a potential reservoir of polarized cells: isolation and study of highly differentiated hepatoma-derived hybrid cells able to form functional bile canaliculi in vitro. *J. Cell Biol.* **115**, 1397–1408.

Cassio, F. and Leao, C. (1991b). Low- and high-affinity transport systems for citric acid in the yeast Candida utilis. *Appl. Environ. Microbiol.* **57**, 3623–3628.

Chambers, T. J., Owens, J. M., Hattersley, G., Jat, P. S. and Noble, M. D. (1993). Generation of osteoclast-inductive and osteoclastogenic cell lines from the H-2Kb-tsA58 transgenic mouse. *Proc. Natl Acad. Sci. USA* **90**, 5578–5582.

Chaumontet, C., Mazzoleni, G., Decaens, C., Bex, V., Cassio, D. and Martel, P. (1998). The polarized hepatic human/rat hybrid WIF 12-1 and WIF-B cells communicate efficiently in vitro via connexin 32-constituted gap junctions. *Hepatology* **28**, 164–172.

Chiu, J. H., Hu, C. P., Lui, W. Y., Lo, S. C. and Chang, C. M. (1990). The formation of bile canaliculi in human hepatoma cell lines. *Hepatology* **11**, 834–842.

Decaens, C., Rodriguez, P., Bouchaud, C. and Cassio, D. (1996). Establishment of hepatic cell polarity in the rat hepatoma-human fibroblast hybrid WIF-B9. A biphasic phenomenon going from a simple epithelial polarized phenotype to an hepatic polarized one. *J. Cell Sci.* **109**, 1623–1635.

De Leon, J. R., Federoff, H. J., Dickson, D. W., Vikstrom, K. L. and Fishman, G. I. (1994). Cardiac and skeletal myopathy in beta myosin heavy-chain simian virus 40 tsA58 transgenic mice. *Proc. Natl Acad. Sci. USA* **91**, 519–523.

Dennis, J. E. and Caplan, A. I. (1996). Differentiation potential of conditionally immortalized mesenchymal progenitor cells from adult marrow of a H-2Kb-tsA58 transgenic mouse. *J. Cell Physiol.* **167**, 523–538.

Efrat, S., Linde, S., Kofod, H., Spector, D., Delannoy, M., Grant, S., Hanahan, D. and Baekkeskov, S. (1988). Beta-cell lines derived from transgenic mice expressing a hybrid insulin gene-oncogene. *Proc. Natl Acad. Sci. USA* **85**, 9037–9041.

Efrat, S., Surana, M. and Fleischer, N. (1991). Glucose induces insulin gene transcription in a murine pancreatic beta-cell line. *J. Biol. Chem.* **266**, 11141–11143.

Ehler, E., Jat, P. S., Noble, M. D., Citi, S. and Draeger, A. (1995). Vascular smooth muscle cells of H-2Kb-tsA58 transgenic mice. Characterization of cell lines with distinct properties. *Circulation* **92**, 3289–3296.

Evans, G. S., Flint, N., Somers, A. S., Eyden, B. and Potten, C. S. (1992). The development of a method for the preparation of rat intestinal epithelial cell primary cultures. *J. Cell Sci.* **101**, 219–231.

Ferrary, E., Cohen-Tannoudji, M., Pehau-Arnaudet, G., Lapillonne, A., Athman, R., Ruiz, T., Boulouha, L., El Marjou, F., Doye, A., Fontaine, J. J., Antony, C., Babinet, C., Louvard, D., Jaisser, F. and Robine, S. (1999). In vivo, villin is required for Ca2+ dependent F-actin disruption in intestinal brush-borders. *J. Cell Biol.* **146**, 819–829.

Fogh, J. and Trempe, G. (1975). New human tumor cell lines. *Human Tumor Cells In Vitro*. ed. J. Fogh. New York: Plenum, 115–141.

Groves, A. K., Entwistle, A., Jat, P. S. and Noble, M. (1993). The characterization of astrocyte cell lines that display properties of glial scar tissue. *Dev. Biol.* **159**, 87–104.

Hanahan, D. (1988). Dissecting multistep tumorigenesis in transgenic mice. *Ann. Rev. Genet.* **22**, 479–519.

Huet, C., Sahuquillo-Merino, C., Coudrier, E. and Louvard, D. (1987). Absorptive and mucus-secreting subclones isolated from a multipotent intestinal cell line (HT-29) provide new models for cell polarity and terminal differentiation. *J. Cell Biol.* **105**, 345–357.

Ihrke, G., Neufeld, E. B., Meads, T., Shanks, M. R., Cassio, D., Laurent, M., Schroer, T. A., Pagano, R. E. and Hubbard, A. L. (1993). WIF-B cells: an in vitro model for studies of hepatocyte polarity. *J. Cell Biol.* **123**, 1761–1775.

Inui, K., Quaroni, A., Tillotson, L. G. and Isselbacher, K. J. (1980). Amino acid and hexose transport by cultured crypt cells from rat small intestine. *Am. J. Physiol.* **239**, C190–C196.

Itoh, M., Hotta, H. and Homma, M. (1998). Increased induction of apoptosis by a Sendai virus mutant is associated with attenuation of mouse pathogenicity. *J. Virol.* **72**, 2927–2934.

Jat, P. S., Noble, M. D., Ataliotis, P., Tanaka, Y., Yannoutsos, N., Larsen, L. and Kioussis, D. (1991). Direct derivation of conditionally immortal cell lines from an H-2Kb-tsA58 transgenic mouse. *Proc. Natl Acad. Sci. USA* **88**, 5096–5100.

Kedinger, M., Simon-Assmann, P. M., Lacroix, B., Marxer, A., Hauri, H. P. and Haffen, K. (1986). Fetal gut mesenchyme induces differentiation of cultured intestinal endodermal and crypt cells. *Dev. Biol.* **113**, 474–483.

Kershaw, T. R., Rashid-Doubell, F. and Sinden, J. D. (1994). Immunocharacterization of H-2Kb-tsA58 transgenic mouse hippocampal neuroepithelial cells. *Neuroreport* **5**, 2197–2200.

Ketterer, M. R., Shao, J. Q., Hornick, D. B., Buscher, B., Bandl, V. K. and Apicella, M. A. (1999). Infection of primary human bronchial epithelial cells by *Haemophilus influenzae*: Macropinocytosis as a mechanism of airway epithelial cell entry. *Infection and Immunity* **67**, 4161–4170.

Mahida, Y. R., Makh, S., Hyde, S., Gray, T. and Borriello, S. P. (1996). Effect of Clostridium difficile toxin A on human intestinal epithelial cells: induction of interleukin 8 production and apoptosis after cell detachment. *Gut* **38**, 337–347.

Martin, F., Knobel, S., Martin, M. and Bordes, M. (1975). A carcinofetal antigen located on the membrane of cells from rat intestinal carcinoma in culture. *Cancer Res.* **35**, 333–336.

Martin, M. S., Caignard, A., Hammann, A., Pelletier, H. and Martin, F. (1987). An immunohistological study of cells infiltrating progressive and regressive tumors induced by two variant subpopulations of a rat colon cancer cell line. *Int. J. Cancer* **40**, 87–93.

Mellon, P. L., Windle, J. J., Goldsmith, P. C., Padula, C. A., Roberts, J. L. and Weiner, R. I. (1990). Immortalization of hypothalamic GnRH neurons by genetically targeted tumorigenesis. *Neuron.* **5**, 1–10.

Morgan, J. E., Beauchamp, J. R., Pagel, C. N., Peckham, M., Ataliotis, P., Jat, P. S., Noble, M. D., Farmer, K. and Partridge, T. A. (1994). Myogenic cell lines derived from transgenic mice carrying a thermolabile T antigen: a model system for the derivation of tissue-specific and mutation-specific cell lines. *Dev. Biol.* **162**, 486–498.

Neutra, M. and Louvard, D. (1989). Differentiation of intestinal cells in vitro. *Functional Epithelial Cells in Culture* Alan R. Liss, Inc., 363–398.

Noble, M., Groves, A. K., Ataliotis, P., Ikram, Z. and Jat, P. S. (1995). The H-2KbtsA58 transgenic mouse: a new tool for the rapid generation of novel cell lines. *Transgenic Res.* **4**, 215–225.

O'Donnell, R. W. and Cockerell, G. L. (1981). Establishment and biological properties of a guinea pig colonic adenocarcinoma cell line induced by N-methyl-N-nitrosourea. *Cancer Res.* **41**, 2372–2377.

Paradis, K., Le, O. N., Russo, P., St-Cyr, M., Fournier, H. and Bu, D. (1995). Characterization and response to interleukin 1 and tumor necrosis factor of immortalized murine biliary epithelial cells. *Gastroenterology* **109**, 1308–1315.

Pujuguet, P., Hamman, A., Moutet, M., Samuel, J. L., *et al.* (1996). Expression of fibronectin ED-A$^+$ and ED-B$^+$ isoforms by human and experimental colorectal cancer. Contribution of cancer cells and tumor-associated myofibroblasts. *Am. J. Path.* **148**, 579–592.

Rogler, G., Daig, R., Aschenbrenner, E., Vogl, D., Schlottmann, K., Falk, W., Gross, V., Schölmerich, J. and Andus, T. (1998). Establishment of long-term primary cultures of small and large intestinal epithelial cells. *Laboratory Investigation* **78**, 889–890.

Ronco, P. M., Prie, D., Piedagnel, R. and Lelongt, B. (1994). Oncogene-transformed renal cell lines: physiological and oncogenetic studies. *NIPS* **9**, 208–214.
Seglen, P. O. (1976). Preparation of rat liver cells. *Methods Cell. Biol.* **13**, 29–83.
Sirica, A. and Gainey, T. W. (1997). A new rat bile ductular epithelial cell culture model characterized by the appearance of polarized bile ducts *in vitro*. *Hepatology* **26**, 537–549.
Shanks, M. R., Cassio, D., Lecoq, O. and Hubbard, A.L. (1994). An improved polarized rat hepatoma hybrid cell line. Generation and comparison with its hepatoma relatives and hepatocytes in vivo. *J. Cell Sci.* **107**, 813–825.
Smith, J. J., Travis, S. M., Greenberg, E. P. and Welsh, M. J. (1996). Cystic fibrosis airway epithelia fail to kill bacteria because of abnormal airway surface fluid. *Cell* **85**, 229–236.
Walther, N., Jansen, M., Ergun, S., Kascheike, B. and Ivell, R. (1996). Sertoli cell lines established from H-2Kb-tsA58 transgenic mice differentially regulate the expression of cell-specific genes. *Exp. Cell Res.* **225**, 411–421.
Whitehead, R. H. and Joseph, J. L. (1994). Derivation of conditionally immortalized cell lines containing the Min mutation from the normal colonic mucosa and other tissues of an "Immortomouse"/Min hybrid. *Epithelial Cell Biol.* **3**, 119–125.
Whitehead, R. H., VanEeden, P. E., Noble, M. D., Ataliotis, P. and Jat, P. S. (1993). Establishment of conditionally immortalized epithelial cell lines from both colon and small intestine of adult H-2Kb-tsA58 transgenic mice [published erratum appears in *Proc. Natl Acad. Sci. USA* 1993 July 15; **90**(14): 6894]. *Proc. Natl Acad. Sci. USA* **90**, 587–591.
Wuthrich, R. P., Glimcher, L. H., Yui, M. A., Jevnikar, A. M., Dumas, S. E. and Kelley, V. E. (1990). MHC class II, antigen presentation and tumor necrosis factor in renal tubular epithelial cells. *Kidney Int.* **37**, 783–792.
Yokomuro, S., Tsuji, H., Lunz III, J. G., Sakamoto, T., Ezure, T., Murase, N. and Demetris, A. (2000). Growth control of human biliary epithelial cells by Interleukin 6, Hepatocyte growth factor, Transforming growth factor β1 and Activin 1: comparison of a cholangiocarcinoma cell line with primary cultures of non-neoplastic biliary epithelial cells. *Hepatology* **32**, 26–35.
Zweibaum, A., Laburthe, M., Grasset, E. and Louvard, D. (1991). Use of cultured cell lines in studies of intestinal cell differentiation and function. *Handbook in Physiology, the gastrointestinal system. In intestinal absorption and secretion.* eds M. Field and R. A. Frizzell, pp. 223–255.

6 Light Microscopy Techniques for Bacterial Cell Biology

Petra Anne Levin
Department of Biology, Washington University, St. Louis, MO 63130, USA

CONTENTS

♦♦♦♦♦♦ INTRODUCTION

Bacteria have typically been viewed as poor candidates for the techniques employed by eukaryotic cell biologists to localize subcellular factors. At a practical level, their small size (1 to 2 μm on average) makes bacteria less than ideal subjects for light microscopy. Furthermore, in the absence of membrane-bound organelles, prokaryotes have often been portrayed as 'sacs of enzymes' exhibiting little if any organization within the confines of their plasma membrane.

Electron microscopy did much to dispel the myth that bacterial cells are intrinsically uninteresting at the subcellular level. Gifted electron microscopists, such as Eduard Kellenberger and Antoinette Ryter, used transmission electron microscopy to create exquisite images of bacterial cells during growth and differentiation. Their work provided fundamental insights into the subcellular organization of the bacterial cell, including the nature of the bacterial nucleoid, the structure of the bacterial cell wall, and the morphological changes *Bacillus subtilis* cells undergo during spore development (Kellenberger and Ryter, 1958; Robinow and Kellenberger, 1994; Ryter, 1964). Immunoelectron microscopy, which uses antibodies conjugated to colloidal gold particles to localize factors of interest in thin sections of cells prepared for EM, also advanced our understanding of bacterial cells. For instance, immunoelectron microscopy first revealed cell type-specific gene expression in developing *B. subtilis* cells, the polar localization of the chemoreceptor MCP in *Escherichia coli*, and the ring structure formed by the bacterial cell division protein FtsZ (Bi and Lutkenhaus, 1991; Maddock and Shapiro, 1993; Margolis *et al.*, 1991).

METHODS IN MICROBIOLOGY, VOLUME 31
ISBN 0-12-521531-2

Ultimately, it was the adaptation of immunofluorescence microscopy (IFM)—a technique routinely used for localizing proteins within eukaryotic cells and tissues—for prokaryotic systems that overcame the notion that bacterial cells are too small to be practical candidates for light microscopy. In contrast to immunoelectron microscopy, which requires highly specialized expertise and equipment, IFM is accessible to any microbiologist seeking to determine the subcellular localization pattern of a protein of interest. Consequently, IFM has been used in a wide range of applications, for instance localizing cell type determinants in *Caulobacter crescentus*, elucidating genetic hierarchies in the localization of cell division proteins to the nascent septal site in *E. coli*, and pinpointing the timing of gene expression in individual *B. subtilis* cells in response to developmental signals (Shapiro and Losick, 2000).

Once the presumption that bacterial cells were too small was disproven, other light microscopy techniques developed for use in eukaryotic cells were rapidly adapted for bacterial systems. These techniques included the use of BrdU (a fluorescent nucleotide analog) to label replicating DNA, and fluorescent *in situ* hybridization (FISH) (Lewis and Errington, 1997; Niki and Hiraga, 1997). The most successful of these adaptations has been the use of the green fluorescent protein (GFP) from the jelly fish *Aequoria victoria* in bacterial systems. GFP can be used both as a transcriptional reporter and as a tag for subcellular localization of both proteins and chromosomal and plasmid DNA (Belmont and Straight, 1998; Tsien, 1998). Since it was first used in bacterial cells, GFP has been widely employed to confirm and expand upon data originally obtained by immunoelectron microscopy or IFM (Margolin, 2000; Webb *et al.*, 1995). Moreover, in contrast to immunoelectron microscopy and IFM, GFP permits the real-time localization of proteins of interest and even regions of the bacterial chromosome in live cells. As a result of this property, GFP fusions have been the basis for experiments demonstrating the rapid movement of newly replicated chromosomal origins and plasmids to opposite poles of the cell during the course of the cell cycle as well as the pole-to-pole oscillations of the bacterial cell division protein MinD (Raskin and de Boer, 1999; Webb *et al.*, 1998).

Together IFM and GFP have fundamentally changed our view of bacterial cells. Far from amorphous bags of DNA and protein, bacteria—with their own highly organized genetic material and cytoskeletal structures—are now considered to have a level of subcellular organization approaching that of eukaryotic cells (Shapiro and Losick, 2000). The question of whether any subcellular ultrastructure exists has been answered. Accordingly, bacterial cell biologists have shifted the focus of their research to investigate how bacterial cells maintain such an elaborate level of organization without membrane-bound organelles and in the apparent absence of motor proteins and cytoskeletal-based trafficking systems.

This chapter provides an overview of the uses of immunofluorescence microscopy and GFP in bacterial systems. The techniques presented require a microscope equipped for epifluorescence and a high power (60× or 100×) DIC objective with a high numerical aperture. Depending on the brightness of the sample, images can be captured either with a standard 35 mm camera mounted on the microscope or with a high resolution, high sensitivity CCD (charged coupled device) camera in conjunction with a digital imaging system. A list of sources of materials and reagents as well as a table of filter sets for visualizing the different fluorophores are provided at the end of the chapter. Although these protocols were originally developed for use in *E. coli* and *B. subtilis*, with some small modifications they should be easily adapted for use with other bacterial species.

◆◆◆◆◆◆ IMMUNOFLUORESCENCE MICROSCOPY

Although it provides neither the resolution of electron microscopy, nor information about protein localization in real time like GFP, IFM offers several advantages over both techniques. First, due to the relatively gentle fixation required to prepare samples and the use of whole cells instead of thin sections, IFM is significantly more sensitive than immunoelectron microscopy. It is possible to visualize proteins that are present at as few as 100 molecules per cell and perhaps those of even lower abundance using IFM (Lemon and Grossman, 1998; Weiss *et al.*, 1997). Moreover, the wide range of available fluorophores facilitates the simultaneous localization of as many as four or five factors of interest in a single cell.

The preparation of cells for immunofluorescence microscopy requires three basic steps: fixation, permeabilization, and staining. Fixation is the means by which growing cells are preserved for IFM. Once fixed, cells are stable for up to a week, allowing analysis at a time that is convenient to the investigator. Permeabilization makes cells porous enough to allow antibodies access to cytoplasmic components. After permeabilization, antibodies conjugated to fluorophores are used to stain cells for factors of interest. If desired, cell wall material and the bacterial nucleoid can also be labeled during this third and final step using wheat germ agglutinin (lectin) conjugated to fluorophore and a nucleic acid stain such as 4′,6-Diamidino-2-phenylindole (DAPI).

Fixation

Although a multitude of variations exist, there are essentially two methods for fixing cells prior to IFM—methanol (Angert, personal communication; Hiraga *et al.*, 1998) (Protocol 1) and gluteraldehyde/paraformaldehyde (Addinall *et al.*, 1996; Harry *et al.*, 1995; Levin and Losick, 1996; Pogliano *et al.*, 1995) (Protocol 2). Each method has its own advantages and disadvantages. For example, although methanol fixation is more gentle, it does not preserve cell wall material as well as gluteraldehyde/paraformaldehyde. The choice of fixative is, therefore, best made empirically.

When using gluteraldehyde/paraformaldehyde, cells grown in rich medium [such as Luria-Bertani (LB) broth] typically require a higher concentration of gluteraldehyde than those grown in defined minimal medium. However, it should be noted that the concentration of gluteraldehyde used in Protocol 2 can effect the sensitivity of the technique. At higher concentrations of gluteraldehyde, background fluorescence increases and antibody-antigen recognition decreases, seriously compromising the sensitivity of the technique. For these reasons it is very important to do a thorough titration of gluteraldehyde when optimizing the fixation protocol. Finally, the repeated centrifugations required to remove fixative can lead to the mislocalization of certain subcellular factors (Lemon and Grossman, 1998). If this is a concern, Protocol 3 provides an alternative means to wash cells following fixation.

Protocol 1: Methanol fixation

1. Drop 1 ml of culture into 10 ml of ice-cold 80% MeOH.
2. Let stand for 60 min at room temperature.
3. Add 200 µl of 16% paraformaldehyde (this step fixes the nucleoids so that they remain intact during permeabilization and staining).
4. Let stand for 5 min at room temperature.
5. Spin down cells at relatively low speed (2500×*g*) and resuspend gently (avoid vortexing... finger flicking is best) in 1 ml of ice–cold 80% MeOH. The cells will tend to clump together at this point. This is not a problem as the clumps will disperse when the cells are adhered to the slide. Cells will keep for up to 1 week in methanol at 4°C.

Solutions

80% MeOH –20°C
16% Paraformaldehyde

Protocol 2: Gluteraldehyde/paraformaldehyde fixation

1. While cells are growing prepare fixative. The gluteraldehyde concentration needs to be optimized for each protein of interest, however, 1.5 to 3 µl of 25% gluteraldehyde per ml of 16% paraformaldehyde is typical. Store on ice.
2. Aliquot 20 µl of 1 M $NaPO_4$ pH 7.4 into 1.5 ml microfuge tubes and label.
3. Immediately prior to sampling add 100 µl of fix into each prepared tube.
4. Take 500 µl aliquots of each culture and add to appropriately labeled tube. Invert tubes one or two times.
5. Incubate for 15 min at room temperature followed by 30 min on ice.
6. Wash cells three times by pelleting in microfuge at approximately 20 000× *g* and resuspending in 1 ml PBS. After the final PBS

wash, resuspend pellets in GTE to a final OD of approximately A600 = 0.200.

Solutions

16% paraformaldehyde
25% gluteraldehyde
Phosphate Buffered Saline (PBS)
50 mM Glucose 25 mM Tris 8.0 10 mM EDTA 8.0 (GTE)
1 M $NaPO_4$ pH 7.4

Protocol 3: Filter wash

1. Proceed with fixation according to either protocol 1 or 2 up to point of spin.
2. Pre-wet filter in holder with PBS. Handle filter with forceps. Do not let filter or cells dry out.
3. Instead of spinning to remove fix from cells and concentrate them, filter the cells through a low protein binding filter using a 5 ml syringe and filter holder. Be gentle when pushing the fixed cells, and later the wash, through the filter.
4. Wash cells on filter with 3 ml of PBS.
5. Resuspend cells in GTE by placing filter in microfuge tube with 200–400 µl of GTE and flicking the tube. Pull out the filter after resuspending cells.

Note: Fixed and filtered cells are best for immunofluorescence when used the same day. However, if necessary, they can be stored for up to 3 days at 4°C.

Equipment

Filter Holder 13 mm SST Swinney Syringe Holder
0.1 µm VVPP Durapore Membrane Filters

Solutions

PBS
GTE

Permeabilization and staining

The permeabilization step is critical to the success of IFM. It is important to achieve adequate breakdown of the cell wall to allow entry of antibodies, while at the same time maintaining cell ultrastructure. *B. subtilis* and *E. coli* cells are adhered to the microscope slide and permeabilized using an isotonic lysozyme solution. In general, *E. coli* require significantly less lysozyme for permeabilization than *B. subtilis* (Addinall *et al.*, 1996) and growth conditions and strain background can also affect the length of time required for effective permeabilization. Thus, titrating both the concentration of lysozyme (or other lytic agent) and/or the incubation time to optimize permeabilization is strongly advised.

While lysozyme is sufficient to permeabilize *E. coli* and *B. subtilis* cells, other species may require additional treatment. Esther Angert found that *Metabacterium polyspora* cells are resistant to standard lysozyme treatment. She, therefore, developed a two-step permeabilization protocol in which the *M. polyspora* cells were subjected to two, 2-hour incubations—first in mutanolysin and then in lysozyme (Angert and Losick, 1998).

In the protocol for IFM presented below (Protocol 4), staining is a two-step process requiring both primary antisera against the protein of interest [or an epitope tag such as c-Myc, FLAG, or hemagglutinin (HA)] and a secondary antibody conjugated to a fluorophore. It is possible to eliminate the second staining step if primary antibodies conjugated to fluorophores are available. The steps at which cell wall and nucleic acid stains can be added are noted in parentheses. If polyclonal antibodies are used, affinity purification of antisera (Protocol 5) can significantly reduce background staining. Protocol 6 is a simple method for fixing and staining bacterial nucleoids that does not require permeabilization.

Protocol 4: Immunofluorescence microscopy

1. Prepare 15- or 8-well slides. Put 10 μl of 1% poly-L-lysine solution into each well. Let stand 5 minutes and aspirate off. Wash once with distilled sterile water and let air dry. (All washes are done with a Pasteur pipette that has been hooked up to an aspirator with a trap. It is important not to touch the wells with the pipette tip during all washes.) Store slides in empty petri dishes to protect them from dust, etc. Use slides within 1 hour of poly-L-lysine treatment for best results.
2. While slides are drying prepare 2 mg ml^{-1*} lysozyme solution in GTE and store on ice.
3. Drop ~10 μl of fixed cells onto well. Let stand 2 min. Aspirate and wash once with 10 μl of PBS. Let air dry. It is important NOT to aspirate too forcefully or touch the wells with the pipette tips.
4. Re-wet cells for 5 min with 10 μl of PBS.
5. Aspirate off PBS and put 10 μl of lysozyme solution onto each well. Incubate 1 to 7 min at room temperature (incubation time may vary depending on strain background and growth conditions).
6. Wash wells once with PBS.
7. Add 10 μl of 2% BSA solution to each well as a blocking agent. Incubate for 10 min at room temperature.
8. Meanwhile dilute primary antibody appropriately in PBS-2% BSA. For untested sera, a titration of between 1:50 to 1:10 000 is generally a good place to start.
9. Aspirate off BSA solution (do not let wells dry!) and add 10 μl of diluted antibody to each well. The incubation period varies but it should be at least 1 hour. If incubating overnight or longer it is best to put the slide at 4°C. To prevent drying, place the slide and a wet piece of tissue (not touching the slide) in a petri dish and seal with parafilm.

10. After primary incubation, wash wells 10 to 25 times with PBS (increasing the number of washes can decrease background) and apply secondary, fluorophore conjugated antibody. Secondary antibodies should be diluted in 2% BSA solution.[†] Incubate for 30 min to 1 h at room temperature. [Wheat germ agglutinin (lectin) conjugated to the fluorophore of choice can be added at this point to stain cell wall material.]
11. Wash wells 10 to 25 times with PBS. Apply one drop of Slow Fade equilibration buffer from Slow Fade kit (Molecular Probes) to each well and let stand 5 min. (1 $\mu g\,ml^{-1}$ DAPI in Slow Fade equilibration buffer can be added at this point to stain nucleoids.)
12. Aspirate off and apply one drop of Slow Fade in glycerol to each well in the uppermost row. Carefully put on cover slip, angling it so that it covers first the top wells and then the middle and bottom wells. Make sure not to get any Slow Fade on top of the cover slip and avoid creating bubbles in the wells below. Aspirate off any extra Slow Fade, as it will make the slide hazy when viewed through an oil objective. Cover slip can be replaced with a fresh one if necessary.
13. The slide is now ready to be examined. Store slides in foil wrapped petri dishes at −20°C.

Note: This procedure can be done on a coverslip instead of an 8- or 15-well slide.

*For *E. coli* much less lysozyme is needed. Typically 1–10 $\mu g\,ml^{-1}$ is sufficient, however, a lysozyme titration is recommended to account for differences between strains.
[†]Titrating fluorophore-conjugated secondary antibodies will ensure optimal results. However, a 1:100 to 1:500 dilution of a 1 $mg\,ml^{-1}$ solution of fluorophore-conjugated secondary antisera is typical.

Solution

PBS
GTE
2% BSA in PBS
1% poly-L-lysine
2 $mg\,ml^{-1}$ lysozyme in GTE
1 $mg\,ml^{-1}$ DAPI stock solution

Supplies

Slow Fade kit
8- or 15-well slides
large cover slips
FITC, Cy3, Cy5, etc. conjugated secondary antibodies

Protocol 5: Affinity purification of polyclonal antisera

1. Run denaturing polyacrylamide gel with 10 μg to 100 μg of the purified protein used to generate antisera. Typical load is 1–10 μg per well or 10 to 100 μg total in one large well.

2. Transfer the protein to PVDF membrane using standard procedure for immunoblotting.
3. Stain the membrane with Ponceau S. Rinse with ddH_2O to destain. Using a razor blade or X-acto knife cut out the region of the membrane containing the protein trying to get as thin a strip as possible.
4. Block the membrane strip with 5% dry milk in PBS for 20 min at RT.
5. Wash membrane strip twice for 2 min in PBS. Cut strip in half. *Note*: if affinity purifying $\leq$300 μl of sera it is possible to use only half of the membrane and save half, tightly wrapped in plastic, at −20°C. Re-wet frozen membrane in MeOH and rinse in PBS before using.
6. Incubate the strip of membrane containing protein with 300 μl of antisera for 1 h at room temperature. It is helpful to cut the strip of membrane into smaller pieces and put them into a microfuge tube for incubation purposes.
7. After incubation, pull off the depleted antisera and set aside. Antisera may retain a relatively high titer of antibodies and can be subjected to further affinity purification should serum be in short supply.
8. Wash the membrane twice for 15 min in 5 ml of PBS.
9. Meanwhile, aliquot 100 μl of $NaPO_4$ pH 8.0 into three microfuge tubes.
10. To strip membrane add 300 μl of 5 mM glycine 150 mM NaCl pH 2.4.
11. Incubate for 30 s at room temperature.
12. Pull off supernatant and put into microfuge tube with $NaPO_4$ solution to neutralize.
13. Repeat steps 10–12 two more times.
14. Combine all three tubes to equilibrate the antibody levels and then divide into fresh aliquots. Generally, dialysis of affinity purified antisera does not appear to be necessary. Store the affinity purified antisera at 4°C as freezing sometimes damages the antibody. Sodium azide can be added to 1 mM.

Note: Yields vary so it is important to titer the antibody after each affinity purification.

Solutions

PBS

Ponceau S

(10× solution: 2% Ponceau S in 30% trichloroacetic acid, 30% sulfosalicylic acid)

5% dry milk in PBS

Stripping solution

(5 mM glycine, 150 mM NaCl pH 2.4)

1 M $NaPO_4$

Supplies

polyacrylamide gel

PVDF membrane

Protocol 6: Quick method for visualizing bacterial nucleoids

1. Immediately prior to fixing cells prepare a poly-L-lysine coated slide:
   - Put 10–20 μl drops of poly-L-lysine into each well and let stand for 2 min.
   - Wash once with ddH_2O.
   - Aspirate to dry completely.
2. Add culture to each well for 30 s to 2 min. Time varies depending on the density of the cell culture. Aspirate and dry with vacuum.
3. Overlay cells with EtOH for 1 to 2 min. Aspirate.
4. Wash three times in 10 to 20 μl ddH_2O. Aspirate and dry.
5. Add DAPI (1 $\mu g\,ml^{-1}$) and coverslip.

Supplies and Solutions

8- or 15-well glass slides
1% poly-L-lysine solution
1 $\mu g\,ml^{-1}$ DAPI solution in ddH_2O
100% EtOH
ddH_2O

◆◆◆◆◆◆ THE GREEN FLUORESCENT PROTEIN

In contrast to IFM, in which sample preparation involves both a fixation step and a lengthy staining procedure, the preparation of cells expressing GFP for microscopy is trivial. At its most basic, sample preparation requires only a glass slide, a coverslip, and a microscope equipped with the appropriate filter set (see Table 6.1). Furthermore, because GFP permits the visualization of proteins of interest in unfixed cells, it can be used for real-time studies of live cells both in culture and, in the case of pathogens and plant symbionts, in host tissues (Cheng and Walker, 1998; Kohler *et al.*, 2000). Finally, in contrast to other fluorescent

Table 6.1 Filter set specifications. Filter sets are available from several sources, however, Chroma is one of the largest suppliers and has excellent service and support

Primary use	Excitation	Beamsplitter	Emission	Chroma #
DAPI	D360/40×	400DCLP	D460/50 m	31000
BFP	D380/30×	420DCLP	D460/50 m	31041
CFP	D436/20×	455DCLP	D480/40 m	31044 v. 2
GFP	HQ480/40×	Q505LP	HQ510LP	41012
FITC	HQ480/40×	Q505LP	HQ535/50 m	41001
YFP	HQ500/20×	Q515LP	HQ520LP	41029
Rhodamine	HQ545/30×	Q570LP	HQ620/60 m	41002c

proteins, GFP fluoresces without the addition of a substrate or co-factor (Tsien, 1998). For these reasons, GFP has become a favorite tool of bacterial cell biologists.

GFP was originally isolated as a protein that co-purified with the chemiluminescent protein aequorin from the jelly fish *Aequoria victoria* (Tsien, 1998). However, because of its inability to fold well at temperatures over 25°C, native GFP proved to be less than ideal for use in most eukaryotic and prokaryotic systems. Moreover, the peak excitation frequency of 395 nm for native GFP (in the UV range) was damaging both to the specimen and to the investigator in the absence of proper eye protection or a UV absorbing screen. To circumvent these problems, several labs developed allelic variants of GFP that increased the folding efficiency of the chromophore at higher temperatures and altered the excitation spectra of the molecule, shifting it towards a peak of 470 nm (Cormack *et al.*, 1996; Tsien, 1998). These so-called 'red-shifted' variants made it possible to use GFP as a transcriptional reporter in lieu of *lacZ* and have facilitated the use of GFP fusions for the subcellular localization of proteins of interest (Kohler *et al.*, 2000; Lemon and Grossman, 1998). Additional site-directed mutagenesis of the chromophore led to the development of blue, cyan, and yellow variants of GFP (BFP, CFP, and YFP respectively); these provide a means of localizing two factors of interest simultaneously in the same cell (see below) (Tsien, 1998).

General considerations

Whether using GFP as a transcriptional reporter or as a tag for the subcellular localization of a target protein, maximizing expression is central to the success of the experiment. For this reason, when making an N-terminal GFP fusion, and in situations where GFP will be used as a transcriptional reporter, it is important to ensure that the Shine-Dalgarno sequence is optimized for the system in which the GFP moiety will be expressed. Codon usage is also an important consideration. Many of the commercially available GFP molecules have been optimized for expression in mammalian cells (so-called 'humanized' GFP). However, trial and error indicates that the 'native' versions of GFP are expressed significantly better in many bacterial systems than the commercially available GFP variants (Lemon, personal communication).

GFP fusion proteins

At 22 kD, the addition of a GFP moiety can significantly alter protein folding and activity *in vivo*. To circumvent problems associated with constructing a functional fusion protein, the GFP moeity can be fused to either the N-terminus or C-terminus of the protein of interest. Additionally, the length and composition of the polypeptide linker that joins the native protein to GFP can also have a surprisingly large effect on protein function (P.A.L. unpublished data; Lindow, personal communication).

Table 6.2 Suppliers

Product	Supplier	Telephone in US
Filter Sets for Fluorescence Microscopy	Chroma	800-824-7662
15- or 8-Well Slides	ICN	800-854-0530
16% Paraformaldehyde	Electron Microscopy Sciences	800-523-5874
25% Gluteraldehyde	Electron Microscopy Sciences	800-523-5874
1% Poly-L-lysine	Sigma	800-325-3010
Slow Fade Kit	Molecular Probes	541-465-8300
FM 4-64	Molecular Probes	541-465-8300
TMA-DPH	Molecular Probes	541-465-8300
DAPI	Molecular Probes	541-465-8300
Conjugated antibodies	Jackson Immunoresearch	800-367-5296
Low protein binding filters	Millipore	800-645-5476
Filter holder	Millipore	800-645-5476
Polyclonal anti-GFP serum	Clontech	800-662-2566

Because of the difficulties associated with making a functional GFP fusion, it is advisable to determine if a GFP fusion protein is able to complement a null mutation to ensure that protein function is not disturbed. Similarly, whenever possible, the localization pattern of a GFP fusion protein should be confirmed by immunofluorescence microscopy. Polyclonal sera against GFP is available commercially from several sources should it be required (see Table 6.2).

If a GFP fusion protein is incapable of complementing a null mutation, it is possible to place the fusion under the control of an inducible/repressible promoter in the presence of the wild type allele (Levin *et al.*, 1999). In this configuration, the fusion protein is expressed at low levels and is used to 'trace' the location of the target molecule without disturbing its activity.

Colored variants of GFP and vital stains for cell membrane

Although the number of colors available for GFP fusions is limited compared to IFM, mutagenesis of the chromophore has led to the development of cyan, blue, and yellow variants (CFP, BFP, and YFP) (Tsien, 1998). To avoid overlapping emission and excitation spectra, these variants can be used in only one of two combinations: BFP and GFP; or CFP and YFP. As with GFP, codon usage is an important consideration when selecting one of the colored variants for expression in bacterial cells. Although colored variants are available commercially, if the non-humanized version is preferred, it is possible to use site-directed

Table 6.3 Key mutations affecting GFP folding, excitation and emission spectra. This is a table of those GFP variants that have been shown to fold and be expressed well in *E. coli* and *B. subtilis*. For a more detailed discussion of the classes of GFP molecules see Tsien (1998). Amino acids that are identical to wild type GFP are left blank

Position	64	65	66	67	68	69	70	71	62	203	Reference
Wild type	Phe	Ser	Tyr	Gly	Val	Gln	Cys	Phe	Ser	Thr	(Tsien, 1998)
mut1	Leu	Thr									(Cormack *et al.*, 1996)
mut2		Ala			Leu				Ala		(Cormack *et al.*, 1996)
mut3		Gly							Ala		(Cormack *et al.*, 1996)
BFPmut2		Ala	His		Leu				Ala		(Levin *et al.*, 1999; Tsien, 1998)
CFPmut2		Ala	Trp		Leu				Ala		(Gueiros-Filho, personal communication; Tsien, 1998)
YFPmut2		Ala			Leu				Ala	Tyr	(Lemon and Grossman, 2000; Tsien, 1998)

This table is based on Table 1 from Cormack *et al.* (1996).

mutagenesis to make the changes in key codons of GFP necessary to generate BFP, CFP or YFP (see Table 6.3) (Levin *et al.*, 1999; Lemon and Grossman, 2000; Gueiros-Filho, personal communication). Although a red fluorescent protein (DsRed) has recently become available commercially, experimental and anecdotal evidence indicates that its folding time is too long to be useful for bacteria with shorter generation times (Baird *et al.*, 2000; Hahn, personal communication).

Vital membrane stains provide a convenient way to visualize the periphery of live cells in the presence of a GFP fusion. The red membrane stain, N-(3-triethylammoniumpropyl)-4-(6-(4-(diethylamino)phenyl) hexatrienyl) pyridinium dibromide (FM 4-64, Molecular Probes), which has an absorbance of 543 nm, has been used in a wide variety of studies to determine cell boundaries and to track membrane dynamics during sporulation in *B. subtilis* (King *et al.*, 1999; Lemon and Grossman, 1998; Pogliano *et al.*, 1999). FM 4-64 is visible in the filter set most commonly used for GFP fusions and, therefore, may not be the best choice of stain for fainter GFP fusions that localize to the periphery of the cell. The blue membrane stain, 1-(4-trimethylammoniumphenyl)-6-phenyl-1,3,5-hexatriene p-toluenesulfonate (TMA-DPH Molecular Probes), which absorbs at 355 nm, can be used as an alternative (although not in conjunction with the blue nucleic acid stain DAPI) (Gueiros-Filho, personal communication).

Other applications

In addition to serving as a marker for protein localization and as a transcriptional reporter, technology developed by Andrew Belmont and

Aaron Straight permits GFP to be used to localize plasmids and regions of chromosomal DNA. Belmont and Straight tagged a region of yeast chromosomal DNA with a tandem array of lactose operator sites and visualized this region by fluorescence microscopy in live cells using GFP fused to the lactose repressor protein (Straight *et al.*, 1996; Belmont and Straight, 1998). This technology has subsequently been adapted for bacterial cells, where it has been used to visualize the position of chromosomal origin and termini, to measure the timing of origin separation during the cell cycle in *B. subtilis*, and to trace the localization of plasmids in *E. coli* cells (Gordon *et al.*, 1997; Webb *et al.*, 1997). A similar technology has also been developed that takes advantage of the tetracycline repressor and its binding site (Michaelis *et al.*, 1997).

The colored variants of GFP provide a means of exploring protein–protein interactions *in vivo* using fluorescence resonance energy transfer (FRET). Due to overlapping emission and excitation spectra, CFP can, following excitation, transfer its energy to a YFP molecule provided that the donor and recipient proteins are in close proximity (10–50 Å). The degree of FRET between two fusion proteins may be measured either *in vitro* with a fluorimeter or *in vivo* with a fluorescence microscope (Heim and Tsien, 1996; Pollok and Heim, 1999; Tsien, 1998). [Although it is possible to use BFP and GFP as a FRET pair, CFP and YFP are strongly preferred due to the relative sensitivity of BFP to photobleaching (Tsien, 1998).] While this technique has not yet been applied to bacterial cells, FRET has been used to determine the oligomeric state of the G-protein-coupled receptor for yeast alpha-factor and to explore interactions between nuclear pore proteins in *Saccharomyces cerevisiae* (Overton and Blumer, 2000; Damelin and Silver, 2000). These results in a similarly sized organism support the idea that FRET will be a powerful method with which to investigate dynamic protein interactions in bacterial cells.

Optimizing growth conditions for GFP

While bright GFP fusions are visible under most growth conditions, fainter fusions may require special attention. In general, it is best to grow cells in minimal defined medium whenever possible to keep background low. Cells grown in Luria-Bertani broth or other rich medium tend to have a faint green background even in the absence of a GFP fusion. This background can interfere with the signal from weaker GFP fusion proteins, making them extremely difficult to see. Temperature is also an important consideration, as even those variants of GFP that have been optimized for higher temperature applications are brighter when cells are grown at 30°C or below. Finally, oxygen is required for maturation of the GFP chromophore (Tsien, 1998). Thus, adequate aeration during growth is critical for maximizing the intensity of any given GFP fusion. (For this reason, it is unlikely that the GFP chromophore would be able to develop under anaerobic conditions, making GFP impractical for use in obligate anaerobes.)

Methods for visualizing GFP fusions in live cells

Protocol 7 presents two methods for preparing live cells for fluorescence microscopy. The first method is the simplest—requiring only pressure to immobilize the cells on the slide. In the second method, an agarose pad is used to provide a solid surface for bacterial adhesion. The agarose pad stabilizes the cells to some degree, making it the method of choice for cells that will be examined over any length of time. In addition, different reagents—isopropylthio-beta-D-galactoside (IPTG), antibiotics, vital dyes, etc—can be added to the agarose pad during the course of the experiment; this allows the investigator to examine the effect of different chemicals on gene expression patterns or protein localization in real time.

Protocol 8 is to be used when staining live cells with vital membrane dyes and nucleic acid stains (typically DAPI, which effectively stains the nucleoids of live bacterial cells). As discussed above the red membrane stain, FM 4-64, which emits both strong red and faint green fluorescence, is visible in the filter set most commonly used for GFP. To circumvent this problem, the filter set recommended for fluorescein-isothiocyanate (FITC), which eliminates the red fluorescence, can be used instead of the GFP filter set. It should be noted, however, that the FITC filter set is not as sensitive as the long-pass GFP filter set, which may be a problem with dimmer fusions. Alternatively, the FM 4-64 solution can be added to the agarose pad after the GFP image has been collected (Sharp and Pogliano, 1999).

In general it is best to look at unfixed, untreated cells to ensure that the GFP fusion is as bright as possible. Although some fusions are visible after fixation, both methanol and gluteraldehyde/paraformaldehyde significantly lower the brightness of most GFP fusion proteins. Poly-L-lysine, the reagent used to adhere fixed cells to the slide for immunofluorescence, can also have a deleterious effect on the brightness and localization pattern of GFP.

Microscopy

A drawback of GFP and its variants is their relatively high sensitivity to photobleaching by the high intensity light required for excitation. Though bright fusions and fusions to highly expressed proteins are less susceptible than their dimmer counterparts, exposure time should be minimized. An electronic shutter system, in which the shutter is controlled by a computer instead of manually, can significantly shorten the period in which cells are exposed to high intensity light. In the same vein, when taking multiple images of the same field of cells using different filter sets, it is best to start looking at fluorophores that excite at the longer (red) wavelengths and work towards the lower end of the spectrum (blue). This approach avoids photobleaching those fluorophores that excite at the shorter wavelengths (CFP and BFP), which tend to be the most sensitive.

To maximize the brightness of a GFP fusion it is best to use a microscope equipped with a high power DIC objective. The polarizing filter can be left in for focusing on a plane of cells and then pulled out to collect GFP images. Phase contrast objectives, although sufficient for the brightest GFP fusions, absorb too much light to be useful for visualizing fusions of below average brightness.

Protocol 7: Preparing live cells for fluorescence microscopy

A. Untreated slides

1. Drop 5 μl of culture onto the middle of a glass slide.
2. Carefully put a coverslip on top of the drop. Avoid creating bubbles if at all possible.
3. Using a tissue or a paper towel carefully press down on the cover slip to remove extra liquid.
4. Examine cells as soon as possible after preparing the slide.

B. Slide with agarose pad

1. Make an agarose pad by dropping ~20 to 50 μl of 1% agarose onto a glass slide. To make surface flat, use two slides, one untreated and one wrapped in tape at either end and coated with rain-X or other siliconizing substance. Drop agarose on treated slide between taped region. Cover with second, untreated slide. Let harden. Lift off taped slide.
2. Drop 10 to 20 μl of cell culture onto an agarose pad. Let it stand for 5 min.
3. Aspirate excess liquid.
4. Place coverslip gently over agarose pad.

Protocol 8: Staining live cells with DAPI and vital membrane dyes

1. Dilute FM 4-64 or TMA-DPH 1 : 100 in PBS or dH_2O.
2. Make an agarose pad as described in Protocol 7.
3. Incubate 200 μl of fresh cells with 2–10 μl of membrane dye for 2 min.
4. Spot cells onto agarose pad.
5. Leave for 5 to 10 min at room temperature.
6. Aspirate excess liquid.
7. If using FM 4-64, a small quantity of DAPI solution in GTE, dH_2O, or minimal medium can be added at this point to stain the nucleoids. Cover with cover slip. Remove any excess liquid.

Note: Cells must be growing prior to adding the stain to ensure that it is incorporated efficiently. The stain of interest can also be added to the media in which the cells are growing 30 min prior to microscopy at a 1 : 10 000 dilution. Finally, if the membrane stain masks the GFP signal it is possible to add the dye after collecting GFP images. See Sharp and Pogliano (1999).

Solutions

1% agarose in minimal media or GTE
1 μg ml^{-1} DAPI in PBS
FM 4-64 1 mg ml^{-1} in ddH_2O
TMA-DPH 1 mg ml^{-1} in DMF

◆◆◆◆◆◆ CONCLUSION

The protocols presented here represent the collected efforts of investigators at all levels working on a diverse array of biological problems. Largely through these efforts bacteria are no longer viewed as the proverbial swimming pool of enzymes; instead they are seen as possessing a level of subcellular organization as complex as that of any eukaryotic cell. These labors, therefore, have been the basis for what can be viewed as a revolution in bacterial cell biology.

While each method has been presented in as complete a manner as possible, the protocols should not be taken as gospel. Instead, these protocols are better seen as simply a place to start. It is likely that modifications will be required in order to optimize each protocol for both the model system and for the application. Do not be afraid to experiment!

Acknowledgments

As it is with good cooks, it is the nature of good biologists to tweak methods in order to streamline the protocol and improve the final product. Accordingly, the protocols presented in this chapter represent the work of many individuals. The author is particularly indebted to Kit Pogliano at UCSD. Kit has been at the forefront of developing technology for bacterial cell biology and has been extremely generous with both her time and her expertise. In addition the author is grateful for the advice of current and former members of the Losick and Grossman laboratories including; Elizabeth Harry, Antje Hofmeister, Esther Angert, Chris Webb, Aurelio Teleman, Frederico Gueiros-Filho, Nicole King, Daniel Lin, Katherine Lemon, Janet Lindow and Iren Kurtser. Their willingness to experiment with reagents and techniques is ultimately responsible for the development of the protocols in this chapter. Finally, the author is grateful to Janet Lindow, Katherine Bacon Schneider, Bradley Smith, and Andrew Erdmann for careful reading of this manuscript prior to publication.

References

Addinall, S. G., Bi, E. and Lutkenhaus, J. (1996). FtsZ ring formation in *fts* mutants. *J. Bacteriol.* **178**, 3877–3884.
Angert, E. R. (personal communication).

Angert, E. R. and Losick, R. M. (1998). Propagation by sporulation in the guinea pig symbiont *Metabacterium polyspora*. *Proc. Natl Acad. Sci. USA* **95**, 10218–10223.
Baird, G. S., Zacharias, D. A. and Tsien, R. Y. (2000). Biochemistry, mutagenesis, and oligomerization of DsRed, a red fluorescent protein from coral. *Proc. Natl Acad. Sci. USA* **97**, 11984–11989.
Belmont, A. S. and Straight, A. F. (1998). In vivo visualization of chromosomes using lac operator-repressor binding. *Trends Cell Biol.* **8**, 121–124.
Bi, E. and Lutkenhaus, J. (1991). FtsZ ring structure associated with division in *Escherichia coli*. *Nature* **354**, 161–164.
Cheng, H. P. and Walker, G. C. (1998). Succinoglycan is required for initiation and elongation of infection threads during nodulation of alfalfa by *Rhizobium meliloti*. *J. Bacteriol.* **180**, 5183–5191.
Cormack, B. P., Valdivia, R. H. and Falkow, S. (1996). FACS-optimized mutants of the green fluorescent protein (GFP). *Gene* **173**, 33–38.
Damelin, M. and Silver, P. A. (2000). Mapping interactions between nuclear transport factors in living cells reveals pathways through the nuclear pore complex. *Mol. Cell* **5**, 133–140.
Gordon, G. S., Sitnikov, D., Webb, C. D., Teleman, A., Straight, A., Losick, R., Murray, A. W. and Wright, A. (1997). Chromosome and low copy plasmid segregation in *E. coli*: visual evidence for distinct mechanisms. *Cell* **90**, 1–20.
Gueiros-Filho, F. J. (personal communication).
Hahn, J. (personal communication).
Harry, E. J., Pogliano, K. and Losick, R. (1995). Use of immunofluorescence to visualize cell-specific gene expression during sporulation in *Bacillus subtilis*. *J. Bacteriol.* **177**, 3386–3393.
Heim, R. and Tsien, R. Y. (1996). Engineering green fluorescent protein for improved brightness, longer wavelengths and fluorescence resonance energy transfer. *Curr. Biol.* **6**, 178–182.
Hiraga, S., Ichinose, C., Niki, H. and Yamazoe, M. (1998). Cell cycle-dependent duplication and bidirectional migration of seqA- associated DNA-protein complexes in *E. coli*. *Mol. Cell* **1**, 381–387.
Kellenberger, E. and Ryter, A. (1958). Cell wall and cytoplasmic membrane of *E. coli*. *J. Biophys. Biochem. Cytol.* **4**, 323–326.
King, N., Dreesen, O., Stragier, P., Pogliano, K. and Losick, R. (1999). Septation, dephosphorylation, and the activation of sigmaF during sporulation in *Bacillus subtilis*. *Genes Dev.* **13**, 1156–1167.
Kohler, R., Bubert, A., Goebel, W., Steinert, M., Hacker, J. and Bubert, B. (2000). Expression and use of the green fluorescent protein as a reporter system in *Legionella pneumophila*. *Mol. Gen. Genet.* **262**, 1060–1069.
Lemon, K. P. (personal communication).
Lemon, K. P. and Grossman, A. D. (1998). Localization of bacterial DNA polymerase: evidence for a factory model of replication. *Science* **282**, 1516–1519.
Lemon, K. P. and Grossman, A. D. (2000). Movement of replicating DNA through a stationary replisome. *Mol. Cell* **6**, 1321–1330.
Levin, P. A., Kurtser, I. G. and Grossman, A. D. (1999). Identification and characterization of a negative regulator of FtsZ ring formation in *Bacillus subtilis*. *Proc. Natl. Acad. Sci. USA* **96**, 9642–9647.
Levin, P. A. and Losick, R. (1996). Transcription factor Spo0A switches the localization of the cell division protein FtsZ from a medial to a bipolar pattern in *Bacillus subtilis*. *Genes & Dev.* **10**, 478–488.
Lewis, P. J. and Errington, J. (1997). Direct evidence for active segregation of *oriC* regions of the *Bacillus subtilis* chromosome and co-localization with the Spo0J partitioning protein. *Mol. Microbiol.* **25**, 945–954.

Lindow, J. (personal communication).
Maddock, J. R. and Shapiro, L. (1993). Polar location of the chemoreceptor complex in the *Escherichia coli* cell. *Science* **259**, 1717–1723.
Margolin, W. (2000). Green fluorescent protein as a reporter for macromolecular localization in bacterial cells. *Methods: Companion Meth. Enzymol.* **20**, 62–72.
Margolis, P., Driks, A. and Losick, R. (1991). Establishment of cell type by compartmentalized activation of a transcription factor. *Science* **254**, 562–565.
Michaelis, C., Ciosk, R. and Nasmyth, K. (1997). Cohesins: chromosomal proteins that prevent premature separation of sister chromatids. *Cell* **91**, 35–45.
Niki, H. and Hiraga, S. (1997). Subcellular distribution of actively partitioning F plasmid during the cell division cycle in *E. coli*. *Cell* **90**, 951–957.
Overton, M. C. and Blumer, K. J. (2000). G-protein-coupled receptors function as oligomers in vivo. *Curr. Biol.* **10**, 341–344.
Pogliano, J., Osborne, N., Sharp, M. D., Abanes-De Mello, A., Perez, A., Sun, Y. L. and Pogliano, K. (1999). A vital stain for studying membrane dynamics in bacteria: a novel mechanism controlling septation during *Bacillus subtilis* sporulation. *Mol. Microbiol.* **31**, 1149–1159.
Pogliano, K., Harry, E. and Losick, R. (1995). Visualization of the subcellular location of sporulation proteins in *Bacillus subtilis* using immunofluorescence microscopy. *Mol. Microbiol.* **18**, 459–470.
Pollok, B. A. and Heim, R. (1999). Using GFP in FRET-based applications. *Trends Cell Biol.* **9**, 57–60.
Raskin, D. and de Boer, P. (1999). Rapid pole-to-pole oscillation of a protein required for directing division to the middle of *Escherichia coli*. *Proc. Natl Acad. Sci. USA* **96**, 4971–4976.
Robinow, C. and Kellenberger, E. (1994). The bacterial nucleoid revisited. *Microbiol. Rev.* **58**, 211–232.
Ryter, A. (1964). Etude morphologique de la sporulation de *Bacillus subtilis*. *Annales de l'Institut Pasteur.* **108**, 40–60.
Shapiro, L. and Losick, R. (2000). Dynamic spatial regulation in the bacterial cell. *Cell* **100**, 89–98.
Sharp, M. D. and Pogliano, K. (1999). An in vivo membrane fusion assay implicates SpoIIIE in the final stages of engulfment during *Bacillus subtilis* sporulation. *Proc. Natl Acad. Sci. USA* **96**, 14553–14558.
Straight, A. F., Belmont, A. S., Robinett, C. C. and Murray, A. W. (1996). GFP tagging of budding yeast chromosomes reveals that protein-protein interactions can mediate sister chromatid cohesion. *Curr. Biol.* **6**, 1599–1608.
Tsien, R. Y. (1998). The green fluorescent protein. *Annu. Rev. Biochem.* **67**, 509–544.
Webb, C. D., Decatur, A., Teleman, A. and Losick, R. (1995). Use of green fluorescent protein for visualization of cell-specific gene expression and subcellular protein localization during sporulation in *Bacillus subtilis*. *J. Bacteriol.* **177**, 5906–5911.
Webb, C. D., Graumann, P. L., Kahana, J. A., Teleman, A. A., Silver, P. A. and Losick, R. (1998). Use of time-lapse microscopy to visualize rapid movement of the replication origin region of the chromosome during the cell cycle in *Bacillus subtilis*. *Mol. Microbiol.* **28**, 883–892.
Webb, C. D., Teleman, A., Gordon, S., Straight, A., Belmont, A., Lin, D. C., Grossman, A. D., Wright, A. and Losick, R. (1997). Bipolar localization of the replication origin regions of chromosomes in vegetative and sporulating cells of *B. subtilis*. *Cell* **88**, 667–674.
Weiss, D. S., Pogliano, K., Carson, M., Guzman, L. M., Fraipont, C., Nguyen-Distéche, M., Losick, R. and Beckwith, J. (1997). Localization of the *Escherichia coli* cell division protein Ftsl (PBP3) to the division site and cell pole. *Mol. Microbiol.* **25**, 671–681.

7 Host–Pathogen Interactions: Structure and Function of Pili

Michelle M Barnhart[1], Joel D Schilling[1], Fredrik Bäckhed[2], Agneta Richter Dahlfors[2], Staffan Normark[2] and Scott J Hultgren[1]
[1]*Department of Molecular Microbiology, Washington University School of Medicine, St. Louis, MO, USA;* [2]*Microbiology and Tumorbiology Center, Karolinska Institute, Stockholm, Sweden*

♦♦

CONTENTS

♦♦♦♦♦♦ INTRODUCTION

Among the earliest events in many bacterial infections are the molecular interactions that occur between the pathogen and host epithelial cells. These interactions are typically required for extracellular colonization or internalization to occur and involve a complex cascade of molecular cross-talk at the host–pathogen interface. Colonization of host tissues is usually mediated by adhesins on the surface of the microbe that are responsible for recognizing and binding to specific receptor moieties present on host cells. The binding of the adhesin to the receptor can activate complex signal transduction cascades in the host cell that can have diverse consequences including the activation of innate host defenses or the subversion of cellular processes facilitating bacterial colonization or invasion. In addition, the binding event can also activate the expression of new genes in the microbe that are important in the pathogenic process. Bacterial pathogens produce a multitude of adhesins and related structures that are expressed by organisms associated with a broad range of diseases. Examples of adhesive structures assembled by bacteria

METHODS IN MICROBIOLOGY, VOLUME 31
ISBN 0-12-521531-2

are pili, which are proteinaceous, filamentous polymeric organelles expressed on the surface of bacteria.

Bacterial attachment to the bladder mucosa is critical for establishing a urinary tract infection (UTI). Pilus-mediated attachment to the host epithelium by uropathogenic *E. coli* (UPEC), the organism most commonly associated with urinary tract infections (UTIs), initiates a complex web of events, including signaling in both the bacterium and host, that influences the outcome of the infection. Several different adhesive organelles are produced by UPEC including S pili, Dr family adhesins, P pili, and type 1 pili, which allow the bacteria to attach to different host receptors (Johnson, 1991). Type 1 and P pili play key roles in the pathogenesis of UTIs. The focus of this chapter will be on methods used to study the biogenesis and roles of type 1 and P pili in pathogenesis. The *E. coli* P pilus, encoded by the *pap* gene cluster (*papA–K*) (Figure 7.1), binds to receptors in the kidney epithelium containing the Galα(1-4)Gal disaccharide (Bock *et al.*, 1985; Hull *et al.*, 1981; Kuehn *et al.*, 1992; Leffler and Svanborg-Eden, 1980; Stromberg *et al.*, 1992, 1990) and is critical for the development of pyelonephritis (Roberts *et al.*, 1994; Wullt *et al.*, 2000). The type 1 pilus of *E. coli*, encoded by the *fim* gene cluster (*fimA–H*) (Figure 7.1), binds to mannose moieties present in the bladder epithelium and has been shown

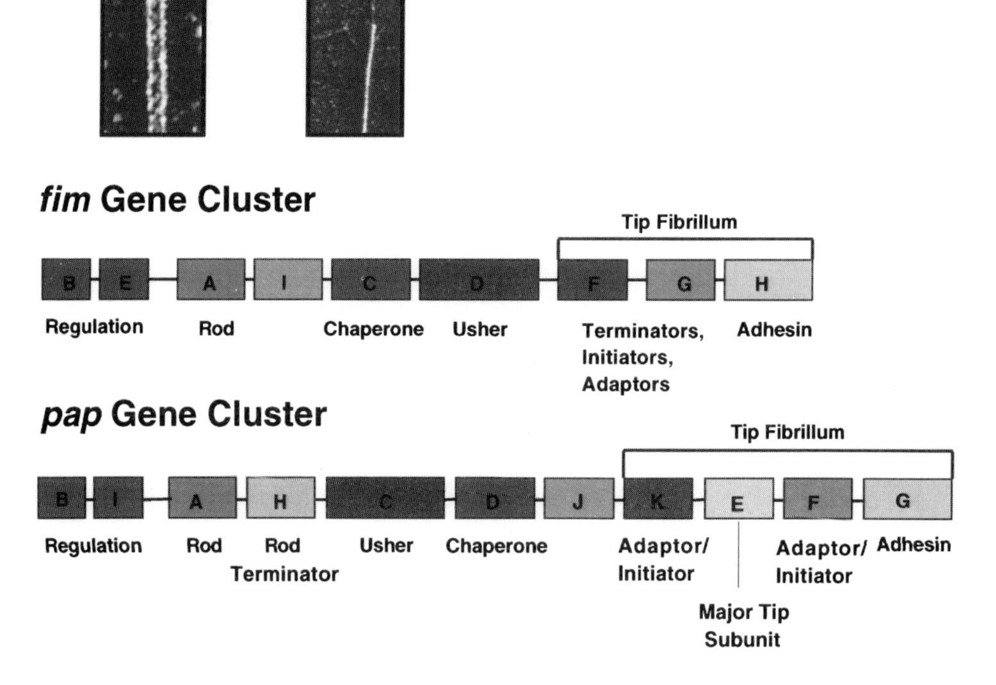

Figure 7.1. High-resolution electron micrographs of a type 1 pilus (A) and a P pilus (B). Below the micrographs are schematic diagrams of the *fim* and *pap* gene clusters. Genes with comparable functions are shaded similarly. Note the similarity in the genetic organization of the two gene clusters.

to mediate attachment to the uroplakin-covered superficial cells at the lumenal surface of the bladder (Connell *et al.*, 1996; Langermann *et al.*, 1997; Mulvey *et al.*, 1998; Thankavel *et al.*, 1997; Wu *et al.*, 1996).

Chaperone/Usher pathway

The assembly of P and type 1 pili proceeds by the highly conserved chaperone–usher pathway, which participates in the biogenesis of at least thirty diverse surface organelles in Gram-negative bacterial pathogens (Figure 7.2) (Holmgren *et al.*, 1992; Hung *et al.*, 1996). Pilus subunits enter the periplasm through the Sec apparatus. The periplasmic chaperone—PapD in the *pap* system, FimC in the *fim* system—interacts with each newly translocated subunit, facilitating its release from the cytoplasmic membrane in a process that may be driven by the folding of the subunit directly on the chaperone template (Jones *et al.*, 1997, 1993; Lindberg *et al.*, 1989; Soto *et al.*, 1998). The chaperone remains bound to the folded

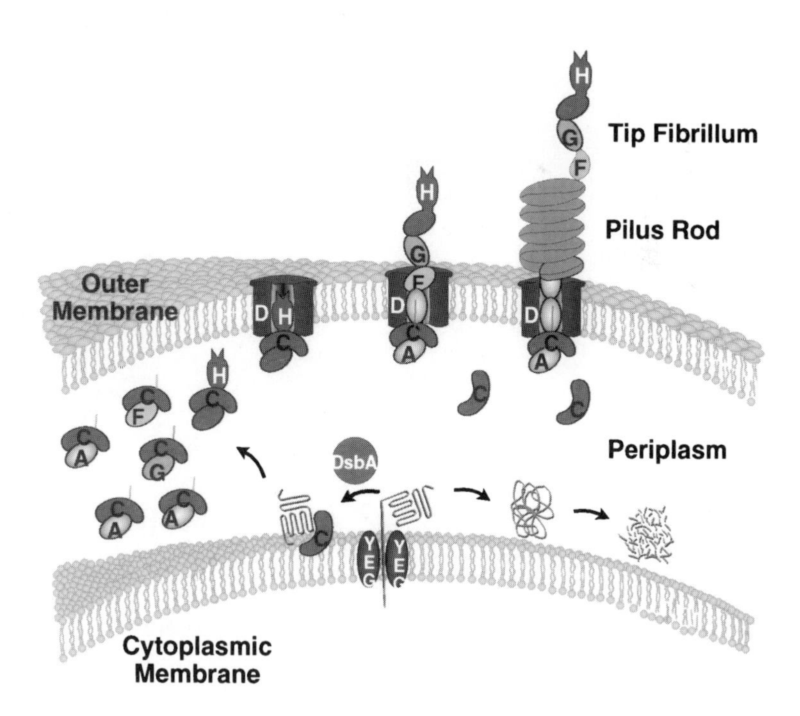

Figure 7.2. A model for type 1 pilus biogenesis via the chaperone-usher pathway. Pilus subunits (FimA, F, G, and H) are translocated from the cytoplasm to the periplasm via the Sec machinery. Upon entry into the periplasm, the subunits interact with the chaperone FimC, which facilitates the folding of the subunits and prevents non-productive interactions among subunits in the periplasm. Subunits that mis-fold are targeted for degradation by periplasmic proteases. The chaperone–subunit complexes (FimC–FimA, FimC–FimF, FimC–FimG, and FimC–FimH) are targeted to the outer membrane usher FimD, which forms a pore in the outer membrane to translocate the pilus subunits to the bacterial cell surface. The chaperone–subunit complexes are targeted to the outer membrane usher in a specific order reflecting their final order in the pilus.

subunit, stabilizing it and capping an interactive surface, thus preventing premature aggregation in the periplasm (Bullitt *et al.*, 1996; Jones *et al.*, 1997; Kuehn *et al.*, 1991, 1993). In the absence of the chaperone the subunits are targeted for degradation by periplasmic proteases such as DegP (Jones *et al.*, 1997). Recently, the crystal structures of the chaperone–adhesin complex FimC–FimH and the chaperone–subunit complex PapD–PapK were solved (Choudhury *et al.*, 1999; Sauer *et al.*, 1999). It was shown that the chaperone completed the fold of the subunits in a mechanism termed donor strand complementation (Figure 7.3) (See the chaperone–subunit

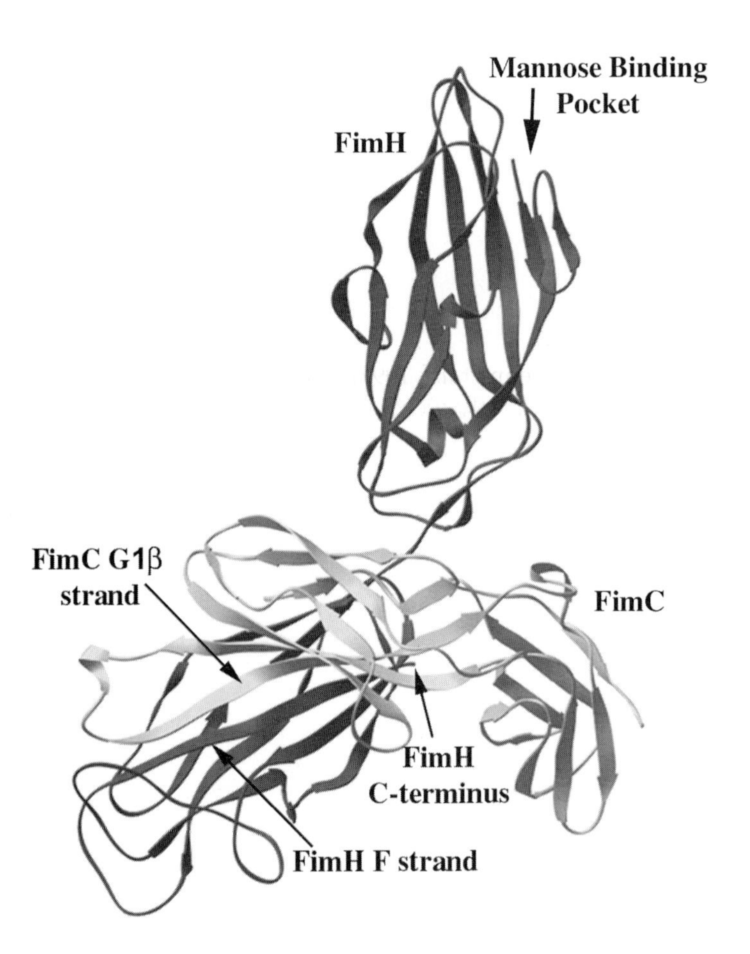

Figure 7.3. Crystal structure of the FimC–FimH complex. Both FimC (gray) and FimH (black) are two domain proteins. The chaperone FimC is composed of two Ig-like folds, which form a boomerang shaped molecule. The adhesin FimH contains an amino terminal receptor binding domain and a carboxyl terminal pilin domain. The pilin domain is also an Ig-like fold. However, this fold is incomplete. It lacks the seventh and carboxyl terminal β strand present in a canonical Ig-fold. In complex with FimC, the Ig-fold of FimH is completed by the G1β strand of FimC in a mechanism termed donor strand complementation. The carboxyl terminal domain of FimH is anchored to the cleft of FimC via the invariant Arg8 and Lys112 residues.

interactions section of the chapter for more details.) Chaperone–subunit complexes are specifically targeted to the outer-membrane usher—PapC in the *pap* system and FimD in the *fim* system (Dodson *et al.*, 1993; Saulino *et al.*, 1998; Thanassi *et al.*, 1998). The PapC usher has been shown to form a channel 2–3 nm in diameter, large enough to allow the passage of pilus subunits (Thanassi *et al.*, 1998). In the *fim* and *pap* systems, the chaperone–adhesin complex binds more rapidly and tightly to the usher than do other chaperone–subunit complexes, an interaction that is thought to initiate pilus assembly. Formation of a chaperone–adhesin–usher ternary complex induces a conformational change in the usher, to an assembly-competent form, that is maintained throughout pilus assembly (Saulino *et al.*, 1998). The usher is thought to facilitate the uncapping of subunits, exposing their interactive surfaces and driving their assembly into the pilus. The diameter of the usher pore suggests that the pilus grows through the usher as a linear fiber and only packages into its final quaternary structure upon reaching the bacterial surface (Thanassi *et al.*, 1998).

The P pilus consists of a thin, flexible tip fibrillum at the distal end of a thicker, rigid pilus rod (Figure 7.1B) (Baga *et al.*, 1984; Bullitt and Makowski, 1995; Gong and Makowski, 1992). The tip fibrillum contains primarily PapE subunits arranged in an open helical conformation (Kuehn *et al.*, 1992); the rod consists of PapA subunits arranged to form a tightly-wound, hollow, right-handed helical structure with a diameter of 70 Å and 3.3 subunits per turn (Baga *et al.*, 1984; Bullitt and Makowski, 1995; Gong and Makowski, 1992). The PapG adhesin, which binds sugars containing the Galα(1-4)Gal disaccharide, is at the distal end of the tip fibrillum (Hultgren *et al.*, 1989; Kuehn *et al.*, 1992; Lund *et al.*, 1987). The PapF and PapK pilus subunits are thought to link the PapG adhesin to the tip fibrillum and the tip fibrillum to the rod, respectively (Jacob-Dubuisson *et al.*, 1993). The type 1 pilus also has a composite structure, with a rod whose dimensions are similar to that of the P pilus joined to a short, stubby tip fibrillum (Figure 7.1A) (Brinton Jr., 1965; Jones *et al.*, 1995). The rod consists primarily of FimA subunits arranged to form a hollow right-handed helix (Jones *et al.*, 1995). The FimH adhesin, which binds mannose, is found in the tip fibrillum, while the FimG subunit is thought to link the FimH adhesin to the rest of the pilus (Abraham *et al.*, 1988; Jones *et al.*, 1995).

The biogenesis of type 1 and P pili on the surface of the bacterium represents a model system for studying protein synthesis, folding via chaperones, secretion, and assembly. Type 1 pili also provide a model system for studying adherence to host cells using both *in vivo* and *in vitro* model systems. This chapter describes techniques that have been used to analyze the biogenesis of pili via the chaperone usher pathway and model systems to study host–pathogen interactions mediated by type 1 piliated *E. coli*. Past reviews have also described methodologies used to analyze pilus biogenesis and are also good references for techniques (Kuehn *et al.*, 1994; Thanassi and Hultgren, 2000).

◆◆◆◆◆◆ IDENTIFICATION OF PILUS GENE FUNCTION

After a pilus gene cluster has been identified, analyzing the function of each of the genes in the cluster is required. A search of gene and protein databases (such as GenBank: www.ncbi.nlm.nih.gov) can be done with the cloned and sequenced genes to identify homology to other genes in the database. This information can be helpful to at least identify the putative functions of each of the genes in the operon. The putative chaperone and usher of the gene cluster can be identified using this strategy as there is high similarity among these genes. However, it must be proven that these proteins do function as chaperones and ushers using techniques that are described in this chapter. To facilitate the analysis of a pilus gene cluster, each individual gene and the entire pilus gene cluster can be cloned into a plasmid with an inducible promoter such as *lac*, *tac*, or *ara*. This makes analysis of the gene cluster easier because the exact conditions to induce the expression of the pilus under the natural promoter may be difficult to obtain in the laboratory. Also, each gene should be mutagenized to gain insight into gene function and to provide a resource for the additional studies described in this chapter. Genes encoding chaperones, ushers, and subunits can be mixed and matched to analyze protein–protein interactions. The roles of each of the individual genes in the gene cluster can be examined by a variety of biochemical, genetic, electron microscopic, and X-ray crystallographic techniques that will be described throughout this chapter.

◆◆◆◆◆◆ HEMAGGLUTINATION ASSAY

Bacterial adhesion can be measured using a hemagglutination assay (HA) (Hultgren *et al.*, 1990). This assay takes advantage of the fact that most bacteria expressing pili have the ability to agglutinate erythrocytes. For example, bacteria expressing type 1 pili are able to agglutinate guinea pig red blood cells via the FimH adhesin (Krogfelt *et al.*, 1990; Old, 1972). An HA titer can be measured by serially diluting a suspension of bacteria in microtiter wells and incubating it with erythrocytes to determine the minimum concentration of bacteria required to agglutinate the erythrocytes (Hultgren *et al.*, 1990; Kuehn *et al.*, 1994). To perform an HA, bacteria are prepared by growing under the proper conditions to induce the expression of the adhesive organelle. The bacteria are diluted to A_{600} of 1.0 in phosphate buffered saline (PBS) and 1 ml of this is spun down in a microfuge for 1 minute. The bacterial cells are resuspended in 100 μl of PBS. The erythrocytes are prepared by washing several times in PBS to get rid of lysed cells. The erythrocytes are diluted to A_{640} of 1.8–2.0. 25 μl of PBS is added to each well of a V-bottom well microtiter plate. To the first well 25 μl of bacteria are added and mixed. 25 μl of this is added to the second well and mixed. This process is repeated across the microtiter plate. Then 25 μl of erythrocytes are added to each well and the plates are incubated at 4°C for 1 h to overnight. The HA titer is read by determining the maximum dilution of bacteria that yields 50% hemagglutination. In the absence of hemagglutination the red blood cells pellet to the bottom of the well and form a dot. In the presence of hemagglutination the red blood cells and the bacteria form a diffuse sheet of clumped cells over the entire well. The host receptor that the adhesin binds can be identified by inhibiting hemagglutination using specific chemically defined carbohydrates (Kihlberg *et al.*, 1989) or using erythrocytes that are deficient in specific blood group antigens. Three different alleles of PapG (PapGI–PapGIII) were identified and the different binding specificities of these alleles were determined by comparing HA titers using different types of erythrocytes (Stromberg *et al.*, 1990). Another method used to determine the host receptor of PapG was an overlay assay where bacteria were incubated with glycolipids that had been separated using thin layer chromatography (Stromberg *et al.*, 1990, 1991).

◆◆◆◆◆◆ IDENTIFICATION OF A PILUS ADHESIN

The identity of the adhesin can be determined by mutating various genes in the pilus gene cluster and determining the effect of the mutation on HA titer since adhesin mutant gene clusters will not agglutinate erythrocytes. This can be problematic because mutations in some of the minor pilins, chaperone, or usher can also result in a negative HA titer (Jacob-Dubuisson *et al.*, 1993; Klemm and Christiansen, 1990; Lindberg *et al.*, 1989; Norgren *et al.*, 1987). An analysis of the genes in the operon can give insight into which gene encodes the adhesin. The adhesins are typically two domain proteins composed of an amino terminal receptor binding domain and a carboxyl terminal pilin domain (Choudhury *et al.*, 1999). In contrast, the pilus subunits contain only a pilin domain and therefore the only homology between the adhesin and the subunit is in the carboxyl terminal domain of the adhesin. The PapG adhesin was identified using a transcomplementation assay where PapG was co-expressed with a related pilus gene cluster *prs*, which has a different receptor binding specificity (Lund *et al.*, 1987). The expression of PapG *in trans* with a *prsG* mutant gene cluster resulted in a strain that now had PapG receptor binding specificity, thus genetically identifying PapG as the adhesin.

◆◆◆◆◆◆ PURIFICATION AND ARCHITECTURE OF PILI

Pili can be purified from the bacterial cell surface to analyze the effect of mutations in certain genes in the operon on the biogenesis pathway (Kuehn *et al.*, 1994; Thanassi and Hultgren, 2000). Bacteria grown to produce pili are harvested and resuspended in 0.5 mM Tris, 75 mM NaCl (bacteria cells can be equilibrated by OD or weight). Pili can be sheared off of the bacteria in a blender or the bacteria can be heated to 65°C for 30 min. After blending or heating the samples are spun at 6000*g* for 10 min to pellet cells. The supernatant is spun a second time at 21 000*g* for 30 min to remove remaining cells and debris. The supernatant is collected and NaCl and $MgCl_2$ are slowly added to final concentrations of 300 mM and 100 mM, respectively. This is rocked at room temperature for two hours to precipitate the pili followed by spinning at 21 000*g* for 40 min. The pili are present in the pellet, which is resuspended in 0.5 mM Tris (pH8) and rocked overnight at 4°C. The next day a final spin at 21 000*g* for 30 min is performed and the purified pili are present in the supernatant. The purified pili can be analyzed on SDS-PAGE gels. Pili are difficult to dissociate and require treatment with urea or acid in order to run into SDS-PAGE gels. The major subunit of the pilus is usually visible on a Coomassie-stained gel; however, the minor pilus subunits such as the adhesin require Western blotting for identification. (Methods for generating pilus antisera are described in Kuehn *et al.*, 1994 and Thanassi and Hultgren, 2000.)

Purified pili can be examined using high resolution electron microscopy (EM) in order to examine the architecture of the pilus (various EM techniques are described in another chapter of this book) (Kuehn *et al.*, 1992). Pili can be visualized on the bacterial cell surface using negative stained EM. However, the architecture of the pilus cannot be determined using this technique. High-resolution EM was essential in determining the molecular architecture of both type 1 and P pili. This technique showed that both type 1 and P pili were composite structures composed of a tip fibrillum and a pilus rod (Figure 7.1A and B) (Jones *et al.*, 1995; Kuehn *et al.*, 1992). The tip fibrillum of the type 1 pilus was short and stubby while the tip fibrillum of the P pilus was longer. Examination of mutant pili by high resolution EM is also useful in studying the architecture of the pilus. Purified *papE* mutant pili were shown to lack a tip fibrillum, which could be restored by expressing *papE in trans* indicating that PapE was the major component of the tip fibrillum (Kuehn *et al.*, 1992). Immunogold EM can also be used to determine the location of a protein in the pilus as it was used to show that PapG localized to the distal end of the tip fibrillum (Kuehn *et al.*, 1992).

◆◆◆◆◆◆ CHAPERONE USHER STRUCTURE FUNCTION

Periplasmic preparations

The periplasmic chaperone is absolutely required for the production of pili on the surface of the bacterium (Lindberg *et al.*, 1989). Pilus subunits expressed in the absence of the chaperone aggregate and are proteolytically degraded by periplasmic proteases such as DegP (Jones *et al.*, 1997). Thus, the pilus subunits require the co-expression of the chaperone in order to be produced and purified from the periplasm. In order to examine or purify a chaperone or chaperone–subunit complexes periplasmic extracts need to be made. Periplasms are prepared by growing the appropriate strain of bacteria expressing a plasmid(s) that encodes a chaperone and/or subunit. The cells are harvested and resuspended per gram of cells in 2 ml of 20% sucrose, 20 mM Tris (pH8), 100 μl of 0.1 M EDTA, and 20 μl of 10 mg ml^{-1} lysozyme. This is incubated on ice for 40 min. Then 80 μl of 0.5 M $MgCl_2$ is added and spheroplasts are pelleted at 15 000*g* for 20 min at 4°C. The supernatant contains the periplasmic extract.

Identification of a chaperone

A putative chaperone can be identified in a pilus gene cluster by looking for homology to other chaperones by searching GenBank. After the putative chaperone has been identified, it should be mutated to determine the effect on subunit synthesis and pilus production by preparing periplasms, pili, and HA titers as in the absence of the chaperone the subunits will be degraded and no pilus formed. The mutated gene cluster should then be complemented with a copy of the chaperone *in trans* to show that the subunits are now stable in the periplasm and pili are made. Additionally, it is important to show that the chaperone interacts with the subunits in the periplasm using co-purification techniques, which are described below.

Purification and structure of the chaperone

The chaperones PapD and FimC have been purified from the periplasm using cation-exchange chromatography as the PapD-like family of chaperones have a characteristic high pI (Kuehn *et al.*, 1994). The chaperone can be purified by preparing periplasmic extracts from a bacterial strain expressing the chaperone. The periplasm is precipitated with 30% ammonium sulfate and spun at 16 000*g* for 30 min. The pellet is resuspended in 20 mM KMES (pH 6.5) followed by dialysis in the same buffer. The dialyzed material is spun at 22 000*g* for 30 min to pellet debris. The supernatant is filtered and applied to a mono S cation exchange column. The chaperone is eluted with a 0 to 0.15 M KCl gradient at a rate of 25 mM per 10 ml. The fractions are monitored for the chaperone by SDS-PAGE and Western blotting. Once a chaperone has been purified, the protein can be crystallized in order to obtain structural information about the chaperone. Methods for X-ray crystallography are beyond the scope of this chapter, however PapD will be used as an example to show the important information that can be obtained from the crystal structure of a protein. The crystal structure of PapD showed that it was a two-domain protein composed of two immunoglobulin (Ig)-like folds oriented towards each other forming a boomerang shaped molecule (Holmgren and Brändén, 1989). This structural information was used to make an alignment of all of the chaperones in the PapD family. The chaperones contained a conserved hydrophobic core and highly conserved residues in the cleft. Site-directed mutations were made in highly conserved surface exposed residues of PapD and the effect of these mutations on pilus biogenesis was tested (Hung *et al.*, 1999; Slonim *et al.*, 1992). Similar mutations could be made in newly identified chaperones by aligning the new chaperone with PapD. The effect of these mutations on pilus assembly can be assayed by expressing the mutant chaperone with the rest of the gene cluster and monitoring pilus production by HA titer and pilus prep. Additionally, the ability of the mutant chaperone to form chaperone–subunit complexes can be determined by co-expressing the mutated chaperones with pilus subunits and monitoring the production of the subunit in the periplasm. Mutational analysis of PapD showed that invariant cleft residues including Arg8 and Lys112 were critical in the ability of the chaperone to bind subunits and mediate their assembly into pili (Slonim *et al.*, 1992).

The chaperone alignment also revealed that there were two subfamilies of chaperones, the FGS (this includes PapD and FimC) and FGL subfamilies (Hung *et al.*, 1996). The two chaperone families are classified based on the length of the loop between the F1 and G1 β strands of the chaperone termed the F1–G1 loop. The FGS family has a short F1–G1 loop and the FGL family has a long F1–G1 loop. The structures assembled by the two sub-families differ in that the structures assembled by the FGS family of chaperones are pili and the structures assembled by the FGL family are non-pilus associated adhesins. The two families also differ in the conserved carboxyl terminal motif present in the subunits that is recognized by the chaperone. The FGS family has a conserved motif of alternating hydrophobic residues that has been termed the β zipper motif (described below). The FGL family has a slightly different motif and this may play a role in the assembly of the structure and may explain the slightly different architecture of the FGL assembled adhesins. The chaperone of a newly identified gene cluster should be examined to see if it is an FGL or FGS chaperone because this information could give insight as to whether the structure assembled by the chaperone is a pilus or non-pilus adhesin.

Purification of chaperone–subunit complexes

Chaperone–subunit complexes can be purified from the periplasm using ion exchange chromatography and hydrophobic interaction columns (Kuehn *et al.*, 1994). The chaperone–subunit complexes can also be purified by expressing subunits with a chaperone containing a poly-histidine tag (or other type of tag) on its carboxyl terminus and using nickel chromatography to purify the complex (Saulino *et al.*, 1998; Thanassi and Hultgren, 2000). This is done by purifying periplasmic extracts from strains expressing a histidine tagged chaperone and non-tagged subunit. The periplasm is dialyzed into PBS in order to remove the EDTA, which interferes with nickel binding. The dialyzed periplasms are incubated with Ni-NTA beads (Qiagen) for 60 min with rocking at room temperature. The beads are washed several times with PBS and the protein complex is eluted from the beads using imidazole or EDTA. The problem with this procedure is that free chaperone is purified with the chaperone–subunit complexes necessitating a second chromatographic step to remove the free chaperone. The reverse purification scheme can also be performed where the histidine tag has been placed on the amino terminus of the subunit and co-expressed with a non-tagged chaperone (Thanassi and Hultgren, 2000). This method is advantageous as a one to one molar ratio of chaperone to subunit can be purified. Chaperone–adhesin complexes can be purified by taking advantage of the specificity of the receptor binding domain of the adhesin. Affinity chromatography has been used to purify both the FimC–FimH and PapD–PapG chaperone–adhesin complexes (Hultgren *et al.*, 1989; Jones *et al.*, 1993). FimC–FimH complexes have been purified from the periplasm using mannose-sepharose chromatography followed by elution with methyl α-D-mannopyranoside. The PapD–PapG complex was similarly purified from the periplasm using Galα(1-4)Gal sepharose chromatography. The chaperone–adhesin complexes can also be purified using ion exchange and hydrophobic interaction columns (Barnhart *et al.*, 2000). Once a chaperone–adhesin complex has been purified it is important to determine that the adhesin has the same receptor binding activity in complex with the chaperone as the adhesin does in the context of the pilus. This was done for the purified PapD–PapG complex (Hultgren *et al.*, 1989). A series of galabioside analogues had been tested for their ability to inhibit hemagglutination of erythrocytes by bacteria expressing P pili. The purified PapD–PapG complex was bound to Galα(1-4)Gal sepharose beads and the ability of some of these analogs to elute the complex from the beads was tested. The galabioside analogues, which were good or poor inhibitors of hemagglutination were also good or poor at eluting the PapD–PapG complex from the Galα(1-4)Gal sepharose beads. Thus, PapG in complex with the chaperone obtained the same receptor binding activity as it did in the context of the P pilus.

Analysis of chaperone–subunit interactions

In order to obtain information about the interaction of the chaperone with subunits the chaperone can be co-crystallized with peptides corresponding to the carboxyl terminus of subunits. The co-crystallization of PapD with peptides corresponding to the carboxyl terminal domain of either PapG or PapK revealed a β zippering interaction between the peptide and the alternating hydrophobic residues of the G1β strand of PapD (Kuehn *et al.*, 1993; Soto *et al.*, 1998). The very carboxyl terminus of the peptide was anchored to the chaperone via the invariant Arg8 and Lys112 residues. The carboxyl terminal domain of subunits is composed of alternating hydrophobic residues with a conserved glycine at position fourteen from the carboxyl terminus and a penultimate aromatic residue (Kuehn *et al.*, 1993). Subunits in a new pilus system can be analyzed to look for a similar or unique motif in their carboxyl terminus that the chaperone may be binding. Site-directed mutagenesis can be performed on these conserved residues to assess their effect on chaperone–subunit and subunit–subunit interactions. Mutations were made in the carboxyl terminal domain of PapG and the ability of the mutants to interact with the chaperone PapD and the subunit PapF, which links PapG to the tip fibrillum of the pilus was analyzed (Soto *et al.*, 1998). The mutant PapG proteins were expressed *in trans* with a *papG⁻pap* gene cluster and the stability of the PapG–PapF interaction was examined. Purified pilus tips were heated to different temperatures to determine the temperature at which PapF and PapG dissociated from one another giving a relative indication of stability. The carboxyl terminal PapG mutants effectively weakened subunit–subunit interactions necessary to join PapG to the tip of the pilus. The interaction of the mutant PapG proteins with PapD was analyzed using an ELISA assay and these mutants were shown to have a defect in chaperone–subunit interactions as well. Thus, the conserved carboxyl terminal motif present in pilus subunits is required for both chaperone–subunit and subunit–subunit interactions. These studies were the first to give insight into the mechanism by which PapD coupled folding with the simultaneous capping of interactive surfaces on the pilus subunits.

After a chaperone–subunit complex has been purified X-ray crystallography can provide valuable information regarding the interactions between the two proteins and the assembly of the adhesive organelle. For example, the crystal structures of the chaperone–subunit complex PapD–PapK and the chaperone–adhesin complex FimC–FimH revealed a possible mechanism for pilus biogenesis (Choudhury *et al.*, 1999; Sauer *et al.*, 1999). The FimH adhesin is a two domain protein composed of an amino terminal receptor binding domain and a carboxyl terminal pilin domain (Figure 7.3), while PapK consists only of a pilin domain. PapK and the pilin domain of FimH have Ig-like folds. However, they lack the seventh and carboxyl terminal β strand present in canonical Ig folds. The absence of the seventh strand results in a deep groove or scar present in the pilin domain exposing the hydrophobic core of the protein. In complex with the chaperone, the Ig-like fold of the subunits is completed by the G1β strand of the chaperone in a mechanism termed donor strand complementation (Figure 7.3). The chaperone completes the fold of the subunit atypically by inserting its G1β strand parallel rather than anti-parallel to the F β strand of the subunit. In the PapD–PapK structure the amino terminal extension of PapK was disordered and previous work had shown that this conserved region is important in subunit–subunit interactions (Sauer *et al.*, 1999; Soto *et al.*, 1998). The amino terminal extensions of the subunits are composed of alternating hydrophobic residues similar to the G1β strand of the chaperone. Thus, it was hypothesized that the highly conserved amino terminal extension of one subunit would displace the G1β strand of the chaperone from its neighboring subunit in a mechanism termed donor strand exchange (Choudhury *et al.*, 1999; Sauer *et al.*, 1999). The amino terminal extension is thought to insert anti-parallel to the F β strand of the subunit, so the mature pilus would consist of an array of perfectly canonical Ig domains, each of which contributes a strand to the fold of its neighboring subunit. When crystallizing a chaperone–adhesin complex, it is often useful to include the receptor or a receptor analog in the crystallization conditions in order to determine the receptor binding pocket of the adhesin

(Figure 7.3) (Dodson *et al.*, 2001; Choudhury *et al.*, 1999). The crystallization conditions for the FimC–FimH complex required the presence of the compound C-HEGA to produce useful crystals (Choudhury *et al.*, 1999). In the structure the C-HEGA molecule bound in a pocket that could potentially represent the mannose binding site.

Construction of adhesin truncates

The structure of FimH showed that it was a two domain protein with the adhesin domain separated from the pilin domain, which is the part of the protein that interacts with the chaperone (Figure 7.3) (Choudhury *et al.*, 1999). This suggested that it could be possible to express the adhesin domain as an independent moiety. The adhesin domain of FimH was cloned with a histidine tag on its carboxyl terminus (Schembri *et al.*, 2000). This protein was stable in the periplasm and purified using nickel chromatography. The receptor binding capability of the purified protein was verified by demonstrating its ability to bind mannose. This technique was also used for the analysis of the PapG adhesion (Dodson *et al.*, 2001). Thus, the generation of receptor binding truncates is a useful way to characterize the molecular basis of adhesin–receptor interactions.

Construction of dsc subunits

The PapD–PapK and FimC–FimH structures revealed a possible mechanism for constructing pilus subunits that are stable in the periplasm in the absence of the chaperone by making donor strand complemented (dsc) subunits (Choudhury *et al.*, 1999; Sauer *et al.*, 1999). Dsc subunits can be made by adding a seventh and C-terminal β strand onto their carboxyl terminus. The seventh strand that is added to the subunit corresponds to the amino terminus of its neighboring subunit. This technology was used to construct donor strand complemented FimH (dscFimH) (Barnhart *et al.*, 2000). FimG and FimH have been shown to interact with each other in the pilus (Jones *et al.*, 1995). The donor strand exchange hypothesis predicts that the amino terminal extension of FimG should complete the Ig fold of the FimH pilin domain (Choudhury *et al.*, 1999; Sauer *et al.*, 1999). Thus, the DNA sequence encoding the first thirteen amino acids of FimG (referred to as the donor strand sequence) was provided to FimH *in cis*, by fusing it to the 3′ end of *fimH* to create dscFimH. A hairpin loop region present in PapD consisting of Asp-Asn-Lys-Gln was inserted upstream of the donor strand to allow the donor strand to fold back into the groove of the FimH pilin domain to complete the Ig fold of FimH. *dscfimH* was cloned into an expression vector and expressed. The production of dscFimH was monitored by preparing periplasmic extracts and performing Western blots using anti-FimCH antisera. DscFimH was a stable protein in the periplasm suggesting that this technology could be applied to other subunits in the pilus in order to study the subunits independently from the chaperone and gain insight into pilus biogenesis (Barnhart *et al.*, 2000). The receptor binding activity of dscFimH was tested by incubating periplasmic extracts with mannose-sepharose beads and eluting the protein off of the beads using methyl α-D-mannopyranoside. The elutions were run on SDS-PAGE and Western blots performed using anti-FimCH antibodies. The results showed that dscFimH bound to and eluted off of the mannose beads suggesting that the protein was properly folded *in vivo*. Donor strand complemented adhesins can be purified for further analysis using receptor binding or ion-exchange and hydrophobic interaction columns (Barnhart *et al.*, 2000). Besides constructing dsc subunits, a method of purifying subunits away from the chaperone has been developed (Pellecchia *et al.*, 1998). This has been successfully performed using the FimC–FimH complex, but is applicable to other chaperone–subunit complexes (Barnhart *et al.*, 2000). First, FimC–FimH complexes were purified from periplasmic extracts. The purified FimC–FimH complex was then incubated in 3 M urea, which dissociates the two proteins. Pure FimH was then collected from the flow through of a Source 15S column (Pharmacia). FimH was stable in 3 M urea presumably because the urea protects the hydrophobic core of the protein, which is normally occupied by the G1 β strand of the FimC chaperone. This has been a useful technique in studying the folding of pilus subunits.

Pilus subunit folding

Pilus subunit folding can be studied by comparing the folding of dsc subunits to subunits that have been separated from the chaperone using the urea separation assay (Barnhart *et al.*, 2000). Denaturation curves for the proteins can be determined by incubating either protein with increasing concentrations of urea starting at 3 M and going up to 9.5 M urea. Denaturation of the proteins is monitored by measuring changes in tyrosine fluorescence. Fluorescence is measured using an excitation wavelength of 290 nm with emission at 350 nm on an AlphaScan PTI fluorometer. The information obtained from the denaturation curves can be used to develop a refolding assay, which uses circular dichroism (CD) to measure protein folding. The CD spectra of native, denatured, and refolded proteins is measured using a JASCO J715 spectropolarimeter. This type of assay was performed with FimH and dscFimH. Both proteins in their native state showed a characteristic β sheet structure. However, when the two proteins were denatured by 9 M urea they lost their characteristic β sheet structure. Refolding of dscFimH and FimH was measured by taking the 9 M urea denatured proteins and diluting away the denaturant to 0.45 M urea. The CD spectra of the proteins were measured after dilution. FimH completely aggregated upon dilution of the denaturant and elicited no CD spectrum. In contrast, dscFimH refolded into its native β sheet conformation. FimH only refolded when the denaturant was diluted away in the presence of FimC suggesting that the chaperone facilitates the folding of pilus subunits.

Identification of the usher

After a putative usher has been identified by homology, it should be mutated and the effect on pilus biogenesis assessed. The ability of the usher mutants to form pili should be analyzed by HA titer, negative stained EM, and pilus purification. Usher mutants, like chaperone mutants, are unable to form pili and are HA negative (Klemm and Christiansen, 1990; Lindberg *et al.*, 1989; Norgren *et al.*, 1987). The usher mutants should also be analyzed for the presence of chaperone–subunit complexes in the periplasm because in the absence of the usher chaperone–subunit complexes accumulate in the periplasm (Klemm and Christiansen, 1990; Norgren *et al.*, 1987). The mutated gene cluster should then be complemented *in trans* with a copy of the usher to show that this restores piliation. The usher should also be expressed and outer membrane preparations (see below) should be made to show that the usher is localized to the outer membrane.

Purification of the usher

The purification of usher proteins requires some special techniques as the ushers are integral outer membrane proteins. The presence of detergents in the purification buffers is required to stabilize membrane proteins. Poly-histidine tagged PapC and FimD can be purified from the outer membrane using nickel chromatography (Saulino *et al.*, 1998; Thanassi *et al.*, 1998). The histidine tagged usher proteins can be cloned into a plasmid with an inducible promoter. The over-expression of an outer membrane protein can be toxic to the host. Thus, it is important to determine the appropriate concentration of inducer to use in order to maintain cell growth. This can be done by varying the concentration of inducer used and monitoring cell growth to determine the maximum concentration of inducer that can be used without being toxic to the host cell. After induction the bacteria are harvested and resuspended in 20 mM Tris (pH 8) plus protease inhibitors such as Complete Protease Inhibitor (Boehringer-Mannheim, Indianapolis, IN) or PMSF (phenylmethylsulfonyl fluoride). After harvesting the cells all of the steps should be carried out at 4°C unless otherwise stated. The cells are then lysed by passage two times through a French press at 14 000 psi. Unbroken cells are removed by centrifugation at 3000*g* for 15 min. N-lauroylsarcosine (Sarkosyl) is added to 0.5% to the supernatant and incubated at room temperature for 5 min. The outer membrane is pelleted by centrifugation at 100 000 *g* for 1 h. The outer membrane pellet is resuspended in 20 mM Hepes (pH 7.5), 0.3 M NaCl. Dodecyl maltoside is added to 0.5% and rocked at room temperature for 1 hour to solubilize the usher. Other non-denaturing detergents such as octyl glucoside, elugent, or Zwittergent 3–14 can be used instead of dodecyl maltoside. The detergent should be tested for efficacy before using. A final spin at 100 000 *g* for 30 min is performed to pellet insoluble material. The usher can then be purified in batch by adding Ni-NTA beads (Qiagen) and rocking overnight at 4°C. After overnight incubation, the beads are washed with buffer A (20 mM Hepes (pH 7.5), 0.3 M NaCl, 0.1% dodecyl maltoside) containing 0–20 mM imidazole. The usher can be eluted from the nickel beads with buffer A plus 30 mM EDTA. The eluate should contain highly purified usher protein, which can be concentrated using a concentration device and/or dialyzed to remove EDTA and salt.

Usher structure

Structural information about integral membrane proteins can be difficult to obtain due to the fact that the proteins tend to aggregate and precipitate during crystallization. Thus, quick-freeze, deep-etch EM can be used to examine purified usher proteins. This technique was used with PapC and showed that PapC appeared to form an oligomeric ring approximately 2–3 nm in diameter suggesting that PapC may form a pore in the outer membrane to translocate the subunits to the cell surface (Thanassi *et al.*, 1998). The ability of usher proteins to form pores can be tested using liposome swelling assays, which have been used to characterize the outer membrane porins as well as the PapC usher (Nikaido *et al.*, 1991; Thanassi *et al.*, 1998). Usher proteins are reconstituted into multilamellar liposomes, which are diluted into isoosmotic solutions containing various sized solutes, usually sugars. The ability of a protein to form a pore is determined by the ability of the solute to enter the liposomes. The solute enters the liposomes down its concentration gradient. If the solute enters the liposome, then swelling occurs as water along with the solute enters the liposomes and this results in a decrease in optical density. The initial rate of the decrease in optical density is measured as the pore-forming or swelling activity of the protein. The size of the pore can be estimated by comparing the swelling activity of different sized solutes. The appropriate controls should be performed with this assay to be assured that the results are reliable. This technique showed that PapC formed a pore 2 nm in diameter, which was in agreement with the data obtained from EM. The diameter of the pilus rod is 6.8 nm and would be unable to fit through the usher fully folded. However, the diameter of the usher is large enough to allow the tip fibrillum to pass through. EM has shown that under certain conditions the pilus rod can be unraveled into its linear fiber and this linear fiber could fit through the usher pore (Thanassi *et al.*, 1998). This data along with the predicted diameter of the usher suggested that the pilus travels through the usher as a linear fiber and once it reaches the cell surface it adopts its final helical structure. The winding of the rod into a helix may facilitate the translocation of the pilus across the outer membrane as pilus assembly has been shown to function independently of cellular energy.

Chaperone–subunit interactions with the usher

The interaction of the usher with various chaperone–subunit complexes can be studied using ELISA assays and far Western blots (Dodson *et al.*, 1993; Saulino *et al.*, 1998). Purified usher is adhered to a microtiter plate or run on SDS-PAGE gel and blotted onto PVDF membranes. Various combinations of purified chaperone–subunit complexes are incubated with the usher coated wells or the membrane. The ELISA or far Western is developed using either anti-chaperone or anti-subunit antibodies. The results from these kind of experiments in the type 1 and P pilus systems showed that the chaperone–adhesin complex had the highest affinity for the usher. This data may provide insight into the mechanism of the localization of the adhesins to the distal end of the tip fibrillum. The other chaperone–subunit complexes bound to a low-level to the usher. Surface plasmon resonance technology can be used to determine both kinetic and equilibrium association and dissociation constants of the chaperone–subunit complexes for the usher (Karlsson *et al.*, 1991; Saulino *et al.*, 1998). These assays are performed using a Biacore 2000 (Pharmacia Bosensor, Piscatawa, NJ) instrument. The usher is covalently linked to a biosensor chip and varying concentrations of chaperone–subunit complexes are injected over the usher-coated chip. The binding of a complex to the usher-coated chip results in a change in refractive index in the solution near the chip surface that is measured as a change in the angle of light reflected off of the chip. These experiments were performed in both the type 1 and P pilus systems confirming the ELISA and far Western data showing that the chaperone–adhesin complexes bound best to the usher (Saulino *et al.*, 1998). It was shown that the association rate of the chaperone–adhesin complexes to the usher for both type 1 and P pili were much greater than the chaperone–subunit complexes. Thus, in the type 1 system the FimC–FimH complex would bind to the usher FimD first followed by FimC–FimG, FimC–FimF, and finally FimC–FimA. The subsequent order of subunit assembly is presumably determined by the donor strand exchange mechanism. The dissociation rate for all of the chaperone–subunit complexes for the usher was slow implying that once a chaperone–subunit complex has bound to the usher it is destined to be incorporated into the pilus.

Purification of assembly intermediates

Assembly intermediates with the usher have been purified in the type 1 system (Saulino *et al.*, 2000). Assembly intermediates can be purified by expressing various chaperone–subunit complexes with a poly-histidine tagged usher. Outer membranes are prepared and nickel chromatography is performed as described previously in this chapter. The elutions are run on an SDS-PAGE gel and analyzed by Western blotting. This assay was performed with cells expressing FimC, FimH, and FimD and showed that the FimC–FimH complex could be co-purified with the FimD usher. The co-expression of FimC–FimG, FimC–FimF, or FimC–FimA with FimD did not result in their co-purification with the usher. However, if and only if FimC–FimH was co-expressed, FimG, FimF, and FimA could be co-purified with the usher (Saulino *et al.*, 2000). Taken together this data suggested that the binding of FimC–FimH could potentially prime the usher into an active form for pilus assembly. Thus, protease susceptibility tests were performed to see if the binding of the FimC–FimH complex to FimD resulted in changes in proteolytic sensitivity of FimD (Saulino *et al.*, 1998). The assay described here is for the type 1 system, but it could be applied to other pilus systems. Whole cells expressing FimD, FimC, and FimH are harvested and resuspended in PBS. The cells are incubated with 100 μg ml^{-1} of exogenously added trypsin in PBS or PBS alone for 2 h at 37°C with rocking. After incubation with trypsin the whole cells are washed one time in PBS and whole cell lysates are run on an SDS-PAGE gel followed by Western blotting with anti-FimD antibodies. This assay showed that when FimC and FimH were co-expressed with the usher there was an enhancement of a trypsin-resistant fragment of FimD. This suggested that the binding of FimC–FimH to the usher FimD resulted in a conformational change in the usher into an assembly competent state.

◆◆◆◆◆◆ HOST–PATHOGEN INTERACTIONS AND CONSEQUENCES

There are several experimental assays that can be done to evaluate bacterial interactions with host cells. These include assessing bacterial adherence, invasion, and intracellular survival, as well as host parameters such as the induction of anti-microbial defenses.

Adherence assays

Type 1 piliated *E. coli* have been shown to bind to bladder epithelial cells using a standard adherence assay (Figure 7.4C) (Elsinghorst, 1994; Martinez *et al.*, 2000). In this assay, host cells are grown to confluence in a 24-well tissue culture plate. During this same interval bacterial cultures are grown under the conditions that induce expression of the desired adhesive organelle(s). When the cells are ready, two sets of triplicate wells are infected at a multiplicity of infection (MOI) of 5–10 bacteria per host cell. After a 2 h incubation at 37°C, one set of triplicate wells should be lysed by adding 10% Triton X-100 to a final concentration of 0.1%. To quantify the bacteria in these lysates, representing the total number of extra- and intracellular bacteria, serial dilutions of the lysate are plated onto the appropriate agar medium. To assess bacterial adherence, wash the second set of triplicate wells five times with PBS (with Mg^{2+}, Ca^{2+}) and subsequently add 1 ml of 0.1% Triton X-100 to lyse the host cells. To quantify the total number of cell associated bacteria, serial dilutions of these lysates are plated onto the appropriate agar plates. Adherence frequencies can be represented as cfu ml^{-1} or as a percentage by dividing the number of adherent bacteria by the total number of bacteria present in each well.

In situ binding assays

In situ binding assays have also been used to monitor bacterial adherence to host cells as well as to determine the host–receptor binding specificity

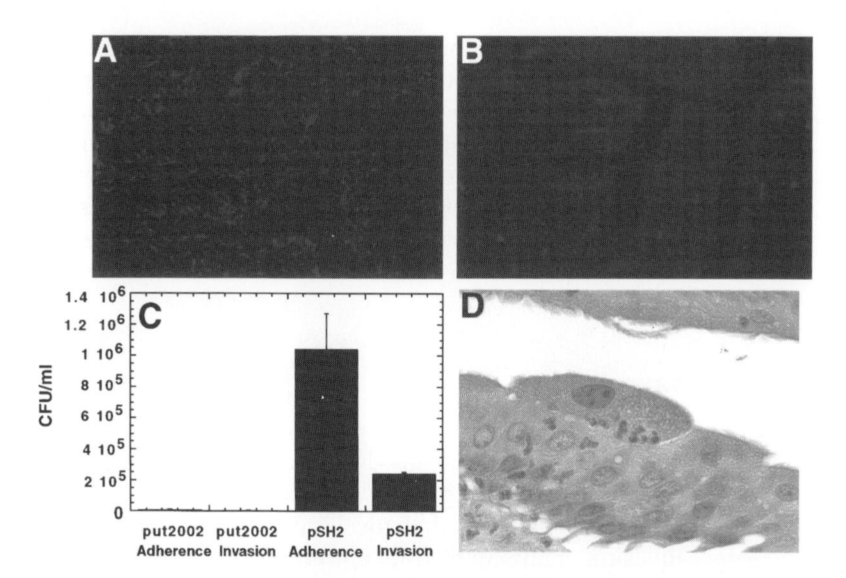

Figure 7.4. Adherence and invasion of bladder epithelial cells mediated by type 1 piliated *E. coli*. An *in situ* binding assay showing that *E. coli* expressing type 1 pili are able to bind to mouse bladder epithelial cells (A) while an isogenic *fimH*$^-$ mutant strain is unable to bind to the mouse bladder epithelium (B). Adherence and invasion of 5637 human bladder epithelial cells by the *E. coli* K12 strain AAEC185 expressing the plasmid pSH2 (*fimH*$^+$) or put2002 (*fimH*$^-$) (C). Bacteria expressing *fimH*$^+$ type 1 pili were able to both adhere to and invade the tissue culture cells while *fimH*$^-$ strains were unable to bind to or invade the bladder epithelial cells. An H & E stained mouse bladder tissue section after infection with a cystitis isolate (D). The cystitis isolate invaded a superficial bladder cell and has started replicating. (This figure is also reproduced in colour between pages 276 and 277.)

and receptor distribution on the tissue (Figure 7.4A and B) (Falk, 1993; Langermann *et al.*, 1997; Striker *et al.*, 1995). This assay has been used to show that bacteria expressing the PapGII adhesin, but not the PapGI or PapGIII adhesin are able to bind to human kidney tissue sections (Striker *et al.*, 1995). *In situ* binding assays have also been used to determine the receptor for *Helicobacter pylori* (Falk, 1993). In these assays, bacteria expressing the appropriate adhesin are labeled with fluorescein isothiocyanate (FITC) by incubating the bacteria with 10 μl of a fresh 10 mg ml^{-1} solution of FITC in dimethyl sufloxide (DMSO) for 1 h. The bacteria are washed to remove excess FITC. The appropriate types of tissue sections are deparaffinized with xylene and isopropanol followed by blocking in a buffer containing PBS, 0.5% tween, and 0.2% bovine serum albumin (BSA). The labeled bacteria are diluted twentyfold into blocking buffer and overlayed onto the tissue sections, which are incubated in a humidified chamber in the dark for 1–5 h. The tissue sections are washed several times with PBS and bacterial binding visualized using a fluorescence microscope. Bacteria can be pre-incubated with glycoproteins or free oligosaccharides prior to tissue binding to determine if these substances can inhibit bacterial adherence. In addition, tissue sections can be treated with proteases or enzymes that cleave specific carbohydrate moieties in order to determine receptor binding specificity. Co-localization experiments can also be performed where a fluorescently labeled lectin or antibody along with FITC labeled bacteria are co-incubated with the tissue sections to determine if they localize to the same portion of the tissue section. Bacteria expressing a mutated pilus gene cluster can also be used in this assay in order to identify the pilus adhesin.

Invasion assays

Some adhesive structures are able to mediate not only adherence to, but also invasion of host cells (Isberg *et al.*, 1987; Mengaud *et al.*, 1996). *E. coli* type 1 pili are an example of one such organelle (Figure 7.4C). In fact, it has been demonstrated that the FimH adhesin is sufficient to mediate the invasion of human bladder epithelial cells (Martinez *et al.*, 2000). To assess bacterial invasion, an additional set of triplicate wells is added to the protocol described for an adherence assay (Elsinghorst, 1994; Isberg *et al.*, 1987). After a 2-h incubation at 37°C, the cells are washed with PBS (with Mg^{2+} Ca^{2+}) and fresh medium is added containing 100 μg ml^{-1} of the host cell-impermeable antibiotic gentamicin. The presence of gentamicin in the medium will kill all of the extracellular bacteria, but not bacteria that have invaded the host cell. To determine the number of intracellular bacteria, lyse the cells and plate out serial dilutions of the lysates. It is often useful to express bacterial invasion as a percentage of adherent bacteria or total bacteria present at the time of the addition of gentamicin. To gain entry into host cells, most invasive bacterial pathogens exploit pre-existing host signal transduction cascades (Bliska *et al.*, 1993; Falkow *et al.*, 1992; Finlay and Falkow, 1997; Finlay *et al.*, 1991). To characterize the signaling molecules involved in an invasion pathway, inhibitors of host cell signal transduction cascades can be added to host cells prior to the addition of bacteria. Initial investigations in the type 1 pilus system have demonstrated that FimH-mediated bacterial invasion of bladder epithelial cells requires the activation of protein tyrosine kinases, phophoinositide-3 kinase (PI 3-kinase), and localized actin cytoskeletal rearrangements (Martinez *et al.*, 2000). To assess intracellular survival or replication following bacterial invasion the host cells can be incubated for longer periods of time in medium containing 15 μg ml^{-1} of gentamicin. Following this incubation the number of intracellular bacteria can be determined as described above and compared to the initial invasion titer. Increases in the intracellular titer would be expected for a bacterium that is able to multiply in an intracellular environment. Using this assay it has been shown that uropathogenic isolates can replicate intracellularly, whereas type 1-piliated *E. coli* K12 strains, although able to invade bladder epithelial cells, cannot multiply within cells (Mulvey *et al.*, 2001). Such observations suggest that intracellular growth is not a general property of all *E. coli* strains, but may require additional virulence factors unique to uropathogenic isolates.

Activation of innate host defenses

Bacteria often activate host defenses when they interact with host cells including the production of cytokines, chemokines, and anti-microbial peptides (Kagnoff and Eckmann, 1997). The ability of bacteria to bind to or invade host cells has been shown to modify the production of inflammatory molecules and thus it is often useful to investigate the contribution of these events to epithelial activation (Backhed *et al.*, 2001; Frendeus *et al.*, 2001; Hedges *et al.*, 1992; Hedlund *et al.*, 2001; Schilling *et al.*, 2001). The mechanism by which type 1 pili augment IL-6 production in response to *E. coli* has been elucidated. Bacterial invasion mediated by type 1 pili enhances epithelial cell responses to this bacteria through an LPS-dependent mechanism (Schilling *et al.*, 2001). As cytokines and chemokines are easily detected in host cell supernatants following exposure of the cells to bacteria, their production serves as an excellent way to quantify epithelial activation. To perform these assays prepare a 24-well tissue culture plate via the same protocol described for adherence and invasion assays. Following the addition of bacteria, however, it is generally recommended that the infection be continued for greater than 2 h to allow for sufficient accumulation of the cytokine/chemokine in the supernatant. A kinetic study is often useful to determine the time at which the cytokine or chemokine is induced maximally in a particular system. After the infection, the supernatants are harvested and the samples are centrifuged to remove the bacteria and cell debris. The samples are frozen at −20°C or −80°C until analysis. A sandwich ELISA can be used to determine the concentration of cytokine or chemokine produced by the cells if one has a monoclonal antibody (capture) and a biotinylated polyclonal antibody (detection) against the cytokine/chemokine of interest (many of these are commercially available). A 96-well ELISA plate is coated with the monoclonal capture antibody and incubated overnight at 4°C. The following day, the plate is washed with PBS, 0.05% Tween and blocked with PBS containing 1% BSA, 5% sucrose and 0.05% NaN_3. Following the block, the sample and standard curve dilutions are added into the appropriate wells. The detection antibody is added and incubated at room temperature for 1.5–2 h after which a strepavidin-HRP conjugate is added to the wells and incubated for approximately 20 min at room temperature avoiding direct light. After the 20-min incubation, the plate is washed and substrate added to the wells (tetramethylbenzidine-hydrochloride (TMB-HCL) in a phosphocitrate buffer). The reaction is developed in the dark for 15–30 min, the reaction stopped with 2 N sulfuric acid and the absorbency of the samples read at 450 nm (subtract 570 nm).

Cystitis mouse model system

Bacterial adherence to and invasion of host tissue can be studied *in vivo* using an animal model system. A mouse model has been developed to study cystitis (Langermann *et al.*, 1997; Mulvey *et al.*, 1998). To illustrate the importance of type 1 pili in the establishment of a bladder infection a cystitis isolate and an isogenic *fimH* mutant were tested using this model system. The bacteria were grown in static broth to induce type 1 pilus production and were subsequently pelleted and resuspended in PBS at a concentration of 2×10^9 colony forming units (cfu) ml^{-1}. Eight- to fifteen-week-old female C57BL/6 mice were inoculated transurethrally with 1×10^8 cfu in a volume of 50 μl. At various time points after infection the mice were sacrificed and the bladders harvested, homogenized and plated to determine the number of cfu/bladder. These experiments demonstrated that mice infected with the *fimH* mutant strain had 99% fewer bacteria in their bladder as compared to the wild-type strain, indicating that type 1 pili are critical in the colonization of the bladder (Langermann *et al.*, 1997). In addition to determining bacterial titers in bladder tissue, mouse bladders can also be analyzed by various imaging techniques such as hematoxylin and eosin (H & E) microscopy (Figure 7.4D), scanning EM, transmission EM and freeze-fracture deep-etch EM (Mulvey *et al.*, 1998; Mulvey *et al.*, 2001; Schilling *et al.*, 2001). Following infection, mouse bladders are harvested at various time points post-infection and prepared for paraffin sections, frozen sections or EM. Studies using EM have demonstrated that the surface of the bladder is coated with protein complexes known as uroplakins (Sun *et al.*, 1996). The uroplakins are composed of four integral membrane proteins (1A, 1B, 2, and 3) and form a hexagonal array on the surface of superficial bladder epithelial cells. *In vitro* studies have shown that FimH can bind to uroplakins 1A and 1B (Wu *et al.*, 1996). However, the first *in vivo* images of type 1 pili interacting with uroplakins came from a high resolution EM study of mouse bladders during the acute phase of a bladder infection (Mulvey *et al.*, 1998). In addition to EM, confocal scanning laser microscopy (CSLM) can be used to study host–pathogen interactions (Richter-Dahlfors *et al.*, 1997; Uhlen *et al.*, 2000).

The fact that adherence mediated by type 1 pili appeared to be a critical event in the establishment of cystitis lead to the hypothesis that a vaccine based on the FimH adhesin would be a successful means to prevent an infection (Langermann *et al.*, 1997). Previous attempts to vaccinate mice with whole pili had not been successful, presumably because FimH is a minor component of the pilus and/or the high degree of antigenic variation among FimA proteins, which compose the bulk of the type 1 pilus. To test this hypothesis mice were vaccinated with varying concentrations (0.6–30 μg) of either truncated FimH protein or the FimC–FimH complex. Nine weeks after vaccination, the mice were challenged with 5×10^7 cfu of type 1 piliated *E. coli* and 2 days after inoculation bacterial persistence in the bladder was determined as described above. As predicted, the mice that received FimH based vaccines were protected from infection with the uropathogen again showing the importance of adherence in establishing cystitis (Langermann *et al.*, 1997).

In vivo invasion assays have also been developed to show that type 1 piliated bacteria are able to invade not only tissue culture cells, but also bladder epithelial cells *in vivo* (Mulvey *et al.*, 1998). To perform this assay, mouse bladders that had been transurethrally inoculated with type 1 piliated *E. coli* or the isogenic *fimH* mutant were removed and bisected, at various time points post infection. Each bladder was placed in ringer solution (155 mM NaCl, 3 mM HCl, 2 mM $CaCl_2$, 1 mM $MgCl_2$, 3 mM NaH_2PO_4, 10 mM glucose, and 5 mM Hepes [pH 7.4]) and gently rinsed. One-half of each bladder was incubated in ringer solution with or without 100 $\mu g\ ml^{-1}$ of gentamicin for 90 min at room temperature. The bladder halves were washed, weighed, and homogenized in Triton X-100 in PBS and the surviving bacteria were plated. Results from these experiments showed that by 48 h after infection the majority of the bacteria remaining in the bladder were intracellular (Mulvey *et al.*, 1998), indicating that the bacteria, which had invaded the epithelium, had a survival advantage during an acute UTI.

◆◆◆◆◆◆ CONCLUSIONS

The techniques described in this chapter will give investigators insights into ways to characterize new pilus gene clusters and analyze the interactions of new adhesive organelles with host cells. A variety of biochemical, genetic, electron microscopic, and X-ray crystallographic techniques have been described to identify the function of each of the genes in a pilus gene cluster. Additionally methodologies were described to investigate adherence to the host and the consequences of this event using both *in vivo* and *in vitro* model systems.

References

Abraham, S. N., Sun, D., Dale, J. B. and Beachey, E. H. (1988). Conservation of the D-mannose-adhesion protein among type 1 fimbriated members of the family Enterobactericeae. *Nature* **336**, 682–684.

Backhed, F., Soderhall, M., Ekman, P., Normark, S. and Richter-Dahlfors, A. (2001). Induction of innate immune responses by Escherichia coli and purified lipopolysaccharide correlate with organ- and cell-specific expression of Toll-like receptors within the human urinary tract. *Cell Microbiol.* **3**, 153–158.

Baga, M., Normark, S., Hardy, J., O'Hanley, P., Lark, D., Olsson, O., Schoolnik, G. and Falkow, S. (1984). Nucleotide sequence of the papA gene encoding the Pap pilus subunit of human uropathogenic *Escherichia coli*. *J. Bacteriol.* **157**, 330–333.

Barnhart, M. M., Pinkner, J. S., Soto, G. E., Sauer, F. G., Langermann, S., Waksman, G., Frieden, C. and Hultgren, S. J. (2000). PapD-like chaperones provide the missing information for folding of pilin proteins. *Proc. Natl Acad. Sci. USA* **97**, 7709–7714.

Bliska, J. B., Galan, J. E. and Falkow, S. (1993). Signal transduction in the mammalian cell during bacterial attachment and entry. *Cell* **73**, 903–920.

Bock, K., Breimer, M. E., Brignole, A., Hansson, G. C., Karlsson, K.-A., Larson, G., Leffler, H., Samuelsson, B. E., Strömberg, N., Svanborg-Edén, C. and Thurin, J. (1985). Specificity of binding of a strain of uropathogenic *Escherichia coli* to Galα(1-4)Gal-containing glycosphingolipids. *J. Biol. Chem.* **260**, 8545–8551.

Brinton, Jr., C. C. (1965). The structure, function, synthesis, and genetic control of bacterial pili and a model for DNA and RNA transport in gram negative bacteria. *Trans. N.Y. Acad. Sci.* **27**, 1003–1165.

Bullitt, E., Jones, C. H., Striker, R., Soto, G., Jacob-Dubuisson, F., Pinkner, J., Wick, M. J., Makowski, L. and Hultgren, S. J. (1996). Development of pilus organelle subassemblies *in vitro* depends on chaperone uncapping of a beta zipper. *Proc. Natl Acad. Sci. USA* **93**, 12890–12895.

Bullitt, E. and Makowski, L. (1995). Structural polymorphism of bacterial adhesion pili. *Nature* **373**, 164–167.

Choudhury, D., Thompson, A., Sojanoff, V., Langermann, S., Pinkner, J., Hultgren, S. J. and Knight, S. (1999). X-ray structure of the FimC-FimH chaperone-adhesin complex from uropathogenic *Escherichia coli*. *Science* **285**, 1061–1065.

Connell, H., Agace, W., Klemm, P., Schembri, M., Marild, S. and Svanborg, C. (1996). Type 1 fimbrial expression enhances *Escherichia coli* virulence for the urinary tract. *Proc. Natl. Acad. Sci. USA* **93**, 9827–9832.

Dodson, K. W., Jacob-Dubuisson, F., Striker, R. T. and Hultgren, S. J. (1993). Outer membrane PapC usher discriminately recognizes periplasmic chaperone-pilus subunit complexes. *Proc. Natl. Acad. Sci. USA* **90**, 3670–3674.

Dodson, K. W., Pinkner, J. S., Rose, T., Magnusson, G., Hultgren, S. J. and Waksman. (2001). Structural basis of the interaction of the pyelonephritic *E. coli* adhesion to its human kidney receptor. *Cell* **105**, 733–743.

Elsinghorst, E. A. (1994). Measurement of invasion by gentamicin resistance. *Methods Enzymol.* **236**, 405–420.

Falk, P., Roth, K. A., Borén, T., Westblom, T. U., Gordon, J. I. and Normark, S. (1993). An *in vitro* adherence assay reveals that *Helicobacter pylori* exhibits cell lineage-specific tropism in the human gastric epithelium. *Proc. Natl. Acad. Sci. USA* **90**, 2035–2039.

Falkow, S., Isberg, R. R. and Portnoy, D. A. (1992). The interaction of bacteria with mammalian cells. *Annu. Rev. Cell Biol.* **8**, 333–363.

Finlay, B. B. and Falkow, S. (1997). Common themes in microbial pathogenicity revisited. *Microbiol. Molec. Biol. Rev.* **61**, 136–169.

Finlay, B. B., Ruschkowski, S. and Dedhar, S. (1991). Cytoskeletal rearrangements accompanying *Salmonella* entry into epithelial cells. *J. Cell Sci.* **99**, 283–296.

Frendeus, B., Wachtler, C., Hedlund, M., Fischer, H., Samuelsson, P., Svensson, M. and Svanborg, C. (2001). *Escherichia coli* P fimbriae utilize the Toll-like receptor 4 pathway for cell activation. *Mol. Microbiol.* **40**, 37–51.

Gong, M. and Makowski, L. (1992). Helical structure of Pap adhesion pili from *Escherichia Coli*. *J. Mol. Biol.* **228**, 735–742.

Hedges, S., Svenson, M. and Svanborg, C. (1992). Interleukin-6 response of epithelial cells to bacterial stimulation *in vitro*. *Infect. Immun.* **60**, 1295–1301.

Hedlund, M., Frendeus, B., Wachtler, C., Hang, L., Fischer, H. and Svanborg, C. (2001). Type 1 fimbriae deliver an LPS- and TLR4-dependent activation signal to CD14-negative cells. *Mol. Microbiol.* **39**, 542–552.

Holmgren, A. and Brändén, C. (1989). Crystal structure of chaperone protein PapD reveals an immunoglobulin fold. *Nature* **342**, 248–251.

Holmgren, A., Kuehn, M. J., Brändén, C.-I. and Hultgren, S. J. (1992). Conserved imunoglobulin-like features in a family of periplasmic pilus chaperones in bacteria. *EMBO J.* **11**, 1617–1622.

Hull, R. A., Gill, R. E., Hsu, P., Minshaw, B. H. and Falkow, S. (1981). Construction and expression of recombinant plasmids encoding type 1 and D-mannose-resistant pili from a urinary tract infection *Escherichia coli* isolate. *Infect. Immun.* **33**, 933–938.

Hultgren, S. J., Duncan, J. L., Schaeffer, A. J. and Amundsen, S. K. (1990). Mannose-sensitive hemagglutination in the absence of piliation in *Escherichia coli*. *Mol. Microbiol.* **4**, 1311–1318.

Hultgren, S. J., Lindberg, F., Magnusson, G., Kihlberg, J., Tennent, J. M. and Normark, S. (1989). The PapG adhesin of uropathogenic *Escherichia coli* contains separate regions for receptor binding and for the incorporation into the pilus. *Proc. Natl Acad. Sci. USA* **86**, 4357–4361.

Hung, D. L., Knight, S. D. and Hultgren, S. J. (1999). Probing conserved surfaces on PapD. *Mol. Microbiol.* **31**, 773–783.

Hung, D. L., Knight, S. D., Woods, R. M., Pinkner, J. S. and Hultgren, S. J. (1996). Molecular basis of two subfamilies of immunoglobulin-like chaperones. *EMBO J.* **15**, 3792–3805.

Isberg, R. R., Voorhis, D. L. and Falkow, S. (1987). Identification of invasin: a protein that allows enteric bacteria to penetrate cultured mammalian cells. *Cell* **50**, 769–778.

Jacob-Dubuisson, F., Heuser, J., Dodson, K., Normark, S. and Hultgren, S. J. (1993). Initiation of assembly and association of the structural elements of a bacterial pilus depend on two specialized tip proteins. *EMBO J.* **12**, 837–847.

Johnson, J. R. (1991). Virulence factors in *Escherichia coli* urinary tract infection. *Clin. Microbiol. Rev.* **4**, 80–128.

Jones, C. H., Danese, P. N., Pinkner, J. S., Silhavy, T. J. and Hultgren, S. J. (1997). The chaperone-assisted membrane release and folding pathway is sensed by two signal transduction systems. *EMBO J.* **16**, 6394–6406.

Jones, C. H., Pinkner, J. S., Nicholes, A. V., Slonim, L. N., Abraham, S. N. and Hultgren, S. J. (1993). FimC is a periplasmic PapD-like chaperone that directs assembly of type 1 pili in bacteria. *Proc. Natl Acad. Sci. USA* **90**, 8397–8401.

Jones, C. H., Pinkner, J. S., Roth, R., Heuser, J., Nicholoes, A. V., Abraham, S. N. and Hultgren, S. J. (1995). FimH adhesin of type 1 pili is assembled into a fibrillar tip structure is the *Enterobacteriaceae*. *Proc. Natl Acad. Sci. USA* **92**, 2081–2085.

Kagnoff, M. F. and Eckmann, L. (1997). Epithelial cells as sensors for microbial infection. *J. Clin. Invest* **100**, 6–10.

Karlsson, R., Michaelsson, A. and Mattsson, L. (1991). Kinetic analysis of monoclonal antibody-antigen interactions with a new biosensor based analytical system. *J. Immunol. Methods* **145**, 229–240.

Kihlberg, J., Hultgren, S. J., Normark, S. and Magnusson, G. (1989). Probing of the combining site of the PapG adhesion of uropathogenic *Escherichia coli* bacteria by synthetic analogues of galabiose. *J. Amer. Chem. Soc.* **111**, 6364–6368.

Klemm, P. and Christiansen, G. (1990). The *fimD* gene required for cell surface localization of *Escherichia coli* type 1 fimbriae. *Mol. Gen. Genet.* **220**, 334–338.

Krogfelt, K. A., Bergmans, H. and Klemm, P. (1990). Direct evidence that the FimH protein is the mannose specific adhesin of *Escherichia coli* type 1 fimbriae. *Infect. Immun.* **58**, 1995–1999.

Kuehn, M. J., Heuser, J., Normark, S. and Hultgren, S. J. (1992). P pili in uropathogenic *E. coli* are composite fibres with distinct fibrillar adhesive tips. *Nature* **356**, 252–255.

Kuehn, M. J., Jacob-Dubuisson, F., Dodson, K., Slonim, L., Striker, R. and Hultgren, S. J. (1994). Genetic, biochemical, and structural studies of biogenesis of adhesive pili in bacteria. *Methods Enzymol.* **236**, 282–306.

Kuehn, M. J., Normark, S. and Hultgren, S. J. (1991). Immunoglobulin-like PapD chaperone caps and uncaps interactive surfaces of nascently translocated pilus subunits. *Proc. Natl Acad. Sci. USA* **88**, 10586–10590.

Kuehn, M. J., Ogg, D. J., Kihlberg, J., Slonim, L. N., Flemmer, K., Bergfors, T. and Hultgren, S. J. (1993). Structural basis of pilus subunit recognition by the PapD chaperone. *Science* **262**, 1234–1241.

Langermann, S., Palaszynski, S., Barnhart, M., Auguste, G., Pinkner, J. S., Burlein, J., Barren, P., Koenig, S., Leath, S., Jones, C. H. and Hultgren, S. J. (1997). Prevention of mucosal *Escherichia coli* infection by FimH-adhesin-based systemic vaccination. *Science* **276**, 607–611.

Leffler, H. and Svanborg-Eden, C. (1980). Chemical identification of a glycosphingolipid receptor for *Escherichia coli* attaching to human urinary tract epithelial cells and agglutinating human erythrocytes. *FEMS Microbiol. Lett.* **8**, 127–134.

Lindberg, F., Tennent, J. M., Hultgren, S. J., Lund, B. and Normark, S. (1989). PapD, a periplasmic transport protein in P-pilus biogenesis. *J. Bacteriol.* **171**, 6052–6058.

Lund, B., Lindberg, F., Marklund, B. I. and Normark, S. (1987). The PapG protein is the alpha-D-galactopyranosyl-(1-4)-beta-D-galactopyranose-binding adhesin of uropathogenic *Escherichia coli*. *Proc. Natl Acad. Sci. USA* **84**, 5898–5902.

Martinez, J. J., Mulvey, M. A., Schilling, J. D., Pinkner, J. S. and Hultgren, S. J. (2000). Type 1 pilus-mediated bacterial invasion of bladder epithelial cells. *EMBO J.* **19**, 2803–2812.

Mengaud, J., Ohayon, H., Gounon, P., Mege, R. M. and Cossart, P. (1996). E-cadherin is the receptor for internalin, a surface protein required for entry of *L. monocytogenes* into epithelial cells. *Cell* **84**, 923–932.

Mulvey, M. A., Lopez-Boado, Y. S., Wilson, C. L., Roth, R., Parks, W. C., Heuser, J. and Hultgren, S. J. (1998). Induction and evansion of host defenses by type1-piliated uropathogenic *Escherichia coli*. *Science* **282**, 1494–1497.

Mulvey, M. A., Schilling, J. D. and Hultgren, S. J. (2001) Establishment of a persistant *E. coli* reservoir during the acute phase of a bladder infection. *Infect. Immun.* **69**, 4572–4579.

Nikaido, H., Nikaido, K. and Harayama, S. (1991). Identification and characterization of porins in *Pseudomonas aeruginosa*. *J. Biol. Chem.* **266**, 770–779.

Norgren, M., Baga, M., Tennent, J. M. and Normark, S. (1987). Nucleotide sequence, regulation and functional analysis of the *papC* gene required for cell surface localization of Pap pili of uropathogenic *Escherichia coli*. *Mol. Microbiol.* **1**, 169–178.

Old, D. C. (1972). Inhibition of the interaction between fimbrial hemagglutinatinins and erythrocytes by D-mannose and other carbohydrates. *J. Gen. Microbiol.* **71**, 149–157.

Pellecchia, M., Guntert, P., Glockshuber, R. and Wuthrich, K. (1998). NMR solution structure of the periplasmic chaperone FimC. *Nat. Struct. Biol.* **5**, 885–890.

Richter-Dahlfors, A., Buchan, A. M. J. and Finlay, B. B. (1997). Murine salmonellosis studied by confocal microscopy: *Salmonella typhimurium* resides intracellularly inside macrophages and exerts a cytotoxic effect on phagocytes in vivo. *J. Exp. Med.* **186**, 569–580.

Roberts, J. A., Marklund, B.-I., Ilver, D., Haslam, D., Kaack, M. B., Baskin, G.,, Louis, M., Mollby, R., Winberg, J. and Normark, S. (1994). The Gal α(1-4) Gal-specific tip adhesin of *Escherichia coli* P-fimbriae is needed for pyelonephritis to occur in the normal urinary tract. *Proc. Natl Acad. Sci. USA* **91**, 11889–11893.

Sauer, F. G., Futterer, K., Pinkner, J. S., Dodson, K. W., Hultgren, S. J. and Waksman, G. (1999). Structural basis of chaperone function and pilus biogenesis. *Science* **285**, 1058–1061.

Saulino, E. T., Bullitt, E. and Hultgren, S. J. (2000). Snapshots of usher-mediated protein secretion and ordered pilus assembly. *Proc. Natl Acad. Sci. USA* **97**, 9240–9245.

Saulino, E. T., Thanassi, D. G., Pinkner, J. and Hultgren, S. J. (1998). Ramifications of kinetic partitioning on usher-mediated pilus biogenesis. *EMBO J.* **17**, 2177–2185.

Schembri, M. A., Hasman, H. and Klemm, P. (2000). Expression and purification of the mannose recognition domain of the FimH adhesin. *FEMS Microbiol. Lett.* **188**, 147–151.

Schilling, J. D., Mulvey, M. A., Vincent, C. D., Lorenz, R. G. and Hultgren, S. J. (2001). Bacterial invasion augments epithelial cytokine responses to *Escherichia coli* through a lipopolysaccharide-dependent mechanism. *J. Immunol.* **166**, 1148–1155.

Slonim, L. N., Pinkner, J. S., Branden, C. I. and Hultgren, S. J. (1992). Interactive surface in the PapD chaperone cleft is conserved in pilus chaperone superfamily and essential in subunit recognition and assembly. *EMBO J.* **11**, 4747–4756.

Soto, G. E., Dodson, K. W., Ogg, D., Liu, C., Heuser, J., Knight, S., Kihlberg, J., Jones, C. H. and Hultgren, S. J. (1998). Periplasmic chaperone recognition motif of subunits mediates quaternary interactions in the pilus. *EMBO J.* **17**, 6155–6167.

Striker, R., Nilsson, U., Stonecipher, A., Magnusson, G. and Hultgren, S. J. (1995). Structural requirements for the glycolipid receptor of human uropathogenic *E. coli*. *Mol. Microbiol.* **16**, 1021–1030.

Stromberg, N., Hultgren, S. J., Russell, D. G. and Normark, S. (1992). In: *Encyclopedia of Microbiology.* (J. Lederberg, ed.), Microbial attachment, molecular mechanisms Academic Press, New York.

Stromberg, N., Marklund, B. I., Lund, B., Ilver, D., Hamers, A., Gaastra, W., Karlsson, K. A. and Normark, S. (1990). Host-specificity of uropathogenic *Escherichia coli* depends on differences in binding specificity to Galα(1-4)Gal-containing isoreceptors. *EMBO J.* **9**, 2001–2010.

Stromberg, N., Nyholm, P.-G., Pascher, I. and Normark, S. (1991). Saccharide orientation at the cell surface affects glycolipid receptor function. *Proc. Natl Acad. Sci. USA* **88**, 9340–9344.

Sun, T. T., Zhao, H., Provet, J., Aebi, U. and Wu, X. R. (1996). Formation of asymmetric unit membrane during urothelial differentiation. *Mol. Biol. Rep.* **23**, 3–11.

Thanassi, D. G. and Hultgren, S. J. (2000). Assembly of complex organelles: pilus biogenesis in gram-negative bacteria as a model system. *Methods* **20**, 111–126.

Thanassi, D. G., Saulino, E. T., Lombardo, M.-J., Roth, R., Heuser, J. and Hultgren, S. J. (1998). The PapC usher forms an oligomeric channel: implications for pilus biogenesis across the outer membrane. *Proc. Natl Acad. Sci. USA* **95**, 3146–3151.

Thankavel, K., Madison, B., Ikeda, T., Malaviya, R., Shah, A. H., Arumugam, P. M. and Abraham, S. N. (1997). Localization of a domain in the FimH adhesin of *Escherichia coli* type 1 fimbriae capable of receptor recognition and use of a domain-specific antibody to confer protection against experimental urinary tract infection. *J. Clin. Invest.* **100**, 1123–1136.

Uhlen, P., Laestadius, A., Jahnukainen, T., Soderblom, T., Backhed, F., Celsi, G., Brismar, H., Normark, S., Aperia, A. and Richter-Dahlfors, A. (2000). Alpha-haemolysin of uropathogenic *E. coli* induces Ca2+ oscillations in renal epithelial cells. *Nature* **405**, 694–697.

Wu, X. R., Sun, T. T. and Medina, J. J. (1996). In vitro binding of type 1-fimbriated *Escherichia coli* to uroplakins Ia and Ib: relation to urinary tract infections. *Proc. Natl Acad. Sci. USA* **93**, 9630–9635.

Wullt, B., Bergsten, G., Connell, H., Rollano, P., Gebretsadik, N., Hull, R. and Svanborg, C. (2000). P fimbriae enhance the early establishment of *Escherichia coli* in the human urinary tract. *Mol. Microbiol.* **38**, 456–464.

8 Measuring and Analysing Invasion of Mammalian Cells by Bacterial Pathogens: The *Listeria monocytogenes* System

J Pizarro-Cerdá, M Lecuit and P Cossart
Unité des Interactions Bactéries-Cellules, Institut Pasteur, Paris, France

◆◆

CONTENTS

List of abbreviations

BHI	Brain heart infusion
BSA	Bovine serum albumin
CFU	Colony forming units
DDSA	Dodecenyl succinic anhydride
DMEM	Dulbecco's modified Eagle medium
DMSO	Dimethyl sulfoxide
FCS	Fetal calf serum
FITC	Fluorescein thiocianate
IPTG	Isopropyl-β-D-thiogalactopyranoside
MOI	Multiplicity of infection
NMA	Nadic methyl anhydride
PBS	Phosphate buffered saline
PCR	Polymerase chain reaction
PO	Propylene oxide
TEM	Transmission electron microscopy

METHODS IN MICROBIOLOGY, VOLUME 31
ISBN 0-12-521531-2

◆◆◆◆◆◆ INTRODUCTION

Bacterial pathogens have developed diverse strategies to associate to eucaryotic cells, subverting or taking advantage of normal cellular functions in order to proliferate and disseminate in the host tissues (Finlay and Cossart, 1997). *Listeria monocytogenes*, a Gram-positive organism responsible for severe food-borne infections in humans and other mammals, is capable of inducing its own internalization into non-professional phagocytes such as epithelial cells, interacting intimately with the plasma membrane of these cells by means of surface invasion proteins that are recognized by specific cellular receptors (Cossart and Lecuit, 1998; Cossart and Bierne, 2001).

Two bacterial proteins, InlA (internalin) and InlB, are known to mediate the internalization process of *Listeria* in *in vitro* cell culture systems. InlA promotes entry into epithelial cells such as the enterocyte-like epithelial cell line Caco-2 (Gaillard *et al.*, 1991) whereas InlB is needed for entry into cultured hepatocytes and some epithelial or fibroblast-like cell lines including Vero, HeLa and CHO cells (Dramsi *et al.*, 1995). The cell adhesion molecule E-cadherin, a transmembrane protein normally involved in homophilic cell/cell interactions, has been identified as the receptor for InlA (Mengaud *et al.*, 1996; Lecuit *et al.*, 1999). Two different cellular receptors have been distinguished for InlB: the receptor of the globular part of the complement component C1q (gC1q-R) (Braun *et al.*, 2000) and also c-Met, the receptor for the hepatocyte growth factor (Shen *et al.*, 2000). Interaction of InlA and InlB with their respective cellular receptors induces recruitment to the sites of bacterial entry of intracellular molecules which are supposed to mediate cytoskeletal rearrangement, necessary to induce bacterial invasion (Ireton *et al.*, 1996, 1999; Lecuit *et al.*, 2000).

In this chapter, we will summarize methods that have been used to study the invasion process of bacterial pathogens, focusing on the *Listeria* system to illustrate these techniques.

◆◆◆◆◆◆ DIRECT ANALYSIS OF BACTERIAL INVASION

The gentamicin survival assay

One of the most widely used methods to study the capacity of a pathogen to invade and to proliferate into host cells is the gentamicin survival assay (Vaudaux and Waldvogel, 1979). This aminoglycoside antibiotic is only very weakly internalized by host cells, allowing the killing of extracellular bacteria—not able to invade host cells—and the survival of intracellular bacteria—protected by the plasma membrane of the host cell from the drug action.

Method

Professional phagocytes (macrophages) as well as non-professional phagocytes (fibroblasts, hepatocytes or epithelial cells) can be used in the gentamicin survival assay. In the *Listeria* system, green monkey kidney cells (Vero) are comfortable cells to work with and suitable to study the InlB-dependent internalization pathway.

Vero cells (ATCC CCL-81) are grown on Dulbecco's modified Eagle medium (DMEM) supplemented with 10% fetal calf serum (FCS) and 2 mM glutamine without antibiotics at 37°C in a 10% CO_2 atmosphere. The day before the assay, cells are plated on 24-well plates at a concentration of 4.5×10^4 cells per well and cultured overnight. A colony of *Listeria monocytogenes* wild type EGD strain (serotype 1/2a) is also grown overnight in 5 ml of brain heart infusion (BHI) broth at 37°C with agitation. The next day, 100 μl of the bacterial overnight culture are diluted in 5 ml of fresh BHI and bacteria are grown at 37°C with agitation to an OD_{600} of 0.8 (approximately 8×10^8 bacteria ml^{-1}). *Listeria* are washed in PBS by two serial centrifugation steps (in order to eliminate proteins from the broth medium such as the listeriolysin, that otherwise would damage the inoculated monolayers), then bacteria are diluted in DMEM without FCS to a concentration of 10^7 bacteria ml^{-1}, and 500 μl of the bacterial dilution are added to each well containing Vero cells (corresponding approximately to a multiplicity of infection—MOI—of 50 bacteria per cell). To synchronize the infection, bacteria can be centrifuged onto the cells at 1000*g* for 1 min (this treatment will also increase, by around 20%, the number of bacteria associated to the cells). Entry can also be synchronized by incubating initially cells and bacteria at 4°C, and then transferring them to 37°C to start the invasion. Inoculated cells are incubated for 1 h at 37°C in a 10% CO_2 atmosphere to allow bacterial entry. Serial dilutions of the bacterial-containing medium are performed in phosphate buffered saline and plated in BHI agar plates in order to assess the exact number of bacteria that were inoculated to the monolayers. After the 1 h incubation period, cells are washed twice with PBS and overlaid with fresh DMEM containing gentamicin at a concentration of 10 μg ml^{-1}, in order to kill the extracellular bacteria. Vero cells are further incubated at 37°C for 2 h, and monolayers are then washed twice with PBS and lysed by the addition of 1 ml of 0.1% Triton X-100 in distilled water for 5 min. The number of viable bacteria released from the cells is assessed by performing serial dilutions of the cell lysates in PBS and plating the dilutions on BHI agar plates: colony forming units (CFU) are counted the next day, and the percentage of bacteria that were able to invade cells is evaluated as the ratio of bacteria which survived to the gentamicin treatment divided by the total number of bacteria that were added to the monolayers (Figure 8.1).

Variation from one experiment to another can be high due to parameters such as differences in the exact number of cells plated each time, differences in the exact number of bacteria inoculated to the monolayers, differential loss of cells during washes, etc. In a single experiment (using the same batch of cells and the same batch of bacteria), it is advisable to perform each treatment (different strains, different target cells, different MOI, etc.) in triplicate to calculate the average number of viable bacteria in each case and to always add internal negative and positive controls (such as known invasive and non-invasive bacteria).

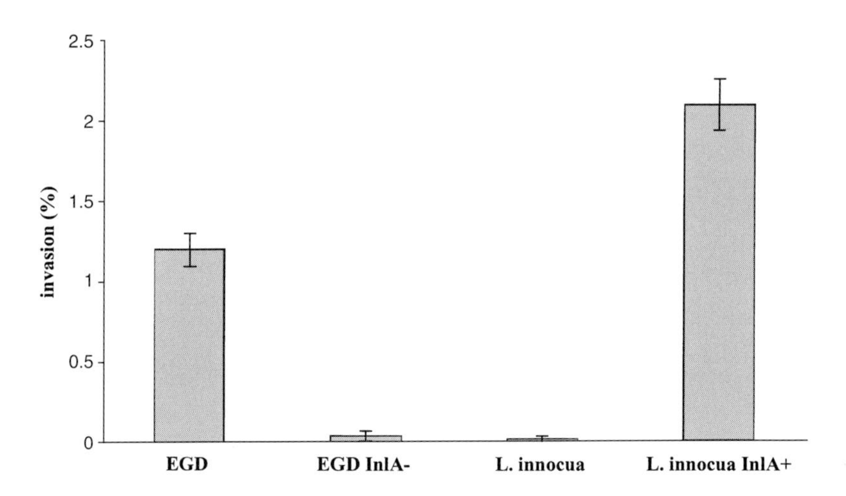

Figure 8.1. Gentamicin survival assay. Caco-2 were infected, as described in the text, with the wild type strain *Listeria monocytogenes* EGD, a mutant EGD strain lacking InlA, wild type *Listeria innocua* and a mutant *L. innocua* strain expressing InlA. Monolayers were then washed and incubated with gentamicin ($10\,\mu g\,ml^{-1}$) for 2 h. Finally, cells were lysed and CFU were estimated by plating on agar dishes serial dilutions of the lysates. Results are expressed as percentage of the inoculum that survived the gentamicin test.

Kinetics of intracellular proliferation can also be calculated if several sets of cells are inoculated at the same time, but lysed at different times after the gentamicin treatment. Before or/and during the infection step, pharmacological inhibitors can be added, such as those described in the final section of this chapter, to analyse cytoskeletal function and signalling cascades during invasion.

Differential immunofluorescence labelling

Another quantitative assay used routinely to assess invasion of many intracellular pathogens is the differential immunofluorescence labelling. This approach relies on the staining of the studied pathogen with a specific antiserum before permeabilization of the host cell plasma membrane, and a second staining after membrane permeabilization. Using secondary antibodies with two different fluorochromes before and after the permeabilization step allows discrimination between adherent extracellular bacteria (or other inoculated particles) and those that have been internalized (Braun *et al.*, 1998; Lecuit *et al.*, 1999).

Method

The enterocyte-like cell line Caco-2 (ATCC HTB-37) can be used to study the internalization of *Listeria* through the InlA-dependent pathway, since this cell line expresses the InlA receptor human E-cadherin (Mengaud *et al.*, 1996). Caco-2 cells are grown on Eagle's Minimal Essential Medium

(MEM) supplemented with non-essential aminoacids, Earl's Salts, 2 mM glutamine and 20% FCS. Caco-2 cells are more tricky to work with than Vero cells, and plating should be done at least 2 days before the labelling is performed to allow cells to spread correctly over the glass substrate. During the trypsinization step, if cellular aggregates do not dissociate easily, cells can be pelleted-down (1200 rpm for 5 min) and re-trypsinized. Cells are plated on 12 mm glass coverslips loaded on 24-well plates. Bacteria are grown the day before the experiment as described for the gentamicin survival assay, and the inoculation of Caco-2 cells is performed as for Vero cells (see above).

One hour after inoculation, monolayers are washed twice with PBS to eliminate non-adherent bacteria, and cells are fixed with 3% paraformaldehyde in PBS for 20 min at room temperature. Cells are then washed once with PBS (at this step, and if the experiment cannot be continued immediately after fixation, fixed cells can be stored at 4°C, preferably not for more than 24 h), and free aldehyde groups are quenched with 50 mM NH_4Cl in PBS for 10 min. Cells are then washed once with PBS and blocked with a non-specific protein (10% mouse serum or 1% bovine serum albumine [BSA] in PBS) for 30 min.

Extracellular bacteria are then detected by incubating cells for 30 min with a specific serum against the analysed pathogen (the serum will be diluted in 5% mouse serum or 0.5% BSA in PBS), washing the cells carefully twice with PBS, and revealing the primary antibodies with secondary antibodies coupled to a specific fluorochrome (for example, if primary antibodies were developed in rabbit, secondary antibodies should be anti-rabbit ones developed in a different species [such as goat, mouse or rat] and labelled with a fluorochrome such as Texas Red). After 20 to 60 min incubation with secondary antibodies (diluted in 5% mouse serum or 0.5% BSA in PB), monolayers are washed three times with PBS. Cells are then permeabilized with an adequate detergent such as Triton X-100 or saponin: if Triton X-100 is used, cells are incubated for 4 min with 0.1% Triton in PBS, then washed, and further incubations can be performed in buffers without the detergent; if saponin is used, all subsequent incubations (and washes) have to be performed in the presence of 0.05% saponin. Intracellular bacteria are then revealed using the primary antibodies that allowed recognition of extracellular bacteria. After 30 min incubation with primary antibodies, monolayers are washed twice with PBS and a 20 to 60 min incubation is performed with secondary antibodies coupled to a fluorochrome different from the one used at the first step (if the first fluorochrome used was Texas Red, the second one could be fluorescein thiocinate [FITC]). Finally, monolayers are washed once with PBS, once with distilled water, and coverslips are mounted on 4 μl of Mowiol (Calbiochem) over a glass slide.

Analysis with a fluorescence microscope will permit the differential detection of extracellular and intracellular bacteria: extracellular bacteria will be doubly labelled and, if Texas Red and FITC were used, they will appear yellow if the FITC and Texas Red channels are overlaid, whereas intracellular bacteria recognized only by antibodies after the permeabilization step, will consequently be labelled only with the second fluorochrome (if it was FITC, intracellular bacteria will be labelled in green).

After the permeabilization step, additional information can be gained by incubating with antibodies directed against intracellular components, structures or compartments, that may be in turn be labelled with a third fluorochrome (such as Cy-5): this would allow the identification of subcellular structures associated with the intracellular pathogen.

Transmission electron microscopy

Another quantitative powerful technique to gain insight into the invasive properties of a pathogen is transmission electron microscopy (TEM). This technique not only allows the direct visualization of micro-organisms inside cells, but also permits one to obtain morphological information concerning the subcellular structures associated with the intracellular parasite (Kocks *et al.*, 1991; Lecuit *et al.*, 2000).

Method

Target cells are infected as described in the previous sections, using 35 mm plastic petri dishes instead of 24-well dishes. Once the inoculation period is achieved, cells are washed twice with PBS to remove non-adherent parasites, and monolayers are fixed for 2 h with 2% glutaraldehyde in 0.2 M sodium cacodylate buffer pH 7.4 at 4°C. The fixing solution is changed once for 1 h, and then cells are washed overnight with sodium cacodylate buffer in order to eliminate any free unreacted glutaraldehyde that remains associated to the monolayer. Aldehydes remaining from the primary fixation are oxidized with two 1 h washes of 1% osmium tetroxide in sodium cacodylate buffer.

Samples are then dehydrated, replacing water with a fluid that acts as a solvent between the aqueous environment of the monolayer and the hydrophobic embedding media; ethanol is the most widely used dehydration agent. Accordingly, cells are incubated progressively with increasing concentrations of ethanol to gradually remove water: first, a 5 min incubation with 30% ethanol is performed; then 15 min incubations are done with 50%, 70% and 95% ethanol, respectively; another 15 min 95% ethanol incubation is realized, and finally two 20 min incubations with absolute ethanol.

The next step involves the replacement of the dehydration solution by another intermediary solvent that is highly miscible with the plastic embedding medium that will be used for the final infiltration. The standard solvent used is propylene oxide (PO), while the Epoxy 812 resin is one of the most popular plastic embedding media. To prepare the Epoxy 812 embedding medium, 50 g of resin are thoroughly mixed with 25 g of dodecenyl succinic anhydride (DDSA) and 25 g of nadic methyl anhydride (NMA), which are hardeners; then, 2 ml of the accelerator 2,4,6-tridimethylamino methyl phenol (DMP-30) are mixed with the Epoxy/DDSA/NMA mixture to complete the embedding media. Samples are first washed three times for 10 min each time with PO only; then, subsequent 2 h incubations are performed with one part of PO and one part of the embedding media, one part of PO and two parts of embedding media, one part of PO and three parts of embedding media, one part of PO and four parts of embedding media, and finally pure embedding media, respectively. The specimens are transferred to gelatin capsules, and the final step in embedding involves the polymerization of the epoxy mixture, achieved by placing the capsules at 60°C for 72 h.

Samples are then trimmed and cut into thin sections in an ultramicrotome, mounted into grids, contrasted with 5% uranyl acetate in methanol for 10 min, washed with 100% methanol, dryed, and analyzed in a transmission electron microscope. In the case of intracellular parasites, different steps of the entry process and of the intracellular replication cycle can be precisely dissected using this potent resolutive technique.

◆◆◆◆◆◆ IDENTIFYING PROTEINS INVOLVED IN BACTERIAL INVASION

Generation and screening of mutant libraries

Disruption of genes by transposon mutagenesis is, when technically possible, the most direct technique used to identify gene products involved in invasion: transposon insertions are obtained in different sites of the genome of the virulent pathogen, and non-invasive clones are identified by use of the simplest final readout to allow efficient screening of a high number of different mutants, such as a simplified version of the gentamicin survival assay.

Method

In the case of *Listeria*, the conjugative transposon Tn*1545* has been successfully used to identify the *inlAB* locus necessary for invasion in non-professional phagocytes (Gaillard *et al.*, 1991). The transposon Tn*1545* is a 26-kilobase DNA encoding resistance to kanamycin, tetracycline and erythromycin (Caillaud *et al.*, 1987). Transposon mutagenesis was achieved by transferring Tn*1545* from *L. monocytogenes* BM4140 serovar 7 (the transposon-bearing strain that received the transposon from *Streptococcus pneumoniae* [Caillaud *et al.*, 1987], where Tn*1545* was originally identified) to *L. monocytogenes* EGD-SmR strain (a spontaneous mutant resistant to streptomycin). Donor and recipient bacteria are grown overnight in BHI broth, then mixed at a ratio of 1 : 10 on a membrane filter, and incubated for 18 h at 37°C on BHI agar. Insertional mutants are selected on Tryptic Soy agar supplemented with 100 μg ml^{-1} of streptomycin, and 10 μg ml^{-1} of tetracycline.

To screen for non-invasive mutants, a modified gentamicin survival assay is used. Caco-2 cells are plated on 96-well plates at a concentration of 5×10^3 cells per well in 200 μl of cell culture medium for at least 48 h. The insertion mutants are grown overnight in Tryptic Soy broth at 37°C also in

96-well plates, and 10 μl of each bacterial culture are added to the Caco-2 monolayers. A 1 h incubation is performed at 37°C, after which monolayers are washed twice with PBS and 200 μl of cell culture medium supplemented with gentamicin (10 μg ml^{-1}) are added to each well. After 2 h of incubation, cells are washed twice with PBS and lysed with 200 μl of 0.1% Triton X-100 in distilled water for 5 min. Then, 10 μl of each lysate are plated on BHI agar to test for bacterial growth. Insertional mutants that do not grow in Caco-2 cells can be tested for a quantitative gentamicin survival assay as described in the first part of this chapter. Cloning of regions of insertion of the Tn*1545* transposon of non-invasive mutants should allow the identification of genes required for invasion of cultured cells.

One problem raised by the use of Tn*1545* is the fact that this transposon induces multiple insertions. An alternative and convenient transposon is Tn*917* which can be delivered to the chromosome after transformation with a thermosensitive plasmid carrying the transposon (Cossart *et al.*, 1989).

Expression of identified candidates in heterologous systems

Once a candidate gene for invasion has been identified, this gene product has to be characterized at the transcriptional and the translational levels. The full length gene, preferably with its own promoter region can be subcloned in an expression vector which can be then transferred into non-invasive bacteria, in order to demonstrate that it is not only necessary for invasion, that it complements the non-invasive mutant, and also may render invasive non-invasive bacteria, demonstrating that it may be sufficient to allow invasion. This has been done for the two *L. monocytogenes* invasion genes *inlA* and *inlB*.

A plasmid carrying the *inlA* gene under the control of its own promoter or a heterologous promoter such as the *spa* promoter from *S. aureus* can be transferred by conjugation from a donor *E. coli* strain to a recipient strain as described in the previous section, or by electroporation using standard protocols into *L. monocytogenes inlA* mutant or in non-invasive bacteria such as the non-pathogenic non-invasive *Listeria innocua* or the more distantly related *Enterococcus faecalis* (Lecuit *et al.*, 1997). This allows the study of an 'isolated' invasion protein, independently of other invasion systems such as *inlB* in the case of *L. monocytogenes*. Such artificial and apparently reductionist approaches have been very instrumental in deciphering step by step the so-called InlA and InlB invasion pathways.

In the case of surface proteins from Gram-positive bacteria such as *L. monocytogenes*, the universal sortase-mediated anchoring system to the petidoglycan can be used to express invasion proteins on the bacterial surface of heterologous Gram-positive hosts (Cossart and Jonquières, 2000). This has been achieved for InlA, which can be successfully expressed on the surface of *Listeria*, *Staphylococcus* and *Enterococcus* species (Lecuit *et al.*, 1997). In the case of proteins associated to the surface by more specific systems, such as the so-called GW-motifs which associates loosely

InlB to the surface of some but not all *Listeria* isolates, another approach can be followed, that is generation of a chimeric protein, fusing the invasion amino-terminus motif of InlB to the carboxy-terminus of either InlA or another cell-wall anchored protein such as staphylococcal protein A (Braun *et al.*, 1999).

Candidate invasion genes can also be subcloned in expression vector such as pET28 in *E. coli*, allowing their biochemical purification following standard procedures such as the His-tag purification technique (Jonquières *et al.*, 1999). Alternatively, modified version of invasion genes lacking their anchoring domain and thus totally released in the broth media can be generated to purify invasion proteins from their natural host. This approach has been followed not only to demonstrate that InlA has to be associated to the bacterial cell surface to mediate entry but also to purify a functional protein, which has been used to identify the InlA receptor (Lebrun *et al.*, 1996; Mengaud *et al.*, 1996).

Purification of identified proteins

Protein purification can be carried out taking advantage of several parameters such as charge, size or activity. One of the most frequently used chromatographic methods is ion exchange due to its ease of use, its wide applicability, and its low cost in comparison to other separation methods. Ion exchange of proteins involves their adsorption to the charged groups of a solid support followed by their elution with fractionation and/or concentration in an aqueous buffer of higher ionic strength. This technique has been commonly used in our laboratory to purify InlA, coupled to a gel filtration step (Mengaud *et al.*, 1996), or to purify InlB (Braun *et al.*, 1997; Ireton *et al.*, 1999; Jonquières *et al.*, 1999).

Method

Production of a recombinant InlA protein without its cell-wall anchor (which allows for protein secretion in the bacterial culture medium) was developed by introducing through electroporation the plasmid pKSV7-2 into the wild type strain EGD, and selecting for gene replacement by homologous recombination (Lebrun *et al.*, 1996; Mengaud *et al.*, 1996). The presence of the truncated form of the *inlA* gene in place of the wild type *inlA* gene was confirmed by Southern blotting, or by PCR analysis using oligonucleotides flanking the 3′ end of *inlA* (5′-CACAACAAGTACAAT GAA-3′ [position 3778–3795] and 5′-TCGTTTCCGCTTTA-3′ [position 5182–5195]).

Bacteria are grown for 24 h in 800 ml of trypticase broth (30 g trypticase, 20 g yeast extract, 0.7 g H_2KPO_4 and 3.7 g Na_2HPO_4 in 1 litre of water) supplemented with erythromicin (final concentration: 8 μg ml^{-1}). Then, the bacterial culture is centrifuged at 5000 rpm for 15 min at 4°C, the supernatant is recovered and a first round of protein precipitation is obtained by incubating the supernatant with 45% of ammonium sulfate for 1 h at 4°C with stirring. The mixture is centrifuged at 5000 rpm for

30 min at 4°C, and InlA is precipitated from the supernatant by incubation with 60% of ammonium sulfate for 1 h at 4°C with stirring. The mixture is centrifuged once again at 5000 rpm for 30 min at 4°C, the supernatant is discarded, and the protein pellet is resuspended in 10 ml of a Tris buffer (50 mM, pH 7.5), and the solution is desalted by dialysis against 4 litres of Tris buffer using a dialysis bag with a cutoff value of 30 000. The desalted solution is loaded on a 5 ml Hitrap Q column (Pharmacia) at 5 ml min^{-1} and InlA is eluted with a 100 ml salt gradient from 0 to 0.2 M NaCl in Tris buffer. Fractions of 1.5 ml are collected and those containing InlA are pooled and concentrated by ultrafiltration (concentrator with a cutoff value of 50 000). Pooled fractions are loaded on a Sephacryl S300 high resolution column (26 mm in diameter and 55 cm in length) at 4 ml min^{-1}. Fractions of 2 ml are collected and InlA-containing fractions are pooled and concentrated by ultrafiltration. From 1 litre of bacterial culture, 4 mg of protein can be obtained (Mengaud *et al.*, 1996).

Production of a recombinant InlB protein coupled to a His-Tag (that can be subsequently removed through trombin cleavage) was achieved by first constructing an expression plasmid by amplification of the *inlB* coding sequence from the plasmid pPE13 (Gaillard *et al.*, 1991) using the primers 5′-CTAGCTAGCGAGACTATCACCGTGTCA-3′ and 5′-CCCAAGCTTGTTGTAGCTTTTTCGTAGG-3′. The amplified product was digested and cloned between the *Nhe*I and *Hind*III sites in the expression vector pET28 (Novagen), allowing the expression of a recombinant form of InlB that contains a 6 × His-Tag at the NH_2 terminus of the protein. For expression of the recombinant protein, the vector was transformed into *E. coli* strain BL21 lambda-DE (Novagen) (Ireton *et al.*, 1999).

The BL21 strain containing the vector is grown at 37°C for 3 h (until early log-phase) in the presence of kanamycine (30 μg ml^{-1}), and expression of the recombinant InlB protein is induced by addition of 1 mM isopropyl-β-D-thiogalactopyranoside (IPTG). After induction for 3 h, the bacterial culture is centrifuged at 4000 rpm for 20 min at 4°C (bacterial pellets can be frozen at −80°C). Bacteria are resuspended in a buffer containing 5 mM imidazole, 500 mM NaCl and 50 mM HEPES pH 7.9 (binding buffer), and are disrupted with a French press. The lysate is centrifuged at 10 000 rpm for 1 h at 4°C and loaded on a nickel column. The column is washed with binding buffer, and then sequentially washed with a similar HEPES/NaCl buffer containing first 20 mM imidazole, second 40 mM imidazole, and finally 60 mM imidazole. Fractions of 1.5 ml are recovered during the 40 mM and 60 mM imidazole elutions, and those containing InlB are pooled and dialysed against 4 litres of a buffer containing 200 mM NaCl/50 mM HEPES (pH 7.6) using a bag with a cutoff value of 30 000. Pooled fractions are then concentrated by ultrafiltration (concentrator cutoff value 50 000) and loaded onto a Poros HS20 column (PerSeptive Biosystems). Fractions of 1.5 ml are collected and InlB is eluted by a continuous salt gradient from 0.2 to 1 M NaCl. InlB-containing fractions are pooled by ultrafiltration, and the 6 × His-Tag is removed by cleavage with thrombin (1,2 thrombin units for 0.3 mg InlB) for 16 h at 16°C in a buffer containing 20 mM Tris pH 8.4, 0.15 M NaCl and 2.5 mM

$CaCl_2$. Cleaved InlB is separated from uncut protein by passage over a nickel column, and thrombin is removed by three successive incubations with an agarose resin coupled to the protease inhibitor *p*-aminobenzamidine. From 1.2 litre of induced bacterial culture, 4 mg of protein can be obtained (Ireton *et al.*, 1999).

Adherence of cells to purified proteins: a quantitative assay

Interaction of the purified microbial proteins with host cells can easily be tested *in vitro* by coating micro-well plates with the protein(s) of interest, and then analyzing the adherence of target cells to the wells: if the identified molecule is recognized by a cellular ligand at the host cell plasma membrane, cells should attach to the wells in a concentration-dependent manner. The amount of attached cells can be determined by quantifying the presence of an endogenous cellular enzyme such as hexosaminidase (Mengaud *et al.*, 1996).

Method

96-well plates (Nunc Immunoplate Maxisorp) are coated with 100 µl per well of the purified protein solution at various concentrations ranging from 0 to 1.0 $\mu g\,ml^{-1}$ in PBS without Ca^{2+} and Mg^{2+}. Incubation is performed overnight at 4°C. The next day, the protein solution is removed and wells are incubated with 200 µl of PBS/BSA 1% without Ca^{2+} and Mg^{2+} for 2 h at 37°C to block non-coated sites. Then, wells are washed three times with 250 µl per well of PBS.

Cells are resuspended in cell culture medium without FCS at a concentration of 1×10^6 cells ml^{-1}, and 50 µl of the cell suspension is distributed in each coated well (controls are obtained by distributing 50 µl of cell culture medium without cells in a parallel series of coated wells). Incubation is performed for 1 h at 37°C in a 10% CO_2 atmosphere. Wells are then washed three times with cell culture medium, and 60 µl of a hexosaminidase substrate solution is added to each well (the substrate stock solution contains 0.03 g of p-nitrophenyl-n-acetyl-β-D-glucosamide in 20 ml of citrate buffer 50 mM pH 5.0/Triton X-100 0.25% and is stored at −20°C). Incubation is carried out for 3 h at 37°C in a 10% CO_2 atmosphere, and the reaction is stopped by adding 90 µl of a solution containing 50 mM glycine/5 mM EDTA pH 10.4. The reaction product is read at an optical density (OD) of 405 nm (Figure 8.2).

Covalent coupling of purified proteins with latex beads

Coating of latex microspheres with identified proteins thought to play a role in invasion may lead to the internalization of these particles into target cells, which otherwise will poorly interact with beads coated with non-specific molecules such as BSA. Although passive absorption of proteins to beads is the simplest method of coating, covalent coupling to a modified chemical group is preferred, and carboxylate-modified latex beads have

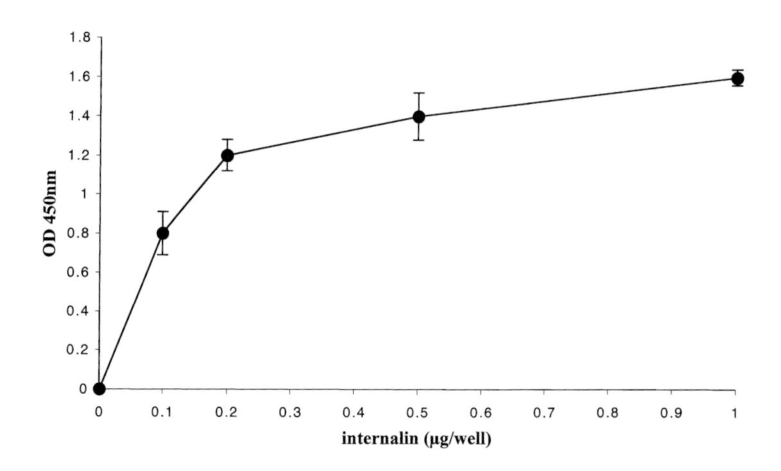

Figure 8.2. Adhesion of Caco-2 cells to purified InlA. A Caco-2 cell suspension was added to a 96-well plate coated with different concentrations of InlA, cells were washed and the number of attached cells was quantified by determining the activity of the lysosomal hexosaminidase. The number of attached cells is proportional to the amount of InlA coated to the wells.

been used successfully to couple the *Listeria* invasion proteins InlA and InlB (Braun *et al.*, 1997; Lecuit *et al.*, 1997).

Method

The candidate protein is concentrated in 15 mM sodium acetate buffer pH 5.0 to approximately 5 mg ml^{-1}. Fluorescent 1 µm carboxylate-modified latex beads (Molecular Probes, 2% aqueous suspension) are sonicated briefly before the coupling in order to correctly disperse the particles. In a 1.5 ml eppendorf tube are mixed 200 µl of latex beads and 80 µl (0.4 mg) of the concentrated protein, and the mixture is incubated for 15 min at room temperature. Then, the surface carboxyl groups are activated by adding 1.6 mg of 1-ethyl-3-(3-dimethylaminopropyl)-carbodiimide (EDAC), and the tube is vortexed. The pH is adjusted to 6.5 by adding 4.7 µl of 0.1 N NaOH and the reaction mixture is incubated on a rocker for 2 h at room temperature (or overnight at 4°C if desired). Finally, to stop the reaction, 11.4 µl of glycine 2.5 M are added and the mix is incubated for 30 min at room temperature. To separate the beads from unreacted protein, the reaction mixture is centrifuged at 2000 rpm for 20 min, and the supernantant is removed (the amount of protein in the supernantant should be measured in order to estimate substractively the degree of covalent coupling). Beads are resuspended in 200 µl of PBS, and centrifuged again at 2000 rpm for 20 min. A similar wash is repeated twice, and finally beads are resuspended in PBS containing 1% BSA to block the remaining hydrophobic sites.

Latex beads covalently coupled to InlA or InlB can be tested for internalization in Caco-2 cells or Vero cells, respectively. 2 µl of latex beads are resuspended in 500 µl of cell culture medium, and cells are incubated with the beads for 1 h at 37°C. After this time period, beads are washed off with PBS, cells are fixed with 3% paraformaldehyde and quenched with 50 mM NH_4CL (as described earlier for the differential immunofluorescence labelling), and a 30 min incubation is performed with antibodies against the target protein, followed by a 20 min incubation with secondary antibodies coupled to a fluorochrome different from the one that characterizes the beads. Cells do not need to be permeabilized, and can be mounted immediately with Mowiol: in a fluorescence microscope, intracellular beads are going to be detected with the fluorochrome intrinsic to the beads, while extracellular beads will be detected by a combination of the intrinsic fluorochrome and the one coupled to the secondary antibodies.

◆◆◆◆◆◆ TESTING CYTOSKELETAL REARRANGEMENTS AND SIGNALLING EVENTS DURING BACTERIAL INVASION

Inhibition of cytoskeleton function

Rearrangements of the cytoskeleton (predominantly microfilaments constituted of polymerized actin) are observed during invasion of host cells by intracellular parasites. Several pathogens, such as *Salmonella typhimurium* or *Shigella flexneri*, induce important cystokeletal modifications during entry that can be easily assessed by immunofluorescence microscopy after labelling of the polymerized filaments with an appropriate probe. *Listeria* induce more subtle local rearrangements of cytoskeleton that are more difficult to analyse by microscopy (Lecuit *et al.*, 2000), but which can also be evaluated using drugs to inhibit the cytoskeletal polymerization and thus blocking its function. For more detailed information concerning this topic, readers can consult the chapter by Boujemaa-Paterski, Wiesner and Carlier in this volume.

Method

Cytochalasins are a family of drugs that can block actin polymerization, and cytochalasin D is one of the most specific and potent drugs among this family of molecules. Testing of cytochalasin D activity is coupled to an invasion assay (such as the gentamicin survival assay) to determine microbial entry in the presence of the drug. A cytochalasin D stock is made in dimethyl sulphoxide (DMSO) at $1\,mg\,ml^{-1}$ and is stored at $-20°C$. Before infection, target cells are incubated for 30 min with cell culture medium containing $1\,\mu g\,ml^{-1}$ of cytochalasin D; then, the gentamicin survival assay is performed with cytochalasin D kept in the cell culture medium (with the bacteria during inoculation) for the whole assay. A parallel immunofluorescence experiment should be accomplished to confirm that only invasion (and not adhesion) of bacteria is prevented (differential immunofluorescence labelling of extracellular and intracellular bacteria can provide this information) and also to confirm that the actin cytoskeleton polymerization has been perturbed (this could be achieved through labelling of the actin cytoskeleton with the fungal drug phalloidin coupled to a fluorochrome).

Inhibition of protein kinase function

In order to trigger the cytoskeletal rearrangements that will lead to pathogen's uptake, many intracellular parasites exploit host cell signal transduction pathways. Several of these pathways use protein or lipid kinases to transmit signals, and in the case of *Listeria*, it has been shown that inhibition of the phosphoinositide 3-kinase (PI 3-K) blocks bacterial entry (Ireton *et al.*, 1996, 1999), highlighting the important role of this enzyme during the invasion process. Tyrosine kinases have also been shown to be involved in the production of phosphoinositides by PI 3-K and thus affect entry during interaction of *Listeria* with host cells (Ireton *et al.*, 1996, 1999). The study of signalling cascades will be approached in more detail in the chapter by Dixon in this volume.

Method

Two structurally unrelated compounds, wortmannin and LY294002 inhibit specifically the PI 3-K activity by different mechanisms (Vlahos *et al.*, 1994; Ui *et al.*, 1995). Stock solutions of these two drugs (50 μM for wortmannin and 50 mM for LY294002) are prepared in DMSO. Cells are incubated 20 min before infection with cell cultured medium supplemented with 50 nM wortmannin or 50 μM LY294002, and then monolayers are infected with bacteria in the presence of these PI 3-K inhibitors for 60 min. A gentamicin survival assay is performed in order to quantify the effect of the drug treatment in microbial invasion; a range of different drug dilutions should be tested to assess more precisely the activity of these compounds during the entry step (Figure 8.3). The protein tyrosine kinase

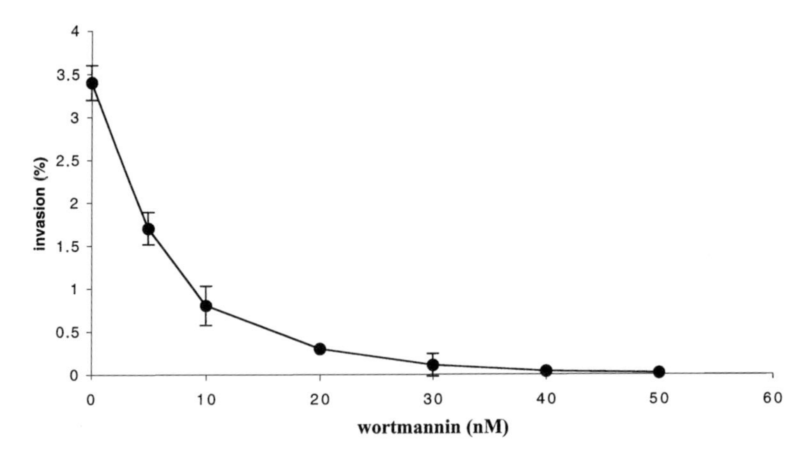

Figure 8.3. Inhibition of entry of *Listeria* to HeLa cells using wortmannin. HeLa cells were infected with *Listeria monocytogenes* strain EGD for 1 h in the presence of different concentrations of wortmannin, and gentamicin survival assays were performed to quantify the percentage of invasion in each case. There is a dose-dependent inhibition of *Listeria* entry by wortmannin.

inhibitor genistein is stored as a 250 mM stock solution in DMSO, and is added to the target monolayers in a final concentration of 250 μM in cell culture medium. The experiment is performed as it is described for the PI 3-K inhibitors.

◆◆◆◆◆◆ CONCLUSIONS

Bacterial pathogens represent excellent tools to address specific cell biology questions, and *Listeria monocytogenes* has been a key model to analyse important phenomena such as actin polymerization during bacterial intracellular movement, as well as subversion of host cell signalling cascades during bacterial entry. In this chapter we have addressed general methods that can be used to characterize the internalization of diverse parasites into host cells, as well as more specific methods related to the *Listeria* model, but that can be adapted to other microbial systems in order to dissect their particular adaptation strategies to their host environment.

References

Braun, L., Dramsi, S., Dehoux, P., Bierne, H., Lindahl, G. and Cossart, P. (1997). InlB: an invasion protein of Listeria monocytogenes with a novel type of surface association. *Mol. Microbiol.* **25**, 285–294.

Braun, L., Ghebrehiwet, B. and Cossart, P. (2000). gC1q-R/p32, a C1q-binding protein, is a receptor for the InlB invasion protein of Listeria monocytogenes. *EMBO J.* **19**, 1458–1466.

Braun, L., Nato, F., Payrastre, B., Mazie, J.-C. and Cossart, P. (1999). The 213-amino-acid leucine-rich repeat region of the Listeria monocytogenes InlB protein is sufficient for entry into mammalian cells, stimulation of PI 3-kinase and membrane ruffling. *Mol. Microbiol.* **34**, 10–23.

Braun, L., Ohayon, H. and Cossart, P. (1998). The InlB protein of Listeria monocytogenes is sufficient to promote entry into mammalian cells. *Mol. Microbiol.* **27**, 1077–1087.

Caillaud, F., Carlier, C. and Courvalin, P. (1987). Physical analysis of the conjugative shuttle transposon Tn1545. *Plasmid* **17**, 58–60.

Cossart, P., Vicente, M. F., Mengaud, J., Barquero, F., Parez-Diaz, J. C. and Berche, P. (1989). Listeriolysin O is essential for the virulence of Listeria monocytogenes: direct evidence obtained by gene complementation. *Infect. Immun.* **57**, 3629–3636.

Cossart, P. and Bierne, H. (2001). The use of host cell machinery in the pathogenesis of Listeria monocytogenes. *Curr. Opin. Immunol.* **13**, 96–103.

Cossart, P. and Jonquiéres, R. (2000). Sortase, a universal target for therapeutic agents against Gram-positive bacteria? *Proc. Natl Acad. Sci. USA* **97**, 5013–5015.

Cossart, P. and Lecuit, M. (1998). Interactions of Listeria monocytogenes with mammalian cells during entry and actin-based movement: bacterial factors, cellular ligands and signalling. *EMBO J.* **17**, 3797–3806.

Dramsi, S., Biswas, I., Maguin, E., Braun, L., Mastroeni, P. and Cossart, P. (1995). Entry of L. monocytogenes into hepatocytes requires expression of

InlB, a surface protein of the internalin multigene family. *Mol. Microbiol.* **16**, 251–261.

Finlay, B. B. and Cossart, P. (1997). Exploitation of mammalian host cell functions by bacterial pathogens. *Science* **276**, 718–725.

Gaillard, J. L., Berche, P., Frehel, C., Gouin, E. and Cossart, P. (1991). Entry of L. monocytogenes into cells is mediated by internalin, a repeat protein reminiscent of surface antigens from Gram-positive cocci. *Cell* **65**, 1127–1141.

Ireton, K., Payrastre, B., Chap, H., Ogawa, W., Sakaue, H., Kasuga, M. and Cossart, P. (1996). A role for phosphoinositide 3-kinase in bacterial invasion. *Science* **274**, 780–782.

Ireton, K., Payrastre, B. and Cossart, P. (1999). The Listeria monocytogenes protein InlB is an agonist of mammalian phosphoinositide 3-kinase. *J. Biol. Chem.* **274**, 17025–17032.

Jonquiéres, R., Bierne, H., Fiedler, F., Gounon, P. and Cossart, P. (1999). Interaction between the protein InlB of Listeria monocytogenes and lipoteichoic acid: a novel mechanism of protein association at the surface of Gram-positive bacteria. *Mol. Microbiol.* **34**, 902–914.

Kocks, C., Hellio, R., Gounon, P., Ohayon, H. and Cossart, P. (1992). *Listeria monocytogenes*-induced actin assembly requires the *actA* gene product, a surface protein. *Cell* **68**, 521–531.

Lebrun, M., Mengaud, J., Ohayon, H., Nato, F. and Cossart, P. (1996). Internalin must be on the bacterial surface to mediate entry of Listeria monocytogenes into epithelial cells. *Mol. Microbiol.* **21**, 579–592.

Lecuit, M., Dramsi, S., Gottardi, C., Fedor-Chaiken, M., Gumbiner, B. and Cossart, P. (1999). A single amino acid in E-cadherin responsible for host specificity towards the human pathogen Listeria monocytogenes. *EMBO J.* **18**, 3956–3963.

Lecuit, M., Hurme, R., Pizarro-Cerdá, J., Ohayon, H., Geiger, B. and Cossart, P. (2000). A role for alpha- and beta-catenins in bacterial uptake. *Proc. Natl Acad. Sci. USA* **97**, 10008–10013.

Lecuit, M., Ohayon, H., Braun, L., Mengaud, J. and Cossart, P. (1997). Internalin of Listeria monocytogenes with an intact leucine-rich repeat region is sufficient to promote internalization. *Infect. Immun.* **65**, 5309–5319.

Mengaud, J., Ohayon, H., Gounon, P., Mére, R.-M. and Cossart, P. (1996). E-cadherin is the receptor for internalin, a surface protein required or entry of L. monocytogenes into epithelial cells. *Cell* **84**, 923–932.

Shen, Y., Naujokas, M., Park, M. and Ireton, K. (2000). InlB-dependent internalization of Listeria is mediated by the Met receptor tyrosine kinase. *Cell* **103**, 501–510.

Ui, M., Okada, T., Hazeki, K. and Hazeki, O. (1995). Wortmannin as a unique probe for an intracellular signalling protein, phosphoinositide 3-kinase. *Trends Biochem. Sci.* **20**, 303–307.

Vaudaux, P. and Waldvogel, F. A. (1979). Gentamicin antibacterial activity in the presence of human polymorphonuclear leukocytes. *Antimicrob. Agents Chemother.* **16**, 743–749.

Vlahos, C. J., Matter, W. F., Hui, K. Y. and Brown, R. F. (1994). A specific inhibitor of phosphatidylinositol 3-kinase, 2-(4-morpholinyl)-8-phenyl-4H-1-benzopyran-4-one (LY294002). *J. Biol. Chem.* **269**, 5241–5248.

List of Suppliers

Amersham Pharmacia Biotech, Inc.
800 Centennial Avenue
P.O. Box 1327
Piscataway, NJ 08855-1327 USA
tel: +1 732 457 8000
fax: +1 732 457 0557
http://www.apbiotech.com

Calbiochem-Novabiochem Corporation
10394 Pacific Center Court
San Diego, CA 92121 USA
tel: +1 858 450 9600
fax: +1 858 453 3552
http://www.calbiochem.com

Molecular Probes, Inc.
4809 Pitchford Avenue
Eugene, OR 97405-0469, USA
tel.: +1 541 465 8300
fax: +1 541 344 6504
http://www.probes.com

Nalge Nunc International Corporation
75 Panorama Creek Drive
P.O. Box 20365
Rochester, NY 14602–0365
tel.: +1 800 625 4327
fax: +1 716 586 8987
http://www.nalgenunc.com

Novagen
601 Science Drive
Madison, WI 53711 USA
tel.: +1 608 238 6100
fax: +1 608 238 1388
http://www.novagen.com

PerSeptive Biosystems
500 Old Connecticut Path
Framingham, MA 01701 USA
tel.: +1 508 383 7700
fax: +1 800 899 5858
http://www.pbio.com

9 Measurement of Bacterial Uptake by Cultured Cells

Amit Srivastava[1] **and Ralph R Isberg**[1,2]
[1]Department of Molecular Biology and Microbiology, [2]Howard Hughes Medical Institute, Tufts University School of Medicine, 136 Harrison Avenue, M & V 409 Boston MA 02111, USA

◆◆

CONTENTS

Bacterial Uptake by Cultured Cells

◆◆◆◆◆◆ INTRODUCTION

A number of bacterial pathogens (e.g. *E. coli, Shigella, Salmonella, Listeria* and *Yersinia*) have evolved a variety of strategies to penetrate their respective target host cells. Internalization by host cells serves a number of roles in bacterial pathogenesis such as immune evasion, initiation of intracellular replication and furnishing a portal into deeper tissue. Elucidation of the entry process is a critical prerequisite for understanding the virulence of a number of bacterial pathogens (Falkow *et al.*, 1992; Finlay and Falkow, 1997). In this chapter, we describe procedures used routinely in our laboratory to dissect uptake of the gut pathogen *Yersinia pseudotuberculosis* by non-phagocytic cultured cells. These approaches may be easily adapted for studying other pathogens with a similar proclivity for promoting internalization by cultured cells. Based on the prevalent understanding of mechanisms of pathogen–host interaction, we consider it appropriate to use the term 'uptake' to specify bacterial entry at a cellular level and 'invasion' to indicate deep tissue incursion by the pathogen. We certainly appreciate the irony of the fact that the first factor demonstrated to promote bacterial entry into cultured cells was called 'invasin' (Isberg *et al.*, 1987). We believe that over the course of time and extensive investigations of bacterial pathogenesis it has become indisputable that the terms 'uptake' and 'invasion' should be reserved for very different processes.

METHODS IN MICROBIOLOGY, VOLUME 31
ISBN 0-12-521531-2

◆◆◆◆◆◆ GENERAL PRINCIPLES

Two primary strategies have been employed to examine bacterial uptake by cultured cells: (i) the gentamicin protection assay and (ii) the differential immunostaining assay. The gentamicin protection assay relies on quantification of viable bacteria protected from externally added antibiotic by virtue of being internalized by the host mammalian cell. The differential immunostaining assay involves antibody labeling of the internalized and external (adherent) bacteria followed by fluorescence microscopic observation. While the gentamicin protection assay remains a reliable workhorse for estimating bacterial ingestion by host cells (Mandell, 1973), it is able to provide only a gross and terminal picture of the uptake process. On the other hand, differential immunostaining affords several distinct advantages: (i) the ability to observe uptake at the level of a single mammalian cell; (ii) the facility to carry out kinetic analyses of uptake relative to the rate of adhesion; (iii) the ability to monitor the participation of host and/or bacterial proteins in the course of uptake in a qualitative and quantitative fashion; and (iv) the ability to analyze different subpopulations of mammalian cells, which is critical for analysis of transfection experiments.

◆◆◆◆◆◆ DIFFERENTIAL IMMUNOSTAINING ASSAY

Since antibody molecules are unable to penetrate the plasma membrane of the mammalian cell, only extracellular bacteria (or exposed parts thereof) are stained during the first round of immunostaining. A second round of staining carried out after the host cells have been treated with a permeabilizing agent allows the antibodies access to the ingested bacteria. This results in differentially stained bacteria that can be easily distinguished from each other. Therefore, it is of utmost importance that the permeabilization agent do no harm to the antigenicity of the bacteria.

Commonly used cultured cell lines (e.g. COS-1, COS-7, HeLa, NIH and Swiss 3T3 cells etc.) are amenable to this uptake assay. Typically mammalian cells are seeded onto sterilized circular (12 mm diameter) glass cover-slips (Fisher Scientific, Pittsburgh PA) at a density of 10–20 000 cells per coverslip in a 24-well tissue culture dish (Falcon®, Becton Dickinson Labware, Franklin Lakes, NJ). Cultured cell lines that have undergone more than 15 passages should be avoided. Glass coverslips may be sterilized by autoclaving or by ethanol flaming before being placed individually in the 24-well dishes; the dishes are then UV sterilized for at least 15 min. The requisite amount of the cell suspension is added to the wells and incubated (6 h to overnight) to allow the cells to adhere to the coverslip and form a monolayer. In the case of *Yersinia pseudotuberculosis*, a bacterial culture in the mid-logarithmic growth phase ($OD_{600} = 0.7$ i.e., 10^8 cfu ml^{-1}) is then diluted in cell culture medium according to the desired MOI (multiplicity of infection). The bacterial suspension is added to the mammalian cell monolayer to initiate the infection. Incubation time with bacteria and the MOI are inversely related and should be adjusted accordingly; 80 min incubation at MOI = 10 and a 20 min incubation at MOI = 150 are examples of infection regimens that we have employed fruitfully. Choice of the incubation time is often governed by the time required to observe a perceptible phenotype, relative to the controls, after perturbation of uptake due to chemical inhibitors or overproduced proteins encoded by transfected DNA, for instance. At the end of the selected infection period, non-adherent bacteria are washed off using Phosphate or Tris Buffered Saline (PBS or TBS) and the cells are processed for immunofluorescence microscopy.

Host cells with adherent and ingested bacteria are fixed using buffered 3–5% (w/v) Formaldehyde solution. It is vital that the fixatives do not distort the cellular architecture. We have found that the fixative, PLP Sucrose (Swanson and Isberg, 1995) is quite satisfactory for fixing both non-phagocytic and phagocytic (e.g. macrophages) cells. For 10 ml PLP Sucrose Fixative: dissolve 21.4 mg of sodium periodate in 7.5 ml Stock A solution and add 2.5 ml of 8% paraformaldehyde. 5 ml aliquots of filter sterilized 8% paraformaldehyde solution may be stored in the dark at −20°C until used (paraformaldehyde should be handled in the fume hood!). Stock A solution (25 ml): 12.5 ml 0.2 M lysine, 3 ml 40% sucrose and 9.5 ml 0.1 M sodium phosphate Buffer pH 7.4, filter sterilized. For best results, fresh PLP Sucrose fixative should be reconstituted and used immediately. Fixation is carried out at 37°C for at least 1 h and may be extended to a few hours without detriment.

All immunostaining may be carried out on parafilm stretched over an inverted 24-well tissue culture dish; not more than 30 μl of diluted antibody solution is required per coverslip. The coverslips bearing cells are placed face down on the droplets of antibody solution using a pair of sharp forceps. Incubation with antibodies can be carried out at room temperature for 45 min to 1 h on the benchtop. Antibodies should be diluted in the buffer used for blocking to reduce non-specific reactivity. Choice of the antibody recognizing the bacteria is very important in that the whole bacterium should be labeled by the antibody. We have obtained satisfactory results with polyclonal antibodies generated in rabbits or rats immunized with killed whole bacteria as the antigen (using standard protocols described in Harlow and Lane, 1988). With respect to fluorophore labeled secondary antibodies, we prefer using the donkey serum based reagents (Jackson Immunoresearch, West Grove, PA) for their high sensitivity and minimal cross-reactivity. As usual, the appropriate antibody dilutions to be used must be determined empirically for each antiserum with due attention to the manufacturers' suggestions. In general, we use secondary antibodies at dilutions ranging from 1/400 to 1/800.

A variety of permeabilization agents may be used to allow antibodies to gain access to the interior of the host cell. The most commonly used permeabilization buffers include a mild detergent such as Triton X-100 (0.1–0.5%). Our method of choice is exposure to cold methanol (−20°C) for exactly 10 seconds per coverslip. Permeabilization is carried out by lifting the coverslip with forceps and immersing it in the methanol followed by immediate immersion in PBS to remove the methanol rapidly. This permeabilization regimen appears to cause minimal disruption of the intracellular architecture; it does not influence the immunoreactivity of bacterial antigens or disrupt existing antigen–antibody complexes.

After immunostaining is complete, the coverslips can be stored in the 24-well dishes immersed in 3% glycerol in PBS for a week or more at 4°C. We mount the coverslips on slides using 3–4 μl of Fluoroguard® Antifade Reagent (Bio-Rad, Hercules, CA) to protect against dehydration of the sample, UV-photobleaching, and quenching. The coverslip is then sealed at the edges using nail polish (choice of color is at user's discretion).

An extremely useful variation of this procedure is to combine analysis of host cell molecules during the bacterial uptake reaction. Host cell molecules may be examined by immunostaining with cognate antibodies after permeabilization of cells. Furthermore, effects of ectopic expression of pertinent host cell molecules on bacterial uptake can be assessed by transfection of the cell monolayer with the relevant plasmid DNA. While most standard transfection protocols and reagents should suffice, we get the best results with Lipofectamine® transfection reagent (Life Technologies, Gaithersburg, MD) and use it according to the manufacturer's instructions. Ectopically expressed molecules may be detected using specific antibodies and/or attached fluorescent tags (e.g. GFP) and detection by either regimen is necessary to distinguish reliably between transfected and untransfected cells. Analysis of transfected cells helps shed light on the role of particular host proteins on bacterial ingestion and may be assessed in terms of: (i) efficiency of bacterial uptake or (ii) co-localization of the overproduced protein with the phagocytic cup. Untransfected cells on the same coverslip constitute the best possible controls for comparison. See Figure 9.1 for an example of a typical result of this assay.

Quantitative assessment of bacterial uptake

The differential immunostaining assay yields multiple parameters that prove extremely useful for the quantitative description of the process of bacterial entry into host cells. Differential staining allows enumeration of adherent bacteria per host cell and internalized bacteria per host cell. Bacteria stained before permeabilization (Red fluorophore) are external and staining after permeabilization labels all cell-associated bacteria (Blue fluorophore) i.e., total bacteria. The number of internalized bacteria can be obtained by subtracting the number of external bacteria from the total. 'Uptake efficiency' or 'percent uptake' is defined as the number of internalized bacteria relative to the total number of cell associated bacteria. These parameters allow one to assess the phenotype of a certain bacterial mutant with respect to entry and/or the contribution of a particular host molecule in bacterial uptake. Ectopic expression of certain host molecules can exert a quantifiable effect on uptake efficiency—a dominant negative mutant may depress uptake while a constitutively active form might stimulate uptake (Alrutz and Isberg, 1998; Alrutz *et al.*,

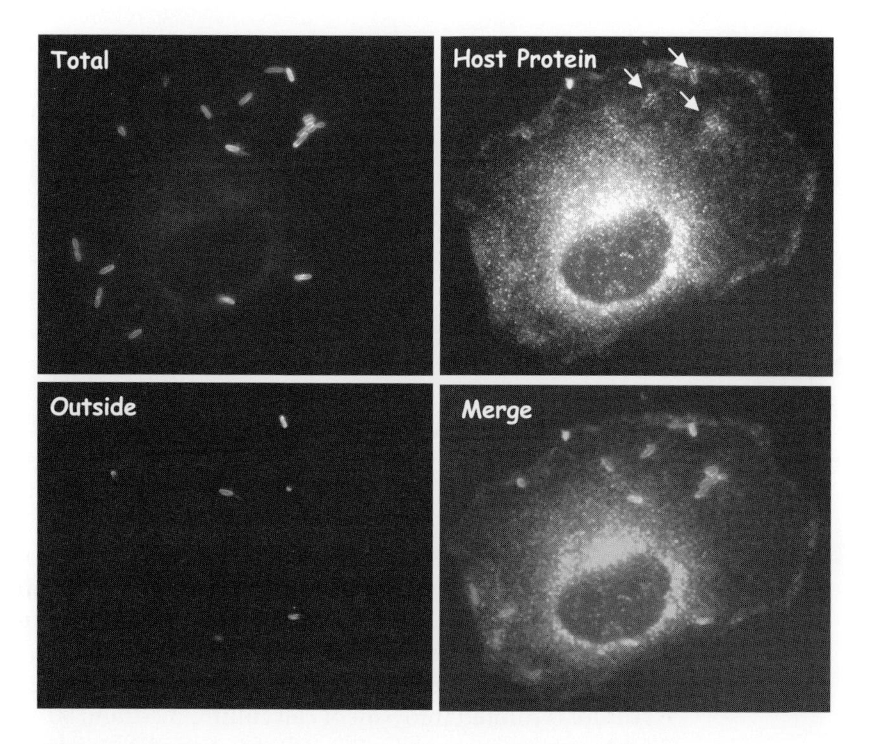

Figure 9.1. Image of *Yersinia pseudotuberculosis* bacteria captured in the dynamic process of entry into a COS-1 cell. Merged image: note the distinct staining of the internalized (Blue) and external (Red) portions of the 'half-in/half-out' bacteria. Localization of a Arp2/3 protein complex during bacterial uptake was examined by immunostaining with an anti Arp3 antibody (Green) after permeabilization of cells (Alrutz *et al.*, 2001). The intense staining around nascent phagosomes strongly suggests that this protein participates in the uptake process; quantitation of this phenotype would reinforce the inference. (This figure is also reproduced in colour between pages 276 and 277.)

2001). Standard statistical methods must be rigorously followed in order to draw a reliable correlation between data sets.

Differential immunostaining can also be used to follow the participation of host cell factors in uptake by following the endogenous proteins or transfected tagged clones thereof. Recruitment of specific molecules to adherent and/or internalized bacteria or a nascent phagosome containing a bacterium can be determined in this fashion. Comparison with the untransfected control cells affords the extraction of statistically significant quantitative data. Furthermore, kinetic experiments of this type can even provide adequate information to order the participation of various host molecules in uptake.

Suggested protocol

Day 1—Seeding coverslips

1. Seed 1×10^4 or 2×10^4 cells per coverslip (in a well of a 24-well tissue culture dish) in a volume of at least 300 μl cell culture

medium. Incubate the dish overnight to allow the cells to adhere to the coverslips.

2. Start a pre-culture of *Yersinia pseudotuberculosis* Y137 (Inv^+ Plasmid cured) using a single colony from a fresh plate (less than 2 weeks old) in 3–5 ml LB medium. Grow overnight with aeration at 28°C.

For experiments involving ectopic expression of host cell molecules, Day 2 will be required for transfection of host cell monolayers with plasmid DNA.

Day 2—Infection

3. Using the overnight pre-culture, inoculate fresh 50 μl, 100 μl and 150 μl into three 5 ml LB tubes. Grow in a water-bath shaker at 28°C for at least 3 h to $OD_{600} = 0.7$. Measure the OD of a 1 : 10 dilution (i.e., expected OD ≈ 0.07).
4. Depending on the MOI desired, dilute the requisite amount of the bacterial culture into 5 ml of pre-warmed cell culture medium to prepare the innoculum. Ensure that the innoculum volume added per well does not exceed 100–150 μl. Mix well.
5. Add the innoculum into each well. 150 μl of mid-logarithmic phase culture is diluted into 5 ml of cell culture medium; 30 μl of this diluted bacterial suspension ($\approx 2 \times 10^5$ cfu) added to each well containing 2×10^4 cells yields an MOI of 10. Rock gently to mix well and place in CO_2 incubator at 37°C for the desired infection time. Infection may also be initiated by brief centrifugation (5 min at 600*g*) allowing bacteria and cells to make contact with each other.
6. After the incubation period, aspirate out the medium and gently wash the coverslips twice with PBS to remove the unbound bacteria.
7. Immerse each coverslip with 300 μl of freshly prepared PLP Sucrose fixative. Rock gently to mix. Place in the 37°C incubator for at least 1 h.

Day 2 and 3—Immunostaining

8. Wash the coverslips three times with PBS. Block with 4% serum (Goat/Donkey) for 45 min on bench-top.
9. Wash three times with PBS. Stain with primary antibody (anti-*Yersinia*).
10. Wash three times with 1 × PBS. Stain with secondary antibody (Texas Red/Rhodamine).
11. Wash three times with PBS. Permeabilize the cells by carefully dipping the coverslips in ice-cold 100% methanol for exactly 10 seconds. Wash three times with PBS.
12. Stain with anti-*Yersinia* primary antibody and against any other protein of interest using antibodies raised in different species than the anti-*Yersinia* antibody (if assessing recruitment to the nascent phagosome). Wash three times with PBS.

13. Stain with secondary antibodies linked to Cascade Blue/Coumarin and FITC/HRP respectively. Wash three times with 1 × PBS. Carry out signal amplification if warranted.
14. Store coverslips under 3% Glycerol in PBS in at 4°C in the dark (wrap the 24-well dish in aluminum foil) or mount in Fluoroguard® Antifade Reagent, seal coverslip with nail polish and observe immediately.

◆◆◆◆◆◆ GENTAMICIN PROTECTION ASSAY

Since this reliable and simple technique has been thoroughly described in the literature (Mandell, 1973; Elsinghorst, 1994) we will provide only a succinct account of the assay. The standard gentamicin protection assay involves seeding cultured cells in a 24-well culture dish at 90–95% confluency. Setting up samples in triplicate at least is necessary to obtain statistically reliable data. An overnight pre-culture of bacteria is used to inoculate multiple dilutions in fresh medium, as described above, to obtain bacteria in the mid-logarithmic phase of growth. The cultured cell monolayer is washed with buffer or growth medium and an aliquot of mid-logarithmic phase bacteria is added (according to a pre-selected MOI) in order to initiate the infection/uptake. A sample is removed for quantitative estimation of the number of bacteria introduced into the assay by viable plate count. After the designated bacteria incubation time, the monolayer is washed with buffer or culture medium to remove external, non-adherent bacteria. Next, the monolayer is incubated in culture medium containing gentamicin (50–100 $\mu g\,ml^{-1}$) to kill off adherent, uninternalized bacteria. At the end of the gentamicin incubation period, the antibiotic is washed off using buffer or culture medium and the cells are lysed with a detergent solution (e.g. 0.1% Triton X-100). Internalized bacteria are discharged from the lysed cells and can now be estimated by plating various dilutions for viable count on appropriate media. Results of the assay are expressed in terms of the percentage of bacteria introduced initially that are protected from subsequent antibiotic killing, by virtue of uptake by cultured cells. Without doubt, it is essential to include the appropriate control bacterial strains—invasive and non-invasive—in each assay to validate its integrity and reliability.

Figure 9.2 displays an example for the use of the gentamicin protection assay to examine the role of focal adhesion kinase (FAK) in invasin-promoted bacterial uptake. Overexpression of FAK stimulates uptake, whereas overexpression of a C-terminal fragment of FAK, FRNK, inhibits uptake, suggesting that it exerts a dominant negative action. Two kinase-defective mutant forms (D562A and K454R) apparently do not influence uptake. However, the autophosphorylation defective mutant Y397F FAK depresses bacterial uptake, suggesting a requirement for tyrosine phosphorylated FAK in invasin-mediated uptake of *Yersinia pseudotuberculosis* (Alrutz and Isberg, 1998).

The gentamicin protection assay is highly sensitive, in that it can detect the internalization of small numbers of bacteria. However, it has some

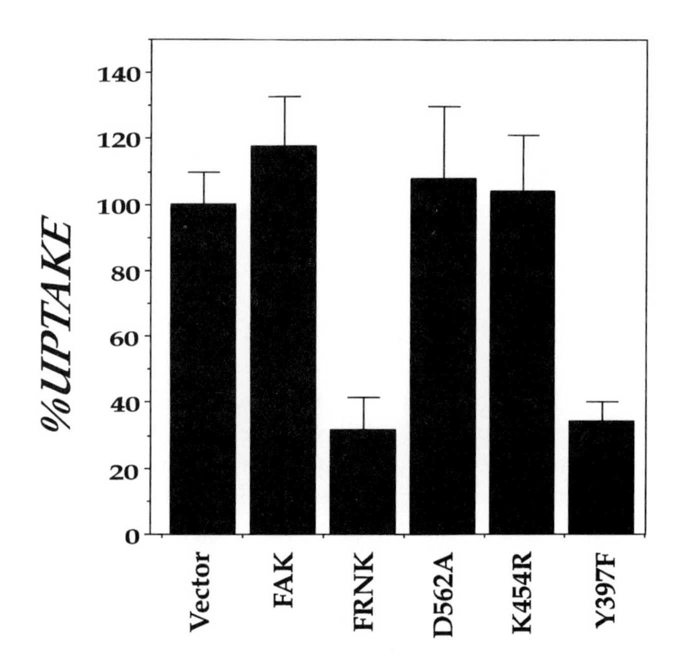

Figure 9.2. Effect of various isoforms of focal adhesion kinase (FAK) on bacterial uptake. Bacterial internalization was determined for Chick Embryo Fibroblasts (CEF) transfected with vector, FAK, FRNK (C-terminal fragment), kinase-defective D562A FAK, kinase-defective K454R FAK and autophosphorylation defective Y397F FAK. Error bars indicate SE for triplicate samples. %UPTAKE refers to amount of internalization observed relative to the vector transfected control estimated by gentamicin protection assay (adapted from Alrutz, 1999).

major shortcomings. First, under certain conditions internalized bacteria may be susceptible to gentamicin action. This may result from experimental treatments that compromise mammalian cell integrity or the innate permeability of certain cell lines. Second, the extracellular medium may not be optimal for antibiotic action resulting in inefficient and incomplete killing of uninternalized bacteria. Finally, incipient and incompletely formed phagosomes can render bacteria vulnerable to gentamicin action. It is often of interest to quantitate the number of nascent phagosomes that form under particular experimental conditions. This susceptibility of bacteria can skew the uptake efficiency to an inaccurate lower number. The differential immunostaining assay is able to account for these dynamic stages of uptake and is therefore able to provide a higher fidelity picture of the uptake process, especially since it facilitates quantitation of nascent phagosomes.

Suggested protocol

Day 1

1. Using a diluted suspension of cultured cells without transfection or after transfection with plasmid DNA, seed 1×10^5 cells per well (in

triplicate for each sample) of a 24-well tissue culture dish in 1 ml volume. Incubate overnight to allow a near-confluent monolayer to form.
2. Start a pre-culture of *Yersinia pseudotuberculosis* Y137 (Inv^+ Plasmid cured) using a single colony from a fresh plate (less than 2 weeks old) in 3–5 ml LB medium. Grow overnight with aeration at 28°C.

Day 2—Infection

3. As described in the previous protocol, set up multiple dilutions of the overnight pre-culture to obtain bacteria in the mid-logarithmic growth phase. Depending on the MOI desired, prepare the innoculum by diluting the requisite amount of bacterial suspension in pre-warmed cell culture medium. Mix well.
4. Remove the culture medium and immerse the monolayer in each well with 1 ml of the bacterial innoculum to initiate the infection/uptake. Rock the dish to mix well and place in CO_2 incubator at 37°C for 1 h. Infection may also be initiated by brief centrifugation (5 min at 600*g*) allowing bacteria and cells to contact each other.
5. Set aside a small amount of the innoculum and carry out a viable plate count to enumerate the bacteria introduced into the assay system. This can be done during the infection period.
6. At the end of the incubation period, aspirate out the medium and gently wash the coverslips three times with PBS to remove the unbound and uningested bound bacteria.
7. Flood each well with 500 μl of pre-warmed culture medium containing 50 $\mu g\, ml^{-1}$ Gentamicin. Rock gently to mix. Place in the 37°C incubator for at least 1 h to kill any remaining external bacteria.
8. Wash the wells three times with PBS and add 500 μl of 0.1% Triton X-100 to each well to lyse the mammalian cells. Mix well by gently pipetting the solution up and down 3–4 times (use a fresh micropipette tip for each well!). Incubate for 5 min with gentle shaking to allow for complete lysis.
9. Enumerate the internalized bacteria liberated by detergent-mediated cell lysis by a viable plate count. Multiple serial dilutions may be necessary to obtain measurable counts and usually range from undiluted to 10^{-6}.
10. Calculate the results as a percentage: (number of cfu recovered after gentamicin killing/total cfu in the initial innoculum) $\times$ 100. Use data from the triplicate samples for statistical correlation.

References

Alrutz, M. A. (1999). *Identification of factors required for invasin-mediated uptake.* PhD thesis, Department of Molecular Biology and Microbiology, Tufts University, Boston, MA.

Alrutz, M. A. and Isberg, R. R. (1998). Involvement of focal adhesion kinase in invasin-mediated uptake. *Proc. Natl Acad. Sci. USA* **95**, 13658–13663.

Alrutz, M. A., Srivastava, A., Wong, K., D'souza-Schorey, C., Tang, M., Ch'ng, L.-E., Snapper, S. B. and Isberg, R. R. (2001). Efficient uptake of *Yersinia pseudotuberculosis* via integrin receptors involves a Rac1-Arp2/3 pathway that bypasses N-WASp function. *Mol. Micro.* (in press).

Elsinghorst, E. A. (1994). Measurement of invasion by gentamicin resistance. In *Bacterial Pathogenesis* (V. L., Clark and P. M., Bavoil, eds), Academic Press, 667–682.

Falkow, S., Isberg, R. R. and Portnoy, D. A. (1992). The interaction of bacteria with mammalian cells. *Annu. Rev. Cell. Biol.* **8**, 333–363.

Finlay, B. B. and Falkow, S. (1997). Common themes in microbial pathogenicity revisited. *Microbiol. Mol. Biol. Rev.* **61**, 136–139.

Harlow, E. and Lane, D. (1988). *Antibodies: a Lab Manual.* Cold Spring Harbor Laboratory Press, Cold Spring Harbor, NY.

Isberg, R. R., Voorhis, D. L. and Falkow, S. (1987). Identification of invasin: a protein that allows enteric bacteria to penetrate cultured mammalian cells. *Cell* **50**, 769–778.

Mandell, G. L. (1973). Interaction of intraleukocytic bacteria and antibiotics. *J. Clin. Invest.* **52**, 1673–1679.

Swanson, M. S. and Isberg, R. R., (1995). Association of *Legionella pneumophila* with the macrophage endoplasmic reticulum. *Infect. Immun.* **63**, 3609–3620.

10 Membranolytic Toxins

F Gisou van der Goot, Marc Fivaz and Laurence Abrami
Dept. Biochemistry, University of Geneva, 30 quai E. Ansermet, 1211 Geneva 4, Switzerland

◆◆◆

CONTENTS

List of abbreviations

BHK	Baby Hamster Kidney
CHO	Chinese hamster ovary cells
DiS-$C_3(5)$	Dye 3, 3′-dipropyl-thiodicarbocyanine iodide
GMEM	Glasgow minimal essential medium
GPI	Glycosylphosphatidyl inositol
IM	Incubation medium
LM	Loading medium
PBFI	K^+-binding benzofuran isophthalate dye
PBS	Phosphate saline buffer
SPQ	6-methoxy-N-(3-sulfopropyl)quinolinium

◆◆◆◆◆◆ INTRODUCTION

Membranolytic toxins (van der Goot, 2001) are produced by a wide range of bacteria both Gram-negative (*Aeromonas* sp., Abrami *et al.*, 2000) and Gram-positive (*Staphylococcus aureus*, Bhakdi and Tranum-Jensen, 1991), both invading (*Listeria monocytogenes*, Scheffer *et al.*, 1988) and extracellular (*Staphylococcus aureus*). Although their exact role in infection is largely unknown, they appear to contribute to spreading of the bacteria. In the case of certain invading bacteria, it has been shown that they mediate release of the micro-organism from the phagocytic vacuole into the cytoplasm of the host cell (Cossart and Lecuit, 1998). In this chapter,

METHODS IN MICROBIOLOGY, VOLUME 31
ISBN 0-12-521531-2

methods for studying the mode of action of pore-forming toxins on mammalian cells are discussed. Although lipases are an important class of membranolytic bacterial virulence factors, they will not be covered by this review.

We will first describe *in vitro* systems to study pore-formation and then focus on methods to study the interaction of pore-forming toxins with target mammalian cells. Most protocols are described for aerolysin from *Aeromonas hydrophila*, based on the expertise of the authors, but are applicable to the toxin of interest.

◆◆◆◆◆◆ MEMBRANE PERMEABILIZATION

Upon identification of a toxic activity that is thought to be membranolytic, three assays are generally performed to show that the toxin has a capacity to form pores: (1) hemolysis; (2) release of dyes or ions from artificial liposomes; and (3) planar lipid bilayers. Only the last method enables one to discriminate between a pore-forming toxin and a lipase.

Hemolytic activity

Time-dependent hemolysis can easily be monitored by following the loss of turbidity of a red blood cell sample as cells are being lysed. Erythrocytes of the desired species must be washed three times1 with PBS and resuspended in this buffer at a concentration of 0.4% (0.4 ml of packed cells per 100 ml of PBS). The turbidity of this sample can then be followed by measuring the absorption at 600 nm using an absorption photometer equipped with thermostated cuvette holder. Kinetics of lysis should be temperature and toxin concentration dependent.

Hemolysis can also be conveniently measured using a 96-well plate. This method is particularly useful to compare wild types to mutants, or to study the effects of the buffer or divalent ions. 100 μl of toxin in PBS is added to the first well of a row. A double dilution down the plate is performed. 100 μl of erythrocytes at a concentration of 0.8% in PBS is then added to each well and the plate is incubated at 37°C. The number of wells lysed per row can be monitored as a function of time.

Liposome release assay

The principle of this method is to encapsulate a dye and/or ions within vesicles, this dye being absent from the outside medium, and to monitor the release of the dye or the ions upon perturbation of the lipid bilayer by the toxin of interest. It is important to note that although this method is extremely useful, it does not provide absolute proof of pore formation and misinterpretation of the data is not uncommon. Since the volume encapsulated in liposomes is very small, transitory destabilization of the membrane by the addition of a membrane interacting protein might empty the liposomes of its dye in the absence of bona fide pore formation. Inactive mutants, if available, should be used as controls.

Many methods exist to make liposomes of different sizes, these include freeze thawing, extrusion and sonication. For all methods, the desired mixture of lipids is prepared in a glass flask from stock solutions in chloroform. The solvent is then evaporated, the desired buffer is added and the flask is vortexed. Briefly, freeze-thawing then consists in submitting the lipid suspension to 3–5 cycles of snap freezing in liquid nitrogen and then thawing at room temperature or 37°C. This method leads to rather large vesicles (>100 nm), inhomogeneous in size. Small liposomes can be obtained by sonicating the lipid suspension using a tip sonicator, on ice, until the solution is clarified. Intermediate size liposomes can be obtained by extrusion. This method consists in repeatedly passing the lipid suspension through polycarbonate filters with calibrated pore sizes by applying pressure (using nitrogen). This method requires the use of an extruder.

We here describe in detail the reverse phase method (Papadopoulos *et al.*, 1990) that might seem tedious and old-fashioned but which provides liposomes that are large and homogeneous in size. The choice of lipids to be used is vast. If one is not searching for lipid specificity, egg yolk lipids which contain a mixture of different chain lengths and saturation is very useful.

In a 25 ml round glass flask, mix 2 ml of egg phosphatidyl choline (from a 25 mg ml^{-1} stock solution in chloroform) and 0.5 ml egg phosphatidic acid (from 2.5 mg ml^{-1} stock). The presence of at least a low percentage of charged lipids is important to prevent the formation of multilamelar, onion like, liposomes thanks to the charge repulsion between the bilayers. Dry the sample using a rotary evaporator. Resuspend the lipids in 2 ml H_2O saturated ether and then add 0.5 ml of buffer containing 100 mM KCL, 10 mM MES, pH 7.4 and the dye of interest, such as, for example, the chloride sensitive dye: 6-methoxy-N-(3-sulfopropyl) quinolinium (SPQ, Molecular Probes) (1.5 mg ml^{-1}). Sonicate the mixture with a tip sonicator on ice for 2 min and submit it to slow evaporation under controlled vacuum in order to slowly progress through the phase reversion. The flask should be immerged in a water bath at approximately 15°C; rotation should be slow and the vacuum should not be lower than 600 bars until the mixture has reached a gel phase that sticks to the bottom of the flask. A controlled vacuum can be obtained by flushing nitrogen into the flask that is under vacuum. When the gel phase has been passed and the sample has become fluid again, add 1 ml buffer with dye and evaporate sample for another 25 min increasing the vacuum in a stepwise manner. In order to calibrate the vesicle size, filter the sample using polycarbonate filters with calibrated pores (Vecsey-Semjen *et al.*, 1996). Remove the external SPQ by gel filtration.

In order to perform the chloride efflux experiments, 5–20 μl of liposomes are diluted into 3 ml of the chloride-free buffer containing 100 mM KNO_3, 10 mM MES, pH 7.4. in a quartz cuvette containing a magnetic stirrer. The sample is exited at 350 nm and emission measured at 422 nm. The change in fluorescence induced by the addition of toxin is measured as a function of time (Vecsey-Semjen *et al.*, 1996).

Planar lipid bilayers

The principle of this method is to create two aqueous compartments separated by a microscopic hole across which is spanned a single lipid bilayer as described in detail by Menestrina and co-workers (Dalla Serra and Menestrina, 2000). After construction of the lipid bilayer, the toxin is added to one of the compartments and the electrical current flowing from one side of the membrane to the other is measured using two electrodes. Upon formation of a channel, an increase of current is observed. This method enables the analysis of single channel events as well as the kinetics of multiple channel formation.

Plasma membrane perforation on nucleated cells

We will describe two techniques that enable one to measure channel formation in the plasma membrane of nucleated cells in culture, the ease of which depends on whether cells are adherent or grow in suspension.

Note that these methods will only detect the formation of rather large pores such as those formed by aerolysin (10–30 Å) or cholesterol binding toxins such as streptolysin O (150 Å) but not very small channels such as those formed by *Helicobacter pylori* vacA. Detection of vacA channels required more sensitive electrophysiological methods such as patch clamp (Tombola *et al.*, 1999).

Adherent cells: kinetics of potassium efflux

Cells are grown to confluency in 6-well plates in order to perform each measurement in triplicate. Cells are first incubated for 30 min at 37°C in incubation medium (IM) containing Glasgow minimal essential medium (GMEM, Sigma, St Louis) buffered with HEPES, pH 7.4. and then for various times1 in the presence of toxin. Cells are then washed rapidly five times1 with ice-cold K^+ and Na^+ free choline medium containing citric acid 5 mM, glucose 5.6 mM, NH_4Cl 10 mM, $MgCl_2$ 0.8 mM, $CaCl_2$ 1.5 mM, H_3PO_4 5 mM, pH 7.4 and solubilized on ice for 15 min with 0.5% Triton X-100 (Ultrapure, Pierce) (0.5 ml per well). Cell debris are further homogenized by passage through a blue tip. After appropriate dilution, the detergent lysat is submitted to a flame emission analysis at 766.5 nm. Such a measurement of potassium efflux upon addition of aerolysin to Baby Hamster Kidney (BHK) cells is illustrated in Figure 10.1.

Non-adherent cells: kinetics of pore formation probed by fluorescence techniques

Determination of K^+ release by flame emission photometry requires rapid washing steps, which are tedious to do on non-adherent cells. Cells growing in suspension are more amenable to quantitative fluorescence studies. Here, we describe three fluorescence assays that allow kinetic measurements of pore formation. Protocols are described to follow the

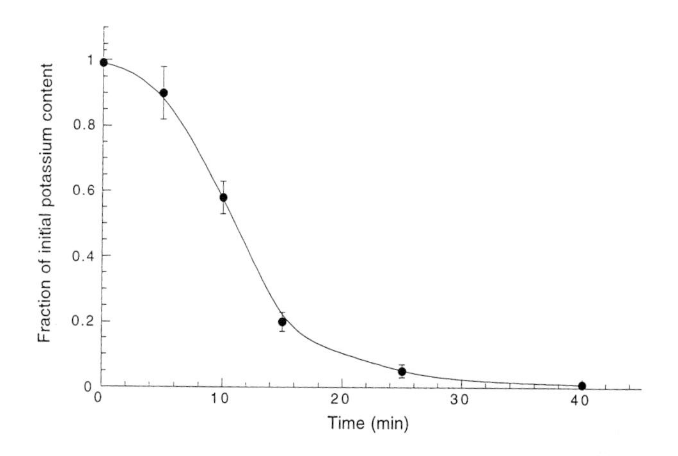

Figure 10.1. Toxin-induced potassium efflux from BHK cells. Cells were incubated with 0.4 nM proaerolysin at 37°C for various times1 and the potassium contents were determined by flame photometry. Experiments were done in triplicate and the standard deviations were calculated.

effects of aerolysin on HL-60 granulocytes. These experiments require a fluorimeter equipped with a thermostated cuvette holder (37°C) and a stirring device.

Measurement of intracellular K^+ using the fluorescent indicator PBFI

Variations of intracellular K^+ concentrations upon channel formation in the plasma membrane can be determined using the K^+-sensitive benzofuran isophthalate dye, PBFI. The cell-permeant acetoxymethyl ester (AM) form of the dye is loaded into cells where it is esterified and binds to K^+ with an *in vivo* K_d of about 100 mM resulting in an increase of dye fluorescence. PBFI-AM and pluronic acid can be purchased from Molecular Probes (Eugene, OR). PBFI stock solutions of 1 mM in DMSO are kept at −20°C, protected from light.

HL-60 granulocytes are washed once and resuspended to a final density of 2×10^7 cells ml^{-1} in loading medium (LM), containing 20 mM Hepes pH 7.4, 5.6 mM glucose, 143 mM NaCl, 6 mM KCl, 1 mM $MgSO_4$, 1 mM $CaCl_2$, containing 0.5% BSA and 0.25 mM sulfinpyrazone, an inhibitor of organic anion transporters, which prevents to some extent release of the dye, once loaded into the cell. PBFI-AM pre-mixed with pluronic acid in DMSO is added to cells to final concentrations of 5 μM PBFI-AM and 0.02% pluronic acid. The dye is allowed to incorporate for 30 min at 37°C plus another 30 min at room temperature. Cells are washed once and resuspended in LM to a final density of 2×10^6 cells ml^{-1}.

The fluorescence of PBFI-loaded cells is acquired in a time-based mode, with excitation and emission wavelengths of 343 nm (5 nm slit) and 460 nm (5 nm slit) respectively. Upon addition of aerolysin, PBFI undergoes a decrease in fluorescence intensity in a dose and time-dependent manner (Krause *et al.*, 1998) (Figure 10.2a). Variations of $[K^+]_i$ are expressed as a fraction of the maximal PBFI intensity. Interestingly,

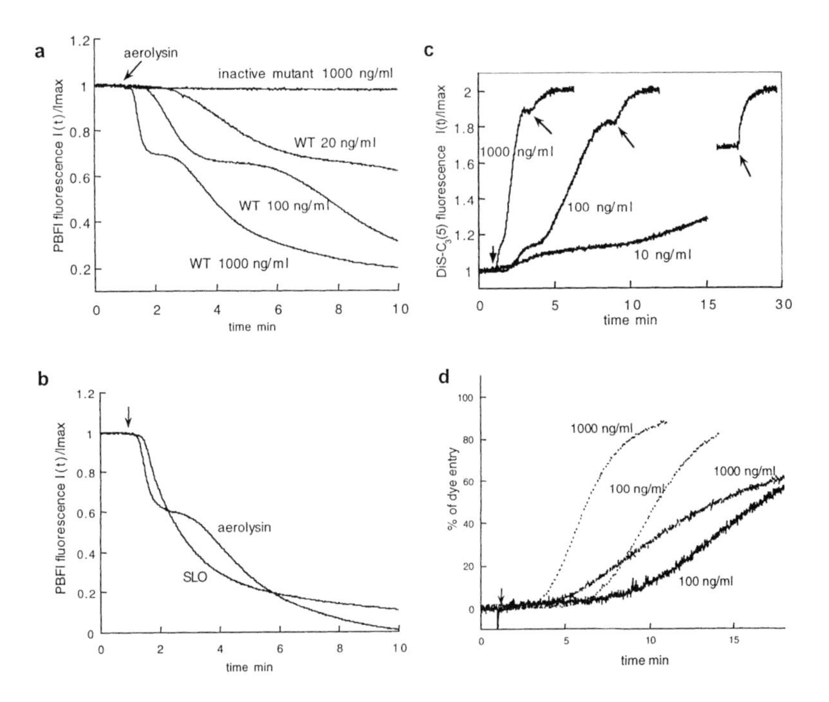

Figure 10.2. (a,b) Kinetics of toxin-induced K^+ release, monitored by the K^+-sensitive dye PBFI. Trypsin-activated aerolysin was added to PBFI-loaded HL-60 granulocytes, at the time indicated by an arrow. Wild type (WT) aerolysin induced a time- and concentration-dependent decrease in PBFI fluorescence, indicating release of intracellular K^+. An inactive mutant of aerolysin, that binds to the cell surface, but is unable to form a pore, had no effect on PBFI fluorescence (a). (b) Kinetics of K^+ efflux obtained with SLO ($30\,ng\,ml^{-1}$) and aerolysin ($1000\,ng\,ml^{-1}$) are compared. Note that kinetics of K^+ efflux induced by WT aerolysin are clearly biphasic, in marked contrast to that caused by streptolysin O, which shows a monophasic behavior. (c) Cell depolarization measured by the voltage-sensitive dye DiS-C_3(5). Increasing concentrations of trypsin-activated aerolysin were added to HL-60 cells, that had been previously loaded with DiS-C_3(5). A time- and dose-dependent biphasic increase of the dye fluorescence occurs, signifying a decrease in the membrane potential. 1% Triton X-100 was added at the end of each trace in order to obtain maximal depolarization. (d) Aerolysin induced dye influxes. Trypsin-activated aerolysin induced a dose-dependent influx of ethidium bromide and ethidium homodimer-1. Aerolysin was added at the time indicated by an arrow (figures reproduced with permission from Krause *et al.*, 1998).

depolarization of HL-60 cells by aerolysin appears to be a biphasic process (Figure 10.2a,b). The origin of this stepwise decrease of the membrane potential remains unknown. We can, however, exclude that this phenomenon is inherent to the K^+ indicator, since streptolysin O (SLO), another pore-forming toxin, caused a monophasic depolarization of HL-60 cells (Figure 10.2b).

Plasma membrane depolarization

Permeabilization of the plasma membrane by a pore-forming toxin generally leads to membrane depolarization, which can be followed using the membrane voltage-sensitive fluorescent cyanine dye 3,3′-dipropyl-thiodicarbocyanine iodide (DiS-C_3(5)) (Waggoner, 1979). DiS-C_3(5) is taken up in a membrane potential-dependent manner. Dye accumulation results in fluorescence quenching. Subsequent depolarization of the cell causes release of the dye and concomitant recovery of fluorescence. Stock solutions of DiS-C_3(5) (Molecular Probes) are prepared in DMSO (100 μM) and kept at −20°C, protected from light. The dye is excited at 625 nm (10 nm slit) and emission is recorded at 670 nm (10 nm slit).

HL-60 granulocytes are washed once and resuspended in LM, to a final density of 3×10^6 cells ml^{-1}. DiS-C_3(5) is added to a final concentration of 200 nM. Membrane incorporation of the dye (corresponding to a decrease in fluorescence intensity) is monitored spectrofluorimetrically and takes about 3 to 5 min. When a steady-state fluorescence level is reached, the toxin is added. Maximal depolarization is obtained at the end of each experiment by adding pre-mixed valinomycin and nigericin, to final concentrations of 2 and 5 μM respectively, to abolish the membrane potential (Kasner and Ganz, 1992). Fluorescent traces are expressed as the ratio $I(t)/I_{max}$, i.e. fluorescence intensity at a given time over maximal fluorescence intensity. As illustrated Figure 10.2c, aerolysin induces a dose and time-dependent depolarization of the cell (Krause *et al.*, 1998).

Measurements of Ethidium bromide and Ethidium homodimer-1 entry

Channel formation at the plasma membrane can also be monitored by looking at the kinetics of entry of cell-impermeant, nucleic acid-binding dyes, such as Ethidium bromide (394 Da) or Ethidium homodimer-1 (857 Da). Both dyes enter cells at breaches in the plasma membranes and undergo a drastic enhancement of fluorescence (20-fold and 40-fold for EtBr and EthD-1 respectively), upon binding to nucleic acids. One important concern when interpreting the data of such an experiment, is to make sure that the observed kinetics of entry reflect translocation of the dye through the toxin pore (i.e. is not due to an overall destabilization of the plasma membrane). The use of dyes of different sizes may help confirm specific translocation of the dye through the pore and also provide information as to the size of the channel.

EtBr and EthD-1 (Molecular Probes) are stored in water (10 mg ml^{-1}) and DMSO/water 1 : 4 (2 mM) respectively. All steps were performed at 37°C. HL-60 granulocytes are washed once and resuspended in LM at a final density of 2×10^7 cells ml^{-1}. EtBr or EthD-1 are added to a final concentration of 100 μM and 6 nM respectively. EtBr and EthD-1 are excited at 340 and 500 nm respectively, and emission is measured at 600 nm. Fluorescent traces are normalized to maximal fluorescence intensity obtained by the addition of 1% Triton X-100. As shown in Figure 10.2d, aerolysin induced a dose-and time-dependent entry of EtBr as well as EthD-1. Faster kinetics of entry are measured for the smaller dye, indicating that a sieving mechanism is taking place and suggest that the dyes enter via the aerolysin channel.

◆◆◆◆◆◆ INVOLVEMENT OF SPECIFIC RECEPTORS

Pore-forming toxins can kill target cells at concentrations in the subnanomolar range. This high sensitivity is generally due to the presence of surface molecules that act as specific receptors. Identified receptors for pore-forming toxins include GPI-anchored proteins (Ricci *et al.*, 2000; Gordon *et al.*, 1999; Abrami *et al.*, 1998; Cooper *et al.*, 1998; Diep *et al.*, 1998; Nelson *et al.*, 1997), sphingomyelin (Yamaji *et al.*, 1998; Lange *et al.*, 1997) and cholesterol (Tweten *et al.*, 2001), i.e. proteins and lipids.

The existence of specific receptors is generally suggested by the observation that very low doses of toxin are required and/or that there is a species or cell type specificity of the toxic activity. Arguments that suggest the existence of a specific protein receptor include the following: binding is saturable, binding of a labeled toxin can be competed by the unlabeled toxin, binding is sensitive to proteolysis (if it is a protein). However, if these criteria are not fulfilled, the existence of a receptor cannot necessarily be ruled out, as recently shown for vacA (Ricci *et al.*, 2000).

Saturation of binding and competition

The easiest manner to quantify binding is to label the toxin of interest with ^{125}I using Iodogen reagent (Pierce) according to the manufacturer's recommendations. Competition experiments can then be performed as follows (Abrami *et al.*, 1998; Escuyer and Collier, 1991). Prepare about ten dishes of cells and incubate them with a mixture of ^{125}I-toxin and unlabelled toxin for 1 h at 4°C, in order to prevent possible internalization of the toxin. This mixture should contain a fixed concentration of ^{125}I-toxin (corresponding preferably to the lowest concentration at which the toxin is effective on the cell type analyzed) and increasing amounts of unlabeled toxin, which can go up to 10^3–10^4 times1 the ^{125}I-toxin. It is important that the labeled and unlabeled toxins are added to the cells simultaneously. Cells should then be extensively washed, scraped from the dish, recovered by centrifugation and radioactivity should be counted. If competition occurs, the number of counts should diminish as the concentration of unlabeled toxin is increased. The sigmoidal competition curve that is thus obtained provides an apparent binding K_d.

To confirm that binding is specific, one should investigate whether binding is saturable. Again five to ten dishes of cells should be prepared and incubated with increasing concentrations of ^{125}I-toxin for 1 h at 4°C. In parallel, dishes should be incubated with the same concentrations of ^{125}I-toxin as well as an excess of unlabeled toxin. In this latter situation, competition occurs and aspecific binding of the toxin, which increases with increasing concentration, can be estimated. The amount of toxin bound per cell should then be plotted as a function of toxin concentration. The contribution of aspecific binding can then be removed for each concentration. The final curve should indicate whether binding is or not saturable.

Sensitivity of binding to proteolysis, to removal of carbohydrates or to PI-PLC

If the receptor for the toxin of interest is thought to be proteinaceous, binding should be sensitive to proteolysis. Binding of ^{125}I-toxin should then be tested on cells pre-treated or not with proteases in the 1–3 mg ml^{-1} range for 30 min at room temperature. Proteases can include trypsin, chymotrypsin, pronase, papain, proteinase K, thermolysin, carboxypeptidase Y, clostripain, elastase and collagenase. Alternatively protease treatment can be performed on cells fixed for 20 min at room temperature with 4% paraformaldehyde and toxin binding can be analyzed by immunofluorescence. Protease sensitivity does not necessarily mean that the toxin binds directly to the protein. Binding could occur via carbohydrate moieties. To address this issue cells can be treated for 30 min at room temperature with glycosidases (1–3 mg ml^{-1}) such as *C. perfringens* neuraminidase, β-N-acetylglucosaminidase, baker's yeast α-glucosidase, jack bean α-mannosidase.

As mentioned, certain pore-forming toxins were found to bind to GPI-anchored proteins. Binding is therefore sensitive to treatment with a phosphatidyl inositol specific phospholipase C. Cells are incubated in IM with 0.2–6 U ml^{-1} of this enzyme for 1 h at 4 or 37°C in the presence of 10 μg ml^{-1} cycloheximide.

Identification of putative protein receptors

Toxin overlay assays

When searching for a toxin receptor, it is worth trying to identify toxin binding proteins using a toxin overlay assay. The principle is to run a membrane protein extract on an SDS gel, to blot it onto a nitrocellulose membrane and to incubate this membrane with the toxin in the hope that the toxin will find its receptor and bind to it. The toxin is then revealed either by autoradiography or using a primary and a secondary antibody. This method is, however, based on the assumption that the receptor has refolded in the blot and/or recovered a conformation compatible with binding. This is, however, not always the case. Therefore the absence of result by overlay does not rule out the existence of a proteinaceous receptor.

A membrane extract is run on an SDS gel. Gels with different concentrations of acrylamide or acrylamide gradient gels should be tried. The gel is then soaked in 50 mM Tris-HCl pH 7.5/20% glycerol twice for 10 min and transferred onto a nitrocellulose membrane using a wet transfer chamber. Transfer is performed for 16–18 h at 100 mM at 4°C in a buffer containing 10 mM $NaHCO_3$, 3 mM Na_2CO_3. Transfer can also be performed using a Genie blotter for 1 h using the manufacturer's instructions. Semi-dry transfers however did not allow receptor identification for aerolysin. Immediately after transfer, the nitrocellulose membrane is incubated for 20–30 min in binding buffer containing 50 mM NaH_2PO_4 pH 7.5, 0.3% Tween 20. Finally the membrane is incubated with the toxin of interest (for example 1–3 nM of aerolysin) in binding buffer for 2 h at room temperature, then washed six times1 for 5 min in binding buffer. The toxin is then revealed either by radiography or by Western blotting.

The result of such an experiment is illustrated Figure 10.3, where the presence of aerolysin receptors was analyzed after fractionation of BHK cells.

Cell surface cross-linking

If overlay experiments fail, toxin interacting proteins can be identified by performing cell surface cross-linking. Incubate cells at 4°C with radio-labeled toxin at a concentration where binding is specific. After extensive washing on ice, incubate cells with a chemical cross-linker. Various cross-linkers should be tried with different size arms, at various concentrations,

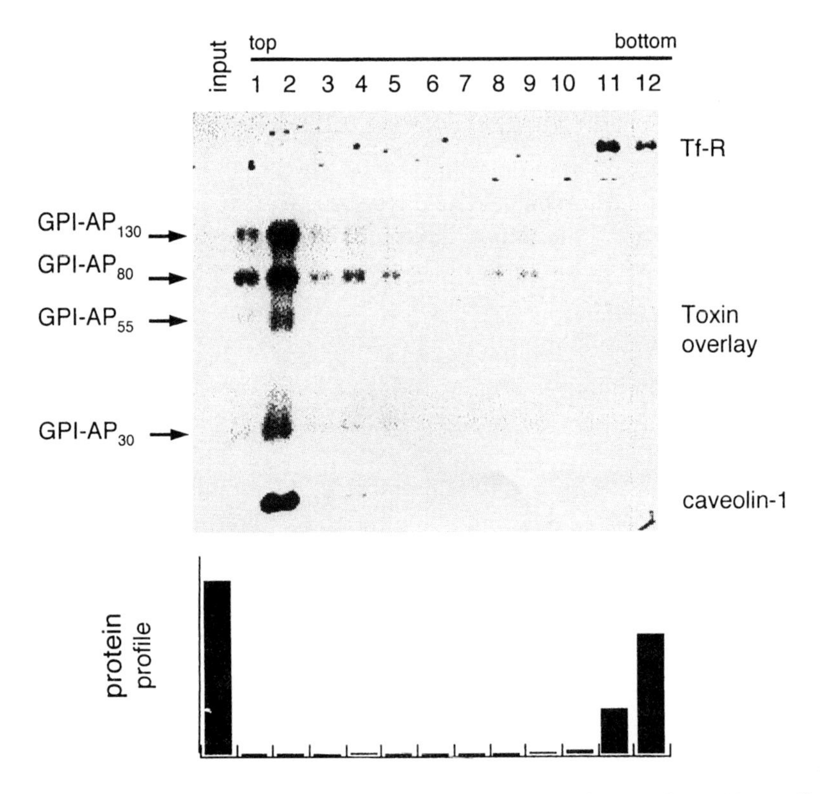

Figure 10.3. Characterization of detergent resistant membranes from BHK cells. BHK cells were extracted in 1% Triton X-100 in the cold and further fractionated on a sucrose gradient (the top of the gradient corresponds to low density fractions). The percentage of proteins contained in each of these sucrose gradient fractions (relative to the input material) is shown at the bottom. The distribution of the transferrin receptor (Tf-R) and that of caveolin-1 was revealed by Western blotting. Aerolysin receptors were revealed by toxin overlay. Four major GPI-anchored receptors can be seen that co-fractionate with caveolin-1 in detergent-insoluble fractions.

for various times1. The use of cleavable (for example thiol cleavable) cross-linkers is useful since it enables the addition of useful controls. This method has been useful to identify the tetanus toxin receptor (Herreros *et al.*, 2000).

Generation of toxin resistant cell lines

To further investigate the binding requirements for a membranolytic toxin, to identify the receptor and/or to confirm its identity, toxin-resistant cell lines can be generated. We describe here how aerolysin-resistant cell lines were generated (Abrami *et al.*, 2001). CHO cells were used for this study since they appear to be functionally haploid for many loci (Adair and Siciliano, 1986).

CHO-K1 cells were maintained in 100 mm diameter dishes containing 10 ml of F-12 medium supplemented with 10% fetal calf serum and 2 mM L-glutamine under standard tissue culture conditions. Cells were mutagenized with ethyl methanesulfonate (400 $\mu g\,ml^{-1}$) at 37°C for 24 h, as described in Esko and Raetz (1978). After incubation in the complete culture medium for 4 days, the mutagenized cells were harvested, re-seeded at 400 colonies per dish and incubated with 0.4 nM proaerolysin for 2 days. Seven days later, the surviving colonies were trypsinized with a filter paper, transferred to 24-well plates and cultured in normal culture medium for several days. Cells were re-seeded and subjected to two other cycles of proaerolysin selection. A total of 314 separated colonies of surviving cells were isolated and purified by limited dilution.

Many of these clones were deficient in toxin binding. By recomplementation of several of the mutants, we could show that GPI biosynthesis was affected proving that GPI anchored proteins are crucial for the high sensitivity of the cells to the toxin. For the identification of the modified genes in the mutant cell lines the readers are referred to the somatic cell genetic work performed, for example, by Kinoshita and co-workers (Watanabe *et al.*, 1998; Kinoshita *et al.*, 1997; Nakamura *et al.*, 1997) or Krieger and co-workers (Guo *et al.*, 1996; Hobbie *et al.*, 1994).

◆◆◆◆◆◆ INVOLVEMENT OF LIPID RAFTS

Most pore-forming toxins require circular polymerization in order to form channels (van der Goot, 2001; Lesieur *et al.*, 1997). Since cells are generally sensitive to very low doses of toxins, it is reasonable to believe that some type of concentration device is available at the target cell surface that enables local concentration of the toxin, a step necessary for encounter between monomers during oligomerization. One possible concentration device would be the association with lipid rafts, which are specialized microdomains of the cell surface that are rich in cholesterol and glycosphingolipids (Brown and London, 1998; Harder and Simons, 1997; Simons and Ikonen, 1997). Interestingly the receptors for a variety of pore-forming toxins are known raft components (for review see Fivaz *et al.*, 1999). Streptolysin O and related toxins require cholesterol for channel formation; lysenin, an earthworm toxin, was shown to bind sphingomyelin and *Vibrio cholera* cytolysin requires both cholesterol and sphingomyelin for channel formation. Aerolysin, *Helicobacter pylori* vacA, *Clostridium septicum* alpha toxin and insecticidal Cry toxins utilize GPI-anchored proteins as cell surface receptors. The association of membranolytic toxins with rafts and the role of rafts in favoring oligomerization has at present only been shown for aerolysin but this property is likely to be common to many pore-forming toxins (Fivaz *et al.*, 1999).

Association with lipid rafts

In order to show association of a toxin with lipid rafts, the following biochemical property of these microdomains can be used: insolubility in non-ionic detergents such as Triton X-100 at 4°C. Because of their high lipid content, these detergent-resistant membranes (DRMs) float to a low density during gradient centrifugation. The method (adapted from Fra *et al.*, 1994; Brown and Rose, 1992) will be described for BHK cells but is applicable to any cell type.

Confluent monolayer of BHK cells (diluted 1/5, 16 h before) are washed three times1 with cold PBS containing 1 mM $CaCl_2$ and 1 mM $MgCl_2$ (PBS 2^+) and then incubated at 4°C with the toxin of interest (for example 0.4 nM proaerolysin or ^{125}I-proaerolysin, Abrami and van der Goot, 1999; Abrami *et al.*, 1998) in IM for 1 h. Cells are then washed three times1 for 5 min with PBS 2^+ at 4°C, further incubated at 37°C in IM for various times1, subsequently washed with ice-cold PBS 2^+, scraped from the dish, and collected by centrifugation at 1500 rpm for 5 min at 4°C. Cells collected from one dish (approx. 2×10^7 cells) are resuspended in 0.5 ml of cold buffer containing 25 mM tris-HCl pH 7.5, 150 mM NaCl, 5 mM EDTA and 1% Triton X-100, with a tablet of Complete, a cocktail of protease inhibitors (Boehringer Mannheim Corp). Membranes are solubilized by rotary shaking at 4°C for 30 min. The detergent-insoluble membranes are obtained either by high-speed centrifugation (30 min, 4°C at 55 000 rpm in TLS 55 Beckman rotor) or purified on a sucrose density gradient as described below. The solubilized cell lysat is adjusted to 40.7% sucrose (in 10 mM tris-HCl, pH 7.4), loaded at the bottom of a SW40 Beckman tube, overlaid with 8 ml of 35% sucrose, topped up with 15% sucrose and centrifuged for 18 h at 35 000 rpm at 4°C. 12 fractions of 1 ml are collected. For each fraction the protein content and the radioactivity are determined. Each fraction is precipitated with 6% trichloroacetic acid in the presence of sodium deoxycholate as a carrier. The same amount of protein (20 μg) of each fraction is analyzed by SDS-PAGE, followed by western blot analysis. As shown in Figure 10.3, the low density fractions containing DRMs correspond to a minor percentage of total cellular proteins, are highly enriched in the caveolar marker caveolin-1 (Harder and Simons, 1997) and are devoid of transferin receptor. These fractions are also highly enriched in the GPI-anchored aerolysin receptors as shown by aerolysin overlay. Using this fractionation method, we could also show that aerolysin prior and after oligomerization was highly enriched in DRMs (Abrami and van der Goot, 1999).

Effect of cholesterol affecting drugs on toxin polymerization

In order to investigate whether rafts play a role in channel formation by pore-forming toxins and more specifically the oligomerization step, the effect of raft disrupting agents can be analyzed. These include cholesterol-removing agents such as methyl-β-cyclodextrin (MCD) or cholesterol binding agents such as filipin or saponin. Saponin or filipin bind to cholesterol and sequester it away from other interactions but do not extract it from the membrane in contrast to MCD.

We will here describe treatments with saponin, filipin and MCD (Abrami and van der Goot, 1999). Cells are washed three times1 with PBS 2^+ and treated with either 0.4% saponin in PBS 2^+ for 1 h at 4°C, 5 μg ml^{-1} filipin for 1 h at 37°C in IM or with 10 mM MCD for 1 h at 37°C in IM (cells must be rocked during this procedure). Under these conditions, MCD removed more than 50% of the total cellular cholesterol from BHK cells, whereas the cholesterol content of saponin-treated cells was the same as that of control cells. It is important to check that the treatment does not affect the binding efficiency of the toxin of interest.

Using these above described procedures we could show that DRM association of aerolysin was not affected by cholesterol extraction from living cells using MCD but that association was abolished after treating cells with saponin. This later treatment led to a redistribution of the receptor bound to toxin all over the plasma membrane and to a dramatic inhibition of the oligomerization kinetics. The above observation led to the hypothesis that raft association favors oligomerization because these microdomains act as concentration platforms.

◆◆◆◆◆◆ INTRACELLULAR EFFECTS

Upon exposure to low doses of pore-forming toxins, cells can remain alive for several hours and for certain toxins, cells even appear to recover (Valeva *et al.*, 2000). Prior to cell death, a number of intracellular effects have recently been documented. These include activation of signaling pathways (Krause *et al.*, 1998; Grimminger *et al.*, 1997, 1991; Bhakdi *et al.*, 1989), apoptosis (Nelson *et al.*, 1999; Jonas *et al.*, 1994) and changes in intracellular morphology and in membrane trafficking. Only this latter aspect will be discussed here. It has been recently observed that vacA (Reyrat *et al.*, 1999), aerolysin (Abrami *et al.*, 1998), *Vibrio cholera* cytolysin (Coelho *et al.*, 2000; Mitra *et al.*, 2000) and *Serratia marcescens* hemolysin (Hertle *et al.*, 1999) lead to the appearance of large vacuoles in the cytoplasm of target mammalian cells. The compartment undergoing vacuolation has only been identified for vacA, which leads to vacuolation of a late endosomal compartment, and for aerolysin, which triggers vacuolation of the endoplasmic reticulum. Vacuoles triggered by *Vibrio cholera* cytolysin and *Serratia marcescens* hemolysin were not identified but were not acidic. Two interpretations are possible: (1) vacuoles do not originate from endocytic compartments; (2) vacuoles do originate from an endocytic compartment that is no longer acidic due to the presence of toxin pores. In the case of vacA, vacuoles are still acidic presumably because the vacA pore are extremely small or do not allow passage of protons. Phase contrast microscopy often reveals changes in cellular morphology such as the appearance of large translucent vacuoles in the cytoplasm, membrane blebing or the presence of apoptotic bodies. To observe living cells by phase contrast microscopy, one can either grow cells on plastic dishes and use a water immersion lens or a long focal distance lens. Alternatively cells can be grown on glass-bottomed microwell dishes (Mattek, Ashland, MA).

To identify the origin of the vacuolating compartment, the distribution of markers for different intracellular organelles should be investigated in toxin treated vs. control cells. We suggest the transferin receptor (Trowbridge *et al.*, 1993), EEA1 (Mu *et al.*, 1995) or 5 min of internalized FITC dextran for early endosomes, the small GTPase rab7 (Chavrier *et al.*, 1990), the lipid lysobisphosphatidic acid (Kobayashi *et al.*, 1998) or the lysosomal glycoprotein lamp1 (Aniento *et al.*, 1993) for late endosomal compartments. Along the biosynthetic pathway, we suggest the following markers: the transmembrane molecular chaperone calnexin for the endoplasmic reticulum (ER) (Wada *et al.*, 1991), the KDEL receptor ERD2 for the intermediate compartment between the ER and the Golgi as well as for the cis-Golgi (Griffiths *et al.*, 1994; Lewis and Pelham, 1992), mannosidase II for the cis-medial-Golgi (Narula *et al.*, 1992), C6-NBD-ceramide (Lipsky and Pagano, 1985) for the Golgi and TGN38 for the trans-Golgi-network.

◆◆◆◆◆◆ CONCLUSION

The protocols described in this chapter will enable investigators to launch studies on the interaction of novel pore-forming toxins with target mammalian cells. This area of research is, however, still in its infancy. Many receptors remain to be identified and the mechanisms by which pore-forming toxins can trigger signaling cascades and vacuolation of specific compartments still remain a mystery. Fascinating structural aspects of these proteins, such as their ability to exist in a soluble and later in a transmembrane configuration have not been addressed here and the reader is referred to a recent book containing reviews on the subject (van der Goot, 2001).

References

Abrami, L. and van der Goot, F. G. (1999). Plasma membrane microdomains act as concentration platforms to facilitate intoxication by aerolysin. *J. Cell Biol.* **147**, 175–184.

Abrami, L., Fivaz, M. and van der Goot, F. G. (2000). Adventures of a pore-forming toxin at the target cell surface. *Trends Microbiol.* **8**, 168–172.

Abrami, L., Fivaz, M., Glauser, P.-E., Parton, R. G. and van der Goot, F. G. (1998). A pore-forming toxin interact with a GPI-anchored protein and causes vacuolation of the endoplasmic reticulum. *J. Cell Biol.* **140**, 525–540.

Abrami, L., Fivaz, M., Kobayashi, T., Kinoshita, T., Parton, R. G. and van der Goot, F. G. (2001). Cross-talk between caveolae and glycosylphosphatidylinositol-rich domains. *J. Bio. Chem.* **276**, 30729–30736.

Adair, G. M. and Siciliano, M. J. (1986). Functional hemizygosity for the MDH2 locus in Chinese hamster ovary cells. *Somat. Cell Mol. Genet.* **12**, 111–119.

Aniento, F., Emans, N., Griffiths, G. and Gruenberg, J. (1993). Cytoplasmic dyne-independent vesicular transport from early to late endosomes. *J. Cell Biol.* **123**, 1373–1387.

Bhakdi, S. and Tranum-Jensen, J. (1991). Alpha-toxin of *Staphylococcus aureus*. *Microbiol. Rev.* **55**, 733–751.

Bhakdi, S., Muhly, M., Korom, S. and Hugo, F. (1989). Release of interleukin-1beta associated with potent cytocidal action of staphylococcal alpha-toxin on human monocytes. *Infect. Immun.* **57**, 3512–3519.

Brown, D. A. and Rose, J. K. (1992). Sorting of GPI-anchored proteins to glycolipid-enriched membrane subdomains during transport to the apical cell surface. *Cell* **68**, 533–544.

Brown, D. A. and London, E. (1998). Functions of lipid rafts in biological membranes. *Annu. Rev. Cell Dev. Biol.*. **14**, 111–136.

Chavrier, P., Parton, R. G., Hauri, H. P., Simons, K. and Zerial, M. (1990). Localization of low molecular weight GTP binding proteins to exocytic and endocytic compartments. *Cell* **62**, 317–329.

Coelho, A., Andrade, J. R., Vicente, A. C. and Dirita, V. J. (2000). Cytotoxic cell vacuolating activity from Vibrio cholerae hemolysin. *Infect. Immun.* **68**, 1700–1705.

Cooper, M. A., Carroll, J., Travis, E. R., Williams, D. H. and Ellar, D. J. (1998). Bacillus thuringiensis Cry1Ac toxin interaction with Manduca sexta aminopeptidase N in a model membrane environment. *Biochem. J.* **333**, 677–683.

Cossart, P. and Lecuit, M. (1998). Interactions of Listeria monocytogenes with mammalian cells during entry and actin-based movement: bacterial factors, cellular ligands and signaling. *EMBO J.* **17**, 3797–3806.

Dalla Serra, M. and Menestrina, G. (2000). Characterization of molecular properties of pore-forming toxins with planar lipid bilayers. *Meth. Mol Biol.* **145**, 171–188.

Diep, D. B., Nelson, K. L., Raja, S. M., McMaster, R. W. and Buckley, J. T. (1998). Glycosylphosphatidylinositol anchors of membrane glycoproteins are binding determinants for the channel-forming toxin Aerolysin. *J. Biol. Chem.* **273**, 2355–2360.

Escuyer, V. and Collier, R. J. (1991). Anthrax protective antigen interacts with a specific receptor on the surface of CHO-K1 cells. *Infect. Immun.* **59**, 3381–3386.

Esko, J. D. and Raetz, C. R. (1978). Replica plating and in situ enzymatic assay of animal cell colonies established on filter paper. *Proc. Natl Acad. Sci. USA* **75**, 1190–1193.

Fivaz, M., Abrami, L. and van der Goot, F. G. (1999). Landing on lipid rafts. *Trends Cell Biol.* **9**, 212–213.

Fra, A. M., Williamson, E., Simons, K. and Parton, R. G. (1994). Detergent-insoluble glycolipid microdomains in lymphocytes in the absence of caveolae. *J. Biol. Chem.* **269**, 30745–30748.

Gordon, V. M., Nelson, K. L., Buckley, J. T., Stevens, V. L., Tweten, R. K., Elwood, P. C. and Leppla, S.H. (1999). Clostridium septicum alpha toxin uses glycosylphosphatidylinositol-anchored protein receptors. *J. Biol. Chem.* **274**, 27274–27280.

Griffiths, G., Ericsson, M., Krijnse, L. J., Nilsson, T., Goud, B. and Soling, H. D., *et al.* (1994). Localization of the Lys, Asp, Glu, Leu tetrapeptide receptor to the Golgi complex and the intermediate compartment in mammalian cells. *J. Cell Biol.* **127**, 1557–1574.

Grimminger, F., Sibelius, U., Bhakdi, S., Suttorp, N. and Seeger, W. (1991). Escherichia coli hemolysin is a potent inductor of phosphoinositide hydrolysis and related metabolic responses in human neutrophils. *J. Clin. Invest.* **88**, 1531–1539.

Grimminger, F., Rose, F., Sibelius, U., Meinhardt, M., Potzsch, B. and Spriestersbach, R., *et al.* (1997). Human endothelial cell activation and mediator release in response to the bacterial exotoxins Escherichia coli hemolysin and staphylococcal alpha-toxin. *J. Immunol.* **159**, 1909–1916.

Guo, Q., Penman, M., Trigatti, B. L. and Krieger, M. (1996). A single point mutation in epsilon-COP results in temperature-sensitive, lethal defects in membrane transport in a Chinese hamster ovary cell mutant. *J. Biol. Chem.* **271**, 11191–11196.

Harder, T. and Simons, K. (1997). Caveolae, DIGs, and the dynamics of srhingo-lipidcholesterol microdomains. *Curr. Opin. Cell Biol.* **9**, 534–542.

Herreros, J., Lalli, G., Montecucco, C. and Schiavo, G. (2000). Tetanus toxin fragment C binds to a protein present in neuronal cell lines and motoneurons. *J. Neurochem.* **74**, 1941–1950.

Hertle, R., Hilger, M., Weingardt-Kocher, S. and Walev, I. (1999). Cytotoxic action of Serratia marcescens hemolysin on human epithelial cells. *Infect. Immun.* **67**, 817–825.

Hobbie, L., Fisher, A. S., Lee, S., Flint, A. and Krieger, M. (1994). Isolation of three classes of conditional lethal Chinese hamster ovary cell mutants with temperature-dependent defects in low density lipoprotein receptor stability and intracellular membrane transport. *J. Biol. Chem.* **269**, 20958–20970.

Jonas, D., Walev, I., Berger, T., Liebetrau, M., Palmer, M. and Bhakdi, S. (1994). Novel path to apoptosis: small transmembrane pores created by staphylococcal alpha-toxin in T lymphocytes evoke internucleosomal DNA degradation. *Infect. Immun.* **62**, 1304–1312.

Kasner, S.E. and Ganz, M.B. (1992). Regulation of intracellular potassium in mesangial cells: a fluorescence analysis using the dye, PBFI. *Amer. J. Physiol.* **262**, F462–F467.

Kinoshita, T., Ohishi, K. and Takeda, J. (1997). GPI-anchor synthesis in mammalian cells: genes, their products, and a deficiency. *J. Biochem.* **122**, 251–257.

Kobayashi, T., Stang, E., Fang, K. S., de Moerloose, P., Parton, R. G. and Gruenberg, J. (1998). A lipid associated with the antiphospholipid syndrome regulates endosome structure and function [see comments]. *Nature* **392**, 193–197.

Krause, K. H., Fivaz, M., Monod, A. and van der Goot, F. G. (1998). Aerolysin induces G-protein activation and Ca2+ release from intracellular stores in human granulocytes. *J. Biol. Chem.* **273**, 18122–18129.

Lange, S., Nussler, F., Kauschke, E., Lutsch, G., Cooper, E. L. and Herrmann, A. (1997). Interaction of earthworm hemolysin with lipid membranes requires sphingolipids. *J. Biol. Chem.* **272**, 20884–20892.

Lesieur, C., Vecsey-Semjn, B., Abrami, L., Fivaz, M. and van der Goot, F. G. (1997). Membrane insertion: the strategy of toxins. *Mol. Memb. Biol.* **14**, 45–64.

Lewis, M. J. and Pelham, H. R. (1992). Ligand-induced redistribution of a human KDEL receptor from the Golgi complex to the endoplasmic reticulum. *Cell* **68**, 353–364.

Lipsky, N. G. and Pagano, R. E. (1985). A vital stain for the Golgi apparatus. *Science* **228**, 745–747.

Mitra, R., Figueroa, P., Mukhopadhyay, A. K., Shimada, T., Takeda, Y., Berg, D. E. and Nair, G. B. (2000). Cell vacuolation, a manifestation of the El tor hemolysin of vibrio cholerae [In Process Citation]. *Infect. Immun.* **68**, 1928–1933.

Mu, F. T., Callaghan, J. M., Steele-Mortimer, O., Stenmark, H., Parton, R. G., Campbell, P. L., *et al.* (1995). EEA1, an early endosome-associated protein. EEA1 is a conserved alpha-helical peripheral membrane protein flanked by cysteine "fingers" and contains a calmodulin-binding IQ motif. *J. Biol. Chem.* **270**, 13503–13511.

Nakamura, N., Inoue, N., Watanabe, R., Takahashi, M., Takeda, J., Stevens, V. L., and Kinoshita, T. (1997). Expression cloning of PIG-L, a candidate N-acetylglucosaminyl-phosphatidylinositol deacetylase. *J. Biol. Chem.* **272**, 15834–15840.

Narula, N., McMorrow, I., Plopper, G., Doherty, J., Matlin, K. S., Burke, B. and Stow, J. L. (1992). Identification of a 200-kD, brefeldin-sensitive protein on Golgi membranes. *J. Cell Biol.* **117**, 27–38.

Nelson, K. L., Raja, S. M. and Buckley, J. T. (1997). The GPI-anchored surface glycoprotein Thy-1 is a receptor for the channel-forming toxin aerolysin. *J. Biol. Chem.* **272**, 12170–12174.

Nelson, K. L., Brodsky, R. A. and Buckley, J. T. (1999). Channels formed by subnanomalar concentrations of the toxin aerolysin trigger apoptosis of T lymphomas. *Cell Microbiol.* **1**, 69–74.

Papadopoulos, G., Dencher, N. A., Zaccai, G. and Büldt, G. (1990). Water molecules and exchangeable hydrogen ions at the active centre of bacteriorhodopsin localized by neutron diffraction. *J. Mol. Biol.* **214**, 15–19.

Reyrat, J. M., Pelicic, V., Papini, E., Montecucco, C., Rappuoli, R. and Telford, J. L. (1999). Towards deciphering the Helicobacter pylori cytotoxin. *Mol. Microbiol.* **34**, 197–204.

Ricci, V., Galmiche, A., Doye, A., Necchi, V., Solcia, E. and Boquet, P. (2000). High cell sensitivity to helicobacter pylori VacA toxin depends on a GPI-anchored protein and is not blocked by inhibition of the clathrin- mediated pathway of endocytosis. *Mol. Biol. Cell* **11**, 3897–3909.

Scheffer, J., Konig, W., Braun, V. and Goebel, W. (1988). Comparison of four hemolysin-producing organisms (Escherichia coli, Serratia marcescens, Aeromonas hydrophila, and Listeria monocytogenes) for release of inflammatory mediators from various cells. *J. Clin. Microbiol.* **26**, 544–551.

Simons, K. and Ikonen, E. (1997). Functional rafts in cell membranes. *Nature* **387**, 569–572.

Tombola, F., Oregna, F., Brutsche, S., Szabo, I., Del Giudice, G. and Rappuoli, R., *et al.* (1999). Inhibition of the vacuolating and anion channel activities of the VacA toxin of Helicobacter pylori. *FEBS Lett.* **460**, 221–225.

Trowbridge, I. S., Collawn, J. F. and Hopkins, C. R. (1993). Signal-dependent membrane protein trafficking in the endocytic pathway. *Annu. Rev. Cell Biol.* **9**, 129–161.

Tweten, R. K., Parker, M. W. and Johnson, A. E. (2001). The cholesterol-dependent cytolysins. In *Pore-forming Toxins.* Goot, F.G.v.d. (ed.) Berlin Heidelberg: Springer Verlag, *Curr. Top. Microbiol. Immunol.* **257**, 1533.

Valeva, A., Walev, I., Gerber, A., Klein, J., Palmer, M. and Bhakdi, S. (2000). Staphylococcal alpha-toxin: repair of a calcium-impermeable pore in the target cell membrane. *Mol. Microbiol.* **36**, 467–476.

van der Goot, F. G. (2001). *Pore-forming Toxins.* Berlin Heidelberg, Springer Verlag.

Vecsey-Semjen, B., Möllby, R. and van der Goot, F. G. (1996). Partial C-terminal unfolding is required for channel formation by staphylococcal alpha-toxin. *J. Biol. Chem.* **271**, 8655–8660.

Wada, I., Rindress, D., Cameron, P. H., Ou, W. J., Doherty, J. D. and Louvard, D., *et al.* (1991). SSR alpha and associated calnexin are major calcium binding proteins of the endoplasmic reticulum membrane. *J. Biol. Chem.* **266**, 19599–19610.

Waggoner, A. S. (1979). The use of cyanine dyes for the determination of membrane potentials in cells, organelles, and vesicles. *Methods Enzymol.* **55**, 689–695.

Watanabe, R., Inoue, N., Westfall, B., Taron, C. H., Orlean, P., Takeda, J. and Kinoshita, T. (1998). The first step of glycosylphosphatidylinositol biosynthesis is mediated by a complex of PIG-A, PIG-H, PIG-C and GPI1. *EMBO J.* **17**, 877–885.

Yamaji, A., Sekizawa, Y., Emoto, K., Sakuraba, H., Inoue, K., Kobayashi, H. and Umeda, M. (1998). Lysenin, a novel sphingomyelin-specific binding protein. *J. Biol. Chem.* **273**, 5300–5306.

11 Cell Transfection, Permeabilization and Microinjection as Means to Study *Shigella*-induced Cytoskeletal Reorganization

Guillaume Duménil[1,*], Laurence Bougnères[1], Philippe Sansonetti[1] and Guy Tran Van Nhieu[1,*]

[1]*Unité de Pathogénie Microbienne Moléculaire, Institut Pasteur, 28 rue du Dr Roux, 75724 Paris, Cedex 15, France*

◆◆

CONTENTS

List of abbreviations

Ab	Antibody
BSA	Bovine serum albumin
DMEM	Dulbecco's modified Eagle's medium
DTT	Dithiothreitol
EDTA	Ethylenediamine tetraacetic acid
FCS	Fetal calf serum
FITC	Fluorescein isothiocyanate
HEPES	N-(2-hydroxyethyl)piperazine-N′-(2-ethanesulfonic acid)
HRP	Horse radish peroxidase
Ig	Immunoglobulin
PFA	Paraformaldehyde

**Present address*: Department of Molecular Biology and Microbiology, Tufts University School of Medicine, 136, Harrison Avenue, Boston, MA 02111, USA

METHODS IN MICROBIOLOGY, VOLUME 31
ISBN 0-12-521531-2

◆◆◆◆◆◆ GENERAL CONSIDERATIONS ON THE ANALYSIS OF *SHIGELLA* ENTRY EFFECTORS

Various gram-negative pathogens modify the cell cytoskeleton by means of bacterial products injected in the cell cytosol through a type III secretion system (Hueck, 1998). For *Shigella*, such products determine bacterial invasion in normally non-phagocytic cells. These effectors induce bacterial internalization by inducing cytoskeletal changes resulting in the formation of cell extensions, that dynamically organize to engulf the bacterium in a large vacuole (Tran Van Nhieu *et al.*, 2000). The formation of these cell extensions require actin polymerization, and the activation of the small GTPases Cdc42 and Rac, whereas the later stages of the bacterial entry process require the activation of the GTPase Rho (Duménil *et al.*, 2000). The Src tyrosine kinase play a dual role during the entry process, by favoring actin polymerization in concert with the activation of Cdc42 and Rac, while down-regulating Rho and actin polymerization during the late stages of entry (Duménil *et al.*, 2000). This indicates that bacterial entry depends on a finely tuned set of responses, which may be induced by the concerted action of various bacterial determinants. Bacterial genetic studies aiming at identifying such products are limited by the fact that entry defective mutants will not only consist of effector mutants, but also of any mutants that are defective for the type III secretory apparatus. An approach based on analyzing the effect on the cell cytoskeleton of individual bacterial products that translocate through the type III secretion apparatus has proven successful in identifying several effectors of bacterial invasion (Galan and Zhou, 2000; Hayward and Koronakis, 1999; Tran Van Nhieu *et al.*, 2000). This can be achieved by various means, from the expression by transfection, to microinjection or semi-permeabilization of the bacterial effectors in cell lines. In parallel to *in vitro* assays using single determinants, we have used independent approaches to analyze signaling pathways that regulate cytoskeletal reorganization induced by *Shigella*. In this chapter, we will discuss experience gained in using these various means to analyze effectors of *Shigella* entry inside cells.

◆◆◆◆◆◆ ANALYSIS OF *SHIGELLA*-INDUCED CYTOSKELETAL REARRANGEMENTS DURING ENTRY INTO EPITHELIAL CELLS

HeLa cells are cells routinely used to analyze entry of *Shigella* into epithelial cells. Although the reasons for this are mostly historical, the spectacular projections that this bacterium induces during the entry process, as well as the distinct phases of the entry structure that can be distinguished in this type of cell are probably the best rationale for this model. Because *Shigella*-induced cytoskeletal changes are highly dynamic, it is important to synchronize the infection so as to compare foci at the same stage of development. As *Shigella* shows little cell-binding activity, it is rendered adhesive to cells by the expression of the *E. coli* Afa E adhesin

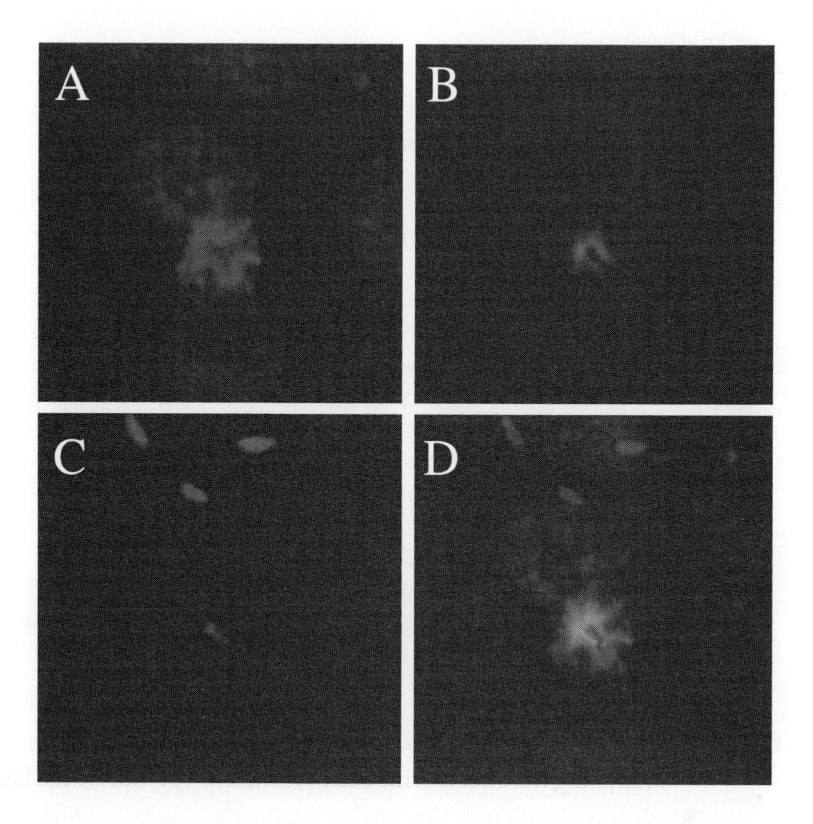

Figure 11.1. Immunofluorescence analysis of ezrin in a *Shigella* focus of internalization. HeLa cells were challenged with *Shigella* for 15 min at 37°C. Samples were fixed and stained for ezrin (Panel A), actin (Panel B) or *Shigella* LPS (Panel C), and analyzed by confocal laser microscopy (LSM510, Zeiss). Images were obtained by reconstruction from sections that do not include the basal cell surface. Panel D shows the superimposition of the triple staining. Ezrin labels the tip of *Shigella*-induced cell extensions where little F-actin is detected. Scale bar: 5 μm. (This figure is also reproduced in colour between pages 276 and 277.)

(Labigne-Roussel *et al.*, 1984). Bacterial binding to cells is performed at 22°C and cytoskeletal changes induced by *Shigella* entry are triggered by placing the samples at 37°C. F-actin staining is used to identify the cytoskeletal rearrangements induced by *Shigella*. Because the projections that *Shigella* induces during entry can reach up to 10 microns in length, confocal laser microscopy may be required in double-labeling procedures to precisely characterize the recruitment of a cytoskeletal protein at the level of the entry structure. Figure 11.1 shows an entry structure that has been induced by *Shigella* on the surface of HeLa cells and that was stained for the cytoskeletal linker ezrin (Panel A) and F-actin (Panel B). Ezrin is recruited at the tip of F-actin rich cell projections, where little F-actin is observed. Bacteria are stained with an anti-LPS antibody (Panel C).

Protocol—Day 1

- HeLa cells are plated onto 24 × 24 mm coverslips in a 35 mm dish, and grown in DMEM containing 10% FCS.

Day 2

- Cells are washed once with DMEM without serum and incubated with bacteria grown in mid-exponential phase and suspended in DMEM containing 50 mM HEPES pH 7.3.
- To prepare the bacterial suspension, *Shigella* strains carrying the AfaE encoding plasmid are grown in TCSB containing spectinomycin at 100 $\mu g\ ml^{-1}$ final concentration until mid-exponential phase (OD 600 nm = 0.3).
- Bacteria are diluted in DMEM-HEPES medium to give an MOI of 10–50 bacteria per cell (0.003 < OD 600 nm < 0.006). 1 ml of the bacterial suspension is added to the well and samples are incubated at room temperature for 15 min to allow bacterial attachment to the cell surface.
- Samples are then shifted at 37°C by floating on a water bath. To avoid problems linked to floating samples, a metal plate can be immersed in the 37°C water bath, so as to leave sufficient water above the plate to allow efficient immersion of the samples.
- After various periods of time, samples are fixed in PBS containing paraformaldehyde at 3.7% final concentration for 20 min at RT.
- Samples are washed three times in PBS. Samples are processed for immunofluorescence staining using standard procedures.

Determination of bacterial entry by differential inside/out immunostaining

Internalized bacteria can be distinguished from extracellular bacteria because they are not accessible to antibodies unless cells have been permeabilized. This makes it possible to differentially stain the extracellular bacteria prior to sample permeabilization with a given fluorochrome, and stain the total bacteria after permeabilization with a different fluorochrome. To avoid cross-reactivity with the secondary Ab, staining is performed with different antibodies (i.e. rabbit polyclonal and mouse monoclonal Abs, or different subclass of monoclonal Abs, with the corresponding secondary Abs). If only one Ab is available, it is possible to covalently link the fluorochromes to the Ab using commercial reagents (i.e. Molecular probes, Calbiochem). Alternatively, the utilization of a bacterial strain that express the green fluorescent protein (GFP) (Rathman *et al.*, 2000) can simplify the procedure, as only the labeling of the extracellular bacteria with a fluorochrome that emits in a different spectrum than the GFP (i.e. rhodamine) is necessary.

Protocol

- Fixed samples are blocked in DMEM containing 10% FCS for at least 30 min.
- Extracellular bacteria are stained with an anti-LPS mAb, followed by an anti-mouse Ig Ab coupled to FITC.
- Samples are permeabilized by incubating for 4 min in PBS containing 0.1% Triton X-100.
- Samples are washed three times in PBS.
- Total bacteria are stained with a anti-LPS rabbit polyclonal Ab, followed by an anti-rabbit Ig Ab coupled to rhodamine.

- Samples are washed three times in PBS and mounted onto slides using 50% glycerol and DABCO at a final concentration of 10 mg ml^{-1} to prevent bleaching of the samples.

If a triple labeling is performed, a UV light excited fluorochrome such as Cascade blue can be used. Because this latter type of fluorescence is more difficult to detect than red or green-emitting fluorochromes, we usually reserve this UV light excited fluorochromes for the labeling giving the strongest and less ambiguous signal, i.e. bacterial LPS labeling. Also, because UV light excitation promotes more sample bleaching than red or green lights, it is preferable to acquire images for quantification purposes. For example, to analyze bacterial internalization in transiently transfected cells, fields containing transfectants will be selected. To limit photobleaching, images corresponding to the UV-excited fluorophore are acquired last.

Analysis of components of *Shigella*-induced entry foci

Shigella entry structure are identified with F-actin staining. Because these structures are readily distinguishable from other cellular structures, *Shigella*-induced foci of actin polymerization can be scored using a computer dedicated program according to the shape and the fluorescence intensity of the entry structure (Duménil *et al.*, 1998). When performing kinetics to localize more precisely a cytoskeletal component within the entry structure, bacteria need to be stained.

Triple-labeling

Samples are fixed and permeabilized. F-actin is stained with Bodipy-linked to phalloidin, the cytoskeletal component to be analyzed is stained with Ab followed by rhodamine-linked secondary Ab. For direct observation, bacteria are labeled with anti-LPS followed by Cascade blue linked Ab. To prepare samples for confocal microscopy analysis, bacteria are labeled with anti-LPS followed by CY-5 linked Ab. Although CY5 cannot easily be distinguished from red-emitting light fluorochrome using commonly used filters, it is readily distinguishable from rhodamine when using the appropriate laser wavelengths, and has the advantage of an excitation light that is less damaging to the sample.

◆◆◆◆◆◆ PURIFICATION OF *SHIGELLA* TYPE III SECRETION EFFECTORS FROM SECRETION MUTANT STRAINS

It is, in general, not a problem to express *Shigella* proteins under a recombinant form in *E. coli*. Several proteins that are secreted via the type III secretion apparatus have been obtained after fusion to GST or tagged with a poly-histidine epitope (Chen *et al.*, 1996; Niebuhr *et al.*, 2000).

Shigella Ipa proteins that are fused to the GST by their N-terminus, are not secreted by the type III secretion apparatus because the GST moiety probably interferes with the secretion signal. Histidine tagged proteins may be still secreted and have been shown in some instances to functionally complement the invasion defect in the corresponding *Shigella* mutant (Niebuhr *et al.*, 2000). Because such technologies are now widely used, we will not discuss technical considerations, but there are two types of limitations to recombinant protein technology that may not be specific for *Shigella* secreted proteins: first, many *Shigella* recombinant proteins are not soluble and tend to form inclusion bodies when overexpressed in *E. coli* (De Geyter *et al.*, 1997; Picking *et al.*, 2001). Purification in this case, implies solubilization from inclusion bodies, using agents such as guanidine hydrochloride or urea. Thus, particular care should be taken to ensure that proteins that are purified this way remain functional or do not show altered properties, specially for those proteins whose activity is likely to be regulated. An alternative to recombinant protein technology, is the purification of proteins from *Shigella* strains, that are expressed under the control of their endogenous promoter. This is feasible for the *Shigella* Ipa proteins, because they are abundant and because secretion provides a means to fractionate proteins that are secreted via the type III secretion apparatus (Bourdet-Sicard *et al.*, 1999; Tran Van Nhieu *et al.*, 1999). After concentration of secreted proteins from *Shigella* culture supernatants, proteins are fractionated by FPLC using a combination of ion-exchange chromatography procedures.

◆◆◆◆◆◆ CHOICE OF THE STRAIN AND GROWTH CONDITIONS

Shigella mutants *ipaB*, *ipaD* or deleted for the *ipa* operon for which the Mxi-Spa apparatus shows constitutive activity (Parsot *et al.*, 1995), are used for these purposes. The use of a mutant strain can also simplify the purification steps. For example, IpaB has been shown to form a stable complex with IpaC after secretion (Ménard *et al.*, 1994); *Shigella ipaD* strain may be used for the isolation of the IpaB-C complex, whereas purification of IpaC from a *Shigella ipaB* strain circumvents the problem of dissociating the IpaB-C complex. For proteins other than the Ipa proteins, that are secreted via the Mxi-Spa apparatus, a *Shigella* strain for which the *ipa* operon is deleted can be used to avoid contamination by these abundant proteins. Strains are grown in trypticase soya or 2 × YT broth. Although casein products present in this medium tend to 'stick' to the chromatography matrix, the yields obtained using TCS broth are significantly higher than those obtained 2 × YT.

Purification procedure—Day 1

- Inoculate a preculture of the *Shigella ipaB* mutant strain.

Day 2

- Inoculate a culture with a 1 : 100 dilution of the overnight preculuture. In general, for IpaA or IpaC, it is possible to obtain about 200 μg of protein starting from a liter of culture of the *ipaB* mutant strain, using the respective endogenous promoters. This yield can be increased by several-folds when using strains transformed with the cloned gene of interest under the control of the Plac promoter. Overexpression, however, appears to alter the solubility of the proteins (not shown).
- Grow on a rotary shaker at 37°C, 250 rpm until mid-exponential phase ($0.6 < OD\ 600 < 1.0$). The recovery of Ipa proteins from culture supernatant at later stages of growth decreases because they tend to become insoluble. The flasks containing the culture are then chilled on ice, and all the purification steps described below are performed at 4°C, unless otherwise stated.
- Centrifuge the bacterial culture at 7000 rpm for 30 min.
- Transfer the supernatant in a beaker, and weigh the appropriate amounts of ammonium sulfate to perform a precipitation at 50% final concentration. Ensure the pH of the sample rapidly neutralized after addition of the ammonium sulfate, and stir samples for at least 2 h.
- Centrifuge samples at 7000 rpm for 30 min and discard supernatant.
- Resuspend pellet in about 1 : 20 of the initial volume of culture in buffer A1 containing: 0.1% Nonidet-P40; 25 mM Tris-HCl pH 7.5; 25 mM NaCl; 0.1 mM EDTA; 1 mM DTT, and a mixture of protease inhibitors. At this stage, the pellet has a dark brown color from the casein precipitates, but it readily comes into solution. Dialyze extensively three times against at least 20 volumes of buffer A1 for at least 2 h for each dialysis batch.

Day 3

- Perform FPLC using a 1 ml monoQ anion exchange column. Proteins are eluted using a 20 ml s NaCl linear gradient with concentrations ranging from 25 mM to 500 mM. IpaA typically elutes in two peaks; a near homogeneous fraction at around 80 mM NaCl, and another fraction contaminated with proteins that migrate at ca. 60 kDa, which probably correspond to IpaH proteins (Figure 11.2A, arrow).
- For IpaC purification, the flow-through of the monoQ column is collected and dialyzed extensively against buffer A2 containing 0.1% Nonidet-P40; 25 mM HEPES pH 7.5; 0.1 mM EDTA; 1 mM DTT, and a mixture of protease inhibitors (complete TM, Pharmacia).
- Proteins are fractionated using a monoS cation exchange column and a 20 ml s NaCl linear gradient with concentrations ranging from 25 mM to 450 mM. Two main peaks are detected that correspond to proteins migrating at 43 kDa, IpaC typically elutes at an NaCl concentration of about 100 mM (Figure 11.2B, arrow). SepA, a secreted protein that shares homology with serine protease, forms a peak that elutes at concentrations slightly inferior to those for IpaC elution, and may contaminate the IpaC containing fractions (Figure 11.2B).
- Purified samples are dialyzed against microinjection buffer containing 25 mM Tris-HCl pH 7.3, 100 mM KCl, 5 mM $MgCl_2$, 1 mM EGTA, 0.1 mM DTT, and protease inhibitors, and concentrated using amicon filters C3000 (Millipore corp.) to a final concentration of about 1 mg ml^{-1}. Aliquots are flash frozen and stored at −20°C.

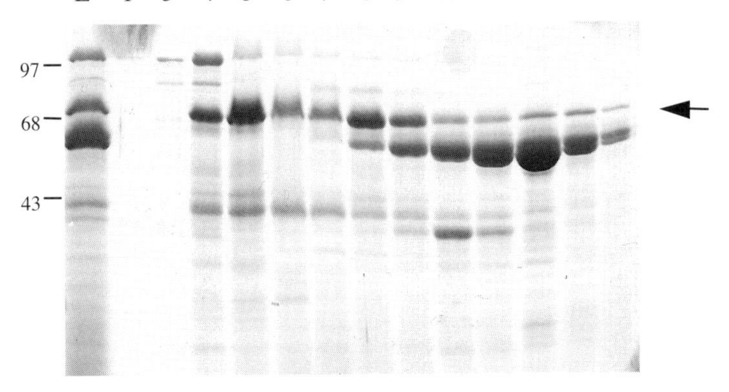

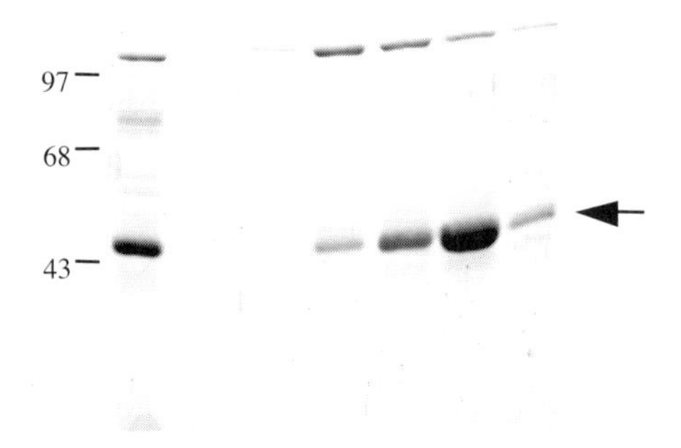

Figure 11.2. Purification of IpaA and IpaC from *Shigella* secretion constitutive mutant strains. Concentrated supernatant of a *Shigella ipaB* mutant strain that shows constitutive secretion, was subjected to FPLC (flow pressure liquid chromatography) on a monoQ anion-exchange column (Panel A). Fractions were eluted with a 20 ml s linear NaCl gradient, with concentrations ranging from 20 mM to 500 mM. The flow-through of the monoQ chromatography was run on a MonoS cation-exchange column (Panel B), and fractions were eluted with a 20 ml s linear NaCl gradient, with concentrations ranging from 0 mM to 450 mM. Panels A and B: E is loaded extract. The numbers above the lanes represent the fraction number from the start of the gradient. The arrow points to IpaA (Panel A) and IpaC (Panel B).

◆◆◆◆◆◆ MICROINJECTION AS A MEANS TO STUDY THE EFFECTS OF BACTERIAL PRODUCTS ON THE ACTIN CYTOSKELETON OF EUKARYOTIC CELLS

Microinjection has an advantage over the transfection procedure, in the sense that it can allow the analysis of short-term effects linked to the microinjected proteins, by fixing samples within minutes after microinjection and performing immunofluorescence microscopy analysis. Microinjection also allows the use of videomicroscopy to analyze the effects of the injected product on the formation of cell projections.

Concerning the choice of the cell line, Swiss 3T3 cells present several advantages over other cell types for the study of cytoskeletal rearrangements (Hall, 1998). These cells show a well-characterized actin cytoskeleton that can be controlled under various cell culture conditions, and there is little heterogeneity in the cell sample (Nobes and Hall, 1995). Furthermore, many studies on the signaling to the actin cytoskeleton have been performed in these cells, and the hierarchy of key regulation events is usually admitted to occur in this cell model. The effects on the cytoskeleton of a particular toxin, or bacterial effector can be visualized best by treating the cells to place them in the optimized responsive state. For example, to study the effects of effectors on depolymerization of actin filaments, cells are usually cultivated in the presence of serum to maximize the presence of stress fibers (Nobes and Hall, 1995). In the case of actin polymerization induced by the IpaC protein, cultivation in the presence of serum appears to interfere with the formation of actin-rich extensions at the cell periphery induced by IpaC (Tran Van Nhieu *et al.*, 1999). Thus, cells need to be cultivated for at least 48 h in the absence of serum.

Protocol—Day 1

- Swiss 3T3 cells between passage 7 and 15 are grown in DMEM containing 10% FCS in a 37°C incubator supplemented with 10% CO_2. Cells are seeded in 24-well plates onto 13-mm-diameter coverslips, that were previously acid-washed and sterilized, at a density of 5×10^4 cells per well.

Day 2

- Cells are washed once in DMEM, resuspended in DMEM and incubated for 48 h in the absence of serum.

Day 4

- Cells are processed for microinjection in a chamber supplemented with 10% CO_2. The microinjection time should not exceed 10 min, this period determining the limit in the amounts of cells being microinjected.
- After microinjection, cells are returned to the incubator in DMEM containing 10% FCS for 5 to 20 min.
- Cells are fixed in PFA and processed for staining of the actin cytoskeleton.

◆◆◆◆◆◆ SEMI-PERMEABILIZATION PROCEDURE TO STUDY THE EFFECTS OF TYPE III SECRETION EFFECTORS ON THE CELL CYTOSKELETON

As an alternative to microinjection, the effects of IpaC on the cytoskeleton can be visualized by permeabilizing cells with trace amounts of the

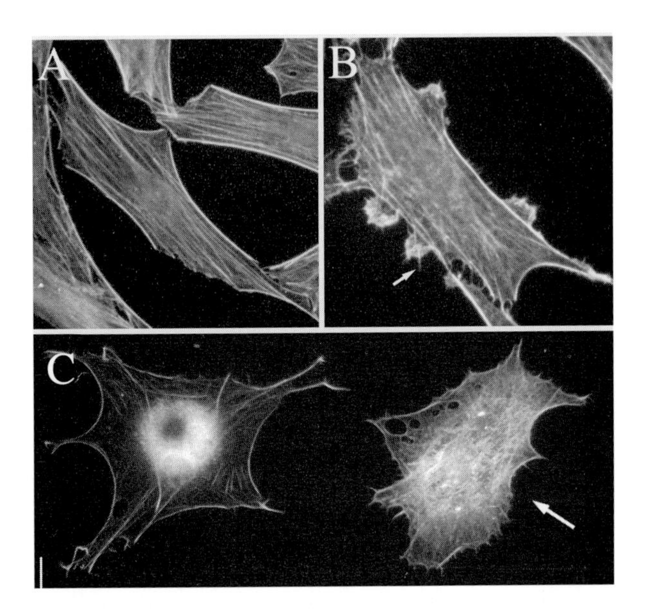

Figure 11.3. Permeabilization versus microinjection of IpaC. Swiss 3T3 cells were permeabilized in the presence of buffer alone (Panel A) or purified IpaC (Panel B) (see text). Samples were fixed and stained for F-actin. IpaC induces the formation of actin-rich filopodial structures that fill in into leaflets, at the cell periphery. Swiss 3T3 cells were microinjected with purified IpaC (Panel C, arrow). The IpaC-microinjected cell shows diffuse actin polymerization, and the formation of numerous microspikes over the cell surface.

detergent saponin (Tran Van Nhieu *et al.*, 1999). Although it is not clear to what extent permeabilization applies to all cytoskeletal effectors, it provides a powerful means to visualize cytoskeletal reorganization because the procedure is less cumbersome than microinjection. Also, because the effects are homogenous in cells from within a sample, quantification is rendered easier. In the case of IpaC, permeabilization leads to the formation of actin-rich filopodial extensions that rapidly fill in to form lamellipodial extensions at the cell periphery (Figure 11.3B). This is in contrast to IpaC microinjection, which leads to diffuse actin polymerization and the formation of numerous thin microspikes on the cell surface (Figure 11.3C, arrow). The reasons for this difference are unclear, but could be linked to leakage of cell components regulating cytoskeletal changes during the permeabilization procedure, or alternatively, a difference in the mode of presentation of IpaC in the two techniques.

Swiss 3T3 cells are grown in 10% FCS-DMEM 1000 mg ml^{-1} glucose in a 37°C incubator supplemented with 10% CO_2, and used between passage 7 and 15. Serum starvation is in DMEM 1000 mg ml^{-1} glucose containing 50 mM HEPES pH 7.3.

Protocol—Buffers used

- UB buffer: 50 mM HEPES pH 7.3, 100 mM Kcl, 3 mM $MgCl_2$, 0.1 mM DTT, 0.2% BSA.

- Permeabilization buffer: 0.003% saponin, 50 mM HEPES pH 7.3, 100 mM KCl, 3 mM $MgCl_2$, 0.1 mM DTT, 0.2% BSA, 1 mM ATP, 100 μM GTP, 100 μM UTP.

Day 1

- Semi-confluent Swiss 3T3 cells (use between passages 7 and 12) are split and plated at 4×10^4 cells/13 mm diameter coverslip or 2×10^5 cells/24 mm × 24 mm coverslip.

Day 2

- Serum starved for 30–48 h.

Day 4

- Wash 1 × PBS, 2 × UB
- Dilute protein (1 mg ml^{-1}) onto parafilm, at least 1 : 10 in permeabilization buffer on parafilm parafilm piece. Use: 40 μl for 13 mm diameter coverslip or 150 μl for 24 × 24 mm coverslip.
- Incubate cells with samples by putting the coverslip face down onto the droplet on parafilm.
- Incubate at 37°C in humid chamber. 20 min on prewarmed metal plate.
- Transfer cells to 3.7% PFA for 30 min at 22°C.
- Process for fluorescence staining of the actin cytoskeleton.

◆◆◆◆◆◆ CELL EXPRESSION OF *SHIGELLA* EFFECTORS BY TRANSIENT TRANSFECTION

The approach consists of cloning the bacterial gene of interest on an eukaryotic expression vector, and introducing the recombinant vector by transfection inside epithelial cells. Although straightforward because it does not require purification of the bacterial product, this approach may be limited by a few considerations that are discussed below. First, any visible effects have to correspond to those occurring at concentrations of protein that allow immunodetection, without warranty that these relate to effects occurring at physiological concentrations. Furthermore, transfection requires incubation, usually from a few hours to a few days, to allow expression and immunodetection of the transfected protein. Thus, the changes observed on the cytoskeleton result from an equilibrium between direct effects of transfected protein and secondary effects that this protein induces. If the direct effects are predominant over secondary effects, transient transfection may be an adequate technique to analyze the effects of a protein on the cytoskeleton. If the direct effects are subtle , or very transient, as one may suspect for the bacterial effectors of entry, the changes observed on the cytoskeleton may correlate only remotely with the real effects of the protein. This is well exemplified by activated Cdc42, which induces filopodia and microspike formation when microinjected

(Nobes and Hall, 1995), whereas it induces stress fiber formation after transient transfection. In this latter situation, the formation of stress fibers probably results from the activation of Rac and Rho GTPases, downstream of Cdc42 (Hall, 1998). For these reasons, transfection should be used as a means to detect obvious effects on the cytoskeleton, to confirm a phenotype, or in conjunction with other pieces of evidences to assess a precise function of a protein.

Choice of the eukaryotic expression vector

- It is convenient to use vectors that also carry a prokaryotic promoter to check for in-frame cloning and expression of the protein to be transfected. Also, many commercially available vectors allow insertion of an epitope tag at one extremity of the protein coding sequence that will facilitate the immunodetection of the protein.

Transfection procedure—Day 1

- Cells are seeded at a density of 5×10^4 cells on a 13 mm diameter coverslip and incubated overnight in DMEM containing 10% FCS.

Day 2

- Cells are transfected using the Fugene reagent according to the manufacturer's instructions, and incubated from 6 h to 3 days.

Analysis of the transfectants

Cells are fixed with PFA and processed for staining of F-actin and to detect expression of the transfected protein. In the case of IpaC cloned into pCDNA.3, immunoanalysis of the transfectants indicate the formation of filopodial structures at the cell periphery, as well as of a reorganization of actin fibers into a branched meshwork of thinner actin cables (Tran Van Nhieu *et al.*, 1999). The changes that are observed on the organization of actin cables of the transfectants are likely to be secondary effects due to the expression of IpaC, as those are not visible during short-term kinetics of cells that have been microinjected with the purified IpaC protein (Figure 11.2B).

◆◆◆◆◆◆ ESTABLISHING SRC STABLE TRANSFECTANTS OF HELA CELLS TO STUDY BACTERIAL-INDUCED SIGNALING TO HOST CELLS

For biochemical purposes, microinjection or transient transfection techniques are not suitable. To analyze association of proteins, or regulation of cytoskeletal organization during bacterial entry, the use of stable transfectants may be required. Although obtaining stable transfectants may not always be possible for proteins that interfere with cell growth, and although it requires significant investment in terms of effort and time, the establishment of stable cell lines clearly opens the way for analysis that cannot be achieved with transient transfection. It allows the obtention of cells that express relatively homogeneous levels of the transfected proteins. Also, it then becomes possible to study a combination of effects, for example, by microinjection or super transfection.

The procedure to obtain stable transfectant is similar to the transient transfection protocol, except that selective medium is added to the cells, usually 24 to 48 h after transfection. In the case of HeLa cells, clones growing as individual clusters of about 200 cells are detectable after 10 days. These clusters should be subcloned at least twice by the limiting dilution or cloning ring technique, to limit the mixture of cell clones. Because not all clones that are resistant to the selective marker will express the protein of interest, it is necessary to analyze several clones growing in the selective medium. In the case of Src constructs, the screening of expressing clones was performed by Western blot on cell lysates corresponding to 5×10^5 to 5×10^6 cells, using the ECL chemiluminescent substrate (Duménil *et al.*, 1998). Although cumbersome, this was necessary because the levels of expression of the transfected Src were too low to detect by immunofluorescence microscopy. Clones expressing levels of the kinase deficient Src could be readily obtained with levels above ten-fold those of endogenous Src, whereas clones expressing constitutively active Src showed only moderate expression of the kinase, with at best a two-fold expression over the endogenous kinase. These latter clones showed a strongly altered morphology, with a disappearance of stress fibers and a decrease in focal adhesions (Duménil *et al.*, 2000). The various Src transfectants are stable and can be transiently transfected with Rho GTPases constructs to study the hierarchy between Src and these GTPases during *Shigella* entry (Duménil *et al.*, 2000).

◆◆◆◆◆◆ INVESTIGATING THE ROLE OF A CYTOSKELETAL REGULATOR BY TRANSIENT TRANSFECTION OR MICROINJECTION

Considerable progress has been made in the knowledge of signaling pathways that govern actin dynamics in recent years, and many key regulators have been identified and characterized in terms of molecular structure and partners. This is opening up the possibility to directly interfere with the function of defined molecules by the introduction of dominant-negative forms that will specifically inhibit or disconnect a given pathway. With the reserves of technical considerations (i.e. difficulties to transfect or to microinject), it is logical to use the same cell model as is commonly used for the bacterial entry process, although one has to be careful about drawing conclusions from what has been established for the implication of a given molecule in another cell type.

We have used microinjection and transfection of dominant-negative forms of Rho GTPases to study the role of these GTPases in *Shigella* entry. Both techniques gave similar results in that inhibition of Cdc42 or Rac-inhibited actin polymerization induced by *Shigella* at the site of entry, whereas inhibition of Rho did not inhibit actin polymerization but the recruitment of ezrin and the Src tyrosine kinase at entry foci (Duménil *et al.*, 2000). Transfection is a straightforward approach because it does not involve protein purification and large-scale microinjection. When expressing a dominant interfering by transfection, however, there are constraints linked to the period of time allowed to visualize expression of the construct that ranges from a minimum of 6 h to several days. This incubation time may lead to the accumulation of pleiotropic effects, and may lead to inhibitory results that are not directly linked to the protein analyzed. Although more cumbersome, microinjection allows one to control the concentration of the inhibitor used, and also to study the effects of the inhibitor a few min after its introduction into the cell cytosol.

Analysis of the expression of dominant-interfering forms of Rho GTPases on *Shigella* entry by transient transfection—Day 1

- HeLa cells are seeded at a density of 2×10^5 cells per 24 mm × 24 mm coverslip in a 35 mm dish, in Dubellco Modified Earl's Medium (Gibco BRL) containing 10% fetal calf serum, in a 37°C incubater supplemented with 10% CO_2.

Day 2

- Cells are transfected with the corresponding construct using the Fugene reagent (Boerhinger Mannheim) according to the manufacturer's instructions. When performing immunofluorescence analysis, transfection using Ca_2PO_4 procedures needs to be avoided because of the interference of Ca_2PO_4 precipitates with the immunodetection procedure.

Day 3–5

- Cells are challenged with *Shigella* and the formation of foci of actin polymerization is analyzed by immunofluorescence as described in earlier.

The expression of the construct is monitored by immunofluorescence analysis. Because most recombinant proteins are tagged with an exogenous epitope, the antibodies and the conditions used for detection of the recombinant proteins are in general well characterized. In theory, it is also possible to interpolate the levels of molecule expressed in the transfectants, by performing the proper calibrations. In practice, however, the levels of inhibitor that should be analyzed are those which lead to inhibition of specific pathways, without causing gross alterations of the cytoskeleton. In this respect, it is important to define the antibody probing conditions so as to observe a range of expression levels related to the fluorescence intensity in the various transfectants. For incubation periods that exceed 24 h after transfection, transfectants are usually observed that expressed weak, intermediate, or high levels of the transfected construct. In the case of dominant interfering N17 Cdc42, N17 Rac or N19 Rho GTPases, transfectants that express high levels of the construct should be omitted from the analysis because of their profoundly altered cytoskeleton. The analysis is performed on cells that express low or intermediate levels of the transfected construct on a statistically significant number of cells from at least three independent experiments.

Analysis of the expression of dominant-interfering forms of Rho GTPases on *Shigella* entry by microinjection

- Purification of dominant-interfering forms of Rho GTPases using recombinant GST fusions in *E. coli* has been described elsewhere (Self and Hall, 1995). After purification, protein samples are dialyzed against microinjection buffer (25 mM Tris-HCl pH 7.3, 100 mM KCl, 5 mM $MgCl_2$, 1 mM EGTA, 0.1 mM DTT and protease inhibitors) and concentrated by centrifugation using Amicon microconcentraters to obtain a protein concentration in the order of 1 mg ml^{-1}. Samples are stored aliquoted at −20°C. The day of microinjection, samples are thawed and mixed with FITC-Dextran (70 kDa, Molecular Probes) previously dialyzed in microinjection buffer, to give a final concentration of 0.2 mg ml^{-1}. Samples are centrifuged for 5 min at 13 K just prior to injection to pellet potential aggregates and the supernatant is used to load microinjection capillaries.

Day 1

- Cells are seeded the day before on 13 mm diameter coverslips in a 24-well plate at a density of 5×10^4 cells per well.

Day 2

- Coverslips are kept in DMEM containing 10% FCS in the 37°C incubator supplemented with 10% CO_2. Just prior to microinjection, samples are transferred to a 60 mm diameter dish in DMEM containing 25 mM HEPES pH 7.3.
- Microinjection is performed on a maximal number of cells for 10 min. With practice, it is usually possible to microinject as many as 50 to 100 cells per sample with a high recovery rate. Samples are returned to the incubator in DMEM containing 10% FCS for 30 min.
- Samples are challenged with *Shigella* and fixed in PFA as described above.
- For IF analysis, microinjected cells are distinguished by the FITC fluorescence. To analyze entry foci induced by *Shigella*, it is convenient to perform staining of bacteria as well as of the component of the foci to be analyzed (i.e. actin) in a manner that is compatible with direct microscopy observation. This can be performed by using red-light emitting fluorophores such as rhodamine, combined with UV light such as Cascade blue-linked antibodies.

◆◆◆◆◆◆ CONCLUDING REMARKS

Because the action of the various *Shigella* effectors, as well as the responses they induce, are finely regulated during the entry process, it is clear that none of the *in vitro* approaches described in this chapter can reproduce what the bacterium achieves during entry. This is well-illustrated in the case of the IpaA protein, which favors the transformation of cell extensions into a structure that is proficient for bacterial uptake. IpaA carries two activities. IpaA binds to the focal adhesion protein vinculin and stimulates its binding to F-actin; this may account for the formation of a focal adhesion-like structure at the intimate contact site between the bacterium and the host cell membrane. The IpaA–vinculin complex also carries an F-actin depolymerizing activity that may account for the transition from filopodial to leaflet-like extensions (Bourdet-Sicard *et al.*, 1999). It is likely that these activities are dynamically regulated during *Shigella* entry, either by the cellular environment, or by the concerted action of other bacterial effectors. Such a type of regulation probably applies to the various effectors of entry, and it will be important to develop approaches that integrate an activity characterized *in vitro* for a purified protein in the bacterial entry process.

References

Bourdet-Sicard, R., Ruediger, M., Jockusch, B., Sansonetti, P. J. and Tran Van Nhieu, G. (1999). Binding of the *Shigella* IpaA protein to vinculin induces actin depolymerization. *EMBO J.* **18**, 5853–5862.

Chen, Y., Smith, M. R., Thirumalai, K. and Zychlinkski, A. (1996). A bacterial invasin induces macrophage apoptosis by binding directly to ICE. *EMBO J.* **15**, 3853–3860.

De Geyter, C., Vogt, B., Benjelloum-Touimi, Z., Sansonetti, P. J., Ruysschaert, J.-M., Parsot, C. and Cabiaux, V. (1997). Purification of IpaC, a protein involved in entry of *Shigella flexneri* into epithelial cells and characterization of its interaction with lipid membranes. *FEBS Lett.* **400**, 149–154.

Duménil, G., Olivo, J.-C., Pellegrini, S., Fellous, M., Sansonetti, P. J. and Tran Van Nhieu, G. (1998). Interferon-alpha inhibits a Src-mediated pathway necessary for Shigella-induced cytoskeletal rearrangements during entry into epithelial cells. *J. Cell Biol.* **143**, 1003–1012.

Duménil, G., Sansonetti, P. and Tran Van Nhieu, G. (2000). Src inhibits stress fiber formation and Rho-dependent responses during *Shigella* entry into epithelial cells. *J. Cell Sci.* **113**, 71–80.

Galan, J. and Zhou, D. (2000). Striking a balance: modulation of the actin cytoskeleton by Salmonella. *Proc. Natl Acad. Sci. USA* **97**, 8754–8761.

Hall, A. (1998). Rho GTPases and the actin cytoskeleton. *Science* **279**, 509–514.

Hayward, R. and Koronakis, V. (1999). Direct nucleation and bundling of actin by the **SipC** protein of invasive *Salmonella*. *EMBO J.* **18**, 4926–4934.

Hueck, C. J. (1998). Type III secretion systems in bacterial pathogens of animals and plants. *Microbiol. Mol. Biol. Rev.* **62**, 379–433.

Labigne-Roussel, A. F., Lark, L., Schoolnik, G. and Falkow, S. (1984). Cloning and expression of an afimbrial adhesin (AFA) responsible for P blood group-independent mannose-resistant hemagglutination from a pyelone-phritic *E. coli* strain. *Infect. Immun.* **46**, 251–259.

Ménard, R., Sansonetti, P. J., Parsot, C. and Vasselon, T. (1994). Extracellular association and cytoplasmic partitioning of the IpaB and IpaC invasins of *Shigella flexneri*. *Cell* **79**, 515–525.
Niebuhr, K., Jouihri, N., Allaoui, A., Gounon, P., Sansonetti, P. and Parsot, C. (2000). IpgD, a protein secreted by the type III secretion machinery of *Shigella flexneri*, is chaperoned by IpgE and is implicated in entry focus formation. *Mol. Microbiol.* **38**, 8–19.
Nobes, C. D. and Hall, A. (1995). Rho, Rac, and Cdc42 GTPases regulate the assembly of multimolecular focal complexes associated with actin stress fibers, lamellipodia, and filopodia. *Cell* **81**, 53–62.
Parsot, C., Ménard, R., Gounon, P. and Sansonetti, P. J. (1995). Enhanced secretion through the *Shigella flexneri* Mxi-Spa translocon leads to assembly of extracellular proteins into macromolecular structures. *Mol. Microbiol.* **16**, 291–300.
Picking, W., Coye, L., Osiecki, J., Baronski Serfis, A., Schaper, E. and Picking, W. (2001). Identification of functional regions within invasion plasmid antigen C (IpaC) of *Shigella flexneri*. *Mol. Microbiol.* **39**, 100–111.
Rathman, M., Jouirhi, N., Abdelmounaaim, A., Sansonetti, P., Parsot, C. and Tran Van Nhieu, G. (2000). The development of a FACS-based strategy for the isolation of *Shigella felxneri* mutant that are deficient in intercellular spread. *Mol. Microbiol.* **35**, 974–990.
Self, A. J. and Hall, A. (1995). Purification of recombinant Rho/Rac/G25K from *Escherichia coli*. *Meth. Enzymol.* **256**, 3–10.
Tran Van Nhieu, G., Bourdet-Sicard, R., Duménil, G., Blocker, A. and Sansonetti, P. (2000). Bacterial signals and cell responses during *Shigella* entry into epithelial cells. *Cell Microbiol.* **2**, 187–193.
Tran Van Nhieu, G., Caron, E., Hall, A. and Sansonetti, P. (1999). IpaC induces actin polymerization and filopodia formation during *Shigella* entry into epithelial cells. *EMBO J.* **18**, 187–193.

12 Modifications of Small GTP-binding Proteins by Bacterial Protein Toxins

Patrice Boquet
INSERM U452, Faculté de Médecine, 28 avenue de Valombrose, Nice, France

◆◆

CONTENTS

◆◆◆◆◆◆ INTRODUCTION

Rho GTP-binding proteins belong to the p21 Ras superfamily which is divided into five main branches: Ras, Rho, Rab, Sar/Arf and Ran. Like other small GTPases Rho are molecular switches playing the role of intracellular timers in signal transduction cascades. Bound to GTP they are in their active form and upon hydrolysis of the nucleotide into GDP they return back to their inactive state. Their ability to hydrolyze GTP into GDP, in conjunction with a helper protein (GTPase-activating-protein GAP), has led to them being named small GTPases. Members of the Rho subfamily of GTPases are clearly preferred targets of several bacterial toxins and virulence factors which can manipulate them to either activate or inhibit these proteins permanently. The *Escherichia coli* cytotoxic necrotizing factor (CNF1) and dermonecrotizing toxin (DNT) from *Bordetella* spp. block the GTP hydrolyzing activity of Rho, keeping the molecule permanently active whereas exoenzyme C3 of *Clostridium botulinum* and *Clostridium difficile* toxins A and B, by modifying covalently the Rho effector domain, inactivate the GTPases. *Clostridium sordellii* lethal toxin, a molecule related to *C. difficile* toxins A and B, also acts on the Ras

METHODS IN MICROBIOLOGY, VOLUME 31
ISBN 0–12–521531–2

subfamily of proteins. Rho-interacting bacterial toxins are, to date, pivotal tools in cell biology to probe the cellular effects of these small regulatory proteins.

◆◆◆◆◆◆ ACTIVITIES OF RHO GTPASES

The amino acid sequences of the different members of the Ras superfamily have shown a high conservation in their primary structure (30 to 55% of homologies one to each other) (Valencia *et al.*, 1991). This is mainly because small GTP-binding proteins have consensus sequences responsible for specific interaction with GDP/GTP and for GTPase activity. They contain in addition a particular domain for their interaction with their downstream effectors (see below). Finally, small GTPases display specific polypeptidic sequences at their carboxy-termini modified by lipids which allow them to stably associate with membranes (Zhang and Casey, 1996).

Small GTPases are molecular switches and more precisely molecular timers (Takai *et al.*, 2001) due to their ability to bind either GDP or GTP. When a signal is detected there is first dissociation of GDP from the small GTP-binding protein in its resting state. Dissociation of GDP is provoked by a guanine exchange factor (GEF). When the GDP bound is released, the empty nucleotide domain of the GTPase is almost immediately refilled by GTP since this nucleotide is about ten times more abundant than GDP in the cytosol. Bound to GTP there is a change of conformation of the so-called effector domain of the GTPase (the switch 1) which can now interact and stimulate a downstream effector. To turn down the activated GTPase, GTP associated with the regulatory protein must be hydrolyzed with release of inorganic phosphate and generation of GDP. Without an additional protein, named the GTPase activating protein (GAP), hydrolysis of GTP into GDP by the small GTP-binding protein is extremely slow. The GAP protein accelerates GTP hydrolysis several thousand-fold, rapidly turning the GTPase back to its inactive state. Different from Ras, Ral and Rap, the Rho and Rab subfamily molecules are regulated in addition by another protein, named the guanine dissociation inhibitor (GDI). Rho GDI and Rab GDI inhibit the basal and the GEF-stimulated dissociation of GDP, maintaining cytosolic the small GTPase in its GDP-bound form. Another role of GDI is to remove the small GTP-binding from the membrane. We have represented in Figure 12.1 the cycle of activation/deactivation of members of the Rho GTPases together with the points of impact of bacterial toxins. This cycle is particularly important to understand how bacterial toxins interfere with these small regulatory proteins.

The Rho GTPases subfamily, which is the most common target of bacterial toxins, encompasses, in higher eukaryotic cells, at least 16 members RhoA, B, C, D, G, Rac1, Rac2, Cdc42, TC10, Rnd1, Rnd2, Rnd3, RhoE, Rho6, Rho7 and Rho8 (Takai *et al.*, 2001). These regulatory molecules are mainly involved in dynamic rearrangements of the actin cytoskeleton implicated in different fundamental cellular processes such as polarization, cytokinesis or cell migration (Hall, 1998).

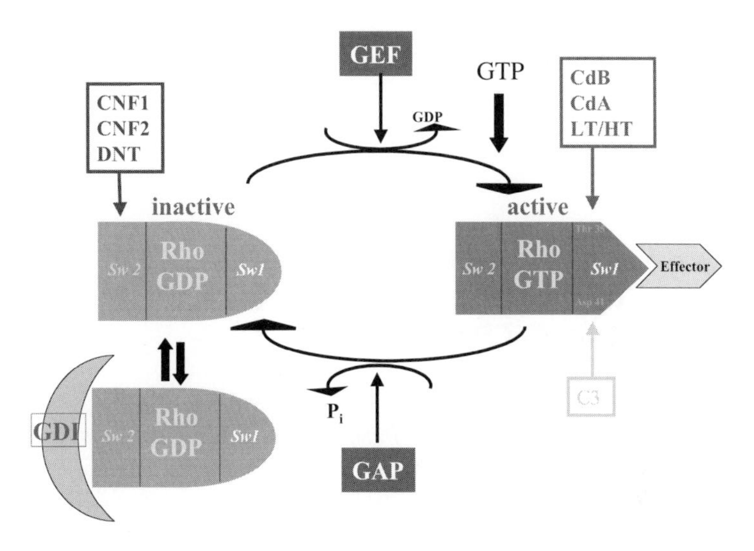

Figure 12.1. The Rho activation/deactivation cell cycle. Rho in its inactive form (associated with GDP) is brought to the membrane where the guanine exchange facture (GEF) allows removal of GDP with exchange with GTP, turning Rho into the active form (bound to GTP). Rho docks to its effector and activates it. To turn off Rho activity, the GTPase activating protein (GAP) introduces an arginine residue in close proximity to Rho glutamine 63, facilitating the hydrolysis of GTP into GDP thereby releasing the GTPase in its inactive conformation. Rho GDP is stocked in the cytosol associated with the protein GDI (guanine dissociation inhibitor). Interaction of bacterial toxins with Rho are indicated by colored arrows. CdB: *Clostridium difficile* toxin B. CdA: *Clostridium difficile* toxin A. LT: *Clostridium sordellii* lethal toxin. HT: *Clostridium sordellii* hemorrhagic toxin. CNF1/2: cytotoxic necrotizing facture 1/2. DNT: dermonecrotic toxin from *Bordetella* sp. (This figure is also reproduced in colour between pages 276 and 277.)

Rho regulates formation of stress fibers and focal adhesion points (Ridley and Hall, 1982). Rac regulates formation of lammellipodia (also named membrane ruffling) (Ridley *et al.*, 1992). Cdc42 controls the formation of filopodia (Nobes and Hall, 1995). These GTPases play also other additional roles and particularly are involved in the control of genes transcription (Minden *et al.*, 1995).

A large number of downstream effectors of Rho proteins have been identified (Takai *et al.*, 2001) but their relative importance in the signalling pathways controlled by Rho GTPases are still not fully defined except for the Rho kinase (ROK or ROCK), the p21 activated kinase (PAK) or the N-Wiskott Aldrich syndrome protein (N-WASP) which seems to represent the main targets for respectively Rho, Rac or Cdc42.

A hierarchy for Rho GTPases activation in Swiss 3T3 cells has been established; Cdc42 driving activation of Rac which finally stimulates Rho (Ridley *et al.*, 1992). If this cascade of Rho GTPases activation seems to apply to the Swiss 3T3 cell model it is clear to date that other cell species do not follow this cascade of activation. In different cell types Rac and Cdc42 down regulate the activation of Rho (Leeuwen *et al.*, 1997) whereas stimulation of Rho inhibits Rac and Cdc42 (Sander *et al.*, 1999). This might be important for the mode of action of the C3 family of exoenzymes which, by their specific inhibitory effects on Rho, may up-regulate Rac and Cdc42 in certain cell types.

◆◆◆◆◆◆ A GENERAL DESCRIPTION OF BACTERIAL TOXINS ACTIVATING OR INHIBITING RHO GTPASES

Before examining molecular and cellular effects of bacterial toxins modifying small GTPases it seems to be important to briefly recapitulate the different members of these proteins isolated to date.

Toxins active on small GTPases can be divided into three families: The C3 family of exoenzymes, the large clostridial family of cytotoxins and the dermonecrotic family of toxins.

The C3 family of exoenzymes is composed of *Clostridium botulinum* exoenzyme C3, *Clostridium limosum* exoenzyme C3, *Bacillus cereus* exoenzyme C3, epidermal differentiation inhibitor (EDIN) from *Staphylococcus aureus* and C3 (Stau) also from *Staphylococcus aureus*. These proteins are not *bona fide* toxins since they do not contain a cell binding/membrane translocating domain. To overcome this problem, C3 has been fused to the cell entry machinery of other toxins (Aullo *et al.*, 1993; Barth *et al.*, 1998) or to membrane translocating polypeptide (Sebbagh *et al.*, 2000; Coleman *et al.*, 2001).

The large clostridial toxins group contains *Clostridium difficile* toxin A, B, *Clostridium novyi* α toxin, lethal toxin (LT) and hemorrhagic (HT) toxins from *Clostridium sordellii*.

The dermonecrotic family of toxins encompasses the cytotoxic necrotizing factor 1 and 2 (CNF1, CNF2) from pathogenic *Escherichia coli*, the dermonecrotic toxin from *Bordetella bronchiseptica* (a CNF1 gene has been recently identified in the genome of *Yersinia pestis* (http://www.sanger.ac.uk/projects/Ypestis.)) and the *Pasteurella multocida* toxin.

The catalytic domain of two toxins (C3 and CNF1) active on small GTP-binding protein has been crystallized (Han *et al.*, 2001; Buetow *et al.*, 2001) Their structures are very informative and we will describe them briefly below.

◆◆◆◆◆◆ EXOENZYME C3 AND THE FAMILY OF EXOENZYMES ACTIVE ON RHO GTPASES

Exoenzyme C3 from *Clostridium botulinum* C and D serotypes is the paradigm of the family of C3 ADP-ribosyltransferases which covalently modify the Rho subfamily of GTPases (Table 12.1). Exoenzyme C3

Table 12.1 Targets of Rho-modifying bacterial toxins

Toxins	Rho	Rac	Cdc42	RhoG	TC10	RhoD	RhoE	Ras	Rap	Ral
C3*	+	−	−	−	−	−	−	−	−	−
C3 Stau*	+	−	−	−	−	−	+	−	−	−
CdtA*	+	+	+	+	+	−	−	−	+	−
CdtB*	+	+	+	+	+	−	−	−	−	−
CsLT*	−	+	−	+	+	−	−	+	+	+
CNF1†	+	+	+	−	−	−	−	−	−	−

C3: exoenzyme C3, C3 Stau, C3 from *Staphylococcus aureus*. CdtA: *Clostridium difficile* toxin A. CdtB: *Clostridium difficile* toxin B. CsLT: *Clostridium sordellii* lethal toxin. CNF1: cytotoxic necrotizing factor 1.
*Toxins inhibiting Rho GTPases.
†Toxin activating Rho GTPases.

catalyses the hydrolysis of cellular NAD^+ into nicotinamide and the concomitant transfer of the ADP-ribose moiety to Rho proteins (Chardin *et al.*, 1989; Rubin *et al.*, 1988; Aktories *et al.*, 1988). The family of exoenzyme C3 encompasses various isoforms of C3 itself (Moriishi *et al.*, 1993) and exoezymes from *Clostridium limosum* (Just *et al.*, 1992), *Bacillus cereus* (Just *et al.*, 1995a), EDIN (Sugai *et al.*, 1992) and Stau (Wilde *et al.*, 2001) from *Staphylococcus aureus* (Table 12.1).

C3 exoenzyme and C3-like exoenzymes are proteins with molecular masses of about 25 kDa. They are lacking membrane binding and translocation domains. These enzymes are strongly positively charged (isoelectric point of about 11) which allows them to stick to the cell membrane (which is negatively charged) and, as classically reported, to enter the cytosol by fluid phase pinocytosis. This idea has been challenged by the recent description that pore-forming toxins such as streptolysin O, elaborated by Gram-positive bacteria, could be a component by which certain of their virulence factors might be translocated into the host cytosol (Madden *et al.*, 2000), thus equivalent to the type III secretion system of Gram negative bacteria (Hueck, 1998). Following this stimulating hypothesis, we could assume that C3 exoenzyme might be translocated into the cytosol through the plasma membrane by a channel formed by the pore-forming botulysin O (an SLO-like pore-forming toxin elaborated by *C. botulinum*).

ADP-ribosylation of Rho A, B and C (and in addition Rnd3, RhoE by Stau (Table 12.1)) occurs at the asparagine 41 residue (N41) located at the border of the effector domain (switch 1) of the GTPases (Sekine *et al.*, 1989). The amino acids sequence $FEN_{41}Y$ was thought to be the only requirement for C3 to recognize RhoA, B and C since the sequence $FDN_{39}Y$, present in Rac and Cdc42, appeared not to allow modification of these GTPases by C3 excepted after their denaturation by low concentration of sodium dodecyl sulfate (Just *et al.*, 1992). Features of RhoA, B and C which make them specific substrates of C3 has recently been studied by making point mutations into Rac to render this GTPase ADP-ribosylable (Wilde *et al.*, 2000). Recognition and catalysis of Rho A by C3 has been achieved also by the recent resolution of the crystal structure of this enzyme (Han *et al.*, 2001)

By constructing a Rac mutant which could be a good substrate for C3, studies have been made concerning which particular amino acids define the substrate specificity of C3 for RhoA (Wilde *et al.*, 2000). This approach led to the conclusion that only the N-terminal 90 residues of Rho A were involved in its specific recognition by C3. In this sequence six residues (Arg 5, Lys 6, Glu 40, Val 43, Glu 47 and Glu 54) present in RhoA and absent in Rac are critical since a mutant Rac harboring these amino acids shows kinetic properties with regard to ADP-ribosylation identical to RhoA. According to this study, RhoA glutamic acid 40 and valine 43 are directly involved in the correct formation of the ternary complex formed by C3, RhoA and NAD^+. Basic RhoA residues arginine 5 and lysine 6, and acidic residues glutamic acids 47 and 54 together with glutamic acid 40 and valine 43 are implicated in the the binding of C3 to RhoA.

The C3 3D structure identifies a bipartite recognition specificity motif for RhoA ADP-ribosylation, termed the ARTT (ADP-ribosylating toxin turn turn) (Han *et al.*, 2001). In its core catalytic domain, C3 contains two antiparallel β-sheets forming a cleft at their interface which accommodates NAD^+. This structure is found in other ADP-ribosylating toxins. However, unlike many ADP-ribosylating toxins, C3 does not contain a loop which acts as an arm to recognize the ADPribose acceptor substrate (EF2 for DT as an example). In contrast, the loop in C3 is replaced by two adjacent protruding turns joining two β plated sheets of the toxin core fold. Turn 1 positions a hydrophobic phenylalanine residue (phe 209) which probably interacts with the hydrophobic region of Rho A, B and C formed by valine 38, phenylalanine 39, valine 43 and tryptophane 58. Turn 2 positions a glutamine residue (gln 212) which recognizes asparagnine 41 of Rho.

We do not know exactly the mechanism by which ADP-ribosylation of Rho A interferes with the effects of the GTPase on its downstream effectors. ADP-ribosylation of RhoA has several effects on the activation/ deactivation cycle of the GTPase: it inhibits the interaction with GEFs thus leaving the protein associated with GDP leading RhoA associated with GDI thus unable to translocate to the cell membrane (Figure 12.2). Stable association of Rho with RhoGDI and impairment of the GTPase to translocate to the cell membrane have been proposed to explain the inhibitory effect of Rho ADP-ribosylation by C3 (Sehr *et al.*, 1998; Fujihara *et al.*, 1997). Inhibition of Rho activation but not inhibition of Rho effector interaction is thus probably the major mechanism underlying inhibition of Rho by C3 ADP-ribosylation (Barth *et al.*, 1999).

◆◆◆◆◆◆ *CLOSTRIDIUM DIFFICILE* TOXIN B AND THE FAMILY OF LARGE CLOSTRIDIAL CYTOTOXINS

Clostridium difficile toxin A (CdtA) and toxin B (CdtB), *Clostridium sordelli* lethal (LT) and haemorrhagic (HT) and *Clostridium novyi* α-toxin have been grouped in the 'large clostridial cytotoxins' family, due to their MW < 250 kDa, and their common enzymatic activity identified as

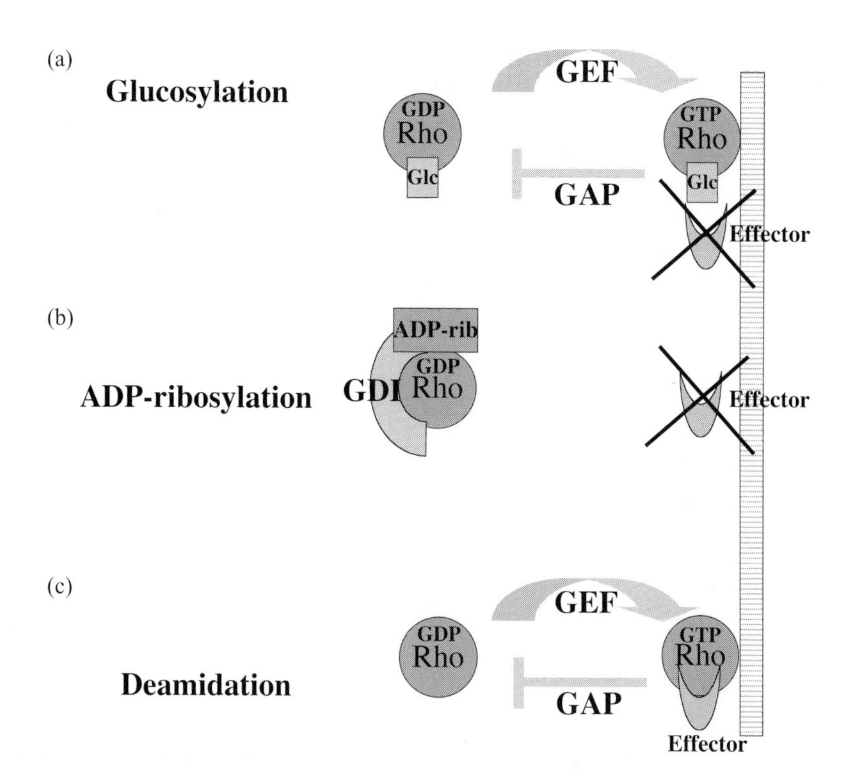

Figure 12.2. Modifications of Rho GTPases by C3, large clostridial toxins and CNF1: consequences on the inhibition/stimulation on the Rho effector. a: Glucosylation by large clostridial cytotoxins inhibits the GAP activity thereby leaving the protein associated with the membrane in its GTP bound form. GDI, which recognizes Rho associated with GDP, is inefficient in removing the GTPase from the membrane. The glucose moiety associated with the effector domain of Rho impairs the docking of the GTPase to its effector. b: C3-ADPribosylation blocks Rho on GDI in its GDP conformation. The GTPase is therefore not translocated to the membrane, cannot be activated by GEF and does not dock on the effector. c: CNF1-deamidation: Rho glutamine 63 is deamidated into glutamic acid rendering the hydrolysis of GTP by the GAP inefficient. The GTPase stays permanently associated with GTP thus stimulating the effector. (This figure is also reproduced in colour between pages 276 and 277.)

mono-glucosyltransferases (Von Eichel-Streiber *et al.*, 1996). Large clostridial cytotoxins are *bona fide* exotoxins which induce morphological changes of cultured target cells due to alteration of the actin microfilament sytem. These cytotoxins are functionally and structurally related. Large clostridial cytotoxins are the major virulence factors of *Clostridium difficile*, a bacteria causative of about 20% of the antibiotic-associated diarrhoea, and systematically of pseudomembranous colitis.

Clostridium difficile toxin B and A are encoded by two genes in tandem on the genome. The bacterium therefore produces the two cytotoxins together. Toxin B is about 100 to 1000 times more toxic than toxin A to cultured cells and has been named cytotoxin. Toxin A, but not toxin B, induces fluid accumulation in animal models and ileal explants and was therefore designated enterotoxin. The cytotoxic effect of both toxins A and B is characterized by rounding up of cells with formation of retraction fibers. By specific staining of the actin cytoskeleton using fluorescent derivatives of phalloidin it can be observed that the cell spanning actin cables (stress fibers) disappear with accumulation of short actin filaments in the perinuclear space (Fiorentini *et al.*, 1990). On polarized epithelial cells (T84) cultivated on filters, application of toxin B opens tight junctions resulting from alteration of the membrane microdomain localization of tight junction proteins (Nusrat *et al.*, 2001).

Toxin A (308 kDa) and toxin B (270 kDa) are intracellularly actin toxins (Von Eichel-Streiber *et al.*, 1996). These molecules are single-chain proteins divided into three domains. The N-terminal 50 kDa domain is the enzymatic part of toxin A and B (Hoffmann *et al.*, 1997) and the receptor binding domain is located in its C-terminal moiety (Sauerbaun *et al.*, 1997). This toxin region is composed of repetitive elements (Von Eichel-Streiber *et al.*, 1996). Cell receptors for toxin A and B are probably carbohydrate structures but their exact composition is still a matter of debate. Toxin A and B enter by receptor-mediated endocytosis and must reach an acidic compartment to translocate their enzymatic activity in the cytosol (Florin and Thelestam, 1986). In this respect it has been demonstrated that toxin B can form channels in artificial lipidic membranes (Barth *et al.*, 2001). Toxin B also can be introduced in the cytosol through the plasma membrane by a low pH pulse (Barth *et al.*, 2001; Qa'Dan *et al.*, 2000). The toxin domain implicated in membrane translocation is most likely located between the enzymatic and the cell binding toxin domains. This toxin domain probably unfolds at acidic pH. It is not yet known which intracellular acidic compartment (early or late endosomes) is the site of toxin B entry into the cytosol.

Toxin B, A, lethal toxin and haemorrhagic toxin have been identified as mono-glucosyltransferases modifying Rho GTPases (Just *et al.*, 1995b,c; Popoff *et al.*, 1996; Genth *et al.*, 1996). The catalytic domain of these toxins transfer a glucose moiety from UDP-glucose (UDP-N-acetyl-glucosamine in the case of *C. novyi* alpha toxin (Selzer *et al.*, 1996)) to the target proteins. In addition to a UDP-sugar, large clostridial cytotoxins require (like many eukaryotic glucosyl transferases) potassium and manganese ions which bind to a DXD motif (Busch *et al.*, 1998). The presence of a conserved tryptophan residue in all the catalytic domain of large clostridial toxins is also required for the full expression of their enzymatic activities (Busch *et al.*, 2000). The target proteins are either Rho or Ras GTPases (Table 12.1).

Large clostridial cytotoxins catalyse the modification of a threonine residue localized in the switch 1 domain of the GTPases of the Rho or Ras subfamilies. In Rho A, B, C threonine 37 is modified. In Rac, Cdc42 or Ras the equivalent threonine 35 is the target of the toxins. Threonine 35/37 is located in the effector domain of the GTPases which couples the GTP-bound regulatory molecule to its downstream effector (Wittinghofer and Pai, 1991). The hydroxyl group of threonine 35/37 is solvent accessible only in the GDP-bound form of the GTPases therefore only this form of Rho or Ras will be modified by large cytotoxins (Just *et al.*, 1995b).

Chimeric toxins associating different parts of the catalytic domains *of* *C. difficile* toxin B and that of the *C. sordellii* lethal toxin have indicated that amino acids 406–468 of toxin B determine the specificity for Rho, Rac and Cdc42 (Hofman *et al.*, 1998). Residues 364–408 of *C. sordellii* lethal toxin determine the specificity for Rac and Cdc42 but not for Rho. The specificity of *C. sordellii* LT toxin for Ras is mediated by residues 408–516. The catalytic domain of large clostridial cytotoxins seems thus organized in modules to recognize their substrates.

Glucosylation of Rho GTPases induces a certain number of modifications of the Rho activation/deactivation cycle (summarized in Figure 12.2). It blocks effector coupling, slows down the release of bound GDP by Rho GEFs (Sehr *et al.*, 1998; Herrmann *et al.*, 1998), and inhibits RhoGAP stimulated hydrolysis of GTP impairing interaction with the GDI leading to permanent entrapment of the glucosylated Rho at the membrane level (Genth *et al.*, 1999). Finally, the 3D structure of Ras glucosylated at threonine 35 by the lethal toxin of *Clostridium sordellii* has been solved (Weller *et al.*, 2000). In this study, it is shown that due to steric constraints, the glucose moiety prevents the formation of the GTP conformation of the effector loop required for binding to the downstream target protein (Raf-kinase for Ras).

It is not clear at the moment why clostridial cytotoxins are required to be so large (>250 kDa) to transfer a catalytic domain of 50 kDa into the cytosol of targets cells. Probably an (or several) additional biological activity (ies) is (are) located in these toxins. In this respect, it has been reported recently that *C. difficile* toxin A was transported to mitochondria and could induce cell death before inducing glucosylation of the Rho GTPases (He *et al.*, 2000).

◆◆◆◆◆◆ THE CYTOTOXIC NECROTIZING FACTOR I (CNF I) FROM UROPATHOGENIC *ESCHERICHIA COLI*: THE PROTOTYPE OF RHO-ACTIVATING TOXINS

Escherichia coli is one of the most important causes of intestinal infections. This bacterium, which is a normal inhabitant of the large bowel, becomes pathogenic upon acquisition of genes encoding virulence factors. Several pathotypes of *E. coli* have been individualized since this bacterium causes a wide variety of extra-intestinal infections such as septicemia, neonatal meningitis and urinary tract infections (UTIs). *E. coli* strains implicated in UTIs often express exotoxins such as hemolysin and the cytotoxic necrotizing factor 1 (CNF1) together with colonization factors. Genes coding for exotoxins and colonizing factors are encoded within regions of chromosomal DNA called pathogenicity islands (PAIs) (Hacker *et al.*, 1997).

CNF1 was first described in the late 1980s as a toxin able to induce an impressive reorganization of cellular actin microfilaments in eukaryotic cells (Fiorentini *et al.*, 1988). It was later demonstrated that CNF1 activates the 21-kDa Rho GTP-binding proteins (Fiorentini *et al.*, 1994, 1995; Oswald *et al.*, 1994) through a novel catalytic activity for bacterial toxins which

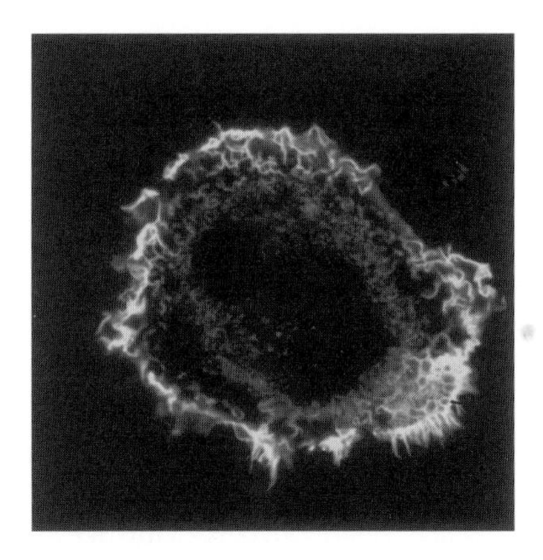
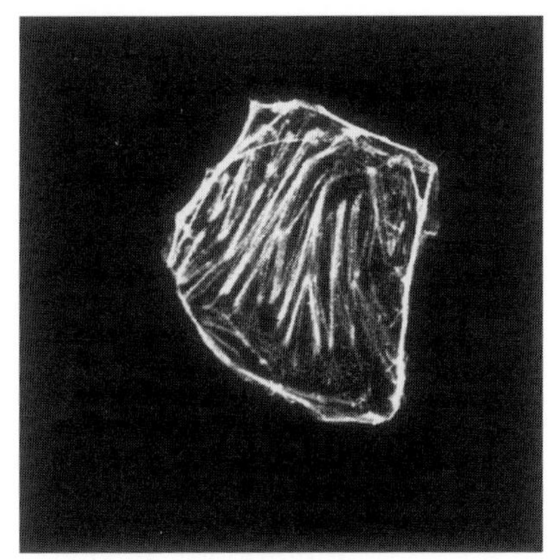

Figure 12.3. Modifications of the actin cytoskeleton induced by CNF1. The actin cytoskeleton is stained by fluorescein isothycyanate-phalloidin. Left panel: HEp-2 cell treated with 10^{-10} M CNF1 for 12 h. Note the membrane ruffling and the formation of pinocytose vesicles induced by CNF1. Right panel: Vero cells treated with 10^{-10} M CNF1 for 12 h. Note the almost exclusive formation by CNF1 of thick stress fibers.

consists in a specific deamidation of a glutamine residue of Rho GTPases (Flatau *et al.*, 1997; Schmidt *et al.*, 1997).

Fibroblastic cells such as Vero exhibit dense actin stress fibers formation whereas epithelial cells such as HEp-2 show mostly lammellipodia and filopodia (Figure 12.3). Spreading of cells follows CNF1 treatment. In HEp-2 cells, formation of large vacuoles is also observed in epithelial cells upon treatment with CNF1 (Figure 12.3). These vacuoles are different from those formed by *Helicobacter pylori* VacA toxin (Reyrat *et al.*, 1999) since they originate from massive pinocytosis triggered by the toxin. In this respect, CNF1 has been shown to induce the non-specific phagocytosis of bacteria and latex beads (Falzano *et al.*, 1993). This toxin effect may be related to the crossing of blood-barrier endothelial cells in meningitis caused by *E. coli* (Wang *et al.*, 2000), and the *cnf1* gene has been detected in *E. coli* K1 strains involved in neonatal meningitis (Bonacorsi *et al.*, 2000; Johnson *et al.*, 2001). In addition, the following activities of CNF1 have been described:

- Inhibition of cytokinesis thereby giving rise to multinucleated cells (Caprioli *et al.*, 1984).
- Activation of the phosphatidylinositol-4-phosphate-5 kinase (PIP5K) associated with the cytoskeleton (Fiorentini *et al.*, 1997a).
- Induction of a decrease in transepithelial electric resistance of Caco 2 cells which is accompanied by an increase paracellular permeability (Gerhard *et al.*, 1998)
- Impairment of CR3 phagocytosis but not the FCγR mediated ingestion of human macrophages (Capo *et al.*, 1998).
- Protection of HEp-2 cells against apoptotic stimuli (such as ultraviolet radiation (UVB)) (Fiorentini *et al.*, 1997b, 1998a). Cell spreading, induced by CNFI, is also involved in protection of apoptosis (Fiorentini *et al.*, 1998b). On the other hand induction of apoptosis in bladder cells by CNFI has been recently

described and suggested to be responsible for exfoliative shedding of cell layers that allow *E. coli* to gain access to deeper tissues in urinary tract infections (Mills *et al.*, 2000).
- Increase of the adherence to epithelia and oxidative burst in polymorphonuclear leukocytes, but decrease of their phagocytosis (Hofman *et al.*, 2000).

The N-terminal domain of CNF1 contains the cell-binding region of the toxin (Lemichez *et al.*, 1997; Fabbri *et al.*, 1999; Meysick *et al.*, 2001). More precisely, this domain has been delimited to the N-terminal first 190 residues of the toxin in which a polypeptide extending from amino acid 53 to 75 plays an important role for cell-binding of CNF1 (Fabbri *et al.*, 1999). The middle domain of CNF, extending from residues 190 to 720, contains hydrophobic structures and could play the role of a membrane translocating domain (T domain). The hydrophobic region 331–414 of CNF1 has been described as a potential membrane-spanning region (Falbo *et al.*, 1993). The hydrophobicity plot of the toxin shows a hydrophilic region (372–392) separating two strongly hydrophobic domains (H1 and H2) (Pei *et al.*, 2001), a feature also found in the diphtheria toxin molecule (DT). A computer search performed with a program analyzing transmembrane domains allows one to predict that this region could be organized in the same manner as the helix-loop-helix motif found in the DT H8-9. H1 helix could extend from 350 to 372, and H2 from 387 to 412, the CNF1-TL loop extending from 373 to 386. These two hydrophobic helices separated by a short hydrophilic loop (H1-2 CNF1 hairpin) appear, like the TH8-9 hairpin of DT (Oh *et al.*, 1999), to be required for the translocation of CNF1 catalytic activity through the membrane (Pei *et al.*, 2001).

The catalytic domain of CNF1 is located at the carboxy-terminus of the toxin (residues 720–1014) (Lemichez *et al.*, 1997; Schmidt *et al.*, 1998). The same architecture of CNF1 is shared by the other members of the dermonecrotic group of toxins: CNF2, the dermonecrotic toxin (DNT) from *Bordetella* and the *Pasteurella multocida* toxin (PMT)). However, the enzymatic activity of PMT, which has been demonstrated to be located in the C-terminal part of the molecule (Busch *et al.*, 2001) remains unknown.

CNF1 tightly binds to HEp-2 cells ($K_d = 20$ pM) (Contamin *et al.*, 2000). The nature and number of CNF1 cell-receptors are still unknown. After binding, CNF1 is endocytosed by both clathrin-dependent and -independent mechanisms (Contamin *et al.*, 2000). Endocytosed CNF1 is routed to the degradative pathway and joins the late endosome compartment. There, the toxin transfers its catalytic activity into the cytosol by an acidic-dependent membrane translocation mechanism reminiscent of that of diphtheria toxin. As for diphtheria toxin, the transfer of the deamidase activity of CNF1 into the cytosol can be induced directly across the membrane of cells by a brief exposure at pH 5.2 (Contamin *et al.*, 2000).

CNF1 provokes the deamidation of glutamine 63 into glutamic acid of Rho (Flatau *et al.*, 1997; Schmidt *et al.*, 1997) or glutamine 61 into glutamic acid of Rac and Cdc42 (Lerm *et al.*, 1999b). CNF1, CNF2 and the CNF1-related dermonecrotizing toxin of *Bordetella* sp. (DNT) act on Rho-GTPases by the identical activities (Sugai *et al.*, 1999; Horiguchi *et al.*, 1997). Glutamine 63 of Rho (or 61 of Rac and Cdc42) is essential for the intrinsic and GTPase-activating protein (GAP)-stimulated GTPase activity of Rho GTP binding proteins (Rittinger *et al.*, 1997). Therefore toxin-modified Rho GTPases are locked in their GTP-bound activated state (Figure 12.2). Deamidation of glutamine residues is observed with various transglutaminases (such as coagulation factor XIII or GMP synthetase). CNF1 also possesses a transglutaminase activity but, different from those enzymes, it selectively modifies Rho glutamine 63 (Schmidt *et al.*, 1998). The transglutaminase activity of CNF1 is however weak and most likely does not play a role in CNF1 activity in opposition to that elicited by the dermonecrotic toxin of *Bordetella pertussis* (DNT) toward Rho GTPases (Schmidt *et al.*, 1999; Masuda *et al.*, 2000). Residues Cys-866 and His-881 of CNF1 are required for deamidase activity (Schmidt *et al.*, 1998). These residues identify CNF1 as a member of the catalytic triad superfamily of deamidases/transglutaminases (Dodson and Wlodawer, 1998). Cystein 866 in CNF1 is involved in the nucleophilic attack of Rho glutamine 63. The activity of cystein-866 is increased by hydrogen bonding to histidine 881. The third residue of the deamidase/transglutaminase is valine 833 (Buetow *et al.*, 2001).

Specificity of CNF1 for the Rho subfamily of small GTP-binding proteins appears to be due, at least partly, to the highly conserved amino-acid sequence of the switch II domain, near glutamine 63 (or 61), of Rho, Rac and Cdc42 (Lerm *et al.*, 1999a; Flatau *et al.*, 2000). The switch II domain of the Rho GTPase (extending from residues 59 to 78) contains glutamine 63. A polypeptide covering aspartate 58 to aspartate 78 of RhoA was a substrate for CNF1 (Lerm *et al.*, 1999a). The minimal peptide shown to be a substrate for the deamidating activity of CNF1 was found to extend from aspartate 59 to leucine 69 of RhoA (Flatau *et al.*, 2000). Arginine 68 appears an important residue for the deamidation of RhoA (Lerm *et al.*, 1999a; Flatau *et al.*, 2000). Interestingly, this residue is also a key amino-acid for the binding of Rho GTPases to the regulatory domain of the guanosine dissociation inhibitor protein (GDI) (Hoffman *et al.*, 2000) which maintains cytosolic Rho GTPases in their GDP-bound state. This suggests that CNF1 does not modify Rho-GTPases when they are associated with Rho GDI.

Recently the 3D structure of the CNF1catalytic domain has been solved (Buetow *et al.*, 2001). The structure of the catalytic region of CNF1 is composed of a central core made by 13 pairs of anti-parallel β-plated sheet sandwiches that is surrounded by helices and extensive loop regions. This enzyme possesses a novel fold forming a pocket in which the catalytic triad cystein (cys 866), histidine (his 881) and valine (val 833) is buried. This pocket is deep and narrow, creating a filter that sterically excludes access of random substrates to the catalytic triad of amino acids. Modeling shows that this filter, in its rigid conformation, does not fit perfectly with the RhoA switch 2 loop region. Thus conformational changes in either the GTPase or the CNF1 or of both seems required to allow full deamidation of Rho glutamine 63.

All activities of CNF1, on cultured cells, can be explained by the stimulating effects of the toxin on Rho GTPases. Formation of actin stress fibers, lamellipodia and filopodia together with constitution of focal adhesion points are well-known actin structures induced and controlled by Rho, Rac and Cdc42 (Ridley and Hall, 1992). Pinocytosis is induced by activation of Rac (Ridley *et al.*, 1992). Cytokinesis is a process controlled by the small GTPases of the Rho family (Propenko *et al.*, 2000). Phagocytosis requires also the activation of Rho, Rac and Cdc42 (May and Machewsky, 2001; Patel *et al.*, 2000). Activation of the oxydative burst in polymorphonuclear leukocytes involves Rac (Abo *et al.*, 1991). Control of tight-junction permeability of polarized epithelial cells depends on the actin cytoskeleton hence on Rho GTPases (Nusrat *et al.*, 1995; Takaishi *et al.*, 1997). The role of Rho GTPases in the control of apoptosis is however less clear. Expression of the anti-apoptotic protein Bcl-2 may be modulated by Rho A (Gomez *et al.*, 1997). On the other hand, induction of apoptosis by activated Rho has been reported, although requiring complementary signals such as those inducing ceramide production (Lacal, 1997). In addition, stimulation of the JNK pathway by CNF1-stimulated Rac and Cdc42 may induce apoptosis by the mitochondrial pathway (Davis, 2000).

Recently it has been observed that activation of Rho GTPases by CNF1 in certain cell types (especially in primary cell cultures) is only transient, due to the subsequent specific degradation of the activated GTPases by a yet unidentified mechanism of proteolysis (Lemichez and Boquet, manuscript in preparation).

◆◆◆◆◆◆ CONCLUDING REMARKS

The discovery of *Clostridium botulinum* exoenzyme C3 has led to the elucidation of the role of the Rho GTPases in the control of the cytoskeleton organization therefore allowing us to understand the molecular mechanism of large clostridial cytotoxins together with the discovery of a new covalent modification of target proteins by bacterial toxins. C3 again was later used to unravel the mode of action of the cytotoxic necrotizing factor CNF1. To date, utilization of exoenzyme C3 (and the chimeric toxins derived from it), large clostridial toxins and CNF1 have allowed a flurry of data concerning the role of Rho GTPases in cell regulations.

It is curious that a physiological regulation of small GTPases of the Rho or Ras subfamilies in eukaryotic cells via endogenous mono-ADP-ribosylation or glucosylation does not seem to take place. On the other hand, recent data indicate that deamidation/transamidation (as those carried out by CNF1 or DNT) could be new post-translational modification used by cells for their regulation (Pepperkok *et al.*, 2000; Singh *et al.*, 2001). Thus the field of bacterial toxins associated with cell biology is still a fertile ground for new discoveries.

References

Abo, A., Pick, E., Hall, A., Totty, N., Teahan, C. G. and Segal, A. W. (1991). The small GTP-binding protein Rac is involved in the activation of the phagocyte NADPH oxidase. *Nature* **353**, 668–670.

Aktories, K., Rosener, S., Blaschke, U. and Chhatwal, G. S. (1988). Botulinum ADP-ribosyltransferase C3. Purification of the enzyme and characterization of the ADP-ribosylation reaction in platelet membranes. *Eur. J. Biochem.* **172**, 445–450.

Aullo, P., Giry, M., Olsnes, S., Popoff, M. R., Kocks, C. and Boquet, P. (1993). A chimeric toxin to study the role of the p21 kDa GTP-binding protein Rho in the control of microfilament assembly. *EMBO J.* **12**, 921–931.

Bar-Sagi, D. and Hall, A. (2000). Ras and Rho GTPases: a family reunion. *Cell* **103**, 227–238.

Barth, H., Hofmann, F., Olenik, C., Just, I. and Aktories, K. (1998). The N-terminal part of the enzyme component (C2I) of the binary *Clostridium botulinum* C2 toxin interacts with the binding component C2II and functions as a carrier system for a Rho ADP-ribosylating C3-like fusion toxin. *Infect. Immun.* **66**, 1364–1369.

Barth, H., Olenik, C., Sehr, P., Schmidt, G., Aktories, K. and Meyer, D. K. (1999). neosynthesis and activation of Rho by *Escherichia coli* cytotoxic necrotizing factor (CNF1) reverse cytopathic effects of ADP-ribosylated Rho. *J. Biol. Chem.* **274**, 27407–27414.

Barth, H., Pfeifer, G., Hofmann, F., Marer, E., Benz, R. and Aktories, K. (2001). Low-pH-induced formation of ion channels by *Clostridium difficile* toxin B in target cells. *J. Biol. Chem.* **276**, 10670–10676.

Bonacorsi, S. P., Clermont, O., Tinsley, C., Le Gall, I., Beaudoin, J. C., Elion, J., Nassif, X. and Bingen, E. (2000). Identification of regions of the *Escherichia coli* chromosome specific for neonatal meningitis-associated strains. *Infect. Immun.* **68**, 2096–2101.

Buetow, L., Flatau, G., Chiu, K., Boquet, P. and Ghosh, P. (2001). Structure of the Rho-activating domain of *E. Coli* cytotoxic necrotizing factor 1. *Nature Struct. Biol.* **8**, 584–588.

Busch, C., Hofmann, F., Selzer, J., Munro, J., Jeckel, D. and Aktories, K. (1998). A conserved motif of eukaryotic glycosyl-transferases is essential for the enzymatic activity of large clostridial cytotoxins. *J. Biol. Chem.* **273**, 19566–19572.

Busch, C., Hofmann, F., Gerhard, R. and Aktories, K. (2000). Involvement of a conserved tryptophan residue in the UDP-glucose binding of large clostridial cytotoxins glycosyltransferases. *J. Biol. Chem.* **275**, 13228–13234.

Busch, C., Orth, J., Djouder, N. and Aktories, K. (2001). Biological activity of a C-terminal fragment of *Pasteurella multocida* toxin. *Infect. Immun.* **69**, 3628–3634.

Capo, C., Sangueldoce, M. V., Meconi, S., Flatau, G., Boquet, P. and Mége, J. L. (1998). Effect of cytotoxic necrotizing factor 1 on actin cytoskeleton in human monocytes: role in the regulation of integrin-dependent phagocytosis. *J. Immunol.* **161**, 4301–4308.

Caprioli, A., Donelli, G., Falbo, V., Possenti, L., Roda, G., Roscetti, G. and Ruggeri, F. M. (1984). A cell division active protein from *Escherichia coli. Biochem. Biophys. Res. Commun.* **118**, 587–593.

Chardin, P., Boquet, P., Madaule, P., Popoff, M. R., Rubin, E. J. and Gill, D. M. (1989). The mammalian G protein Rho C is ADP-ribosylated by *Clostridium botulinum* C3 and affects actin microfilaments in Vero cells. *EMBO J.* **8**, 1087–1092.

Coleman, M. L., Sahai, E. A., Yeo, M., Bosch, M., Dewar, M. and Olson, M. F. (2001). Membrane blebbing during apoptosis results from caspase-mediated activation of ROCK-1. *Nature Cell Biol.* **3**, 339–345.

Contamin, S., Galmiche, A., Doye, A., Flatau, G., Benmerah, A. and Boquet, P. (2000). The p21-Rho-activating toxin cytotoxic necrotizing factor 1 is endocytosed by a clathrin-independent mechanism and enters the cytosol by an acidic dependent membrane translocation step. *Mol. Biol. Cell* **11**, 1775–1787.

and Davis, R. G. (2000). Signal transduction by the JNK group of MAP kinases. *Cell* **103**, 239–252.

Dodson, G. and Wlodawer, A. (1998). Catalytic triads and their relatives. *Trends Biochem. Sci.* **19**, 15–18.

Fabbri, A., Gauthier, M. and Boquet, P. (1999). The 5′ region of *cnf1* harbours a translational regulatory mechanism for CNF1 synthesis and encodes the cell-binding domain of the toxin. *Mol. Microbiol.* **33**, 108–118.

Falbo, V., Pace, T., Picci, L. and Caprioli, A. (1993). Isolation and nucleotide sequence of the gene encoding cytotoxic necrotizing factor type I. *Infect. Immun.* **9**, 1247–1254.

Falzano, L., Fiorentini, C., Donelli, G., Michel, E., Kocks, C., Cossart, P., Cabanié, L., Oswald, E. and Boquet, P. (1993). Induction of phagocytic behaviour in human epithelial cells by *Escherichia coli* cytotoxic necrotizing factor type I. *Mol. Microbiol.* **9**, 1061–1070.

Fiorentini, C., Arancia, G., Caprioli, A., Falbo, V., Ruggeri, F. M. and Donelli, G. (1988). Cytoskeletal changes induced in HEp-2 cells by the cytotoxic necrotizing factor of *Escherichia coli*. *Toxicon.* **26**, 1047–1056.

Fiorentini, C., Malorni, W., Paradisi, S., Giuliano, M., Mastrantonio, P. and Donelli, G. (1990). Interaction of *Clostridium difficile* toxin A with cultured cells: cytoskeletal changes and nuclear polarization. *Infect. Immun.* **58**, 2329–2336.

Fiorentini, C., Giry, M., Donelli, G., Falzano, L., Aullo, P. and Boquet, P. (1994). *Escherichia coli* cytotoxic necrotizing factor 1 increases actin assembly via the p21 Rho GTPase. *Zentralbl. Bakteriol. Suppl.* **24**, 404–405.

Fiorentini, C., Donelli, G., Matarrese, P., Fabbri, A., Paradisi, S. and Boquet, P. (1995). *Escherichia coli* cytotoxic necrotizing factor type I: evidence for induction of actin assembly by constitutive activation of the p21 Rho GTPase. *Infect. Immun.* **63**, 3936–3944.

Fiorentini, C., Fabbri, A., Flatau, G., Donelli, G., Matarrese, P., Lemichez, E., Falzano, L. and Boquet, P. (1997a). *Escherichia coli* cytotoxic necrotizing factor 1 (CNF1) a toxin that activates the Rho GTPase. *J. Biol. Chem.* **272**, 19532–19537.

Fiorentini, C., Fabbri, A., Matarrese, P., Falzano, L., Boquet, P. and Malorni, W. (1997b). Hinderance of apoptosis and phagocytic behaviour induced by *Escherichia coli* necrotizing factor (CNF1): two related activities in epithelial cells. *Biochem. Biophys. Res. Comm.* **241**, 341–346.

Fiorentini, C., Matarrese, P., Straface, E., Falzano, L., Fabbri, A., Donelli, G., Cossarizza, A., Boquet, P. and Malorni, W. (1998a). Toxin-induced activation of the Rho-GTP binding protein increases Bcl-2 expression and influences mitochondrial homeostasis. *Exp. Cell Res.* **242**, 341–350.

Fiorentini, C., Matarrese, P., Straface, E., Falzano, L., Donelli, G., Boquet, P. and Malorni, W. (1998b). Rho-dependent cell spreading activated by *E. coli* cytotoxic necrotizing factor 1 hinders apoptosis in epithelial cells. *Cell Death Diff.* **5**, 720–728.

Flatau, G., Landraud, L., Boquet, P., Bruzzone, M. and Munro, P. (2000). Deamidation of RhoA glutamine 63 by the *Escherichia coli* CNF1 toxin requires a short sequence of the GTPase switch 2 domain. *Biochem. Biophys. Res. Comm.* **267**, 588–592.

Flatau, G., Lemichez, E., Gauthier, M., Chardin, P., Paris, S., Fiorentini, C. and Boquet, P. (1997). Toxin-induced activation of the G-protein p21 Rho by deamidation of glutamine. *Nature* **387**, 729–733.

Florin, I. and Thelestam, M. (1986). Lysosomal involvement in cellular intoxication with *Clostridium difficile* toxin B. *Microb. Pathog.* **1**, 373–385.

Fujihara, H., Walker, L. A., Gong, M. C., Lemichez, E., Boquet, P., Somlyo, A. V. and Somlyo, A. P. (1997). Inhibition of RhoA translocation and calcium sensitization by *in vivo* ADP-ribosylation of Rho with the chimeric DC3B. *Mol. Biol. Cell.* **8**, 2437–2447.

Genth, H., Aktories, K. and Just, I. (1999). Monoglucosylation of RhoA at threonine 37 blocks cytosol-membrane cycling. *J. Biol. Chem.* **274**, 29050–29056.

Genth, H., Hofmann, F., Selzer, J., Rex, G., Aktories, K. and Just, I. (1996). Difference in protein substrate specificity between hemorrhagic toxin and lethal toxin from *Clostridium sordellii*. *Biochim. Biophys. Res. Commun.* **229**, 570–574.

Gerhard, R., Schmidt, G., Hofmann, F. and Aktories, K. (1998). Activation of Rho GTPases by *Escherichia coli* cytotoxic necrotizing factor 1 increases intestinal permeability in caco-2 cells. *Infect. Immun.* **66**, 5125–5131.

Gomez, J., Carlos-Martinez, A., Giry, M., Garcia, A. and Rebollo, A. (1997). Rho prevents apoptosis through Bcl-2 expression: implications for interleukin-2 receptor signal transduction. *Eur. J. Immunol.* **27**, 2793–2799.

Hacker, J., Blum-Oehler, G., Muhldorfer, H. and Tschäpe, H. (1997). Pathogenicity islands of virulent bacteria: structure, function and impact on microbial evolution. *Mol. Microbiol.* **23**, 1089–1097.

Hall, A. (1998). Rho GTPases and the actin cytoskeleton. *Science* **279**, 509–514.

Han, S., Arvai, A. S., Clancy, S. B. and Tainer, J. A. (2001). Crystal structure and novel recognition motif of Rho ADP-ribosylating C3 exoenzyme from *Clostridium botulinum*: Structural insights for recognition specificity and catalysis. *J. Mol. Biol.* **305**, 95–107.

He, D., Hagen, S. J., Pothoulakis, C., Chen, M., Medina, M. D., Warny, M. and Lamont, T. J. (2000). *Clostridium difficile* toxin A causes early damage to mitochondria in cultured cells. *Gastroenterology* **119**, 139–150.

Herrmann, C., Ahmadan, H. R., Hofmann, F. and Just, I. (1998). Functional consequences of monoglucosylation of H-Ras at effector domain amino acid threonine 35. *J. Biol. Chem.* **273**, 16134–16139.

Hofman, F., Busch, C. and Aktories, K. (1998). Chimeric clostridial cytotoxins: identification of the N-terminal region involved in protein substrate recognition. *Infect. Immun.* **66**, 1076–1081.

Hofmann, F., Busch, C., Prepens, U., Just, I. and Aktories, K. (1997). Localization of the glucosyl transferase activity of *Clostridium difficile* toxin B to the N-terminal part of the holotoxin. *J. Biol. Chem.* **272**, 11074–11078.

Hofman, P., Le Negrate, G., Mograbi, B., Hofman, V., Brest, P., Alliania-Schmid, A., Flatau, G., Boquet, P. and Rossi, B. (2000). *Escherichia coli* cytotoxic necrotizing factor-1 (CNF1) increases the adherence to epithelia and the oxidative burst of human polymorphonuclear leukocytes but decreases bacteria phagocytosis. *J. Leuk. Biol.* **68**, 522–528.

Hoffman, G. R., Nassar, N. and Cerione, R. A. (2000). Structure of the Rho family GTP-binding protein Cdc42 in complex with multifunctional regulator RhoGDI. *Cell* **100**, 345–356.

Horiguchi, Y., Inoue, N., Masuda, M., Kashimoto, T., Katahira, J., Sugimoto, N. and Matsuda, M. (1997). *Bordetella bronchiseptica* dermonecrotizing toxin induces reorganization of actin stress fibers through deamidation of Gln-63 of the GTP-binding protein Rho. *Proc. Natl Acad. Sci. USA* **94**, 11623–11626.

Hueck, C. J. (1998). Type III secretion systems in bacterial pathogens of animals and plants. *Microbiol. Mol. Biol. Rev.* **62**, 379–433.

Johnson, J. R., Delavari, P. and O'Bryan, T. T. (2001). *Escherichia coli* 018:K1:H7 isolates from patients with acute cystitis and neonatal meningitis exhibit common phylogenic origins and virulence factor profiles. *J. Infect. Dis.* **183**, 425–434.

Just, I., Mohr, C., Schallehn, G., Menard, L., Didsbury, J. R., Vandekerckhove, J. and Van Damme, J. (1992). Purification and characterization of an ADP-ribosyltransferase produced by *Clostridium limosum*. *J. Biol. Chem.* **267**, 10274–10280.

Just, I., Selzer, J., Jung, M., Van Damme, J., Vandekerckhove, J. and Aktories, K. (1995a). Rho ADP-ribosylating exoenzyme from *Bacillus cereus*: purification, characterization and identification of the NAD-binding site. *Biochemistry* **34**, 334–340.

Just, I., Selzer, J., Wilm, M., von Eichel-Streiber, C., Mann, M. and Aktories, K. (1995b). Glucosylation of Rho proteins by *Clostridium difficile* toxin B. *Nature* **375**, 500–503.

Just, I., Wilm, M., Selzer, J., Rex, G., Von Eichel-Streiber, C., Mann, M. and Aktories, K. (1995c). The enterotoxin from *Clostridium difficile* (Toxin A) monoglucosylates the Rho proteins. *J. Biol Chem.* **270**, 13932–13936.

Just, I., Selzer, J., Hofmann, F., Green, G. A. and Aktories, K. (1996). Inactivation of Ras by *Clostridium sordellii* lethal toxin-catalyzed glucosylation. *J. Biol. Chem.* **271**, 10149–10153.

Lacal, J. C. (1997). Regulation of proliferation and apoptosis by Ras and Rho GTPases through specific phospholipid-dependent signaling. *FEBS Lett.* **410**, 73–77.

Leeuwen, F. N., Kain, H. E., Van Der Kammen, R. A., Michiels, F. N., Kranenburg, O. W. and Collard, J. G. (1997). The guanine exchange factor Tiam 1 affects neuronal morphology: opposing roles for the small GTPases Rac and Rho. *J. Cell Biol.* **139**, 797–807.

Lemichez, E., Flatau, G., Bruzzone, M., Boquet, P. and Gauthier, M. (1997). Molecular localization of the *Escherichia coli* cytotoxic necrotizing factor 1 cell-binding and catalytic domains. *Mol. Microbiol.* **24**, 1061–1070.

Lerm, M., Schmidt, G., Goehring, U. M., Schrimer, J. and Aktories, K. (1999a). Identification of the region of Rho involved in substrate recognition by *Escherichia coli* cytotoxic necrotizing factor 1 (CNF1). *J. Biol. Chem.* **274**, 28999–29004.

Lerm, M., Selzer, J., Hoffmeyer, A., Rapp, U. R., Aktories, K. and Schmidt, G. (1999b). Deamidation of Cdc42 and Rac by *Escherichia coli* cytotoxic necrotizing factor 1: activation of C-Jun N-terminal kinase in HeLa cells. *Infect. Immun.* **67**, 496–503.

Madden, J. C., Ruiz, M. and Caparon, M. (2001). Cytolysin-mediated translocation (CMT): a functional equivalent of type III secretion in Gram-positive bacteria. *Cell* **104**, 143–152.

Masuda, M., Betancourt, L., Matsuzawa, T., Kashimoto, T., Takao, T., Shimonishi, Y. and Horiguchi, Y. (2000). Activation of Rho through a cross-link with polyamines catalyzed by Bordetella dermonecrotizing toxin. *EMBO J.* **19**, 521–530.

May, R. C. and Machewsky, L. M. (2001). Phagocytosis and the actin cytoskeleton. *J. Cell Sci.* **114**, 1061–1077.

Meysick, K. C., Mills, M. and O'Brien, A. D. (2001). Epitope mapping of monoclonal antibodies capable of neutralizing cytotoxic necrotizing factor type 1 of uropathogenic *Escherichia coli*. *Infect. Immun.* **69**, 2066–2074.

Mills, M., Meysick, K. C. and O'Brien, A. D. (2000). Cytotoxic necrotizing factor type 1 of uropathogenic *Escherichia coli* kills human uroepithelial 5637 cells by an apoptotic mechanism. *Infect. Immun.* **68**, 5869–5880.

Minden, A., Lin, A., Claret, F. X., Abo, A. and Karin, M. (1995). Selective activation of the JAK signalling cascade and C-Jun transcriptional activity by the small GTPase Rac and Cdc42. *Cell* **81**, 1147–1157.

Moriishi, K., Syuto, B., Saito, M., Oguma, K., Fujii, N., Abe, N. and Naiki, M. (1993). Two different types of ADP-ribosyltransferases C3 from *Clostridium botulinum* type D lysogenized organisms. *Infect. Immun.* **61**, 5309–5314.

Nobes, C. D. and Hall, A. (1995). Rho, Rac and Cdc42 regulate the assembly of multimolecular focal complexes associated with actin stress fibers, lamellipodia and filopodia. *Cell* **81**, 1–20.

Nusrat, A., Giry, M., Turner, J. R., Colgan, S. P., Parkos, S. A., Carnes, D., Lemichez, E., Boquet, P. and Madara, J. L. (1995). Rho protein regulates tight-junction and perijunctional actin organization in polarized epithelia. *Proc. Natl Acad. Sci. USA* **92**, 10629–10633.

Nusrat, A., Von Eichel-Streiber, C., Turner, J. R., Verkade, P., Madara, J. L. and Parkos, C. A. (2001). *Clostridium difficile* toxins disrupt epithelial barrier junctions by altering membrane microdomain localization of tight junction proteins. *Infect. Immun.* **69**, *1329–1336.*

Oh, K. J., Senzel, L., Collier, R. J. and Finkelstein, A. (1999). Translocation of the catalytic domain of diphtheria toxin across planar phospholipid bilayers by its own T domain. *Proc. Natl Acad. Sci. USA* **96**, 8467–8470.

Oswald, E., Sugai, M., Labigne, A., Wu, H. C., Fiorentini, C., Boquet, P. and O'Brien, A. D. (1994). Cytotoxic necrotizing factor type 2 produced by virulent *Escherichia coli* modifies the small GTP-binding protein Rho involved in assembly of actin stress fibers. *Proc. Natl Acad. Sci. USA* **91**, 3814–3818.

Patel, J. C., Hall, A. and Caron, E. (2000). Rho GTPases and macrophage phagocytosis. *Methods Enzymol.* **325**, 462–473.

Pei, S., Doye, A. and Boquet, P. (2001). Acidic residues of the CNF1 T domain are important for cell membrane translocation of the toxin. *Mol. Microbiol.*

Pepperkok, R., Holz-Wagenblatt, A., Konig, M., Girod, A., Bossmeyer, D. and Kinzel, V. (2000). Intracellular distribution of mammalian protein kinase A catalytic subunit altered by conserved Asn 2 deamidation. *J. Cell Biol.* **148**, 715–726.

Popoff, M. R., Chaves, O. E., Lemichez, E., Von Eichel-Streiber, C., Thelestam, M., Chardin, P., Cussac, D., Chavrier, P., Flatau, G., Giry, M., Gunzburg, J. and Boquet, P. (1996). Ras, Rap, and Rac small GTP-binding proteins are targets *for Clostridium sordellii* lethal glucosylation. *J. Biol. Chem.* **271**, 10217–10224.

Propenko, S., Saint, R. and Bellen, H. J. (2000). Untying the Gordian knot of cytokinesis: role of small G proteins and their regulators. *J. Cell Biol.* **148**, 843–848.

Qa'Dan, M., Spyres, L. M. and Ballard, J. D. (2000). pH-induced conformational changes in *Clostridium difficile* toxin B. *Infect. Immun.* **68**, 2470–2474.

Reyrat, J. M., Pelicic, V., Papini, E., Montecucco, C., Rappuoli, R. and Telford, J. L. (1999). Toward deciphering the *Helicobacter pylori* cytotoxin. *Mol. Microbiol.* **34**, 197–204.

Ridley, A. J. and Hall, A. (1992). The small GTP-binding Rho regulates the assembly of focal adhesion and actin stress fibers in response to growth factors. *Cell* **70**, 389–399.

Ridley, A. J., Paterson, H. F., Johnston, C. L., Dickmann, O. and Hall, A. (1992). The small GTP-binding Rac regulates growth factor-induced membrane ruffling. *Cell* **70**, 401–410.

Rittinger, K., Walker, P. A., Eccleston, J. F., Smerdon, S. J. and Gamblin, S. J. (1997). Structure at 1,65 A of RhoA and its GTPase-activiating protein in complex with a transisition state analogue. *Nature* **389**, 753–762.

Rubin, E. J., Gill, D. M., Boquet, P. and Popoff, M. R. (1988). Functional modification of a 21-kilodalton G protein when ADP-ribosylated by exoenzyme C3 of *Clostridium botulinum*. *Mol. Cell Biol.* **8**, 418–426.

Sander, E. E., Ten Klooster, J. P., Van Delft, S., Van der Kammen, R. A. and Collard, J. G. (1999). Rac downregulates Rho activity: reciprocal balance between both

GTPases determines cellular morphology and migration behavior. *J. Cell Biol.* **147**, 1009–1022.

Sauerbaum, M., Leukel, P. and Von Eichel-Streiber, C. (1997). The C-terminal ligand-binding domain of *Clostridium difficile* toxin A (TcdA) abrogates TcdA-specific binding to cells and prevents mouse lethality. *FEMS Microbiol. Lett.* **155**, 45–54.

Schmidt, G., Sehr, P., Wilm, M., Selzer, J., Mann, M. and Aktories, K. (1997). Rho Gln-63 is deamidated by *Escherichia coli* cytotoxic necrotizing factor 1. *Nature* **387**, 725–729.

Schmidt, G., Selzer, J., Lerm, M. and Aktories, K. (1998). The Rho-deamidating cytotoxic necrotizing factor 1 from Escherichia coli posseses transglutaminase activity. *J. Biol. Chem.* **273**, 13669–13674.

Schmidt, G., Goehring, U. M., Schirmer, J., Lerm, M. and Aktories, K. (1999). Identification of the C-terminal part of Bordetella dermonecrotic toxin as a transglutaminase for Rho GTPases. *J. Biol. Chem.* **274**, 31875–31881.

Sebbagh, M., Renvoizé, C., Hamelin, J., Riché, N., Bertoglio, J. and Bréard, J. (2001). Caspase 3-mediated cleavage of Rock 1 induces MLC phosphorylation and apoptotic membrane blebbing. *Nature Cell Biol.* **3**, 346–352.

Sehr, P., Joseph, G., Genth, H., Just, I., Pick, E. and Aktories, K. (1998). Glucosyla-tion and ADP-ribosylation of Rho proteins: effect on nucleotide binding, GTPase activity and effector coupling. *Biochemistry* **37**, 5296–5304.

Sekine, A., Fujiwara, M. and Narumyia, S. (1989). Asparagine residue in the Rho gene product is the modification site for botulinum ADP-ribosyltransferase. *J. Biol. Chem.* **264**, 8602–8605.

Selzer, J., Hofman, F., Rex, G., Wilm, M., Mann, M., Just, I. and Aktories, K. (1996). *Clostridium novyi* α-toxin-catalyzed incorporation of GlcNac into Rho subfamily proteins. *J. Biol. Chem.* **271**, 25173–25177.

Singh, U. S., Kunar, M. T., Kao, Y. L. and Baker, K. (2001). Role of transglutaminase II in retinoic acid-induced activation of RhoA associated kinase-2. *EMBO J.* **20**, 2413–2423.

Sugai, M., Hashimoto, K., Kikuchi, A., Inoue, S., Okumura, H., Matsumoto, K., Goto, Y., Ohgai, H., Moriishi, K. and Syuto, B. (1992). Epidermal cell differentiation inhibitor ADP-ribosylates small GTP-binding proteins and induces hyperplasia of epidermis. *J. Biol. Chem.* **267**, 2600–2604.

Sugai, M., Hatazaki, K., Mogami, A., Ohta, H., Peres, S. Y., Herault, F., Horiguchi, Y., Masuda, M., Ueno, Y., Komatsuzawa, H., Suginata, H. and Oswald, E. (1999). Cytotoxic necrotizing factor type 2 produced by pathogenic *Escherichia coli* deamidates a gln residue in the conserved G-3 domain of the rho family and preferentially inhibits the GTPase activity of RhoA and rac1. *Infect. Immun.* **67**, 6550–6557.

Takai, Y., Sasaki, T. and Matozaki, T. (2001). Small GTP-binding proteins. *Physiological Rev.* **81**, 153–208.

Takaishi, K., Sasaki, T., Kotani, H., Nishioka, H. and Takai, Y. (1997). Regulation of cell-cell adhesion by Rac and Rho small G proteins in MDCK cells. *J. Cell Biol.* **139**, 1047–1059.

Valencia, A., Chardin, P., Wittinghofer, A. and Sander, C. (1991). The Ras protein family: evolutionary tre and role of conserved amino acids. *Biochemistry* **30**, 4637–4648.

Von Eichel-Streiber, C., Boquet, P., Sauerborn, M. and Thelestam, M. (1996). Large clostridial cytotoxin: a family of glycosyl transferases modifying small GTP-binding proteins. *Trends Microbiol.* **4**, 375–382.

Wang, Y., Wass, C. A. and Kim, K. S. (2000). Cytotoxic necrotizing factor 1 promotes *E. coli* K1 invasion of brain microvascular endothelial cells in vitro and in vivo. *In 100thAmerican Society for Microbiology, General Meeting 65*. Los Angeles.

Weller, I. R., Hofmann, F., Wohlgemuth, S., Herrmann, C. and Just, I. (2000). Structural consequences of mono-glucosylation of Ha-Ras by *Clostridium sordellii* lethal toxin. *J. Mol. Biol.* **301**, 1091–1095.

Wilde, C., Genth, H., Aktories, K. and Just, I. (2000). Recognition of RhoA by *Clostridium botulinum* C3 exoenzyme. *J. Biol. Chem.* **275**, 16478–16483.

Wilde, C., Chhatwal, G. S., Schmalzing, G., Aktories, K. and Just, I. (2001). A novel C3-like ADP-ribosyl transferase from *Staphylococcus aureus* modifying RhoE and Rnd3. *J. Biol. Chem.* **276**, 9537–9542.

Wittinghofer, A. and Pai, E. F. (1991). The structure of Ras protein: a model for universal switch. *Trends Biochem. Sci.* **16**, 382–387.

Zhang, F. L. and Casey, P. J. (1996). Protein prenylation: molecular mechanism and functional consequences. *Annu. Rev. Biochem.* **65**, 241–269.

13 Actin-based Motility of *Listeria monocytogenes* and *Shigella flexneri*

Sebastian Wiesner, Rajaa Boujemaa-Paterski and Marie-France Carlier
Dynamique du Cytosquelette, Laboratoire d'Enzymologie et Biochimie Structurale, CNRS, Gif-sur-Yvette, 91198 France

◆◆

CONTENTS

◆◆◆◆◆◆ INTRODUCTION

Spatially controlled assembly of actin filaments, in response to environmental factors, is responsible for the formation of cell protrusions such as lamellipodia, filopodia and pseudopodia. It is by using these extensions that motile cells explore the extracellular space and find their way toward their targets, in chemotactic locomotion, embryonic/metastatic cell migration or in morphogenetic processes like neural cone growth. Actin polymerization is also involved in the movement of endosomes, of cortical actin patches in yeast, and in the formation of phagosomes. Bacterial pathogens, upon entry into or adhesion to epithelial cells, induce local actin polymerization (see Chapter 11). To summarize, actin assembly is the hallmark of the motile response of cells to extracellular signals.

Interestingly, a small number of intracellular pathogens, such as *Listeria monocytogenes*, *Shigella flexneri* or *Rickettsia* (as well as a few viruses like vaccinia virus), following escape from the phagocytic vacuole, induce actin polymerization at their surface by harnessing the host machinery leading to actin assembly in response to signaling (see Higley and Way, 1997; Beckerle, 1998; Dramsi and Cossart, 1998, for reviews). The bacteria continuously generate an actin meshwork or 'tail' that promotes their propulsion in the host cytoplasm. A single

METHODS IN MICROBIOLOGY, VOLUME 31
ISBN 0-12-521531-2

bacterial factor, the *Listeria* protein ActA or the unrelated *Shigella* protein IcsA (also called virG) is required for actin assembly and bacterial movement (Kocks *et al.*, 1992; Domann *et al.*, 1992; Bernardini *et al.*, 1989). Therefore these pathogens have been considered powerful tools to approach the mechanism of actin-based motility. Major progress in this direction was accomplished when actin-based motility of *Listeria* was reconstituted *in vitro* in cell extracts (Theriot *et al.*, 1994). *Xenopus* egg extracts, the conventional tool in studies on microtubule dynamics, were first successfully used, followed by platelet extracts, brain and Hela cell extracts (Welch *et al.*, 1997; Laurent and Carlier, 1998; Theriot and Fung, 1998; Laurent *et al.*, 1999). The *in vitro* motility assay in cell extracts presented several advantages. The extracts were easy to prepare, could be stored at $-80^{\circ}C$ in small aliquots for long periods of time without loss in activity, and could be immunochemically and biochemically manipulated. Depletion and add-back studies of actin regulatory proteins could be performed. Cell extracts were helpful to demonstrate that the protein IcsA was sufficient to induce actin-based motility of *Shigella*, and could induce motility in non-pathogenic *E. coli* (Goldberg and Theriot, 1995; Kocks *et al.*, 1995). Fractionation of platelet extracts led to the identification of Arp2/3 complex as the cellular factor responsible for local activation of actin assembly at the surface of *Listeria* (Welch *et al.*, 1997), and the important discovery that its biological activity required interaction with the *Listeria* protein ActA (Welch *et al.*, 1998). Recruitment and activation of Arp2/3 complex by *Shigella* was also demonstrated using cell extracts (Egile *et al.*, 1999).

Although motility assays of *Listeria* in cell extracts were instrumental in the discovery of essential factors like Arp2/3 complex in actin-based movement, they had many drawbacks. First, only plasmid-transformed bacterial strains expressing high levels of ActA or IcsA were motile in cell extracts. Wild type and mutated standard strains did not move, which limited the functional analysis of the domains of the bacterial protein involved in motility. Second, the active state of all putative essential components of the extracts could not be controlled, resulting in poor reproducibility of the motile behavior from one batch to another. Finally, technical problems arose in depletion studies due to the possible uncontrolled co-removal of ligands of the depleted protein. The effects of the depletion were therefore difficult to attribute to a single factor in a non-ambiguous fashion. Failure to restore 100% of motility upon adding back the pure protein was often difficult to interpret.

The full reconstitution, from a minimum set of pure components, of actin-based motility of *Listeria* or *Shigella* or any particle able to initiate the cascade of events leading to local activation of polarized actin assembly, was a challenging issue. Its achievement, two years ago, has opened up new perspectives both in the experimental approach to the mechanism of production of force by vectorial assembly of a polymer, and in the potential ways to overcome the virulence of these pathogens.

◆◆◆◆◆◆ DEVELOPMENT OF A CONSTITUTIVE MOTILITY MEDIUM FROM PURE PROTEINS

Rationale: functions and players in actin-based motility

Actin treadmilling

Observations of the forward movement of the leading edge of motile cells and of the asssociated dynamics of actin filaments in the extending lamellipodia clearly pointed out to two striking features. Filaments with their barbed ends pointing toward the leading edge were constantly generated and growing at the plasma membrane during movement, while depolymerizing at the rear of the lamellipodium, in a rapid treadmilling process (Wang, 1985; Small, 1995). Lamellipodium extension therefore results from the steady state cycle of actin assembly, the flux of depolymerized actin subunits being used to feed the growth of barbed ends formed at specific sites at the leading edge. Movement of *Listeria* and *Shigella* uses the same mechanism, therefore the length of the actin tail as well as the rate of propulsion are maintained constant due to the constant recycling of actin subunits. Two essential biochemical functions appear necessary to maintain stationary movement: (1) filaments exposing free barbed ends able to grow actively must be maintained at specific sites; (2) growth of these filaments at the specific sites must be made possible by the maintenance of a concentration of monomeric actin well above the critical concentration for barbed end assembly.

Recent progress has been made in the genetic identification of proteins that are involved in cell motility and in the biochemical analysis of proteins that control the dynamics of actin filaments (Carlier, 1998; Carlier *et al.*, 1999). These advances have helped to elaborate the composition of a minimum reconstituted medium that would be made of pure proteins and would support sustained actin-based movement of *Listeria* and *Shigella* (Loisel *et al.*, 1999). Three essential components in the motility medium and their properties are listed below.

Actin depolymerizing factor

Actin depolymerizing factor (ADF)/cofilin is a small (18 kDa), ubiquitous, conserved actin binding protein which is responsible for the rapid turnover of actin filaments in motile regions of the cell. ADF/cofilin interacts with both F- and G-actin in the ADP bound form preferentially. Upon binding to F-ADP-actin, ADF/cofilin modifies the structure and the dynamic properties of the actin filament. ADF-bound filaments are destabilized and become more dynamic due to a large increase in the rate of depolymerization from the pointed end, which is the rate-limiting step in the treadmilling cycle. A higher concentration of monomeric ATP-G-actin is therefore established, so that faster barbed end growth balances ADF-induced faster pointed end depolymerization. Hence ADF contributes to the fast growth of barbed ends at steady state, which powers actin-based motility (see Carlier *et al.*, 1999, for review).

Capping proteins

Capping proteins block actin dynamics at the barbed ends. Barbed end-capped actin filaments can only depolymerize from their pointed ends at steady state. If a large proportion of filaments are capped at their barbed ends, the concentration of monomeric actin settles at a value very close to (slightly lower than) the critical concentration at the pointed end. The few remaining non-capped barbed ends grow so as to balance the flux of depolymerization from all the pointed ends. Treadmilling therefore is funnelled, in that each individual non-capped barbed end grows at a faster rate than it would in the absence of capping proteins (Carlier and Pantaloni, 1997). Hence capping proteins act in an additive fashion with ADF to enhance actin-based motility. Any capping protein (CP, the homolog of CapZ in non-muscle cells, or CapG, the macrophage capping protein, or gelsolin) is equally efficient in the motility medium (Pantaloni *et al.*, 2000).

Arp2/3 complex

Arp2/3 complex is a conserved seven subunit complex comprising in particular the **A**ctin **r**elated **p**roteins Arp2 and Arp3. The Arp2/3 complex is the downstream target of a variety of signaling pathways leading to actin polymerization (see Higgs and Pollard, 1999, for review). It is responsible for the generation of new actin filaments when it interacts with an activator. The *Listeria* protein ActA, which is solely responsible for actin polymerization around the bacterium and for its propulsion, directly activates Arp2/3 complex (Welch *et al.*, 1998). The *Shigella* protein IcsA does not activate Arp2/3 complex, but recruits N-WASp (Suzuki *et al.*, 1998; Egile *et al.*, 1999), the ubiquitous homolog of WASp/Scar family proteins, which are the natural activators of Arp2/3 complex in eukaryotic cells (see Higgs and Pollard, 1999, for review). Hence both *Listeria* and *Shigella* use Arp2/3 complex to generate new actin filaments at their surface upon activation.

A consensus has not been reached yet concerning the mechanism by which actin polymerization is stimulated by Arp2/3 complex. There is general agreement however that Arp2/3 complex multiplies the filaments locally by branching either from their sides (Blanchoin *et al.*, 2000) or from their barbed ends (Pantaloni *et al.*, 2000), and is responsible for the dendritic filament array observed in the lamellipodium (Svitkina and Borisy, 1999). Depletion experiments have demonstrated that the Arp2/3 complex is required for actin-based motility of both *Listeria* and *Shigella* in cell extracts (Welch *et al.*, 1997; Egile *et al.*, 1999). Accordingly, the Arp2/3 complex is the third essential actin regulatory factor in the reconstituted motility medium.

Accessory proteins: VASP, profilin and alpha-actinin

Early genetic studies of ActA together with biochemical manipulation of cell extracts used for motility have demonstrated that a central proline-rich region of ActA plays an important role in movement by interacting with VASP, a focal adhesion protein later shown also to be present at the leading edge of motile cells (Smith *et al.*, 1996; Gertler *et al.*, 1996; Rottner *et al.*, 1999). VASP belongs to a family of proteins that comprises four members: Ena (*Drosophila enabled*, a substrate of the Abl kinase), its murine homolog Mena, VASP and Evl (Ena-VASP-like). Mena and Evl can well substitute for VASP in *Listeria* movement (Laurent *et al.*, 1999). Although the role of VASP in *Listeria* (but not *Shigella*) movement is undisputable, the molecular mechanism by which its function in motility is elicited is not known (see Machesky, 2000, for review).

Profilin is a well-known actin-binding protein which interacts with ATP-G-actin specifically. The profilin–actin complex productively associates with barbed ends exclusively, hence profilin cooperates with ADF to enhance the processivity of the treadmilling of actin filaments (Didry *et al.*, 1998), and improves actin-based motility; however, it is not specifically required.

Alpha-actinin is not required for bacterial actin-based motility, but its ability to cross-link the filaments is important to anchor the actin tail of the bacteria in the background meshwork that forms the medium (Loisel *et al.*, 1999). By preventing the drift of the bacterium + tail, alpha-actinin makes the measurements of bacterial speed easier and more accurate.

Spatially controlled activation of Arp2/3 complex: role of the bacterial surface proteins

Shigella protein IcsA

The *Shigella* protein IcsA is required and sufficient for bacterial motility. IcsA contains in particular a domain made of six glycine-rich repeats which binds N-WASp (but not WASp, the homolog of N-WASp in hematopoietic cells, Egile *et al.*, 1999). The region of N-WASp with which IcsA interacts is not known yet. WASp family proteins (Miki *et al.*, 1998) have a multimodular structural organization. WASp and N-WASp contain in particular a WH1 (WASp homology 1) domain whose function is not known, yet which binds F-actin, a basic region that interacts with PIP2, a GBD (G protein binding domain) that specifically interacts with the small G-protein Cdc42 in the GTP bound form, a proline-rich region that binds SH3 domains of Grb2, Nck and some tyrosine kinases, and a conserved C-terminal region (VCA) which binds one G-actin molecule and Arp2/3 complex. The isolated VCA domain constitutively activates Arp2/3-induced actin polymerization (Machesky *et al.*, 1999; Rohatgi *et al.*, 1999; Egile *et al.*, 1999; Higgs *et al.*, 1999). In the full length N-WASp protein, binding of actin and Arp2/3 complex to VCA is prevented due to an autoinhibited fold of N-WASp (Abdul-Manan *et al.*, 1999; Kim *et al.*, 1999). Binding of PIP2, Cdc42-GTP, Grb2 to their respective targets on N-WASp promotes a structural change of N-WASp, exposing the VCA domain for association with G-actin and Arp2/3 complex, leading to the stimulation of filament branching (Rohatgi *et al.*, 1999, 2000; Carlier *et al.*, 2000; Higgs and Pollard, 2000). IcsA mimics Cdc42 in relieving the autoinhibited fold of N-WASp. The interaction between IcsA and N-WASp is so strong that N-WASp remains associated with the bacterial surface during movement of *Shigella* (Egile *et al.*, 1999). Hence Arp2/3 complex is constitutively activated exclusively at the surface of *Shigella*. Similarly, N-WASp remains associated with the surface of rocketting endosomes during movement (Taunton *et al.*, 2000).

Listeria protein ActA

The *Listeria* protein ActA is also required and sufficient for bacterial motility. Two domains of ActA play defined functional roles in movement. The N-terminal domain (residues 1–160) is sufficient to activate Arp2/3 complex (Skoble *et al.*, 2000). ActA uses the same molecular mechanism as the VCA domain of N-WASp: it binds one G-actin molecule and Arp2/3, the resulting ternary complex stimulates actin polymerization *in vitro* and generates branched filaments (Boujemaa *et al.*, 2001). The central domain consists of four proline-rich repeats and binds the N-terminal EVH1 domain of VASP very tightly. VASP is almost essential in movement (Loisel *et al.*, 1999) and comet tail structural organization. The molecular mechanism by which VASP enhances motility is currently unknown. The C-terminal EVH2 domain of VASP is essential in VASP function. This domain interacts with F-actin with an equilibrium dissociation constant of 2 μM (Laurent *et al.*, 1999). This weak interaction may be involved in the role of VASP in motility.

◆◆◆◆◆◆ METHODOLOGY OF THE MOTILITY ASSAY

As outlined above, the basic motility medium is a solution of actin assembled at steady state at physiological ionic strength in the presence of ATP, and supplemented with ADF, a capping protein, Arp2/3 complex, and preferably profilin, alpha-actinin and VASP (for *Listeria* only). Because the medium is optically much cleaner than the cell extracts, it is not absolutely necessary that actin be traced with rhodamine-actin for fluorescence observation. The phase contrast observation is sufficient to visualize the actin tail and monitor bacterial movement.

Preparation and handling of the proteins that compose the motility assay

Actin

Rabbit muscle actin is used (previous studies in *Xenopus* egg or platelet extracts have established that muscle actin supports motility as well as non-muscle actin). Actin is isolated in the monomeric CaATP-G-actin form by standard procedures, with a final gel filtration step on Sephadex G-200 in G buffer (5 mM Tris-Cl- pH 7.8, 1 mM DTT, 0.1 mM $CaCl_2$, 0.2 mM ATP, 0.01% NaN_3). ATP-G-actin (generally at 50 ± 10 μM) can be kept at

0°C (on ice) in G buffer for about a month. A stock solution of F-actin is prepared by adding 1 mM $MgCl_2$ and 0.1 M KCl to monomeric actin, immediately following its conversion into the physiological MgATP-G-actin form. Conversion of CaATP-G-actin into MgATP-G-actin is conveniently made by addition of one molar equivalent actin + 10 μM of $MgCl_2$ (for instance add 40 μM $MgCl_2$ to a 30 μM CaATP-G-actin solution), followed by 0.2 mM EGTA. Replacement of bound Ca^{++} ion by Mg^{++} ion takes 3 min. The solution of Mg-F-actin can be kept at room temperature for the day.

The actin stock solution may be supplemented with a fraction (10 to 20%) of rhodamine-labeled actin for fluorescence optical microscopy. The rhodamine labeling of actin (in the F-form) by rhodamine-NHSR is described in Isambert *et al.* (1995). Labeling of lysines rather than cysteine 374 is recommended for this assay, since proteins like profilin do not bind well to cysteine 374-derivatized actin (Malm *et al.*, 1984).

The final actin concentration in the motility medium is 6–8 μM.

ADF

Bacterial recombinants of ADFs from different sources (plant, amoeba, vertebrate) have been used successfully in the motility medium. The procedures for purification of recombinant ADF is described (Carlier *et al.*, 1997). ADF can be stored for months at −80°C. Thawed aliquots can be stored on ice for a few weeks without loss of activity.

The optimum concentration of ADF in the motility medium is 50% of the actin concentration, which leads to the maximum increase in treadmilling (Carlier *et al.*, 1997).

Capping protein

Several capping proteins have been used successfully and interchangeably in the motility medium. Gelsolin and CapG (a gift from Dr Helen Yin) can be expressed as bacterial recombinants. Capping protein $\beta 2$, the homolog of muscle CapZ can be purified from bovine erythrocytes (Kuhlmann and Fowler, 1997). Capping proteins can be stored at −80°C. CapG and gelsolin require the presence of free Ca^{++} ions (0.1 mM) in the motility medium. Capping protein and CapG are used at final concentrations of 40–60 nM in the motility medium. Gelsolin is used at optimum gelsolin : actin ratios in the range 1 : 200 to 1 : 500 (Pantaloni *et al.*, 2000).

Arp2/3 complex

Arp2/3 complex can be purified from a variety of sources, including amoeba (Mullins *et al.*, 1998), and vertebrate tissues like thymus (Higgs *et al.*, 1999) or brain (Egile *et al.*, 1999). Several purification protocols have been described. The yield is about 1 mg pure Arp2/3 complex recovered from 1 kg of tissue. One of the fastest and easiest protocols involves a cation exchange step followed by affinity chromatography on a VCA column (Egile *et al.*, 1999). Arp2/3 complex is stored at −80°C at concentrations of 1–2 μM. Thawed solutions kept on ice are used for two days. Arp2/3 complex is used at 70–100 nM in the motility medium.

N-WASp

His-tagged human N-WASp is expressed in Sf9 cells using the baculovirus system (Egile *et al.*, 1999). The protein is purified by DEAE cellulose and Ni-affinity. It is stored at −80°C. Thawed aliquots kept on ice can be used for motility assays for a week. Bacteria expressing IcsA are coated with N-WASp at 0.1 to 0.2 μM for 10 min and washed before being placed in the motility medium.

Profilin

Either recombinant profilin from different sources or profilin purified from bovine spleen (Perelroizen *et al.*, 1996) can be used. Profilin is stored at −80°C. Thawed aliquots can be stored on ice for several weeks. Profilin is used at 1–2.5 μM in the motility assay. Profilin causes a two to three-fold increase in the propulsion rate of *Listeria* or *Shigella*.

VASP

His-tagged human VASP is expressed in insect Sf9 cells using the baculovirus system, and purified as described in Laurent *et al.* (1999). It is stored at −80°C. Thawed aliquots kept on ice can be used in the motility assay (at optimum final concentrations of 0.1 to 0.2 μM) for one week. VASP has no effect on *Shigella* movement, and causes an increase in actin-based movement of *Listeria* of at least 10-fold (movement is barely detectable in the absence of VASP).

Alpha-actinin

Alpha-actinin is commercially available (Sigma). It is used at 0.5 μM in the motility assay.

Bacteria or functionalized beads: protocols for the preparation and use of different motility substrates for the reconstituted motility medium

Listeria monocytogenes

Either wild-type LO28 or ActA-overexpressing Lut12(pActA3) *Listeria* (Kocks *et al.*, 1992) may be used. Wear protective clothing and decontaminate all materials that have been in contact with the pathogen.

Grow *Listeria* at 37 °C in brain–heart infusion medium (BHI, Difco), in the presence of 10 μg ml^{-1} chloramphenicol for strain Lut12(pActA3), to an OD of 1.5 to 2.0. Add glycerol to 30% v/v, aliquot by 200 μl and store at −80°C. Centrifuge thawed aliquots for 5 min at 13 krpm in an eppendorf centrifuge and resuspend in 20 μl of buffer X (for buffer composition see motility assay section) to make 20× concentrated *Listeria* stock. Store on ice and use within one day. Motility assay: both strains require the addition of 0.5 μM VASP to the assay for optimal motility.

Escherichia coli expressing the *Shigella* protein IcsA

The protocol refers to *E. coli* BUG 968 (Kocks *et al.*, 1995) overexpressing the *Shigella flexneri* IcsA protein. Handle *E. coli* with the usual precautions.

Grow *E. coli* at 37°C in 2YT medium with 100 μg ml^{-1} ampicillin to OD 1.5–2.0. Add glycerol to 30% v/v, aliquot by 200 μl and store at −80°C. From thawed aliquots, prepare 20× concentrated stock as for *Listeria*.

Motility assay: incubate 20× concentrated stock with 20–30 nM N-WASp for 15 min at room temperature, centrifuge and resuspend the pellet in 20 μl buffer Xb. Store on ice and use on the same day.

Functionalized polystyrene beads

Polymer beads are a valuable alternative to micro-organisms. Among their advantages are uniform physical properties, easy handling and storage and low health hazard. In selecting beads for a motility experiment, the most important parameters are material, size and surface chemistry. Polystyrene-based (PS) beads are most often used. They allow easy immobilization of proteins via hydrophobic interactions, and their density and refractive index allow fast sedimentation and good detection in phase contrast. PS beads of sizes between 50 nm and 10 μM have been successfully used in reconstituted motility assays (Yarar *et al.*, 1999; Cameron *et al.*, 1999; our results, Bernheim *et al.*, manuscript in preparation); sizes of 0.2–1 μM are most convenient for light microscopy and move fastest. We have found beads from PolySciences Inc. and Bangs Laboratories Inc. to work equally well.

Proteins may be immobilized to beads by adsorption or by covalent coupling. The choice of the appropriate method depends on the protein used; most researchers prefer adsorption to coupling because it is faster and requires less material. However, where a zero level of free protein or a specific orientation is required, coupling may be necessary. Extensive documentation on possible chemistries can be obtained from bead manufacturers. The default technique is to use a water-soluble carbodiimide to couple carboxylated PS beads to proteins via their primary amino groups.

Adsorption of recombinant N-WASp to PS beads is performed as follows. To 0.1% w/v 1 μm diameter polystyrene beads (Bangs Laboratories Inc., Cat. No. PCO3N) in buffer X, add 0.1% BSA and 200 nM recombinant N-WASp. Mix well. Incubate for 15 min on ice, store on ice and use within 1 week.

Motility assay: immediately before use, pellet an aliquot of beads and resuspend in the same volume of buffer X containing 0.1% BSA. Check monodispersity, and sonicate briefly if the beads are aggregated.

Observation of movement, data acquisition and analysis

Sample preparation

Solutions

Buffer X: 10 mM HEPES pH 7.5, 0.1 M KCl, 1 mM $MgCl_2$, 0.1 mM $CaCl_2$, 1 mM ATP. Prepare with distilled water and adjust pH with KOH. Store aliquots at −20°C.

1% methyl cellulose: Heat 1% w/v methyl cellulose in distilled water to 60°C and stir until the solution clears. Store on ice.

Stock solutions of 10% BSA, 0.2 M DTT, 22 mM DABCO (1,4-diazabicyclo[2,2,2]octane from Sigma) are all prepared in distilled water. Store small aliquots at −20°C.

0.2 M ATP (Roche Molecular Biochemicals) is dissolved in distilled water and adjusted to pH 7.0 with 6 N NAOH. Store aliquots at −20°C.

Valap: Mix vaselin, lanolin and solid paraffin in equal amounts and homogenize at 50°C.

All chemicals are reagent grade if not indicated otherwise.

Assay

Prepare Mg-F-Actin (35 to 50 μM) as described in previous section. Polymerize for 1 h at room temperature.

Thaw aliquots of 22 mM DABCO, 0.2 M ATP pH 7.0, 0.2 M DTT, buffer X and proteins. Prepare 30 mM ATP/60 mM $MgCl_2$ stock solution by adding 15 μl of 0.2 M ATP pH 7.0 and 6 μl 1 M $MgCl_2$ to 79 μl of distilled water. Prepare the 'MIX' solution by mixing 20 μl of ATP/$MgCl_2$, 20 μl 2.2 mM DABCO and 10 μl 0.2 M DTT. Store on ice and use within three days.

Prepare 1% BSA in buffer X and dilute protein stocks with this buffer to the desired concentrations. Store on ice.

A standard typical motility assay is made up as described on Chart 1. Because too high a concentration of methyl cellulose can cause excessive filament bundling, all components excluding actin and methyl cellulose are mixed first; F-actin and methyl cellulose are added. When modifying the sample pipetting scheme given below, keep all non-protein components and 10% BSA at constant relative volumes. To make up to final volume, use 1% BSA in buffer X.

Pipet 2–3 μl of the mixture of medium + bacteria or beads on a clean slide and cover with a clean coverslip. Seal the preparation with Valap and observe at room temparature. Actin tails are usually seen after 5–20 min.

Comments

The addition of BSA is important to prevent protein adsorption to slides and coverslips. Methyl cellulose reduces Brownian movement in the assay, but is not essential for the development of tails and does not significantly affect velocity in the concentration range used.

Chart 1. Composition of the motility assay

Solution	Volume	= final concentration
Mg-F-Actin 48 μM	3.8 μl	7.6 μM
BSA 10%	1.2 μl	0.5%
ADF 55 μM	1.6 μl	3.7 μM
Profilin 50 μM	1.2 μl	2.5 μM
Capping Protein 0.5 μM	2.4 μl	50 nM
α-Actinin 5 μM	1.2 μl	0.25 μM
Arp2/3 1 μM	1.8 μl	75 nM
Methyl cellulose 1%	5.4 μl	0.23%
MIX	3.0 μl	
Rhodamin-G-Actin 40 μM	0.6 μl	1 μM
Listeria or *E.coli* (IcsA) or beads	0.6 μl	$2 \cdot 10^8$/ml
VASP 8 μM (for *Listeria*) or Buffer X 1% BSA	1.4 μl	0.47 μM
Final volume	**24.0 μl**	

Microscopy equipment

The actin tails formed in reconstituted motility medium are usually observable in phase contrast alone when optimal conditions are used (Figure 13.1). When working under suboptimal conditions, epifluorescence may be necessary to detect actin tails.

The images shown in this chapter were acquired either in phase contrast with a 40× objective (Nikon inverted microscope) and a Sony Hyper HAD CCD camera, or in epifluorescence with a 100 × 1.3 na Neofluar objective (Zeiss) and a low light camera (LHESA LH 72 LL). Images were improved with an Argus-10 image processor (Hamamatsu) The video signal is digitized by video frame grabber (VG-5 Scion corp). NIH Image 1.62 software was used for image analysis.

Data analysis

The reconstituted motility assay can be used as an analytical tool. Parameters that have been exploited are speed, trajectories and percentage of motile particles.

A straightforward, low-tech way to determine speed is to measure, after calibration, the distance between two points on a trajectory, marked at different time points on the screen with a felt-tip. More exactly, velocities can be obtained on video sequences with 10 seconds intervals, using Macros of NIH image which compute the velocity and average velocity at each time of the differential translation of the XY center of individual bacteria or beads. The quality of the data will of course depend on the number of time/way intervals measured; as a rule of thumb, one should measure about 5 to 15 intervals. During such measurements, it is imperative to avoid image translation by (thermal) movements of the optical system and convection in the sample.

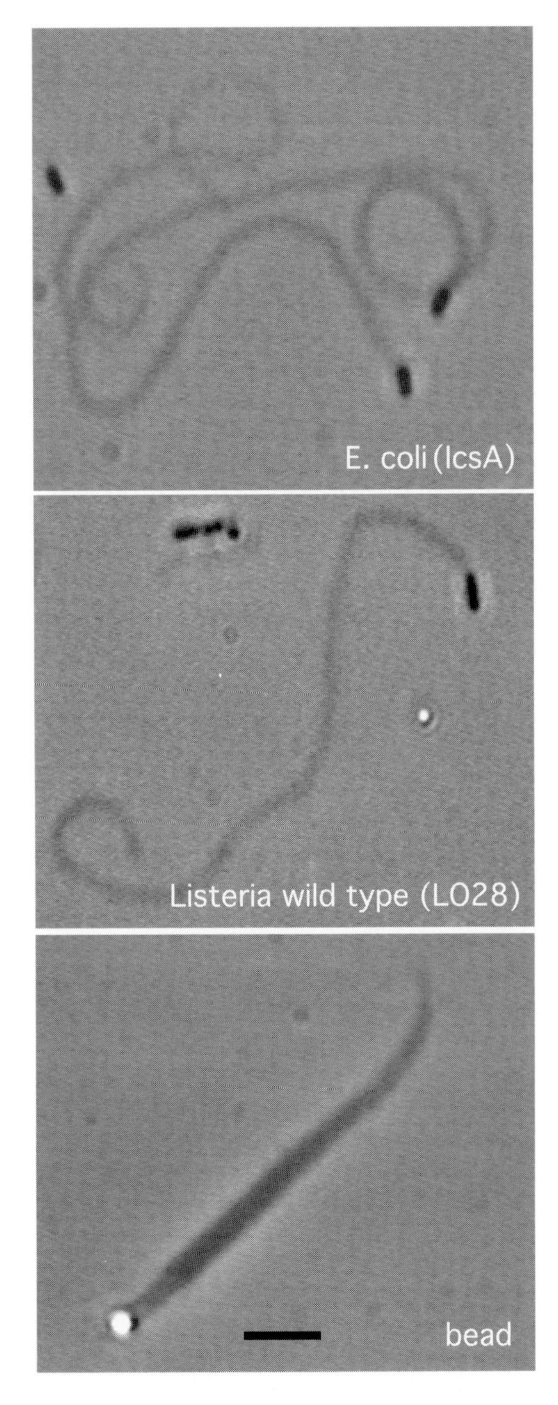

Figure 13.1. Actin-based movement in the reconstituted motility assay. *E. coli* (IcsA) (top panel), *Listeria monocytogenes* (middle panel) and N-WASp-coated beads (2 μm diameter) are observed in phase contrast microscopy following one hour incubation in the standard minimum motility assay described in Chart 1. The bar represents 4 μm.

Velocity measurements have proved a sensitive tool to measure the effect of capping proteins on tail formation (Pantaloni *et al.*, 2000). This point is illustrated in Figure 13.2.

The trajectories (the actin tails) can provide information by their lengths, densities and overall form. For example, it has been shown that the average tail length correlates with ADF concentration (Carlier *et al.*, 1997, and Figure 13.2). Tail densities may be measured by fluorometry (Cameron *et al.*, 1999), and a wealth of 'non-standard' tail forms that can occasionally be observed in motility assays may hold important information and await systematic investigation. (Weisner *et al.*, manuscript in preparation.)

The percentage of motile bacteria/beads is usually determined by manually counting the ratio of motile bacteria/beads in several fields of view and subsequent averaging. As samples may be heterogeneous, especially on the edges and in regions contaminated by dirt, only central, clean regions should be chosen. Typical values under optimal conditions fall between 70 and 80% for living micro-organisms and will be near 100% for polymer beads.

◆◆◆◆◆◆ PERSPECTIVES AND CHALLENGES

Functional analysis of the bacterial surface proteins involved in local activation of Arp2/3 complex. Analysis of mutant strains

One of the great advantages of the reconstituted motility assay, in understanding bacterial virulence, is that it is the first *in vitro* assay enabling to monitor actin-based propulsion of wild-type bacteria. So far wild-type bacteria moved in living cells, not in cell extracts. Only engineered bacteria harboring a high level of plasmidic expression of ActA or IcsA were able to move in cell extracts. The fact that the reconstituted motility medium is optimized for movement, and more controllable than cell extracts may account for this difference. The motility medium therefore can be used in a straightforward fashion for the analysis of mutant bacteria, so as to understand the motility defect in biochemical terms. For instance, a mutation on ActA might lead to a poorly motile *Listeria* because the affinity of the mutated ActA for Arp2/3 is lowered, resulting in a poor stimulation of actin assembly at the surface of the bacteria. Increasing the concentration of Arp2/3 complex in the motility medium in this case is expected to restore the motility by favoring the ActA–Arp2/3 interaction. Apart from this simple case, the biochemical control of the motility medium is potentially useful to analyze the origin of special motility phenotypes exhibited by some mutants, like the *Listeria* showing periodic rocketting (Lasa *et al.*, 1997) or 'skidding' motion (Rafelski *et al.*, 2000). Similarly, mutations in IcsA are expected to be useful to decipher the structure–function relationship in N-WASp, the connector responsible for activation of Arp2/3 complex in eukaryotic cells.

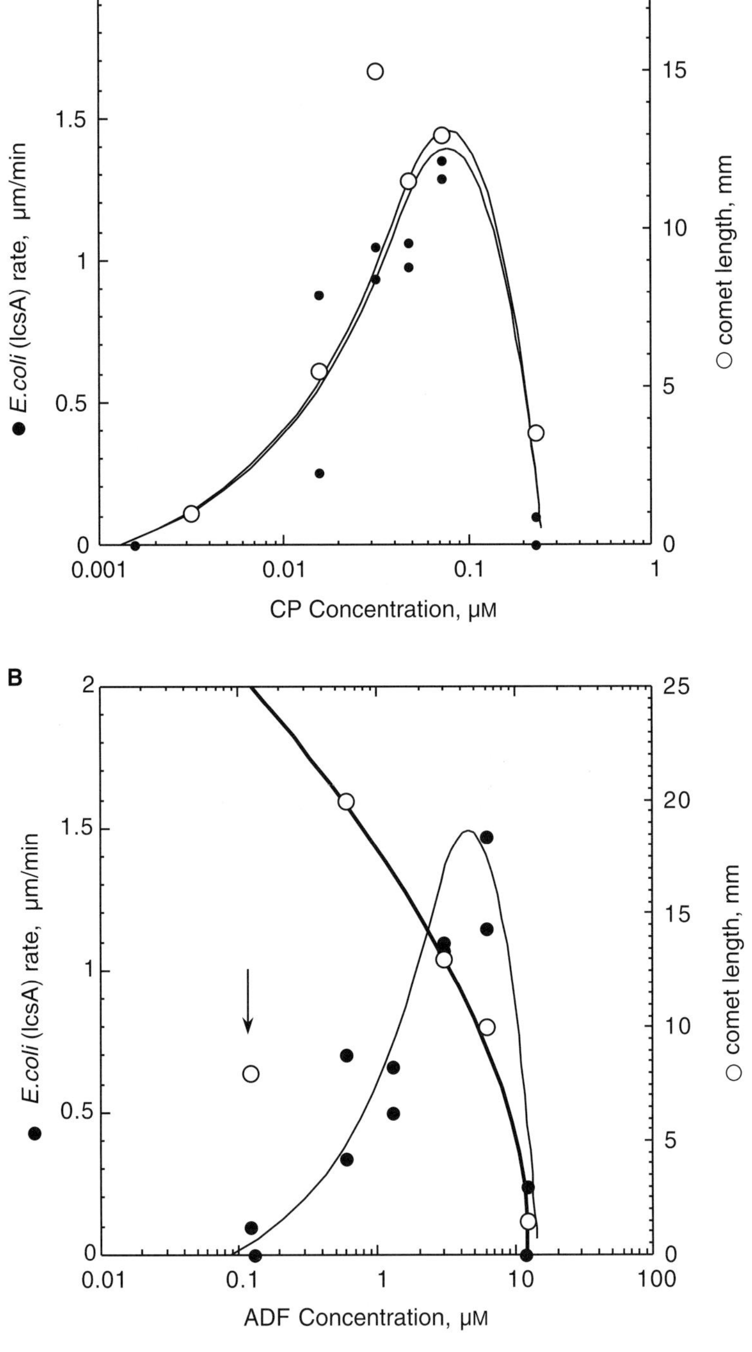

Figure 13.2. Role of two essential proteins, ADF and capping protein, in actin-based movement. Actin-based propulsion of N-WASp-coated *E. coli* (IcsA) was monitored in the reconstituted motility assay containing 75 nM Arp2/3 complex and either (top) 2 µM ADF and capping protein as indicated, or (bottom) 50 nM capping protein and ADF as indicated. The length of the actin tails (open symbols, thick lines) and the rate of bacteria propulsion (closed symbols, thin lines) were recorded. The tail length measurement at 0.1 µM ADF is not taken into account because at this low ADF concentration, the steady-state length was not reached within a reasonable period of time.

Test of the function of cellular regulatory proteins involved in actin-based motile processes

The minimum motility medium has the potential to test the motile function of other proteins. For instance we have tested the function of Ciboulot, a protein of the actobindin family which *in vivo* appears to play an important role in axonal growth during *Drosophila* brain development (Boquet *et al.*, 2000). Ciboulot as well as actobindin appear to enhance motility and efficiently substitute for profilin in the motility medium, confirming independent biochemical evidence (Hertzog *et al.*, manuscript in preparation) that then affect actin dynamics like profilin. The function of many other proteins, suspected to play a regulatory role in diverse motile processes like phagocytosis by interacting with essential partners, can be tested using the motility medium, either to replace one of the essential components or to modulate their activity.

Use of the reconstituted motility medium for cytoskeletal drug discovery

The reconstituted motility assay provides a biochemically well-defined and controllable basis for a screen of putative inhibitors of cell motility, therefore it is expected to facilitate cytoskeletal drug discovery. This perspective is more broadly opened by replacing the bacteria by chemically derivatized particles carrying diverse molecules able to activate Arp2/3 complex at different steps in the signaling pathways. Such molecules could be natural intermediates along the pathways, or synthesized compounds designed to mimic one such intermediate. Drug therapy design therefore is an open avenue for the reconstituted motility assay.

Acknowledgment

This work was realized with financial support from the Ligue Nationale Contre le Cancer and a grant from Human Frontiers in Science (# RG 0227/1998-M).

References

Abdul-Manan, N., Aghazadeh, B., Liu, G. A., Majumdar, A., Ouerfelli, O., Siminovitch, K. A. and Rosen, M. K. (1999). Structure of Cdc42 in complex with the GTPase-binding domain of the 'Wiskott-Aldrich syndrome protein. *Nature* **399**, 379–383.

Beckerle, M. C. (1998). Spatial control of actin filament assembly: Lessons from Listeria. *Cell* **95**, 741–748.

Bernardini, M. L., Mounier, J., d'Hauteville, H., Coquis-Rondon, M. and Sansonetti, P. (1989). Identification of IcsA, a plasmid locus of *Shigella flexneri* that governs bacterial intra and intercellular spread through interaction with F-actin. *Proc. Natl Acad. Sci. USA* **86**, 3867–3871.

Blanchoin, L., Amann, K. J., Higgs, H. N., Marchand, J.-B., Kaiser, D. A. and Pollard, T. D. (2000). Direct observation of dendritic actin filaments networks nucleated by Arp2/3 complex and WASp/Scar proteins. *Nature* **404**, 1007–1011.

Boquet, I., Boujemaa, R., Carlier, M.-F. and Préat, T. (2000). Ciboulot regulates actin assembly during *Drosophila* brain metamorphosis. *Cell* **102**, 797–808.

Boujemaa-Paterski, R., Gouin, E., Hausen, G., Samarin, S., Le Clainche, C., Didry, D., Kocks, C., Cossart, P., Dehoux, P., Carlier, M.-F. and Pantaloni, D. (2000). *Biochemistry* **40**, 11390–11404.

Cameron, L. A., Footer, M. J., Van Oudenaarden, A. and Theriot, J. A. (1999). Motility of ActA protein coated microspheres driven by actin polymerization. *Proc. Natl Acad. Sci. USA* **96**, 4908–4913.

Carlier, M.-F., Ressad, F. and Pantaloni, D. (1999). Control of actin dynamics in cell motility: Role of ADF/cofilin. *J. Biol. Chem* **274**, 33827–33830.

Carlier, M.-F. (1998). Control of actin dynamics. *Curr. Opin. Cell Biol* **9**, 45–51.

Carlier, M.-F. and Pantaloni, D. (1997). Control of actin dynamics in cell motility. *J. Mol. Biol.* **269**, 459–467.

Carlier, M.-F., Laurent, V., Santolini, J., Melki, R., Didry, D., Xia, G. X., Hong, Y., Chua, N. H. and Pantaloni, D. (1997). Actin Depolymerizating Factor (ADF/Cofilin) enhances the rate of filament turnover: Implication in actin-based motility. *J. Cell Biol.* **136**, 1307–1322.

Carlier, M.-F., Nioche, P., Broutin-L'HermitteBoujemaa, R., Le Clainche, C., Egile, C., Garbay, C., Ducruix, A., Sansonetti, P. and Pantaloni, D. (2000). GRB2 links signaling to actin assembly by enhancing interaction of neural Wiskott-Aldrich syndrome protein (N-WASp) with Actin-related protein (ARP2/3) complex. *J. Biol. Chem.* **275**, 21946–21952.

Didry, D., Carlier, M.-F. and Pantaloni, D. (1998). Synergy between Actin Depolymerizing Factor (ADF/cofilin) and profilin in increasing actin filament turnover. *J. Biol. Chem.* **273**, 25602–25611.

Domann, E., Wehland, J., Rhode, M., Pistor, S., Hartl, M., Goebel, W., Leismeister-Wächter, M., Wuenscher, M. and Chakraborty, T. (1992). A novel bacterial virulence gene in *L. monocytogenes* required for host cell microfilament interaction with homology to the proline-rich region of vinculin. *EMBO J.* **11**, 1981–1990.

Dramsi, S. and Cossart, P. (1998). Intracellular pathogens and the actin cytoskeleton. *Ann. Rev. Cell Dev. Biol.* **14**, 137–166.

Egile, C., Loisel, T. P., Laurent, V., Li, R., Pantaloni, D., Sansonetti, P. J. and Carlier, M.-F. (1999). Activation of the Cdc42 effector N-WASP by the *Shigella flexneri* IcsA protein promotes actin nucleation by Arp2/3 complex and bacterial actin-based motility. *J. Cell Biol.* **146**, 1319–1332.

Gertler, F. B., Niebuhr, K., Reinhard, M., Wehland, J. and Soriano, P. (1996). Mena, a relative of VASP and Drosophila Enabled, is implicated in the control of microfilament dynamics. *Cell* **87**, 227–239.

Goldberg, M. C. and Theriot, J. A. (1995). *Shigella flexneri* surface protein IcsA is sufficient to direct actin-based motility. *Proc. Natl Acad. Sci. USA* **92**, 6572–6576.

Higgs, H. N. and Pollard, T. D. (1999). Regulation of actin polymerization by Arp2/3 complex and WASp/Scar proteins. *J. Biol. Chem.* **274**, 32531–32534.

Higgs, H. N., Blanchoin, L. and Pollard, T. D. (1999). Influence of the C-terminus of WASp and the Arp2/3 complex on actin polymerization. *Biochemistry* **38**, 15212–15222.

Higgs, H. N. and Pollard, T. D. (2000). Activation of Cdc42 and PIP2 of Wiskott-Aldrich Syndrome protein stimulates actin nucleation by Arp2/3 complex. *J. Cell Biol.* **150**, 1311–1320.

Higley, S. and Way, M. (1997). Actin and cell pathogenesis. *Curr. Opin. Cell Biol.* **9**, 62–69.

Isambert, H., Venier, P., Maggs, A. C., Fattoum, A., Kassab, R., Pantaloni, D. and Carlier, M. F. (1995). Flexibility of actin filaments derived from thermal fluctuations. Effect of bound nucleotide, phalloidin, and muscle regulatory proteins. *J. Biol. Chem.* **270**, 11437–11444.

Kim, A. S., Kakalis, L. T., Abdul-Manan, N., Liu, G. A. and Rosen, M. K. (1999). Autoinhibition and activation mechanisms of the Wiskott-Aldrich syndrome protein. *Nature* **404**, 151–158.

Kocks, C., Gouin, E., Tabouret, M., Berche, P., Ohayon, H. and Cossart, P. (1992). *Listeria monocytogenes*-induced actin assembly requires the actA gene product, a surface protein. *Cell* **68**, 521–531.

Kocks, C., Marchand, J.-B., Gouin, E., d'Hauteville, H., Sansonetti, P. J., Carlier, M.-F. and Cossart, P. (1995). The unrelated proteins ActA of *Listeria monocytogenes* and IcsA of *Shigella flexneri* are sufficient to confer actin-based motility on *Listeria innocua* and *E. coli* respectively. *Mol. Microbiol.* **18**, 413–423.

Kuhlmann, P. A. and Fowler, V. M. (1997). Purification and characterization of an apha1beta2 isoform of capZ from human erythrocytes: cytosolic location and inability to bind to Mg^{++} ghosts suggests that erythrocytes actin filaments are capped by transducin. *Biochemistry* **36**, 13461–13472.

Lasa, I., Gouin, E., Goethals, M., Vancompernolle, K., David, V., Vandekerckhove, J. and Cossart, P. (1997). Identification of two regions in the amino terminal domain of ActA involved in the actin comet tail formation by *Listeria monocytogenes*. *EMBO J.* **16**, 1531–1540.

Laurent, V. and Carlier, M.-F. (1998). Use of platelet extracts for actin-based motility of Listeria monocytogenes. *Cell Biology: A laboratory handbook.* second edition., Vol. 2, 359–365.

Laurent, V., Loisel, T. P., Harbeck, B., Wehman, A., Gröbe, L., Jockusch, B. M., Wehland, J., Gertler, F. B. and Carlier, M.-F. (1999). Role of proteins of the Ena/VASP family in actin-based motility of *Listeria monocytogenes*. *J. Cell Biol.* **144**, 1245–1258.

Loisel, T. P., Boujemaa, R., Pantaloni, D. and Carlier, M.-F. (1999). Reconstitution of actin-based motility of Listeria and Shigella using pure proteins. *Nature* **401**, 613–616.

Machesky, L. M. (2000). Putting on the brakes: a negative regulatory function for Ena/VASP proteins in cell migration. *Cell* **101**, 685–688.

Machesky, L. M., Mullins, R. D., Higgs, H. N., Kaiser, D. A., Blanchoin, L., May, R. C., Hall, M. E. and Pollard, T. D. (1999). Scar, a WASp-related protein, activates nucleation of actin filaments by the Arp2/3 complex. *Proc. Natl Acad. Sci. USA* **96**, 3739–3744.

Malm, B. (1984). Chemical modification of Cys-374 of actin interferes with the formation of the profilactin complex. *FEBS Lett.* **173**, 399–402.

Miki, H., Sasaki, T., Takai, Y. and Takenawa, T. (1998). Induction of filopodium formation by a WASP-related actin-depolymerizing protein N-WASP. *Nature* **391**, 93–96.

Mullins, R. D., Heuser, J. A. and Pollard, T. D. (1998). The interaction of Arp2/3 complex with actin nucleation: high affinity pointed end capping, and formation of branching networks of actin filaments. *Proc. Natl Acad. Sci. USA.* **95**, 6181–6186.

Pantaloni, D., Boujemaa, R., Didry, D., Gounon, P. and Carlier, M.-F. (2000). The Arp2/3 complex branches filament barbed ends: functional antagonism with capping proteins. *Nature Cell Biol.* **2**, 385–391.

Perelroizen, I., Didry, D., Christensen, H., Chua, N.-H. and Carlier, M.-F. (1996). Role of nucleotide exchange and hydrolysis in the function of profilin in actin assembly. *J. Biol. Chem.* **274**, 12302–12309.

Rafelski, S. M., Lauer, P. and Theriot, J. A. (2000). Skidding and doing donuts: *Listeria monocytogenes* strains carrying mutations within the ActA N-terminal domain exhibit significant alterations in actin-based motility. *Mol. Biol. Cell.* **11**, 85a.

Rohatgi, R., Ho, H. Y. and Kirschner, M. W. (2000). Mechanism of N-WASp activation by Cdc42 and PIP2. *J. Cell Biol.* **150**, 1299–1309.

Rohatgi, R., Ma, L., Miki, H., Lopez, M., Kirschhausen, T., Takenawa, T. and Kirschner, M. W. (1999). The interaction between N-WASp and the Arp2/3 complex links Cdc42-dependent signals to actin assembly. *Cell* **97**, 221–231.

Rottner, K., Behrend, B., Small, J. V. and Wehland, J. (1999). VASP dynamics during lamellipodia protrusion. *Nat. Cell Biol.* **1**, 321–322.

Skoble, J., Portnoy, D. A. and Welch, M. D. (2000). Three regions within ActA promote Arp2/3 complex-mediated actin nucleation and *Listeria monocytogenes* motility. *J. Cell Biol.* **150**, 527–538.

Small, J. V. (1995). Getting the actin filaments straight: nucleation-release or treadmilling. *Trends Cell Biol.* **5**, 52–54.

Smith, G. A., Theriot, J. A. and Portnoy, D. A. (1996). The tandem repeat domain in the *Listeria monocytogenes* ActA protein controls the rate of actin-based motility the percentage of moving bacteria and the localization of VASP and profilin. *J. Cell Biol.* **135**, 647–660.

Suzuki, T., Miki, H., Takenawa, T. and Sasakawa, C. (1998). Neural Wiskott-Aldrich syndrome protein is implicated in the actin-based motility of *Shigella flexneri*. *EMBO J.* **17**, 2767–2776.

Svitkina, T. M. and Borisy, G. G. (1999). Arp2/3 complex and Actin Depolymerizing Factor/cofilin in dendritic organization and treadmilling of actin filament array in lamellipodia. *J. Cell Biol.* **145**, 1009–1026.

Taunton, J., Rowning, B. A., Coughlin, M. L., Wu, M., Moon, R. T., Mitchison, T. J. and Larabell, C. A. (2000). Actin-dependent propulsion of endosomes and lysosomes by recruitment of N-WASP. *J. Cell Biol.* **148**, 519–530.

Theriot, J. A. and Fung, D. C. (1998). *Listeria monocytogenes*-based assays for actin assembly factors. *Methods Enzymol.* **159**, 114–122.

Theriot, J. A., Rosenblatt, J., Portnoy, D. A., Goldschmitt-Clermont, P. J. and Mitchison, T. J. (1994). Involvement of profilin in the actin-based motility of *L. monocytogenes* in cells and cell-free extracts. *Cell* **76**, 505–517.

Wang, Y.-L. (1985). Exchange of actin subunits at the leading edge of living fibroblasts: possible role of "tread milling". *J. Cell Biol.* **101**, 597–602.

Welch, M. D., Iwamatsu, A. and Mitchison, T. J. (1997). Actin polymerization is induced by the Arp2/3 complex at the surface of *Listeria monocytogenes. Nature* **385**, 265–269.

Welch, M. D., Rosenblatt, J., Skoble, J., Portnoy, D. A. and Mitchison, T. J. (1998). Interaction of human Arp2/3 complex with the *Listeria* protein ActA in actin filament nucleation. *Science* **281**, 105–108.

Yarar, D., To, W., Abo, A. and Welch, M . D. (1999). The Wiskott-Aldrich Syndrome protein directs actin-based motility by stimulating actin nucleation with the Arp2/3 complex. *Curr. Biol.* **9**, 555–558.

14 Transport and Intracellular Movement—Protein Translocation via Dedicated Secretion Systems

Sabine Tötemeyer and Guy R Cornelis
Christian de Duve Institute of Cell Pathology, Université catholique de Louvain, Av Hippocrate, 74, B1200 Brussels, Belgium

◆◆

CONTENTS

◆◆◆◆◆◆ INTRODUCTION

In this chapter we discuss the methods used to elucidate the action of the Yop virulon, an archetype of type III secretion/translocation systems on eukaryotic target cells. This 'weapon' is encoded by the pYV plasmid (*Yersinia* virulence) and consists of three major parts: (1) type III secretion machinery (Ysc) that directs Yop proteins over the bacterial membranes; (2) translocator Yops that allow delivery of effector Yops in the cytosol of the eukaryotic cell; and (3) Yop effector proteins that interfere with the eukaryotic host cells (recently reviewed by Cornelis *et al.*, 1998; Cornelis, 2000).

◆◆◆◆◆◆ YOP SECRETION *IN VITRO*

For *Y. enterocolitica*, Yop production and secretion can be artificially triggered *in vitro* by incubation in rich Ca^{2+}-deprived medium at 37°C. This artificial secretion is accompanied by a restriction of growth. These

METHODS IN MICROBIOLOGY, VOLUME 31
ISBN 0–12–521531–2

phenomena are very useful to check *Y. enterocolitica* strains and mutants for their ability to produce and secrete the Yops.

Growth on MOX

When placed at 37°C in rich medium deprived of Ca^{2+}, *Y. enterocolitica* carrying a functional pYV plasmid, secrete Yops and cease growing. This phenomenon, called Ca^{2+}-dependency is generally tested by plating bacteria in parallel on tryptic soy agar (TSA) (Difco) and on tryptic soy agar supplemented with 20 mM $MgCl_2$ and 20 mM sodium oxalate (MOX). *Y. enterocolitica* mutants unable to secrete Yops because of the loss of the pYV plasmid or spontaneous mutations in a *ysc* gene or in the transcription activator gene *virF* can be selected on MOX at 37°C.

Secretion profile of Yops on SDS page

For *in vitro* induction of Yop synthesis, *Y. enterocolitica* are routinely inoculated to an OD_{600} of 0.1 in 10 ml brain–heart infusion (BHI, Difco) supplemented with 20 mM sodium oxalate, 20 mM $MgCl_2$ and 0.4% glucose (BHI-Ox). They are grown at room temperature in a shaking incubator for 2 h, then shifted to a 37°C shaking waterbath and incubated for further 4 h. Bacteria are sedimented by centrifugation and the proteins contained in the supernatant are precipitated overnight at 4°C, either with 10% (w/v) trichloroacetic acid (TCA) or with 3.5 g ammonium sulphate for 9 ml of supernatant. The TCA pellet is washed with acetone and the ammonium sulphate pellet is washed rapidly with water. The proteins are then redissolved in electrophoresis sample buffer (about 50 µl for the Yops contained in 10 ml of culture) and separated by electrophoresis in 12% (w/v) polyacrylamide gels in the presence of SDS (SDS-PAGE) as described by Laemmli (1970). After electrophoresis proteins can be either stained with Coomassie brilliant blue to see all the proteins or transferred by electroblotting to a nitrocellulose membrane to detect a specific Yop (Towbin *et al.*, 1979). Revealing of immunoblots is performed with secondary antibodies conjugated to horseradish peroxidase (1:2000; Dako) before development with supersignal chemiluminescent substrate (Pierce) or α-naphthol solution.

◆◆◆◆◆◆ BACTERIAL GROWTH FOR CELL INFECTION

Y. enterocolitica strains are routinely grown in tryptic soy broth (TSB) and plated on TSA containing the required antibiotics. Prior to cell infection, *Y. enterocolitica* is grown in BHI under selective pressure for maintaining the relevant plasmids. Since the pYV plasmid from many serotypes of *Y. enterocolitica* contains arsenite resistance genes, one can add 1 mM arsenite to the culture medium to exert a selective pressure in favour of the pYV plasmid (Neyt *et al.*, 1997).

The eukaryotic cell types are grown in their respective growth media (summarized in Boyd *et al.*, 2000), eventually supplemented with antibiotics. Adherent cells are seeded into 24-well plates the day before infection. Non-adherent cells are seeded the day of the experiment. When using HUVECs the plates and coverslips are coated with gelatine, except for the cAMP translocation assays. On the day of infection the cells are washed with culture medium without antibiotics. *Y. enterocolitica* are grown overnight in TSB at RT. On the day of the infection, they are inoculated in BHI at $OD_{600} = 0.2$ and grown for 2 h at room temperature. After harvesting and washing in saline, the bacteria are added to the cells at a multiplicity of infection (MOI) depending on the experiment. For infection of SF9 insect cells the bacteria are pre-induced for Yop production by growth at room temperature for 1 h and then at 37°C for 1 h before harvesting. Infections are carried out for a number of hours depending on the experiment at 37°C (except in the case of SF9 insect cells where infection needs to be carried out at 28°C, as SF9 cells do not tolerate 37°C). For longer assays gentamycin is added to stop extracellular bacterial growth but to allow the development of the effects of internalized Yops to be observed.

◆◆◆◆◆◆ STUDY OF THE ACTION OF YERSINIA ON THE CYTOSKELETON

Four different Yop effectors alter the cytoskeleton dynamics: YopE (Rosqvist *et al.*, 1991), YopH (Black and Bliska, 1997; Black *et al.*, 1998; Persson *et al.*, 1997) YopO and YopT (Iriarte and Cornelis, 1998). Only YopE and YopT cause strong morphological changes, generally referred to as cytotoxicity This cytotoxic effect, especially on HeLa cells, has been very important for the discovery of the translocation process (summarized by Cornelis *et al.*, 1998).

Morphological changes analysed by phase contrast microscopy

Following addition of *Y. enterocolitica* to eukaryotic cell monolayers of 30% confluence (MOI = 50–200), the morphology of the cells is observed after 1, 2, 3 and 5 h of infection by phase contrast microscopy. Cytotoxicity manifests by rounding up of the cells and detachment from the extracellular matrix, due to the disruption of the actin cytoskeleton. These morphological changes can be observed in all adherent cell types infected with wildtype *Y. enterocolitica* (HeLa, PU5-1.8, COS, RAT-1 and HUVECs,) except primary neurons (summarized in Table 14.1). Infected neurons do not round up or develop an amorphic appearance under the conditions of the assay (Boyd *et al.*, 2000). This is in agreement with the absence of such a morphological alteration upon treatment of neurons

Table 14.1 Assessment of Yop translocation into various target cell types (data obtained from Boyd *et al.*, 2000)

Cell types	Source	Cytotoxicity (MRS40 (pYV))	Translocation of $YopE_{130}$-Cya (nmole cAMP mg^{-1} protein)	
			Wild type (MRS40 (pYV) (pMSIII))	YopB mutant (MRS40 (pPW401) (pMSIII))
Cell lines				
HeLa	Human epithelial cells	++	8.1 ± 0.8	0.1 ± 0.1
U937	Human monocyte-macrophage		9.8 ± 3.3	0.3 ± 0.2
Jurkat	Human T-cells	N/A	14.2 ± 5.5	0.1 ± 0.1
Kaso	Human B-cells	N/A	8.6 ± 2.2	0.1 ± 0.1
PU5-1.8	Mouse monocyte-macrophage	+	12.0 ± 4.8	0.1 ± 0.1
CHO	Chinese hamster epithelial cells	+	25.7 ± 9.1	0.3 ± 0.2
RAT-1	Rat fibroblast	+++*	15.3 ± 3.6	0.1 ± 0.1
COS	Monkey fibroblast	++	9.8 ± 4.5	0.2 ± 0.1
SF9	Insect ovarian		5.1 ± 1.5	0.1 ± 0.1
Primary cells				
Neurons	Rat	–	9.2 ± 2.4	0.1 ± 0.1
HUVECs	Human umbilical vein endothelial cells	+++*	3.7 ± 1.3	0.6 ± 0.2

* Suitable for actin filament staining.
\+ cytotoxicity visible.
++ cytotoxicity well visible.
+++ cytotoxicity very dramatic.

with the actin filament disruptive factor cytochalasin D (Boyd *et al.*, 2000; Bradke and Dotti, 1999).

Actin filament staining

Actin filament staining with phalloidine-FITC post-infection of HUVECs or RAT-1 fibroblasts is the best way to visualize the action of YopE, YopH, YopO and YopT (Boyd *et al.*, 2000).

Y. enterocolitica are added to 30% confluent cells grown on coverslips at a multiplicity of infection (MOI) of 20–50. After 2–$2\frac{1}{2}$ h infection the first signs of cytotoxicity appear and the cells are fixed with 2% paraformaldehyde for 20 minutes. Cells are washed three times with PBS and cell membranes are permeabilised with 0.1% (v/v) Triton X-100. Actin filaments are stained by adding 5 μg ml^{-1} phalloidin-FITC for 40 min at 37°C. After washing the coverslips are mounted in Mowiol 50% (Polysciences, Inc.) containing 100 mg ml^{-1} diazabicyclo-octane (DABCO, Sigma).

◆◆◆◆◆◆ TRANSLOCATION OF YOP EFFECTORS INTO THE EUKARYOTIC CYTOSOL

Various methods are used to show translocation of bacterial proteins across the eukaryotic cell membrane. To prove translocation it must be demonstrated that the prokaryotic protein is found inside the eukaryotic cell compartment in the absence of phagocytosed bacteria. It is, therefore, important to prevent the uptake of bacteria into the eukaryotic cells, for example by treatment with cytochalasin D (5 μg ml^{-1}). There are three different approaches to demonstrate translocation: fusion of a reporter protein to the potential bacterial effector protein; fractionation of the various compartments by using detergents; and confocal microscopy. One caveat of every method is the use of *yop* genes or *yop* fusion genes carried by multicopy plasmids. Indeed, over-expression of a gene of the *yop* regulon may affect the expression levels of other proteins and, therefore, may affect the secretion/translocation process. Additionally, certain Yop proteins are not stable without their respective chaperone protein (Syc proteins) and this chaperone could be a limiting factor.

Fusion of a reporter protein to the potential *Yersinia* effector protein

To study the translocation process Sory and Cornelis (1994) and Sory *et al.* (1995) developed the cya assay. For this assay recombinant *Y. enterocolitica* are constructed, producing a Yop protein fused to a reporter enzyme. Calmodulin-dependent adenylate cyclase form *Bordatella pertussis* is a particularly suitable reporter enzyme for translocation into eukaryotic cells for the following reasons: (i) translocated adenylate cyclase should lead to the accumulation of cAMP in the eukaryotic cytoplasm; (ii) intra-bacterial adenylate cyclase is inactive since bacteria do not produce calmodulin (for review see Botsford and Harmann, 1992); (iii) adenylate cyclase secreted into the medium is inactive; (iv) it has been shown that infection of a cell monolayer by wildtype *Y. pseudotuberculosis* does not result in enhanced levels of cAMP. Therefore, detection of cAMP after infection of eukaryotic cells by a recombinant *Y. enterocolitica* demonstrates intracellular location of the hybrid protein. This method gives only qualitative information about translocation since the enzyme activity of the reporter is dependent on bacterial expression levels as well as stability and enzyme activity in the eukaryotic interacellular environment.

Hybrid proteins are engineered by fusing codons 2–400 of the *Bordetella pertussis cyaA* gene encoding the catalytic domain of adenylate cyclase in 3′ to a full-length *yop* gene or to a truncated yop gene (Sory and Cornelis,

1994; Sory *et al.*, 1995). The catalytic domain of Cya is unable to penetrate eukaryotic cells by itself.

Bacteria and cells are pre-grown as described earlier. Cytochalasin D ($5\,\mu g\,ml^{-1}$) is added to the cells and 30 min later bacteria are added at a MOI of 20–100. Following infection, cells are washed and then lysed in denaturing conditions (100°C for 5 min in 50 mM HCl, 0.1% Triton X-100). The lysate is neutralized by NaOH and cAMP extracted with ethanol. After centrifugation the supernatant is dried and cAMP is assayed by an enzyme immunoassay (Biotrak, Amersham, Uppsala, Sweden). All experiments should be performed three times with duplicate samples.

So far translocation of *Y. enterocolitica* YopE-cya has been shown in every cell type tested; (i) cell lines: HeLa (human epithelial cells), U937 (human monocyte-macrophage), Jurkat (human T-cells), Kaso (human B-cells), PU5-1.8 (mouse monocyte-macrophage), CHO (Chinese hamster epithelial cells), RAT-1 (rat fibroblasts), COS (monkey fibroblasts), SF9 (insect ovarian); (ii) primary cells: HUVECs (human umbilical vein endothelial cells), rat neurons (summarized in Table 14.1).

A number of other fusion proteins have been used to elucidate effector translocation: Green fluorescent protein (Jacobi *et al.*, 1998), neomycin phosphotransferase (Lee *et al.*, 1998) and Diphteria toxin subunit A (Boyd *et al.*, 2000).

Fractionation of the various compartments by using detergents

This method allows one to distinguish between the inside and outside of the eukaryotic cell and therefore to distinguish between attached bacteria and translocated effector proteins. Great care needs to be taken in the choice of detergent: the detergent needs to be strong enough to permeabilize the eukaryotic cell and to potentially detach the translocated proteins from the membrane but should not affect the integrity of the bacterial cell membrane.

Bacteria and cells are pre-grown as described earlier. Twenty hours before infection, cells are seeded to confluency into 10 cm diameter tissue culture plates. Before infection the cells are washed and covered with medium lacking FBS and antibiotics. Subsequently, cytochalasin D (Sigma) is added 30 min before infection at a final concentration of $5\,\mu g\,ml^{-1}$ (stock solution $2\,mg\,ml^{-1}$ in dimethyl sulphoxide). Cells are then infected with an MOI of 20. After an incubation period of 2 h at 37°C under 5% CO_2, the culture supernatant is carefully removed to analyze bacterial proteins simply released in the medium. The culture supernatant is centrifuged at 5000 rpm for 15 min to sediment detached cells and non-attached bacteria. Proteins in this supernatant are precipitated by the addition of TCA to a final concentration of 10% (w/v). In the mean time, 1 ml of ice-cold extraction buffer [0.1% (v/v) Triton X-100 in PBS (136 mM NaCl, 2.7 mM KCl, 10.1 mM Na_2HPO_4, 1.8 mM KH_2PO_4, pH 7.4), 1 mM phenylmethylsulphonyl fluoride, $16\,mg\,ml^{-1}$ leupeptin (Sigma), $1\,mg\,ml^{-1}$ pepstatin (Sigma), prepared fresh before extraction] was added to the tissue culture plate to lyse the cells. Leave on ice for 15 min. The cell lysate was centrifuged at 13000 rpm for 15 min at 4°C to collect the Triton-insoluble fraction. The supernatant is transferred to a new tube and the centrifugation repeated. After that the supernatant was precipitated with TCA, as described above. All three fractions (culture supernatant, Triton-soluble and Triton-insoluble) are suspended in SDS-PAGE loading buffer, separated by electrophoresis in 12% (w/v) polyacrylamide gels in the presence of SDS (SDS-PAGE). After electrophoresis proteins are transferred by electroblotting to a nitrocellulose membrane (Towbin *et al.*, 1979) and immunoblotting is carried out by using the relevant antibody. Revealing of immunoblots is performed with secondary antibodies conjugated to horseradish peroxidase (1:2000; Dako) before development with supersignal chemiluminescent substrate (Pierce) or α-naphthol solution.

If there is no antibody available against the translocated protein, a tag can be added.

Confocal microscopy

HeLa cells grown on coverslips (0.5×10^5) are infected with an MOI of 40. At various times after infection the cell monolayers are washed twice with PBS and are then fixed in 2% paraformaldehyde, permeabilized with 0.5% Triton X-100 and further processed for indirect immunofluorescence labelling (for details see Rosqvist *et al.*, 1991) using biotine-conjugated anti-rabbit antibodies followed by FITC conjugated strepavidine. In double labelling experiments infected HeLa cells are stained for $1\frac{3}{4}$ h post-infection with wheat germ agglutinin (WGA) conjugated to Texas Red ($25\ \mu g\ ml^{-1}$ for 5 min at room temperature) to stain the cell membrane. The specimens are then fixed, permeabilized and incubated with affinity purified antibodies against the Yop of interest followed by FITC-conjugated antibodies (Rosqvist *et al.*, 1994).

The specimens are analysed using an epifluorescence microscope or a confocal laser scanning microscope equipped with dual detectors and argon-krypton (Ar/Kr) laser for simultaneous scanning of two different fluorochromes (Multiprobe 2001, Molecular Dynamics, Sunnyvale, CA). Confocal microscopy is then used to analyse Yop localization. Sets of fluorescent images are acquired simultaneously for Texas Red and fluorescein-tagged markers. Companion images (30 sections with image size 256×256) are scanned with 0.07 μm pixel size and 0.2 μm step size. The images obtained in the red channel (excitation 568 nm, WGA-Texas Red staining) background noise is substracted. The images are then subjected to 3D reconstruction using surface shading to create an image of the cell membrane. The images obtained in the green channel (excitation 488 nm, Yop staining with FITC conjugated antibodies) are also processed to reduce background noise, and then presented as a look-through projection. By the use of volume visualization techniques (Kaufmann, 1991) it is possible to highlight different aspects of the interaction between the bacteria and the eukaryotic cell surface. A low opacity value for the eukaryotic cell membrane (no look-through) and a low density value for the Yop concentration allows one to show bacteria/parts of bacteria outside the cell. A high opacity value for the eukaryotic cell membrane and a high density value for the Yop concentration can be used to determine the spatial localization of Yops (Rosqvist *et al.*, 1994).

◆◆◆◆◆◆ APOPTOSIS OF MACROPHAGES

Similar to YopE and YopH (as described in IV), the effect of YopP/YopJ on macrophages was discovered before its translocation was shown by any of the methods detailed YopP/YopJ induces apoptosis in the mouse macrophage like cell lines J774A.1 (Mills *et al.*, 1997). The following three methods were used to show apoptosis:

Terminal deoxyribonucleotidyl transferase-mediated dUTP-Dioxigenin nick end-labelling (TUNEL) reaction and epifluorescence microscopy

With this method fragmented genomic DNA (characteristic for apoptosis) is detected by immunofluorescence using the TUNEL reaction followed by the addition of fluorescein isothiocyanate-conjugated anti-digoxigenin antibodies. J774.1 macrophages are seeded at 5×10^5 cells per coverslip in 24-well tissue culture plates 15 h in advance of infection. These cells are infected with an MOI of 50 with *Y. enterocolitica* pre-grown for 2 h at room temperature and 30 min at 37°C. After 30 min infection, cells are incubated with gentamycin at 30 $\mu g\ ml^{-1}$ to kill extra cellular bacteria. After 2–5 h infection (depending on the *Y. enterocolitica* strain) the cells are fixed for 20 min in 2.5% paraformaldehyde, extracted with 2:1 ethanol: acetic acid for 5 min at −20°C, and processed for epifluorescence microscopy using the Apotag In Situ Apoptosis Detection Kit (Oncor Apotag S7110-Kit) according to manufacturer's instructions. Processed cells are evaluated by phase contrast and epifluorescence (480 nm) microscopy of three random fields of view (100–175 cells per view) to determine the percentage of TUNEL-positive nuclei at 400× magnification. That 95–100% of the cells have bacteria associated with them was determined by indirect immunofluorescence staining of *Y. enterocolitica*.

DNA fragmentation

For DNA isolation, 10^7 J774A.1 cells are seeded into 100 mm tissue culture dishes and infected using an MOI of either 10 or 100 as described earlier. J774A.1 DNA is isolated 2 h after infection in 5 ml of lysis buffer (8 mM EDTA, 10 mM Tris, pH 7.2, 0.2% Triton X-100, 15 $\mu g\ ml^{-1}$ proteinase K). DNA fragments are separated by agarose gel electrophoresis (Zychlinsky *et al.*, 1992).

Electron microscopy

This method allows one to visualize structural changes in the eukaryotic cell.

J774A.1 cells (20×10^6) are seeded into 100 mm tissue culture dishes and infected using an MOI of either 10 or 100 as described earlier. Cells are fixed with 1% (w/v) glutaraldehyde and processed for electron microscopy in pellets as described by Nassogne *et al.* (1997). Semi-thin sections are stained with toluidine blue and examined by light microscopy.

Typical structural features of apoptosis include (i) peripheral chromatin condensation in crescents, except in the vicinity of nuclear pores; (ii) bulging of nuclear crescent into the cytoplasm; and (iii) appearance of central clusters of small particles of unknown nature, typical of apoptosis (Kerr *et al.*, 1995).

◆◆◆◆◆◆ PORE FORMATION IN EUKARYOTIC TARGET CELL

Translocation of Yop effectors requires the formation of a pore involving YopB and YopD (Hakansson *et al.*, 1996; Neyt and Cornelis, 1999). This pore formation is analysed either on red blood cells or on nucleated cells like macrophages. It requires the use of a multi-mutant of *Y. enterocolitica* that is not expressing the effector proteins YopH, YopO, YopP, YopE, YopM and YopN (ΔHOPEMN), presumably because the translocated effectors obstruct the translocation channel.

Red blood cell lysis

For the hemolysis assay as described by Hakansson *et al.* (1996) and Neyt and Cornelis (1999), overnight Yersinia cultures are diluted to an OD of 0.1 in BHI-Ox medium supplemented with 20 mM $MgCl_2$ and 0.4% glucose and grown for 1.5 h at room temperature and for 2 h at 37°C. Sheep erythrocytes are washed and centrifuged three times for 5 min at 2000*g* with cold PBS until the supernatant is essentially colourless, counted with a haemocytometer and diluted to 4×10^9 cells ml^{-1}. Erythrocyte suspension (50 μl) is mixed with 10^8 bacteria in a round bottom 96-well plate and, after 10 min centrifugation at 1200*g* (to facilitate contact between red blood cells and bacteria), the plate is incubated for 1.5 h at 37°C. The pellets are then resuspended in PBS to a total volume of 250 μl, and the plate is centrifuged for 15 min at 1200*g*. Supernatants (100 μl) are then transferred to a flat bottom 96-well plate, and the OD at 570 nm is measured. As a positive control erythrocytes lysed with water are used. The percentage of hemolysis is determined by dividing the sample value by the OD_{570} of water-lysed erythrocytes and multiplying by 100.

Assessment of pore formation in macrophages by phase contrast microscopy

After infection of PU5-1.8 macrophages with *Y. enterocolitica* ΔHOPEMN almost all of the cells (90%) flatten except for the nucleus, evoking fried eggs. This phenomenon can already be observed after 1 h of infection if the bacteria have been pre-incubated at 37°C to induce Yop synthesis. In the absence of pre-induction of Yop synthesis, flattened cells are only observed after 5 h infection. In contrast, cells infected with wildtype *Y. enterocolitica* do not flatten, but rather round up under the action of the cytotoxic effectors YopE and YopT (Neyt and Cornelis, 1999).

To ensure that the flattening of macrophages after infection with ΔHOPEMN is the result of pore formation but not cell lysis, one can monitor the release of LDH from infected cells. Indeed, this 135 kDa molecule is not released simply by pore formation. LDH released into the medium can be determined with the CytoTox 96 cytotoxicity assay kit (Promega): 50 μl aliquots of medium are transferred to a 96-well plate. The substrate is then added and, after 20–30 min incubation, the reaction is terminated by the addition of 50 μl stop solution, and the absorbance is measured at 490 nm. As a positive control for total cell associated LDH, macrophages are lysed with 0.09% Triton X-100. As negative control, uninfected cells and cells infected with a *Y. enterocolitica* strain lacking the pYV plasmid (pYV-) were used.

Assessment of pore formation in macrophages by fluorescence microscopy

Pore formation can also be studied by monitoring the permeability towards various fluorescent dyes. Dye exclusion experiments are performed essentially as described by Kirby *et al.* (1998) and the following dyes are used: lucifer yellow CH, lithium salt (625 μg ml^{-1}, Molecular Probes) and Texas red-X phalloidin (1 U per coverslip). Bacteria that are not induced for Yop production (grown as described earlier) are diluted to an OD_{600} of 1.0. Bacterial suspension (100 μl) is added to cells (1.5×10^5 cells per well) grown on coverslips in 1 ml medium, previously washed with RPMI containing 30 mM dextran 6000 (RPMI-dextran). After incubation for 3 h at 37°C, macrophages are washed once with ice-cold PBS-dextran, incubated for 10 min on ice and then incubated with fluorescent dyes in RPMI-dextran for 1 h on ice. Macrophages are then washed seven times with ice-cold PBS-dextran and fixed for 20 min with 3% paraformaldehyde in PBS-dextran. After washing the coverslips, samples are mounted in 50% Mowiol (Polysciences) containing 100 mg ml^{-1} diazabicyclooctane (DABCO; Sigma). Cells are examined under a fluorescence microscope. The *Y. enterocolitica* pore confers permeability towards Lucifer Yellow (443 Da) but not to Texas red-phalloidin (1490 Da) (Neyt and Cornelis, 1999).

Assessment of pore formation in macrophages by BCECF release

The BCECF-AM (623 Da, estimated molecular radius: 0.8 nm) release assay is based on the principle described by Bhakdi *et al.* (1989). Cells and bacteria were grown and prepared as described for LDH assay. Just before infection, cells are washed twice with 1 ml PBS and labelled by incubation with 10 μM BCECF-AM [2′,7′-bis-(2-carboxyethyl)-5-(and-6)-carboxyfluorescein, acetoxymethyl ester; Molecular Probes] for 20 min at 37°C.

Cells are then washed twice with RPMI without phenol red and, after 1 h infection with an MOI of 500 they are pelleted by centrifugation at 250*g* for 4 min. The cell culture supernatant is transferred into 2 ml cuvettes, and fluorescence is measured using a spectofluorimeter (Versafluor; BioRad) with an excitation wavelength of 490 nm and emission at 520 nm. Cells treated with Triton X-100 (0.09%) for 45 min were used as a positive control. As negative control, uninfected cells and cells infected with *Y. enterocolitica* strain lacking the pYV plasmid (pYV-) were used. Percentage of lysis was calculated using the following formula: %lysis = (sample-uninfected) × 100.

In summary, here we have discussed a range of methods used with *Y. enterocolitica* to investigate the process of translocation of effectors into eukaryotic cells. Some methods can easily be used to investigate any type III secretion systems, such as the use of fusion proteins or the assays to show pore formation, while other assays depend on effector activity, such as cytotoxicity and apoptosis assays. Additionally, the following points need to be considered to investigate the translocation of potential effector proteins.

1. Under what experimental conditions are the bacteria able to produce and secrete/translocate the virulence proteins.
2. It is necessary to distinguish between translocated proteins and internalized bacteria.
3. It is also important to distinguish between translocated proteins and proteins inside the bacteria or secreted into the environment.
4. The action of some translocated effectors could interfere with the assay. Therefore, it might be necessary to use mutants devoid of one or more effector proteins to show the translocation or action of other effector proteins, especially if the latter are expressed or translocated at low levels.
5. Controls need to show the difference between the effect of bacterial infection *per se* and translocated effector proteins.
6. Importance of assay detection sensitivity limits and the degree of expression of proteins.
7. The use of genetically modified strains to show specificity—e.g. secretion and translocation mutants.

References

Bhakdi, S. and Greulich, S. *et al.* (1989). Potent leukicidal action of *Escherichia coli* hemolysin mediated by permeabilization of target cell membranes. *J. Exp. Med.* **169**, 737–759.

Black, D. S. and Bliska, J. B. (1997). Identification of p130Cas as a substrate of Yersinia YopH (Yop51), a bacterial protein tyrosine phosphatase the translocates into mammalian cells and targets focal adhesions. *Embo J.* **6**(10), 2730–2744.

Black, D. S. and Montagna, L. G. *et al.* (1998). Identification of an amino-terminal substrate-binding domain in the Yersinia tyrosine phosphatase that is required for efficient recognition of focal adhesion targets. *Mol. Microbiol.* **29**(5), 1263–1274.

Botsford, J. L. and Harmann, J. G. (1992). Cyclic AMP in prokaryotes. *Microbiol. Rev.* **56**(1), 100–122.

Boyd, A. P. and Grosdent, N. *et al.* (2000). Yersinia enterocolitica can deliver Yop proteins into a wide range of cell types: development of a delivery system for heterologous proteins [in process citation]. *Eur. J. Cell Biol.* **79**(10), 659–671.

Bradke, F. and Dotti, C. G. (1999). The role of local actin instability in axon formation. *Science* **283**(5409), 1931–1934.

Cornelis, G. R. (2000). Molecular and cell biology aspects of plague. *Proc. Natl Acad. Sci. USA* **97**(16), 8778–8783.

Cornelis, G. R. and Boland, A. *et al.* (1998). The virulence plasmid of Yersinia, an antihost genome. *Microbiol. Mol. Biol. Rev.* **62**(4), 1315–1352.

Hakansson, S. and Schesser, K. *et al.* (1996). The YopB protein of Yersinia pseudotuberculosis is essential for the translocation of Yop effector proteins across the target cell plasma membrane and displays a contact-dependent membrane disrupting activity. *Embo J.* **15**(21), 5812–5823.

Iriarte, M. and Cornelis, G. R. (1998). YopT, a new Yersinia Yop effector protein, affects the cytoskeleton of host cells. *Mol. Microbiol.* **29**(3), 915–929.

Jacobi, C. A. and Roggenkamp, A. *et al.* (1998). In vitro and in vivo expression studies of yopE from Yersinia enterocolitica using the gfp reporter gene. *Mol. Microbiol.* **30**(4), 865–882.

Kerr, J. F. and Gobe, G. C. *et al.* (1995). Anatomical methods in cell death. *Methods. Cell Biol.* **46**, 1–27.

Kirby, J. E. and Vogel, J. P. *et al.* (1998). Evidence for pore-forming ability by Legionella pneumophila. *Mol. Microbiol.* **27**(2), 323–336.

Laemmli, U. K. (1970). Cleavage of structural proteins during the assembly of the head of bacteriophage T4. *Nature* **227**(259), 680–685.

Lee, V. T. and Anderson, D. M. *et al.* (1998). Targeting of Yersinia Yop proteins into the cytosol of HeLa cells: one-step translocation of YopE across bacterial and eukaryotic membranes is dependent on SycE chaperone. *Mol. Microbiol.* **28**(3), 593–601.

Mills, S. D. and Boland, A. *et al.* (1997). Yersinia enterocolitica induces apoptosis in macrophages by a process requiring functional type III secretion and translocation mechanisms and involving YopP, presumably acting as an effector protein. *Proc. Natl Acad. Sci. USA* **94**(23), 12638–12643.

Nassogne, M. C. and Louahed, J. *et al.* (1997). Cocaine induces apoptosis in cortical neurons of fetal mice. *J. Neurochem.* **68**(6), 2442–2450.

Neyt, C. and Cornelis, G. R. (1999). Insertion of a Yop translocation pore into the macrophage plasma membrane by Yersinia enterocolitica: requirement for translocators YopB and YopD, but not LcrG. *Mol. Microbiol.* **33**(5), 971–981.

Neyt, C. and Iriarte, M. *et al.* (1997). Virulence and arsenic resistance in Yersiniae. *J. Bacteriol.* **179**(3), 612–619.

Persson, C. and Carballeira, N. *et al.* (1997). The PTPase YopH inhibits uptake of Yersinia, tyrosine phosphorylation of p130Cas and FAK, and the associated accumulation of these proteins in peripheral focal adhesions. *Embo J.* **16**(9), 2307–2318.

Rosqvist, R. and Forsberg, A. *et al.* (1991). Intracellular targeting of the Yersinia YopE cytotoxin in mammalian cells induces actin microfilament disruption. *Infect. Immun.* **59**(12), 4562–4569.

Rosqvist, R. and Magnusson, K. E. *et al.* (1994). Target cell contact triggers expression and polarized transfer of Yersinia YopE cytotoxin into mammalian cells. *Embo J.* **13**(4), 964–972.

Sory, M. P. and Boland, A. *et al.* (1995). Identification of the YopE and YopH domains required for secretion and internalization into the cytosol of macrophages, using the cyaA gene fusion approach. *Proc. Natl Acad. Sci. USA* **92**(26), 11998–12002.

Sory, M. P. and Cornelis, G. R. (1994). Translocation of a hybrid YopE-adenylate cyclase from Yersinia enterocolitica into HeLa cells. *Mol. Microbiol.* **14**(3), 583–594.

Towbin, H. and Staehelin, T. *et al.* (1979). Electrophoretic transfer of proteins from polyacrylamide gels to nitrocellulose sheets: procedure and some applications. *Proc. Natl Acad. Sci. USA* **76**(9), 4350–4.

Zychlinsky, A. and Zheng, L. M. *et al.* (1992). Cytolytic lymphocytes induce both apoptosis and necrosis in target cells. *J. Immunol.* **146**(1), 393–400.

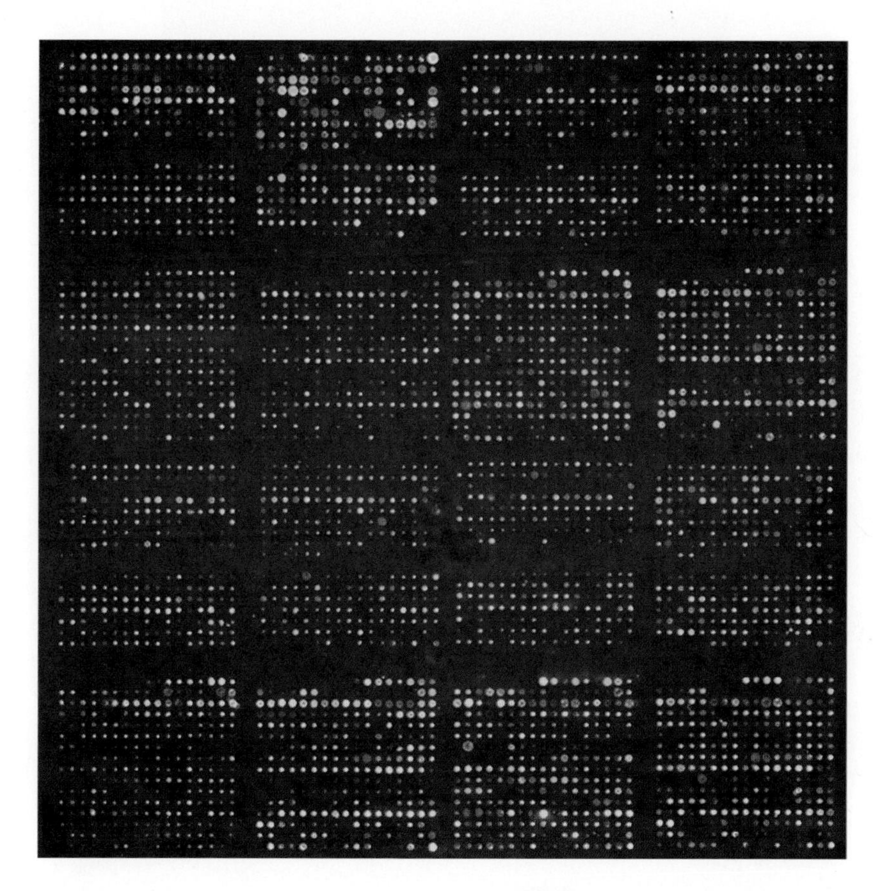

Figure 2.1 DNA labeled with the specified protocol hybridized to a *Helicobacter pylori* whole genome spotted array.

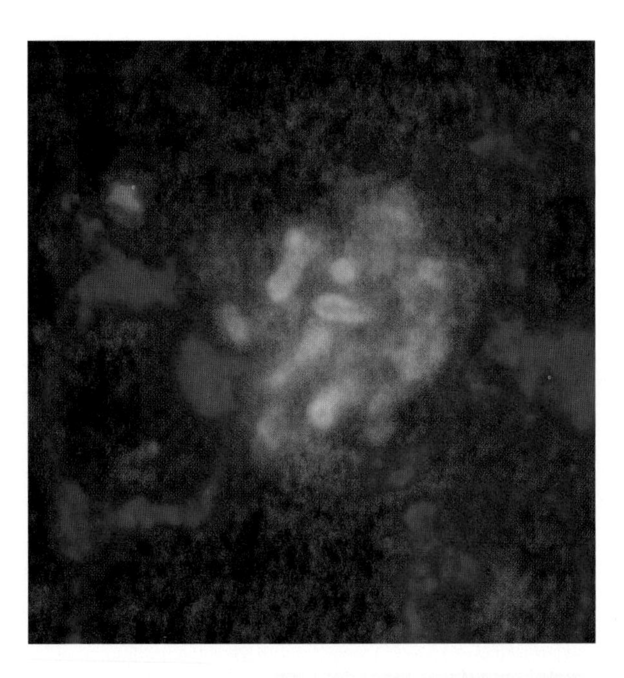

Figure 4.1 Spleen section of BALB/c mouse infected with *S.* Typhimurium. *Salmonella* cells that carry a single copy *ssaG::gfp* transcriptional fusion in the chromosomal *put* locus are detected with an anti-LPS antibody (red fluorescence). An anti-CD18 monoclonal antibody (blue fluorescence) identifies all phagocytic cells. Green fluorescence reflects *ssaG::gfp* expression within phagocytic cells.

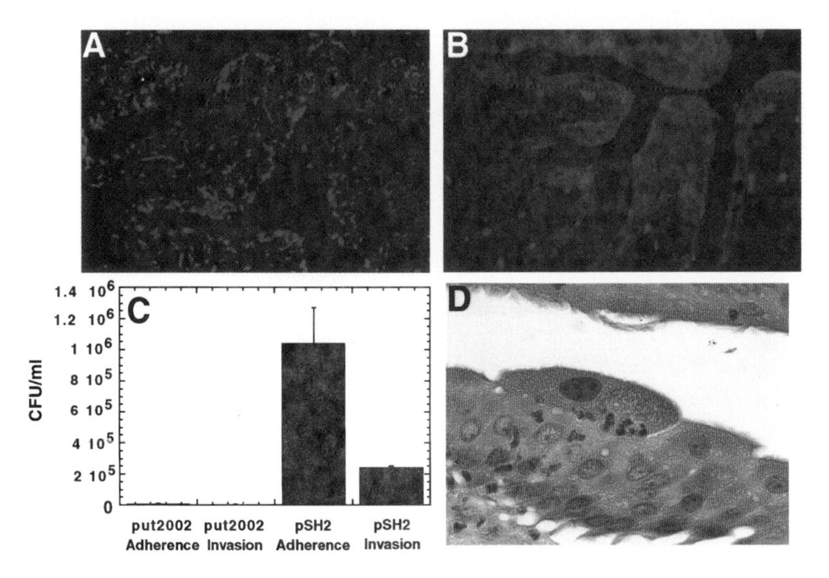

Figure 7.4 Adherence and invasion of bladder epithelial cells mediated by type 1 piliated *E. coli*. An *in situ* binding assay showing that *E. coli* expressing type 1 pili are able to bind to mouse bladder epithelial cells (A) while an isogenic $fimH^-$ mutant strain is unable to bind to the mouse bladder epithelium (B). Adherence and invasion of 5637 human bladder epithelial cells by the *E. coli* K12 strain AAEC185 expressing the plasmid pSH2 ($fimH^+$) or put2002 ($fimH^-$) (C). Bacteria expressing $fimH^+$ type 1 pili were able to both adhere to and invade the tissue culture cells while $fimH^-$ strains were unable to bind to or invade the bladder epithelial cells. An H & E stained mouse bladder tissue section after infection with a cystitis isolate (D). The cystitis isolate invaded a superficial bladder cell and has started replicating.

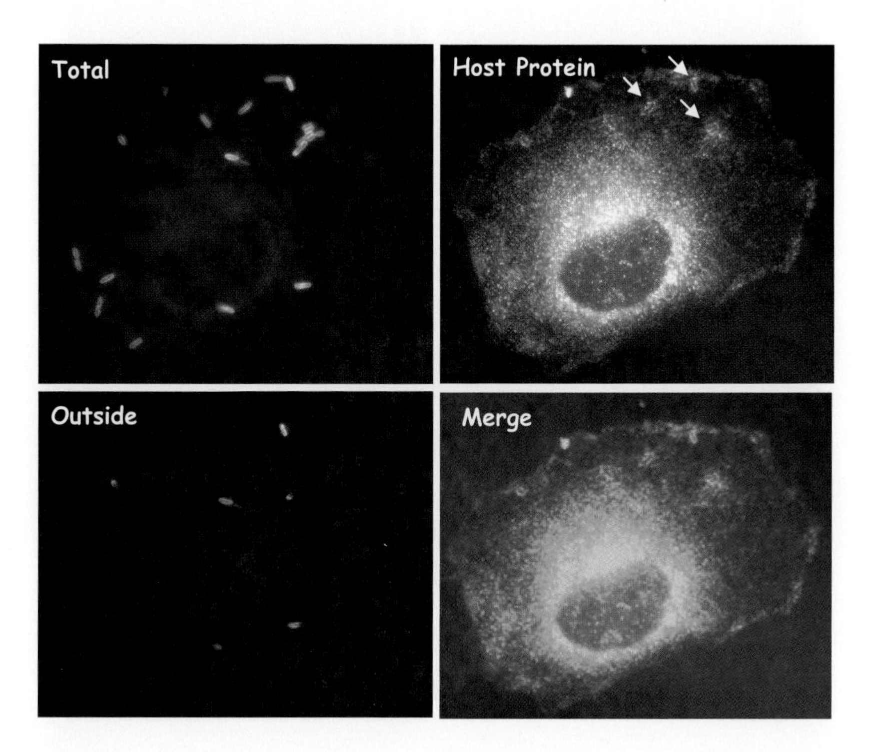

Figure 9.1 Image of *Yersinia pseudotuberculosis* bacteria captured in the dynamic process of entry into a COS-1 cell. Merged image: note the distinct staining of the internalized (Blue) and external (Red) portions of the 'half-in/half-out' bacteria. Localization of a Arp2/3 protein complex during bacterial uptake was examined by immunostaining with an anti Arp3 antibody (Green) after permeabilization of cells (Alrutz *et al.*, 2001). The intense staining around nascent phagosomes strongly suggests that this protein participates in the uptake process; quantitation of this phenotype would reinforce the inference.

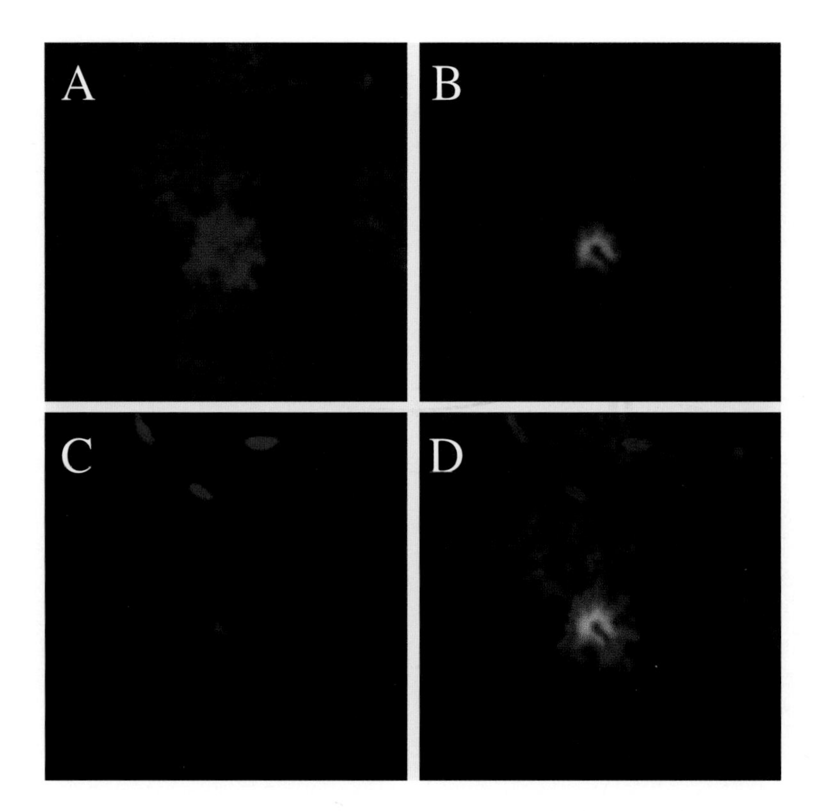

Figure 11.1 Immunofluorescence analysis of ezrin in a *Shigella* focus of internalization. HeLa cells were challenged with *Shigella* for 15 min at 37°C. Samples were fixed and stained for ezrin (Panel A), actin (Panel B) or *Shigella* LPS (Panel C), and analyzed by confocal laser microscopy (LSM510, Zeiss). Images were obtained by reconstruction from sections that do not include the basal cell surface. Panel D shows the superimposition of the triple staining. Ezrin labels the tip of *Shigella*-induced cell extensions where little F-actin is detected. Scale bar: 5 μm.

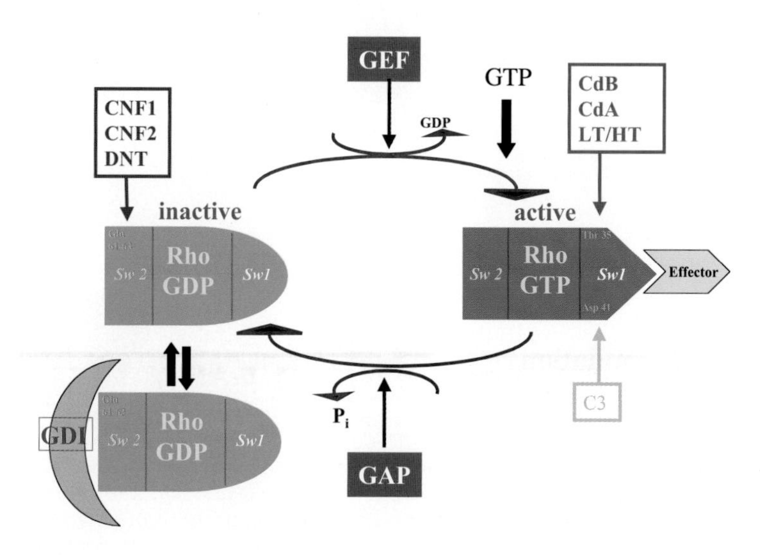

Figure 12.1 The Rho activation/deactivation cell cycle. Rho in its inactive form (associated with GDP) is brought to the membrane where the guanine exchange facture (GEF) allows removal of GDP with exchange with GTP, turning Rho into the active form (bound to GTP). Rho docks to its effector and activates it. To turn off Rho activity, the GTPase activating protein (GAP) introduces an arginine residue in close proximity to Rho glutamine 63, facilitating the hydrolysis of GTP into GDP thereby releasing the GTPase in its inactive conformation. Rho GDP is stocked in the cytosol associated with the protein GDI (guanine dissociation inhibitor). Interaction of bacterial toxins with Rho are indicated by colored arrows. CdB: *Clostridium difficile* toxin B. CdA: *Clostridium difficile* toxin A. LT: *Clostridium sordellii* lethal toxin. HT: *Clostridium sordellii* hemorrhagic toxin. CNF1/2: cytotoxic necrotizing facture 1/2. DNT: dermonecrotic toxin from *Bordetella* sp.

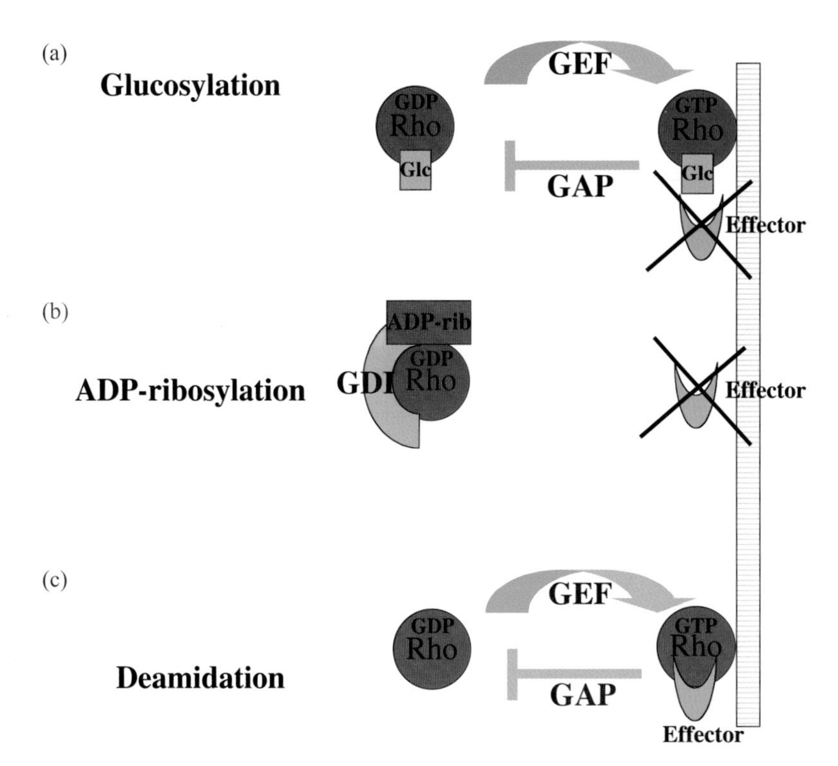

Figure 12.2 Modifications of Rho GTPases by C3, large clostridial toxins and CNF1: consequences on the inhibition/stimulation on the Rho effector. a: Glucosylation by large clostridial cytotoxins inhibits the GAP activity thereby leaving the protein associated with the membrane in its GTP bound form. GDI, which recognizes Rho associated with GDP, is inefficient in removing the GTPase from the membrane. The glucose moiety associated with the effector domain of Rho impairs the docking of the GTPase to its effector. b: C3-ADPribosylation blocks Rho on GDI in its GDP conformation. The GTPase is therefore not translocated to the membrane, cannot be activated by GEF and does not dock on the effector. c: CNF1-deamidation: Rho glutamine 63 is deamidated into glutamic acid rendering the hydrolysis of GTP by the GAP inefficient. The GTPase stays permanently associated with GTP thus stimulating the effector.

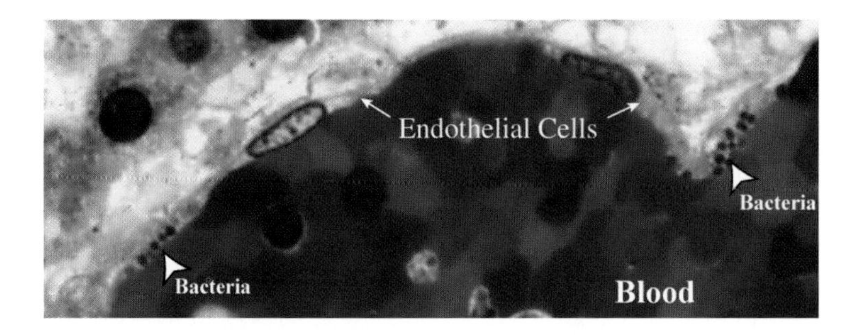

Figure 23.1 Postmortem section of a capillary in the choroid plexus of the brain showing *Neisseria meningitidis* adhering to the endothelial cells.

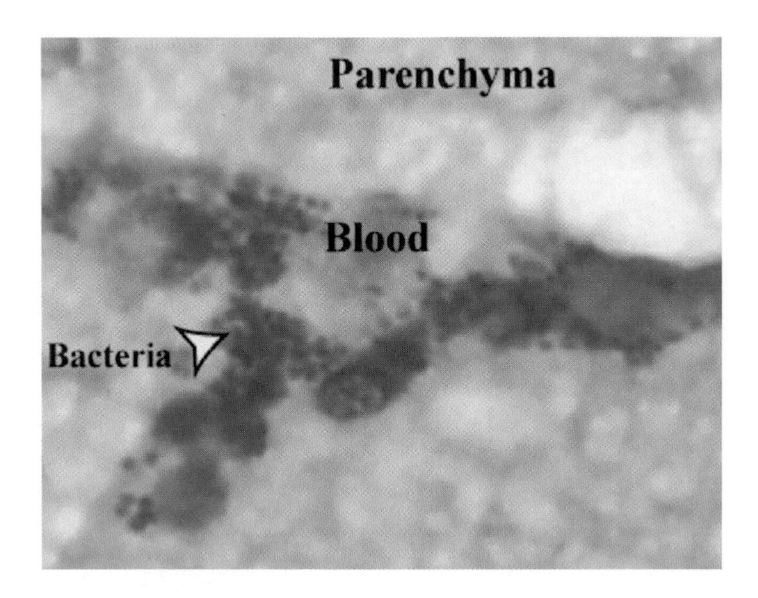

Figure 23.2 Postmortem section of the brain parenchyma showing *Neisseria meningitidis* adhering to brain capillaries.

Figure 24.2 Detection of apoptotic cells using TUNEL. Rabbit ileal loops were inoculated with wild-type *Shigella flexneri*. Sections of Peyer's patches were stained with TUNEL (apoptotic cells appear green) and counterstained with propidium iodide.

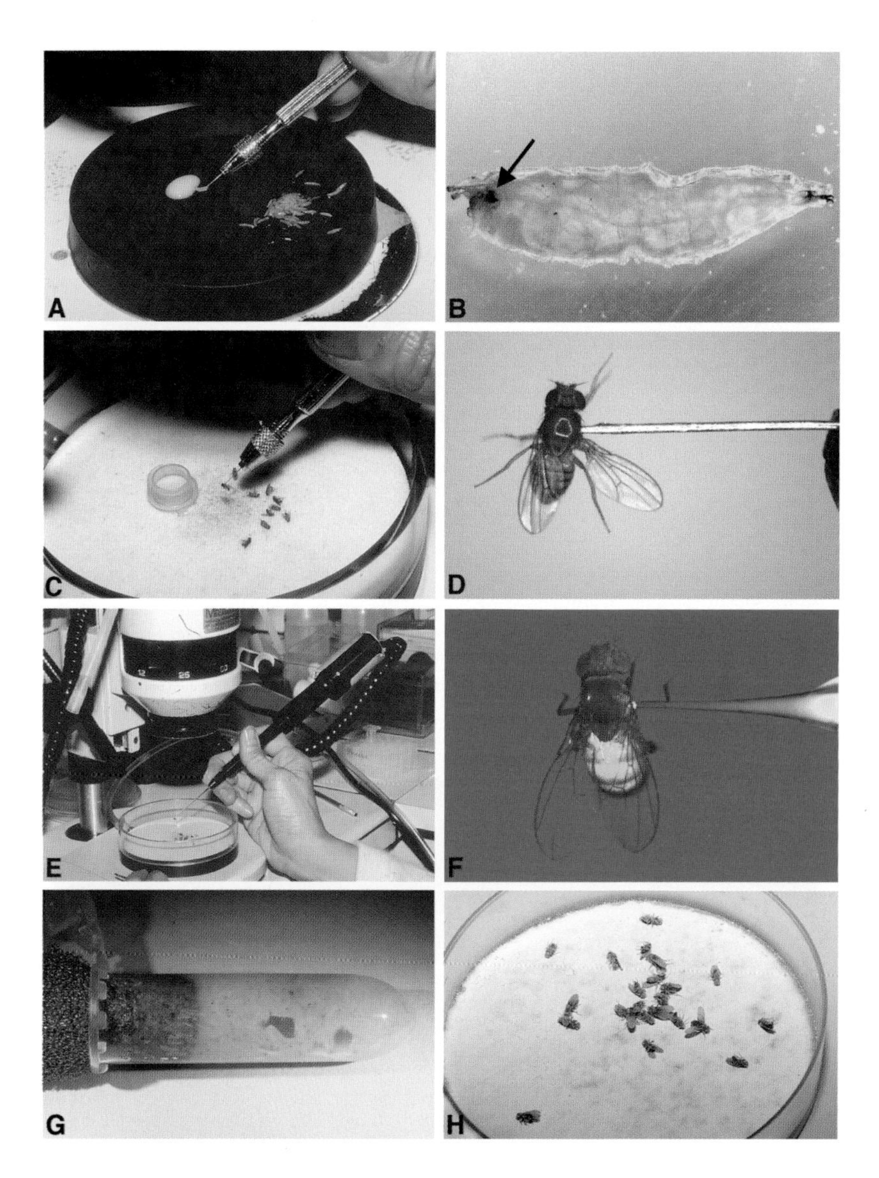

Figure 27.1 Microbial infection of *Drosophila*. (A) Septic injury of larvae is performed with a tungsten needle. A rubber block is employed to protect the needle from damage. A drop of microbial solution is placed on the rubber block, and larvae are placed inside a drop of water. (B) Larva are punctured on their posterior lateral side, triggering a melanization reaction at the injury site (arrow). (C) The needle is mounted on a handle to prick adult flies. The CO_2 pad (Inject + Matic^{TM}, Geneve) provides a convenient way to anesthetize flies. The microbial solutions are concentrated and placed into the cap of a microfuge tube. (D) An adult fly is pricked on the dorsal side of the thorax. (E) Drummond Nanoject injector: a fine capillary tip is backfilled with mineral oil before mounting onto the injector handle, and dipping into microbial solution to load. After specifying the quantity, at each pulse, the injector will release the exact amount (varies from 4 to 73 nl) into the body cavity of the fly. (F) An adult fly being injected with 9.2 nl of GFP expressing bacteria (OD = 100). (G) Natural infection of larvae is performed by mixing the following in a centrifuge vial: crushed banana, bacterial pellet, and third instar larva. After 30 min of incubation at room temperature, the larvae and the bacterial mixture are directly transferred to a standard fly vial. (H) Natural fungal infection is done by covering the flies thoroughly with fungal spores. Flies are anesthetized and shaken on a Petri dish containing a sporulating fungal species.

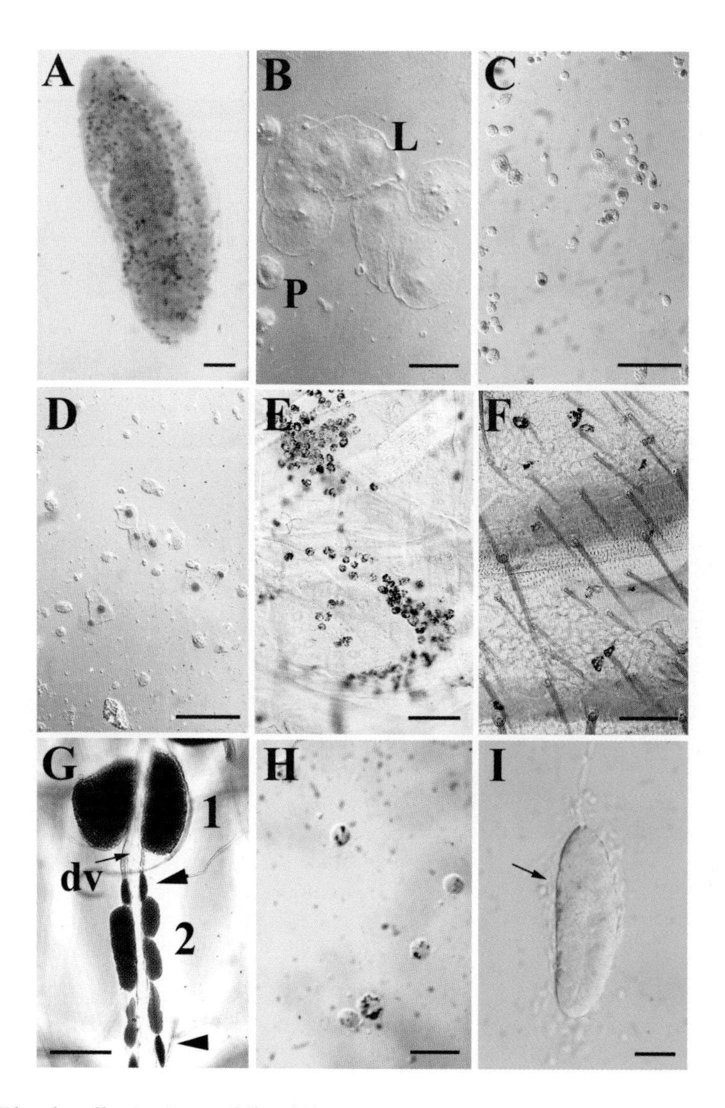

Figure 27.2 Blood cells in *Drosophila*. (A) Distribution of hemocytes in a late embryo, as evidenced by anti-croquemort antibody. Bar: 50 μm. (B) Larval blood cells observed by interference phase contrast microscopy. P: plasmatocyte; L: lamellocyte. Bar: 20 μm. (C, D) *lacZ* expression in larval hemocytes: staining in plasmatocytes in line *l(3)05309* (C) and in lamellocytes in line *l(3)06946* (D) (Braun *et al.*, 1997). Bar: 50 μm. (E, F) Observation of sessile hemocytes through the cuticle after Indian ink injection into a third instar larva (E) or into a *yellow* adult (F). Bar: 50 μm. (G) Dissection of a larval lymph gland attached to the dorsal vessel and stained with osmium tetroxide. Arrow heads: pericardial cells; 1 and 2 designate the first and second lobes of the lymph glands: the second lobes are well developed when larvae are raised at 18°C; dv: dorsal vessel. Bar: 100 μm. (H) Indian ink phagocytosis by larval plasmatocytes observed 2 h after injection. Bar: 20 μm. (I) Encapsulated *L. boulardi* egg in larval hemocoel 24 h after parasitization. The wasp egg is surrounded by lamellocytes (arrow); blackening has not yet occurred. Bar: 50 μm.

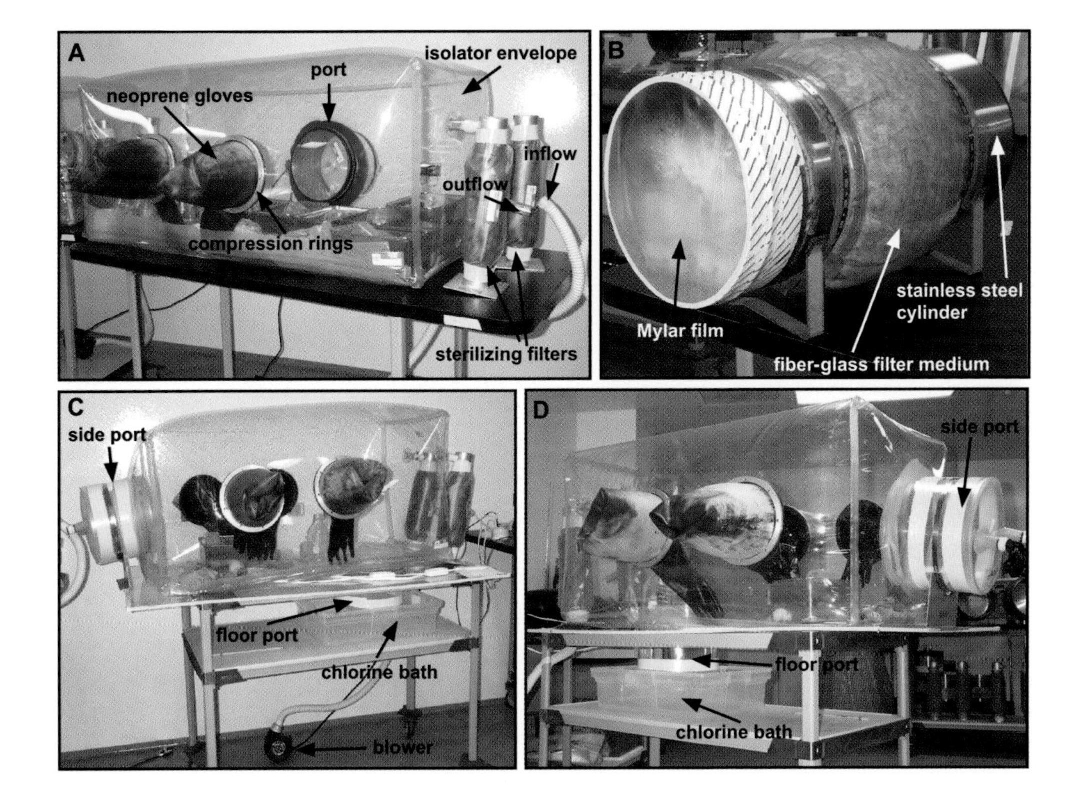

Figure 29.1 Use of flexible film isolators for maintaining and re-deriving germ-free mice. (A) *Flexible film isolator.* A clear vinyl envelope, placed on a cart, completely encloses a sterile space containing cages and supplies. Neoprene gloves attached to the envelope by metal compression rings allow manipulations to be carried out within the sterile workspace. A double-doored port located in the wall of the isolator opposite the gloves is used to import supplies into the isolator. Air is supplied to the isolator by a blower attached to a sterilizing filter by a flexible vinyl hose. Air exits the isolator through an identical filter assembly. (B) *Sterilizing drum.* Sterile supplies such as food, water, and bedding, are autoclaved in a stainless steel cylinder whose diameter is exactly that of the isolator port. Its open end is covered prior to autoclaving with a sheet of Mylar film. After autoclaving, the drum is attached to the isolator port with a plastic transfer sleeve. The sleeve interior is sterilized by fogging with Clidox. The inner port cap is then removed from inside the isolator, the Mylar seal on the mouth of the cylinder is punctured, and the supplies are imported into the isolator. (C and D) *Two views of a surgical isolator for Cesarean re-derivation.* This isolator differs from the normal flexible film isolator by having two sets of gloves and an additional port in the floor. Prior to re-derivation, the outer floor port cap is removed and the mouth of the port is submerged in a chlorine bath. The uterus containing the live pups is removed aseptically from the female, and each horn of the uterus is clamped using sterile hemostats. The uterus is passed into the chlorine bath, and brought up into the germ-free isolator. Pups are immediately removed from the uterus and fostered to germ-free lactating females.

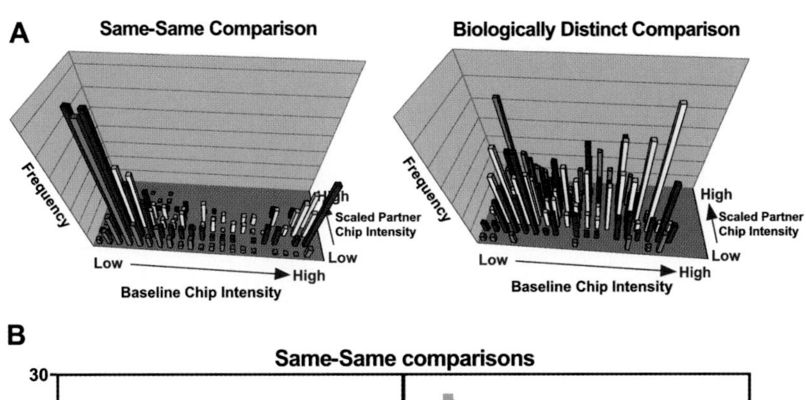

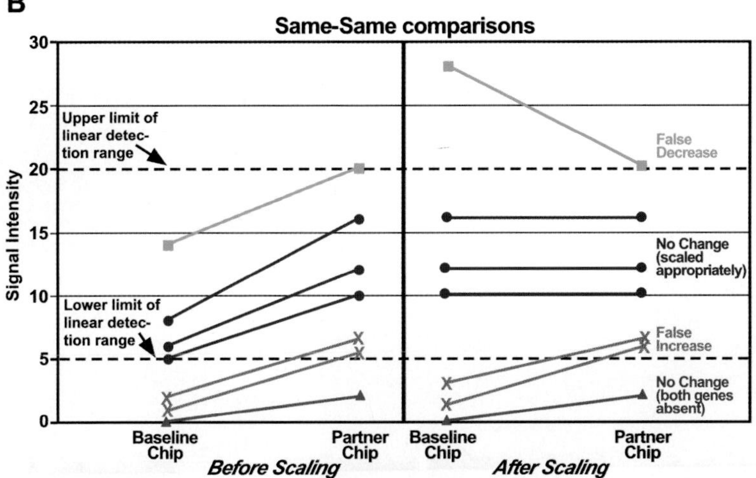

Figure 29.3 Strategy for eliminating noise from GeneChip datasets. (A) *Distributions of Partner vs. Baseline chip intensities in Same–Same GeneChip comparisons differ from comparisons of biologically distinct RNA*. Examples of two types of comparisons are depicted: (1) Same–Same, where chips are hybridized to cRNAs prepared from the same starting RNA and (2) Biologically Distinct, where the two chips in the comparison are hybridized to cRNAs prepared from different starting populations of RNA. The entire range of baseline chip signal intensities ('Average Differences') is represented by the horizontal axis. The corresponding entire range of partner chip intensities is represented by the axis going into the plane of the page. The frequency of probe sets having specified combinations of Baseline and Partner chip intensities is represented by the vertical axis. Only the probe sets called Increased or Decreased are plotted. Note how Same–Same comparisons result in probe sets at the extremes of Baseline chip intensity and at the low end of Partner chip intensity, whereas Biologically Distinct comparisons result in probe sets in the intermediate Baseline chip intensity values and/or high Partner chip intensity values. (B) *Microarray scaling as a source of false-positive changes in gene expression*. Affymetrix GeneChips typically have a false positive rate of 1 to 2% (Lipshutz *et al*., 1999; Lee *et al*., 2000). Given the scale of a typical chip experiment, even such a low error rate can produce a large proportion of genes falsely called Increased or Decreased. A significant source of false positive errors is global scaling, in which the signal intensities of all probe sets on both chips are multiplied by a scaling factor so that the average chip-wide signal intensities are equal. Scaling corrects for chip-wide differences in signal intensity due to variations in hybridization conditions or target labeling. Following scaling, the GeneChip software can compare individual genes across multiple chips and determine whether there are changes in their expression levels. In this example, signal intensities for specific probe sets from a comparison of two microarrays probed with duplicate cRNA targets (a 'Same-Same' comparison) are represented, both before and after scaling. Although in practice both partner and baseline chips are scaled to an arbitrary chip-wide target intensity, to simplify the example we have assumed here that the partner chip needs no scaling. Because the targets are derived from the same tissue RNA sample, the signal outputs from each of the probe sets theoretically should be equal between the two chips following scaling. However, in practice, global scaling works well only for genes whose expression levels are in the linear range of signal detection (represented by blue lines). Artefacts may be created at the extremes of chip signal detection. This can happen for genes having low Fold Changes or low Average Differences (gray lines), or for genes whose signal intensity is so strong that it is near maximal on both chips (green lines). Adapted from Mills and Gordon, 2001.

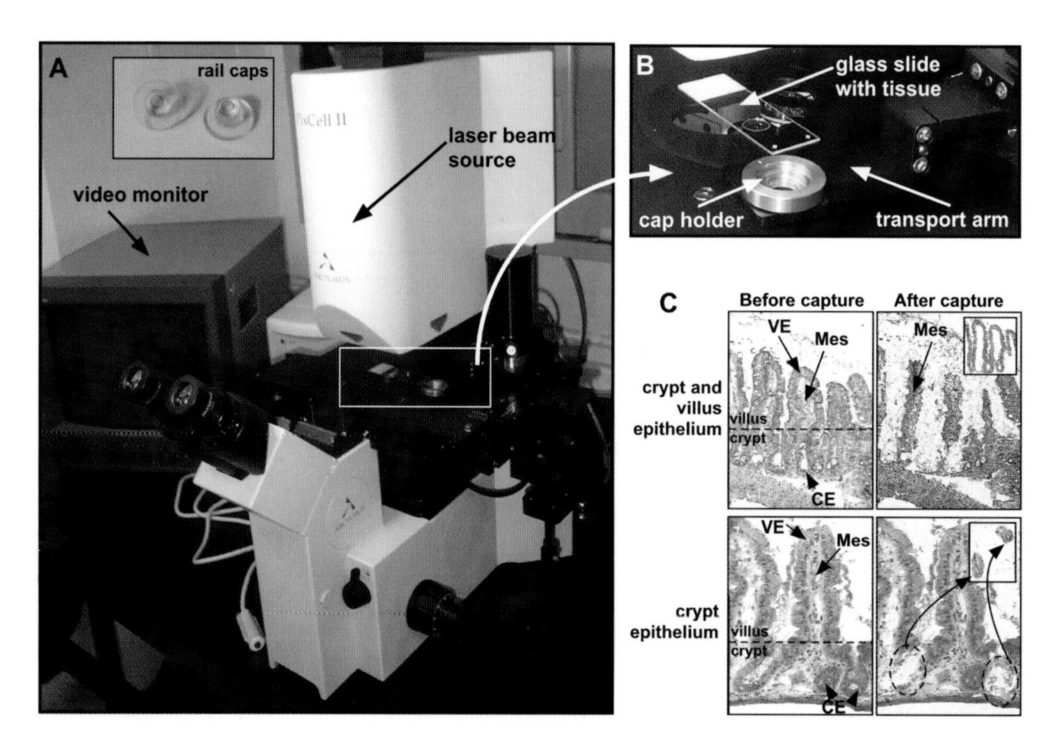

Figure 29.5 Laser capture microdissection. (A and B) *Laser capture microscope*. Frozen tissue sections are placed on the microscope stage and can be viewed on the monitor screen. A 'rail cap' is placed over the tissue and activated laser pulses result in the transfer of selected cells to the cap. (C) *Example of laser capture microdissection of small intestinal epithelial cells*. Sections are stained with nuclear fast red. Insets show captured villus and crypt epithelial populations as they appear on the cap. VE = villus epithelium; CE = crypt epithelium; Mes = mesenchyme.

15 Membrane Translocation by Bacterial AB Toxins

Kamran Badizadegan[1,4], R John Collier[2] and Wayne I Lencer[3,5,6]
[1]Departments of Pathology, [2]Microbiology and Molecular Genetics, and [3]Pediatrics, Harvard Medical School, Boston, MA 02115; [4]Department of Pathology and [5]the Combined Program in Pediatric Gastroenterology, Children's Hospital, Boston, MA 02115, and [6]the Harvard Digestive Diseases Center, Boston, MA 02115

◆◆

CONTENTS

◆◆◆◆◆◆ INTRODUCTION

Many toxins of bacterial origin act by catalytic modification of cytosolic substrates. To gain access to these substrates, the toxins must somehow cross the plasma membrane of their target cells. For most toxins of this type, which are collectively referred to as AB toxins, the catalytically active domain (component A) resides on a different polypeptide from the domain responsible for toxin binding to the cell surface (component B) (Gill, 1978). The AB toxins can further be divided into two general groups based on the intracellular site of membrane translocation of the A subunit. In the first group, exemplified by anthrax and diphtheria toxins, the site of membrane translocation is the early endosomal compartment. For these toxin, translocation of the catalytic unit depends on acidification of the endosomal compartment (Koehler and Collier, 1991; Miller *et al.*, 1999). In the second group, exemplified by cholera and related *E. coli* enterotoxins, shiga toxin, ricin, and *Pseudomonas* exotoxin A, toxicity requires retrograde trafficking into the Golgi complex and the endoplasmic reticulum (ER) of the host cell (Bastiaens *et al.*, 1996; Hazes and Read, 1997; Lencer *et al.*, 1995a; Majoul *et al.*, 1996). Here, membrane translocation of the catalytic unit is thought to depend on ER lumenal chaperones and the protein conducting pore, sec61p (Koopmann *et al.*, 2000; Schmitz *et al.*, 2000; Simpson *et al.*, 1999; Wesche *et al.*, 1999). In what follows, we have

Membrane Translocation by Bacterial AB Toxins

METHODS IN MICROBIOLOGY, VOLUME 31
ISBN 0-12-521531-2

focused mainly on anthrax and cholera toxins as important examples of AB toxins that translocate from the endosome and the ER, respectively.

Anthrax toxin: entry by protein translocation across endosomal membranes

Anthrax toxin is a collective term that refers to a set of three proteins secreted by *Bacillus anthracis*, the causative agent of the disease anthrax. The three proteins are: (i) the protective antigen (PA); (ii) the edema factor (EF); and (iii) the lethal factor (LF). While each of these three proteins is non-toxic by itself, the combination of PA and EF is an edema toxin (EdTx) causing edema in experimental animals, and the combination of PA and LF is a lethal toxin (LeTx) causing death in experimental animals (Leppla, 1982). EF and LF are the catalytic or A subunits of edema toxin and lethal toxin, respectively. PA serves as a common B subunit for binding and intracellular delivery of each of the catalytic subunits (Leppla, 1982).

EdTx and LeTx are assembled at the surface of target cells from their component parts (Figure 15.1). PA, the toxin's B-subunit, is an 83 kDa protein, which mediates receptor binding, self assembly, and translocation of the catalytic unit to the cytosol (Benson *et al.*, 1998; Miller *et al.*, 1999; Wesche *et al.*, 1998). PA binds to a cell surface receptor that has a protein component and appears to be saturable (Escuyer and Collier, 1991). The cell surface receptor for PA is not yet identified, but it appears to be present on all cell types tested. In the human intestinal T84 cell line, the PA receptor is confined to the basolateral membranes (Beauregard *et al.*, 1999). Although the exact sequence of events is unknown, after binding to its cell surface receptor, PA is thought to be cleaved by furin or a related protease into an N-terminal 20 kDa peptide (PA_{20}) and a C-terminal 63 kDa

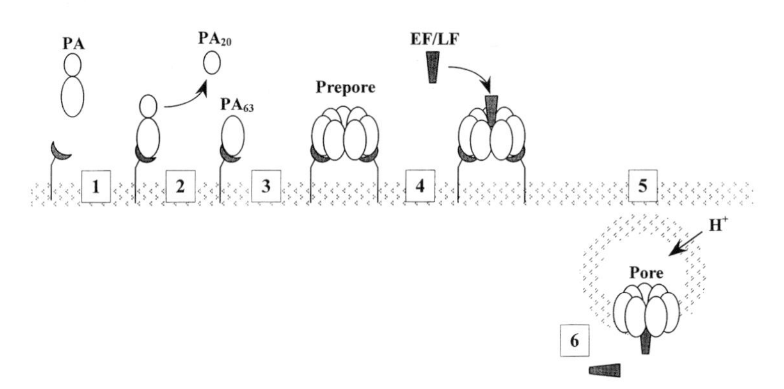

Figure 15.1. Proposed model of anthrax toxin binding and entry into the cell (based on Miller *et al.*, 1999). (1) Binding of PA to its membrane receptor; (2) proteolytic activation of PA and dissociation of PA_{20}; (3) self-association of monomeric PA_{63} to form the heptameric prepore; (4) binding of the enzymatically active component (EF or LF) to the prepore; (5) endocytosis and acidification of the early endosome; and (6) pH-dependent conversion of the prepore to pore with subsequent translocation of the toxin to the cytosol.

fragment (PA_{63}) (Klimpel *et al.*, 1992; Singh *et al.*, 1989). After PA_{20} is released into the medium, PA_{63} remains bound to the cell surface and spontaneously oligomerizes to form a heptameric, ring-shaped prepore that binds the A-moieties, EF and LF (Benson *et al.*, 1998; Miller *et al.*, 1999; Petosa *et al.*, 1997).

The EF/LF-PA_{63}-receptor complexes enter the cell by endocytosis where the low lumenal pH of the endosomal compartment triggers prepore to pore conversion of the PA_{63} heptamer. It has been proposed that the process of prepore to pore conversion involves conformational rearrangement of a disordered amphipathic loop (D2L2; residues 302–325), in which loops from the seven protomers combine to form a transmembrane 14-stranded β barrel (Benson *et al.*, 1998). The pore formed conducts solutes and mediates translocation of EF and LF to the cytosol, although the precise mechanism of protein translocation remains unknown (Benson *et al.*, 1998; Miller *et al.*, 1999; Petosa *et al.*, 1997).

EF is an 89 kDa potent calmodulin-dependent adenylyl cyclase that when delivered to the cytosol catalyzes the formation of cAMP (Leppla, 1982, 1984). LF, the A-subunit of *B. anthracis* lethal toxin is a metalloprotease that cleaves the amino terminus of mitogen-activated protein kinase kinases 1 and 2 (MAPKK1 and MAPKK2). This cleavage inactivates MAPKK1 and inhibits the MAPK signal transduction pathway, leading to destruction and death of macrophages (Duesbery *et al.*, 1998).

The EF and LF bind to the same site on PA_{63}, and their entry into host cells follows the same sequence of events (Figure 15.1). Thus, translocation of both EF and LF depends critically on the function of PA. PA binds a membrane receptor targeted for endocytosis, oligomerizes into a transmembrane solute conducting pore, and somehow mediates protein translocation across the limiting membrane of the endosomal compartment.

Cholera toxin: entry by translocation across the ER membranes

Cholera toxin (CT) produced by *Vibrio cholerae* is the toxin responsible for induction of the massive secretory diarrhea of Asiatic cholera. CT is comprised of a pentameric B-subunit (CTB; 55 kDa) that binds ganglioside G_{M1} at the cell surface, and an enzymatic A-subunit (CTA; 27 kDa) that translocates to the cytosol of host cells and activates adenylyl cyclase by catalyzing the ADP-ribosylation of the regulatory GTPase Gsα (Holmes *et al.*, 1995, 1990; Spangler, 1992). In the small intestinal crypts, CT elicits a cAMP-dependent chloride secretory response fundamental to the pathogenesis of secretory diarrhea.

Entry of CT into the intestinal epithelial cell occurs by receptor-mediated endocytosis and retrograde transport into the Golgi and endoplasmic reticulum (Figure 15.2) (Bastiaens *et al.*, 1996; Lencer *et al.*, 1995a, 1993, 1992; Majoul *et al.*, 1996; Orlandi, 1997; Orlandi and Fishman, 1993). We have recently obtained direct evidence that the intact holotoxin traffics into the ER (unpublished data, Fujinaga and Lencer) where the A-subunit is unfolded and dissociated from the B-pentamer by a redox-sensitive chaperone (Tsai *et al.*, 2001). Available evidence indicates that the unfolded A-subunit likely enters the cytosol through the ER membranes

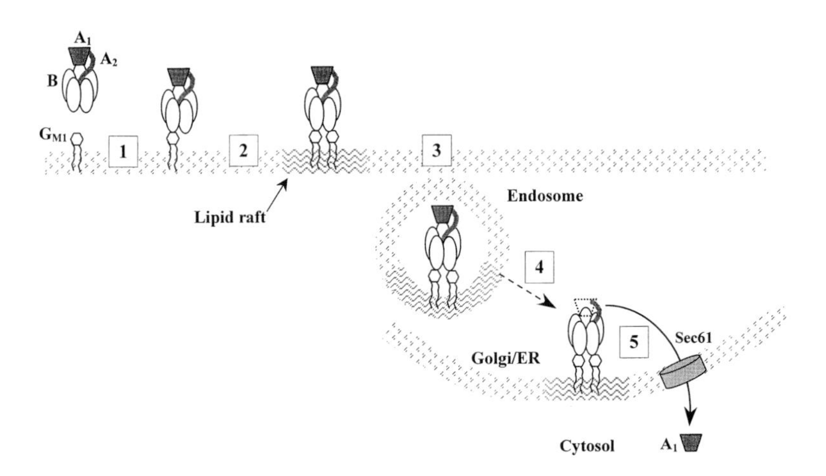

Figure 15.2. Proposed model of cholera toxin binding and entry into the cell (based on Lencer *et al.*, 1999). (1) Binding of the holotoxin to its receptor ganglioside G_{M1} on the cell surface; (2) clustering of the toxin-receptor complex in lipid rafts; (3) endocytosis of the toxin receptor complex (it is not known if the toxin continues to remain associated with lipid rafts as shown here); (4) retrograde transport of the endocytic vesicle to Golgi and ER cisternae; (5) unfolding and translocation of the enzymatically active A_1 peptide into the cytosol via Sec61.

by 'dislocation' through the sec61 complex (Koopmann *et al.*, 2000; Schmitz *et al.*, 2000; Simpson *et al.*, 1999; Wesche *et al.*, 1999). Thus, unlike anthrax and diphtheria toxins, translocation of CT to the cytosol of host cells depends critically on trafficking into the ER where the A-subunit exploits the machinery for membrane translocation of proteins endogenous to the host cell itself.

Retrograde trafficking into the ER is essential for CT action. Movement into the ER depends on the function of the B-subunit, which binds ganglioside G_{M1} on the cell surface, and on the ER-targeting KDEL motif at the C-terminus of the A-subunit (Lencer *et al.*, 1995a, 1999; Wolf *et al.*, 1998). The specificity of ganglioside G_{M1} function in toxin action correlates with the ability of G_{M1} to partition CT into detergent-insoluble membrane microdomains that display some of the biochemical characteristics of caveolae (Badizadegan *et al.*, 2000; Orlandi and Fishman, 1998; Wolf *et al.*, 1998). We have proposed that CT binding to G_{M1} at the cell surface represents a form of protein acylation that acts as the dominant sorting motif for entry into this retrograde pathway, and that depends on association with detergent-insoluble membranes (Lencer *et al.*, 1999).

Detergent-insoluble glycosphingolipid-rich membranes (DIGs) are distinct membrane structures rich in cholesterol and glycolipids, which function in various cell types as membrane organizing centers for signal transduction, protein and lipid sorting, endocytosis, and transcytosis (reviewed in Anderson, 1998 and Brown and London, 2000). DIGs can be isolated from essentially all mammalian cell types by virtue of their insolubility in non-ionic detergents such as Triton X-100 at 4°C and their relatively light buoyant density. Although the terminology has been the subject of debate, plasma membrane microdomains that conform to the functional definition of DIGs can be classified into two categories: classical caveolae, which are limited to some cell types and exhibit a characteristic ultrastructural morphology including a cytoplasmic coat rich in caveolin-1 (Peters *et al.*, 1985; Rothberg *et al.*, 1992; Yamada, 1955), and non-caveolar DIGs, which can be isolated from essentially all cell types including the intestinal epithelia. Given the paucity of classical caveolae in the intestinal epithelia, non-caveolar DIGs are most likely involved in CT binding and action in the intestinal epithelia.

The lag phase: time required for toxin entry into the cell

As described above, neither anthrax nor cholera toxins act by receptor-mediated signal transduction from the cell surface of sensitive cells. Such a mechanism of action does occur in nature as exemplified by *E. coli* heat-stable toxin. *E. coli* heat-stable toxin mimics the intestinal paracrine factor guanylin and acts immediately upon binding guanylate cyclase at the apical membrane. CT and anthrax toxins, however, exhibit a characteristic delay between binding cell surface receptors and bioactivity. This 'lag phase' corresponds to the time required for endocytosis and trafficking the toxin into specific subcellular compartments necessary for membrane translocation of the A-subunit. Thus, the essential rate limiting reactions for AB toxins occur in the time interval between binding to receptors at the cell surface and delivery of the A-subunit to the cytosol.

In this chapter we present some of the basic experimental assays that have been developed or used in our laboratories for the study of cholera and related *E. coli* heat labile enterotoxins, diphtheria toxin, and anthrax toxins. We emphasize the importance of experimental systems that provide a high degree of temporal resolution when measuring toxin function. For most of the assays described below, we have made an attempt to provide as much experimental detail as possible, but essentially all of the following techniques must be individually optimized for the specific research question(s) in hand.

◆◆◆◆◆◆ CHARACTERIZATION OF BINDING TO THE CELL SURFACE

Quantification of cell surface binding

We have utilized a modified ELISA to assess binding of toxins to the cell surface at 4°C. T84 or other appropriate cells are grown to confluency in 96-well culture plates (Costar, Cambridge, MA), washed in a physiological buffer, and incubated with varying concentrations of the toxin for 1 h at 4°C. Monolayers are then washed and incubated with a primary anti-toxin antibody (1 h), followed by additional washes and incubation with an HRP-labeled secondary antibody (1 h). Cells are washed again and the amount of bound HRP is quantitated spectrophotometrically after development with an appropriate chromogen. This assay is well suited to comparative studies of cell surface binding between different toxins or variants of the same toxin. However, if absolute quantitation of binding kinetics of toxins is needed, the ligand must be directly labeled with biotin and/or radioactive markers. This will allow estimation of the molar quantities of toxin bound. In order to show specificity of binding, competition with unlabeled ligands must be included in such experiments.

Fractionation with lipid rafts

Fractionation with lipid rafts at the cell surface is thought to be required for toxicity of CT (Orlandi and Fishman, 1998; Wolf *et al.*, 1998) but not for the anthrax toxins (Beauregard *et al.*, 1999). To determine fractionation with lipid rafts, confluent monolayers of human intestinal T84 cells are grown on 45 cm^2 Transwell inserts (Costar) as previously described (Badizadegan *et al.*, 2000; Dharmsathaphorn and Madara, 1990). All reagents and buffers are kept at 4°C. Monolayers are rinsed by gentle immersion in Hank's Balanced Salt Solution (HBSS, Sigma Chemical Co., St. Louis, MO), and then equilibrated for 15 min at 4°C with 10 ml apical and 10 ml basolateral HBSS. The apical and basolateral buffers are removed and replaced with 10 ml each of fresh buffer containing a saturating concentration of toxins. Toxins are allowed to bind for 45–60 min, and monolayers rinsed three times in 10 ml of apical and 10 ml of basolateral HBSS for 5 min each. The inserts are then removed from the holder plates, and gently placed in the bottom half of a standard 45 cm^2 culture dish (the membrane should make contact with the dish in standard 45 cm^2 plates). At this time, 1.5–2 ml of a detergent extraction buffer such as 1% Triton X-100 in 10 mM Tris-HCl, 150 mM NaCl, pH 7.4, containing an appropriate amount of protease inhibitors (such as Complete protease inhibitor tablets, Boehringer-Mannenheim, Indianapolis, IN) is added to the apical chamber of the insert. The cells are then gently scraped off the membrane using a large cell-scraper (Costar), and transferred to a Dounce homogenizer with the tight-fitting piston B. Cells are homogenized approximately 20 times over a period of 10–15 min. Nuclei and other large debris are pelleted at 1000*g* for 5 min. The 1000*g* supernatant is well mixed with an equal volume of 80% sucrose in detergent extraction buffer, placed at the bottom an ultracentrifuge tube, and layered with a continuous or step gradient from 30% sucrose to 5% sucrose in detergent extraction buffer. The material is centrifuged at an average of 100 000*g* for 3 h to overnight, after which a floating membrane fraction should be visible at 15–25% sucrose. The tube contents are fractionated and analyzed for toxin and other relevant proteins by Western blotting. Results of a typical experiment with intestinal epithelial cells are presented in Figure 15.3.

Microscopy on polarized monolayers

Microscopic techniques for non-polarized cells and monolayers grown on glass or plastic are well described. For studies of protein binding, sorting,

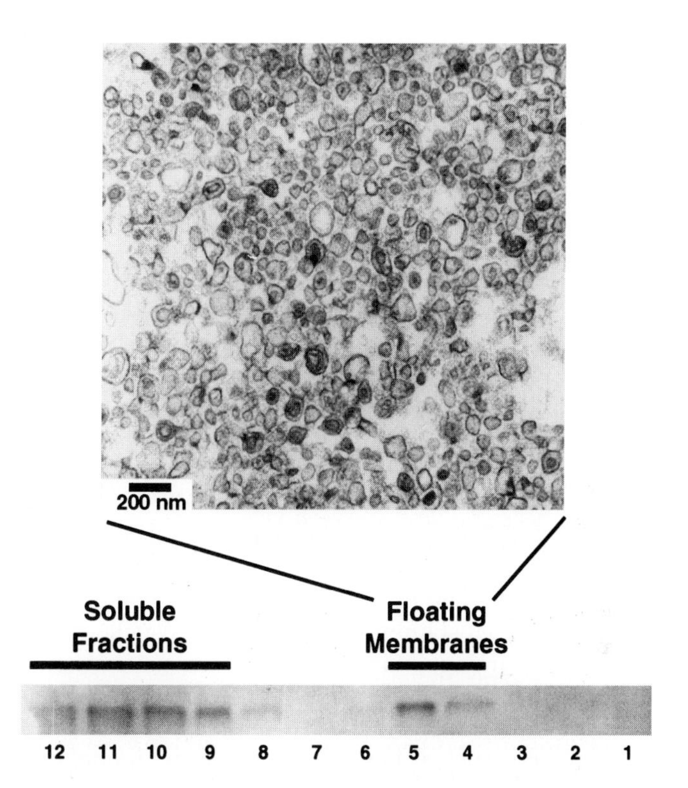

Figure 15.3. Fractionation with lipid rafts in T84 cells. Polarized T84 monolayers grown on 45 cm^2 Transwell inserts were scraped and homogenized in 1% Triton X-100 in 10 mM Tris buffer (pH 7.5) containing 150 mM NaCl and protease inhibitors. The homogenate was brought to 40% sucrose and layered under a continuous gradient of 30–10% sucrose in the same buffer. After centrifugation at 39 000 rpm in an SW41 swinging bucket rotor (Beckman Instruments) for 18 h, the entire sample was collected in 1 ml fractions and blotted for caveolin-1 (bottom panel). A sample of floating membranes from fractions 4–5 was processed for transmission electron microscopy and shows vesicular membrane fragments typical of lipid rafts.

or intracellular trafficking in polarized epithelial cells, morphological examination of polarized monolayers grown on permeable supports is essential. Here, we briefly discuss the principles of microscopy on polarized monolayers for study of toxin binding or internalization.

Polarized cell monolayers grown on 0.33 cm^2 Transwell inserts (Costar) are used for microscopy. Detailed protocols for cryosectioning of polarized monolayers grown on permeable supports have been described previously (Kendall *et al.*, 1992). In general, monolayers are washed in phosphate buffered saline (PBS) and embedded in OCT compound (Tissue-Tek, Torrance, CA) for snap-freezing and sectioning on a cryomicrotome (such as Leica CM3050, Nussloch, Germany). Monolayers may be fixed prior to embedding in OCT, or post-fixed after sectioning. The choice of fixative must be empirically determined, but we have found that fixation of monolayers for 10–15 min in a 3–4% solution of paraformaldehyde in PBS is suitable for most applications. Cryoprotective agents are not absolutely necessary, but may be beneficial (Kendall *et al.*, 1992). Five-micron frozen sections are air-dried at room temperature, post-fixed in 4% paraformaldehyde in PBS if necessary, washed in PBS, and blocked in an appropriate blocking solution such as 10% normal goat serum (Zymed Laboratories, South San Francisco, CA). A mild detergent such as 0.01% Tween-20 may be added to the wash buffers to reduce non-specific binding. Sections are stained with primary antibodies diluted to 2–2.5 $\mu g\, ml^{-1}$ in the blocking solution, and detected with a fluorophore-conjugated secondary antibody for fluorescent microscopy. We routinely mount all fluorescent microscopic preparations in ProLong anti-fade reagent according to manufacturer's instructions (Molecular Probes, Eugene, OR). Frozen sections can be examined by epifluorescence or laser confocal microscopes.

As an alternative approach to frozen sectioning, intact monolayers may be fixed, stained, and examined by a confocal microscope. When intact monolayers are stained for intracellular antigens, however, a suitable permeabilization step must be added after fixation to allow antibody access to the intracellular compartment. Permeabilization in 0.1–0.2% Triton X-100 in PBS for 5 min at room temperature is commonly used, but for optimal staining other detergents such as saponin may be necessary. After staining, intact monolayers are cut into small pieces (approximately 0.2 cm in each dimension), placed with cells facing up on a microscope slide, and mounted in an antifade medium. For staining cell surface antigens such as bound toxins in intact monolayers, the permeabilization step is not necessary. Whole-mount sections prepared as such can be examined by epifluorescence, or preferably by laser confocal microscopy.

◆◆◆◆◆◆ DIRECT MEASURES OF TOXIN INTERNALIZATION AND TRANSLOCATION

In vitro translocation assay

When cells with surface-bound EdTx are exposed to an acidic buffer (pH 4.8), elevations in intracellular cAMP can be observed, indicating translocation of EF into the cytoplasm (Gordon *et al.*, 1988; Milne and Collier, 1993). In order to directly demonstrate translocation of EF under these conditions, we have utilized a protease-protection assay for translocation of a radioactively labeled EF into intact cells (Wesche *et al.*, 1998). Rat myoblast L6 cells obtained from the American Type Culture Collection (ATCC, Rockville, MD) are grown in DMEM with 5% fetal calf serum. Cells are chilled to 4°C and incubated with PA or PA_{63} prepore in buffered medium for 2 h. Cells are washed and incubated for 2 h with *in vitro* transcribed and translated LFn (N-terminal 255 residues of LF) that is internally labeled with ^{35}S-methionine (Promega, following kit instructions). After another washing step, the extracellular pH is lowered to 4.8 at 37°C for 30 s, and the cells are either lysed or treated with Pronase E (a non-specific protease) and then lysed. Cells treated similarly at neutral pH are used as controls. Proteins are precipitated from lysates by incubation with 5% trichloroacetic acid, followed by ether washing to remove detergent.

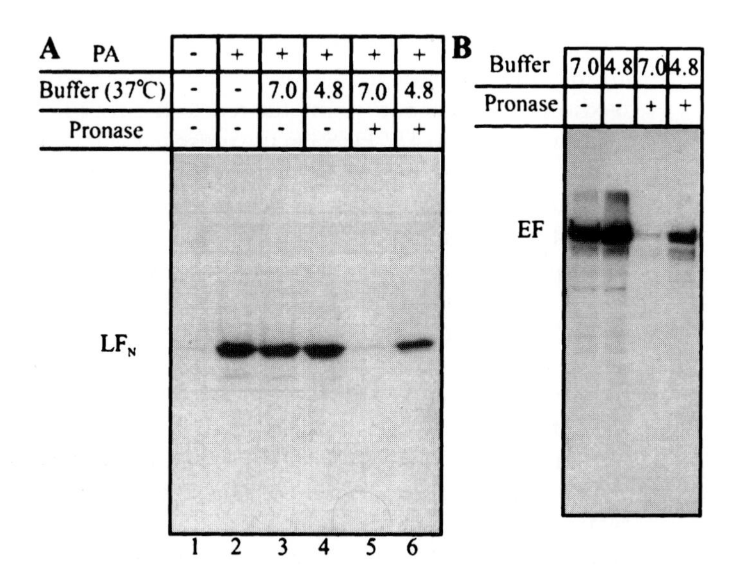

Figure 15.4. Pronase protection assay for translocation. (A) L6 cells were incubated with (lanes 2–6) or without (lane 1) trypsin-nicked PA for 2 h at 4°C, washed, and then incubated with radiolabeled LF_N for 2 h at 4°C. The cells were then either lysed directly (lanes 1–2), or briefly incubated at 37°C with an acidic (lanes 4 and 6) or a neutral buffer (lane 3 and 5). In some cases (lanes 5 and 6), cells were subsequently treated with Pronase. In all cases, the cells were lysed, the nuclei were removed, and the proteins were TCA-precipitated and analyzed by SDS-PAGE followed by fluorography. (B) CHO-K1 cells were incubated with nicked PA and EF and treated as in (A). (Adapted from Wesche *et al.*, 1998.)

Protein pellets are solubilized, and examined by SDS-PAGE followed by phosphorimager analysis. The fraction of LFn that translocates across the plasma membrane during the low-pH pulse is protected from degradation by Pronase, and is visible on the gels as a radioactive band. Radioactive bands can be quantified and compared to samples not treated with Pronase for a measure of translocation efficiency (Figure 15.4).

Endocytosis assay

Endocytosis of cell surface bound toxin can be observed by morphology (Lencer *et al.*, 1993) or measured by quantitative assays. The basic principle in quantitative assays of endocytosis is to bind a known quantity of toxin to the cell surface, then physically or biochemically separate internalized ligands from the residual cell surface ligands as a function of time. We have taken two experimental approaches to achieve this, each of which has its specific limitations.

The first approach is based on a method previously published for internalization via the coated pits (Schmid and Smythe, 1991). Purified toxin is labeled with a disulfide-linked biotin and applied to the cell monolayers grown on 0.33 cm^2 Transwell inserts or in 96-well culture plates (Costar) for 30 min at 4°C in HBSS to allow receptor binding. Excess toxin is removed by washing in excess HBSS at 4°C, and the monolayers are rapidly brought to the test conditions of temperature and culture medium. For each data point, the reaction is rapidly stopped by returning the monolayers to 4°C HBSS in the presence of the membrane impermeant reducing agent 2-mercaptoethanesulfonic acid (MESNA). MESNA will remove the biotin tag of the residual cell surface toxins, but will not affect any internalized biotinylated toxin. The reducing potential of MESNA is then quenched with iodoacetimide followed by N-ethylmaleimide. The monolayer is then rinsed in 4°C HBSS, entirely solubilized, and analyzed by SDS-PAGE and avidin blot to measure the relative quantity of biotin left compared with the baseline (total bound biotinylated toxin at 4°C) after subtraction of the background signal (total bound biotinylated toxin at 4°C that remains after treatment with MESNA). Specificity of toxin internalization via the cell surface receptor is demonstrated by competition with non-biotinylated toxin at any given time point. In our experience, the main drawbacks of this approach for CT have been failure to completely strip biotin from the cell surface bound toxin, and the non-specific release of bound toxin into the culture medium during the transition from 4°C to 37°C. Since the assay depends on an indirect measurement of internalized toxin, the data may be confounded by (and must be corrected for) the cell surface toxin released into the medium.

In order to avoid confounding by the potential release of the cell-surface ligand into the medium, internalized toxin can also be directly quantitated. Here, the monolayers are incubated with radioactively labeled toxin and the reaction quenched at any given time point by rapid transfer to 4°C buffer as described above. The residual toxin on the cell surface is then stripped by repeated washes at low pH (2.5–3.0), and the internalized toxin quantitated by gamma counting. Specificity of toxin internalization via its cell surface receptor is demonstrated by competition with non-radioactive toxin at any given time point. With this approach, the optimal cycles of low-pH washes and the background signal which includes cell surface toxin still remaining after repeated washes must be characterized for each experimental system by loading the cell surface with radioactive toxin at 4°C, removing the unbound toxin, and plotting the residual radioactivity after each subsequent low-pH wash.

Transcytosis assay

Available evidence from *in vivo* studies indicate that CT may breech epithelial barriers to act as a mucosal vaccine adjuvant by directly affecting the function of sub-epithelial cells in the lamina propria of the intestine or respiratory tract. To test this idea, we adapted methods of selective cell surface biotinylation as originally described by Sargiacomo and Rodriguez-Boulan (Sargiacomo *et al.*, 1989). In these studies, we examined the movement of CT across polarized T84 cell monolayers by vesicular traffic, a process termed transcytosis (Lencer *et al.*, 1995b).

Since the reactive agent used in selective cell surface biotinylation, sulfo-NHS-biotin (Pierce, Rockford, IL), is impermeable to cell membranes and intact tight junctions, it is possible to selectively label proteins containing free amino groups on either the apical or basolateral cell surface of polarized monolayers. Thus, if toxin bound to apical receptors enters the transcytotic route, the B-subunit should become accessible to sulfo-NHS-biotin applied to basolateral cell surfaces after toxin transit through the cell. This can then be detected by solubilizing the cell, immunoprecipitating the toxin, and examining the immunoprecipitate by SDS-PAGE and Western and ligand blot using HRP-labeled avidin.

In a typical experiment shown in Figure 15.5, CT (20 nM) is applied apically to T84 monolayers and incubated for various times at 37°C or 4°C (a temperature which completely inhibits vesicular traffic and CT-induced signal transduction). The monolayers are then returned to 4°C and biotinylated either at the apical (lane 1) or basolateral (lanes 2–5) cell surface. The large avidin-peroxidase signal (lane 1, lower panel) demonstrates that CT B-subunit can be labeled with biotin while bound to G_{M1} at the cell surface. In the absence of vesicular traffic (at 4°C), however, the CT B-subunit was not labeled by applying biotin to basolateral reservoirs (lane 2). Lane 4 shows that in monolayers exposed to apical CT and incubated at 37°C for 2 h, basolaterally applied biotin has now labeled a fraction of the B-subunit at the basolateral membrane, indicating transcytosis to the basolateral membrane. In contrast (lane 3), the CT B-subunit was not labeled in monolayers incubated at 37°C for only 30 min, a point in the lag phase where a CT-induced secretory response is not yet detectable.

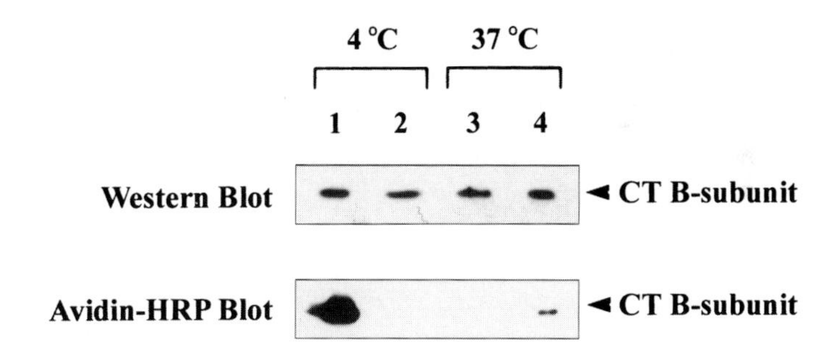

Figure 15.5. Transcytosis assay. CT (20 nM) was applied apically to T84 monolayers and incubated for 120 min at 4°C, a temperature which inhibits toxin internalization and trafficking. Some monolayers (lanes 3 and 4) were further incubated at 37°C for 30 min (lane 3) or 120 min (lane 4). The monolayers were then biotinylated either at the apical (lane 1) or the basolateral (lanes 2–4) cell surface. Western blotting for the CT B-subunit shows the presence of equal amounts of bound CT in all samples. The large avidin-peroxidase signal (lane 1, lower panel) demonstrates that CT B-subunit can be labeled with biotin while bound at the cell surface. In the absence of vesicular trafficking at 4°C, however, the CT B-subunit was not labeled by applying biotin to basolateral compartment (lane 2). Lane 4 shows that in monolayers exposed to apical CT and incubated at 37°C for 2 h, basolaterally applied biotin has now labeled a fraction of the B-subunit at the basolateral membrane indicating transcytosis of CT from the apical to the basolateral membranes. In contrast, lane 3 shows that the CT B-subunit was not biotinylated in monolayers incubated at 37°C for only 30 min, a point in the lag phase characteristic of CT action (see Figure 15.6C). (Modified from Lencer *et al.*, 1995b.)

In describing this method, we refer the reader to studies that define the experimental limitations of this approach (Gottardi *et al.*, 1995) and emphasize two important details that led to successful adaptation of this technology. First, it is essential that all cellular proteins (and internalized toxin) are fully solubilized. For CT, this required boiling the epithelial monolayers in SDS and then reducing the cell lysate into a mixed micelle buffer for subsequent immunoprecipitations. We found, however, that such severe conditions for cell lysis caused the reactive sulfo-NHS-biotin to non-specifically label proteins (including CT) in the cell lysate. This occurred even after quenching the reactive group and extensive washing to remove the reagent from the experimental system prior to solubilization. Thus to control for this non-specific background, we introduced a second critical modification by selectively quenching all available reactive sites with sulfo-NHS-acetate on the contralateral apical membrane prior to selective biotinylation of the basolateral membrane.

ER/Golgi transport assays

It has been known for several years that CT and certain other non-pore-forming toxins such as ricin, *Pseudomonas* exotoxin A, and shiga toxin, must enter the ER to unfold and translocate to the cytosol by 'dislocation' through the translocon sec61p (Hazes and Read, 1997). This idea arises from recent evidence that the biosynthetic pathways of eukaryotic cells are endowed with the ability to identify and eliminate misfolded, unassembled, or aberrantly modified membrane or secreted proteins by proteosome-dependent degradation in the cytosol (for reviews see (Bonifacino and Weissman, 1998; Brodsky and McCracken, 1997; Kopito, 1997; Plemper and Wolf, 1999)).

To test these ideas for ricin, the ER specific transfer of N-linked oligosaccharides and the *trans* Golgi specific transfer of sulfate have been exploited (Huttner, 1982, 1988; Rapak *et al.*, 1997; Wesche *et al.*, 1999). These approaches have also been used successfully for studies on shiga toxin (Johannes *et al.*, 1997). We have recently applied this technology to studies on trafficking of CT (unpublished, Fujinaga and Lencer). Both oligosaccharide transferase and tyrosylprotein sulfotransferase are ubiquitously expressed *in vivo* and in cultured cell lines (Huttner, 1988; Silberstein and Gilmore, 1996). Thus, recombinant toxin variants can be engineered to contain one or more consensus N-glycosylation (N-X-T/S, Pless and Lennarz, 1977; Silberstein and Gilmore, 1996) and tyrosine sulfation motifs (Bundgaard *et al.*, 1997; Huttner, 1988). The placement of such motifs must be carefully considered and proven to not affect toxin function.

N-glycosylation of ^{35}S-sulfated toxin-subunits are assessed by a shift in molecular mass detected by SDS-PAGE and fluorography after immunoprecipitation against toxin subunits, or after dual immunoprecipitation first with anti-toxin antibodies and then with the lectin concanavalin A. Each immunoprecipitated sample is further divided into two samples. One of each pair is treated with PGNase F to digest N-linked oligosaccharides and thus confirm evidence for N-glycosylation. The results of such studies can be verified and extended using brefeldin A to block retrograde vesicular transport (Lencer *et al.*, 1993), tunicamycin to block N-glycosylation by oligosaccharide transferase, and Endo-H to provide evidence for the ER- and Golgi-processed forms of N-glycosylated toxin subunits.

◆◆◆◆◆◆ INDIRECT MEASURES OF TOXIN TRANSLOCATION

Bacterial AB toxins are designed to translocate a functional enzyme into the cytosol. Thus, the functional signal induced by membrane translocation of the enzymatic toxin-subunit is magnified tremendously over the signal that can be obtained by direct measurement of the mass of translocated protein. This is one of the key advantages in using bacterial toxins to study membrane translocation and vesicular transport in intact cells.

Measurement of cAMP-mediated chloride secretion by electrophysiology

All reagents and buffers are kept at 37°C. HBSS with 10 mM HEPES buffer (pH 7.4) is used for these assays. T84 cells are grown to confluency on collagen-coated 0.33 cm^2 Transwell inserts (Costar) as previously described (Lencer *et al.*, 1995a). Monolayers are drained of media by inverting and rinsed by gentle immersion in a large beaker of HBSS (care must be taken to minimize fluid sheer in this and all subsequent steps to avoid disruption of the monolayer and loss of electrical resistance). Inserts are then placed in a clean 24-well plate (Costar) placed on a 37°C plate-warmer containing 1 ml of HBSS in the lower chamber, and 200 µl of HBSS is added to the upper chamber by gently sliding the buffer down the side of the well. Electrical currents and potentials are measured using a commercial voltage clamp (Iowa Dual Voltage Clamps, University of Iowa Bioengineering) interfaced with a pair of calomel electrodes submerged in saturated KCl and a pair of Ag–AgCl electrodes submerged in HBSS without glucose. Electrode pairs are in turn interfaced with the monolayers using a set of four 20-cm-long agar bridges (Figure 15.6A). Agar bridges are prepared ahead of time by suctioning heated 6% agar dissolved in HBSS without glucose into 1-mm-bore polypropylene tubing, and stored submerged in HBSS without glucose. The agar bridges are held in place above and below each monolayer using a custom-made bridge-holding device that positions two bridges above and two below each monolayer as shown in Figure 15.6A. For each data point, two successive measurements are made: (1) the spontaneous transepithelial potential (E_0), and (2) the instantaneous potential generated by passing 20 µA of current (E_{25}). The transepithelial resistance (R) and transepithelial short-circuit current (I_{sc}) are then calculated using Ohm's law. Monolayers can be followed for at least 4 h under these conditions, with measurements taken every few minutes to characterize the cAMP-mediated chloridogenic short-circuit currents induced by the bacterial toxins. Data from up to 24 individual

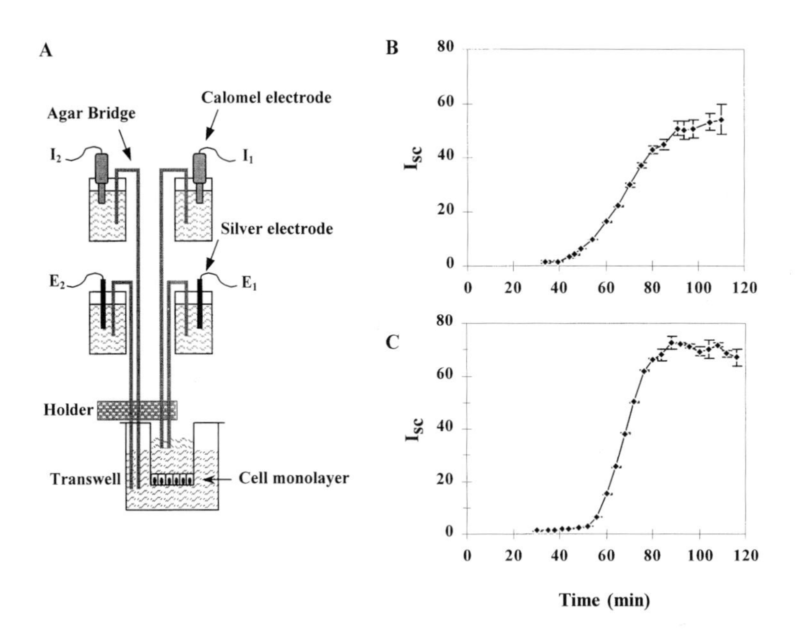

Figure 15.6. (A) Schematic representation of the set up for electrophysiological measurements using 0.33 cm^2 Transwell inserts. A voltage clamp apparatus is used to measure the resting transepithelial potential (ΔE_0), as well as the instantaneous potential when a current of 25 μA is applied across the monolayer (ΔE_{25}). The transepithelial resistance (R) and the short circuit current (I_{sc}) are then calculated using Ohm's law. Typical sets of data plotting short-circuit currents generated as a function of time by (B) basolateral PA/EF or (C) basolateral CT demonstrate the high degree of temporal resolution enabling detailed analysis of the kinetics of toxin action.

monolayers can be sequentially tabulated in a single experiment. Figures 15.6B and 15.6C show typical sets of data for short-circuit currents induced by PA/EF and CT, respectively.

Toxin translocation across planar bilayers

This assay originally developed for the study of diphtheria toxin translocation (Oh *et al.*, 1999; Senzel *et al.*, 2000), may well be applicable to the study of other pH-dependent AB toxins such as anthrax toxin. Diphtheria toxin (DT) is a classic AB toxin synthesized as a single polypeptide containing three folding domains: (i) the amino terminal catalytic domain C; (ii) the intermediate transmembrane domain T; and (iii) the carboxyl terminal receptor binding domain R. After binding to its cell-surface receptor via the R domain, DT is proteolytically broken into two disulfide linked peptides, one corresponding to the C domain and the other corresponding to domains T and R. This complex is endocytosed into an acidic compartment where a conformational change allows insertion of the T domain into the membrane followed by translocation of the C domain into the cytosol. Once in the cytosol, the C domain inhibits protein synthesis in the cell by catalyzing the

ADP-ribosylation of elongation factor 2 (for selected reviews see Collier, 1975, 1988; Olsnes *et al.*, 1988; Wilson and Collier, 1992).

The basic property of DT underlying this assay, which is described in detail elsewhere (Oh *et al.*, 1999; Senzel *et al.*, 2000), is that the T domain of DT forms ion-conducting channels, which can be electrically monitored in planar phospholipid bilayers. Planar lipid bilayers made from lecithin type IIS (Sigma Chemical Co.) from which neutral lipids are removed (Lien and Racker, 1971) are formed at room temperature across a 100 mm hole as previously described (Qiu *et al.*, 1996; Silverman *et al.*, 1994; Wonderlin *et al.*, 1990). The solutions on both sides of the bilayer membrane contains 1 M KCl, 2 mM $CaCl_2$, with or without 1 mM EDTA. The *cis* solution (to which T domain is added) is buffered to acidic pH by 30 mM Mes, pH 5.3, while the *trans* solution contains 50 mM HEPES, pH 7.2. Both solutions are stirred by small magnetic bars. After a given amount of toxin is added to the *cis* solution, a known voltage is applied across the lipid bilayer and the current monitored as previously described (Jakes *et al.*, 1990).

Experiments with T-domain, whole DT, or DT with truncated R domain were conducted with H6 histidine tags at the amino terminus of the polypeptides (Oh *et al.*, 1999). In the presence of amino-terminal H6 tags, channels formed by these polypeptides rapidly close when a negative voltage is applied across the lipid bilayer. Because of the high affinity of nickel for histidine tags, membrane impermeant Ni^{2+} can be selectively added to *cis* or *trans* compartments to localize the histidine tag. Under the above experimental conditions, addition of Ni^{2+} to the *trans* compartment prevented channel closure in response to an applied negative voltage, while addition of Ni^{2+} to the *cis* compartment had no effect (Oh *et al.*, 1999). These and additional data (Oh *et al.*, 1999; Senzel *et al.*, 2000) directly demonstrated translocation of the amino-terminal regions, including in the case of whole DT, the catalytic domain, from the *cis* compartment into the *trans* compartment. Self-translocation of the DT catalytic domains of the toxin have also been demonstrated in another experimental system developed by London *et al.* (Jiang *et al.*, 1991; Sharpe *et al.*, 1999; Sharpe and London, 1999).

Direct cAMP measurements

As described above, translocation of CTA and LF into the cytosol leads to formation of intracellular cAMP. Direct measurement of intracellular cAMP levels can therefore be used as an indirect measure of toxin translocation. With the advent of new cAMP ELISA assay kits (Amersham-Pharmacia Biotech, Piscataway, NJ), these measurements can be made rapidly and without the need for radioactive substrates. Briefly, T84 monolayers grown to confluency on 0.33 cm^2 Transwell inserts are incubated with the appropriate toxin(s). For every data point, selected monolayers are rapidly cooled by immersion in ice-cold buffer containing 1 mM 3-isobutyl-1-methylxanthine (IBMX, Sigma) to prevent further generation or degradation of intracellular cAMP. Monolayers are then cut out of the inserts, and the entire monolayer (including the plastic support) placed in

20 μl of the lysis buffer provided by the manufacturer. The cAMP elevations induced by as little as 2-4 nM CT are readily detectable in one 0.33 cm^2 monolayer.

Protein synthesis inhibition assay

Diphtheria toxin A subunit (DTA) and *Pseudomonas* exotoxin A both inhibit protein synthesis in their target cells by ADP-ribosylation of elongation factor 2. As a result, the extent to which cellular protein synthesis is inhibited can be used as an indirect assay of toxin translocation. This assay cannot only be used for DTA and exotoxin A, but also for the shiga toxin family, the ricin family of plant toxins, and hybrid toxins. For example, for a hybrid containing DTA fused to LF_N, the N-terminal PA_{63}-binding domain of LF, the assay is performed as previously described (Blanke *et al.*, 1996; Wesche *et al.*, 1998). Briefly, CHO-K1 hamster ovary cells are obtained from ATCC and propagated in HAM's F12 medium (Sigma) supplemented with 10% fetal calf serum. Cells are plated at a density of approximately 4×10^4 cells per well in 24-well plates (Costar) 18 h prior to the addition of test proteins. Cells are incubated with approximately 10^{-8} M trypsin-nicked PA (or a similar pore-forming unit) and approximately 10^{-9} M LF-DTA (or a similar toxin) for 24 h. Cells are then washed and incubated in leucine-free medium containing 1 μCi ml^{-1} of ^{3}H-leucine for 1 h. Cells are washed with cold buffer and the total cellular protein is precipitated with 10% TCA. The extent of protein synthesis is measured by the level of tritium incorporation into TCA-insoluble materials by scintillation counting, and reported as percentage of radioactivity incorporated into cells untreated with DTA.

◆◆◆◆◆◆ DISCUSSION OF UNKNOWNS

The technologies described in this chapter have been used to identify the mechanisms of internalization and intracellular compartments from which AB-toxins gain entry to the cytosol. We still do not understand, however, the molecular mechanisms of toxin unfolding and membrane translocation. These will be fundamentally different for anthrax and cholera toxins. We also do not understand fully the sorting motifs opportunistically used by these toxins for trafficking into specific organelles of host cells. These membrane dynamics harnessed by AB toxins to enter mammalian cells, move retrograde into the endosome or ER, and then in some cases cross the cell by transcytosis are fundamental to the structure and function of eukaryotic cells themselves and of central importance to cell biology and microbial pathogenesis.

References

Anderson, R. G. (1998). The caveolae membrane system. *Annu. Rev. Biochem.* **67**, 199–225.

Badizadegan, K., Dickinson, B. L., Wheeler, H. E., Blumberg, R. S., Holmes, R. K. and Lencer, W. I. (2000). Heterogeneity of detergent insoluble membranes from

human intestine containing caveolin-1 and ganglioside GM1. *Am. J. Physiol.* **278**, G895–914.
Bastiaens, P. I., Majoul, I. V., Verveer, P. J., Soling, H. D. and Jovin, T. M. (1996). Imaging the intracellular trafficking and state of the AB5 quaternary structure of cholera toxin. *Embo. J.* **15**, 4246–53.
Beauregard, K. E., Wimer-Mackin, S., Collier, R. J. and Lencer, W. I. (1999). Anthrax toxin entry into polarized epithelial cells. *Infect. Immun.* **67**, 3026–30.
Benson, E. L., Huynh, P. D., Finkelstein, A. and Collier, R. J. (1998). Identification of residues lining the anthrax protective antigen channel. *Biochemistry* **37**, 3941–8.
Blanke, S. R., Milne, J. C., Benson, E. L. and Collier, R. J. (1996). Fused polycationic peptide mediates delivery of diphtheria toxin A chain to the cytosol in the presence of anthrax protective antigen. *Proc. Natl Acad. Sci. USA* **93**, 8437–42.
Bonifacino, J. S. and Weissman, A. M. (1998). In *Annual Review of Cell and Developmental Biology* (ed. J. A. Spudich), Annual Reviews, Palo Alto, Vol. 14, pp. 19–57.
Brodsky, J. L. and McCracken, A. A. (1997). ER-associated and proteosome-mediated protein degradation: how two topologically restricted events came together. *Trends Cell Biol.* **7**, 151–6.
Brown, D. A. and London, E. (2000). Structure and function of sphingolipid- and cholesterol-rich membrane rafts. *J. Biol. Chem.* **275**, 17221–4.
Bundgaard, J. R., Vuust, J. and Rehfeld, J. F. (1997). New consensus features for tyrosine O-sulfation determined by mutational analysis. *J. Biol. Chem.* **272**, 21700–5.
Collier, R. J. (1975). Diphtheria toxin: mode of action and structure. *Bacteriol. Rev.* **39**, 54–85.
Collier, R. J. (1988). Structure-activity relationships in diphtheria toxin and Pseudomonas aeruginosa exotoxin A. *Cancer Treat. Res.* **37**, 25–35.
Dharmsathaphorn, K., and Madara, J. L. (1990). Established intestinal cell lines as model systems for electrolyte transport studies. *Methods Enzymol.* **192**, 354–89.
Duesbery, N. S., Webb, C. P., Leppla, S. H., Gordon, V. M., Klimpel, K. R., Copeland, T. D., Ahn, N. G., *et al.* (1998). Proteolytic inactivation of MAP-kinase-kinase by anthrax lethal factor. *Science* **280**, 734–7.
Escuyer, V. and Collier, R. J. (1991). Anthrax protective antigen interacts with a specific receptor on the surface of CHO-K1 cells. *Infect. Immun.* **59**, 3381–6.
Gill, D. M. (1978). In *Bacterial Toxins and Cell Membranes* (eds J. Jeljaszewicz and T. Wadström), Academic Press, London and New York, pp. 291–332.
Gordon, V. M., Leppla, S. H. and Hewlett, E. L. (1988). Inhibitors of receptor-mediated endocytosis block the entry of Bacillus anthracis adenylate cyclase toxin but not that of Bordetella pertussis adenylate cyclase toxin. *Infect. Immun.* **56**, 1066–9.
Gottardi, C. J., Dunbar, L. A. and Caplan, M. J. (1995). Biotinylation and assessment of membrane polarity: caveats and methodological concerns. *Am. J. Physiol.* **268**, F285–95.
Hazes, B. and Read, R. J. (1997). Accumulating evidence suggests that severalAB-toxins subvert the endoplasmic reticulum-associated protein degradation pathway to enter target cells. *Biochemistry* **36**, 11051–4.
Holmes, R. K., Jobling, M. G. and Connell, T. D. (1995). In *Bacterial Toxins and Virulence Factors in Disease* (eds J. Moss, B. Iglewski, M. Vaughan and A. T. Tu), Marcel Dekker, Inc., New York, Vol. 8, pp. 225–55.
Holmes, R. K., Twiddy, E. M., Pickett, C. L., Marcus, H., Jobling, M. G. and Petitjean, F. M. J. (1990). In *Symposium on Molecular Mode of Action of Selected*

Microbial Toxins in Foods and Feeds (eds A. E., Pohland, V. R., Dowell and J. L., Richard), Plenum Press, New York, pp. 91–102.

Huttner, W. B. (1982). Sulphation of tyrosine residues – a widespread modification of proteins. *Nature (Lond.)* **299**, 273–6.

Huttner, W. B. (1988). Tyrosine sulfation and the secretory pathway. *Ann. Rev. Physiol.* **50**, 363–76.

Jakes, K. S., Abrams, C. K., Finkelstein, A. and Slatin, S. L. (1990). Alteration of the pH-dependent ion selectivity of the colicin E1 channel by site-directed mutagenesis. *J. Biol. Chem.* **265**, 6984–91.

Jiang, J. X., Chung, L. A. and London, E. (1991). Self-translocation of diphtheria toxin across model membranes. *J. Biol. Chem.* **266**, 24003–10.

Johannes, L., Tenza, D., Antony, C. and Goud, B. (1997). Retrograde transport of KDEL-bearing B-fragment of shiga toxin. *J. Biol. Chem.* **272**, 19554–61.

Kendall, D., Lencer, W. I. and Matlin, K. S. (1992). Cryosectioning of epithelial cells grown on permeable supprots. *J. Tiss. Cult. Meth.* **14**, 181–6.

Klimpel, K. R., Molloy, S. S., Thomas, G. and Leppla, S. H. (1992). Anthrax toxin protective antigen is activated by a cell surface protease with the sequence specificity and catalytic properties of furin. *Proc. Natl Acad. Sci. USA.* **89**, 10277–81.

Koehler, T. M. and Collier, R. J. (1991). Anthrax toxin protective antigen: low-pH-induced hydrophobicity and channel formation in liposomes. *Mol. Microbiol.* **5**, 1501–6.

Koopmann, J. O., Albring, J., Huter, E., Bulbuc, N., Spee, P. and Neefjes, J., Hammerling, G. J., *et al.* (2000). Export of antigenic peptides from the endoplasmic reticulum intersects with retrograde protein translocation through the Sec61p channel. *Immunity* **13**, 117–27.

Kopito, R. R. (1997). ER quality control: the cytosolic connection. *Cell* **88**, 427–30.

Lencer, W. I., Constable, C., Moe, S., Jobling, M. G., Webb, H. M., Ruston, S., Madara, J. L. *et al.* (1995a). Targeting of cholera toxin and Escherichia coli heat labile toxin in polarized epithelia: role of COOH-terminal KDEL. *J. Cell Biol.* **131**, 951–62.

Lencer, W. I., de Almeida, J. B., Moe, S., Stow, J. L., Ausiello, D. A. and Madara, J. L. (1993). Entry of cholera toxin into polarized human intestinal epithelial cells. Identification of an early brefeldin A sensitive event required for A1-peptide generation. *J. Clin. Invest.* **92**, 2941–51.

Lencer, W. I., Delp, C., Neutra, M. R. and Madara, J. L. (1992). Mechanism of cholera toxin action on a polarized human intestinal epithelial cell line: role of vesicular traffic. *J. Cell Biol.* **117**, 1197–1209.

Lencer, W. I., Hirst, T. R. and Holmes, R. K. (1999). Membrane traffic and the cellular uptake of cholera toxin. *Biochim. Biophys. Acta* **1450**, 177–90.

Lencer, W. I., Moe, S., Rufo, P. A. and Madara, J. L. (1995b). Transcytosis of cholera toxin subunits across model human intestinal epithelia. *Proc. Natl Acad. Sci. USA* **92**, 10094–8.

Leppla, S. H. (1982). Anthrax toxin edema factor: a bacterial adenylate cyclase that increases cyclic AMP concentrations of eukaryotic cells. *Proc. Natl Acad. Sci. USA* **79**, 3162–6.

Leppla, S. H. (1984). Bacillus anthracis calmodulin-dependent adenylate cyclase: chemical and enzymatic properties and interactions with eucaryotic cells. *Adv. Cyclic Nucleotide Protein Phosphorylation Res.* **17**, 189–98.

Lien, S. and Racker, E. (1971). Partial resolution of the enzymes catalyzing photophosphorylation. 8. Properties of silicotungstate-treated subchloroplast particles. *J. Biol. Chem.* **246**, 4298–307.

Majoul, I. V., Bastiaens, P. I. and Soling, H. D. (1996). Transport of an external Lys-Asp-Glu-Leu (KDEL) protein from the plasma membrane to the endoplasmic reticulum: studies with cholera toxin in Vero cells. *J. Cell Biol.* **133**, 777–89.

Miller, C. J., Elliott, J. L. and Collier, R. J. (1999). Anthrax protective antigen: prepore-to-pore conversion. *Biochemistry* **38**, 10432–41.

Milne, J. C. and Collier, R. J. (1993). pH-dependent permeabilization of the plasma membrane of mammalian cells by anthrax protective antigen. *Mol. Microbiol.* **10**, 647–53.

Oh, K. J., Senzel, L., Collier, R. J. and Finkelstein, A. (1999). Translocation of the catalytic domain of diphtheria toxin across planar phospholipid bilayers by its own T domain. *Proc. Natl Acad. Sci. USA* **96**, 8467–70.

Olsnes, S., Moskaug, J. O., Stenmark, H. and Sandvig, K. (1988). Diphtheria toxin entry: protein translocation in the reverse direction. *Trends Biochem. Sci.* **13**, 348–51.

Orlandi, P. A. (1997). Protein-disulfide isomerase-mediated reduction of the A subunit of cholera toxin in a human intestinal cell line. *J. Biol. Chem.* **272**, 4591–9.

Orlandi, P. A. and Fishman, P. H. (1993). Orientation of cholera toxin bound to target cells. *J. Biol. Chem.* **268**, 17038–44.

Orlandi, P. A. and Fishman, P. H. (1998). Filipin-dependent inhibition of cholera toxin: evidence for toxin internalization and activation through caveolae-like domains. *J. Cell Biol.* **141**, 905–15.

Peters, K. R., Carley, W. W. and Palade, G. E. (1985). Endothelial plasmalemmal vesicles have a characteristic striped bipolar surface structure. *J. Cell Biol.* **101**, 2233–8.

Petosa, C., Collier, R. J., Klimpel, K. R., Leppla, S. H. and Liddington, R. C. (1997). Crystal structure of the anthrax toxin protective antigen. *Nature* **385**, 833–8.

Plemper, R. K. and Wolf, D. H. (1999). Retrograde protein translocation: Eradication of secretory proteins in health and disease. *Trends Biol. Sci.* **24**, 266–70.

Pless, D. D. and Lennarz, W. J. (1977). Enzymatic conversion of proteins to glycoproteins. *Proc. Natl Acad. Sci. USA* **74**, 134–138.

Qiu, X. Q., Jakes, K. S., Kienker, P. K., Finkelstein, A. and Slatin, S. L. (1996). Major transmembrane movement associated with colicin Ia channel gating. *J. Gen. Physiol.* **107**, 313–28.

Rapak, A., Falnes, P. Ø. and Olsnes, S. (1997). Retrograde transport of mutant ricin to the endoplasmic reticulum with subsequent translocation to the cytosol. *Proc. Natl Acad. Sci. USA* **94**, 3783–8.

Rothberg, K. G., Heuser, J. E., Donzell, W. C., Ying, Y. S., Glenney, J. R. and Anderson, R. G. (1992). Caveolin, a protein component of caveolae membrane coats. *Cell* **68**, 673–82.

Sargiacomo, M., Lisanti, M., Graeve, L., Le Bivic, A. and Rodriguez-Boulan, E. (1989). Integral and peripheral protein composition of the apical and basolateral domains in MDCK cells. *J. Membrane Biol.* **107**, 277–86.

Schmid, S. L. and Smythe, E. (1991). Stage-specific assays for coated pit formation and coated vesicle budding in vitro. *J. Cell Biol.* **114**, 869–80.

Schmitz, A., Herrgen, H., Winkeler, A. and Herzog, V. (2000). Cholera toxin is exported from microsomes by the Sec61p complex. *J. Cell Biol.* **148**, 1203–12.

Senzel, L., Gordon, M., Blaustein, R. O., Oh, K. J., Collier, R. J. and Finkelstein, A. (2000). Topography of diphtheria Toxin's T domain in the open channel state [see comments]. *J. Gen. Physiol.* **115**, 421–34.

Sharpe, J. C., Kachel, K. and London, E. (1999). The effects of inhibitors upon pore formation by diphtheria toxin and diphtheria toxin T domain. *J. Membr. Biol.* **171**, 223–33.

Sharpe, J. C. and London, E. (1999). Diphtheria toxin forms pores of different sizes depending on its concentration in membranes: probable relationship to oligomerization. *J. Membr. Biol.* **171**, 209–21.

Silberstein, S. and Gilmore, R. (1996). Biochemistry, molecular biology, and genetics of the oligosaccharyltransferase. *FASEB J.* **10**, 849–858.

Silverman, J. A., Mindell, J. A., Zhan, H., Finkelstein, A. and Collier, R. J. (1994). Structure-function relationships in diphtheria toxin channels: I. Determining a minimal channel-forming domain. *J. Membr. Biol.* **137**, 17–28.

Simpson, J. C., Roberts, L. M., Romisch, K., Davey, J., Wolf, D. H. and Lord, J. M. (1999). Ricin A chain utilises the endoplasmic reticulum-associated protein degradation pathway to enter the cytosol of yeast. *FEBS Lett.* **459**, 80–4.

Singh, Y., Chaudhary, V. K. and Leppla, S. H. (1989). A deleted variant of Bacillus anthracis protective antigen is non-toxic and blocks anthrax toxin action in vivo. *J. Biol. Chem.* **264**, 19103–7.

Spangler, B. D. (1992). Structure and function of cholera toxin and the related Escherichia coli heat-labile enterotoxin. *Microbiol. Rev.* **56**, 622–47.

Tsai, B., Rodighiero, C., Lencer, W. I. and Rapoport, T. A. (2001). Protein disulfide isomerase acts as a redox-dependent chaperone to unfold cholera toxin. *Cell* **104**, 937–48.

Wesche, J., Elliott, J. L., Falnes, P. O., Olsnes, S. and Collier, R. J. (1998). Characterization of membrane translocation by anthrax protective antigen. *Biochemistry* **37**, 15737–46.

Wesche, J., Rapak, A. and Olsnes, S. (1999). Dependence of ricin toxicity on translocation of the toxin A-chain from the endoplasmic reticulum to the cytosol. *J. Biol. Chem.* **274**, 34443–9.

Wilson, B. A. and Collier, R. J. (1992). Diphtheria toxin and Pseudomonas aeruginosa exotoxin A: active-site structure and enzymic mechanism. *Curr. Top. Microbiol. Immunol.* **175**, 27–41.

Wolf, A. A., Jobling, M. G., Wimer-Mackin, S., Ferguson-Maltzman, M., Madara, J. L., Holmes, R. K. and Lencer, W. I. (1998). Ganglioside structure dictates signal transduction by cholera toxin and association with caveolae-like membrane domains in polarized epithelia. *J. Cell Biol.* **141**, 917–27.

Wonderlin, W. F., Finkel, A. and French, R. J. (1990). Optimizing planar lipid bilayer single-channel recordings for high resolution with rapid voltage steps. *Biophys. J.* **58**, 289–97.

Yamada, E. (1955). The fine structure of the gall bladder epithelium of the mouse. *J. Biophys. Biochem. Cytol.* **1**, 445–458.

16 Bacterial Toxins: Intracellular Trafficking and Target Identification

Marina de Bernard, Ornella Rossetto and Cesare Montecucco
CNR, Centro CNR Biomembrane and Dipartimento di Scienze Biomediche Sperimentali, Via G. Colombo 3, 35121 Padova, Italia

◆◆◆

CONTENTS

List of abbreviations

ER	Endoplasmic reticulum
HEPES	(N-[2-hydroxyetil]piperazine-N′-[2-ethanesulfonic acid])
DMEM	Dulbecco's modified Eagle's medium
MDCK	Madin–Darby canine kidney
V-ATPase	Vacuolar ATPase
BSA	Bovine serum albumin
FITC	Fluorescein Isothiocyanate
PBS	Phosphate buffered saline
SNARE	SNAP receptor
TGN	Trans Golgi network
ERAD	ER-associated protein degradation
SDS-PAGE	Sodium dodecyl sulfate polyacrylamide gel electrophoresis
HRP	Horseradish peroxidase
PNS	Post nuclear supernatant
PFA	Paraformaldehyde
EGF	Epidermal growth factor
CI-MPR	Catione independent mannose-6-phosphate receptor
GFP	Green fluorescence protein
LF	Lethal factor
IEF	Isoelectric focusing

METHODS IN MICROBIOLOGY, VOLUME 31
ISBN 0–12–521531–2

◆◆◆◆◆◆ INTRODUCTION

A large variety of protein toxins made by bacteria, plants and animals are known (Rappuoli and Montecucco, 1997) and many more are still to be discovered. In principle, a toxin can act on a selected protein of the cell surface (for example ion channel blockers) or on the plasma membrane or inside the cell after translocation into the cytosol. Many bacterial toxins act by damaging or altering the plasma membrane (Alouf and Freer, 1999). They will not be dealt with in this chapter which is dedicated to the trafficking and target identification of some bacterial toxins acting in the cytosol (for actinomorphic toxins see Chapter 12 by Boquet).

The majority of plant and bacterial toxins which act on mammalian cells are referred to as AB-toxins because of their structural organization consisting of an A (active) part endowed with catalytic activity and a B (binding) moiety responsible for toxin binding to cell-surface receptors, and in some cases also in the entry of A into the cytosol. After binding, these toxins are internalized inside endocytic vesicles which merge, within a few minutes, into early endosomes. Most of the early endosomes recycle to the plasma membrane, whereas a minor part forms carrier vesicles that fuse with late endosomal compartments (within 15–30 min). Here, the hydrolytic degradation of endocytosed molecules begins and is completed inside lysosomes. Along this route, endocytosed molecules encounter progressively more acidic pH values.

Several toxins have taken advantage of this acidity and they change conformation at low pH becoming capable of translocating the A part across the endosomal membrane into the cytosol, where it acts. Other toxins traffic from endosomes to the Golgi apparatus and then to the ER, wherefrom they cross the membrane and enter the cytosol (Figure 16.1). It cannot be excluded that some toxins will be found to proceed from the ER into the nucleus. Retrograde transport of toxins through the Golgi apparatus to the ER may take place via different mechanisms. Some toxins have a KDEL sequence (e.g. cholera toxin, *Pseudomonas exotoxina* A), and binding to KDEL receptors could facilitate retrograde transport by a COPI-dependent mechanism (Sandvig and van Deurs, 1996; Jackson *et al.*, 1999). However, other toxins do not have such a sequence (e.g. ricin, shiga toxin) and may use a different, perhaps Rab6-dependent, retrograde pathways to the ER (Sandvig and van Deurs, 1996; Girod *et al.*, 1999).

A molecular understanding of the mechanism of cell intoxication by a given toxin requires both the determination of the trafficking route into the cell and the identification of target and of the type of biochemical modification elicited by the toxin A moiety. The present chapter provides a practical guide to the strategies used for the identification of the intracellular routing of a toxin and to the identification of toxin targets and target modifications, excluding the most common toxin target, i.e. actin, which is dealt with in Chapter 12.

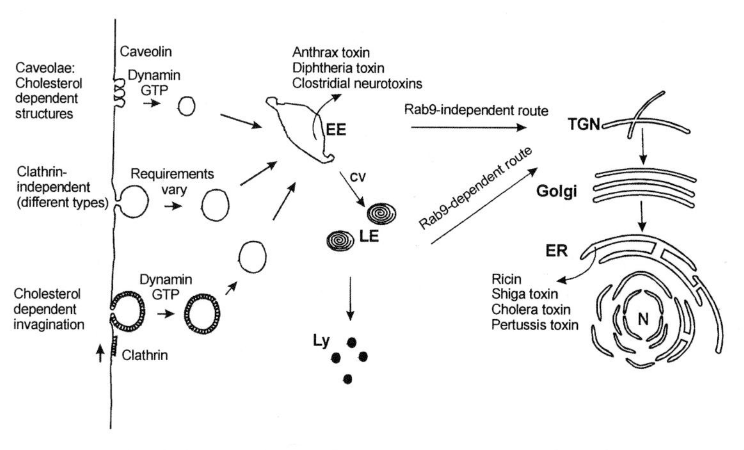

Figure 16.1. Intracellular transport of toxins. Toxins are taken up by endocytosis and transported through the endocytic pathway. Diphtheria toxin, Anthrax toxins and Clostridial neurotoxins are able to translocate from endosomes whereas other toxins such as ricin, Shiga toxin, Cholera toxin and Pertussis toxin, are transported to the Golgi apparatus and further to the ER where translocation takes place. Elements of plasma membrane involved in endocytosis are also indicated in the figure. EE, early endosomes; CV, carrier vesicles; LE, late endosomes; Ly, lysosomes; ER, endoplasmic reticulum.

◆◆◆◆◆◆ INTRACELLULAR ROUTING OF BACTERIAL PROTEIN TOXIN

Endocytosis may take place via the specialized mechanism involving coated pits and coated vesicles, which form after binding of adaptor molecules to plasma membrane protein clusters, via the less characterized route involving small membrane caveolae (Sorkin, 2000) and via micro- and macro-pinocytosis (Figure 16.1). Diphtheria toxin, Shiga toxin, *Pseudomonas aeruginosa* exotoxin A and *Clostridium difficile* toxin bind to receptors which enter into clathrin-coated vesicles (Alouf and Freer, 1999). Such a type of uptake can be distinguished from the others because it is inhibited by selected treatments: (a) exposure to hypertonic medium (Daukas and Zigmond, 1985) or depletion of cell potassium (Larkin *et al.*, 1983); (b) adjustment of the cytoplasmic pH to one pH unit lower than that of the extracellular medium (Sandvig *et al.*, 1987); (c) expression of dominant negative dynamin mutants which block the formation of coated vesicles (Damke *et al.*, 1994). These treatments were reported to inhibit diphtheria toxin toxicity, while they had little effect on that of ricin (Moya *et al.*, 1985; Simpson *et al.*, 1998), indicating that the productive uptake of ricin does not, to a large extent, implicate the coated vesicles route of endocytosis. Recently, methods have been developed to clarify the different modes of clathrin-independent endocytosis (Sandvig and van Deurs, 1999).

Methods for discriminating between clathrin-dependent and clathrin-independent toxin endocytosis

Hypotonic shock/K^+ depletion

Cells are grown in 24-well plates until reaching confluence. The medium is discarded and each monolayer is washed twice with 0.5 ml of 50 mM NaHEPES, pH 7.4, containing 100 mM NaCl. The cells are hypotonically shocked for 5 min by incubation in 0.5 ml of DMEM/water (1 : 1), followed by incubation for 5–15 min in isotonic K^+-free medium (50 mM Na-HEPES, pH 7.4, 100 mM NaCl). The toxin is added at various concentrations for 30 min at 37°C, and the cells are then washed with K^+free medium to remove unbound toxin. The washing medium must also be capable of removing non-endocytosed surface-bound toxin (e.g. by including pronase 10/mg ml^{-1} for 30 min). For different toxins, this result can be achieved in different ways: for example, in the case of ricin, a K^+ free buffer containing 100 mM lactose is able to remove toxin molecules bound to the cell surface. After toxin molecules are removed by washing, cells are incubated in complete medium for a period of time varying with the type of toxin analysed before proceeding to the analysis of the cytopathic effect. It is important to note that various cell lines respond differently to the action of intracellular K^+ depletion, and the actual extent of depletion has to determined by atomic adsorption or radioactive tracer methods. It may also be useful to perform different type of approaches in parallel.

Treatment of cells in hypertonic medium

The extent of the inhibition of coated vesicle-dependent endocytosis in hypertonic medium, depends on the tonicity: partial inhibition occurs in 0.45 and 0.6 osmolar medium, while maximal inhibition occurs in 0.75 osmolar medium (2.5 times normal tonicity). The actual degree of inhibition is rather independent of the chemical nature of the solute used to increase the tonicity: sodium chloride (0.225 M NaCl), sucrose (0.45 M sucrose), or lactose (0.6 M) inhibits uptake to a similar extent. The inhibition is rapid, the maximal effect being achieved within 5 min of placing cells in hypertonic medium, and is fully reversible upon returning cells to isotonic medium. Briefly, cells grown in 24-well plates for 18 h, are incubated for 10 min at 37°C in hypertonic medium; toxin is added at various concentrations for a further 30 min. Cells are then washed with the same medium to remove bound and unbound toxin (following the outlines described above: see *hypotonic shock method*) and incubated in normal medium (Daukas and Zigmond, 1985). Inhibition of endocytosis can be monitored by measuring cell associated radioactivity after a pulse of I^{125}-Tfn (Hopkins and Trowbridge, 1983).

Acidification of the cytosol

Different methods to acidify the cytosol have been reported, and the one described here is well tolerated by cells (Sandvig *et al.*, 1987). Cells grown for 18 h are incubated for 30 min at 37°C in HEPES medium, pH 7.5 containing 25–40 mM NH_4Cl. The medium is removed and substituted with a buffer containing 20 mM Na-HEPES pH 7.0, 0.13 M KCl, 2 mM $CaCl_2$, 1 mM $MgCl_2$, 1 mM amiloride. After a further 5 min incubation, toxin is added at various concentrations for 30 min at 37 °C. Cells are then washed with the same medium in order to remove bound and unbound toxin (see *hypothonic shock method*) and incubated in complete medium. The cytosolic pH can be determined from the distribution of the weak acid [^{14}C]DMO as described by Deutsch *et al.* (1979).

Overexpression of dynamin mutants

The GTPase dynamin is essential for chlatrin-mediated endocytosis and its involvement has been studied in detail (Sever *et al.*, 2000). Cells stably transformed with tetracycline-inducible plasmids containing the cDNAs encoding for dyn^{wt} and dyn^{K44A} (Damke *et al.*, 1994) are seeded with tetracycline and treated with the toxin 48 h later. The effect of the toxin will be evaluated after an appropriate time. However, dynamin appears to be involved also in caveolae-dependent endocytosis (Oh *et al.*, 1998), and thus additional mutants which inhibit specifically the formation of clathrin-coated vesicles can later be used. They include (1) a dominant negative mutant (Edelta95/295) of the Eps15, a protein recently identified as a constituent of plasma membrane clathrin-coated pits and required for the earliest steps of clathrin-dependent endocytosis (Benmerah *et al.*, 1999); (2) the SH3 domain of intersectin, which inhibits the intermediate events leading to the formation of constricted coated pits (Simpson *et al.*, 1999a). These mutants were recently used to study the mechanism of cell entry of the *Helicobacter pylori* vacuolating cytotoxin (Ricci *et al.*, 2000).

Clathrin-independent endocytosis includes events originating from caveolae, small cholesterol-rich invaginations rich in endothelial cells and some other cell types (Parton *et al.*, 1994; Oh *et al.*, 1998), and events independent of caveolae (Fra *et al.*, 1994). A highly regulated form of clathrin-independent endocytosis takes place on the apical side of polarized MDCK cells (Holm *et al.*, 1995), whose caveolae are localised on the basolateral side (Vogel *et al.*, 1998). Methods have been introduced to assay for a toxin entering cells via caveolae based on their cholesterol dependence. However, invagination of clathrin-coated pits is also dependent on cholesterol (Rodal *et al.*, 1999) (Figure 16.1) and, once more, the use of more than one protocol is recommended.

Extraction of cholesterol with methyl-β-cyclodextrin

Cells seeded 15–18 h before experiment, are incubated with 10 mM MβCD (Sigma) for 15 min at 37 °C. Toxin is added at various concentrations for further 30 min, and then cells are washed with the same medium to remove external toxin (see *hypotonic shock method*) and incubated in normal medium without MβCD. The endocytosed toxin can be monitored directly by measuring cell associated radioactivity after administering a pulse of labelled toxin or, indirectly, by following the specific cellular alteration caused by the toxin.

Treatment with cholesterol-complexing drugs

The involvement of cholesterol-rich microdomains in cytotoxicity can also be studied by treating cells with nystatin or filipin, drugs which complex cholesterol thus altering the structure and function of glycolipid microdomains and caveolae (Skretting *et al.*, 1999; Orlandi and Fishman, 1998). Cells are preincubated for 30 min with 5 μM filipin or 25 μg ml^{-1} nystatin before the addition of the toxin and incubation for 30 min at 37 °C. Cells are then washed with the same medium to remove external toxin (see *hypotonic shock method*) and incubated in normal medium. The proportion of endocytosed toxin can be determined as described above.

While toxin entry from the plasma membrane has not yet demonstrated for any bacterial toxin, and all of them require endocytic uptake, toxins differ with respect to their requirement for a passage through an acidic intracellular compartment to enter the cytosol. The simplest indication that this is an essential step of the intoxication process comes by observing the effect of NH_4Cl, similarly to that done for the first time by Kim and Groman (1965) in the case of diphtheria toxin. This method can still be used, but more specific procedures involving ionophores capable of preventing acidification of intracellular compartments (Sandvig and Olsnes, 1982), or bafilomycins, which are specific inhibitors of the vacuolar-type H^+-ATPase, are now available (Bowman *et al.*, 1988; Umata *et al.*, 1990; Papini *et al.*, 1993; Williamson and Neale, 1994; Barth *et al.*, 2000). The methods described in the following section allow one to investigate the role of the endosomal acidic pH in the intoxication process.

Methods to evaluate the relevance of endo-lysosomal acidification in cell intoxication

Cells are plated 15–18 h before the experiment, and are pre-incubated with either 100 mM ammonium chloride or various concentrations in the 5–100 nM range of bafilomycin A1 (Sigma) for 60 min at 37 °C. The toxin is then added without removing the pH interfering agents from the medium because their effects are reversible. The specific effect of the toxin under study is then evaluated in accordance with the known activity and time course of each toxin. Alternatively, the carboxylic ionophore monensin which exchanges K^+/H^+ can be used (Ohkuma and Poole, 1978). Cells are seeded as described above and incubated with monensin (suggested range of concentration between 100 nM and 1 μM). After 15 min incubation at 37 °C, the toxin is added. The ionophore should be maintained for the entire duration of the experiment because of the reversibility of its action.

To monitor the activity of the V-ATPase or more in general to check the real increase of the endosomal pH after these treatments, it is possible to measure the acidity of intracellular compartments with two different assays. Qualitative methods are based on the use of dyes such as neutral red, acridine orange or quinacrine, which are membrane permeant weak bases. In their unprotonated form such dyes can cross the cell membranes, while they become relatively membrane-impermeable once they are protonated inside an acidic compartment. Cells seeded on glass coverslips and pre-treated with the specific agent, are further incubated for 20 min with the dye. The cells are washed, mounted and observed under a fluorescence microscope. Strong granular fluorescence should be observed in the control cells, whereas such fluorescence should be lowered or absent in treated cells.

Quantitative methods able to estimate the pH value of the endocytic compartments are also available. Briefly, cells are seeded on glass coverslips and after pre-incubation with the specific agent they are treated for 60 min at 37°C in DMEM, 20 mM HEPES, pH 7.4, 0.2% BSA containing 10 mg ml^{-1} FITC-dextran, washed, and further incubated for 15 min in the same medium without FITC-dextran. Coverslips are mounted in a thermostatic chamber (37°C), overlaid with 0.5 ml PBS, 5.6 mM glucose and placed in the stage of an inverted microscope. Single cell fluorescence values are determined with excitation wavelengths of 494 and 450 nm and a fluorescein emission filter, with the aid of a computerized image analysis system. pH values are obtained by comparing the ratio of the two wavelength fluorescence values, corrected for cell intrinsic fluorescence, with those of FITC-dextran dissolved in internalization media of different pH (Maxfield, 1982).

While several toxins enter the cytosol from acidic endosomes following a low pH induced conformational change which makes them capable of translocating the organelle limiting membrane, others have to be transported to the Golgi apparatus and the ER, before translocation into the cytosol may take place (Figure 16.1). Advancements in the comprehension of the machinery which mediates transport along the endocytic pathway are rapidly being made. For example, the two SNARE proteins VATIP8 and Syntaxin 7 were recently shown to be involved in carrier vesicles–late endosomes fusion as well as in late endosomes homotypic fusion (Mullock *et al.*, 2000; Antonin *et al.*, 2000). Mutant forms of these proteins might be used in order to infer from which specific organelle a toxin translocates into the cytosol. It is also clear that small GTP binding proteins of the Rab family control different steps of endocytosis, and that the use of dominant negative Rabs may provide insights on the particular step of intracellular trafficking having a primary role for toxin entry in the cytosol (Papini *et al.*, 1997). In addition, there is growing evidence for a major role in trafficking being played by the cell cytoskeleton.

Methods to assess the involvement of cytoskeletal components in the intoxication process

Microtubules are involved in the transport of vesicles from early to late endosomal compartments (Gruenberg *et al.*, 1989) and actin filaments are reported to have a role in the cargo delivery from late endosomes to lysosomes (van Deurs *et al.*, 1995; Barois *et al.*, 1998). To test any influence of cytoskeletal components on toxin uptake, microtubules and actin filaments can be disassembled with nocodazole and cytochalasin D, respectively. Cells, seeded 15–18 h before the experiment, are pre-incubated for 2 h at 37°C with a medium containing 20 μg ml^{-1} nocodazole or 0.5 μg ml^{-1} cytochalasin D before adding the toxin and incubation is prolonged in the same medium for 30 min at 37°C. Cells are then washed and incubated in normal medium. This method was used to demonstrate that the *C. botulinum* C2 toxin is released in the cytosol from early endosomes (Barth *et al.*, 2000). It is noteworthy that actin also plays a role in the very preliminary phases of the endocytosis process (Lamaze *et al.*, 1997; van Deurs *et al.*, 1995; Lamaze and Schmid, 1995), and cell treatment with cytochalasin D should be performed after toxin binding and endocytosis.

Methods to monitor the entry of protein toxins via the Golgi apparatus

There appear to be more than one route from endosomes to the Golgi apparatus. One of them is controlled by the small GTP binding protein Rab9 (Riederer *et al.*, 1994), but ricin and shiga toxin reach the Golgi apparatus by another route, possibly directly from the perinuclear recycling compartment to the TGN without any involvement of late endosomes (Simpson *et al.*, 1995; Mallard *et al.*, 1998). The first possibility may be probed by the inducible expression of a dominant negative rab9 mutant (Riederer *et al.*, 1994). A simple way to assess the toxin entry via the Golgi apparatus is that of using the specific drug brefeldin A. This is a fungal drug known to have several effects on cells, the most notable one being that of disrupting the Golgi stack in many cell types (Klausner *et al.*, 1992). Cells seeded 15–18 h before experiment, are pre-incubated in normal medium containing 5 μg ml^{-1} brefeldin A for 30 min at 37°C. Toxin is then added to the medium in the presence of the drug for a period of time which varies according to the type of toxin analysed before proceeding to the analysis of the specific cellular effect induced by each toxin.

Methods to monitor the retrograde transport of a toxin to the endoplasmic reticulum

To demonstrate whether a toxin is transported retrogradely from the trans-Golgi network to the ER, a tyrosine sulfation site alone or in combination with an N-glycosylation site is introduced in the gene encoding for the toxin under study, which is then expressed as a recombinant protein (Rapak *et al.*, 1997). Since addition of a sulfate group takes place in the Golgi apparatus, the labelling of the toxin molecule with radioactive sulphate is diagnostic of its passage through this compartment. Similarly, N-glycosylation is diagnostic of the protein toxin transport to the ER.

Identification of the translocation mechanism from the ER to the cytosol

How protein toxins translocate into the cytosol from the ER lumen is not clear, though it was known that proteins failing to fold or oligomerize correctly were targeted for degradation by a non-lysosomal process (Klausner and Sitia, 1990). Some aspects of this ER-associated protein degradation (ERAD) pathway have now been elucidated. Aberrant proteins are exported from the ER lumen into the cytosol where most ERAD substrates are degraded by the cytosolic ubiquitin/proteasome system. It has been proposed that toxins such as ricin A chain might also gain access to the cytosol from the ER by masquerading as ERAD substrates (Lord, 1996; Hazes and Read, 1997). Studying the translocation of a toxin in mammalian cells may be difficult due to the high toxicity and the very low proportion of the added toxin which actually exerts the toxic effect on the cell.

Recently a yeast-based model system was introduced to assess the possibility that a protein toxin is exported from the ER by ERAD. The ERAD systems of mammalian cells and yeast appear to be mechanistically similar sharing the Sec61p translocon. Sec61p mutant yeast cells can be directly intoxicated or transfected with a construct encoding for the toxin and the rate of its proteasomal degradation can be determined (Simpson *et al.*, 1999b).

Methods to monitor the translocation to the cytosol

Determination of the toxic effect caused by a toxin on cells is usually sufficient to monitor the actual entry of the A moiety of the toxin in the cytosol. However, to study the efficiency of translocation and to test the possibility that other portions of the toxin, in addition to A, are translocated, it is necessary to use more direct methods. If this is required, the toxin gene can be modified to include a tag which can be specifically modified by enzymes or other molecules exclusively present in the cytosol. The detection of the modified toxin molecule in the cell is diagnostic of its cytosolic localization. The protein can be supplied with a C-terminal farnesylation signal, a CAAX (C, cysteine; A, an aliphatic amino acid; X, frequently methionine) box (Falnes *et al.*, 1995; Wiedlocha *et al.*, 1995). If the four C-terminal amino acids of a cytosolic protein form a CAAX box, enzymes in the cytosol will link a farnesyl group on to the cysteine residue with subsequent removal of the three-terminal amino acids and carboxylmethylation of the exposed terminal cysteine residue. Farnesylating enzymes are present in the cytosol of eukaryotic cells and the modification can be assessed in different ways. One system consists of labelling the cells with radioactive mevalonic acid, a precursor of the farnesyl group and the translocated protein can be visualized by subsequent solubilization of the cells and immunoprecipitation with specific anti-toxin antibodies, followed by SDS-PAGE. The modified protein usually migrates slightly more rapidly in SDS-PAGE than the unmodified molecule (Falnes *et al.*, 1995).

Subcellular fractionation and microscopic analysis of cells intoxicated with labelled-toxins

These methods are based on the availability of toxins that preserve activity after labelling to follow their intracellular routing via microscopic analysis and/or subcellular fractionation. These methods combined with the use of appropriate drugs (e.g. cytochalasin D, nocodazole, Brefeldin A) were very useful to demonstrate that shiga toxin reaches the Golgi apparatus directly from early/recycling endosomes without involvement of late endosomal compartments (Mallard *et al.*, 1998).

Subcellular fractionation of cells incubated with a iodine-labelled toxin (Fraker and Speck, 1978)

Cells seeded 15–18 h before experiment in 12-well plates, are incubated for 1 h at 37°C with ^{125}I-toxin (at least 10 000 cpm ng^{-1}). HRP (3 mg ml^{-1}) is usually added to label endosomal/lysosomal compartments. The incubation is terminated by cooling the cells and all subsequent steps are carried out at 4°C. Cells are washed 3 × 10 min in cold PBS/0.2% BSA. 0.5 ml of homogenation buffer (H-buffer: 0.3 M sucrose, 3 mM Imidazole, pH 7.4) are added to each well and the cells are scraped off with a rubber policeman. Each well is subsequently scraped once more and washed with 2 × 1 ml of H-buffer. Cells from 3–6 wells may be pooled for each experimental point. The pooled cells are pelletted by centrifugation for 10 min at 100 g. 1 ml of H-buffer is added and cells are resuspended and homogenized by passing 10 times up and down through a 1 ml blue tip on a pipette, followed by six times through a 1 ml syringe with a 22 G $1\frac{1}{4}$ needle. The homogenate is centrifuged at 2500 rpm for 10 min in 1.5 ml tubes (Brinkman Instruments Inc., Westbury, NY) in a centrifuge (model 5415; Beckman Instruments, Inc.) to obtain a nuclear pellet and a postnuclear supernatant (PNS). The PNS is subjected to discontinuous gradient centrifugation: in the bottom of SW40 tubes, gradients are made of 4.5 ml light solution (1.15 M sucrose, 15 mM CsCl) and 1.5 ml heavy solution (1.15 M sucrose, 15 mM CsCl, 15% Nycodenz [w/v]). The gradients are made in a Biocomp Gradient Master (Nycomed, Oslo, Norway) (angle 74, speed 16, time 2 min 45 s). PNS (5.6 parts) is mixed with 2 M sucrose, 10 mM CsCl (4.4 parts), usually a total of 1.5 ml, and layered on top of the gradient. This is again overlaid with 3 ml 0.9 M sucrose, and finally 1–2 ml of 0.3 M sucrose. After 4.5 h at 33 000 rpm, the gradients are fractionated (25–30 fractions) and analysed with respect to marker distribution: HRP for endosomes and lysosomes, esterase for ER, galactosyl- and sialyltransferase for Golgi apparatus will be used as markers (Sandvig *et al.*, 1991). See as example, Figure 16.2.

Immunofluorescence studies

Cells, seeded on coverslips are placed on ice, fluorophore-labelled toxin is added in HEPES-containing medium and then incubated with the cells for 30 min. The cells are then washed three times with ice-cold culture medium and shifted to 37°C in a CO_2 incubator. At different time points

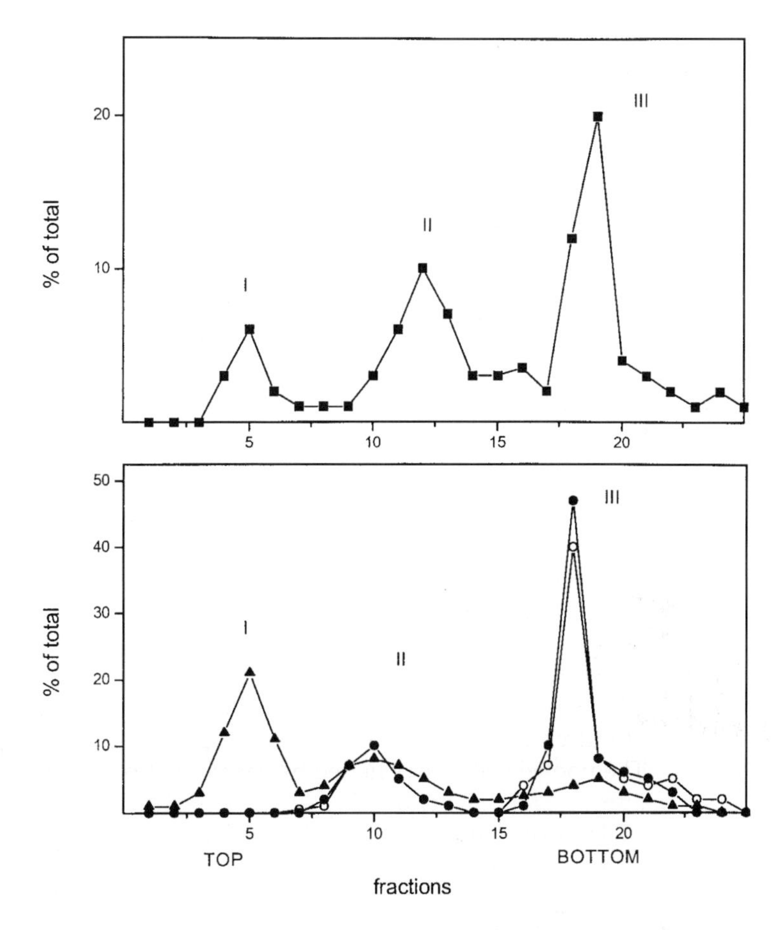

Figure 16.2. Distribution of Shiga toxin, HRP, and organelle markers on a discontinuous sucrose/Nycodenz gradient system. Filter-grown MDCK cells were incubated apically with ^{125}I-Shiga toxin and HRP for 60 min at 37°C. (■), ^{125}I-Shiga; (▲), UDP-galactosyltransferase; (●), β-N-Acetylglucosaminidase; (○), HRP. Peak I corresponds to the Golgi apparatus; peak II corresponds to the part of the gradient containing markers released from disrupted endosomes and lysosomes during homogenization as well as plasma membrane-bound toxin. Peak III contains a number of organelles that move downward into the heavy gradient. Endosomes, lysosomes, and ER are found in this peak. Figure modified from Sandvig *et al.* (1991).

cells are washed three times with PBS, fixed with 3% PFA for 10 min, permeabilized with saponin, stained with primary and secondary antibodies, and then mounted and analysed by confocal microscope. For the specific organelle identification the following marker proteins can be used: ferro-transferrin and EGF fluorophore-labelled (Molecular Probes) and antibodies anti rab7, anti TGN 38 and anti CI-MPR. The same approach can be used on living cells by following the route of a fuorophore-labelled toxin applied on cells seeded on coverslides in a thermostatic chamber (37°C) on the stage at a confocal microscope. Also in this case, the organelle identification can be done with specific labelled protein markers as indicated previously.

Electron microscopy

Cells incubated with HRP-labelled toxin (see van Deurs *et al.*, 1986, for the labelling method) for 30 min at 4°C and further incubated at 37°C for different times are processed for electron microscopy. Various methods can be used and one of them is summarized below. Cells are fixed with 2% glutaraldehyde in 0.1 M Na-cacodylate buffer, pH 7.2, for 60 min at room temperature. The cells are then carefully washed with PBS and incubated with diaminobenzidine-H_2O_2 as described elsewhere (van Deurs *et al.*, 1986). The cells are post-fixed with OsO4, treated with 1% uranyl acetate in distilled water, embedded in Epon, cut at 50 nm, and examined in an electron microscope.

◆◆◆◆◆◆ TARGET IDENTIFICATION

An increasing number of bacterial toxins is found to exert their action on targets exposed to the cell cytosol. The different toxins catalytically modify very different cytosolic components and the consequences for the cell vary from minor effects on cell physiology to cell death.

The identification of the molecular targets of toxins is essential in the molecular understanding of the pathogenesis of the disease in which they are involved. Moreover, it frequently provides major advancements in the knowledge of fundamental cellular processes. This is the result of long-term evolutionary processes that have 'shaped' the activity of each toxin around key physiological processes of the host whose modification/alteration leads to substantial evolutionary advantages to the toxigenic bacteria. Examples of very important molecules whose role was discovered via toxins are the signalling trimeric G inhibitory proteins target of pertussis toxin (Rappuoli and Pizza, 1991) and the SNARE proteins VAMP/synaptobrevin, SNAP-25 and syntaxin identified as essential components of neurotransmitter release machinery because they are targets of tetanus and botulinum neurotoxins (Schiavo *et al.*, 2000).

Unfortunately, there is no single straightforward method or rule to identify the biochemical activity and target of an intracellular acting toxin that can be summarized as defined protocols. Therefore, in this second part of the chapter, we present the general modern approaches as to what can be done to orient the search.

Toxin sequence analysis

If the toxin sequence is available, it should be analysed with appropriate programmes for the presence of specific sequence motifs and for similarities with proteins or protein domains of known function. In many cases, such analysis has provided fundamental clues on the possible biochemical activity, which is essential to proceed further with the identification of the target. Several primary sequence databases are now available, including the well-annotated SWISS-PROT. The computer programs BLAST and FASTA are available from several web sites and are commonly used for sequence similarity searches. In addition to the numerous primary databases, there are many secondary databases (e.g. PROSITE, PRINTS), so-called because they contain the result of analysis of the sequences in the primary sources. The rationale behind their developments was that protein families could be simply and effectively characterized by the single most conserved motif observable in a multiple alignment of known homologues, such motifs usually encoding key biological functions (e.g. enzyme active sites, ligand or metal binding sites, etc.). Searching such a database helps in determining to which protein family a new sequence might belong, or which domain(s) or functional site(s) it might contain. If the structure and function of the family are known, searches of secondary databases thus offer a fast track to the inference of biological function. Because secondary databases are derived from multiple sequence information, searches of them are often better able to identify distant relationship than are corresponding searches of the primary databases. However, none of the secondary database is yet complete; they should therefore be used only to implement primary database searches, rather than to replace them.

Analysis of toxin-induced cellular alterations

The information derived from sequence studies can then be coupled to cell biology or physiological data as to which cellular/physiological process is altered/impaired by the toxin. As an example, the sequence analysis of tetanus neurotoxin and of botulinum neurotoxins led to the identification of the common presence of the HExxH motif characteristic of metalloproteases (Schiavo *et al.*, 1992a and b). Coupled with the well-known fact that these neurotoxins block neurotransmitter release, this led to the search for protein(s) present on neurotransmitter containing synaptic vesicles that could be proteolytically cleaved by these neurotoxins (Schiavo *et al.*, 1992c). Other examples of biochemical experiments oriented by knowledge of cellular alterations are provided in Chapter 12 by Boquet.

If the analysis of the toxin sequence does not provide a clue about its function, one should follow the line indicated by the specific cellular alteration induced by the toxin under study. As an example, the *Clostridium botulinum* C2 toxin and the *Clostridium perfringens* iota toxin were known to alter actin polymerization in cells. By analogy with the well-established ADP-ribosyltransferase activity of several bacterial toxins including diphtheria toxin and cholera and pertussis toxins (Alouf and Freer, 1999), it was soon found also that these clostridial toxins were ADP-ribosyltransferases. Then isolated actin was incubated with the toxins in the presence of radioactive NAD and the labelled actin was fragmented. The isolated radiolabelled fragment was sequenced and it was thus found that Arg-177 of actin was specifically modified and that this sole modification was responsible for its altered polymerization properties (Aktories *et al.*, 1989).

In less fortunate cases, the supposed identification of the biochemical activity is not sufficient to allow one to design experiments aimed at identification of the target(s). An example of such an occurrence is provided by the lethal factor of *B. anthracis*, which was identified as a metalloprotease (Klimpel *et al.*, 1994; Kochi *et al.*, 1994). However, the only cellular effect known to be caused by this toxin was the induction of macrophage death (Friedlander, 1986). When no clue on how to proceed is provided by the cellular effect(s) of the toxin, one has to perform 'unbiased' approaches, i.e. screening type of experiments that do not rely on any previous knowledge or preliminary assumptions or hypothesis. We provide here a general discussion of such non-hypothesis driven approaches.

Yeast two-hybrid system

The yeast two-hybrid system is an extremely powerful method for detecting protein–protein interactions and provides a relatively straightforward approach to understanding protein function (Drees, 1999). In this technique a protein is expressed in yeast as a fusion to the DNA-binding domain of a transcription factor lacking a transcription activation domain. The DNA-binding fusion protein is generally called the 'bait'. The yeast strain also contains one or more reporter genes with binding sites for the DNA-binding domain. To identify proteins that interact with the bait, a plasmid library that expresses cDNA-encoded proteins fused to a transcription activation domain is introduced into the strain. Interaction of a cDNA-encoded protein with the bait results in activation of the reporter genes, allowing cells containing the interactors to be identified. Therefore this approach should, in principle, be a suitable method to screen for potential toxin substrates. A yeast two-hybrid system was applied successfully in our laboratory to the identification of the cytosolic target(s) of the metallo-proteolytic activity of the anthrax lethal factor. Using this approach the MAPKKs Mek1 and Mek2 were identified as lethal factor substrates (Vitale *et al.*, 1998). A Hela c-DNA library was screened using as bait an LF mutated at the glutamic acid of the zinc binding motif (LF^{E687A}), because mutation of this residue is known to abolish the proteolytic activity of metallo-proteases. The use of an inactive mutant overcomes the possibility that putative preys are lost because the interaction with the prey is terminated following its biochemical modification by the bait.

Here we describe the protocol used to identify the substrate of the lethal factor (LF) of *B. anthracis* (Vitale *et al.*, 1998).

Experimental procedure

The yeast reporter strain L40 (Vojtek *et al.*, 1993) was transformed with the plasmid encoding a fusion between LF^{E687A} and the bacterial protein LexA (LexA::LF^{E687A}), which recognizes specific DNA sequences upstream of the two reporter genes HIS3 and lacZ, using a lithium-acetate-based method (Schiestl and Giest, 1989) and grown on synthetic medium lacking tryptophan. This screening strain was subsequently transformed with a plasmid cDNA library (MATCHMAKER HeLa cell oligo-dT-primed library in pGADGH, Clontech) encoding C-terminal fusion proteins with the transcriptional activation domain of Gal4 (GAD). The transformants were plated on synthetic medium lacking histidine, leucine and tryptophan. Colonies were picked 4 days after plating and tested for β-galactosidase activity, using a replica filter assay (Vojtek *et al.*, 1993). Library plasmid from his-βGal positive clones were rescued into E. coli and subsequently analysed by re-transformation test and DNA sequencing. One clone showed specific interaction with LF^{E687A} but not with the control bait encoding a LexA fusion with LaminC and its sequence revealed a reading frame infusion with GAD encoding for amino acid 31–400 of the MAPKK Mek2.

Mammalian two-hybrid system

Some interactions between a bacterial protein and its mammalian target may not occur inside yeast cells because of the possible lack of associating factors, protein modifications (such as signal-induced phosphorylation), or correct protein folding. To overcome these limitations, mammalian cell two-hybrid screening systems were recently developed (Luo *et al.*, 1997). In a commercially available one (Promega CheckMate), the pBIND vector contains the yeast GAL4 DNA binding domain upstream of a multiple cloning region and the pACT vector contains the herpes simplex virus VP16 activation domain upstream of the multiple cloning region. In addition, the pBIND vector expresses the *Renilla reniformis* luciferase, which allows the user to normalize the transfection efficiency. The two genes encoding the two potentially interacting proteins of interest are cloned into pBIND and pACT vectors to generate fusion proteins with the DNA binding domain of GAL4 and the activation domain of VP16, respectively. The interaction provides functional transcription activation from RNA polymerase II basal promoters with upstream GAL4 binding site (Sadowski *et al.*, 1988). The pG25*luc* vector contains five GAL4 binding sites upstream of a minimal TATA box which, in turn, is upstream of the firefly luciferase gene (*luc*$^+$) providing a sensitive and quantitative reporter system for functional assessment of reconstituted GAL4 : VP16 activity. The pGAL4 and the pVP16 fusion constructs are transfected along with pG5*luc* vector into the mammalian cells of choice. Two to three days after transfection the cells are lysed, and the amount of Renilla luciferase and firefly luciferase are quantitated using the Dual-Luciferase Reporter Assay System (Promega). Interaction between the two test proteins, as GAL4 and VP16 fusion construct, results in an increase in firefly luciferase expression over the negative controls.

Variations of this mammalian two-hybrid system have been described by other authors who use different bait–prey pairs as model interactors and detect the two-hybrid interaction by expression of the GFP reporter gene (Shioda *et al.*, 2000).

Screening cells with chemical libraries

If the toxin under study has a defined effect on cell lines that can be adapted to an automated or semi-automated readout, one can screen a chemical library comprising many thousands of chemical compounds for their effect(s) on the same cell lines. Cells are seeded in 96-well plates and a range of concentration of each chemical is added. After an appropriate incubation time, the plate is scanned and the effect/dose profile is determined. Such a screening will provide a profile of activity of the various chemical compounds that can be compared with the toxin activity. Those chemical compounds with effect similar to that of the toxin will be further considered. They may have a selected target within the cell or a defined mechanism of action thus orienting further research on the toxin activity itself. In this way it was found that LF had overlapping activity with PD98059, a specific inhibitor of MAPK kinases, and this information led to the identification of the target of the proteolytic activity of LF (Duesbery *et al.*, 1998).

Identification of targets by Proteomics

'Proteomics' is the large-scale screening of the proteins of a cell, organism or biological fluid. This process requires stringently controlled steps of sample preparation, 2-D electrophoresis, image detection and analysis, spot identification, and database searches. The most essential and at the same time weakest part of the proteomics approach is separation of the proteins of the complex sample by 2-D electrophoresis. The weak point resides especially in obtaining a clear and reproducible pattern. Correct sample preparation is very important and would ideally result in the complete solubilization, disaggregation, denaturation and reduction of proteins in the sample. If any of these steps is not optimized for a particular sample, separations may be incomplete or distorted and information may be lost. At the present time, no other technique capable of simultaneously resolving thousands of proteins is available. Proteins are separated according to two independent properties in two discrete steps: the first-dimension step is isoelectric focusing (IEF) which separates proteins according to their isoelectric points (pI); the second-dimension step is SDS-polyacrylamide gel electrophoresis (SDS-PAGE) which separates proteins according to their molecular weights. Ideally, each spot on the resulting two-dimensional array corresponds to a single molecular entity, i.e. different proteins or the same protein carrying different chemical modifications (phosphorylations, glycosilations etc. including the toxin-induced modification). Even complex 2-D patterns can be analysed with currently available software and compared with 2-D databases, allowing identification of spots by their position. Immunoblotting provides a rapid spot identification if appropriate specific antibodies are available. In addition, the molecular weight can be precisely estimated by mass spectrometry. Microsequencing of the spot can be performed providing a further means of spot identification and pieces of sequence information that open the way to cloning the gene encoding for unknown proteins. In the attempt to identify a new toxin target, all intracellular proteins of a cell or of a subcellular compartment (e.g., nuclei, mitochondria, plasma membrane) can be analysed by 2-D electrophoresis before and after treatment with the toxin. Then estimation of the molecular weight of the target molecule, or of its fragments, provides insightful information

References

Alouf, J. E. and Freer, J. H., (eds) (1999). *The Comprehensive Sourcebook of Bacterial Toxins*. Academic Press, London.

Aktories, K. and Wegner, A. (1989). ADP-ribosylation of actin by clostridial toxins. *J. Cell Biol.* **109**, 1385–1387.

Antonin, W., Holroyd, C., Fasshauer, D., Pabst, S., Fisher von Mollard, G. and Jahn, R. (2000). A SNARE complex mediating fusion of late endosomes defines conserved prpoerties of SNARE structure and function. *EMBO J.* **19**, 6453–6464.

Barois, N., Forquet, F. and Davoust, J. (1998). Actin microfilaments control the MHC class II antigen presentation pathway in B cells. *J. Cell Sci.* **111**, 1791–1800.

Barth, H., Blocker, D., Behlke, J., Bergsma-Schutter, W., Brisson, A., Benz, R. and Aktorie, K. (2000). Cellular uptake of Clostridium botulinum C2 Toxin requires oligomerization and acidification. *J. Biol. Chem.* **275**, 18704–18711.

Benmerah, A., Bayrou, M., Cerf-Bensussan, N. and Dautry-Varsat, A. (1999). Inhibition of clathrin-coated pit assembly by an Eps15 mutant. *J. Cell Sci.* **112**, 1303–1311.

Bowman, E. J., Siebers, A. and Altendorf, K. H. (1988). Bafilomycins: A class of inhibitors of membrane ATPases from microorganisms, animal cells and plant cells. *Proc. Natl Acad. Sci. USA* **85**, 7972–7976.

Damkè, H., Baba, T., Warnock, D. E. and Schmid, S. L. (1994). Induction of mutant dynamin specifically blocks endocytic coated vesicle formation. *J. Cell Biol.* **127**, 915–934.

Daukas, A. and Zigmond, S. H. (1985). Inhibition of receptor-mediated but not fluid-phase endocytosis in polymorphonuclear leukocytes. *J. Cell Biol.* **101**, 1673–1679.

Deutsch, C. J., Holian, A., Holian, S. K., Daniele, R. P. and Wilson, D. F. (1979). Transmembrane electrical and pH gradients across human erythrocytes and human peripheral lymphocytes. *J. Cell Physiol.* **99**, 79–94.

Drees, B. L. (1999). Progress and variations in two-hybrid and three-hybrid technologies. *Curr. Opin. Chem. Biol.* **3**, 64–70.

Duesbery, N. S., Webb, C. P., Leppla, S. H., Gordon, V. M., Klimpel, K. R., Copeland, T. D., Ahn, N. G., Oskarsson, M. K., Fukasawa, K., Paull, K. D. and Vande Woude, G. F. (1998). Proteolytic inactivation of MAP-kinase-kinase by anthrax lethal factor. *Science* **280**, 734–737.

Falnes, P. O., Wiedlocha, A., Rapak, A. and Olsnes, S. (1995). Farnesylation of CaaX tagged diphtheria toxin A-fragment as a measure of transfer to the cytosol. *Biochemistry* **34**, 11152–11159.

Fra, A. M., Williamson, E., Simons, K. and Parton, R. G. (1994). Detergent-insoluble glycolipid microdomains in lymphocytes in the absence of caveolae. *J. Biol. Chem.* **269**, 30745–30748.

Fraker, P. J. and Speck, J. C. (1978). Protein and cell membrane iodinations with a sparingly soluble chloroamide, 1,3,4,6-tetrachloro-3a,6a-diphrenylglycoluril. *Biochem. Biophys. Res. Commun.* **80**, 849–857.

Friedlander, A. M. (1986). Macrophages are sensitive to anthrax lethal toxin through an acid-dependent process. *J. Biol. Chem.* **261**, 7123–7126.

Girod, A., Storrie, B., Simpson, J. C., Johannes, L., Goud, B., Roberts, L. M., Lord, J. M., Nilsson, T. and Pepperkok, R. (1999). Evidence for a COP-I-independent transport route from the Golgi complex to the endoplasmic reticulum. *Nat. Cell Biol.* **1**, 423–430.

Gruenberg, J., Griffiths, G. and Howell, K. H. (1989). Characterization of the early endosome and putative endocytic carrier vesicles in vivo and with an assay of vesicle fusion in vitro. *J. Cell Biol.* **108**, 1301–1316.

Hazes, B. and Read, R. J. (1997). Accumulating evidence suggests that several AB-toxins subvert the endoplasmic reticulum-associated protein degradation pathway to enter target cells. *Biochemistry* **36**, 11051–11054.

Holm, P. K., Eker, P., Sandvig, K. and van Deurs, B. (1995). Phorbol myristate acetate selectively stimulates apical endocytosis via protein kinase C in polarized MDCK cells. *Exp. Cell Res.* **217**, 157–168.

Hopkins, C. R. and Trowbridge, I. S. (1983). Internalization and processing of transferrin and the transferrin receptor in human carcinoma A431 cells. *J. Cell Biol.* **97**, 508–521.

Jackson, M. E., Simpson, J. C., Girod, A., Pepperkok, R., Roberts, L. M. and Lord, J. M. (1999). The KDEL retrieval system is exploited by Pseudomonas exotoxin A, but not by Shiga-like toxin-1, during transport from the Golgi complex to the endoplasmic reticulum. *J. Cell Sci.* **112**, 467–475.

Kim, K. and Groman, N. B. (1965). In vitro inhibition of diphtheria toxin action by ammonium salts and amines. *J. Bacteriol.* **90**, 1552–1556.

Klausner, R. D. and Sitia, R. (1990). Protein degradation in the endoplasmic reticulum. *Cell* **62**, 611–614.

Klausner, R. D., Donaldson, J. G. and Lippincot-Schwartz, J. (1992). Brefeldin A: Insights into the control of membrane traffic and organelle structure. *J. Cell Biol.* **116**, 1071–1080.

Klimpel, K. R., Arora, N. and Leppla, S. H. (1994). Anthrax toxin lethal factor contains a zinc metalloprotease consesus sequence which is required for lethal toxin activity. *Mol. Microbiol.* **13**, 1093–1100.

Kochi, S. K., Schiavo, G., Mock, M. and Montecucco, C. (1994). Zinc content of the Bacillus anthracis lethal factor. *FEMS Microbiol. Lett.* **124**, 343–348.

Lamaze, C. and Schmid, S. L. (1995). The emergence of clathrin independent pinocytic pathways. *Curr. Opin. Cell Biol.* **7**, 573–580.

Lamaze, C., Fujimoto, L. M., Yin, H. L. and Schmid, S. L. (1997). The actin cytoskeleton is required for receptor-mediated endocytosis in mammalian cells. *J. Biol. Chem.* **272**, 20332–20335.

Larkin, J. M., Brown, M. S., Goldstein, J. L. and Anderson, R. G. W. (1983). Depletion of intracellular potassium arrests coated pit formation and receptor-mediated endocytosis in fibroblasts. *Cell* **33**, 273–285.

Lord, J. M. (1996). Go outside and see the proteasome. Protein degradation. *Curr. Biol.* **6**, 1067–1069.

Luo, Y., Batalao, A., Zhou, H. and Zhu, L. (1997). Mammalian two-hybrid system: a complementary approach to the yeast two-hybrid system. *BioTechniques* **22**, 350–352.

Mallard, F., Antony, C., Tenza, D., Salamero, J., Goud, B. and Johannes, L. (1998). Direct pathway from early/recycling endosomes to the Golgi apparatus revealed through the study of shiga toxin B-fragment transport. *J. Cell Biol.* **143**, 973–990.

Maxfield, F. R. (1982). Weak bases and ionophores rapidly and reversibility raise the pH of endocytic vesicles in cultured mouse fibroblasts. *J. Cell Biol.* **95**, 676–681.

Moya, M., Dautry-Varsat, A., Goud, B., Louvard, D. and Boquet, P. (1985). Inhibition of coated pits formation in Hep2 cells blocks the cytotoxicity of diphtheria toxin but not that of ricin toxin. *J. Cell Biol.* **101**, 548–559.

Mullock, B. M., Smith, C. W., Ihrke, G., Bright, N. A., Lindsay, M., Parkinson, E. J., Brooks, D. A., Parton, R. G., James, D. E., Luzio, J. P. and Piper, R. C. (2000). Syntaxin 7 is localised to late endosome compartments, associate with Vamp8, and is required for late endosome-lysosome fusion. *Mol. Biol. Cell* **11**, 3137–3153.

Oh, P., McIntosh, D. P. and Schnitzer, J. E. (1998). Dynamin at the neck of caveolae mediates their budding to form transport vesicles by GTP-driven fission from the plasma membrane of endothelium. *J. Cell Biol.* **141**, 101–104.

Ohkuma, S. and Poole, B. (1978). Fluorescence probe measurement of the intralysosomal pH in living cells and the perturbation of pH by various agents. *Proc. Natl Acad. Sci. USA* **75**, 3327–3331.

Orlandi, P. A. and Fishman, P. H. (1998). Filipin-dependent inhibition of cholera toxin: evidence for toxin internalization and activation through caveolae-like domains. *J. Cell Biol.* **141**, 905–915.

Papini, E., Bugnoli, M., de Bernard, M., Figura, N., Rappuoli, R. and Montecucco, C. (1993). Bafilomycin A1 inhibits Helicobacter pylori-induced vacuolization of HeLa cells. *Mol. Microbiol.* **7**, 323–327.

Papini, E., Satin, B., Bucci, C., de Bernard, M., Telford, J. L., Manetti, R., Rappuoli, R., Zerial, M. and Montecucco, C. (1997). The small GTP binding protein rab7 is essential for cellular vacuolation induced by Helicobacter pylori cytotoxin. *EMBO J.* **16**, 15–24.

Parton, R. G., Joggerst, B. and Simons, K. (1994). Regulated internalization of caveolae. *J. Cell Biol.* **127**, 1199–1215.

Rapak, A., Falnes, P. and Olsnes, S. (1997). Retrograde transport of mutant ricin to the endoplasmic reticulum with subsequent translocation to cytosol. *Proc. Natl Acad. Sci. USA* **94**, 3783–3788.

Rappuoli, R. and Montecucco, C. (eds) (1997). *Protein Toxins and Their Use in Cell Biology.* Oxford: University Press.

Rappuoli, R. and Pizza, M. G. (1991). Structural and evolutionary aspects of ADP-ribosylating toxins. *In Sourcebook of Bacterial Protein Toxins.* Aloufs, J. E., and Freer, J. H., (eds) pp. 1–23. Academic press, London.

Ricci, V., Galmiche, A., Doye, A., Necchi, V., Solcia, E. and Boquet, P. (2000). High cell sensitivity to Helicobacter pylori VacA toxin depends on a GPI-anchored protein and is not blocked by inhibition of the clathrin-mediated pathway of encocytosis. *Mol. Biol. Cell* **11**, 3897–3909.

Riederer, M. A., Soldati, T., Shapiro, A. D., Lin, J. and Pfeffer, S. R. (1994). Lysosomes biogenesis requires Rab9 function and receptor recycling from endosomes to the trans-Golgi network. *J. Cell Biol.* **125**, 573–582.

Rodal, S. K., Skretting, G., Garred, O., Vilhardt, F., van Deurs, B. and Sandvig, K. (1999). Extraction of cholesterol with methyl-beta-cyclodextrin perturbs formation of clathrin-coated endocytic vesicles. *Mol. Biol. Cell* **10**, 961–974.

Sadowski, I., Ma, J., Triezenberg, S. and Ptashne, M. (1988). GAL4-VP16 is an unusually potent transcriptional activator. *Nature* **335**, 563–564.

Sandvig, K. and Olsnes, S. (1982). Entry of the toxin proteins abrin, modeccin, ricin, and diphtheria toxin into cells. II. Effect of pH, metabolic inhibitors, and ionophores and evidence for toxin penetration from endocytotic vesicles. *J. Biol. Chem.* **257**, 7504–7513.

Sandvig, K. and van Deurs, B. (1996). Endocytosis, intracellular transport, and cytotoxic action of Shiga toxin and ricin. *Physiol. Rev.* **76**, 949–966.

Sandvig, K. and van Deurs, B. (1999). Endocytosis and intracellular transport of ricin: recent discoveries. *FEBS Lett.* **452**, 67–70.

Sandvig, K., Olsnes, S., Petersen, O. W. and van Deurs, B. (1987). Acidification of the cytosol inhibits endocytosis from coated pits. *J. Cell Biol.* **108**, 1331–1343.

Sandvig, K., Prydz, K., Ryd, M. and van Deurs, B. (1991). Endocytosis and intracellular transport of the glycolipid-binding ligand Shiga toxin in polarized MDCK cells. *J. Cell Biol.* **113**, 553–562.

Schiavo, G., Benfenati, F., Poulain, B., Rossetto, O., Polverino De Laureto, P., DasGupta, B. and Montecucco, C. (1992a). Tetanus and Botulinum-B neurotoxins block neurotransmitters release by proteolytic cleavage of synaptobrevin-2. *Nature* **359**, 832–835.

Schiavo, G., Poulain, B., Rossetto, O., Benfenati, F., Tauc, L. and Montecucco, C. (1992b). Tetanus toxin is a zinc protein and its inhibition of neurotransmitter release and protease activity depend on zinc. *EMBO J.* **11**, 3577–3583.

Schiavo, G., Rossetto, O., Santucci, A., DasGupta, B. and Montecucco, C. (1992c). Botulinum neurotoxins are zinc proteins. *J. Biol. Chem.* **267**, 23479–23483.

Schiavo, G., Matteoli, M. and Montecucco, C. (2000). Neurotoxins affecting neuroexocytosis. *Physiol. Rev.* **80**, 717–766.

Schiestl, R. H. and Gietz, R. D. (1989). High efficiency transformation of intact yeast cells using single stranded nucleic acids as a carrier. *Curr. Genet.* **16**, 339–46.

Sever, S., Damke, H. and Schmid, S. L. (2000). Dynamin: GTP controls the formation of constricted coated pits, the rate limiting step in clathrin-mediated endocytosis. *J. Cell Biol.* **150**, 1137–1148.

Shioda, T., Andriole, S., Yahata, T. and Isselbacher, K. J. (2000). A green-fluorescent protein-reporter mammalian two-hybrid system with extrachromosomal maintenance of a prey expression plasmid: Application to interaction screening. *Proc. Natl Acad. Sci. USA* **97**, 5220–5224.

Simpson, J. C., Dascher, C., Roberts, L. M., Lord, J. M. and Balch, W. E. (1995). Ricin cytotoxicity is sensitive to recycling between the endoplasmic reticulum and the Golgi complex. *J. Biol. Chem.* **270**, 20078–20083.

Simpson, J. C., Smith, D. C., Roberts, L. M. and Lord, J. M. (1998). Expression of mutant dynamin protects cells against diphtheria toxin but not against ricin. *Exp. Cell Res.* **239**, 293–300.

Simpson, F., Hussain, N. K., Qualmann, B., Kelly, R. B., Kay, B. K., McPherson, P. S. and Schmid, S.L. (1999a). SH3-domain-containing proteins function at distinct steps in clathrin-coated vesicle formation. *Nat. Cell Biol.* **1**, 119–124.

Simpson, J. C., Roberts, L. M., Romisch, K., Davey, J., Wolf, D. H. and Lord, J. M. (1999b). Ricin A chain utilises the endoplasmic reticulum-associated protein degradation pathway to enter the cytosol of yeast. *FEBS Lett.* **459**, 80–84.

Skretting, G., Torgersen, M. L., van Deurs, B. and Sandvig, K. (1999). Endocytic mechanisms responsible for uptake of GPI-linked diphtheria toxin receptor. *J. Cell Sci.* **112**, 3899–3909.

Sorkin, A. (2000). The endocytosis machinery. *J. Cell Sci.* **113**, 4375–4376.

Umata, T., Moriyama, Y., Futai, M. and Mekada, E. (1990). The cytotoxic action of diphtheria toxin and its degradation in intact Vero cells are inhibited by bafilomycin A1, a specific inhibitor of vacuolar-type H(+)-ATPase *J. Biol. Chem.* **265**, 21940–21945.

Van Deurs, B., Tonnessen, T. I., Petersen, O. W., Sandvig, K. and Olsnes, S. (1986). Routing of internalized ricin and ricin conjugates to the Golgi complex. *J. Cell Biol.* **102**, 37–47.

Van Deurs, B., Holm, P. K., Kayser, L. and Sandvig, K. (1995). Delivery to lysosomes in the human carcinoma cell line Hep2 involves an actin filament-facilitated fusion between mature endosomes and preexisting lysosomes. *Eur. J. Cell Biol.* **66**, 309–323.

Vitale, G., Pellizzari, R., Recchi, C., Napolitani, G., Mock, M. and Montecucco, C. (1998). Anthrax lethal factor cleaves the N-terminus of MAPKKs and induces tyrosin/threonine phosphorylation of MAPKs in cultured macrophages. *Biochem. Biophys. Res. Commun.* **248**, 706–711.

Vogel, U., Sandvig, K. and van Deurs, B. (1998). Expression of caveolin-1 and polarized formation of invaginated caveolae in Caco-2 and MDCK II cells. *J. Cell Sci.* **111**, 825–832.

Vojtek, A. B., Hollenberg, S. M. and Cooper, J. A. (1993). Mammalian Ras interacts directly with the serine/threonine kinase Raf. *Cell* **74**, 205–214.

Wiedlocha, A., Falnes, P. O., Rapak, A., Klingenberg, O., Munoz, R. and Olsnes, S. (1995). Translocation to cytosol of exogenous CAAX-tagged acidic fibroblast growth factor. *J. Biol. Chem.* **270**, 30680–30685.

Williamson, L. C. and Neale, E. A. (1994). Bafilomycin A1 inhibits the action of tetanus toxin in spinal cord neurons in cell culture. *J. Neurochem.* **63**, 2342–2345.

17 Flow Cytometric Analysis of *Salmonella*-containing Vacuoles

Stéphane Méresse[1], B Brett Finlay[2] and Jean-Pierre Gorvel[1]
[1] *Centre d'Immunologie INSERM-CNRS-Univ. Med. de Marseille-Luminy, Marseille France;*
[2] *Biotechnology Laboratory and Departments of Biochemistry and Microbiology, University of British Columbia, Vancouver, British Columbia, Canada*

Analysis of *Salmonella*-containing Vacuoles

◆◆

CONTENTS

List of abbreviations

cfu	Colony forming unit
Cy5	Cyanin 5
DMEM	Dulbeco's minimal essential medium
EEAI	Early endosomes associated protein I
ER	Endoplasmic reticulum
FACS	Fluorescence-activated cell sorting
FCS	Fetal calf serum
FITC	Fluorescein isothiocyanate
GFP	Green fluorescent protein
HB	Homogenization buffer
Lampl	Lysosomal membrane glycoprotein I
LB	Luria-Bertani
PBS	Phosphate saline buffer
PE	Phycoerythrin
PNS	Post-nuclear supernatant
SCV	*Salmonella*-containing vacuole
SPI	*Salmonella* pathogenicity island

METHODS IN MICROBIOLOGY, VOLUME 31
ISBN 0-12-521531-2

◆◆◆◆◆◆ INTRODUCTION

Salmonella enterica Serovar Typhimurium (*S. typhimurium*) is an enteric pathogen that causes gastroenteritis in humans. In mice, it is the etiologic agent of a systemic infection similar to typhoid fever. Thus it is used as a model to study the pathophysiology of typhoid fever (Jones and Falkow, 1996). Over the past few years, significant information has been obtained about how *Salmonella* interacts with cultured epithelial cells. However, despite this surge in information, there are several key questions that need to be addressed to further understand how this pathogen actually causes disease. A critical stage of *Salmonella* infections is its life within host cells, both epithelial and macrophage. Little is known about the intracellular environment, and how this pathogen manages to survive within this niche. We also know little about which bacterial virulence factors are expressed within host cells. Therefore it is important to characterize the nature of the *Salmonella*-containing vacuole (SCV). This can be done by indirect methods such as gene fusions with reporters to monitor gene expression. However, to directly characterize the host and bacterial components present in the SCV, isolation of the vacuole is needed. *S. typhimurium* is rapidly internalized into non-phagocytic epithelial cells. The first well-characterized event which occurs upon interaction of Salmonella with an epithelial cell is the formation of membrane ruffles which facilitate the entry of the bacteria and involves extensive cytoskeletal rearrangements within the host cell (Finlay *et al.*, 1991). Ruffle formation is mediated by the delivery of virulence factors directly into the host cytosol, triggering host cell signalling pathways that lead to localized membrane ruffling, macropinocytosis and bacterial uptake. *Salmonella* use a type III secretion system, needle-like complexes spanning bacterial inner and outer membranes, to target effectors into host cells (Kubori *et al.*, 1998). *Salmonella* type III secretion system chromosomal genes are clustered in the Pathogenicity Island-1 (SPI-1), which also encodes secreted effector proteins such as SipA, SptP and SipC. These proteins, together with SopE (a protein not encoded by SPI-1), which serves as a guanine nucleotide exchange factor for activating Rac and Cdc42, are injected in the host cytoplasm to promote cytoskeleton rearrangements and bacterial uptake (Mills *et al.*, 1998).

After uptake, *Salmonella* resides within a vacuole that successively acquires markers of the early endosomes, the recycling compartment and late endocytic compartments (Steele-Mortimer *et al.*, 1999). SCV maturation is characterized by a gradual enrichment in lysosomal membrane proteins, but not mannose-6-phosphate delivered proteins. This specialized routing occurs through the interactions with vesicles enriched in the rab7 small GTPase and lysosomal glycoproteins (Méresse *et al.*, 1999). During the course of infection, a second bacterial type III secretion system (*Spi/Ssa*) encoded by the SPI-2 pathogenicity island appears to function intracellularly and is required for the survival of *Salmonella* in macrophages (Marcus *et al.*, 2000). Recent studies have established that SpiC (Uchiya *et al.*, 1999) and SifA, are exported via the Spi/Ssa type III secretion system into the host cell cytosol and control SCV maturation (Brumell *et al.*, 2001; Beuzon *et al.*, 2000). Methods based on centrifugation through sucrose gradients showed difficulties in the purification of SCVs, due to vacuole instability and perhaps ER contamination, although it was suitable for latex bead and Yersinia vacuole isolation (Mills and Finlay, 1998). We have recently developed a new method for the characterization and purification of SCVs. This method had previously been developed for the isolation of rab5- or rab7-containing vesicles (Chavrier *et al.*, 1997; Méresse *et al.*, 1997) and involves labelling of bacterial surfaces with fluorescein or bacteria expressing GFP, infecting host cells and analysing and isolating SCVs by flow cytometry. The analysis is based on a positive selection for fluorescent bacteria, thereby decreasing potential contamination with other subcellular organelles. In addition, this technology enables the specific sorting of SCVs and can be extended to other intracellular pathogens.

◆◆◆◆◆◆ PREPARATION OF FLUORESCENT BACTERIA FOR INVASION OF PROFESSIONAL OR NON-PROFESSIONAL PHAGOCYTES

Salmonella grown under different conditions are needed whether the invasion of phagocytic or non-phagocytic is being used. As *Salmonella* grown to the log phase induce apoptosis of macrophages (Monack *et al.*, 1996), invasion of this cell type requires bacteria grown to stationary phase. In contrast, maximum invasion efficiency of epithelial cells is obtained when the bacteria are grown to late log phase. Bacteria are both highly motile and invasive at this growth phase. An efficiency of invasion of HeLa cells of between 0.5 and 5% is seen when 200 cfu are added per cell for 5 min at 37°C; thus each cell is infected with approximately 1–10 bacteria (Steele-Mortimer *et al.*, 1999). A rapid and efficient invasion protocol is required to obtain synchronised SCV maturation. Protocols for bacterial growth follow:

For epithelial cell invasion: S typhimurium (strains SL1344 or 12023) are grown overnight at 37°C with shaking and then subcultured at a dilution of 1 : 33 in fresh Luria-Bertani (LB) broth and incubated at 37°C with shaking for 3 h. *For macrophage invasion*: Bacteria are grown overnight at 37°C with shaking.

Bacteria expressing GFP are available for several bacterial pathogens (Valdivia and Falkow, 1997). However, tranformation and GFP expression in some pathogens is problematic, and thus other ways of fluorescently tagging bacteria can be considered. We have found that with *Salmonella*, surface labelling bacteria with FITC has no effect on the viability and invasion capacity of bacteria. The fluorescence of the intracellular bacteria can be assessed visually by fluorescence microscopy. FITC-labelled bacteria are clearly visible in cells for several hours, indicating that the label is fairly stable on the bacterial surface. However, the level of fluorescence decreases with time and bacterial replication.

FITC labelling of bacteria. Bacterial subcultures are pelleted in a microfuge at 10 000 rpm for 2 min and resuspended in PBS. Freshly solubilized FITC mixed isomers are added to a final concentration of 0.5 mg ml^{-1} and the bacteria are incubated at 37°C for 10 min. Free FITC is removed by washing in PBS and the bacteria are either resuspended in Earle's buffered salt solution, pH 7.4 or opsonized in FCS.

◆◆◆◆◆◆ INFECTION AND SUBCELLULAR FRACTIONATION OF CELLS

Optimised cell invasion is necessary for the analysis of fluorescent vacuoles by flow cytometry. Optimal detection is achieved when vacuoles represent more than 5% of particles. This requires cells infected with between 1 and 10 fluorescent bacteria. Subconfluent tissue culture cells must be freshly seeded and grown in the absence of antibiotics. Highly motile and invasive bacteria in log phase growth are added to the non-phagocytic cell culture medium. Infection of macrophages first requires the opsonization of bacteria in FCS and the centrifugation of bacteria onto cell monolayer.

Culture and infection of non-phagocytic cells. Cells are grown in 10 cm dishes at a density of approximately 5×10^6 cells per dish and incubated overnight at 37°C in 5% CO_2. Growing medium is replaced by 10 ml of EBSS containing the bacterial inoculum at a multiplicity of infection of approximately 100 : 1. The infection is carried out at 37°C in 5% CO_2 for 5 min, following which free bacteria are removed by washing three times with PBS. The monolayers are then incubated in growing medium supplemented with 100 μg ml^{-1} gentamicin. For experiments in which cells are incubated for more than 2 hours after infection, the gentamicin concentration can be decreased to 10 μg ml^{-1}.
Culture and infection of macrophages. Raw 264.7 murine macrophages grown in suspension in Petri dishes designed for bacteriology (in which they do not adhere) are seeded at a density of 10^7 cells per 10 cm cell culture dishes. Overnight cultures of *Salmonella* in LB are pelleted and opsonized in DMEM containing FCS and 10% normal mouse serum for 20 min. Bacteria are added to the monolayers at a multiplicity of infection of approximately 100 : 1, centrifuged at 170*g* for 5 min at room temperature and incubated for 25 min at 37°C. The monolayers are then washed three times with PBS and incubated in growing medium supplemented with 100 μg ml^{-1} gentamicin. For experiments in which

cells are incubated for more than 2 hours after infection, the gentamicin concentration can be decreased to 10 μg ml^{-1}.

Subcellular fractionation for flow cytometry analysis consists of the disruption of the infected cell plasma membrane to release intracellular organelles including the SCVs (Steele-Mortimer *et al.*, 1999). Gentle conditions of homogenization are needed to limit possible damages to vacuole membranes. Harsh conditions should be avoided to limit the breakage of both lysosomes and nuclei which cause proteolysis of SCVs and organelle aggregation, respectively. We found that adding skin gelatin during homogenization limits SCV breakage. The cells are released from the dish by scraping with the sharp edge of a rubber policeman. At this stage, breakage should not exceed 5–10% of the cells. After collecting the cells by centrifugation, the cells are resuspended in a small volume. Homogenization is easier at a relatively high density of cells (20–30% v/v). The cells are then homogenized by passage through a needle and the post-nuclear supernatant (PNS) is prepared by centrifugation to remove unlysed cells and large debris.

Homogenization. All manipulations are performed at 4°C and buffers prefiltered (0.45 μM). Two 10 cm dishes of infected or non-infected cells are used for each time point. Cells are washed extensively with PBS containing 0.2% BSA, scraped with a rubber policeman and centrigufed at 100*g* for 5 min. 3 ml of homogenization buffer (HB : 250 mM sucrose, 3 mM imidazole, pH 7.4, 0.1% gelatin, 0.5 mM EGTA, 1 μg ml^{-1} antipain, 1 μg ml^{-1} pepstatin, 1 μg ml^{-1} leupeptin, 15 μg ml^{-1} benzamidin) are added to the top of cell pellets and recentrifuged at 1800*g* for 5 min. Cell pellets are then resuspended in 400 μl HB and gently homogenized by 3–6 passages through a 22G needle fitted to a 1 ml syringe. Care must be taken to avoid bubbles. Homogenates are diluted to 2 ml with HB, and post-nuclear supernatants (PNS) are prepared by three successive centrifugations at 100*g* for 5 min to eliminate nuclei and cell debris.

◆◆◆◆◆◆ FLOW CYTOMETER SETTINGS AND ANALYSIS OF VACUOLE

Flow cytometry is usually used as a tool for analysis and/or sorting of mammalian cells. The recent development of subcellular particle-specific probes combined with gentle subcellular fractionation techniques allow investigators to now study intracellular particles by flow cytometry (Figure 17.1). This method has been successively used for the sorting of individual chromosomes (Fawcett *et al.*, 1994) or the analysis and the sorting of Rab5- and Rab7-containing endocytic compartments (Chavrier *et al.*, 1997; Méresse *et al.*, 1997). This method is particularly well adapted

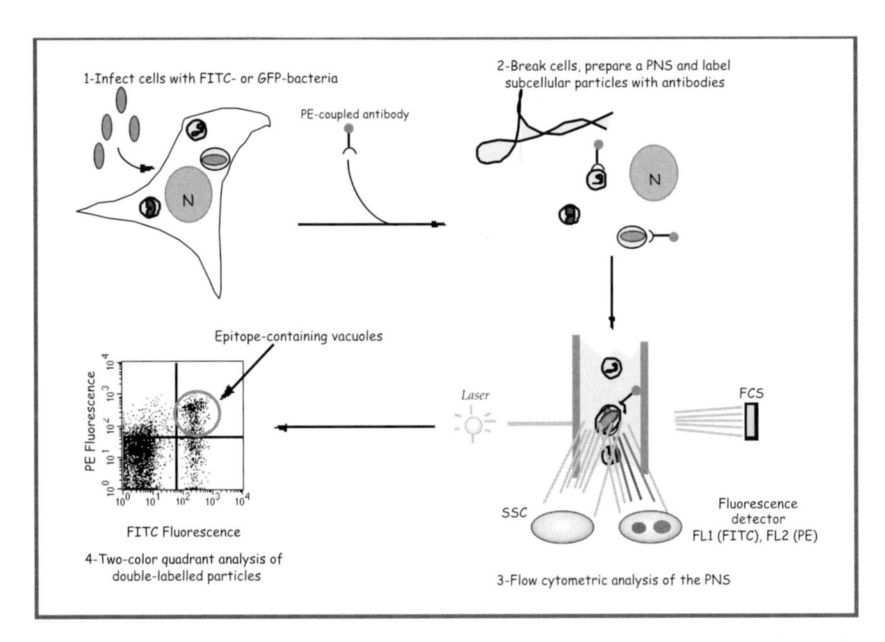

Figure 17.1. Flow cytometric analysis of pathogen-containing vacuoles. Host cells are infected with FITC- or GFP-bacteria and submitted to subcellular fractionation. A PNS is prepared and incubated with directly or indirectly PE-coupled antibodies specific for a given epitope. As they pass in front of the laser beam, each particle is analysed for its forward and side scattered light (FSC and SSC) and fluorescence emissions (FITC and PE). Two-colour quadrant analysis allows the detection of double-labelled particles, which represent the epitope-containing vacuoles.

to the study of vacuoles as their detection is based on the selection of fluorescent pathogens inside membrane compartments. Flow cytometer settings useful for the analysis of SCVs are described below.

Flow cytometry allows the rapid statistical analysis of a large number of particles. For example, 40 analyses can be performed with 2 ml of PNS (see above). 50 μl of PNS in which SCVs represent 5% of total particles, are used to analyse 20 000 particles at a speed of 1000 events s^{-1} enabling the study of 1000 SCVs in 20 s. In a fluorescence versus side scatter dot-plot (Figure 17.2), fluorescent particles are well discriminated from other particles (compare the analysis of PNSs of non-infected in Figure 17.2A and infected cells in Figure 17.2B). It is then important to check that fluorescent particles detected by the cytometer correspond to SCVs and not to free bacteria released during homogenization. For this, free fluorescent bacteria can be added to PNSs from non-infected and infected cells. As shown in Figures 17.2C and D, free fluorescent *Salmonella* display a very homogeneous distribution and are separated from SVCs both in the fluorescence and the side scatter channels.

During the course of infection, SCVs undergo a maturation process characterized by the successive enrichment and release of endocytic markers. The distribution of these markers can be analysed by confocal microscopy and by flow cytometry using specific antibodies. As PNS contains non-fixed and non-permeabilized particles, immunolabelling

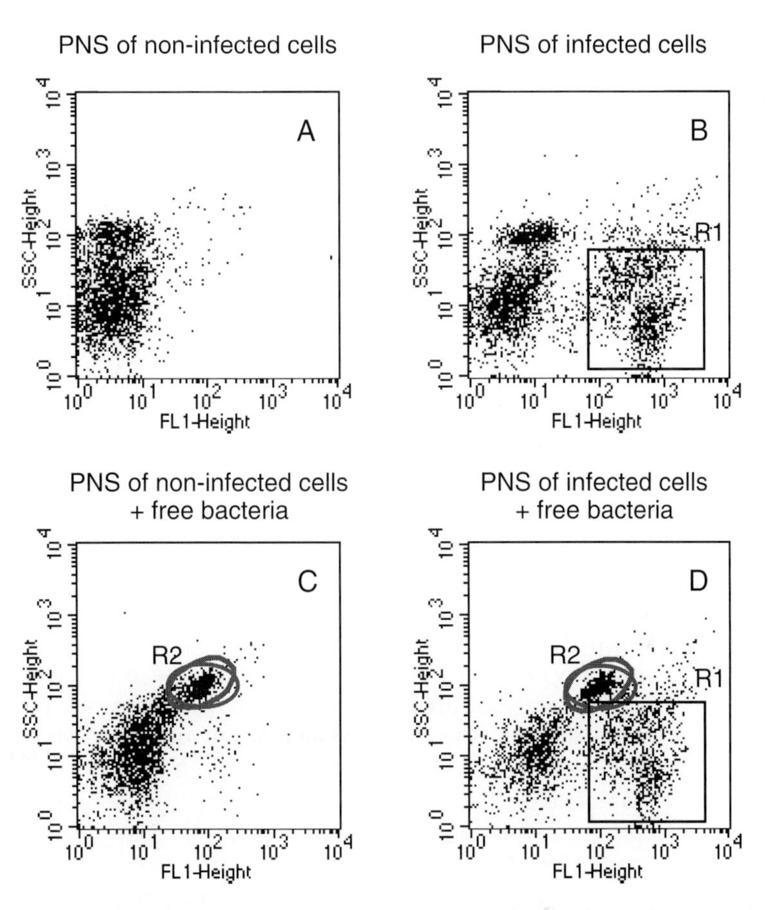

Figure 17.2. Flow cytometry discriminates between free and membrane-enclosed fluorescent bacteria. FITC fluorescence versus SSC analysis of PNS of *Salmonella*-infected (B, D) or non-infected (A, C) cells in which free fluorescent bacteria were added (C, D) or not (A, B). Free bacteria and SCVs are identified on the basis of FITC fluorescence (FL1-Height) and light scattering signal (SSC-Height). The R1 window corresponds to SCVs which display a strong fluorescence signal and a less homogeneous distribution than free bacteria (R2 window).

is restricted to epitopes located at the cytoplasmic surface of SCVs. Immunolabelling of cytoplasmic domains of integral proteins (transferrin receptor, lysosomal glycoproteins, mannose-6-phosphate receptor) and of lipid-anchored proteins (small GTPases of the Rab family and their effectors such as EEA1) has been successfully performed (Méresse *et al.*, 1999; Steele-Mortimer *et al.*, 1999)

Immunolabelling. 50 µl of PNS are mixed gently with 50 µl of PBS containing 5% normal goat serum and an appropriate dilution of primary antibodies and incubated for 10 min on ice. PE- and Cy5-conjugated secondary antibodies diluted in 50 µl of PBS are then added for 10 min on ice. Samples are diluted with 1 ml PBS and immediately analysed.

Flow cytometer settings. Becton Dickinson FACStar Plus flow cytometer and CellQuest software (Becton Dickinson, Franklin Lakes, NJ) displaying filter settings of 530/30 nm, 575/26 nm and 660/13 nm band-pass for FITC (FL1), PE (FL2) and Cy5 (FL4), respectively are well adapted for SCV fluorescence detection and analysis. An analysis gate containing most of the FITC-positive particles is set using the SSC (log)/FCS (linear) dot-plot and used for double fluorescence quadrant analysis.

Immunofluorescence microscopic analysis of infected cells allows the direct localization of a given epitope on SCVs. The main benefit of the flow cytometric method described above is that it brings subtle information about the level of epitope expression on SCVs, and allows epitope quantitation, which is difficult using fluorescence microscopy. Double fluorescence quadrant analysis allows the determination of both the mean and standard deviation of fluorescence intensity for a given epitope on SCVs. These data directly reflect the amount and the dispersion of the analysed marker on vacuole. For example, confocal microscopic observations show that most SCVs contain lamp1 from 30 min of infection onwards. We learned from the flow cytometric analysis of the same cells that the number of lamp1 molecules per SCV increases progressively to reach a maximum at 3 h post-infection (Steele-Mortimer *et al.*, 1999) (Figure 17.3).

By combining both confocal microscopy and flow cytometry it has been possible to correlate morphological observation and statistical analysis of key marker distribution in *Salmonella* infected cells and to refine our understanding of SCV maturation.

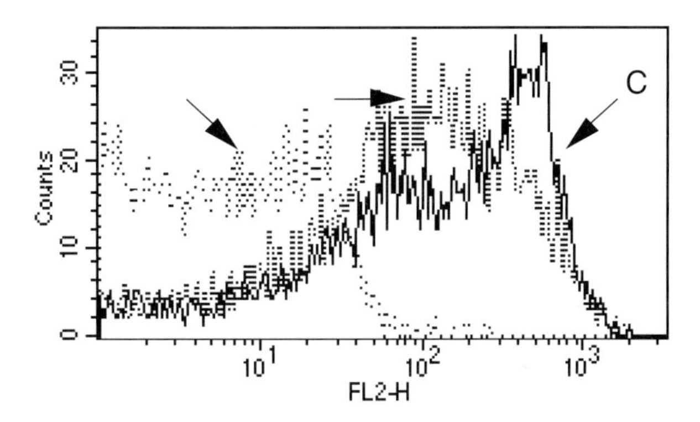

Figure 17.3. Time course FACS analysis of SCVs unravels a progressive enrichment in lamp1. Fluorescence (FL2-H) histograms of 20 min (B) or 60 min late (C) SCVs immunolabelled with a rabbit antibody specific for lamp1 and PE-coupled anti-rabbit IgGs. As a control, the primary antibody was omitted (A).

◆◆◆◆◆◆ FLOW CYTOMETRIC SORTING OF VACUOLES. PRESENT AND FUTURE

Setting parameters defined for the flow cytometric analysis can then be used to sort SCVs from PNS. After passage in front of the laser beam, the fluid jet is broken up into droplets. The sort decision that is formulated electronically is executed by charging the droplets containing the fluorescent SCV. These droplets are deflected as they pass the deflection plate and collected. Sorting of FITC-positive particles in a PNS of infected cells should allow the collection of a homogeneous population of SCV devoid of contamination by other subcellular organelles. Sorting can be based on multiple parameters analysis of immmunolabelled SCV; for example SCV that are positive for EEA1 and negative for lamp1. This procedure permits the isolation of SCVs subpopulation corresponding to different steps of the vacuole maturation.

A conventional sorter, running at a maximum sorting speed of 1000 events per second for 4 h would permit the isolation of about 1.5×10^6 SCVs from a PNS in which vacuoles represent 10% of the define analysis gate (see *Flow cytometer settings*). In terms of protein, sorted SCV represent a very limited amount (less that 1 μg). Such a preparation is convenient for further Western-blotting analysis (Méresse *et al.*, 1997) but not for the determination of SCV polypeptide map by a proteomic approach.

High-speed sorters were originally designed for the isolation of chromosomes. Commercial high-speed sorters have become available with sorts at rates up to 20 000 events s^{-1} and increased droplet-sorting yield. In conclusion, both the quantity and the quality of sorted material are improved.

◆◆◆◆◆◆ CONCLUSION

Many bacterial pathogens reside within host cells, either transiently or throughout an infection, and this intracellular lifestyle is often a key component to disease. Immunofluorescence microscopy has been used to follow routing of pathogens inside host cells, but does not allow marker quantitation, nor biochemistry. The ability to enrich and study vacuoles containing individual organisms generates many opportunities to better characterize processes which are currently poorly understood. The methods described here for the analysis of *Salmonella*-containing vacuoles may be applied to other intracellular pathogens which are transiently found in vacuoles (*Listeria, Shigella*) or reside in a vacuole (*Brucella, Mycobacterium, Toxoplasma, Leishmania, Legionella*). By using FACS techniques, the kinetics of vacuole maturation can be followed as the vacuole matures. There are several significant techniques that are rapidly being developed which may be applied to sorted vacuoles. The ability to utilize 2D-gel analysis of purified vacuoles and lysosomes presents wonderful opportunities to obtain a proteomic analysis of isolated vacuoles. However, sample quantity is a limiting factor, and this may be possible only for pathogens whose vacuoles are relatively stable. Similarly, the genomes of many pathogens are completed, and mouse and human genomes are rapidly being finished. Array analysis of both host and bacterial components of genes being expressed within a vacuole would significantly enhance our knowledge of the intracellular lifestyle of various pathogens. However, sample quantity will remain an issue to be addressed prior to successful application of arrays. Biochemical studies on vesicle fusion events have been very successful in defining how vesicles fuse. Since many pathogens appear to alter fusion capabilities, isolation of vacuoles will permit such studies to be performed with pathogen vacuoles. These studies will be essential to define the molecular mechanisms used by pathogens to alter their intracellular routing within host cells. They will also add further information to our current understanding of normal fusion events.

Although bacterial uptake (invasion and phagocytosis) has been fairly well studied, little is known about the events that follow pathogen uptake into host cells. The ability to enrich and isolate pathogen-containing vacuoles such as SCVs provides many additional opportunities to understand these later events. It is clear that the next few years will provide unprecedented opportunity to understand intracellular parasitism, and FACS analysis will be central to these developments.

Acknowledgments

This work was supported by operating grants from the Medical Research Council of Canada (BBF) and CNRS, INSERM (S.M. and J.-P.G.).

References

Beuzon, C. R., Méresse, S., Unsworth, K. E., Ruiz-Albert, J., Garvis, S., Waterman, S. R., Ryder, T. A., Boucrot, E. and Holden, D. W. (2000). Salmonella maintains the integrity of its intracellular vacuole through the action of SifA. *EMBO J.* **19**, 3235–49.

Brumell, J. J., Rosenberger, C. M., Gatto, G. T., Mareus, S. L. and Finlay, B. B. (2001). SifA permits survival and replication of Salmonella typhimurium in murine macrophages. *Cell Microbiol.* **3**(2), 75–84.

Chavrier, P., Van der Sluijs, P., Mishal, Z., Nagelkerken, B. and Gorvel, J. P. (1997). Early endosome membrane dynamics characterized by flow cytometry. *Cytometry* **29**, 41–9.

Fawcett, J. J., Longmire, J. L., Martin, J. C., Deaven, L. L. and Cram, L. S. (1994). Large-scale chromosome sorting. *Methods Cell Biol.* **42**, 319–330.

Finlay, B. B., Ruschkowski, S. and Dedhar, S. (1991). Cytoskeletal rearrangements accompanying salmonella entry into epithelial cells. *J Cell Sci.* **99**, 283–296.

Jones, B. D. and Falkow, S. (1996). Salmonellosis: host immune responses and bacterial virulence determinants. *Annu. Rev. Immunol.* **14**, 533–561.

Kubori, T., Matsushima, Y., Nakamura, D., Uralil, J., Lara-Tejero, M., Sukhan, A., Galan, J. E. and Aizawa, S. I. (1998). Supramolecular structure of the Salmonella typhimurium type III protein secretion system. *Science* **280**, 602–605.

Marcus, S. L., Brumell, J. H., Pfeifer, C. G. and Finlay, B. B. (2000). Salmonella pathogenicity islands: big virulence in small packages. *Microbes Infect.* **2**, 145–156.

Méresse, S., André, P., Mishal, Z., Barad, M., Brun, N., Desjardins, M. and Gorvel, J. P. (1997). Flow cytometric sorting and biochemical characterization of the late endosomal rab7-containing compartment. *Electrophoresis* **18**, 2682–2688.

Méresse, S., Steele-Mortimer, O., Finlay, B. B. and Gorvel, J.-P. (1999). The rab7 GTPase controls the maturation of Salmonella typhimurium-containing vacuole in HeLa cells. *EMBO J.* **18**, 4394–4403.

Mills, S. D. and Finlay, B. B. (1998). Isolation and characterization of Salmonella typhimurium and Yersinia pseudotuberculosis-containing phagosomes from infected mouse macrophages: Y. pseudotuberculosis traffics to terminal lysosomes where they are degraded. *Eur. J. Cell Biol.* **77**, 35–47.

Mills, S. D., Ruschkowski, S. R., Stein, M. A. and Finlay, B. B. (1998). Trafficking of porin-deficient Salmonella typhimurium mutants inside HeLa cells: ompR and envZ mutants are defective for the formation of Salmonella-induced filaments. *Infect. Immun.* **66**, 1806–1811.

Monack, D. M., Raupach, B., Hromockyj, A. E. and Falkow, S. (1996). Salmonella typhimurium invasion induces apoptosis in infected macrophages. *Proc. Natl Acad. Sci. USA* **93**, 9833–9838.

Steele-Mortimer, O., Meresse, S., Toh, B. H., Gorvel, J. P. and Finlay, B. B. (1999). Biogenesis of Salmonella typhimurium-containing vacuoles in epithelial cells involves interactions with the early endocytic pathway. *Cell Microbiol.* **1**, 33–49.

Uchiya, K., Barbieri, M. A., Funato, K., Shah, A. H., Stahl, P. D. and Groisman, E. A. (1999). A Salmonella virulence protein that inhibits cellular trafficking. *EMBO J.* **18**, 3924–3933.

Valdivia, R. H. and Falkow, S. (1997). Fluorescence-based isolation of bacterial genes expressed within host cells. *Science* **277**, 2007–2011.

18 Proteomic Analysis of Phagosomes

Michel Desjardins, Sophie Duclos and Jean-François Dermine
Département de pathologie et biologie cellulaire, Université de Montréal, C.P. 6128, Succ. Centre ville, Montréal, QC, Canada, H3C 3J7

◆◆◆

CONTENTS

List of abbreviations

2-D gel	Two-dimensional gel electrophoresis
ER	Endoplasmic reticulum
EST	Expression sequence tag
FACS	Fluorescence-activated cell sorter
FAOS	Fluorescence-activated organelle sorting
GFP	Green fluorescent protein
IPG	Immobilized pH gradient
LAMP	Lysosomal-associated membrane protein
LPG	Lipophosphoglycan
MALDI-TOF MS	Matrix-assisted laser desorption/ionization time-of-flight mass spectrometry
PBS	Phosphate buffered saline
PNS	Post-nuclear supernatant
RPM	Rotation per minute
SNAREs	Soluble N-ethylmaleimide-sensitive factor attachment protein receptors

◆◆◆◆◆◆ INTRODUCTION

The vast sum of information generated by the analysis of the genome of many organisms is going to have major impacts on many aspects of biology, as it will allow one to find the differences associated to particular

METHODS IN MICROBIOLOGY, VOLUME 31
ISBN 0–12–521531–2

diseases, the differences between host and tumor cells and the reason why we are predisposed to various genetic diseases. However, DNA sequence data reveal little about the level of expression of proteins, their function, the regulation of their function, and their subcellular localization. Indeed, it is apparent that genomic information is the starting point of the more challenging aim of understanding how cells work (Pennington *et al.*, 1997). In the last few years, the field of genomic research has contributed a new discipline whose task is to systematically identify and study all proteins of a given organism, cell or organelle. This new field, 'proteome research' (for all the PROTEins expressed by the genOME) (Wilkins *et al.*, 1996), is likely to become an important tool of biological science, as proteins are the direct effectors of cell functions. The data generated by proteome research will transform fundamentally our understanding of the molecular composition and function of cells in health and diseases.

Application of proteomic approaches to the study of host–pathogen interaction has already given valuable information on how microorganisms can alter the function of host cells. By comparing phagosomes containing living or dead mycobacteria, Pieters and collaborators (Ferrari *et al.*, 1999) were able to identify a coat protein, named TACO, involved in inhibition of intracellular trafficking of mycobacteria to phagolysosomes. According to their model, TACO represents a component of the phagosome coat that is normally released prior to phagosome fusion with lysosomes. Inhibition of phagosome–lysosome fusion appears to be a fundamental way by which intracellular pathogens survive in their host cells. In our laboratory, we were able to demonstrate that the lipophosphoglycan (LPG) of *Leishmania*, the major surface glycolipid of this parasite, is responsible for the inhibition of fusion occurring between *Leishmania*-containing phagosomes and late endocytic organelles at the onset of infection of J774 macrophages (Desjardins and Descoteaux, 1997; Scianimanico *et al.*, 1999; Dermine *et al.*, 2000). Purification of phagosomes containing wild type *Leishmania* (non-fusogenic phagosomes) or a mutant lacking LPG (present in fusogenic phagosomes), followed by their analysis by two-dimensional gel electrophoresis (2-D gel) clearly indicated the presence of specific proteins associated to each type of phagosome and likely to confer specific functions to these organelles (Dermine, Huber and Desjardins, unpublished results).

◆◆◆◆◆◆ PHAGOSOME FORMATION AND ISOLATION

Phagosomes can be formed by feeding cells with latex particles, a simple procedure allowing isolation of this compartment by flotation on sucrose step gradients. Latex beads are phagocytosed by a variety of cells when added to the culture medium. A procedure used in the past to isolate phagosomes from a variety of cells is the following.

Latex beads are diluted in culture medium and cells are incubated in this medium for the desired time at 37°C in a CO_2 incubator. After bead internalization, cells are washed 3×5 min with cold phosphate buffered saline (PBS). This eliminates most of the non-internalized particles. Cells are then scraped with a rubber policeman in cold PBS and pelleted at low speed. The cell pellet is washed twice in 1 ml of homogenization buffer (250 mM sucrose, 3 mM imidazole, pH 7.4) containing protease inhibitors, and then broken gently by pipeting it through a Pasteur pipette 10 to 12 times. Cells are further broken by passages in a syringe with a $22\frac{1}{2}$G needle. This procedure is monitored under the light microscope after each set of five strokes and continued until most of the cells are broken without major breakage of the nucleus. The homogenate is centrifuged in a 15 ml Falcon tube at 2000 rotations per minute (rpm) for 5 min to get rid of the unbroken cells and nucleus. After centrifugation, the post-nuclear supernatant containing the phagosomes is present in the upper phase. This part is recovered and mixed with a 62% sucrose solution to bring it to a final concentration of 42% sucrose. Phagosomes are then isolated on a step sucrose gradient consisting of the following steps in an SW41 Beckman tube: a cushion of 1 ml of 62% sucrose; the phagosome sample at 42% sucrose (about 2 ml); 2 ml of 35% sucrose; 2 ml of 25% sucrose; and 2 ml of 10% sucrose. This gradient is centrifuged at 24 000 rpm for 60 min at 4°C. After centrifugation, phagosomes are recovered in a single band at the 10–25% sucrose interface. Phagosomes are then suspended in 10 ml of cold PBS and pelleted in the same kind of tube by centrifugation at 15 000 rpm for 15 min at 4°C. The phagosome pellet is then resuspended in an appropriate buffer for further analyses.

◆◆◆◆◆◆ ISOLATION OF PHAGOSOMES CONTAINING MICRO-ORGANISMS

The method described above is simple, and efficiently provides pure preparations of phagosomes containing latex beads. Because these phagosomes are fusogenic both *in vivo* and *in vitro* and move along microtubules (Desjardins *et al.*, 1994; Jahraus *et al.*, 1998), and contain several of the molecules required for protein degradation and antigen processing (Garin *et al.*, 2001), they are well suited to the analysis of membrane organelle functions. However, an understanding of the mechanisms and molecules altered in phagosomes housing various micro-organisms requires the purification of pathogen-containing compartments. Unfortunately, these organelles cannot be simply isolated by flotation on sucrose gradients. Instead, sedimentation procedures are used (Chakraborty *et al.*, 1994; Alvarez-Dominguez *et al.*, 1996). Conventional approaches involve the use of sucrose and/or Ficoll gradients. Prior to the isolation procedure, great care has to be taken during cell lysis and several conditions have to be tested in order to break host cells without damaging phagosomes. This can be particularly difficult when micro-organisms reside in spacious vacuoles. In any case, isolation on step gradients is likely to generate phagosome preparations displaying significant levels of contamination by other host cell organelles.

Faced with these limitations, we have used a different method to isolate *Leishmania*-containing phagosomes. This method, referred to as

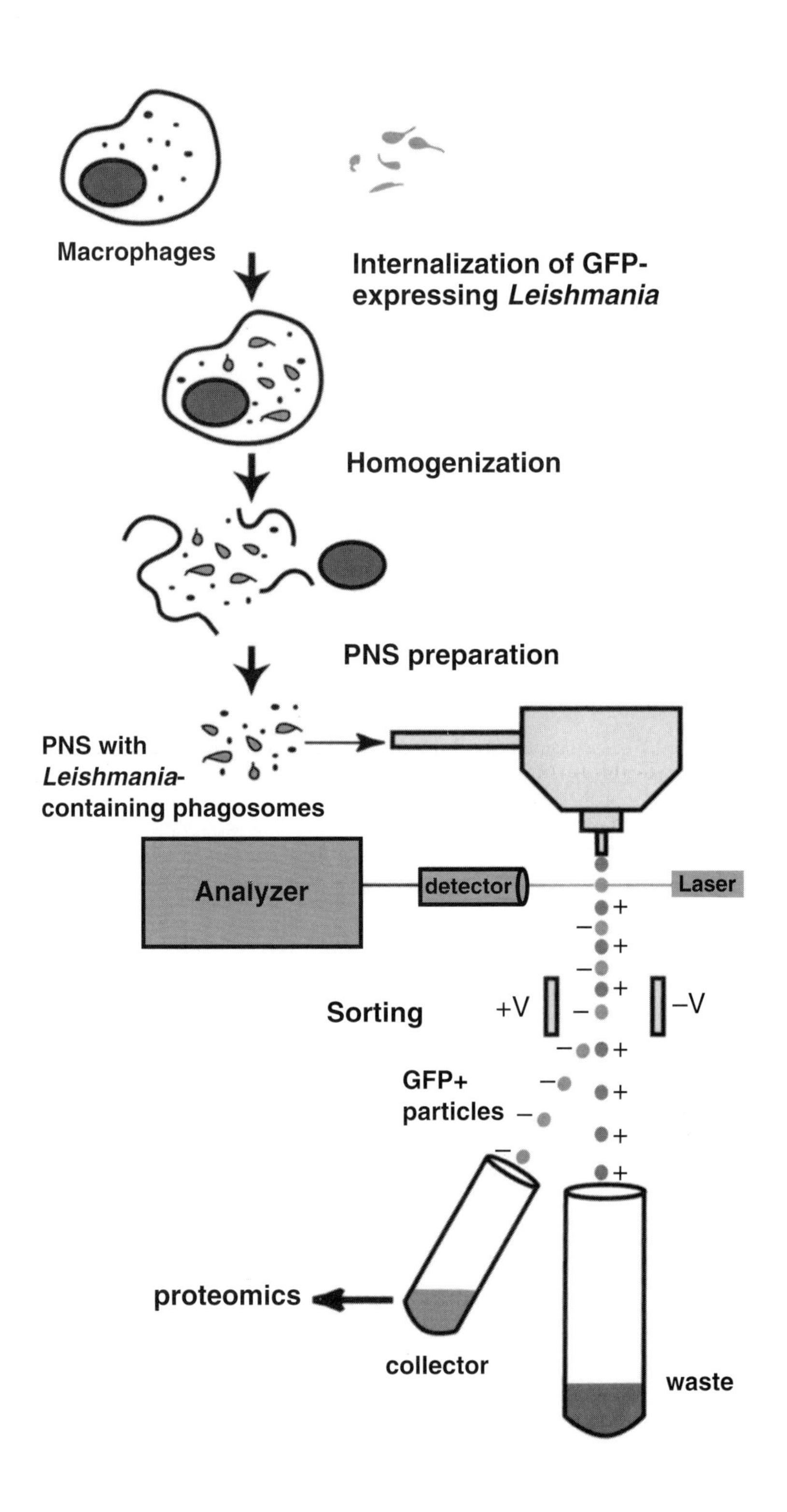

Macrophages
Internalization of GFP-expressing *Leishmania*
Homogenization
PNS preparation
PNS with *Leishmania*-containing phagosomes
Analyzer
detector
Laser
Sorting
+V
–V
GFP+ particles
proteomics
collector
waste

fluorescence-activated organelle sorting (FAOS) (Böck *et al.*, 1997) takes advantage of the possibility to sort and separate fluorescent particles from heterogeneous samples using a fluorescence-activated cell sorter (FACS). This approach requires that phagosomes display a fluorescent signal. This can be done in several ways. Micro-organisms can be modified to express the green fluorescent protein (GFP) or can be opsonized with fluorescent molecules. The former approach has the advantage that the surface of the micro-organism is not modified in any way that could alter its binding to host cell surface receptors. The cytoplasmic side of phagosomes can also be made fluorescent by binding fluorescent antibodies directed against phagosome proteins like lysosomal-associated membrane protein1 (LAMP1) or LAMP2. A limitation of this approach is that there is no protein marker specific to phagosomes, as all of their molecules identified to date are also present on other organelles such as early and late endosomes, lysosomes or the plasma membrane. Thus, the use of GFP-expressing micro-organisms appears to be the method of choice.

In this case, phagosomes are formed by infecting cells with GFP-micro-organisms for various periods of time depending on the pathogen studied. In many cases, longer periods of incubation than those needed for latex beads have to be used in order to form enough phagosomes. After infection, cells are washed to remove free micro-organisms and detached from Petri dishes for homogenization. Homogenization can be performed in the buffer described above. Cells can be broken by passages through a syringe with the appropriate needle ($22\frac{1}{2}$G can be used). Extreme care should be taken to avoid pricking. The number of strokes needed has to be determined empirically in order to break most of the cells without breaking the phagosome membrane. After centrifugation at low speed to get rid of unbroken cells and nuclei, phagosomes can be separated from the other host cell organelles directly in the FACS (see Figure 18.1).

Figure 18.1. Isolation and purification of phagosomes containing micro-organisms (*Leishmania donovani*). Analysis of the proteome of phagosomes containing various micro-organisms requires procedures allowing purification of these phagosomes from other cellular organelles. To isolate phagosomes containing pathogens such as *Leishmania donovani*, these micro-organisms can be engineered to express the green fluorescent protein (GFP). For micro-organisms where expression of foreign proteins is not possible, opsonization of their surface with fluorescent probes can be performed. Macrophages are allowed to internalize the GFP-expressing *Leishmania* for a certain period of time. Cells are then homogenized in a pH 7.4 buffer made of 250 mM sucrose, 3 mM imidazole and protease inhibitors by passages through a $22\frac{1}{2}$G needle. After centrifugation, a post-nuclear supernatant (PNS) is obtained, which contains the phagosomes housing the GFP-expressing *Leishmania* parasites. Phagosomes are then separated from the other host cell organelles in the FACS. In this instrument, the cell suspension passes through a vibrating nozzle and single particles emerge surrounded by a buffer stream. Each of the particles emits scatter and fluorescent light signals when passing through a focused laser beam. Various optical devices detect these light signals. Droplets containing phagosomes of *Leishmania* are given, for example, a negative charge, while the non-fluorescent droplets will acquire a positive charge. The different droplets will then be deflected by an electric field into collector tubes according to their charge. The highly enriched preparation of *Leishmania*-containing phagosomes is then ready for proteomic analysis.

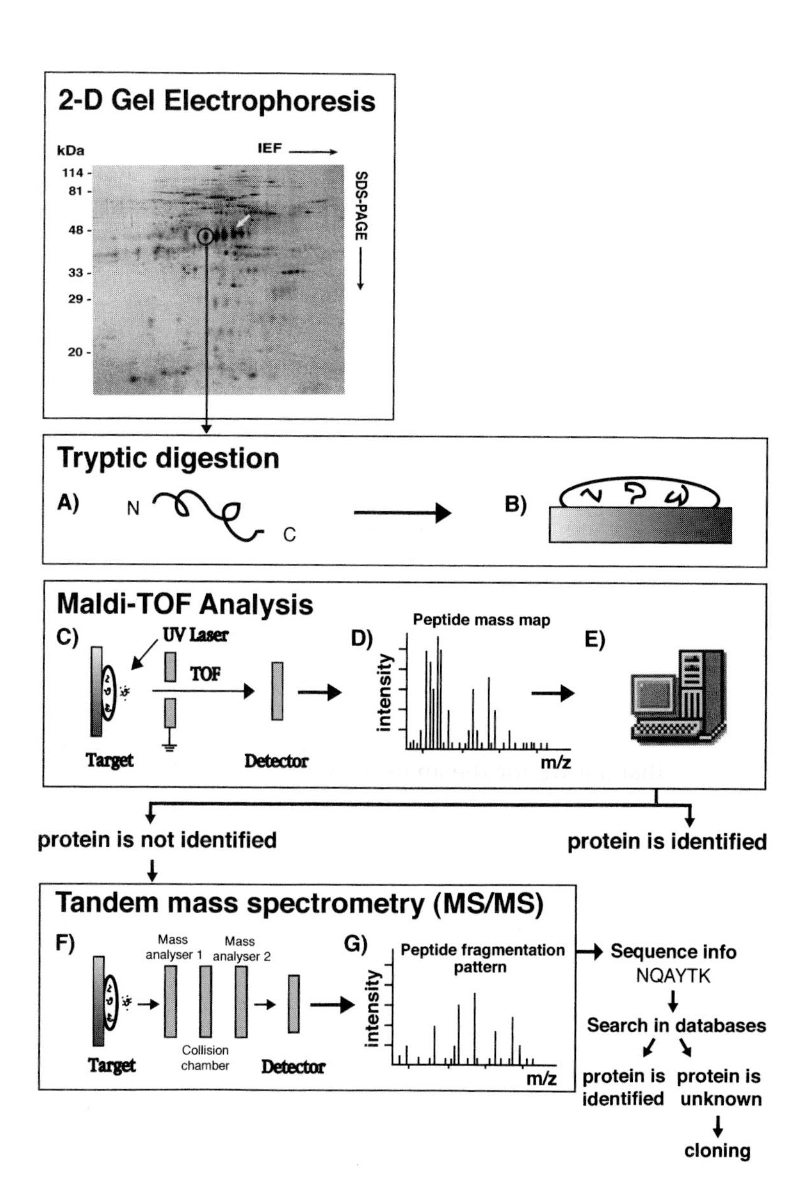

Figure 18.2. Proteomic analysis. To uncover the phagosome proteome, phagosomal proteins isolated from J774 macrophages are first separated according to their isoelectric point, using immobilized pH gradient (IPG) strips. The second dimension is performed on an SDS acrylamide gel, by standard electrophoresis techniques. The complexity of phagosomes is illustrated by the demonstration that hundreds of protein spots are associated to this organelle (top panel; the white arrow indicates the position of actin on the gel). The following panels illustrate the steps of a typical approach leading to the identification of proteins by matrix-assisted laser desorption/ionization time-of-flight mass spectrometry (MALDI-TOF MS) or tandem mass spectrometry (MS/MS) analysis. First, each spot corresponding to a protein is cut out, washed and dehydrated (A). The protein is then 'in gel' digested with trypsin, an enzyme that specifically cuts after the arginine and lysine residues, generating peptides of different masses. This mixture of peptides is incorporated in a thin layer of a UV-absorbing matrix on the mass spectrometer probe tip (B). The target (peptides + matrix) is struck by a UV laser, which converts molecules to gas-phase ions. Once ions are created, individual

Settings have to be determined empirically depending on the microorganism studied and the type of sorter used. Increasing sample-flow rates will decrease the time needed to generate enough materials for biochemical analysis but will increase the potential impurity of the samples. To decrease the level of contamination, phagosomes can be enriched first on a sucrose gradient and then sorted in the FACS. Additional steps will, however, decrease the yield. The methods described above should allow the isolation of highly enriched phagosome preparations suited for various biochemical analysis such as high resolution 2-D gel electrophoresis.

◆◆◆◆◆◆ PROTEOMIC ANALYSIS

The emergence of proteome research relies on the recent advent of powerful new methods to separate and analyze rapidly and efficiently minute amounts of proteins from complex samples. First, technical refinements in 2-D gel electrophoresis now allow one to separate thousands of proteins in a single gel in a highly reproducible manner. Second, mass spectrometry approaches have been developed to a level that allows for the analysis of proteins isolated from very small samples and matching to their corresponding genes in DNA sequence databases using dedicated software algorithms (Lamond and Mann, 1997) (see Figure 18.2). The introduction of immobilized pH gradients for isoelectrofocusing in the first dimension has enabled a great level of reproducibility not delivered with previous gel systems. This enables gels of different samples performed by different persons to be compared, a prerequisite for complex differential display experiments such as comparison of disease states. However, it is still difficult to generate 2-D database that can be screened and used by different investigators, mostly for lack of standard operating protocols. Indeed, in the last few years, several manufacturers have proposed complete proteomic systems of their

---

mass-to-charge ratios (m/z) are separated by a time-of-flight (TOF) mass analyzer: m/z ratios are determined by measuring the time it takes for ions to move from the ion source to the detector (C). The peptide mass fingerprint obtained (D) is then compared, using appropriate software, with theoretical fingerprints of proteins (or translated DNA) from databases available on the Internet, to identify the corresponding protein or expression sequence tag (EST) (E). In the case where the protein cannot be identified by this approach, the MS/MS is used to obtain more information on the peptide. In the MS/MS device, the peptides will be further fragmented into amino acids by progressing through tandem-arranged mass analyzers (F). As each amino acid displays a specific m/z ratio, the peptide fragmentation pattern obtained (G) will allow an amino acid sequence to be deduced. This sequence can then again be compared with sequences available in databases on the Internet, which will lead either to identification of the protein or to the conclusion that a new protein has been found. At that point, the protein can be cloned using the sequence information obtained from the MS/MS analysis.

own for protein separation, protein preparation and mass spectrometry analysis. The fact that several investigators use hybrid or home made systems further limits comparison between laboratories.

Despite the increases in sensitivity of mass spectrometers in recent years, proteomic analysis of subcellular fractions is still difficult. One of the limitations comes from the fact that purifying an organelle is often accompanied by low yield of materials, not allowing spot visualization in 2-D gels by conventional staining methods compatible with mass spectrometry analysis. This is likely to be the case for the isolation of phagosomes containing micro-organisms where purification methods involve several centrifugation steps leading to loss of material and disruption of the phagosomes. To circumvent this problem, it is possible to perform qualitative analysis of very small amounts of isolated phagosomes on mini 2-D gels with silver staining. Although this approach is unlikely to generate enough materials for mass spectrometry analysis, it generally allows one to generate high-resolution minigels that can be used to select the spots of interest. The next step is then to generate enough materials for the mass spectrometry analysis, a procedure that can be done by loading more materials of slightly contaminated samples, knowing the spots of interests. This is best done when narrow ranges of pH are used for the first dimension, an approach allowing the loading of more materials with a higher resolution. Recent results indicate that an approach bypassing the gel separation step for direct analysis of complex samples in the mass spectrometer, using liquid chromatography, could be used with success to analyze organelle preparations (Wu *et al.*, 2000). Indeed, gel-less approaches would limit the loss of materials in the gel matrix and during the various steps of the in-gel degradation procedure, allowing one to identify proteins from an even lower amount of materials, thus facilitating the work with subcellular fractions.

◆◆◆◆◆◆ PROTEOMIC APPROACH TO STUDY HOST–PATHOGEN INTERACTION

In recent years, it has become evident that infection by intracellular pathogens is the result of complex interaction and exchanges between the pathogen and its host cells (for a review see Méresse *et al.*, 1999). In many cases, it appears that the strategy used by the pathogen consists in the direct modulation of the host cell phagosome. Analysis of the proteome of latex bead-containing phagosomes provided direct insight into the functions of this organelle in macrophages (Garin *et al.*, 2001), and pointed at potential molecular targets for pathogens. To date, over 200 proteins have been identified on phagosomes, 140 of which were reported in the aforementioned study. Among these, several proteins not previously identified on phagosomes were found, directly indicating new functions of the organelle. For example, several proteins known to be associated to lipid rafts on the plasma membrane were present in our phagosome preparations, suggesting the presence of these structures on phagosomes. Lipid rafts have been postulated to represent a general feature of the plasma membrane of eukaryotic cells (Simons and Ikonen, 1997). Lipid-modified proteins and some transmembrane proteins are concentrated in the rafts whereas other proteins are excluded. Lipid rafts have been implicated in many important cellular processes, such as polarized sorting of apical membrane proteins in epithelial cells and signal transduction. Recent evidence further indicates that a raft-based mechanism might be involved in the sorting of soluble N-ethylmaleimide-sensitive factor attachment protein receptors (SNAREs) to the plasma membrane and in their function in apical membrane docking and fusion events (Lafont *et al.*, 1999). As this is in no way an exhaustive list of the potential functions of lipid rafts, it appears that membrane subdomains represent important sites conferring specialized properties to foci within biological membranes. Using a cell biology and biochemical approaches we were able to demonstrate that flotillin-1-enriched lipid rafts were accumulating on maturing phagosomes, suggesting that specialized functions occur at specific foci on the phagosome membrane (Dermine *et al.*, 2001). Interestingly, the formation of flotillin-1-enriched lipid rafts is inhibited by the intracellular pathogen *Leishmania* indicating that survival of this parasite in macrophage phagosomes depends on the alteration of functions taking place at lipid raft sites. The fact that an LPG-deficient mutant of *Leishmania* can be internalized in phagosomes displaying flotillin-1-enriched lipid rafts further suggests that LPG is used by the parasite to alter lipid raft formation. This hypothesis is currently being studied in our laboratory.

The proteome analysis of phagosomes also pointed to two new features of this organelle. First, we were able to show the presence on phagosomes of several molecules normally associated to the endoplasmic reticulum (ER). Although this might represent a contamination, further data suggested that ER elements could interact directly with phagosomes (Gagnon and Desjardins, unpublished results). The presence of ER molecules on phagosomes is not surprising. Recent studies have shown that micro-organisms such as *Legionella* and *Brucella* reside within their host cells in compartments displaying ER features (see Méresse *et al.*, 1999). In J774 macrophages, it is not uncommon to observe cells that had engulfed more than 25 latex beads of 1 μm in size. Despite the important surface of plasma membrane needed for the engulfment of these particles, the cell does not consume itself but rather maintains a relatively stable size. This suggests that while plasma membrane recycles rapidly back to the cell surface, membrane from an intracellular source has to be recruited to keep the particles within closed compartments, our results suggesting that the ER could be used for this task. Second, a series of molecules involved in apoptosis, as well as anti-apoptotic molecules, were identified on phagosomes. This suggests that macrophages may be able to monitor the state of the internal lumen of their phagosomes and trigger an apoptotic signal in threatening conditions. By sacrificing itself, the macrophage would limit pathogen replication and further infection.

◆◆◆◆◆◆ CONCLUDING REMARKS

Proteomics is obviously a powerful new approach for the study of host–pathogen interaction. The global analysis of phagosomes containing various micro-organisms will allow us to compare the protein content of all these organelles and identify the specific features associated to the properties of each of them. This approach should indicate why some phagosomes are fusogenic while others do not fuse with endocytic organelles, and why some phagosomes can fuse with organelles like ER and avoid the degradative pathway. Isolation of phagosomes at different times after their formation should further indicate the precise mechanisms involved in these processes by providing a constant state of the phagosome composition, and this in various cellular conditions or after various stimuli. Proteomic analysis of pathogens isolated from infected cells should provide the counter image and indicate how micro-organisms adapt and react to the hostile environment encountered in the host cells. Understanding how pathogens modulate their own composition in host cells should provide a series of potential targets for the development of new drugs or vaccines.

References

Alvarez-Dominguez, C., Barbieri, A. M., Berón, W., Wandiger-Ness, A. and Stahl, P. D. (1996). Phagocytosed live *Listeria monocytogenes* influences Rab5-regulated *in vitro* phagosome-endosome fusion *J. Biol. Chem.* **271**, 13834–13843.

Böck, G., Steinlein, P. and Huber, L. A. (1997). Cell biologists sort things out: analysis and purification of intracellular organelles by flow cytometry. *Trends Cell Biol.* **7**, 499–503.

Chakraborty, P., Sturgill-Koszycki, S. and Russell, D. G. (1994). Isolation and characterization of pathogen-containing phagosomes. *Methods Cell Biol.* **45**, 261–276.

Dermine, J.-F., Scianimanico, S., Privé, C., Descoteaux, A. and Desjardins, M. (2000). Leishmania promastigotes require lipophosphoglycan to actively modulate the fusion properties of phagosomes at an early step of phagocytosis. *Cell Microbiol.* **2**, 115–126.

Dermine, J.-F., Duclos, S., Garin, J., St-Louis, F., Rea, S., Parton, R. G. and Desjardins, M. (2001). Flotillin-1-enriched lipid raft domains accumulate on maturing phagosomes. *J. Biol. Chem.* **276**, 18507–18512.

Desjardins, M. and Descoteaux, A. (1997). Inhibition of phagolysosomal biogenesis by the *Leishmania* lipophosphoglycan *J. Exp. Med.* **185**, 2061–2068.

Desjardins, M., Huber, L. A., Parton, R. G. and Griffiths, G. (1994). Biogenesis of phagolysosomes proceeds through a sequential series of interactions with the endocytic apparatus. *J. Cell Biol.* **124**, 677–688.

Ferrari, G., Langen, H., Naito, M. and Pieters, J. (1999). A coat protein on phagosomes involved in the intracellular survival of mycobacteria. *Cell* **97**, 435–447.

Garin, J., Diez, R., Kieffer, S., Dermine, J.-F., Duclos, S., Gagnon, E., Sadoul, R., Rondeau, C. and Desjardins, M. (2001). The phagosome proteome: insight into phagosome functions. *J. Cell Biol.* **152**, 165–180.

Jahraus, A., Tjelle, T. E., Berg, T., Habermann, A., Storrie, B., Ullrich, O. and Griffiths, G. (1998). In vitro fusion of phagosomes with different endocytic organelles from J774 macrophages. *J. Biol. Chem.* **273**, 30379–30390.

Lafont, F., Verkade, P., Galli, T., Wimmer, C., Louvard, D. and Simons, K. (1999). Raft association of SNAP receptors acting in apical trafficking in Madin-Darby canine kidney cells. *Proc. Natl Acad. Sci. USA* **96**, 3734–3738.

Lamond, A. I. and Mann, M. (1997). Cell biology and the genome projects—a concerted strategy for characterizing multiprotein complexes by using mass spectrometry. *Trends Cell Biol.* **7**, 139–142.

Méresse, S., Steele-Mortimer, O., Moreno, E., Desjardins, M., Finlay, B. and Gorvel, J. P. (1999). Controlling the maturation of pathogen-containing vacuoles: a matter of life and death. *Nat. Cell Biol.* **1**, E183–188.

Pennington, S. R., Wilkins, M. R., Hochstrasser, D. and Dunn, M. J. (1997). Proteome analysis: from protein characterization to biological function. *Trends Cell Biol.* **7**, 168–173.

Scianimanico, S., Desrosiers, M., Dermine, J-F., Méresse, S., Descoteaux, A. and Desjardins, M. (1999). Impaired recruitment of the small GTPase rab7 correlates wth the inhibition of phagosome maturation by *Leishmania donovani* promastigotes. *Cell Microbiol.* **1**, 19–32.

Simons, K. and Ikonen, E. (1997). Protein and lipid sorting from the trans-Golgi network to the plasma membrane in polarized cells. *Semin. Cell Dev. Biol.* **9**, 503–509.

Wilkins, M. R., Sanchez, J. C., Gooley, A. A., Appel, R. D., Humphery-Smith, I., Hochstrasser, D. F. and Williams, K. L. (1996). Progress with proteome projects: why all proteins expressed by a genome should be identified and how to do it. *Biotechnol. Genet. Eng. Rev.* **13**, 19–50.

Wu, C. C., Yates, J. R., 3rdNeville, M. C. and Howell, K. E. (2000). Proteomic analysis of two functional states of the Golgi complex in mammary epithelial cells. *Traffic* **1**, 769–82.

19 Studying Trafficking of Intracellular Pathogens in Antigen-presenting Cells

U E Schaible and S H E Kaufmann
Max-Planck-Institute for Infection Biology, Schumannstr. 21-22, D-10117 Berlin, Germany

◆◆

CONTENTS

◆◆◆◆◆◆ INTRODUCTION

Intracellular bacteria and parasites have developed strategies which allow them to survive and proliferate inside host cells to avoid the consequences of hazardous humoral host responses. The preferred host cells of most of these micro-organisms are macrophages (Mϕ) and—to some extent—dendritic cells (DC) due to their profound capability to phagocytose and their longevity (up to 1 month) compared to granulocytes. Mϕ and DC are not only host cells for these pathogens but are also professional antigen-presenting cells (APC) able to process and present antigens for recognition by T-cells via MHC class I and II and MHC-like molecules such as CD1. DCs are the most potent APC, whereas the major task of Mϕ is the elimination of bacteria and parasites and the maintenance of an ongoing immune response. Only professional APC express MHC class II molecules, whereas MHC class I molecules are found on virtually all nucleated cells in the body.

METHODS IN MICROBIOLOGY, VOLUME 31
ISBN 0-12-521531-2

The group of intracellular pathogens comprise important human pathogens such as the tubercle bacillus, *Mycobacterium tuberculosis*, opportunistic mycobacteria, e.g. the *M. avium* complex, the agents of legionaires disease, *Legionella pneumophila*; listeriosis, *Listeria monocytogenes*; typhoid fever, *Salmonella enterica*; trachoma, *Chlamydia trachomatis* and leishmaniosis, *Leishmania mexicana*, *L. major*, *L. donovani*, and others. These pathogens reside in distinct intracellular compartments: mycobacteria prefer the early, leishmaniae the late endosomal stage of the phagosome, whereas legionellae and chlamydiae direct their phagosomes out of the conventional maturation pathway and listeriae lyse the phagosomal membrane to escape into the cytoplasm (Schaible *et al.*, 1999a). The intracellular localization influences the immune response elicited and principally also the response important for protection. Microorganisms which remain in the phagosome mainly induce MHC class II restricted CD4 positive T-helper cells (Th), whereas the cytoplasm-bound listeriae induce MHC class I restricted CD8 positive cytotoxic T-cells (CTL).

MHC class I antigens are processed to mainly nonameric peptides inside the cytoplasm by proteasomes and loaded onto MHC class I molecules after transfer via transporters-associated-with-antigen- presentation (TAP) into the endoplasmic reticulum (ER). MHC class II molecules are loaded within endosomes in the APC with peptides of 14 to 20 amino acids (AA) generated by endosomal/lysosomal proteases. Similarly, the non-polymorphic group I CD1 molecules in man (CD1a, CD1b, CD1c) and in part also the non-classical MHCIb molecules, are loaded with their antigens inside endosomes. In contrast to MHC molecules, CD1 molecules present glycolipid antigens to T-cells.

When intracellular pathogens dwell inside these APC, they interfere with several functions of these cells to establish their intracellular habitat. At the same time they risk that their antigens are recognized by T-cells. In order to understand which immune response participates in protection against these agents, their intracellular fate and that of their antigens deserves the interest of experimental researchers. This chapter describes experimental approaches to tackle the trafficking of intracellular pathogens using microscopical and biochemical methods.

◆◆◆◆◆◆ ANTIGEN-PRESENTING CELLS

Murine bone marrow derived macrophages (BMMϕ)

Mice are killed by cervical dislocation and bone marrow cells are flushed out of femura and tibiae using a syringe with a 23G gauge (Braun, Melsungen, Germany). Bone marrow cells are cultured in bacterial petri dishes at $3–5 \times 10^6$ cells per dish in 10 ml DMEM (Biochrom KG, Berlin, Germany) containing $3.7\,g\,l^{-1}$ Na_3CO_3, $4.5\,g\,l^{-1}$ glucose, 10 mM Hepes, 10 mM glutamine, 1 mM pyruvate, 10% heat inactivated (h.i.) fetal calf serum (FCS; Biochrome), 5% h.i. horse serum (HS; Biochrome), and 20% culture supernatant (CSN) from L929 cells (L929 CSF cell line derived from ATCC, Manassas, Va). Differentiation takes 6 days at 37°C/7% CO_2.

BMMϕ can be recovered from the dishes by replacing the medium with ice-cold phosphate buffered saline (PBS; Biochrome) and keeping the BMMϕ at 4°C for 30 min. Mϕ can easily be removed by gentle scraping using a rubber policeman or by vigorously flushing the cells off the plate using ice-cold PBS. The cells can then be used for further experiments. It should be pointed out that all media and supplements should be free of lipopolysaccharide (LPS). Cell culture media, FCS, HS and PBS should be chosen or made up to contain as low amounts of LPS as possible. Less than 0.01 pg LPS or endotoxin units ml^{-1} as tested by the Limulus assay (Sigma, St. Louis, MO) are recommended. To handle the cells, only plastic pipettes should be used. We also recommend testing several batches of medium, FCS and HS from various commercial providers since we found marked variability from batch to batch in supporting Mϕ differentiation as well as activation.

Resident and elicited peritoneal macrophages

In order to elicite Mϕ, mice are injected intraperitoneally (i.p.) 4 days before harvest using either one of the following agents: thioglycollate (10 mg ml^{-1} in water, sterile filtered and aged for at least 1month) or sodium periodate (1mg ml^{-1} in 0.5 ml PBS). To harvest peritoneal cells, mice are killed by cervical dislocation and 10 ml of PBS are carefully injected (18G gauge) into the lower peritoneum and recovered using the same gauge from the upper bit of the blown up peritoneum. We recommend using female mice due to the fact that male belly buttons rupture more easily during the washing procedure. Cells are washed and cultured in DMEM containing 10% FCS (7% CO_2/37°C). The phenotypes of resident Mϕ vs. those elicited by the different compounds differ in their phenotype. Elicited macrophages are more activated and respond more promptly by respiratory burst to PMA or bacterial products. For further details see Alford *et al.* (1991) and Gordon (1995).

Human peripheral blood monocytes (PBMC)

Human PBMC are derived from heparinized blood or buffy coats (2× diluted in PBS) from blood donors by consecutive ficoll/percoll gradients. Cells are resuspended in 10 ml of sterile PBS and carefully layered on top of a bed of 15 ml Ficoll (density: 1077; Biochrome) in a 50 ml tube. The gradient is spun at 2000 rpm (800*g*) for 20 min at 18°C with the brakes off (Megafuge; Heraeus-Kendro, Hanau, Germany). The cell-containing interface is removed using a Pasteur pipette, resuspended in 10 ml of PBS and layered on top of 15 ml Percoll (56% in PBS; density 1.124; Biochrome). The gradient is spun at 18°C for 20 min at 2200 rpm (Megafuge, Heraus) with the brakes off. The cell-containing interface is removed using a Pasteur pipette, resuspended in PBS and the Percoll is removed by two washings in PBS. Finally the cells are resuspended in PBS and seeded into petri dishes at 1×10^7 per dish. After 2 h at 37°C, non-adherent cells are removed and RPMI containing 10% FCS is added for further culture at 5% CO_2 at 37°C.

Dendritic cells (DC)

Murine bone marrow cells are seeded in petri dishes in DMEM (10% FCS, glutamin, 20% GMCSF-containing and 10% IL4-containing CSN of NIH cells transfected with the gene for the respective cytokine, non-adherent cells are removed after 2 h and adherent cells are further cultured at 37°C, 7% CO_2 for 3–6 days (Inaba *et al.*, 1992). Human DC are generated from PBMC by culturing them in RPMI containing 10% FCS and 100 U ml^{-1} rhIL-4 and rhGMCSF (Preprotech, Rocky Hill, NJ) (Romani *et al.*, 1996). All DC are analysed by FACS for the expression of MHC class II high (human clone L243, ATCC; mouse clone Tib120; ATCC), CD1a (Okt6; ATCC), CD1b (Serotech; Oxford, GB) and B7.1 or 2 (Pharmingen BD, Heidelberg, Germany) using specific monoclonal antibodies (mAb).

◆◆◆◆◆◆ INFECTION

Infection of APC is performed in normal culture medium as described above. For most pathogens a multiplicity of infection (MOI) of 5 : 1 to 10 : 1 is sufficient for a decent infection rate although in the case of virulent salmonella and *M. tuberculosis*, an MOI of 2 : 1 should not be exceeded to avoid excess killing of host cells by apoptosis. In infection experiments with leishmania, salmonella, listeria and legionella, the extracellular organisms in the cultures are removed 30 min after infection, mycobacteria are left on the APC for at least 2 h to achieve an appropriate infection rate. Apart from parasites and mycobacteria, the addition of 5 μg ml^{-1} gentamycin to the cultures impairs the growth of extracellular bacteria such as salmonellae and listeriae without affecting their intracellular growth. Although leishmania parasites are infective for APC in both the pro- and the amastigote stages, in species where amastigote-like forms can be generated *in vitro* (*L. mexicana*, *L. donovani*), these stages should preferably be used for *in vitro* infections (Hodgkinson *et al.*, 1996). They are generated by culture for at least 2 days in the following UM54 medium: 200 ml M199 (Sigma), 0.5 glucose, 10 g trypticase, 0.15 g glutamine, 1.49 g Hepes, 250 μl hemin, 40 ml FCS. Furthermore, for mycobacteria, the addition of 5% horse serum to the cultures has been found to enhance the infection efficiency up to 95%. This is most probably due to the presence of the complement component C2 in horse serum (Schorey *et al.*, 1997). For virulent salmonellae, culture conditions can influence the type of uptake by the APC: standing overnight cultures produce an invasive phenotype whereas bacteria cultured under shaking conditions are preferentially phagocytosed.

◆◆◆◆◆◆ MICROSCOPY VS. BIOCHEMISTRY

Pathogen-containing compartments can be studied either by various microscopical methods such as confocal or immuno electron microscopy or by biochemical analyses of isolated subcellular fractions of infected APC including purified pathogen-containing phagosomes. The advantage of microscopical over biochemical analyses is that the former can be performed independently of the physical properties of the phagosomes. In contrast, subcellular fractionation allows better quantification. Intact *L. mexicana*-containing phagosomes for example cannot be isolated after infection time has exceeded 10 h due to their spacious morphology whereas microscopy can be performed over several days of infection. Furthermore, using microscopy individual compartments in single cells can be studied whereas subcellular fractionation provides data from enriched vesicle fractions of whole cell populations.

In both approaches the respective compartments are characterized using antibodies for intracellular markers and/or soluble tracers which label specific compartments or vesicular stages. Table 19.1 describes a selection of antibodies and tracers of interest for studies with intracellular pathogens. Using antibodies, the steady-state situation of a cell population can be analysed whereas tracers can be used to study the dynamics of intracellular compartments by employing pulse/chase protocols. Tracers are either fluid phase molecules engulfed by micro-/macropinocytosis or natural ligands for cellular receptors or membrane lipids which are endocytosed, transported to and enriched in specific intracellular compartments. For confocal microscopy tracers are labelled with fluorescent dyes, for immunoelectronmicroscopy and biochemical analysis with biotin or chemical groups which allows the use of specific antibodies for detection such as dinitrophenol (DNP), fluorescein or digoxigenin (Boehringer Mannheim; Molecular Probes, Junction City, OR). It should be pointed out that we found that biotinylated tracers combined with fluorescently labelled streptavidin did not give satisfactory results for fluorescent/confocal microscopy due to high background. All labelled tracers which bind to specific receptors have to be controlled for receptor specificity by incubating cells in a 10 × surplus of unlabelled tracer which should compete for binding and abolish uptake of the labelled tracer. Transferrin labelled with excess dye for example will end up in late instead of in early endosomes. Most tracers are commercially available from several companies (Molecular Probes; Sigma). Testing for receptor specificity is also recommended for these products. We usually generate our own labelled transferrin using the following protocol: Holotransferrin (Calbiochem, Bad Soden, Germany) is dissolved in 1 ml 0.2 M sodium bicarbonate buffer (7 mg ml^{-1}; pH not adjusted, ~pH 8.3) and mixed with 1 mg N-hydroxy-succinimidester derivatives of fluorescein, biotin or digoxigenin (Boehringer Mannheim). The gemisch is incubated for 1 h at room temperature turning head over end, followed by dialysis overnight against 2 l of PBS using a membrane (cut off 10 kDa). Similarly, low-density lipoprotein (LDL), α2 macroglobulin (α2M), ovalbumin or BSA can be labelled. Labelling with Texas Red requires a higher protein to dye ratio (10/1) and labelling should be performed at pH 9.0 in 35 mM borate-saline buffer.

Table 19.1 Tracers and antibodies of interest

Marker	Pulse/chase	Subcellular localization
Tracer		
Transferrin	45 min pulse	Early endosomes
ManBSA, LDL, α2M	1 h pulse	Early to late endosomes
ManBSA, LDL, α2M	1 h/1 h pulse/chase	Late endosomes to lysosomes
Ovalbumin, Dextran, HRP	16 h/4 h pulse/chase	Lysosomes
Antibodies		
LAMP-1		Early endosomes to lysosomes in increasing amounts
Mannose receptor[2]		Early endosomes
Transferrin receptor		Early/recycling endosomes
Early endosomal antigen 1 (EEA1)[1]		Recycling endosomes
MHC class II		Late endosomes, lysosomes, surface
α Mannosidase		Golgi apparatus
Rab 5[1]		Early/recycling endosomes
Rab 7[1]		Late endosomes
Cathepsin D/B		Late endosomes, lysosomes
Calnexin[1]		ER
Protein disulfide isomerase[3]		ER
Dyes		
Hoechst dye[5]		Nuclei
Phalloidin[4]		Actin filaments

[1] Transduction Lab., Lexington, KY; [2] Pharmingen; [3] Stressgene, Victoria, Canada;[4] Molecular Probes; [5] Sigma.

◆◆◆◆◆◆ TRACERS FOR THE ENDOSOMAL SYSTEM

In several studies, we used labelled transferrin to trace early endosomal compartments. Transferrin is engulfed by the transferrin receptor and delivered to the early compartments where transferrin releases iron, separates from the TFR and recycles to the cell surface. Since transferrin is recycled, true enrichment in early endosomes is not achieved and therefore, the cells should be incubated for 5–20 min at a concentration of 5 to 50 μg ml^{-1} depending on the cell type used (Sturgill-Koszycki *et al.*, 1996; Schaible *et al.*, 1998, 2000). Labelling should take place in transferrin-free media. We use Ringer solution for this purpose: 155 mM NaCl, 5 mM KCl, 2 mM $CaCl_2$, 1 mM $MgCl_2$, 2 mM NaH_2PO_4, 10 mM Hepes, 10 mM glucose, pH 7.2.

Late endosomal compartments can be marked using labelled fluid phase molecules such as dextran or ovalbumin in complete medium at a concentration of 100 μg ml^{-1}. The classic tracer to label late endosomes/lysosomes for electron microscopy is horseradish peroxidase (HRP; Sigma) which can easily be employed for biochemical analysis of subcellular fractions using a peroxidase substrate and an ELISA reader or after SDS-PAGE and Western blotting using the ECL system and film. HRP is predominantly taken up via the macrophage mannose receptor (MMR) which trafficks to early endosomes to deliver its ligand into the late endosomal/lysosomal pathway (Lang and DeChastellier, 1985). Mannosylated BSA (Sigma) which is also bound by the MMR can be used in a similar way (Schaible *et al.*, 2000). Both tracers are used at 50 to 100 μg ml^{-1} concentration and labelling can be done in complete medium. Other tracers to label late endosomes/lysosomes include labelled LDL (Calbiochem) and α2M (Calbiochem) which bind to their respective receptors. Which endosomal stage is labelled using these tracers depends on the time used for the pulse/chase periods. Labelling of lysosomes requires at least a 2 h pulse followed by a 2 to 4 h chase period. To get satisfactory enrichment of the tracer in the lysosomes, 4 h pulsing followed by an overnight chase is recommended. Shorter chase times such as 30 min to 1 h still label late endosomes as well whereas no chase covers the full early to late endosomal to lysosomal continuum (Table 19.1).

Cells are either placed on coverslips in a 24- or 6-well plate or in multichamber slides (Nunc, Roskilde, Denmark) for these studies. Upon infection and tracer loading, the cells are washed twice in PBS, fixed using 4% paraformaldehyde in PBS, permeabilized using 0.1% Triton-X100 (Sigma) in PBS for 5 min at 4°C and blocked in PBS containing 5% h.i. goat serum and 0.05% Tween 20 for 30 min at room temperature (r.t.). For immunofluorescence, cells are incubated with the respective primary antibodies for 30 min followed by 10 washings in PBS, and secondary antibodies for another 30 min. It should be noted that many rabbit sera and secondary antibodies from rabbit, goat or donkey are generated using complete Freund adjuvant containing mycobacteria. Similarly, animals used for immunization could be carriers of the pathogens investigated or closely related species. Therefore, these antibody preparations usually contain (myco-)bacteria-specific antibodies. We recommend (i) to use affinity purified rabbit antibodies only and (ii) to absorb polyclonal secondary reagents by incubation with the respective pathogens for 30 min at 4°C followed by a spin at 13 000 rpm in a microfuge to remove bacteria. We generally use secondary reagents from Jackson Immunoresearch Lab. (West Grove, PA) with satisfactory results. As intracellular markers, we generally use antibodies against the transferrin receptor (Mouse, Tib 219; human L5.1, both ATCC), LAMP-1 (mouse ID4B; human H4A3, both from the Developmental Hybridoma Bank, NIH, Bethesda, MD) and cathepsin B or D (Calbiochem).

In order to study the access of plasmamembrane material to the phagosome, the host cell surface can be labelled using either NHS derivatives of fluorescein, digoxigenin or biotin. The culture medium is removed, cells are washed twice in cold Hanks buffered salt solution

(HBSS; Gibco, BRL, Karlsruhe, Germany) and covered with a small amount of cold HBSS containing 200 μg ml^{-1} of the NHS derivative (5 mg ml^{-1} stock in DMSO) and placed on a slowly rocking platform in the cold for 10 to 30 min. Cells are washed twice in HBSS and further cultured in medium for various time periods.

◆◆◆◆◆◆ TRACERS TO STUDY CYTOPLASMIC ACCESS

In order to study the fate of luminal material during phagosome maturation coloading of tracers during infection has been employed. For labelling, the membrane impermeable green fluorescent dyes 8-hydoxy-pyrene1,3,6-trisulfone acid (HPTS; Molecular Probes) and Lucifer Yellow (Molecular Probes) as well as the tracers dextran (Molecular Probes) or ovalbumin labelled with different fluorophores or biotin are used (Beauregard *et al.*, 1997; Schaible *et al.*, submitted). During infection these dyes/tracers are added at 5 mM (HPTS) or 1 mg ml^{-1} (Lucifer Yellow, dextran, ovalbumin). The transfer of the dyes to the cytoplasm is followed by microscopy or dot blot/Western blot upon subcellular fractionation, respectively. Note, that in addition to the phagosomes other endocytic vesicles are labelled during the incubation time.

To study access of cytoplasmic material to phagosomes, tracers can be delivered to the cytoplasm of infected APC. This can be achieved on the one hand by microinjection using a microinjection apparatus (Eppendorf, Hamburg, Germany). This method allows study of only a small number of cells at once. In order to study whole cell populations, loading of tracers into the cytoplasm by ATP, glass beads or scraping/syringe treatment has been employed. Using Lucifer Yellow (Sigma) or labelled dextrans of different sizes (Molecular Probes), these techniques were employed to study the access of cytoplasmic material to endosomes, autophagosomes or parasitophorous vacuoles containing *L. mexicana* (Steinberg *et al.*, 1988; McNeil and Warder, 1987; Schaible *et al.*, 1999b). Lucifer Yellow is a green fluorescent low-molecular weight organic anion (800 Da) which is transported via organic anion transporters present in late endosomes (Steinberg *et al.*, 1988). These transporters which also transfer small peptides, antibiotics and toxins belong to the multi-drug-resistance gene family of ABC-transporters and are inhibited using either Probenecid (5 mM) or Gemfibrocil (0.5 mM; Sigma). Therefore, Lucifer Yellow can be loaded into cells using ATP: 1 mg ml^{-1} dye is incubated together with 5 mM ATP in DMEM at 37°C for 5 min followed by three washing steps. This technique leads to a labelling efficiency of almost 100% and can also be used for non-adherent cells. Using mechanical techniques to inject the dyes via microlesions in the plasma membrane, less cells are loaded and loading efficiency depends on the size of the tracers (Schaible *et al.*, 1999b). For glass bead loading, the tracers are dissolved in cold Ringer solution and placed onto the APC grown either on glass coverslips or in Petri dishes. Around 80% of the cell monolayer is covered with a layer of acid washed and autoclaved glass beads (400 μm diameter; Sigma). The beads are allowed to roll over the cell monolayer five times before the cells are washed ten times in ice-cold Ringer solution (alternatively FCS-free medium) and further incubated in complete culture medium at 37°C for the time period desired. Alternatively, the cells are covered with the tracer solution scraped off the dish using a rubber policeman and passed three times through a 25 gauge tuberculin syringe. Cells are pelleted and washed three times in complete culture medium. The latter method can also be used for non-adherent cells. Small organic anions such as Lucifer Yellow are transported via specific transporters into late endosomes within 30 min whereas larger molecules are collected by autophagosomes within the next 10 to 24 h of incubation (Schaible *et al.*, 1999b). We found the addition of 1% Pluronic F-68 improved the numbers of living cells recovered after loading. Uptake of regurgitated material can be avoided by subsequently changing the medium twice.

◆◆◆◆◆◆ ANALYSES OF PURIFIED INTRACELLULAR COMPARTMENTS

Separation of intracellular vesicles

Various methods have been established to isolate fractions of intracellular compartments such as Golgi, ER membranes, mitochondria or light versus dense endosomal compartments such as lysosomes. There are two principles: the first approach is based on compounds which form layers of defined densities depending on their concentration like sugar polymers such as sucrose or ficoll, metrizamide or silica gels such as percoll and can be used to perform density gradient centrifugations. The second approach takes advantage of the differential electric charges of distinct membrane vesicles such as free flow electrophoresis (FFE). Since FFE depends on elaborate equipment this method will not be discussed here but a method combining the benefits of both techniques called density gradient electrophoresis (DGE) will shortly be described (Lindner, 2001).

Host cells such as Mϕ are grown to a complete monolayer and lysed in 5 ml ice-cold homogenization buffer (20 mM Hepes or 10 mM triethanolamine/acetate buffer both pH 7.4, containing 250 mM sucrose, 0.5 mM EGTA, 0.01% BSA or gelatine; Sigma). To avoid extensive proteolytic degradation these procedures have to be performed on ice. In addition, a mix of the following protease inhibitors is added to the homogenization buffer in a 1 in 1000 dilution TLCK (stock 100 mg ml^{-1}), Pepstatin A (stock 50 mg ml^{-1}), Leupeptin (stock 100 mg ml^{-1}) and E64 (stock 25 mg ml^{-1}; all purchased from Sigma and diluted in DMSO). Lysis is achieved by forcing cells 20–40 times through a 27G gauge syringe or a glass douncer (15 ml vol.; Wheaton, USA) with a tightly fitted pistil. The lysis efficiency has to be monitored microscopically and the procedure should be terminated when > 80% free nuclei are visible. Lysis is followed by three post-nuclear spins at 1000*g* (Heraeus Megafuge) to remove nuclei. The post-nuclear supernatant (PNS) is then further separated. Note the following remarks: EGTA can lyse the nuclear membrane releasing DNA, which is sticky and influences the viscosity of the lysate as well as the motility of the vesicles in the gradient. Treatment of PNS with 1 U DNAse for 30 min at 37°C and another spin at 2200*g* can help in this case. Gelatine or BSA interferes with protein determination or SDS gel electrophoresis if a global protein staining is used such as silver staining or Coomassie. In both situations buffers without these compounds should be applied first.

Percoll density gradients are a simple and, in most cases, sufficient way to separate distinct intracellular vesicle populations (Figure 19.1). Percoll will form a continuous gradient during centrifugation. We use 20% or 30% Percoll in either a triethanolamine or HEPES buffer containing 250 mm sucrose and 0.5 mM EGTA. Figure 19.1a shows a 20% Percoll gradient where each each layer was measured using a refractometer (SPER Scientific). Alternatively, the Percoll density can be tested by density marker beads (Amersham-Pharmacia, Little Chalfont, Buckinghamshire, GB; Figure 19.1c). 2 ml of the lysate is carefully layered on top of 10 ml Percoll solution and the tubes are spun for 21 min at 18 000 rpm (25 000*g*, SS34 rotor; Sorvall RC5B, Kendro, Hanau, Germany) with the brakes off. The fractions are collected from the gradient by carefully removing 1 or 2 ml fractions from the top using a Pasteur pipette. Alternatively, commercially available fraction collectors can be used. Percoll has to be removed from the fractions using a 100 000*g* spin in an ultracentrifuge (Beckman Optima TLX Table Top; Beckman, Palo Alto, CA). The vesicles form a skin-like layer on top of the Percoll pellet, which can be removed by

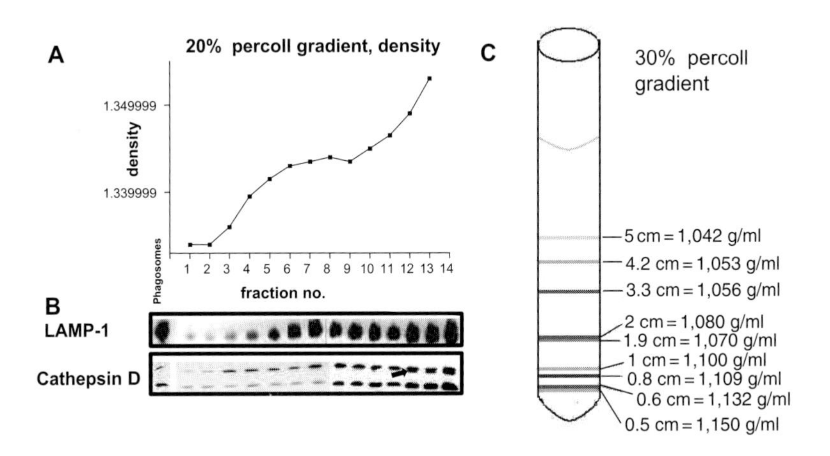

Figure 19.1. (a) Fractions from a 20% Percoll gradient were analysed for their density using a refractometer. (b) These subcellular fractions from mycobacteria-infected APC were tested by Western blot. Phagosomes were isolated from human DC infected with mycobacteria (*M. bovis* BCG) for 2 days using a sucrose step gradient. Subsequently, the 20% Percoll gradient was performed to purify vesicles from the phagosome-deprived lysate. Individual fractions were analysed by SDS PAGE and Western blot using antibodies to LAMP-1 and Cathepsin D (arrow points to the proform of the enzyme). Equal amounts of protein were loaded onto the gel. (**c**) The density distribution in a 30% percoll gradient after a 21 min spin at 25 000 g is shown using density beads (Pharmacia).

careful resuspension in plain buffer. Fractions can be stored at −70°C for further analysis. We recommend freezing down several aliquots of each fraction since more than one analysis has to be performed. We routinely test the fractions for protein content using the BCA assay kit (Pierce; Rockford, IL) and the lysosomal enzymes β galactosidase or β glucosaminidase. Other enzyme tests for plasma membrane (phosphodiesterase), Golgi (α mannosidase) and cytoplasm (lactate dehydrogenase; Sigma) can also be used (Appendix 19.1; Storrie and Madden, 1990).

Density electrophoresis has been used to separate vesicles including phagosomes from infected Mϕ (Beatty and Russell, 2000; Hasan *et al.*, 1997) in one go as well as intracellular compartments such as those involved in MHC class II antigen processing (MIIC) or Golgi apparatus (Lindner, 2001). An electrophoresis chamber is built consisting of two buffer containers which are connected by rubber sealed holes with a diameter to fit 50 ml pipettes. The upper has the anode, the lower the cathode consisting of wire loops around the chamber's inner walls. The PNS is pretreated with DNAse and the protein content is determined. The PNS is treated using soy bean trypsin (25 mg mg^{-1} protein in the PNS) for 5 min at 37°C followed by a four times excess of Soy bean trypsin inhibitor (Sigma). The gradient is prepared in 12 cm pieces cut from 50 ml pipettes which are closed at the bottom using a dialysis membrane (cut off 10 kDa) and placed into the rubber sealed holes of the chamber. The lower half is filled with 10 ml of 12% Ficoll (70 kDa) in homogenization buffer followed by 2 ml of the PNS which is adjusted to 10% Ficoll. This layer is followed by a continuous gradient generated from 8 (5 ml) to 0% (5 ml) Ficoll in ice-cold homogenization buffer using a gradient maker (Pharmacia). Both chambers are filled with ice-cold homogenization buffer in a way that the upper end of the pipette is covered and the lower end reaches into the buffer. To avoid disturbance of the gradient in the upper chamber a filter paper should be placed on top of the tubes. Electrophoresis is performed in a cold room at 5 mA (160–250 V) for 2 h. The chambers should be hooked up to a tubing system connected with a buffer reservoir and a peristaltic pump to allow slow recirculation of the buffer from the lower to the upper chamber. Finally, 1 ml fractions are collected from the top of the gradient using a Pasteur pipette or a fraction collector and the vesicles are enriched by a 1 h spin at 100 000*g* using a table-top ultracentrifuge (Beckmann).

Pathogen-containing phagosomes

Pathogen-containing phagosomes are purified from infected cells using step gradients. We will describe phagosome preparations for mycobacteria, leishmaniae, listeriae and magnetic beads. These protocols can be altered to adjust them to the isolation of different types of pathogens or particles. The most critical parameters for different conditions are listed as following. (i) Infection time and lysing conditions: phagosomes may become more susceptible to mechanical rupture or osmotic changes during infection. Thus *L. mexicana* vacuoles can only be isolated without significant loss of luminal material within the first 10 hours of infection.

It can be extrapolated that the spacious phagosomes of salmonellae or the inclusion bodies of chlamydiae are similarly sensitive in this matter. (ii) Size and density of the pathogens: parameters to be altered include the concentration of the gradient material. In our protocols the diametral characteristics of the large and loose leishmanial vacuoles vs. the small and tight mycobacterial phagosomes represent polar forms and therefore allow determination of appropriate conditions. (iii) Cytoskeleton assembly: some phagosomes induce aggregation of cytoskeleton compounds such as actin around the membrane. Since it is difficult to get rid of this cargo which influences the buoyancy of the phagosomes, trypsin treatment of the sample is probably needed before layering it on the gradient.

Mycobacterial phagosomes have been isolated from Mϕ and DC at time points p.i. as long as 10 days in the case of *M. avium* and BCG phagosomes (Sturgill-Koszycki *et al.*, 1996; Schaible *et al.*, 1998, Schaible, unpublished data; Figure 19.1a,b). In contrast, *M. tuberculosis* phagosomes can only be isolated between 2 and 4 days p.i. (Dietrich *et al.*, 2000). It should also be noted that the procedure to isolate *M. tuberculosis* phagosomes meets proper safety requirements. Thus, isolation has to be done in an S3/P3 facility using S3P3 laminar flow and personal protection to avoid exposure to aerosols, which occur during lysis. To reduce the aerosol release into the flow area we use 50 ml plastic tubes (Falcon) with a hole in the lid and a 20 ml syringe tightly fitted into the hole. In addition, the syringe is fixed to the lid using parafilm. Infected cells are scraped off the flasks in homogenization buffer containing 250 mM sucrose and lysed until around 80% of the nuclei are freed (see above). Lysis is followed by three post-nuclear spins at 200*g* for 8 min in a Megafuge (Heraeus). Note that preparing PNS from infected cells requires a lower *g* force to retain the phagosomes in the PNS. Because mycobacterial phagosomes tend to gather around the nucleus up to 5 additional washings of the nuclear pellet in homogenization buffer are required to free most of the phagosomes. All supernatants are pooled and 3 to 6 ml are layered on top of a step gradient consisting of 2 ml 50% sucrose followed by 2 ml of 12% sucrose in the respective buffer (20 mM Hepes or 10 mM triethanolamine/acetate; see above) using 15 ml conical tubes (Falcon). The gradient is spun at 800*g* for 45 min with the brakes off (Megafuge, Heraeus). After the spin, the 12%/50% interface, visible as a turbid ring containing most of the phagosomes, is retrieved. This is done by sticking a gauge through the tube wall at the 2 ml mark and slowly pulling 2 ml. Alternatively, the upper layers are carefully taken off the gradient down to the 3.5 ml mark using a Pasteur pipette connected to a pump. The phagosomes are collected by diluting the 12%/50% interface 3× in sucrose-free homogenization buffer and spun at 100 000*g* using a table-top ultracentrifuge (Beckmann). For further purification the phagosome-containing fraction is placed in a 15 ml conical tube (Falcon) and under-layered with 10% Ficoll (70 kDa, in 5% sucrose-containing homogenization buffer). This gradient is spun at 1400*g* for 40 min in the cold. The gradient is removed using a pump and the pellet is resuspended in buffer for further analysis. In order to further separate other intracellular compartments from infected cells, the lysate on top of the phagosomal phase is collected with a Pasteur pipette and diluted in homogenization buffer to around 8.55% sucrose. To remove remaining bacteria, the lysate is spun for 15 min at 4–5000 rpm (Heraus Megafuge) The bacteria-free supernatant can then be further separated using a Percoll gradient or density gradient electrophoresis (see above).

Leishmania phagosomes are separated from infected macrophages not later than 10 h post-infection to avoid extensive destruction of the parasitophorous vacuoles (Chakraborty *et al.*, 1994; Schaible *et al.*, 1999b). Infected macrophages are lysed in homogenization buffer by pushing the lysate through a 27G gauge and a syringe. To remove nuclei the lysate is submitted to 3 consecutive post-nuclear spins at 200*g* for 8 min. The resulting PNS is loaded on top of a gradient consisting of 20%, 40% and 60% sucrose in 30 mM Hepes buffer containing 100 mM KCl, 0.5 mM $CaCl_2$, 0.5 mM $MgCl_2$ (pH 6.5) and spun for 25 min at 700*g* with the brake off (Heraeus Megafuge). The phagosomes are collected from the 40% to 60% turbid interface diluted 1 : 2 in buffer before pelleting them by a spin at 12 000 g for 25 min in an ultracentrifuge (Beckman).

Similar to leishmania phagosomes, phagosomes containing listeriae and other intracellular bacteria and particles have been purified using a sucrose step gradient and the reader should refer to the respective publications (Collins *et al.*, 1997; Alvarez-Dominguez *et al.*, 1997; Desjardins *et al.*, 1994a,b). Magnetic beads (Dynatech) have been used to study the general fate of particles upon phagocytosis by Mϕ (Sturgill-Koszycki *et al.*, 1996). Bead phagosomes are prepared and isolated as follows. Tosyl activated magnetic beads (2 µm diameter; Dynal, Lake Success, NY) are washed using a magnet and covalently coated with the proteins of choice for 2 h in PBS turning head over. We usually use human IgG (Sigma) to target the beads rapidly into late endosomal/lysosomal compartments. The beads (less than 3–5 per cell) are placed onto the cells in the cold and incubated for 5 min, washed, warmed to 37°C and further cultured for the time required for the chase. By 1 h most of the beads have entered lysosomes. Subsequently, cells are lysed using a gauge and the supernatant is spun through a 30% Ficoll (400 kDa; Sigma) cushion in homogenization buffer at 300*g* for 10 min with the brake off (Heraeus Megafuge). The supernatant is removed and the pellet containing most of the phagosomes is resuspended in homogenization buffer (250 µl), placed into a flat bottom 2 ml microfuge tube (screw cap, Sarstedt). The supernatant is underlayered using a syringe with 500 µl 30% sucrose in homogenization buffer and the tube is placed on a magnet (Dynal) in ice to suck the phagosomes through the high sucrose cushion (~5 min). This procedure is repeated three times before the phagosomal pellet is further analysed.

◆◆◆◆◆◆ T-CELL ASSAYS

T-cell functions are generally assessed by proliferation assays, cytotoxicity assays and ELIspot or ELISA assays to determine cytokine production.

They are described in detail elsewhere (Coligan *et al.*, 2000; Kaufmann and Kabelitz, 1998). Usually murine Mϕ or DC are used as APC which are seeded into 96-well plates at 1×10^5 per well containing RPMI (5% FCS, 10 mM glutamate, 10 mM mercaptoethanol, 50 mM pyruvate). Cells are either infected for 24 to 48 h or pulsed with purified antigens for 24 h ($< 2\,\mu g\,ml^{-1}$) or antigenic peptides for 2 h ($< 100\,ng\,ml^{-1}$). Spleen cells are isolated from infected or vaccinated animals, single cell suspensions are prepared and T-cells or T-cell subpopulations such CD4 or CD8 T-cells separated using specific antibodies and magnetic bead isolation procedures (Miltenyi, Cologne, Germany; Dynatech). T-cells are seeded onto APC (1×10^4 to 1×10^5 per well) and incubated at 37°C at 5% CO_2. Cultures are observed microscopically each day for cellular activation as shown by the appearance of blast cells in the culture. Similarly, T cells from peripheral blood of patients or blood donors separated by Ficoll gradient as described above are used with APC from the same donor. Upon activation, the supernatant is harvested for analysis of cytokine production using ELISA or bioassays. The supernatant is replaced by medium containing 12 μCi ^{3}H-thymidine, further cultured for 16 h and the cells are harvested onto glass filters and thymidine incorporation is analysed in a beta counter.

A number of drugs can be used which influence intracellular trafficking and/or antigen processing and presentation. These inhibitors are added to the respective T-cell assay and the culture time is usually determined by the toxicity of the compounds. We recommend to first determine the concentration of the respective compound which does not kill the cells. Therefore, the concentrations given are only hints. (i) To raise the intracellular pH, lysosomotrophic bases can be used such as NH_2Cl (3 mM) or chloroquine (20 μM). These compounds block processing and presentation of MHC class II antigens. Similarly, bafilomycin (2 μM) blocks the proton ATPase activity in endosomes and phagosomes with a similar effect on MHC class II antigen presentation. (ii) The cystein protease inhibitor leupeptin ($10\,\mu g\,ml^{-1}$; Sigma) is commonly used to block cathepsins which are involved in invariant chain degradation and/or antigen processing and ultimately in MHC class II presentation. Lactocystin is a potent inhibitor of proteasome activity and therefore of MHC class I processing and presentation of endogenous antigens but it is also rather toxic (10 μM; for < 4 h). (iii) Transport of newly synthesized MHC and other molecules from the Golgi apparatus can be blocked by Brefeldin A ($2.5\,ng\,ml^{-1}$; Sigma) or monensin (2 μM; Sigma) similar to methods used to enrich cytokines in the Golgi apparatus for intracellular cytokine staining (Jung *et al.*, 1993). The cytoskeleton of the APC can be influenced using the following drugs: Cytochalasin D ($1\,\mu g\,ml^{-1}$; Sigma) blocks actin polymerization and nocodazol ($1\,\mu g\,ml^{-1}$; Sigma) inhibits microtubule formation and therefore movement and fusion of intracellular vesicles. Autophagy depends on intact actin polymers and can therefore be blocked using cytochalasin D (Schaible *et al.*, 1999b).

♦♦♦♦♦♦ LABELLING OF PATHOGENS

To follow antigen trafficking inside host cells the respective pathogens can be labelled using several methods. For microscopical analysis, we use fluorescent, biotin or digoxigenin hydrazide or carbazide derivatives to label terminal sugar residues using the Schiff's reaction or N-hydroxysuccinimidylester (NHS) derivatives to label aminogroups of surface proteins. Hydrazide labelling: pathogens are pelleted and resuspended at a concentration of $10^8\,ml^{-1}$ in 100 µl 0.2 M sodium acetate buffer (pH 5.5) containing 15 mM sodium periodate and agitated for 15 min in the dark. The reaction is stopped by 100 µl 15 mM sodium bisulfite for 5 min. Finally, the desired label is added such as Alexa™568-hydrazide (50 µM; Molecular Probes) and the gemisch is turned head over end at room temperature (r.t.) for 3 h (Schaible *et al.*, 2000). NHS labelling: pathogens are diluted in PBS at a concentration of $10^8\,ml^{-1}$ and $1\,mg\,ml^{-1}$ of the N-hydroxysuccinimidester derivatives such as NHS-carboxyfluorescein (stock solution in DMSO; Boehringer Mannheim) are added. The gemisch is agitated for 1 h at r.t. Labelled bacteria are washed five times in PBS and twice in medium before being used for infection. We found that mycobacteria retain approximately 95% viability after this labelling.

Confocal microscopy using sugar labelling has to be performed on live cells cultured on coverslips. For this purpose, cells are kept in Ringer solution at 37°C for the time of observation while the coverslips (round 2 cm diameter, or rectangular 2×4 cm) are sealed with a thin ring of vaseline applied with a syringe.

For biochemical analysis pathogens can be either surface labelled, using either hydrazide or NHS derivatives of biotin or digoxigenin, or radioactively, employing ^{35}S-methionine, ^{14}C-amino acid mixes for proteins, ^{14}C-acetate as a global label, ^{14}C palmitic acids for glycolipids and lipoproteins and ^{14}C-glucose for glycolipids (NEN, Boston, MA). A concentration of the label of $10–50\,\mu Ci\,ml^{-1}$ is usually sufficient. For some studies, the organisms are labelled in their conventional culture medium. To increase incorporation of radioactivity by the organisms by avoiding competition through cold compounds, culture media without or with a lower concentration of the cold component such as methionine-free medium for ^{35}S-methionine labelling is recommended. Although for most bacteria labelling with ^{14}C-compounds and ^{35}S-methionine can be done in 4 hours, mycobacteria, due to their slow replication rate, need to be incubated at least overnight. Labelling time does not only influence the amount of radioactivity incorporated by the organisms but also the pattern of labelled compounds. Therefore, the incubation time has to be adjusted to achieve optimal labelling of the chemical entities of interest.

Acknowledgments

The authors would like to thank Dr Robert Lindner, University of Hannover, Germany, who shared unpublished data and his broad experience in the field of subcellular fractionation methods with us, Dr David G. Russell, Cornell University, Ithaka, NY, USA, for helpful discussions and technical advices, and Kristine Hagens and Jana Enders for expert technical assistance. We also thank the Fonds der Chemischen Industrie and the DFG (SFB 421) for financial support.

◆◆◆◆◆◆ APPENDIX I: ENZYME ASSAYS

We usually take 1 to 10 µl of the samples/fractions to be tested for the following enzyme activities.

1. *β hexosaminidase (lysosomes)*: The substrate 4-methylumbelliferyl-2-acetoamido-2-deoxy-β-D-glucopyranoside (Sigma; stock solution 200 mM in DMSO; working dilution 4 mM in water) is diluted 1:2 in 0.4 M sodium acetate buffer/0.25% Triton X-100. Samples are filled up to 10 µl using PBS, placed in a 96-well black maxisorp F16 plate (475515; Nunc, Roskilde, Denmark) and mixed with 40 µl of the substrate solution. The plate is covered with parafilm and incubated for 60 min at 37°C. Reaction is stopped with 100 µl 0.5 M glycine/sodium carbonate buffer (pH 7) and read in an ELISA plate fluorometer (Spectra Fluor; Tecan, Crailsheim, Germany) using an extinction wavelength of 364 nm and an emission wavelength of 448 nm.
2. *β galactosidase (lysosomes)*: The substrate 4-methylumbelliferyl-β-D-galactopyranoside (Sigma; stock solution 100 mM in DMSO) is directly diluted 1:100 in 0.1 M sodium acetate buffer/0.25% Triton X-100. The assay follows the procedure described above.
3. *α mannosidase (Golgi)*: The substrate 4-methylumbelliferyl-α-D-mannopyranoside (Sigma; stock solution 400 mM in DMSO; working solution 4 mM in water) diluted 1:2 in PBS buffer/0.25 % Triton X-100. The assay follows the procedure described above but using a 30 min incubation time.
4. *Phosphodiesterase (plasmamembrane)*: The substrate sodium-thymidine 5′-monophosphate-p-nitrophenylester (stock solution 100 mM in water; working dilution 1:10 in water) is diluted 1:2 in 0.2 M Tris-HCL pH 9 and further diluted 1:2 in water. Samples are filled up to 10 µl using PBS, placed in a 96-well ELISA plate (Nunc) and mixed with 40 µl of the substrate solution. The plate is covered with parafilm and incubated for 2 h or 18 h at 37°C. Reaction is stopped with 100 µl 0.5 M glycine/sodium carbonate buffer (pH 7) and read in an ELISA plate reader (SpectraMax 250; Molecular Devices, Munich, Germany) at an absorbance of 410 nm.
5. *Lactate dehydrogenase (cytoplasm)*: We employ a ready-to-use kit purchased from Sigma.

References

Alford, C. E., King, T. E. and Campbell, P. A. (1991). Role of transferrin, transferrin receptors, and iron in macrophage listericidal activity. *J. Exp. Med.* **174**(2), 459–66.

Alvarez-Dominguez, C., Roberts, R. and Stahl, P. D. (1997). Internalized Listeria monocytogenes modulates intracellular trafficking and delays maturation of the phagosome. *J. Cell Sci.* **110**(Pt 6), 731–43.

Beatty, W. L. and Russell, D. G. (2000). Identification of mycobacterial surface proteins released into subcellular compartments of infected macrophages. *Infect. Immun.* **68**(12), 6997–7002.

Beauregard, K. E., Lee, K. D., Collier, R. J. and Swanson, J. A. (1997). pH-dependent perforation of macrophage phagosomes by listeriolysin O from Listeria monocytogenes. *J. Exp. Med.* **186**(7), 1159–63.

Chakraborty, P., Sturgill-Koszycki, S. and Russell, D. G. (1994). Isolation and characterization of pathogen-containing phagosomes. *Methods Cell Biol.* **45**, 261–76.

Collins, H. L., Schaible, U. E., Ernst, J. D. and Russell, D. G. (1997). Transfer of phagocytosed particles to the parasitophorous vacuole of Leishmania mexicana is a transient phenomenon preceding the acquisition of annexin I by the phagosome. *J. Cell Sci.* **110**(Pt 2), 191–200.

Coligan, J. E., Kruisbek, A. M., Margulies, D. H., Shevach, E. M. and Strober, W. (2000). (eds) *Current Protocols in Immunology*. John Wiley & Sons.

Desjardins, M., Celis, J. E., van Meer, G., Dieplinger, H., Jahraus, A., Griffiths, G. and Huber, L. A. (1994a). Molecular characterization of phagosomes. *J. Biol. Chem.* **269**(51), 32194–200.

Desjardins, M., Huber, L. A., Parton, R. G. and Griffiths, G. (1994b). Biogenesis of phagolysosomes proceeds through a sequential series of interactions with the endocytic apparatus. *J. Cell Biol.* **124**(5), 677–88.

Dietrich, G., Schaible, U. E., Diehl, K. D., Mollenkopf, H., Wiek, S., Hess, J., Hagens, K., Kaufmann, S. H. and Knapp, B. (2000). Isolation of RNA from mycobacteria grown under in vitro and in vivo conditions. *FEMS Microbiol Lett.* **186**(2), 177–80.

Gordon, S. (1995). The macrophage. *Bioessays* **17**(11), 977–86.

Hasan, Z., Schlax, C., Kuhn, L., Lefkovits, I., Young, D., Thole, J. and Pieters, J. (1997). Isolation and characterization of the mycobacterial phagosome: segregation from the endosomal/lysosomal pathway. *Mol. Microbiol.* **24**(3), 545–53.

Hodgkinson, V. H., Soong, L., Duboise, S. M. and McMahon-Pratt, M. (1996). D Leishmania amazonensis: cultivation and characterization of axenic amastigote-like organisms. *Exp. Parasitol.* **83**(1), 94–105.

Inaba, K., Inaba, M., Romani, N., Aya, H., Deguchi, M., Ikehara, S., Muramatsu, S. and Steinman, R. M. (1992). Generation of large numbers of dendritic cells from mouse bone marrow cultures supplemented with granulocyte/macrophage colony-stimulating factor. *J. Exp. Med.* **176**(6), 1693–702.

Jung, T., Schauer, U., Heusser, C., Neumann, C. and Rieger, C. (1993). Detection of intracellular cytokines by flow cytometry. *J. Immunol. Methods.* **159**(1–2), 197–207.

Kaufmann, S. H. E. and Kabelitz, D. (1998). Immunology of infection. In *Methods in Microbiology*, Vol. 25, 1st edn., Academic Press, San Diego and London.

Lang, T. and de Chastellier, C. (1985). Fluid phase and mannose receptor-mediated uptake of horseradish peroxidase in mouse bone marrow-derived macrophages. Biochemical and ultrastructural study. *Biol. Cell* **53**(2), 149–54.

Trafficking of Intracellular Pathogens

Lindner, R. (2001). One-step separation of endocytic organelles, Golgi/trans-Golgi network and plasma membrane by density gradient electrophoresis. *Electrophoresis* **22**, 386–93.
McNeil, P. L. and Warder, E. (1987). Glass beads load macromolecules into living cells. *J Cell Sci.* **88**(Pt 5), 669–78.
Romani, N., Reider, D., Heuer, M., Ebner, S., Kampgen, E., Eibl, B., Niederwieser, D. and Schuler, G. (1996). Generation of mature dendritic cells from human blood. An improved method with special regard to clinical applicability. *J. Immunol. Methods* **196**(2), 137–51.
Schaible, U. E., Sturgill-Koszycki, S., Schlesinger, P. H. and Russell, D. G. (1998). Cytokine activation leads to acidification and increases maturation of Mycobacterium avium-containing phagosomes in murine macrophages. *J. Immunol* **160**(3), 1290–6.
Schaible, U. E., Collins, H. L. and Kaufmann, S. H. (1999a). Confrontation between intracellular bacteria and the immune system. *Adv. Immunol.* **71**, 267–377.
Schaible, U. E., Schlesinger, P. H., Steinberg, T. H., Mangel, W. F., Kobayashi, T. and Russell, D. G. (1999b). Parasitophorous vacuoles of Leishmania mexicana acquire macromolecules from the host cellcytosol via two independent routes. *J. Cell Sci.* **112**(Pt 5), 681–93.
Schaible, U. E., Hagens, K., Fischer, K., Collins, H. L. and Kaufmann, S. H. (2000). Intersection of group I CD1 molecules and mycobacteria in different intracellular compartments of dendritic cells. *J. Immunol.* **164**(9), 4843–52.
Schorey, J. S., Carroll, M. C. and Brown, E. J. (1997). A macrophage invasion mechanism of pathogenic mycobacteria. *Science* **277**(5329), 1091–3.
Steinberg, T. H., Swanson, J. A. and Silverstein, S. C. (1988). A prelysosomal compartment sequesters membrane-impermeant fluorescent dyes from the cytoplasmic matrix of J774 macrophages. *J. Cell Biol.* **107**(3), 887–96.
Storrie, B. and Madden, E. A. (1990). Isolation of subcellular organelles. *Methods Enzymol.* **182**, 203–25.
Sturgill-Koszycki, S., Schaible, U. E. and Russell, D. G. (1996). Mycobacterium-containing phagosomes are accessible to early endosomes and reflect a transitional state in normal phagosome biogenesis. *EMBO J.* **15**(24), 6960–8.

20 Effector Proteins of Bacterial Type III Protein Secretion Systems: Elucidating their Biochemical Effects on Eukaryotic Signaling Cascades

Kim Orth[1], **Jack E Dixon**[2] **and James B Bliska**[3]
[1]*University of Texas Southwestern Medical Center, Dallas, TX 75390 USA;* [2]*University of Michigan, Ann Arbor MI, USA;* [3]*State University of New York at Stony Brook, New York, USA*

◆◆

CONTENTS

◆◆◆◆◆◆ INTRODUCTION

Bacterial pathogens use a variety of strategies to overcome host barriers to infection (Finlay and Falkow, 1997). A number of pathogens intentionally activate or inhibit eukaryotic signaling cascades to subvert host defenses. For example, bacterial pathogens can control fluid secretion, actin dynamics, vesicle trafficking, cytokine production or host cell survival by manipulating eukaryotic signaling cascades. Type III secretion systems are utilized by many Gram-negative bacteria to manipulate eukaryotic signaling cascades (Hueck, 1998). It is clear that type III secretion systems accomplish this task via the introduction of multiple effector proteins into membranes or interior compartments of host eukaryotic cells (Hueck, 1998). However, it is often not immediately clear how an effector protein functions to co-opt a signaling cascade. This is particularily true when the function of the effector is not easily deduced from inspection of its primary sequence, or when it is difficult to associate a phenotype with the effector in an *in vitro* infection system. Thus, elucidating the biochemical functions of type III effector proteins is, and will continue to be, a major challenge in the

METHODS IN MICROBIOLOGY, VOLUME 31
ISBN 0-12-521531-2

field of bacterial pathogenesis. This chapter reviews several approaches that can be used to uncover the biochemical functions of type III effector proteins and to elucidate their effects on eukaryotic signaling cascades.

◆◆◆◆◆◆ BASIC CHARACTERIZATION OF THE EFFECTOR

Before embarking on a method of analysis for identification of the effector's target, it is helpful to obtain some preliminary characteristics about the effector protein. First, the amino acid sequence of the effector should be studied. The effector sequence should be analyzed by BLAST searches (Basic Local Alignment Search Tool; *ncbi.nlm.nih.gov/BLAST*) (Altschul *et al.*, 1997) to look for molecules that might encode similar primary sequence. In addition, the effector should be analyzed by the SMART program (Simple Modular Architecture Research Tool; *smart.embl-heidelberg.de*) (Schultz *et al.*, 1998; Schultz *et al.*, 2000) to search for functional domains. These searches might provide some important insight into the activity of an effector and give clues about potential interacting molecules. However, one should be cautious not to over-interpret small regions of sequence homology (see Case study).

Second, in the case where the activity of an effector protein can be predicted by secondary structure analysis, mutants harboring mutations at critical residues can be generated. The mutant proteins can then be tested in the assays described below. For effectors where no known host protein appears to be similar, in many cases there are homologues of the effector present in other bacterial pathogens. Therefore, an alignment of these homologous effectors would indicate what domains or individual amino acid residues have been evolutionarily conserved. Mutation of conserved and rare amino acids (charged and/or bulky) in the effector could produce a mutant effector that is no longer active. Unfortunately, some mutants may simply be misfolded molecules that do not retain a native three-dimensional structure. In an attempt to assess whether the effector maintains some resemblance of a native molecule, the mutants can be tested for normal transport through the secretion apparatus during infection. Additional studies with deletion mutants might indicate a region of the effector that is critical for its activity. These mutants are not only valuable for initial characterization of the effector but are also useful tools for future experiments.

Third, the production of a recombinant form of the effector is required for many of the methods described below and, therefore, knowing the expression profile (solubility, stability, yields) for a particular effector is essential information. If possible, it also may be helpful to purify mutant forms of the effector. Described below are three techniques that have been successfully used to identify the targets of various effector proteins. As protocols have been described in great detail elsewhere, this chapter will focus on considerations that are important for designing and implementing these experiments.

◆◆◆◆◆◆ IDENTIFYING TYPE III EFFECTOR TARGETS

Yeast two-hybrid screen

The yeast two-hybrid system is an assay used to detect protein–protein interaction by detecting the activation of transcription of a reporter gene (Chien *et al.*, 1991; Fields and Song, 1989). The system consists of two plasmids. One plasmid (the 'bait vector') encodes a transcription factor DNA binding domain fused to a DNA fragment encoding a test protein, in this case an effector protein. The second plasmid (the 'prey vector') encodes a transcription factor activation domain that is fused to a DNA fragment encoding a random protein sequence generated from a cDNA library. The cDNA library can be derived from virtually any type of cultured cell or tissue. When a bait protein interacts with a protein encoded by the cDNA library, a functional transcription factor is reconstituted. The DNA binding domain is now linked via a protein bridge to the transcription activation domain. Reporter genes that express β-galactosidase or are essential for yeast to grow are turned on by the active transcription factor. Therefore, positive interactions in the yeast two-hybrid system can be isolated based on the yeast ability to grow in a selective media or detected by a colorometric assays using β-galactosidase. Complete protocols for the yeast two-hybrid system can be obtained from a number of sources and a very reliable example of this is one described by Vojtek and Hollenberg (1995). Yeast two-hybrid libraries and systems can be purchased from a number of companies (2000 BioSupplyNet Source Book; SciQuest.com).

A number of important controls must be performed before initiating a yeast two-hybrid screen. After the effector is cloned in frame with the DNA binding domain into the bait vector, the bait vector must then be tested for toxicity in the yeast. Knowing that a number of the effectors disrupt homeostasis in eukaryotic cells, it is not surprising that they might be toxic in yeast. However, this toxicity may be bypassed because the bait fusion protein is targeted at the nucleus. In a case where the bait is toxic, an inactivating mutant of the effector could replace the wild type effector as the bait in a yeast two-hybrid screen. However, the mutant protein must retain a native three-dimensional structure. In addition, the bait vector must be tested for autoactivation. If the effector that is fused to a DNA binding moiety resembles an activation domain, the reporter genes will be activated. In the event that a bait weakly autoactivates the system, the screen could be carried out in the presence of 3-amino triazole (Vojtek and Hollenberg, 1995). However, in cases where the autoactivation is strong, it is recommended that another method for detection of the target be used. Finally, when possible, expression of the bait fusion protein should be confirmed by immunoblotting.

The next step in preparation for a two-hybrid screen is to consider the source of the library to be used in the screen. If possible, the library should be prepared from a cell type that is sensitive to the action of the effector. However, in many cases effectors target ubiquitous cellular machinery and therefore a library derived from most cultured cell lines will work satisfactorily. The cDNA can be prepared using either oligo-dT primers, resulting in longer inserts, or random primers, resulting in short inserts. The advantage and disadvantages of each type of library are discussed in reviews (CLONTECH Lab. Inc., matchmaker.clontech.com; Vojtek and Hollenberg, 1995). In addition, the length of the inserts and the complexity of the libraries may vary. Many commercial libraries are available from a number of sources (2000 BioSupplyNet Source Book; SciQuest.com).

After the primary two-hybrid screen, the strains containing the putative positive prey vectors must be sorted. It is essential that the positive strains be consecutively replated on selective media at least three times. Replating the yeast and allowing for new growth dramatically decreases the chance that a single positive colony will contain more than one prey vector. Next, yeast strains containing the putative positive prey vectors are cured of the bait vector. These cured strains are then mated back to yeast strains containing the original bait vector or a control bait vector (for example a bait vector encoding the DNA binding domain fused to lamin). Colonies that activate the reporter genes when mated to the control vector are eliminated. Strains containing prey vectors that interact positively with the original bait vector are isolated. The prey vectors can then be directly analyzed by DNA sequencing to reveal the identity of the cDNA insert. For assistance with a yeast two-hybrid screen, Erica Golemis's research group has created an extremely helpful and informative WEB site referred to as the 'Interaction Trap at Work' (chaos.fccc.edu/research/labs/golemis/InteractionTrapInWork.html). This WEB site provides insightful information about false positives commonly isolated in two-hybrid screens in addition to helpful suggestion for troubleshooting a two-hybrid screen.

When the putative positive cDNAs are identified, confirmation of the protein–protein interaction between the effector (bait) and the protein encoded by the cDNA (the 'target') is strongly recommended. A common method for confirming the protein–protein interaction *in vitro* utilizes a recombinant GST-effector protein and a radiolabeled *in vitro* translated target protein. GST pull down assay is performed to detect protein–protein interaction. Additionally, the effector and the target can be expressed in mammalian cells and their interaction confirmed by co-immunoprecipitation experiments. An example of a successful two hybrid screen was used for the *Yersinia* effector YopJ that identified members of the mitogen activated protein kinase (MAPK) kinase family as host targets (Orth *et al.*, 1999).

Expression cloning

Expression cloning is a method where a library of cDNA is inserted into a λgt11 or related phage vector and during the lytic phase the recombinant target proteins are produced in large quantities. Phage plaques immobilized on filters can then be screened with the radiolabeled effector for interacting proteins. Expression cloning therefore requires the production of a recombinant radiolabeled effector protein. Phage that produce target proteins that bind to the radiolabeled effector are then purified and the cDNA sequenced. Numerous phage expression libraries are commercially available (2000 BioSupplyNet Source Book; SciQuest.com). In principle this technique appears rather straightforward. However, as is the case with any large screen, thoughtful preparation is essential and success is not guaranteed. Margolis and Young (1995) have written an excellent chapter describing the pros and cons of this screening system. As with the two-hybrid screen, false positives can be misleading and hugely expensive in terms of time. If the primary positive plaques do not amplify in the secondary and tertiary screens, it is a clear indication that these are not real positives and one should move on. If this screen does produce real positives, then you not only have the protein but the DNA that encodes the protein and DNA sequencing will reveal its identity. An example of a successful expression cloning project is described in Hardt *et al.* (1998). They were able to identify Rac1 and Cdc42 as targets of *Salmonella* SopE.

Affinity purification

Affinity purification techniques have not been commonly applied to the identification of type III effector targets, yet this represents one of the most direct approaches to achieve this aim. In addition, techniques for the identification of proteins using very small quantities of material are now available in many universities world-wide. The purified effector protein can be immobilized on a matrix either before or after it is exposed to cellular proteins. For example, an effector protein with a tag (epitope, GST, 6XHIS) can be introduced into cultured cells by transfection. The transfected cells are then lysed and the effector protein is purified using a matrix that binds tightly to the tag. Alternatively, a purified effector can be coupled to a matrix and then incubated in a cell lysate that contains the target. Once the effector–target complexes are formed, the matrix is washed extensively to remove proteins that bind non-specifically. It is important to empirically establish wash conditions that effectively remove non-specific proteins and elution conditions that specifically disrupt target–effector complexes. It is also useful to carry out mock purifications in parallel using a matrix that does not contain the effector. Any proteins that are eluted from both the mock and experimental matrix should be considered non-specific. In pilot experiments eluted target proteins can be detected by: (1) immunoblotting if the identity of the target can be surmized based on its physical properties; (2) silver staining; or (3) autoradiography in cases where lysates from radiolabeled cells are used. Once pilot experiments yield favorable results, the purification scheme can be scaled up to obtain sufficient protein material for protein microsequencing (a minimum of 5 pmols). Host cell targets of the *Yersinia* effector YopH have been identified by affinity purification and protein microsequencing (Black *et al.*, 2000).

Mass spectrometric techniques can be used to identify targets that correspond to proteins in databases. This involves digesting as little as a few hundred femptomoles of gel purified protein with a site-specific protease (i.e. trypsin) and then accurately determining the mass of the resulting peptides. The peptide masses are then submitted to one of several mass database search algorithms that are available on web based servers. These search algorithms generate theoretical mass profiles for each protein in a specified database and compares them with the experimental data to arrive at a statistical correlation.

◆◆◆◆◆◆ STUDYING THE IMPACT OF EFFECTOR PROTEINS ON HOST SIGNALING CASCADES

Introducing the effector into cells

Several different approaches can be used to introduce effector proteins into host cells in order to study their biological effects on host signaling cascades. The easiest and most biologically relevant way to accomplish this is by bacterial infection. However, for a number of reasons it is advisable to introduce effector proteins into host cells using one or more additional methods such as transfection or microinjection. The advantages of using transfection or microinjection include the following. (1) The activity of the effector can be analyzed in the absence of other effectors that might have redundant activity or otherwise mask the activity of the effector. (2) Transfection or microinjection may result in higher levels of intracellular protein than can be achieved by infection. This may yield an informative phenotype, or facilitate determination of the effector's subcellular location. (3) Individual domains of the protein can be separated from secretion and translocation signals and analyzed for function using a transfection or microinjection. Finally, confirming *in vivo* protein–protein interactions (see above) between an effector and a putative target(s) may be facilitated by transfection due to higher protein levels. Artefacts may be obtained using these delivery methods, and one should avoid over-interpretation of such data in the absence of supporting results from infection assays.

The choice of cultured host cell for analysis is also an important consideration. The primary selection criteria should be one of *in vivo* relevance. What type of cell is likely to be the target of the effector during infection of an animal host? A secondary consideration is one of technical practicality. Can the natural target cell be obtained in sufficient quantities for experimentation, and is the target cell amenable to molecular and cellular analysis? In some cases the issue of technical practicality may be overriding. For example, if the target cell is a neutrophil, effector delivery by bacterial infection would likely be the only feasible approach. In addition, with current technology it would be difficult or impossible to introduce altered forms of the effector target into neutrophils. In this case, another type of professional phagocyte that is more amenable to experimental manipulation, such as a macrophage, may be an acceptable substitute. At the extreme, many experiments carried out with cell types chosen strictly on the basis of technical feasibility (e.g. COS cells because of high transfection efficiencies) have yielded extremely useful results.

Transfection

A review of the literature indicates that many type III effector proteins have been sucessfully produced in host cells using transient transfection techniques (for examples see Hardt *et al.*, 1998; Juris *et al.*, 2000; Palmer *et al.*, 1999; Pederson *et al.*, 1999). Typically, the coding region for the effector is inserted into a standard vector (e.g. pCNDA3, Invitrogen) with an epitope tag appended to either the N- or C-terminus of the protein to allow for detection. If a tag is used, one should verify that the tag does not interfere with effector function. For example, addition of a C-terminal tag to *Yersinia* protein kinase A (YpkA) resulted in a non-functional protein (J.E.D. and K.O., unpublished observation). Commercially available reagents (e.g. FuGENE 6, Roche Molecular Biochemicals) can be used to transfect many types of cultured mammalian cells with high efficiency. Expression of the protein can be assessed by immunoblotting or immunofluorescence microscopy using anti-epitope antibody. Optimal conditions for expression of your protein can be determined by varying the concentration of input plasmid DNA or the time of incubation following transfection. Retroviral vectors can be utilized to express effector proteins in host cell types that are difficult to transfect with standard techniques (e.g. primary murine macrophages). In addition, retroviral technology permits transfection of essentially every cell in a population. Retroviral vectors and packaging cell lines are available commercially (Imgenex, Clontech). In addition, a new generation of mono- and bi-cistronic transfection vectors have been developed based on fluorescent protein technology (e.g. pCMS-EGFP; Clontech). These vectors could be used to couple production of type III effectors and a fluorescent protein, or to directly fuse the type III protein to a fluorescent protein tag.

Microinjection

Microinjection offers an alternative to transfection for delivery of effector proteins into host cells. Detailed descriptions of microinjection techniques have been published (Bar-Sagi, 1995). It is possible to microinject either protein (typically into the cytoplasm) or plasmid DNA (typically into the nucleus); only the former will be discussed here. Perhaps the major advantage of protein microinjection is that the response of the host cell to the effector can be monitored within a relatively short period of time (e.g. 1 h or less). For example, Hardt *et al.* (1998) were able to detect membrane ruffling stimulated by *Salmonella* SopE 45 min post-microinjection. Microinjection can approximate bacterial delivery of the effector because the amount of effector that is microinjected can be easily controlled by dilution of the protein sample. In addition, some effector proteins may be too toxic to be produced by transient transfection. The three major disadvantages of microinjection are: (1) the effector has to be in a purified and soluble form; (2) host cell responses can only be monitored at the single-cell level (e.g. by microscopy); and (3) specialized equipment is required (inverted microscope with phase objective, microinjector, micromanipulator and capillary micropipettes).

Several investigators have overcome the purification issue by fusing an effector to the C-terminus of an easily purified protein such as glutathione-*S* transferase (GST) (Fu and Galan, 1998; Hardt *et al.*, 1998; Hersh *et al.*, 1999). In addition to facilitating the purification of the effector, the presence of the GST domain may increase the solubility properties of the protein. Surprisingly, the presence of the GST domain does not appear to interfere with the biological functions of many proteins to which it is attached. Therefore, the GST domain does not necessarily need to be removed before the protein is microinjected. This can be extremely advantageous, as it allows for rapid purification of the fusion protein, detection of the injected fusion protein using anti-GST antibodies, and microinjection of GST alone can be used as an appropriate negative control.

The protein sample to be microinjected is prepared starting at a concentration of approximately 1–5 mg ml^{-1} in an appropriate buffer (Bar-Sagi, 1995). Cultured host cells can be seeded on glass coverslips or on petri dishes; only cells that adhere tightly to the substrate can be easily injected. The use of coverslips with an etched grid facilitates the localization of injected cells at various time points post-injection. Many different types of cultured eukaryotic cells have been successfully microinjected. In general, large flat cells (e.g. epithelial cells or fibroblasts) are relatively easy to microinject, while small round cells (e.g. macrophages) are more difficult. Dishes containing the cells growing in standard serum-containing media are placed on the microscope stage and an appropriate volume of protein solution is injected into a group of cells (Bar-Sagi, 1995). With a high quality microinjection apparatus and practice it is possible to process approximately 100 cells in a 10 min period. Typical survival rates for easily injected cells are in the range of 80–90%. The injected cells are then returned to the tissue culture incubator for varying lengths of time before processing (see below).

Semi-permeable cell assays

Tran Van Nhieu *et al.* (1999) have developed a semi-permeable cell assay that has been used to identify *Shigella* IpaC as an effector protein that can directly trigger actin reorganization in host cells. This assay could be used as a general approach to study effector protein function as long as the effector protein and the cellular response are not perturbed by transient exposure to the permeation conditions. Low concentrations of saponin are used to permeabilize fibroblast cells in the presence of high concentrations of BSA and effector proteins. The detergent is then removed by washing and the cells are allowed to reseal their membranes. At various time points thereafter, cells are fixed and analyzed for cellular response (see below). It appears that cell viability remains high as long as the permeation step is held to less than 20 min.

Introducing modified forms of signaling cascade components into host cells

It is often useful to introduce modified components of signaling cascades into host cells to further probe the mechanism of effector function. For example, if an effector is suspected to have a negative effect on a particular signaling pathway, one can attempt to pinpoint the step at which the effector blocks the pathway via the co-expression of the effector with a constitutively activated signaling component. If an effector is able to block signaling in the presence of the activated component, then it is likely that the effector acts at or below the level of the activated component. This strategy was used to identify the step at which the *Yersinia* YopJ protein blocks MAP kinase signaling pathways (Orth *et al.*, 1999) (see below). Conversely, dominant interfering forms of signaling components can be used to probe mechanisms of effector proteins that have positive effects on signaling cascades. Hardt *et al.* (1998) found that dominant interfering forms of Rac1 or Cdc42 counteracted *Salmonella* SopE activity, indicating that SopE acts at or above the level of Rac1 or Cdc42 to stimulate actin polymerization. This is true because SopE acts directly on Rac1 and Cdc42 to catalyze the GDP to GTP exchange reaction (Hardt *et al.*, 1998). Genetically-modified signaling components and effectors can be introduced simultaneously into host cells using one or more of the techniques described above. Alternatively, some investigators studying type III effectors have used well-characterized bacterial exotoxins (Boquet, Chapter 12 this volume) to activate (e.g. *E. coli* CNF1) or inactivate (e.g. *Clostridium* C3 toxin) endogenous Rho family GTPases (Pederson *et al.*, 1999; Tran Van Nhieu *et al.*, 1999).

Analysis of host cells

Fluorescence microscopy

Fluorescence microscopy is a reliable and highly sensitive technique for accessing (1) the concentration and subcellular location of an effector or target protein in host cells; and (2) morphological changes that may result from the action of a type III effector. Specimens can be analyzed by either light microscopy, or by confocal scanning laser microscopy. The use of epitope tags to label type III effector proteins bypasses the need for production of antibodies, provided that the tag does not interfere with function. In addition, epitope tags are a particularly effective way to distinguish a transfected or microinjected target protein from the endogenous polypeptide. A basic procedure for immunofluorescent labeling of cultured eukaryotic cells is described elsewhere (Harlow and Lane, 1988). The use of appropriate controls to demonstrate specificity of immunolabeling is of critical importance. As a minimum, cells that do not contain the effector protein (or epitope-tagged target protein) should be labeled with primary and fluorochrome-conjugated secondary antibodies and analyzed to establish levels of non-specific staining.

A large variety of commercially available antibodies and other types of fluorescent probes can be used in conjunction with fluorescence microscopy to monitor morphological alterations in host cells resulting from the action of type III effector proteins. Such alterations include actin rearrangements (see Tran Van Nhue, Chapter 11 this volume), cytotoxic or apoptotic changes (see Zychlinsky, Chapter 24 this volume) or changes in vesicle or protein trafficking. A recent example of the application of immunofluorescence microscopy to the study of proinflammatory signaling cascades is provided by the work of Yuk *et al.* (2000). Immunofluorescent labeling of NFκB in epithelial cells infected with *Bordetella bronchiseptica* indicated that one or more type III effectors of *Bordetella* interfere with normal trafficking of NFκB to the nucleus (Yuk *et al.*, 2000).

Assays that directly measure activation of signaling cascade components

There are numerous assays available that can be used to directly measure activation of signaling cascade components. This section will describe only several of the most recently-developed assays that we believe will be highly beneficial in studies of type III effector mechanisms. Typically, following introduction of the effector and the application of appropriate stimuli if needed, host cells or the isolated nuclei derived thereof are lysed in a detergent-containing solution to solubilize components of the signaling cascade under study. In some cases the signaling component can be assayed in a crude lysate while in others it needs to be purified prior to assay. Most assays of this type are developed using SDS/PAGE combined with either immunoblotting or autoradiography.

Many proteins that function at early steps of signaling cascades are regulated by tyrosine (Tyr) phosphorylation. It is therefore not surprising that at least two, and possibly more, type III effectors exhibit tyrosine phosphatase activity (DeVinney *et al.*, 2000). Methods for detecting phospho-Tyr (pTyr), as well as other types of protein modifications (see below), have been revolutionized by the development of specific antibody reagents. Global changes in pTyr are easily measured by immunoblot analysis of lysates using commercially available antibodies (e.g. 4G10, Upstate Biotechnology or P-Tyr-100, Cell Signaling Technology). These antibodies appear to specifically recognize pTyr in a context-independent manner. Certain proteins upon tyrosine phosphorylation enter into relatively insoluble complexes with cytoskeletal elements etc., so it is often useful to analyze detergent-soluble and insoluble fractions of cells by anti-pTyr immunoblotting. Tyrosine phosphorylation of individual proteins can be accessed using immunoprecipitation to isolate the protein from the lysate, followed by immunoblotting with anti-pTyr antibodies.

Ras and the small GTPases of the Rho subfamily (Rac1, Cdc42 and Rho) are key upstream regulators of the mitogen-activated protein kinase (MAPK) cascades. These small GTPases appear to be preferential targets of many bacterial toxins, including those secreted by type III systems (Boquet, 2000). Small GTPases cycle between an active GTP-bound state and an inactive GDP-bound state. When bound to GTP they bind tightly to specific downstream partner proteins. This property can be exploited as a convenient method to measure their activation. For example, the

GTP-bound forms of Rac1 and Cdc42 bind tightly to a distinct domain (known as PBD) in the p21-activated kinase (Pak). A purified GST-PBD fusion protein is incubated in cell lysates and then recovered using glutathione agarose (Benard *et al.*, 1999). After washing, bound GTPase protein is eluted and detected by immunoblotting with antibodies specific for Rac or Cdc42. Similar assays have been developed to measure activation of Rho (Ren *et al.*, 1999) or Ras (Taylor and Shalloway, 1996). All reagents necessary to perform these assays are available commercially (Upstate Biotechnology).

The functional MAPK signaling pathway is a three-tiered cascade. MAPKs are phosphorylated and activated by MAPK-kinases (MKKs), which in turn are phosphorylated and activated by MAPKK-kinases (MKKKs). Phosphorylated MAPKs enter the nucleus and activate transcription factors via phosphorylation. The traditional method for measuring activation of MAPKs utilizes an *in vitro* kinase assay (Hardt *et al.*, 1998; Orth *et al.*, 1999). In the case of ERK MAP kinase, mylein basic protein (MBP) is used as a substrate in place of a transcription factor. MBP is incubated in a reaction containing immunoprecipitated ERK and 32P-ATP (Orth *et al.*, 1999; Sugimoto *et al.*, 1998). Phosphorylation of MBP by activated ERK is then quantified by SDS-PAGE and autoradiography. Alternatively, activation of ERK and other MAPKs can be measured by immunoblotting using phospho-specific antibodies that are available from several companies (e.g. Cell Signaling Technology, Upstate Biotechnology). This type of phospho-specific antibody is raised against a synthetic phosphopeptide that corresponds to the site of dual phosphorylation in an MAPK. Thus, these antibodies are relatively specific for a single signaling component. Phosphospecific immunoblotting bypasses the need to perform kinase reactions with 32P-ATP. In some cases the antibodies are of sufficient specificity to allow analysis of whole cell lysates (Fu and Galan, 1999; Palmer *et al.*, 1999). In other cases it is preferable to first immunoprecipitate the target protein and then perform immunoblotting. Orth and colleagues (1999) used phospho-specific immunoblotting to demonstrate that *Yersinia* YopJ inhibits phosphorylation, and thereby activation, of the MKK for ERK.

It is worth noting that the availability of specific antibody reagents to assay activation of signaling components is increasing rapidly. For example, reagents are now available to directly measure activation of a variety of transcription factors (e.g. STATs and NFκB; Cell Signaling Technology). Due to the 'user friendly' nature of phospho-specific technology, it is likely to have a major impact on future studies of the kind described in this chapter.

Assays that measure gene expression

Nuclear responses triggered by signaling cascades are commonly measured by one of the following: (1) production and release of cytokines or chemokines; (2) production of messenger RNA; or (3) activation of transfected reporter genes. The former two methods require analysis of whole cell populations, and thus far have only been applied to experiments that utilize infection to deliver effector proteins (Hobbie *et al.*, 1997; Palmer *et al.*, 1998; Schesser *et al.*, 1998). Reporter plasmids offer a convenient method for measuring gene expression within a subset of a cell population. Three of the more commonly used reporter molecules are β-galactosidase (LacZ), chloramphenicol acetyltransferase (CAT) or luciferase (Luc). Genes encoding the reporter molecule can be placed under control of a basic promoter element (TATA box) and a multimerized transcription factor binding site (e.g. pNFκB-Luc; Stratagene). A second reporter plasmid under the control of constitutive promoter (e.g. pCMV-LacZ) can be used to control for transfection efficiency between samples. Cultured cells are co-transfected with both plasmids for 1 to 2 days. The cells are then stimulated if necessary and incubated an additional 2 to 4 hours to allow reporter gene expression. The cells are then lysed and the lysates are analyzed for luciferase production using a luminometer. The values obtained are normalized to levels of LacZ determined by a separate assay.

◆◆◆◆◆◆ CASE STUDY

In the last part of this chapter, the characterization of the *Yersinia* YopJ protein will be used as a case study to exemplify how the function of an effector can be uncovered using many of the tools described above. YopJ was initially described as a 30 kD protein that was secreted by *Yersinia pestis* (Straley and Bowmer, 1986). Through infection studies, YopJ was found to be required for suppression of TNFα production by infected macrophages (Boland and Cornelis 1998; Palmer *et al.*, 1998). These studies also indicated that YopJ was required for down regulation of the MAP kinase pathways. Infection studies also demonstrated that YopJ was required for *Yersinia* species to induce program cell death in macrophages (Ruckdeschel *et al.*, 1997; Mills *et al.*, 1997; Monack *et al.*, 1997; Monack *et al.*, 1998) and that signaling by the NFκB pathway was impaired under these conditions (Ruckdeschel *et al.*, 1998; Schesser *et al.*, 1998). Palmer *et al.* (1999) used transfection to demonstrate that YopJ was directly responsible for inhibition of MPK signaling pathways. These experiments required the production of mammalian expression plasmids that produced a protein product that had been shown by infection experiments to be fully active. Palmer *et al.* (1999) fused an epitope tag (M45) to the carboxy terminus of YopJ and demonstrated that this molecule was expressed and translocated by the type III secretion system.

In other studies, YopJ was shown to belong to a family of proteins produced by a wide variety of plant and animal pathogens as well as plant symbiots (Ciesiolka *et al.*, 1999; Freiberg *et al.*, 1997; Hardt and Galan, 1997; Whalen *et al.* 1993). However, the primary sequence of YopJ was not similar to any known eukaryotic protein or motif. Preliminary studies had suggested that YopJ might contain an SH2 domain, however, the homology was very weak (Schesser *et al.*, 1998). A form of YopJ lacking this SH2 domain motif was no longer able to inhibit signaling but it was unclear whether the mutant protein was expressed and translocated at wild-type levels.

In the absence of any obvious sequence similarities, a two-hybrid screen was carried out in an attempt to identify a host cell target (Orth *et al.*, 1999). A two-hybrid LexA-YopJ fusion protein was not toxic to yeast cells and did not autoactivate. The screen identified multiple members of MAPK kinase family as YopJ-interacting proteins. A related kinase responsible for activation of NF-κB, IκB kinase, was also shown to interact with YopJ *in vitro* and *in vivo*. Recombinant YopJ was produced as a GST fusion protein. The interaction between YopJ and various MAPK kinases and IκB kinase was confirmed by GST pull down experiments (Orth *et al.*, 1999). Additionally, co-immunoprecipitation experiments confirmed that YopJ bound to MAPK kinases and IκB kinase in host cells.

Transfection experiments with YopJ were then used to perform a biochemical epistasis experiment (Orth *et al.*, 1999). In essence, the step in the linear MAPK pathway (from receptor to ERK) that was blocked by YopJ was determined. The pathway was activated at various points, in the presence or absence of YopJ, and the production of active ERK was assessed. It was shown that YopJ inhibited the ability of Raf to phosphorylate MEK1 (an MAPK kinase). Thus, the two-hybrid screen and the epistasis experiment both arrived at the conclusion that YopJ was acting at the level of MAPK kinases. Although these studies clarified how one effector could inhibit so many different pathways, it was still unclear as to what molecular mechanism YopJ might use to inhibit these signaling pathways.

Understanding of the molecular mechanism of YopJ came from careful analysis of the secondary structure of YopJ (Orth *et al.*, 2000). The secondary structure of YopJ was compared to all known secondary structures derived from crystallographic data. The programs used to assess similarities were the CPHmodels and Threader programs. From this analysis, YopJ appeared to have a predicted structure similar to the known structure of adenoviral protease. Human proteins that share sequence similarity to the adenoviral protease, are the ubiquitin-like protein proteases. This suggested the possibility that ubiquitin-like proteins are the substrate for YopJ (Orth *et al.*, 2000). For analysis of effectors in this manner, our recommendation is to collaborate with scientists that specialize in this type of structure analysis. Our mutational analysis of YopJ, which was based on alignments of the known YopJ homologues, proved very useful in confirming that YopJ was a protease. Mutations in the predicted catalytic triad of YopJ destroyed its ability to inhibit signaling pathways in transfection and infection studies (Orth *et al.*, 2000). Similar studies were performed with AvrBsT, a homologue of YopJ in the plant pathogen *Xanthomonas campestris*. AvrBsT is responsible for inducing the hypersensitive response in plants. Catalytically inactive forms of AvrBsT were unable to initiate the hypersensitive response in the infected plant leaf. The catalytically-inactive forms of YopJ and AvrBsT were translocated into their respective host cells at wild-type levels.

Collectively, these studies exemplify how a number of protocols described in this chapter were used to identify the target and activity of the *Yersinia* effector YopJ. Infection studies identified a number of signaling systems affected by YopJ including MAP kinase pathways, the NFκB pathway and programmed cell death. A two-hybrid screen identified host cell targets of YopJ as members of the MAPK kinase superfamily. Transfection studies demonstrated that YopJ was necessary for inhibition of the signaling pathways but was not sufficient to induce programmed cell death. By careful analysis of the primary sequence of YopJ, the effector was identified to encode a cysteine protease. This observation was confirmed in both an animal and a plant homologue of YopJ mutagenesis studies. By further sequence analysis, the endogenous activity that YopJ homologues are mimicking was identified as ubiquitin-like protein protease activity. Therefore, utilizing a combination of the described tools has led to a clearer understanding of a family of homologous effector proteins that are found in both plant and animal pathogens as well as plant symbionts.

References

Altschul, S. F., Madden, T. L., Schaffer, A. A., Zhang, J., Zhang, Z., Miller, W. and Lipman, D. J. (1997). Gapped BLAST and PSI-BLAST: a new generation of protein database search programs. *Nucleic Acids Res.* **25**, 3389–3402.

Bar-Sagi, D., ed. (1995). *Methods in Enzymology.* Academic Press, San Diego.

Benard, V., Bohl, B. and Bokoch, G. M. (1999). Characterization of Rac and Cdc42 activation in chemoattractant-stimulated human neutrophils using a novel assay for active GTPases. *J. Biol. Chem.* **274**, 13198–13204.

Black, D. S., Marie-Cardine, A., Schraven, B. and Bliska, J. B. (2000). The Yersinia tyrosine phosphatase YopH targets a novel adhesion-regulated signalling complex in macrophages. *Cell Microbiol.* **2**, 401–414.

Boland, A. and Cornelis, G. R. (1998). Role of YopP in suppression of tumor necrosis factor alpha release by macrophages during Yersinia infection. *Infect. Immun.* **66**, 1878–1884.

Boquet, P. (2000). Small GTP binding proteins and bacterial virulence. *Microbes Infect.* **2**, 837–844.

Chien, C. T., Bartel, P. L., Sternglanz, R. and Fields, S. (1991). The two-hybrid system: a method to identify and clone genes for proteins that interact with a protein of interest. *Proc. Natl Acad. Sci. USA* **88**, 9578–9582.

Ciesiolka, L. D., Hwin, T., Gearlds, J. D., Minsavage, G. V., Saenz, R., Bravo, M., Handley, V., Conover, S. M., Zhang, H., Caporgno, J., Phengrasamy, N. B., Toms, A. O., Stall, R. E. and Whalen, M. C. (1999). Regulation of expression of avirulence gene avrRxv and identification of a family of host interaction factors by sequence analysis of avrBsT. *Mol. Plant–Microbe Interactions* **12**, 35–44.

DeVinney, R., Steele-Mortimer, O. and Finlay, B. B. (2000). Phosphatases and kinases delivered to the host cell by bacterial pathogens. *Trends Microbiol* **8**, 29–33.

Fields, S. and Song, O. (1989). A novel genetic system to detect protein–protein interactions. *Nature* **340**, 245–246.

Finlay, B. B. and Falkow, S. (1997). Common themes in microbial pathogenicity revisited. *Microbiol. Mol. Biol. Rev.* **61**, 136–169.

Freiberg, C., Fellay, R., Bairoch, A., Broughton, W. J., Rosenthal, A. and Perret, X. (1997). Molecular basis of symbiosis between Rhizobium and legumes [see comments]. *Nature* **387**, 394–401.

Fu, Y. and Galan, J. E. (1999). A *Salmonella* protein antagonizes Rac-1 and Cdc42 to mediate host-cell recovery after bacterial invasion. *Nature* **401**, 293–297.

Fu, Y. X. and Galan, J. E. (1998). The Salmonella Typhimurium tyrosine phosphatase Sptp is translocated into host cells and disrupts the actin cytoskeleton. *Mol. Microbiol.* **27**, 359–368.

Hardt, W. D. and Galan, J. E. (1997). A secreted Salmonella protein with homology to an avirulence determinant of plant pathogenic bacteria. *Proc. Natl Acad. Sci. USA* **94**, 9887–9892.

Hardt, W.-D., Chen, L.-M., Schuebel, K. E., Bustelo, X. R. and Galan, J. E. (1998). *S. typhimurium* encodes an activator of Rho GTPases that induces membrane ruffling and nuclear responses in host cells. *Cell* **93**, 815–826.

Harlow, E. and Lane, D. (1988). *Antibodies: A Laboratory Manual.* Cold Spring Harbor, Cold Spring Harbor Laboratory Press, NY.

Hersh, D., Monack, D. M., Smith, M. R., Ghori, N., Falkow, S. and Zychlinsky, A. (1999). The *Salmonella* invasin SipB induces macrophage apoptosis by binding to caspase-1. *Proc. Natl Acad. Sci. USA* **96**, 2396–2401.

Hobbie, S., Chen, L. M., Davis, R. and Galan, J. (1997). Involvement of mitogen-activated protein kinase pathways in the nuclear responses and cytokine production induced by *Salmonella typhimurium* in cultured intestinal epithelial cells. *J. Immunol.* **159**, 5550–5559.

Hueck, C. J. (1998). Type III protein secretion systems in bacterial pathogens of animals and plants. *Microbiol. Mol. Biol. Rev.* **62**, 379–433.

Juris, S. J., Rudolph, A. E., Huddler, D., Orth, K. and Dixon, J. E. (2000). A distinctive role for *Yersinia* protein kinase: actin binding, kinase activation, and cytoskeletal disruption *Proc. Natl Acad. Sci. USA* **97**, 9431–9436.

Margolis, B. and Young, R. A. (1995). Screening l expression libraries with antibody and protein probes, in *DNA Cloning-expression Systems: A Practical Approach.*, Hames, D. and Glover, D. (eds), Oxford University Press, London.

Mills, S. D., Boland, A., Sory, M. P., van der Smissen, P., Kerbourch, C., Finlay, B. B. and Cornelis, G. R. (1997). Yersinia enterocolitica induces apoptosis in macrophages by a process requiring functional type III secretion and translocation mechanisms and involving YopP, presumably acting as an effector protein. *Proc. Natl Acad. Sci. USA* **94**, 12638–12643.

Monack, D. M., Mecsas, J., Ghori, N. and Falkow, S. (1997). Yersinia signals macrophages to undergo apoptosis and YopJ is necessary for this cell death. *Proc. Natl Acad. Sci. USA* **94**, 10385–10390.

Monack, D. M., Mecsas, J., Bouley, D. and Falkow, S. (1998). Yersinia-induced apoptosis in vivo aids in the establishment of a systemic infection of mice. *J. Exp. Med.* **188**, 2127–2137.

Orth, K., Palmer, L. E., Bao, Z. Q., Stewart, S., Rudolph, A. E., Bliska, J. B. and Dixon, J. E. (1999). Inhibition of the mitogen-activated protein kinase kinase superfamily by a *Yersinia* effector. *Science* **285**, 1920–1923.

Orth, K., Xu, Z., Mudgett, M. B., Bao, Z. Q., Palmer, L. E., Bliska, J. B., Mangel, W. F., Staskawicz, B. and Dixon, J. E. (2000). Disruption of signaling by *Yersinia* effector YopJ, a ubiquitin-like protein protease. *Science* **290**, 1594–1597.

Palmer, L. E., Hobbie, S., Galan, J. E. and Bliska, J. B. (1998). YopJ of *Yersinia pseudotuberculosis* is required for the inhibition of macrophage TNFa production and the downregulation of the MAP kinases p38 and JNK. *Mol. Microbiol.* **27**, 953–965.

Palmer, L. E., Pancetti, A. R., Greenberg, S. and Bliska, J. B. (1999). YopJ of *Yersinia* spp. is sufficient to cause downregulation of multiple mitogen activated protein kinases in eukaryotic cells *Infect. Immun.* **67**, 708–716.

Pederson, K. J., Vallis, A. J., Aktories, K., Frank, D. W. and Barbieri, J. T. (1999). The amino-terminal domain of *Pseudomonas aeruginosa* ExoS disrupts actin filaments via small-molecular weight GTP-binding proteins. *Mol. Microbiol.* **32**, 393–401.

Ren, X.-D., Kiosses, W. B. and Schwartz, M. A. (1999). Reguation of the small GTP-binding protein Rho by cell adhesion and the cytoskeleton. *EMBO J.* **18**, 578–585.

Ruckdeschel, K., Roggenkamp, A., Lafont, V., Mangeat, P., Heesemann, J. and Rouot, B. (1997). Interaction of Yersinia enterocolitica with macrophages leads to macrophage cell death through apoptosis. *Infect. Immun.* **65**, 4813–4821.

Ruckdeschel, K., Harb, S., Roggenkamp, A., Hornef, M., Zumbihl, R., Kohler, S., Heesemann, J. and Rouot, B. (1998). Yersinia enterocolitica impairs activation of transcription factor NF-kappaB: involvement in the induction of programmed cell death and in the suppression of the macrophage tumor necrosis factor alpha production. *J. Exp. Med.* **187**, 1069–1079.

Schesser, K., Spiik, A.-K., Dukuzumuremyi, J.-M., Neurath, M. F., Pettersson, S. and Wolf-Watz, H. (1998). The *yopJ* locus is required for *Yersinia*-mediated inhibition of NF-kB activation and cytokine expression: YopJ contains a eukaryotic SH2-like domain that is required for its repressive activity. *Mol. Microbiol.* **28**, 1067–1080.

Schultz, J., Milpetz, F., Bork, P. and Ponting, C. P. (1998). SMART, a simple modular architecture research tool: Identification of signalling domains. *Proc. Natl Acad. Sci. USA* **95**, 5857–5864.

Schultz, J., Copley, R. R., Doerks, T., Ponting, C. P. and Bork, P. (2000). SMART: A Web-based tool for the study of genetically mobile domains. *Nucleic Acids Res.* **28**, 231–234.

Straley, S. C. and Bowmer, W. S. (1986). Virulence genes regulated at the transcriptional level by Ca2+ in Yersinia pestis include structural genes for outer membrane proteins. *Infect. Immun.* **51**, 445–454.

Sugimoto, T., Stewart, S., Han, M. and Guan, K. (1998). The kinase suppressor of Ras (KSR) modulates growth factor and Ras signaling by uncoupling Elk-1 phosphorylation from MAP kinase activation. *EMBO J.* **17**, 1717–1727.

Taylor, S. J. and Shalloway, D. (1996). Cell cycle-dependent activation of Ras. *Curr. Biol.* **6**, 1621–1627.

Tran Van Nhieu, G., Caron, E., Hall, A. and Sansonetti, P. (1999). IpaC induces actin polymerization and filopodia formation during *Shigella* entry into epithelial cells *EMBO J.* **18**, 3249–3262.

Vojtek, A. B. and Hollenberg, S. M. (1995). Ras-Raf interaction: two-hybrid analysis *Meth. Enzymol.* **255**, 331–342.

Whalen, M. C., Wang, J. F., Carland, F. M., Heiskell, M. E., Dahlbeck, D., Minsavage, G. V., Jones, J. B., Scott, J. W., Stall, R. E. and Staskawicz, B. J. (1993). Avirulence gene avrRxv from Xanthomonas campestris pv. vesicatoria specifies resistance on tomato line Hawaii 7998. *Mol. Plant–Microbe Interactions* **6**, 616–627.

Yuk, M. H., Havill, E. T., Coter, P. A. and Miller, J. F. (2000). Modulation of host immune responses, induction of apoptosis and inhibition of NF-kB activation by the *Bordetella* type III secretion system. *Mol. Microbiol.* **35**, 991–1004.

21 Modeling Microbial–Epithelial Interactions in the Intestine

Andrew T Gewirtz, Katherine A Reed, Didier Merlin, Michael Hobert, Andrew S Neish and James L Madara

Epithelial Pathobiology Division, Department of Pathology and Laboratory Medicine, Emory University, Atlanta, GA 30322, USA

◆◆

CONTENTS

◆◆◆◆◆◆ INTRODUCTION

The intestinal epithelium is one of the major interfaces with the outside world. This interface is very heavily colonized with gram-negative bacteria and yet permits absorption of life-sustaining nutrients while protecting the tissues below from microbial onslaught. Some of the mechanisms by which the intestinal epithelium can successfully accomplish such seemingly mutually exclusive functions have begun to be elucidated, in part, due to the development of reductionist cell culture models. Such models allow well-defined manipulation of experimental parameters while maintaining many of the key aspects of intestinal physiology especially with regard to bacterial–epithelial interactions. This review discusses these model systems and outlines many of the techniques that can be used with these systems to study bacterial–epithelial interactions in the intestine.

◆◆◆◆◆◆ PREPARATION OF MODEL EPITHELIA

Any *in vitro* study of bacterial epithelial interactions will, of course, require preparation of model epithelia and like the epithelial surface of the

METHODS IN MICROBIOLOGY, VOLUME 31
ISBN 0–12–521531–2

gut these model epithelia should be highly polarized. The exact type of model epithelia prepared should depend both on its physiological relevancy to the question being asked as well as any technical parameters the methods to be utilized require. Here, we discuss some of the cell lines and configurations that our group has utilized and describe in detail, as a specific example, preparation of our most commonly used type of T-84 model epithelia.

Cell types for model epithelia

Physiologists have long used stripped mucosal layers with the conjecture that, by stripping the mucosal layer from the circular muscle layer below, the nerve plexi were removed thereby eliminating the nervous input to the epithelial cells. However, it has been shown in a number of studies (Bridges *et al.*, 1986; Cooke, 1986) that even in stripped intestinal epithelium, effects from the remaining neurons can be observed. Moreover, the presence of immune effector cells in the mucosa adds to the complexity of the experimental model. Further complexity was recognized by the fact that other non-epithelial cells in the lamina propria may release paracrine signals to the epithelial cells. These confounding elements have been circumvented by the use of intestinal cell lines which can be grown as monolayers on permeable filters to an epithelium-like structure such that it is possible to study properties and responses of epithelial cells without interference of lamina propria cells. Since 1984, the T-84 human colon carcinoma cell line has been used as a model in Ussing chamber studies (Dharmsathaphorn *et al.*, 1984). Although investigators need to remember that all of these cell lines may develop properties not common to normal cells or lose specific features found in native epithelium, their homogeneity as well as the possibility to select clones with specialized functions e.g. electrolyte or mucous secretion, makes these cell lines very important tools in epithelial research. Some of the other commonly used intestinal epithelial cell lines that form polarized monolayers are shown in Table 21.1.

Table 21.1 Properties of some intestinal epithelial cell lines

Cell lines	Cell type and function	*TER* ($\Omega\,cm^2$)	Reference
T84	Colonic-like Cl-secretion	1000 to 2000	Dharmasthaphorn, 1984.
HT29-Cl.19A	Colonic-like Cl-secretion	100 to 200	Augeron and Laboisse, 1984.
HT29-Cl.16E	Colonic-like Cl-secretion and Mucous Secretion	100 to 200	Augeron and Laboisse, 1984. Merlin *et al.*, 1998
Caco2BBE	Small intestine-like Di/tri peptides absorption	100 to 200	Mooseker, 1985

Choosing configuration

There are a variety of commercially available apparati for growing cells on permeable filters. The variables are filter size, pore size, and material. In general, as bigger permeable supports tend to be more susceptible to defects, it is best to use the smallest size filter that will provide enough material for the experiment at hand. For example, a 0.33 cm^2 monolayer of T-84 cells secretes sufficient IL-8 for detection via ELISA but a 5 cm^2 monolayer will be needed to generate enough IL-8 mRNA for detection via Northern blot. Regarding pore size, most cells grow and polarize well on 0.4 μm pore size filters but such filters will prevent neutrophil movement and somewhat retard the movement of bacteria making them inappropriate for some studies. While the most commonly used cell culture filters are made of polycarbonate, transparent polyester filters are also available. Such filters improve the signal to noise ratio in fluorescence microscopy and fluorometric analysis of intracellular cations (specialized 'homemade' configuration described below), although we have found that epithelial cells do not grow quite as well on these materials. Lastly, to study pathogen-induced neutrophil transmigration in the physiologic direction, it is necessary to prepare model epithelia in an inverted configuration (described below).

Example—T84 model epithelia

T84 cells are available from American Type Culture Collection and can be used reliably to at least passage 75. Maintain the cells as confluent monolayers in a 1 : 1 mixture of Dulbecco's Vogt modified Eagle's media and Ham's F-12 medium supplemented with 15 mM N-2-hydroxyethyl-piperazine-N'-2-ethanesulfonic acid (HEPES) buffer (pH 7.5), 14 mM $NaHCO_3$, 40 μg ml^{-1} penicillin, 90 μg ml^{-1} streptomycin, and 5% newborn calf serum. Cells should be subcultured every 7 days. Cells can be subcultured or prepared for plating on permeable supports by trypsinization with 0.1% trypsin and 0.9 mM EDTA in Ca^{2+}/Mg^{2+}-free phosphate buffered saline (PBS). A critical requirement of this cell line is the need for relatively high density. Thus, splits of 1 : 2 are required to prevent loss of the polarized phenotype.

Collagen coat filters by adding 300 μg ml^{-1} rat tail collagen (in 70% ethanol) per square centimeter surface area overnight. Add T-84 cells (200 μl^{-1} of a 7.5×10^6 cells ml^{-1} stock per cm^2, i.e. 67 μl or 5×10^5 cells for a 0.33 cm^2 filter) to the collagen-coated filter. Allow 3 h for the cells to begin adhering and then add media to the underside of the filter, i.e. the basolateral reservoir (see Figure 21.1). T-84 cells cultured in this manner will form tight junctions in around 5 days and can generally be used for studying bacterial–epithelia interaction from 6–14 days without additional changing of media.

Preparation of inverted epithelia (Parkos *et al.*, 1991) requires silicone rings machined to fit over the lower surface of the permeable support surrounding the filter. One source of these rings is the Harvard University machine shop. Glue the rings onto the underside of the cell culture

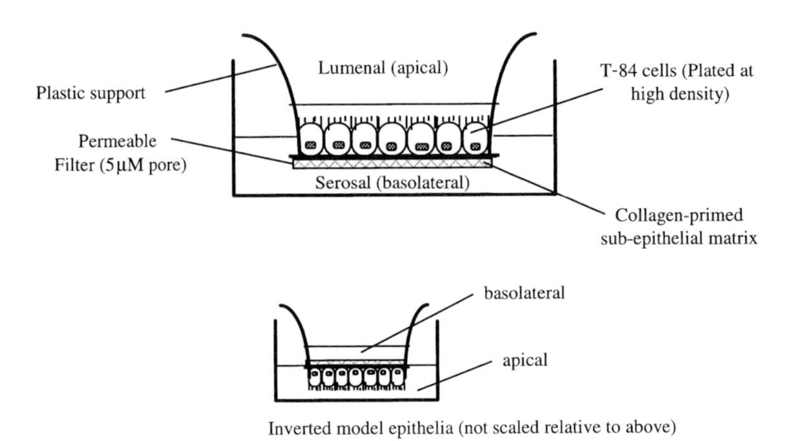

Figure 21.1. Schematic drawing of a model intestinal epithelia. T84 are cultured on permeable supports as described in text.

support with a silicone based adhesive. Next, sterilize the apparatus via brief incubation in 70% ethanol. Place the apparatus ring-side up in a sterile petri dish. Coat with collagen and add T-84 cells as described above. 12–18 h after adding the cells, right the inverted cell culture support and return to a standard 24-well tissue culture plate and add media to both sides of the filter (cells are now on the underside of the filter with the apical surface facing the bottom of the cell culture dish).

◆◆◆◆◆◆ TESTING MODEL EPITHELIA

Following their preparation, model epithelia should be tested to make sure they have the features expected of that cell type and that are important for the planned experiments.

Morphological examination

While not recommended as a routine way to test model epithelia, it is a good idea to initially, and occasionally, examine them structurally by electron microscopy or confocal microscopy (using orienting probes to f-actin and junctional proteins) using relatively standard previously described methods (Madara *et al.*, 1987). At low power, the epithelial cells should appear tightly packed and should completely cover all parts of the filter. The cells should form a single monolayer as opposed to stacked monolayers and there should not be any cells growing out of the underside of the filter. If this is not the case, the growth conditions should be altered or new cells passaged. At high power, intercelluar tight junctions, desmosomes, and adherens junctions should be visible near the apical surface. The monolayer should appear obviously polarized, though the degree of pronouncement of the microvilli (brush border) will vary considerably among model epithelia of different cell types (e.g. much more for Caco-2 BBE than T-84).

Biochemical testing

The apical and basolateral membranes of model, like natural, epithelia express distinct sets of proteins. For example, the apical but not basolateral membrane expresses CFTR while the reverse is true for β1 integrin. These examples are of particular importance to the study of bacterial epithelial interactions as CFTR serves as an entry point for *Salmonella typhi* (Pier *et al.*, 1998) while β1 integrin is an essential ligand for Yersinia spp. (McCormick *et al.*, 1997). Thus, biochemical (or immunohistological) examination of model epithelia will often be important for testing epithelia or interpreting the results of studies of bacterial–epithelial interactions. Examination of the polarity of expression of membrane proteins can be accomplished by immunostaining and confocal microscopy or by specifically biotinylating the apical or basolateral surfaces followed by immunoprecipitation with antibody to protein of interest and then western blotting using streptavidin for detection (Lencer *et al.*, 1995).

Physiological testing of model epithelia

Like native intestinal epithelia, model intestinal epithelia should be a significant barrier to passive transport. An additional property of colonic crypt epithelia is the ability to apically secrete chloride ions that drive the fluid secretion that is the molecular basis of secretory diarrhea. Both barrier function and chloride secretion can be rapidly assessed via electrical measurement (Madara *et al.*, 1992). The former is measured by transepithelial electrical resistance while the latter can be assayed by measuring short-circuit current. Physiologic examination of model epithelia should be carried out at 37°C with best results typically achieved 7–16 days post-plating. First, place model epithelia in a bicarbonate-free buffer such as Hanks buffered salt solution (HBSS). Alternatively, if it is necessary to perform such studies in standard cell culture media taken (i.e. bubbling through CO_2) to stabilize the pH. To determine currents, transepithelial potentials, and conductances, a commercial voltage clamp (Bioengineering Dept., Univ. of Iowa, Iowa City, IA) can be interfaced with equilibrated pairs of calomel electrodes submerged in saturated KCl and with paired Ag-AgCl electrodes. Apical and basolateral experimental medium volume should remain constant during the course of the experiment, and the electrodes should be stably placed on either side of the monolayer (see Figure 21.2). Positive currents correspond to anion secretion/cation absorption, i.e. a lumen-negative potential under open-circuit conditions. Before each experiment, a blank filter can be used to compensate for both the fluid resistance and the resistance of the filter. Transepithelial electrical resistance can be calculated from the measured potential via Ohm's law.

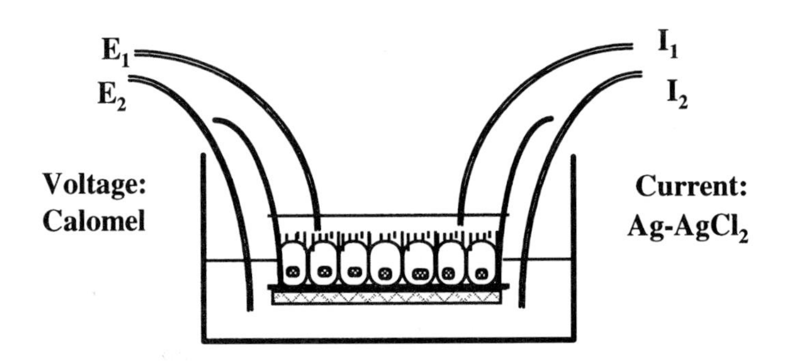

Figure 21.2. Measurement of TER and Isc of model intestinal epithelia. Electrodes are placed in both the apical and basolateral reservoirs to measure transepithelial electrical resistance and chloride secretion (ISC).

In addition to being used to verify the properties of epithelia prior to their use, these techniques can be used to study bacterial–epithelial interactions. For example some bacteria or their products can reduce barrier function as measured by transepithelial resistance. Furthermore, under some conditions bacterial–epithelial interactions lead to the production of mediators that induce epithelial chloride secretion. When measuring these parameters in response to experimental stimulus, it may be desirable to evaluate the ion conductance of the isolated apical plasma membrane. This can be done by adding the polyene ionophore amphotericin B to the basolateral solution at a concentration of 100 μM, the lowest concentration that gives a maximal change in steady-state conductance (Merlin *et al.*, 1998). Under these conditions, only the plasma membrane facing the amphotericin-containing solution (basolateral membrane in this case) incorporates the ionophore, thus electrically isolating the opposing (apical) plasma membrane. To study the Cl^- conductance of the apical plasma membrane, the voltage across the monolayer is sequentially stepped from a holding voltage of 0 mV to values between −40 and +40 mV over a period of ~10 s. Positive currents correspond to anion secretion/cation absorption, i.e. a lumen-negative potential under open-circuit conditions.

◆◆◆◆◆◆ BACTERIAL ADHERENCE AND INVASION

Assessment of bacterial invasion by gentamicin protection

The invasion potential of bacteria can be assessed in polarized epithelia by protection against gentamicin (McCormick *et al.*, 1993), a non-permeant bactericidal antibiotic, following procedures developed by Lee and Falkow (Lee *et al.*, 1992). Duplicate test-infected samples can be used to quantitate both total cell-associated and invaded bacteria during the same experimental period. On completion of the experimental infection period, the bacterial suspension on the apical surface of the monolayer is removed and the cells, grown on permeable supports or plastic, are gently washed six times in ice-cold phosphate buffered saline (PBS) to remove non-adherent bacteria and prevent further internalization. Duplicate monolayers are used in this case, one for total cell-associated bacteria, and the other to quantitate internalized organisms. The monolayers on which internalization will be assessed are placed in empty culture plate wells and 200 µl of pre-warmed PBS containing 500 µg ml^{-1} gentamicin (Sigma Chemical Co., St Louis, MO) is added to the apical surface and incubated at 37°C for 60 min. The duplicate monolayers are placed in the empty wells of a separate culture plate and 100 µl of pre-chilled 1% Triton X-100 (Sigma Chemical Co.) in PBS, placed on the apical surface of each and incubated on a rotating platform for 90 min at 4°C. On completion of the 90 min incubation the whole monolayer/filter-monolayer is disrupted with a sterile pipette tip and 900 µl of sterile Luria broth (LB) added and incubated for a further 10 min at 4°C. The gentamicin-treated monolayers are washed nine times per monolayer in PBS to remove any trace of the antibiotic and treated in the same manner as the total counts. Samples are then serially diluted in PBS and 100 µl of the final two dilutions spread on MaConkey agar plates and incubated overnight at 37°C. Dilution series: for total cell-associated counts 4–5, 1 : 10 dilutions and internalized 2–3, 1 : 10 dilutions. Adherence can be assessed as the total cell-associated counts minus the internalized counts.

Immunocytochemical staining of *Salmonella typhimurium*

(Clark *et al.*, 1996)

On completion of the experimental infection period, the bacterial suspension at the apical surface of the model epithelial monolayer is removed and the cells, grown on permeable supports, are gently washed six times in ice-cold buffer to remove non-adherent bacteria and prevent further internalization. To visualize the bacteria and assess the extent and pattern of bacterial adherence and internalization a dual immunocytochemical labeling technique was employed. The initial stage involved a 20 min reaction with anti-Salmonella antibodies (CSA-1) while the culture inserts were maintained on ice. The antibodies were prepared in goats by immunization with 17 strains of heat killed Salmonella to give a broad specificity, most likely against the 'O' antigens (1 : 200 in phosphate buffered saline (PBS); Kirkegaard & Perry Laboratories Inc., Gaithersburg, MD). The inserts were washed six times in ice-cold PBS and a secondary antibody added for 20 min, rabbit anti-goat IgG conjugated to fluorescein isothiocyanate (FITC), (1 : 100 in PBS; Sigma Chemical Co., St Louis, MO), again while maintained on ice. After a further six washes the culture inserts were placed in ice-cold methanol for 30 min to permeabilize the cell membranes. The inserts were washed in PBS and the staining procedure repeated with the goat anti-Salmonella antibody (CSA-1) (1 : 200 in PBS) for 20 min at room temperature. The secondary on this occasion is a rabbit anti-goat IgG conjugated to tetramethyl rhodamine isothiocyanate (TRITC) (1 : 50 in PBS; Sigma) for 20 min. Finally the inserts were washed and the inorganic membranes with the cells adhered to them removed and mounted with the apical side uppermost on a microscope slide in anti-fading medium (P-phenylenediamine; Sigma Co.). To assess the degree of adherence and internalization of the bacteria, the monolayers can be viewed using a fluorescent microscope with adherent bacteria (FITC-labeled) viewed under a blue filter at 520 nm and total numbers (TRITC-labeled) under a green filter at 576 nm and images collected using a confocal laser scanning microscope (CLSM) with an Argon/Krypton laser.

Bacterial transepithelial migration

The methods described above for quantitating bacteria can also be used to measure how many bacteria cross model epithelia into the basolateral reservoir. Simply add the bacteria to the apical reservoir and quantitate their numbers in the basolateral reservoir. Use a 5 μM pore size filter as smaller ones can somewhat impede bacterial movement. The absolute numbers of bacteria that cross the epithelia will typically be small. For example about 10^4 bacteria out of a 10^9 inoculum of *S. typhimurium* cross T-84 model epithelia within 90 min.

◆◆◆◆◆◆ BACTERIAL-INDUCED SIGNALS IN POLARIZED MODEL EPITHELIA

Bacteria interact quite differently with polarized versus non-polarized epithelial cells. Furthermore, bacteria can interact quite differently with the apical versus the basolateral membranes of polarized epithelia. For example, flagellin (the primary structural component of flagella) from either *E. coli* or Salmonella will activate IL-8 expression in non-polarized epithelia, but only induces this response in polarized epithelia if flagellin can gain access to the basolateral membrane. Thus, although the basic biochemical principles underlying the methods used to study signaling are the same in polarized and non-polarized cells, results can be vastly different with polarized versus nonpolarized cells. For this reason, we have adapted established methods for studying intracellular signaling so that they can be used with polarized model epithelia under configurations that allow selective access to the basolateral and apical compartments.

Measuring intracellular [Ca^{++}] (Gewirtz *et al.*, 2000)

The first step to measure Intracellular [Ca^{++}] is to prepare model epithelia in an appropriate configuration, preferably with the help of a good machine shop. Begin by cutting pieces of clear plastic (polycarbonate works well) such that they fit precisely diagonally across a standard fluorescence cuvette (see Figure 21.3). If possible, corner the side edges and cut an ellipse across the bottom (corresponding to the dimple on most disposable fluorescence cuvette) such that the piece will fit closely enough that, when its edges are greased, fluid added to one side of the cuvette

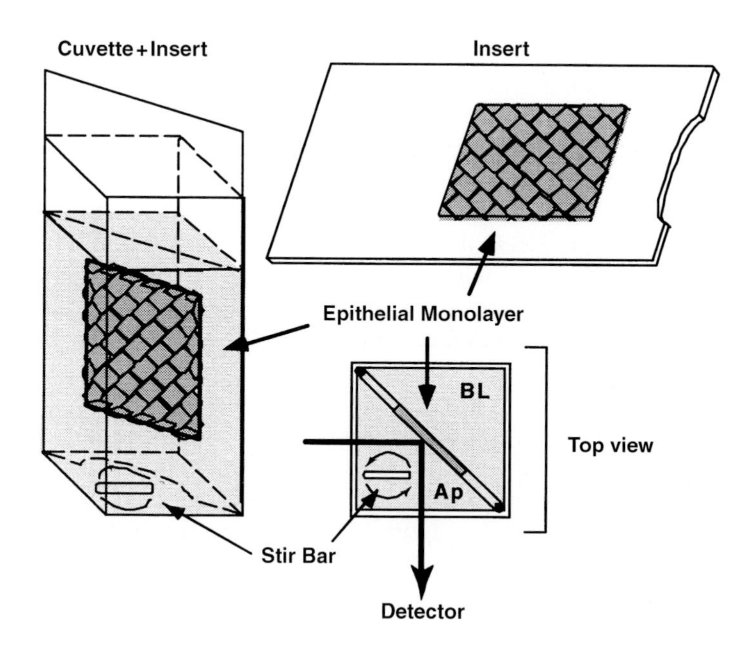

Figure 21.3. Measurement of Ca^{++} mobilization in model epithelia. Cell culture configuration that permits measurement of intracellular Ca^{++} in polarized model epithelia (details in text).

will not pass to the other side. Next cut a window in the plastic that will fully encompass the location where the excitation beam from the fluorimeter contacts. Coat the edges of the window with a silicone based adhesive (e.g. RTV 108, GE silicones, Waterford, NY), and place a transparent polyester filter (Corning Materials, Corning, NY) over the window and lightly press on it with a clean petri dish (or other convenient object) such that the polyester filter adheres fully across the window. After allowing the glue to dry (about 30 min) dip the plastic piece with attached filter in 70% ethanol to sterilize. Then coat the filter with rat tail collagen using 200 µl of a 100 $\mu g\,ml^{-1}$ collagen solution (in ethanol) per cm^2 of filter and allow to dry overnight.

In a petri dish, place gel spacers (or other similarly sized objects) below the plastic (i.e. filter holders) such that the filter is not touching the gel spacers or the petri dish. Add epithelial cells at high density (2×10^5 cells in 200 ml per cm^2 for T-84). Allow 3–4 h for the cells to begin adhering to the filter at which time media should be added to the petri dish such that the lower surface of the filter is in contact with media. This should permit formation of a fully confluent epithelial monolayer within 6–10 days. As the filter is somewhat transparent its confluence can be judged via a light microscope.

Next, wash the cells and filter apparatus (i.e. the model epithelia) three times in HBSS, and place into HBSS containing 5 µM Fura-2-acetoxymethylester (AM) (Molecular Probes, Eugene OR) for 75 min at 37°C, followed by three more washes. Coat the bottom and side edges of the plastic piece with vacuum grease and place into a fluorescence cuvette.

Place 1 ml of HBSS wash into the basolateral reservoir of the model intestinal epithelia. After verifying the integrity of the grease seal by noting that the HBSS did not visibly leak over to the apical aspect, add 1 ml of HBSS to the apical reservoir of this model epithelium. Then incubate the Fura-2-loaded model epithelia for an additional 10 min to permit diffusional washout of Fura-2 sequestered in the extracellular spaces. Aspirate out the HBSS (both apical and basolateral) and add fresh HBSS (1 ml to each aspect) without disturbing the grease seal.

Place the cuvette containing the Fura-2 loaded epithelia into a spectrofluorimeter (thermostated to 37°C) such that the apical surface faces (at a 45° angle) the excitation beam. Place a 4 mm stir bar (Fisher scientific) in the aspect (apical or basolateral) which is to receive addition of an agonist. Read fluorescence emission at 505 nm while the excitation wavelength is changed from 340 nm to 380 nm four times per second, preferably via a controlling computer. After reading fluorescence for 3–5 min, add bacteria or other agonist. To observe a response to *S. typhimurium* we add 10^9 CFU ml^{-1}.

Values of intracellular $[Ca^{++}]$, can be calculated via the Grynciewitz equation $(R - R_{min})/(R_{max} - R)^{*}K_d$ which equals about 2.54×10^{-7} M. R_{max} and R_{min} are measured by adding digitonin (10 μM) and then EGTA (20 mM) respectively to Fura-2-loaded model epithelia. Fura-2 leakage which occurred over the course of the $[Ca^{++}]$ measurement can be assessed and corrected for by adding EGTA (20 mM) at the end of the measurement and subtracting the changes in fluorescence that immediately (within 2 s) result. The autofluorescence of the bacteria as well as changes in autofluorescence of the model epithelia that were induced by the bacteria should also be measured using non-Fura-2-loaded epithelia and subsequently corrected for by subtraction. To increase the signal-to-noise ratio of our measurement the fluorescence values can be averaged to produce one ratio value and thus one value for intracellular $[Ca^{++}]$ every 2 s.

Activation of the NF-κB signaling pathway

Since its discovery in 1988, NF-κB has been recognized as a key signaling system for response to immune and pro-inflammatory stimuli. More recently, the NF-κB pathway has been demonstrated to play a central role in the mucosal response to bacteria. Indeed, a highly homologous pathway is present in insects, where the system functions as a rapidly inducible signaling mechanism against bacteria and other pathogens, and thus constitutes a primitive innate immune system.

Biochemically, NF-κB is a collective term for homo- or heterodimeric combinations of the rel family of transcriptional activators, which include the proteins p50, p65, RelB, and cRel. Various dimeric combinations may be present in specific cell types or different properties in transcriptional activation and DNA recognition. As a classic transcription factor, NF-κB is a sequence specific DNA binding protein, that binds to a characteristic motif (GGGYNNUUCC) present in the promoters of many immune and inflammatory effector genes.

NF-κB is activated by a unique post-translational process. The NF-κB dimer is sequestered in the cytoplasm by the action of a third protein, IκB or inhibitor of κB, of which several isoforms are known. As currently understood, pro-inflammatory stimuli, such as bacterial products or endogenous cytokines, induce a variety of signal transduction cascades that converge on the activation of a large multi-subunit IκB kinase complex. The catalytically active subunits, termed IκKinase α and β, phosphorylate IκB on conserved serine residues. This modification allows phospho-IκB to be recognized by a second multi-subunit complex, designated SCF-β-TrCP. The SCF complex is a ubiquitin ligase or E3 enzyme, which conjugates a series of 76 AA ubiquitin molecules to IkB. Polyubiquitination serves as a signal for recognition and rapid degradation by the 26S proteasome. Once IκB is degraded, the NF-κB dimer can then translocate to the nucleus via regulated transport across the nuclear pore, bind to relevant promoters and activate transcription.

One rapidly upregulated NF-κB responsive gene is IκB-α, the primary inhibitor of NF-κB translocation and binding. In most cells, IκB-α is newly synthesized within 60 min, enters the nucleus and removes DNA bound NF-κB. Thus, resynthesis of IκB forms a negative regulatory feedback loop, allowing the termination of the pro-inflammatory signaling and 'resetting' of the system.

In this section, we will discuss current methods of assaying the NF-κB pathway during microbial–epithelial interactions. Because bacterially elicited signals can be dependent on whether they interact with the apical or basolateral aspect of an epithelial monolayer, we will focus on the analysis of polarized models, as these culture systems more accurately reflect the physiologic epithelia.

EMSA

Perhaps the simplest assay for detection of activated NF-κB (and other DNA binding transcription factors) is the gel shift assay or EMSA (electrophoretic mobility shift assay). The mobility shift assay is based on the observation that DNA–protein complexes migrate through non-denaturing polyacrylamide gels more slowly than unbound DNA fragments. This technique utilizes a double-stranded DNA probe of 20–30 bp that spans a candidate NF-κB binding motif. The radiolabeled probe is incubated with cellular or nuclear extracts to allow nucleic acid/protein complexes to form. The binding reactions are performed in the presence of competitor DNA to suppress non-specific DNA interactions, and empirically established conditions of salt and divalent cations. The binding reactions are resolved on a non-denaturing gel; the free probe migrates rapidly through the gel, while nucleo-protein complexes are retarded in their migration, giving rise to a 'gel shift'. Resolved complexes must further be analyzed by specific and non-specific competition experiments to ensure the observed bands represent sequence specific DNA binding, and addition of antibodies to the binding reaction allows identification of bound proteins ('supershift' analysis).

EMSA is a biochemical population-based assay for NF-κB/DNA interactions that only indirectly correlates with NF-κB's transcriptional stimulating activity. Nevertheless, EMSA is a useful assay to unambiguously demonstrate the presence of NF-κB binding protein and can be used to determine whether a given organism or agonist induces nuclear translocation of DNA binding dimer, for following the kinetics of induction, and for assaying potential inhibitors of the activation process.

Preparation of nuclear extracts (Gewirtz *et al.*, 2000). A variety of protocols for preparation of nuclear and cytoplasmic extracts are available. The use of epithelial cells presents special problems in the generation of nuclear extracts. Because of the attachment of the basolateral aspect of the cells to the porous membrane necessary for polarized cultivation, the usual method of physically scraping the cells off a non-porous plastic substrate will rupture the polarized cells, precluding the preparation of useful nuclear extracts. We perform the hypotonic lysis step *in situ* by flooding the intact washed monolayer with a minimal volume of Buffer A (Hypotonic lysis buffer, 100 $\mu l\ cm^{-2}$ area) and scraping the cells with a rubber policeman. The cell suspension is transferred to a tube on ice, gently pelleted, and the size of the cell pellet estimated. The pellet is resuspended in an equal volume of Buffer A and lysis of the plasma membrane allowed to proceed for 5 min with occasional gentle agitation. Adequate lysis can be monitored by the loss of trypan blue exclusion. The suspension is centrifuged at 14 000 rpm for 20 s and supernatants removed. The nuclei are resuspended in 2/3 packed cell volumes of Buffer B (high salt extraction buffer), and shaken vigorously for 30 min. The mixture is pelleted at 14 000 rpm for 5 min, and the supernatant is removed and added to an equal volume of Buffer C, aliquoted and stored at −70°C.

Buffer A	Buffer B	Buffer C
10 mM Hepes, pH 7.9	20 mM Hepes, pH 7.9	20 mM Hepes, pH 7.9
1.5 mM $MgCl_2$	1.5 mM $MgCl_2$	20% Glycerol
10 mM KCl	420 mM KCl	100 mM KCl
1.0 mM DTT	1.0 mM DTT	1.0 mM DTT
0.5% NP-40	0.2 mM EDTA	0.2 mM EDTA

The best results require use of multiple protease inhibitors. All buffers will contain the following cocktail of protease inhibitors: 0.5 mM PMSF, 10 $mg\ ml^{-1}$ Leupeptin, 1 $mg\ ml^{-1}$ Pepstatin, 10 $mg\ ml^{-1}$ Aprotinin. All operations are performed at 4°C.

EMSA Binding reactions consist of probe, binding solution and nuclear extract. We generally use 2–4 μl of nuclear extract (5–10 μg protein) per reaction. Binding reactions (salt and divalent cation concentrations, pH, protein concentration, type and concentration of bulk carrier DNA) and gel electrophoresis conditions (acrylamide %, ionic strength, acetate vs borate) may need to be optimized empirically. Probes are generally constructed from 15–30 bp complementary oligonucleotides spanning the candidate NF-κB site, and are designed with single stranded overhangs for labeling with Klenow mediated incorporation of radiolabeled dCTP-^{32}P, by standard techniques.

For detection of NF-κB binding in extracts prepared as described, the following conditions generally work well. Binding reactions are 20–30 μl and contain of 5–10 μg protein with 50 000 cpm probe in 25 mM HEPES, pH 7.9, 150 mM NaCl, 0.5 mM DTT, 1 mM EDTA, and 1 μg poly dI·dC. Binding reactions are allowed to proceed for 20 min at room temperature and are loaded onto a 5% Tris/Borate/EDTA gel in 45 mM TBE.

In order to obtain meaningful results when using the mobility-shift assay, it is crucial to demonstrate the appropriate control reactions, including (1) competition of the observed gel shift with a molar excess of unlabeled probe (50–100 fold molar excess); (2) lack of competition with a similar sized, but structurally unrelated probe, or lack of competition with a probe containing a specifically mutated binding site.

Antibodies can be added to a gel shift reaction mixture to 'supershift' observed nucleoprotein complexes if the antibody is specific to the protein binding to the DNA probe. Antibody may be added to the binding reaction prior to the addition of probe, or after DNA–protein complexes have been allowed to form. This variation will allow the identification of the dimeric composition of the bound NF-κB. To demonstrate a specific binding interaction between an antibody and a supershifted band, a control binding reaction containing an equal amount of non-immune rabbit serum, or an antibody to a different protein, must be included in each experiment.

Unstimulated epithelial cells generally have little or no κB-specific DNA binding activity in properly prepared nuclear extracts. Cell stimulation with pro-inflammatory stimuli generally elicits an unambiguous qualitative appearance of DNA-bound NF-κB. In our experience, the translocation event can be detected as rapidly as 20 min with low MW soluble agonists such as TNF, and as long as 2 h with some invasive bacteria such as *S. typhimurium* (Gewirtz *et al.*, 2000). Cell type variation should be expected to influence the kinetics of NF-κB activation. Practically, these observations require that experiments exploring the activation potential of a given candidate agent or organism be studied over time.

Immunofluorescence

A complementary approach is to determine the subcellular location of NF-κB components, such as p65, is by immunohistochemistry or immunofluorescence. These techniques allow individual cell analysis (as opposed to population based techniques) to directly determine cytoplasmic vs nuclear location. As with EMSA analysis, this assay does not directly measure transcriptional activation.

Polarized monolayers grown on polycarbonate filters can be colonized with a given test organism, washed and fixed. We routinely excise the fixed monolayers still attached to the underlying support material, and process the mounted tissue away from the plastic tissue culture supports. Post-fixed monolayers are permeabilized and incubated with test antibody, washed, and followed by fluorescent-labeled secondary Ab. Conditions for fixation, permeabilization, and titration of primary and secondary antibodies have been well described though conditions need to be established for each new antibody. Relevant controls include peptide competition to establish specificity, omission of primary antibody to assess non-specific binding of the fluorescent secondary antibody, omission of secondary antibody to access endogenous autofluoresence.

As an example, we have had success with the anti-65 antibody from Rockland (Gilbertsville, PA) under the following conditions (Neish *et al.*, 2000). Post-treatment, cells are immediately washed in cold HBSS, fixed in 3% paraformaldehyde (20 min) and permeabilized with 0.1% saponin (20 min). Fixed cells are blocked in PBS/5% normal goat serum for 30 min at room temperature and then incubated with primary antibody, also in PBS/5%NGS for 1 h. Cells are washed in PBS and appropriate FITC-conjugated anti-rabbit secondary antibody applied for 30 min. Cells are again washed, treated with L-phenolein antifade, and mounted on glass slides. Fluorescence is viewed by standard fluorescence or laser confocal microscopy.

Control cells and cells exposed to inert bacteria or stimuli will exhibit cytoplasmic p50 and p65 while cells exposed to activating stimuli will exhibit nuclear staining of these proteins. Changes in relative cellular localization of these proteins change over the activation cycle as described above, so evaluation over a time course is essential. Nuclear counterstains such as DAPI, or cytoplasmic stains such as phalloidin may be useful to define nuclear and cytoplasmic boundaries in some cell types. The ability to stain with two antibodies (detected with different fluors) in combination with fluorescent histochemical stains greatly increases the power of this method by allowing the investigator to colocalize proteins and subcellular structures.

Analysis of IκB turnover

A major rate-limiting step in NF-κB activation is the degradation of the inhibitor IκB, an event that immediately precedes the translocation of the active dimer. A simple method to monitor this event in tissues and cultured cells is by immunoblotting cell lysates with antibodies to IκB family members. We routinely prepare whole cell lysates by flooding washed polarized monolayers with denaturing SDS-PAGE lysis buffer. This procedure minimizes potential proteolytic degradation and facilitates analysis of multiple monolayers, such as in a time course experiment. Lysates are analyzed by standard SDS-PAGE and Western blotting with an antibody to the IκB under study.

In response to proinflammatory stimuli such as TNF-α, near total degradation of whole cell levels of IκB-α are observed within 15–30 min. Proinflammatory stimuli such as invasive bacteria may be slower. For example, *S. typhimurium* requires 1–2 h of colonization to observe maximal degradation. As mentioned above, the IκB-α gene is rapidly induced by NF-κB activation. When evaluating the degradation of IκB in response to slow-acting stimuli this resynthesis can be concurrent with degradation, masking any potentially observable degradation and also, presumably, attenuating the amount of translocated NF-κB. The use of the eukaryotic protein synthesis inhibitor cycloheximide may be used to block resynthesis and eliminate such secondary effects (Gewirtz *et al.*, 2000).

The key step in IκB degradation is the induced phosphorylation on two serine residues. Thus, detection of this modified species is a useful marker of induction events proximal to IκKinase activation. Phospho-IκB can be observed in Western blots as a higher molecular weight species seen in lysates from activated cells. However, phospho-IκB is a substrate for the constitutively active E3-SCF complex and is rapidly degraded *in vivo*. The use of cell permeant inhibitors of the proteasome such as MG-132, MG-262, and lactacystin (Calbiochem, San Diego, CA; Affiniti Research Products, Exeter, UK) prior to experimental activation retards IκB degradation and permits visualization of this highly labile intermediate (Neish *et al.*, 2000). The availability of phospho-specific IκB antibodies is also useful to confirm this activation event (New England Biolabs, Beverly, MA).

Transient transfection

The detection of specific NF-κB binding in a nuclear extract, or visualization of nuclear p65, correlates with but does not directly prove transcriptional activation. A functional demonstration of NF-κB activation can be inferred by the transcriptional induction of genes that are known to require that factor. This is easily accomplished with NF-κB as numerous genes require NF-κB binding in their promoters for activation (this in no way implies NF-κB is sufficient), such as IL-8, TNF, and e-selectin, VCAM-1, etc. Another functional technique is the use of reporter gene constructs and transient transfection assays. In this approach, a synthetic promoter comprised of iterated motifs, for example three identical consensus NF-κB motifs, is coupled to an easily assayed 'reporter' gene. The fusion gene is contained in a transfectable plasmid. When introduced into cultured cells and stimulated with an appropriate agonist, activity of the reporter gene is interpreted as an index of promoter function. The relative promoter activities in transfected epithelial cells under both quiescent and infected conditions will provide a functional assay of NF-κB activity.

Numerous methods and commercially available reagents exist for these assays. Many vendors provide plasmid vectors for this purpose, including a choice of reporter gene, such as CAT, GFP, luciferase etc. Stratagene (La Jolla, CA) markets pre-made reporters bearing repeated NF-κB motifs expressly designed for this purpose. In addition there is a wide variety of transfection reagents for cultured cells and the choice is largely one of preference. A limitation of transient transfection assays is the variable 'transfectability' of polarized epithelial derived cell lines. Polarized Caco and HT-29 cells are generally amenable to this technique, however T-84 cell lines are virtually untransfectable by current chemically mediated techniques.

Bacterially induced changes in gene expression in polarized epithelia

Transcriptional control is the process of regulating new transcript synthesis. A given promoter is activated by the action of DNA binding transcription factors, of which NF-κB is a classic example. In the generally accepted models, transcription factors serve to recruit, activate or augment the RNA polymerase machinery that actually catalyzes creation of a nascent mRNA chain. Strictly speaking, transcriptional activation refers only to events that influence these processes. The 'gold standard' method for analysis of this event is the nuclear run-on assay, which measures incorporation of radiolabeled ribo-nucleotides into nascent mRNA chains from freshly prepared isolated nuclei. Alternatively, activation of reporter gene by a physiologic promoter sequence by the method of transient transfection is often used as a substitute to demonstrate transcriptional control of a particular candidate gene.

Changes in gene expression are understood to represent the increase or decrease in total quantity of a given mRNA transcript. Many genes activated by interaction of pathogens with epithelial cells, for instance certain chemokines, are regulated by post-transcriptional mechanisms, commonly, differential control of mRNA half-life. Since most assays of gene expression (Northern blotting, RT-PCR, microarray analysis) are based on the total steady-state level of transcription, post-transcriptional controls register as changes in gene expression, but are not changes in gene transcription. The distinction is not academic. If changes in gene expression are experimentally shown to be mediated by post-transcriptional mechanisms, analysis of promoter structure and DNA binding proteins is not likely to be informative.

In any event, measurement of steady-state mRNA levels is essential to the analysis of gene regulation. Total RNA is easily prepared from epithelial monolayers by washing, aspirating the wash solution, and flooding the cells with a strong denaturant followed by organic phase extraction. We routinely use Trizol reagent (GIBCO-BRL, Gaithersburg, MD). Numerous techniques exist for analysis of RNA. Northern blotting is semi-quantitative, and the electrophoretic separation step allows measurement of transcript size and detection of different splice forms. Qualitative changes in mRNA levels, such as *de novo* activation of IL-8 during infection, is easily detected by this method, especially when hybridization levels are normalized to a 'housekeeping' gene such as GAPDH or actin (Gewirtz *et al.*, 1998). More accurate quantitation has been attempted by reverse transcription PCR, and the recent introduction of 'real time' RT-PCR promises to refine this approach.

The large-scale, parallel analysis of mRNA abundance by microarray analysis is an emerging technique of great promise that has already been applied to the study of host-pathogen interactions (Eckmann *et al.*, 2000). These techniques utilize hundreds to thousands of cDNA or oligonucleotide elements immobilized to a solid substrate to generate an 'array'. Labeled RNA populations from cells under study are hybridized to the arrays and quantitated. While these experiments are conceptually similar to commonly used hybridization based techniques, the vast amounts of data generated presents new problems in interpretation, requiring the use of powerful data management and data-mining software. It is likely that 'in silico' analysis of gene expression will become the norm in the near future.

Polarized cytokine/chemoattractant secretion

It is now well established that the epithelial cells that line the intestine secrete a panel of cytokines and chemokines that regulate mucosal immune cells. While non-polarized epithelial cell lines (e.g. Henle-407) as well as epithelial cells grown in standard tissue culture plates (i.e. non or minimally polarized) also secrete cytokines and chemokines, both the response to stimuli and the chemokine secretion is polarized and thus studies using polarized model epithelia will often yield different and much more physiologically relevant results. Specifically, not only does the ability of a bacteria or its metabolites to activate chemokine secretion depend on what membrane surfaces it accesses, but also the chemokine secretion itself is polarized. While the majority of cytokines (e.g. IL-8, TNF, MCP, ENA-78) are secreted basolaterally, other important cytokines (e.g. IL-6) as well as pathogen-elicited epithelial chemoattractant (PEEC) are secreted apically. Thus, although it is a little more work, important physiological insights can come from studying bacterial-induced cytokine/chemokine secretion in a polarized system.

Here, we describe the basic assay of measuring IL-8 secretion from polarized model epithelia (Gewirtz *et al.*, 1998). Other cytokines could, of course, be measured by simply using a different ELISA. Alternatively, various bacterial strains, mutant epithelial cells, or a variety of pharmacologic inhibitors can be added as is appropriate to the question being asked, with the one caveat being that if the drug or mutant protein has a major effect on transepithelial resistance, this could lead to erroneous conclusions.

First, prepare *S. typhimurium* by placing a colony into 10 ml LB in a 50 ml tube and grow at 37°C while shaking for 6–8 h. Dilute 1 to 1000 and grow overnight (15 h) at 37°C without shaking. Centrifuge at 5000*g* for 10 min, wash in HBSS and recentrifuge, and resuspend pellet so bacteria are concentrated 30-fold from overnight culture (this yields an approximate concentration of 10^{10} organisms ml^{-1}). Next, wash 0.33 cm^2 model epithelia three times in HBSS and place into 300 μl of HBSS, and add 50 ml HBSS to apical reservoir. Allow to equilibrate at 37°C for 30–45 min at which time 25 ml of the concentrated (or diluted as desired) bacterial suspension is added. When adding the bacteria, mix them to disperse evenly throughout the apical reservoir. One hour later, if desired, rinse off non-adherent organisms by dipping into HBSS. Five hours after initially adding bacteria discard the monolayer and add 33 μl of 10× wash buffer (5% Goat Serum, 1% Tween-20 in HBSS) of PBS to basolateral supernatants. These samples can then be stored overnight at 4°C or at −20°C for longer. Measure IL-8 via ELISA.

A variety of reagents for cytokine ELISAs are commercially available. The following 'sandwich' ELISA works well for IL-8. Coat 96-well plates (non-tissue culture treated) overnight with goat anti-human IL-8 (R&D systems) using 100 ml of 8 mg ml^{-1} Ab in 0.1 M $NaHC0_3$ (pH-9.6). Aspirate and wash four times with wash buffer (0.5% Goat Serum, 0.1% Tween-20 in HBSS) and add 100 µl per well samples or standards (0.025–10 ng ml^{-1} recombinant IL-8 in wash buffer). Incubate for 1h at 37°C. Aspirate and wash four times with wash buffer and add 100 µl per well of 8 mg ml^{-1} rabbit anti-human IL-8 (in wash buffer). Incubate for 1h at 37°C. Aspirate and wash four times with wash buffer and add 100 µl per well of 0.15 mg ml^{-1} goat anti-rabbit (HRP coupled, KPL, in wash buffer). Incubate for 1 h at 37°C. Aspirate and wash four times with wash buffer, add HRP substrate (TMB from KPL works well), and read absorbance at appropriate wavelength.

◆◆◆◆◆◆ BACTERIAL-INDUCED NEUTROPHIL MOVEMENT

As a consequence of the epithelial chemokine secretion induced by *S. typhimurium*, neutrophils are recruited to and directed to migrate across the intestinal epithelium resulting in a crypt abscess, which is the defining hallmark of active intestinal inflammation. This PMN movement can be measured across model epithelia and is induced by pathogenic *S. typhimurium* but not by normal gut *E. coli*. Regardless of where *S. typhimurium* is added, the epithelial chemokine solution is such that PMN will migrate from the basolateral to the apical reservoir but will not migrate in the reverse direction without an exogenously applied chemotactic gradient. Thus, use of inverted epithelia is essential to study bacterial-induced neutrophil movement.

Salmonella typhimurium-induced PMN movement can be considered a multi-stage process with two of the steps being movement through the subepithelia or lamina propria (transmatrix migration) followed by migration across the epithelium itself. These events are driven by distinct chemokines and can be measured separately from one another.

PMN transepithelial migration assay (McCormick *et al.*, 1993)

S. typhimurium-induced PMN trasmigration is largely driven by pathogen-elicited chemoattractant (McCormick *et al.*, 1998) and is measured as follows. Place model epithelia in HBSS. Prepare a culture of 10^{10} organisms ml^{-1} of *S. typhimurium* in HBSS as described above. Remove inverted model epithelia from each well and place upside-down in a moist chamber such that the epithelial apical surface faces upward. Layer 25 µl of the bacterial suspension (approximately 1.25×10^8) gently onto the apical surface and incubate for 60 min at 37°C. Remove non-adherent bacteria by washing three times in HBSS buffer. Transfer the inverted model epithelia right-side up back into the 24-well tissue culture tray containing 1.0 ml HBSS buffer in the lower (apical membrane now oriented upward and colonized with *S. typhimurium*) reservoir and add 160 µl to the upper (basolateral) reservoir. To the basolateral reservoir, add 40 µl (1×10^6) of isolated PMN and incubate for 2 h at 37°C. For a positive control, include non-colonized epithelia where 10^{-7} M fMLP is present in the apical reservoir. The number of transmigrated PMN can most easily be quantified by collecting the basolateral reservoir, adding 0.5% Triton X-100, and assaying for the PMN azurophilic granule marker myeloperoxidase using a variety of peroxidase substrates such as ABTS. Prepare lysates from a series of dilutions of PMN (from 1×10^6 to 5×10^4) to serve as standards for quantitation.

PMN transmatrix assay (McCormick *et al.*, 1995)

A variant of the above technique can be used to measure neutrophil movement through the sub-epithelial matrix (corresponding to the lamina propria). Transmatrix migration is driven by secreted IL-8 that imprints the matrix with a long-lasting haptotactic gradient for PMN. To measure transmatrix migration, colonize inverted model epithelia as described above. Allow 3 h after initial bacterial–epithelial contact (2 h after removing non-adherent bacteria) for the secretion of chemokines that will imprint the matrix with a chemotactic gradient to occur. Remove the epithelial cells from their matrix and the filter with two successive 1h incubations with 500 mM EGTA (while swirling). There should be no visible portions of epithelial cells remaining. Return the filters to a 24-well plate, add PMN as described above, and quantitate migration over a 90 min period. Unlike neutrophil transepithelial migration, matrices from uncolonized epithelia will support considerable migration (about 15×10^5 PMN), due to either basal IL-8 secretion or the rate of PMN simply falling through the filter.

References

Augeron, C. and Laboisse, C. L. (1984). Emergence of permanently differentiated cell clones in a human colonic cancer cell line in culture after treatment with sodium butyrate. *Cancer Res.* **44**(9), 3961–3969.

Bridges, R. J., Rack, M., Rummel, W. and Schreiner, J. (1986). Mucosal plexus and electrolyte transport across the rat colonic mucosa. *J. Physiol. (Lond)* **376**, 531–542.

Clark, M. A., Reed, K. A., Lodge, J., Stephen, J., Hirst, B. J. and Jepson, M. A. (1996). Invasion of murine intestinal M cells by Salmonella typhimurium inv mutants severely deficient for invasion of cultured cells. *Infect Immun.* **64**(10), 4363–4368.

Cooke, H. J. (1986). Neurobiology of the intestinal mucosa. *Gastroenterology* **90**(4), 1057–1081.

Dharmsathaphorn, K., Mandel, K. G., Mc Roberts, J., Tisdale, L. D. and Masui, H. (1984). A Human colonic tumor cell line that maintains vectorial electrolyte transport. *Am. J. Physiol.* **246**, 6204–6208.

Eckmann, L., Smith, J. R., Housley, M. P., Dwinell, M. B. and Kagnoff, M. F. (2000). Analysis by high density cDNA arrays of altered gene expression in human intestinal epithelial cells in response to infection with the invasive enteric bacteria Salmonella. *J. Biol. Chem.* **275**(19), 14084–14094.

Gewirtz, A. T., Mc Cormick, B., Neish, A. S., Petasis, N. A., Gronert, K., Serhan, C. N. and Madara, J. L. (1998). Pathogen-induced chemokine secretion from model intestinal epithelium is inhibited by lipoxin A4 analogs. *J. Clin. Invest.* **101**(9), 1860–1869.

Gewirtz, A. T., Rao, A. S., Simon, P. O., Jr., Merlin, D., Carnes, D., Madara, J. L. and Neish, A. S. (2000). Salmonella typhimurium induces epithelial IL-8 expression via Ca(2+)- mediated activation of the NF-kappaB pathway. *J. Clin. Invest.* **105** (1), 79–92.

Lee, C., Jones, B. and Falkow, S. (1992). Identification of a *Salmonella typhimurium* invasion locus by selection for hyperinvasive mutants. *Proc. Natl Acad. Sci. USA* **89**, 1874–1851.

Lencer, W. I., Moe, S., Rufo, P. A. and Madara, J. L. (1995). Transcytosis of cholera toxin subunits across model human intestinal epithelia. *Proc. Natl Acad. Sci. USA* **92**(22), 10094–8.

Madara, J., Stafford, J. and Carlson, S. (1987). Structural analysis of a human intestinal epithelial cell line. *Gastroenterology* **92**, 1133–1145.

Madara, J. L., Colgan, S. P., Nusrat, A., Delp, C. and Parkos, C. A. (1992). A simple approach to measurement of electrical parameters of cultured epithelial monolayers: use in assessing neutrophil epithelial interactions. *J. Tissue Culture Meth.* **15**, 209–16.

McCormick, B. A., Nusrat, A., Parkos, C. A., D'Andrea, L., Hofman, P. M., Carnes, D., Liang, T. W. and Madara, J. L. (1997). Unmasking of intestinal epithelial lateral membrane beta1 integrin consequent to transepithelial neutrophil migration in vitro facilitates inv-mediated invasion by Yersinia pseudotuberculosis. *Infect. Immun.* **65**(4), 1414–1421.

McCormick, B. A., Parkos, C. A., Colgan, S. P., Carnes, D. K. and Madara, J. L. (1998). Apical secretion of a pathogen-elicited epithelial chemoattractant (PEEC) activity in response to surface colonization of intestinal epithelia by Salmonella typhimurium. *J. Immunol.* **160**, 455–456.

McCormick, B. A., Colgan, S. P., Archer, C. D., Miller, S. I. and Madara, J. L. (1993). Salmonella typhimurium attachment to human intestinal epithelial monolayers: transcellular signalling to subepithelial neutrophils. *J. Cell Biol.* **123**, 895–907.

Merlin, D., Augeron, C., Tien, X. Y., Guo, X., Laboisse, C. L. and Hopfer, U. (1994). ATP-stimulated electrolyte and mucin secretion in the human intestinal goblet cell line HT29-C1.16E. *J. Membr. Biol.* **137**(2), 137–149.

Merlin, D., Jiang, L., Strohmeier, G. R., Nusrat, A., Alper, S. L., Lencer, W. I. and Madara, J. L. (1998). Distinct Ca2+- and cAMP-dependent anion conductances in the apical membrane of polarized T84 cells. *Am. J. Physiol.* **275**(2 Pt 1), C484–495.

Mooseker, M. S. (1985). Organization, chemistry, and assembly of the cytoskeletal apparatus of the intestinal brush border. *Annu. Rev. Cell Biol.* **1**, 209–241.

Neish, A. S., Gewirtz, A. T., Zeng, H., Young, A. N., Hobert, M. E., Karmali, V., Rao, A. S. and Madara, J. L. (2000). Prokaryotic regulation of epithelial responses by inhibition of IκB-alpha ubiquitination [see comments]. *Science* **289**(5484), 1560–1563.

Parkos, C. A., Delp, C., Arnaout, M. A. and Madara, J. L. (1991). Neutrophil migration across a cultured intestinal epithelium: dependence on a CD1lb/CD18- mediated event and enhanced efficiency in the physiologic direction. *J. Clin. Invest.* **88**, 1605–1612.

Pier, G., Grout, M., Zaidi, T., Meluleni, G., Mueschenborn, S., Banting, G., Ratcliff, R., Evans, M. and Colledge, N. (1998). Salmonella typhi uses CFTR to enter intestinal epithelial cells. *Nature* **393**, 79–82.

22 Activation of Innate Immune Receptors by Bacterial Products

HD Brightbill and RL Modlin
Department of Microbiology, Immunology and Molecular Genetics, Division of Dermatology, UCLA School of Medicine, University of California, Los Angeles, CA, USA

◆◆◆

CONTENTS

List of abbreviations

APC	Antigen presenting cell
DMEM	Dulbecco's modified Eagle's medium
ELISA	Enzyme linked immunosorbent assay
EMSA	Electrophoretic mobility shift assay
FCS	Fetal calf serum
GM-CSF	Granulocyte macrophage-colony stimulating factor
IL	Interleukin
IFN-γ	Interferon-gamma
IKK	I-κB kinase
I-κB	Inhibitor of NF-κB
iNOS	Inducible nitric oxide synthase
IRAK	IL-1 receptor accessory protein kinase
JNK	Jun N-terminal kinase
LPS	Lipopolysaccharide
NF-κB	Nuclear factor-κB
NIK	NF-κB inducing kinase
NO	Nitric oxide
PBMC	Peripheral blood mononuclear cells
PBS	Phospho-buffered saline
TLR	Toll-like receptor
TNF	Tumor necrosis factor
TRAF	TNF receptor associated factor

METHODS IN MICROBIOLOGY, VOLUME 31
ISBN 0-12-521531-2

◆◆◆◆◆◆ INTRODUCTION

A renewed interest in the innate immune response has emerged with the identification of a family of receptors involved in the response to infection in flies (Chapter 27, this volume), plants and mammals. Such a striking similarity between these organisms suggests that these receptors play a fundamental role in regulating how the immune system responds to pathogenic micro-organisms.

Mammalian Toll-like receptors (TLR) were identified for their homology to the Toll receptor family of *Drosophila* that is involved in the response to fungal and bacterial infection (Belvin and Anderson, 1996). Unlike the adaptive immune system, which generates its receptor repertoire through rearrangements of receptor gene segments, Toll-like receptors have evolved predetermined specificity (Medzhitov and Janeway, 1997). The mammalian Toll-like receptors are composed of a growing number of family members, of which TLR2, TLR4, and the B cell specific, RP-105, are the most widely studied thus far. Also a member of the IL-1R family, TLRs mediate their activity through common signaling pathways, such as the evolutionarily conserved, Rel/NF-κB and MAP kinase pathways. Interestingly, these signaling pathways play a fundamental role in the activation of inflammatory cytokine gene expression (IL-12, TNFα), antimicrobial responses (iNOS, NO) (Chapter 26, this volume) and apoptosis (Chapter 24, this volume) in response to microbial ligands (Yang *et al.*, 1998; Kirschning *et al.*, 1998; Brightbill *et al.*, 1999; Aliprantis *et al.*, 1999).

A wide range of microbial ligands have been identified that activate these innate immune responses, the most well studied being lipopolysaccharide (LPS). The exact mechanism of how TLRs respond to these ligands is the subject of much debate, however studies of TLR deficient mice suggest that TLRs have clear specificity for particular pathogens and their ligands (Takeuchi *et al.*, 1999; Brightbill and Modlin, 2000). Much like the *Drosophila* Toll receptors, TLR specificity may have important implications for which downstream target genes are activated by TLRs in response to different pathogens. This chapter addresses the techniques and reagents available to study Toll-like receptors; specifically, how they are activated by microbial ligands from a wide range of microbial pathogens. In addition, this chapter will serve to also address the molecular regulation of target genes downstream of TLR activation. We feel this to be a future direction of TLR biology and hope this will give insight into the role of TLR signal transduction and downstream gene regulation in determining the influence of these receptors on innate immune responses to infection.

◆◆◆◆◆◆ MICROBIAL LIGANDS

A great number of microbial molecules have been identified that elicit strong responses from cells of the innate immune system. Over the past 2 years these microbial molecules have been shown to activate Toll-like receptor family members (Table 22.1). Of the receptors identified thus far, TLR2 has a varied specificity including molecules from Gram-positive bacteria, yeast, and mycobacteria, such as microbial lipoproteins; while TLR4 and RP105 primarily respond to LPS and Gram-negative bacteria (Hoshino *et al.*, 1999; Takeuchi *et al.*, 1999, 2000; Ogata *et al.*, 2000; Yang *et al.*, 1998; Kirschning *et al.*, 1998; Brightbill *et al.*, 1999; Hirschfeld *et al.*, 1999; Underhill *et al.*, 1999). Interestingly, the TLR4 mutant mouse strains and TLR-deficient mice have implicated a role for TLR specificity in the innate immune response to infection. The microbial ligands and mutant mouse strains are reviewed below.

Microbial ligands and TLR specificity

For many years, a receptor for LPS was sought to explain the responses involved in Gram-negative bacterial sepsis. Following the identification of TLR4, Toll-like receptors were referred to as 'the missing piece of the puzzle' (Wright, 1999). In addition to LPS, other molecules that have been purified from Gram-negative and Gram-positive bacteria, mycobacteria, and yeast that elicit similar inflammatory responses from cells of the innate immune system, such as the macrophage. Although these potential microbial TLR ligands are derived from such a wide variety of organisms,

Table 22.1 TLR family members mediate the activation of microbial ligands

Microbial ligand	Associated TLR family member	Reference
LPS, Lipid A	TLR-4, RP-105 (B cells)	
Gram-negative cell walls	TLR-4	Takeuchi *et al.*, 1999
Gram-positive lipoteichoic acids	TLR-4	Takeuchi *et al.*, 1999
Microbial lipoproteins	TLR-2	Brightbill *et al.*, 1999; Hirschfeld *et al.*, 1999; Takeuchi *et al.*, 2000
Synthetic lipopeptides	TLR-2	Brightbill *et al.*, 1999; Hirschfeld *et al.*, 1999 Aliprantis *et al.*, 1999; Takeuchi *et al.*, 2000
Gram-positive peptidoglycan	TLR-2	Takeuchi *et al.*, 1999
Gram-positive cell walls	TLR-2	Takeuchi *et al.*, 1999
Lipoarabinomannan, avirulent *M. tb*	TLR-2	
Yeast	TLR-2	Underhill *et al.*, 1999
Virus	TLR-2	Kurt-Jones *et al.*, 2000
Bacterial CpG DNA	TLR-9	Hemmi *et al.*, 2000

they all share common characteristics that the innate immune response has evolved to respond to. Due to their distinct structure, microbial ligands are referred to as 'Pathogen Associated Molecular Patterns' (PAMPs). They all share a repetitive structure that is distinct from the host and essential for the survival and pathogenicity of the organism (Medzhitov and Janeway, 1997; Medzhitov *et al.*, 1997). Microbial lipoproteins from spirochete and mycoplasma species, and *Mycobacterium tuberculosis* (*M. tb*) have been purified and shown to induce macrophage activation and cytokine production to similar levels as LPS (Radolf *et al.*, 1995; Takeuchi *et al.*, 2000). These lipoproteins share a common NH_2-terminal lipidation motif that is responsible for its activity (Table 22.2, structure outlined below). Other stimulatory ligands include lipoarbinomannan from *M. tb* (Means *et al.*, 1999), Gram-positive cell walls (Takeuchi *et al.*, 1999), peptidoglycan and lipoteichoic acids from *Listeria monocytogenes* and *Staph aureus* (Flo *et al.*, 2000; Takeuchi *et al.*, 1999; Yoshimura *et al.*, 1999; Schwandner *et al.*, 1999), and rough preparations of Gram-negative bacterial cell walls (Takeuchi *et al.*, 1999), yeast (Underhill *et al.*, 1999) and virus (Kurt-Jones *et al.*, 2000) (Table 22.1).

TLR mutant mice

Early studies of LPS responses identified three mouse strains that were naturally hyporesponsive to LPS stimulation. These include C3H/HeJ, C57BL/10ScCr and C57BL/10ScN strains (Vogel, 1992; Qureshi *et al.*,

Table 22.2 (A) N-terminal lipid (PAM3Cys) motif. (B) Hexapeptide sequences from *Borrelia burgdorferi* and *Treponema pallidum* lipoproteins

(A)

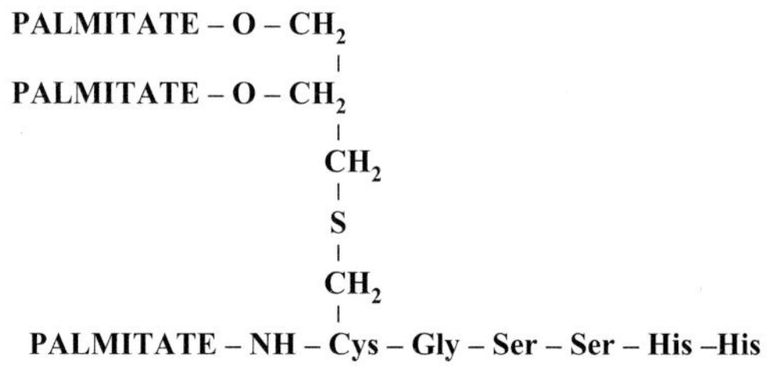

```
PALMITATE – O – CH2
                 |
PALMITATE – O – CH2
                 |
                CH2
                 |
                 S
                 |
                CH2
                 |
PALMITATE – NH – Cys – Gly – Ser – Ser – His –His
```

(B)

Borrelia burgdorferi	
OspA	**Cys-Lys-Gln-Asn-Val-Ser**
OspB	**Cys-Ala-Gln-Lys-Gly-Ala**
Treponema pallidum	
Tp47	**Cys-Gly-Ser-Ser-His-His**

1999a,b). The various mutations in these mice all involve the TLR4 gene (Poltarak *et al.*, 1998a,b; Qureshi *et al.*, 1999a,b). Although these mouse strains are hyporesponsive to high doses of LPS they respond normally to microbial lipoproteins (Vogel *et al.*, 1992; Qureshi *et al.*, 1999a,b; Ma and Weiss, 1993). These early observations suggest that much like in *Drosophila*, multiple mammalian receptors mediate differential responses to microbial infection.

The characterization of TLR2 and TLR4 knockout mice has put these responses into perspective. TLR4-deficient mice fail to respond to LPS. Along with the mutant mouse strains discussed above, TLR4 has been accepted as the primary LPS receptor. Although TLR4−/− macrophages are hyporesponsive to LPS they still are able to respond normally to other ligands, such as microbial lipoproteins. The opposite is true for TLR2-deficient mice, which respond normally to LPS, but are hyporesposive to Gram-positive bacteria and microbial lipoproteins (Table 22.3; Takeuchi *et al.*, 2000).

In contrast to TLR2 and TLR4, which are found on a number of different cell types including macrophages, dendritic cells and endothelial cells, the TLR family member, RP105, appears predominantly on B cells and may serve to mediate B cell responsiveness to LPS. RP105-deficient mice develop their B cell compartment normally, but do not respond to LPS. RP105 deficient B cells exhibit impaired proliferation and antibody responses when stimulated with LPS (Ogata *et al.*, 2000). Interestingly, this study implicated some cooperativity between RP105 and TLR4 in a transfected cell line. Cooperativity between TLR family members has been suggested as a means of mediating pathogen and cell type specificity,

Table 22.3 TLR deficient mice and mutant mouse strains

Factor	Phenotype	Reference
TLR4−/−	Do not respond to LPS	Hoshino *et al.*, 1999
TLR2−/−	Do not respond to Gram-positive bacterial cell walls, peptidoglycan, or microbial lipoproteins	Takeuchi *et al.*, 1999, 2000
RP105−/−	B cells do not respond to LPS	Ogata *et al.*, 2000
CD14−/−	Do not respond to LPS	Haziot *et al.*, 1996
MyD88−/−	Do not respond to LPS	Kawai *et al.*, 1999
C3H/HeJ	TLR4 mutation. LPS hyporesponsive	Vogel, 1992
C57BL10/ScCN	TLR4 mutation. LPS hyporesponsive	Qureshi *et al.*, 1999; Poltorak *et al.*, 1998a,b; Qureshi *et al.*, 1991a,b
Hamster macrophages	TLR2 null mutation	Heine *et al.*, 1999

since heterodimerization has been observed between TLR family members (Ozinsky *et al.*, 2000). A newly identified family member, TLR9, has recently been implicated in macrophage responses to bacterial CpG DNA (Hemmi *et al.*, 2000). The number of TLR family members is growing rapidly and as the specificity of these receptors is characterized, the set of parameters that regulate these responses is more than likely to be very complex indeed.

Synthetic lipopeptides

Microbial lipoproteins are able to activate macrophages by a common N-terminal lipidation modification. Hexapeptides containing amino acid sequences from a number of microbial lipoproteins can be synthesized by peptide synthesizer (Table 22.3; DeOgny *et al.*, 1994). The peptides themselves do not activate via Toll-like receptors, and serve as an appropriate negative control for activation. M.V. Norgard (Univ. Texas Southwestern) using a solid-phase chemistry synthesis procedure (DeOgny *et al.*, 1994), produced tripalmitoyl derivatives (PAM_3Cys) of these peptides that closely resemble the modification of these lipoproteins *in vivo* and are able to activate macrophages in a TLR-dependent manner (Brightbill *et al.*, 1999; Hirschfeld *et al.*, 1999). Synthetic Lipid A derivatives of LPS can also be obtained commercially (Sigma Chemicals) (Ogata *et al.*, 2000; Lien *et al.*, 2000). Although these peptides are synthetic in nature they do provide an important tool to study the activation of TLRs, in that they are more easily obtained than the purified lipoproteins from the organism. In addition, the threat of endotoxin contamination is greatly reduced.

Preparations of synthetic peptides

Lipid-modified hexapeptides are rather insoluble and should be resuspended in sterile water or fetal calf serum. Aliquots of resuspended peptides may be kept in the −70°C freezer. Prior to use, peptides should be vortexed until they are fully resuspended in solution. Non-lipidated peptides should be used as negative controls and titrations should be conducted to determine the optimum concentration to be used (1–10 $\mu g\, ml^{-1}$) (DeOgny *et al.*, 1994).

◆◆◆◆◆◆ ISOLATION OF PRIMARY MACROPHAGES AND COMMONLY USED CELL LINES

Cells of the innate immune system, such as the macrophage, play a fundamental role in host defense and rely on ancient mechanisms of responding to infectious micro-organisms, such as phagocytosis, antimicrobial responses and inflammation. In addition to their role in the innate immune response, these 'antigen presenting cells' (APC) provide the link to the adaptive immune response through antigen presentation and cytokine production. Macrophages and other APCs have been shown to express TLR family members (Medzhitov and Janeway, 1997; Medzhitov *et al.*, 1997). As we discuss later, since TLRs mediate responses to these microbial ligands their downstream targets may play a critical role not only in innate inflammatory responses. In addition, TLR targets influence the adaptive immune response through differential induction of critical cytokines that have been shown to bias immune responses in disease. The following methods outline the isolation and activation of primary macrophages from mouse and human, as well as commonly used cell lines.

Mouse

Mouse macrophages can be isolated by different techniques. Care should be taken to decide which approach is best for the questions being asked in the study since elicited macrophages may have a more activated phenotype than other macrophages due to the reagents used. Mouse primary macrophages can be isolated from peritoneum, bone marrow, thymus and spleen. In addition, macrophages can be elicited to the peritoneum by Thioglycollate (Johnson *et al.*, 1978), Biogel polyacrylamide (Fauve *et al.*, 1983) or BCG organisms (Haworth and Gordon, 1998). Detailed descriptions of these techniques are beyond the scope of this chapter, however these techniques are reviewed in Haworth and Gordon (1998) and in Current Protocols in Immunology (1991). TLR-deficient mouse have predominantly used thioglycolate elicited macrophages. The advantage to elicited macrophages is the higher cell number that can be harvested compared to resident macrophages.

Human

Human monocyte/macrophages can be isolated from peripheral blood and are commonly used to study cytokine induction by microbial ligands (Coligan *et al.*, 1993). Blood can be isolated directly from donors or through Leukopak preparations available from the American Red Cross. Care should be taken to take proper precautions and have authorization from the host institution when conducting experiments with human cells.

Methods

Isolation of human peripheral blood monocytes

1. Dilute blood 1:1 with complete RPMI 1640 media (10% FCS, 2 mM L-glutamine, 50 U ml^{-1} penicillin and streptomycin).
2. Isolate white blood cells by density gradient centrifugation. Layer 30 ml of blood carefully over 15 ml of Ficoll-Hypaque cushion (Pharmacia). Do not mix.
3. Centrifuge at 1600 rpm for 20 min at room temp. Do not use brake.

4. Aspirate supernatant within a few ml of the interface containing white blood cells.
5. Collect interface by pipette (only 15 ml). Two interfaces may be combined, then add RPMI media to bring volume up to 50 ml in a conical tube.
6. Centrifuge at 1800 rpm (550*g*) for 10 min at room temperature. May use brake.
7. Aspirate supernatant and resuspend pellet in a few ml of RPMI and pool common cell suspensions into a 50 ml conical tube bringing the total volume to 35 ml with RPMI.
8. Wash: Centrifuge at 1200 rpm (250*g*) for 10 min and repeat.
9. Aspirate supernatant and resuspend in 5 ml of complete RPMI per original 30 ml volume of blood.
10. Count total white blood cells.
11. Adhere cells for 1–3 h in 1% FCS/RPMI in a culture flask.
12. Aspirate and wash cells vigorously with complete RPMI medium. Monocytes will remain adhered and other lymphocytes will be washed off.
13. Monocytes will make up 10–20% of total cells adhered.
14. Adherent cells can be removed by 100 mM EDTA in 1× PBS for 5 min. Add an equal volume of complete media to do cell count.

Commonly used cell lines

A number of cell lines have been used in the study of macrophages and Toll-like receptors. Macrophage cell lines include RAW264.7 (Brightbill *et al.*, 1999, 2000; Raschke *et al.*, 1978), J774 (Ralph *et al.*, 1975; Sanjabi *et al.*, 2000), p388D1 (Koren *et al.*, 1975) and THP-1 (Libraty *et al.*, 1997). The human embryonic kidney (HEK293T) cell line has been used in transfection studies of TLR family members, receptor mutants, and CD14. HEK293T cells do not express the known TLR family members (Kirschning *et al.*, 1998; Yang *et al.*, 1998). The BaF3 cell line, also TLR-deficient, was used in studies of TLR4 and MD2 (Shimazu *et al.*, 1999), as well as RP105 (Ogata *et al.*, 2000). All cell lines are available through the American Type Culture Collection (ATCC). Cell culture conditions for each cell line are available through their web site, www.atcc.org.

Monitoring the activation of macrophages by microbial ligands

Macrophage activation is commonly monitored by cytokine release by mouse and human macrophages, and NO release by mouse macrophages. A number of pro-inflammatory (IL-12, TNFα) and anti-inflammatory cytokines (IL-10) are produced in response to microbial ligands. Whether the activation of these genes is direct or secondarily through the induction of other factors is under much debate. However, TLR signaling pathways play a fundamental role in providing the initial activating signals.

Methods

Preparation of microbial ligands

Preparations of microbial ligands are often contaminated with bacterial endotoxin. Ligand preparations should be tested for endotoxin contamination by the Limulus Amoebocyte Lysate assay (Whittaker Bioproducts) (Brightbill *et al.*, 1999). Low levels of endotoxin are tolerable if the same amount of LPS does not activate production of the cytokine being studied. If the low levels of endotoxin do not activate, the predominant activation would be from the purified ligand itself. However, if higher levels of endotoxin are detected, endotoxin can be removed by Detoxi-Gel endotoxin removing columns (Pierce). Conversely, lipoprotein contamination of LPS preparations was also found suggesting that the purity of the ligand being studied is crucial to the validity of the resulting data (Hirschfeld *et al.*, 2000).

Activation of monocyte/macrophages by microbial ligands

The activation of primary cells or cell lines can be performed in a 96-well plate format in 200 µl. Following cell harvest from mouse peritoneum or human peripheral blood, plate $\sim 5 \times 10^5$ cells per well in 100 µl of complete RPMI medium. Allow the cells to adhere for 6 h to overnight in the 37°C incubator. Activate the cells with the microbial ligand (100 ng–10 µg) in an additional 100 µl for 16–24 h. Supernatants for cytokine ELISA and NO assays can be stored at −20°C.

IFN-γ

IFN-γ plays an important role in the maturation of monocytes and the stimulation of macrophages. At the level of gene expression, IFN-γ has been shown to enhance or provide a required signal for optimal production of cytokines, such as IL-12. However, IFN-γ treatment can also lead to the downregulation of anti-inflammatory cytokines, such as IL-10 (Libraty *et al.*, 1997). IFN-γ (100 U/ml, Endogen) can be added at the time the cells are plated or concurrently with the microbial ligand.

Blocking antibodies

Blocking antibodies are commonly used in these assays to ascertain the role of different receptors in the activation of macrophages. Few antibodies directed against TLR family members have been produced thus far, however an anti-human TLR2 antibody has been demonstrated to block the activation of macrophages by microbial lipoproteins (Brightbill *et al.*, 1999). Another group demonstrated antibody inhibition of TLR4 activation by LPS and Lipid A (Lien *et al.*, 2000). Studies of RP105 utilized a monoclonal antibody that activates signaling through RP105 (Chan *et al.*, 1998). CD14 has been shown to act as a TLR coreceptor in response to a number of ligands (Kirschning *et al.*, 1998; Yang *et al.*, 1998; Brightbill *et al.*, 1999). In fact, CD14-deficient mice are also hyporesponsive to LPS stimulation (Haziot *et al.*, 1996). Studies using a blocking antibody to CD14 also suggest that CD14 plays an important role in responses of the innate immune system to a wide range of antigens (Wright, 1995). Modifying the protocol outlined above, blocking antibodies (5–10 $\mu g\,ml^{-1}$) should be added approximately 30 min prior to activation by the microbial ligand. Isotype control antibodies should be used in parallel to control for the observed inhibition.

Cytokine ELISA

Cytokine production is commonly measured by an ELISA assay, which quantifies levels of cytokine against a titration of standards. ELISA kits and antibody pairs are widely available commercially for most cytokines. In brief, supernatants from unactivated and microbial ligand activated cells are incubated with a cytokine specific primary antibody adhered to the plastic 96-well plate. After the supernatants have been thoroughly washed off a secondary antibody is added that is linked to horseradish peroxidase (HRP) enzyme. Using a colorimetric substrate, a known titration of cytokine standards can be used to determine relative amounts of the cytokine in each experimental sample. Common macrophage produced cytokines studied include IL-12 (p40 and p70), TNFα, IL-18, IL-6, and IL-10 (Libraty *et al.*, 1997; Netea *et al.*, 2000).

NO assay

Nitric oxide (NO) is the only macrophage mycobactericidal mechanism known (Chan *et al.*, 1995). The iNOS promoter and NO production are induced in response to a number of microbial ligands (Brightbill *et al.*, 1999). NO production can be monitored by nitrite levels using the colorimetric Griess reaction according to Haworth and Gordon (1998).

◆◆◆◆◆◆ MOLECULAR TARGETS OF TLR ACTIVATION BY MICROBIAL LIGANDS

Activation of signaling pathways downstream of toll-like receptors

Much like *Drosophila* Toll, mammalian TLRs have been shown to activate conserved signal transduction pathways involved in the activation of the innate immune system. Signaling downstream of TLRs involves a discrete series of kinases that lead to the activation of gene expression. The following section describes the methods used to study the activation of signaling molecules involved in the NF-κB and MAP kinase pathways.

Methods

Preparation and immunoprecipitation of the whole cell lysate

Primary macrophages or cell line should be activated with the microbial ligand as outlined above for a short time course from 10 min to 1 h.

Following activation, $\sim 10^7$ cells are lysed in ~1 ml ice-cold lysis buffer containing appropriate protease and phosphatase inhibitors. Centrifuge the whole cell lysate at maxium speed (14 000 rpm) in a microcentrifuge. Incubate 100–500 μg of total protein from each sample extract with 0.2–10 μg of the antibody specific for the signaling molecule (i.e. IRAK or JNK, ERK, p38 MAP kinases) rotating at 4°C for 1 h to overnight. The amount of antibody should be titrated initially to determine the best concentration. Following the incubation with the antibody add ~ 20 μl of either Protein A or Protein G-agarose/sepharose beads and repeat the incubation rotating at 4°C. Collect the immune complexes by centrifugation at 2500 rpm for 5 min at 4°C and aspirate. Wash the pellet four times in either lysate buffer or PBS (less stringent). For the kinase assays, do the final wash in the appropriate kinase buffer. Split the each sample in two and resuspend half in either electrophoresis sample buffer for Western blot analysis or in kinase buffer for the *in vitro* kinase assay (Adachi *et al.*, 1998; Chan *et al.*, 1998; Muzio *et al.*, 1998). Western blot analysis should be used to normalize for protein levels used in each reaction.

In vitro *kinase assay*

Antibodies used for the JNK, Erk2, and p38 are available from Santa Cruz Biotech. IRAK antibodies are available from Transduction labs (Takeuchi *et al.*, 1999). Following the isolation of the immunoprecipitates, the pellet should be resuspended in kinase buffer. Kinase reactions should be carried out in 20 μl containing the appropriate kinase buffer, 0.5–1 μg of appropriate substrate, and 10 μCi [γ-^{32}P] ATP for 30 min at 30°C. c-Jun (Santa Cruz Biotechnology) is an AP-1 family transcription factor that is activated by JNK and serves as the substrate for the JNK *in vitro* kinase assay (Adachi *et al.*, 1998; Muzio *et al.*, 1998). The IRAK *in vitro* kinase assay measures IRAK autophosphorylation (Kojima *et al.*, 1998), therefore no additional substrate is necessary. Acid denatured rabbit muscle enolase (Sigma Chemicals) can be used as the substrate for Erk and p38 MAP kinase assays (Chan *et al.*, 1998). Terminate the reactions with electrophoresis sample buffer. Separate proteins by SDS-PAGE and visualize by autoradiography or quantitate by PhosphorImager (Adachi *et al.*, 1998).

Detection of phosphorylated forms by Western

In contrast to the *in vitro* kinase assay, antibodies have been developed to recognize phosphorylated forms of a number of signaling molecules, thus removing the need for the use of radioactivity (Santa Cruz Biotech). Many of the signaling proteins are phosphorylated on tyrosine residues, which can be detected by anti-phosphotyrosine antibodies (Upstate Biotechnology). Using lysates from activated macrophages and TLR transfected cell lines, the microbial ligand induced phosphorylation of these molecules can be detected by standard Western blot analysis techniques (Chan *et al.*, 1998). It is important to normalize these results to a Western blot using standard antibodies directed against the same molecule for the amount of the protein used in each sample.

Dominant negative mutants of the NF-κB signaling pathway

The NF-κB signaling pathway plays an essential role in development and the immune response. Activation of NF-κB involves the recruitment of multiple factors, including MyD88 and IRAK, which through TRAF6, lead to the activation of MAPKKKs, such as NIK. NIK activates IKK family members, which are involved in the phosphorylation of members of the NF-κB inhibitory family, I-κB. Following phosphorylation, I-κB is ubiquitinated resulting in its degradation and the subsequent release of NF-κB to the nucleus (Baeuerle and Baltimore, 1998). Dominant negative and dominant active forms of many of these molecules have been used to dissect the signaling pathways downstream of TLR family members and other receptors, such as TNFR and IL-1R (Muzio *et al.*, 1998; Yang *et al.*, 1999; Zhang *et al.*, 1999). Either through truncation or amino acid substitution, each mutation serves to alter its function to block TLR activation at particular steps in the signaling pathway. Table 22.4 outlines a few of the mutations generated thus far.

Reagents

RIPA lysis buffer for Western blot: 1 × PBS, 1% Nonidet P-40, 0.5% sodium deoxycholate, 0.1% SDS, protease and phosphatase inhibitors.

Lysis buffer for the *in vitro* kinase assay: NP-40 based-kinase assay 20 mM Tris-Cl [pH 8.0], 137 mM NaCl, 5 mM EDTA, 10% glycerol, 1% Triton X-100, 1 mM EGTA, protease and phosphatase inhibitors.

Kinase assay buffer (IRAK, JNK): 25 mM HEPES [pH 7.6], 20 mM $MgCl_2$, 20 mM β-glycerophosphate, 1 mM Na_3VO_4, and 2 mM DTT.

Kinase assay buffer (Erk, p38 MAP kinases): 20 mM Tris (pH 7.2), 10 mM $MgCl_2$, 10 mM $MnCl_2$, 0.1% NP-40, and 1 mM Na_3VO_4.

Protease inhibitors: 1 mM PMSF, 1–20 μg ml^{-1} Aprotinin, 20–50 μg ml^{-1} Leupeptin.

Phosphatase inhibitors: 1 mM sodium vanidate (Na_3VO_4), 1 mM $Na_4P_2O_7$, 10 mM NaF, and 10 mM β-glycerophosphate.

Gene expression downstream of Toll-like receptors

LPS and other microbial ligands induce transcription of a number of genes including cytokines (IL-12, TNF, IL-10) and the antimicrobial gene, iNOS (Libraty *et al.*, 1997; Barnes *et al.*, 1992; Xie *et al.*, 1993). Future studies of Toll-like receptor biology will address the downstream gene targets of these receptors and their ligands. To date only a small number of promoters upstream of these downstream target genes have been characterized, however clear trends in the regulation of these genes have been found. Such transcription factors as NF-κB, C/EBP and AP-1 have been shown to regulate many pro-inflammatory genes activated by microbial ligands presumably downstream of Toll-like receptors (Plevy

Table 22.4 Dominant negative of the NF-κB signaling pathway (Δ = truncation)

Factor	Mutation	Reference
TLR2Δ1	C terminal truncation	Yang *et al.*, 1999, 1998
TLR2Δ2	C terminal truncation	Yang *et al.*, 1999, 1998
TLR4Δ	C terminal truncation	Muzio *et al.*, 1998
ΔMyD88	(152–296)	Wesche *et al.*, 1997; Muzio *et al.*, 1997
ΔIRAK	(1–96)	Muzio *et al.*, 1997
ΔIRAK2	(1–96)	Muzio *et al.*, 1997
ΔTRAF2	(87–522)	Cao *et al.*, 1996; Rothe *et al.*, 1995
ΔTRAF6	(289–522)	Cao *et al.*, 1996
ΔNIK	(KK429–430AA)	Malinin *et al.*, 1997
IKKα	(K44A)	Kirschning *et al.*, 1998
IKKβ	(K44A)	Kirschning *et al.*, 1998

et al., 1997; Brightbill *et al.*, 1999). The following sections will address the study of these factors and how they regulate TLR target genes through promoter regulatory elements.

Methods

Nuclear extract preparation

Following activation of macrophages, as stated previously, $\sim 10^7$ cells should be washed in PBS and then resuspend in 400 μl of cold Buffer A. Following 10–15 min on ice, pellet nuclei by microcentrifugation at full speed for 10 min, thus removing cytoplasmic proteins. Resuspend the nuclei pellet in cold Buffer C, vortex and rotate at 4°C for 30 min. After the nuclear proteins have been eluted, centrifuge again at maximum speed for 10 min and transfer the supernatant to a fresh tube (Adachi *et al.*, 1998; Sanjabi *et al.*, 2000). Store the cytoplasmic and nuclear extracts at −80°C. Macrophage extracts contain high levels of proteases, therefore appropriate protease inhibitors should be used.

NF-κB gel shift

NF-κB family members are held in the cytoplasm in an inactive state by a family of inhibitory proteins, I-κB. Following activation by microbial ligands, the I-κB/NF-κB complex is broken and NF-κB family members translocate to the nucleus where gene expression of a number of cytokine (IL-12, TNFα, IL-6) and antimicrobial (iNOS) genes are affected. Therefore, the translocation of NF-κB to the nucleus is commonly used as a means to monitor the activation of Toll-like receptors by microbial ligands (Kirschning *et al.*, 1998; Yang *et al.*, 1998). The following EMSA protocol measures the binding of NF-κB (or other transcription factors) to an annealed oligonucleotide probe (Table 22.5; Plevy *et al.*, 1997; Sanjabi *et al.*, 2000). Binding of proteins to the ^{32}P-labeled DNA probe makes the complex migrate more slowly than the probe alone giving a distinct pattern for each transcription factor. A comparison of nuclear extracts from unactivated as well as microbial ligand activated cells should result in an increase in NF-κB complex following activation.

1. Add 5 μg of nuclear extract to 2 μg of poly(dI-dC) and binding buffer on ice.
2. Add the ^{32}P-labeled oligonucleotide probe (10^5 cpm) to a total volume of 25 μl.
3. Incubate on ice for 45 min.
4. Separate EMSA products on a 5% polyacrylamide –0.4× TBE gel. Run at room temperature for 1.5 h at 150 V.
5. Dry gel on whatman paper for 1 h. Expose to film at −70 °C with intensifing screen as needed or PhosphorImager for quantification.

Supershift and competitions in the gel shift assay

The specificity of the DNA binding complex can be assessed by transcription factor specific antibodies or unlabeled gel shift probes comprised of consensus transcription factor binding sites. Unlabeled consensus binding site probes should compete binding of NF-κB to the labeled probe composed of sequences from the promoter being studied. A mutated consensus probe should be used as a control that does not compete for binding. Antibodies to NF-κB family members can be used to ensure that the gel shift complex being studied is specific to NF-κB family members. The p65/p50 complex is generally referred to as NF-κB, however there are other NF-κB/Rel family members (RelA, RelB, c-Rel, p50, p52) that are also inducibly activated and expressed in a cell type specific manner (Baeurle and Baltimore, 1996). Addition of the antibody to the binding reaction will either result in a supershift (higher mobility complex) or inhibit binding of NF-κB to the probe altogether. Antibodies to most NF-κB family members are available from Santa Cruz Biotechnology. Prior to adding the labeled probe to the nuclear extract in binding buffer, pre-incubate the nuclear extract with the antibody (1–2 μg) or unlabeled competitor probe (100-fold molar excess) for approximately 30–45 min on ice. The protocol outlined above can then be followed to completion (Plevy, *et al.*, 1997; Sanjabi *et al.*, 2000).

C/EBP and AP-1

C/EBP and AP-1 are also important transcription factors involved in the activation of TLR target genes (Plevy *et al.*, 1997). Although the pathways leading to the activation of these factors downstream of TLR are not as established as NF-κB their activation can also be monitored by the gel shift assay. Gel shift probes for C/EBP, AP-1 and NF-κB are listed in Table 22.5.

Table 22.5 Gel shift probes for NF-κB, C/EBP, and AP-1 EMSA

Probe	Sequence	Reference
NF-κB consensus	5′-CAGAGGGGACTTTCCGAGA-3′	Plevy *et al.*, 1997
C/EBP consensus	5′-AAGCTGCAGATTGCGCAATCTGCAGCTT-3′	Plevy *et al.*, 1997
AP-1 consensus	5′-CGCTTGATGACTCAGCCGGAA	Santa Cruz Biotechnology

Reagents

Buffer A: 10 mM HEPES [pH 7.8], 10 mM KCl, 0.1 mM EDTA, 1 mM DTT, 0.5% Triton X-100 and protease inhibitors.

Buffer C: 50 mM HEPES [pH 7.8], 420 mM KCl, 0.1 mM EDTA, 5 mM $MgCl_2$, 10% Glycerol, 1 mM DTT, and protease inhibitors.

EMSA probes: Annealed oligonucleotide probes (100–200 ng) can be labeled by filling in overhang ends with the Klenow enzyme and [α-^{32}P]GTP or [α-^{32}P] CTP, or label the 5′ end with Polynucleotide kinase and [γ-^{32}P] ATP.

NF-κB EMSA binding buffer: 10 mM Tris [pH 7.5], 50 mM NaCl, 1 mM dithiothreitol, 1 mM EDTA, 5% Glycerol.

TLR activation of gene expression

Promoter driven reporter genes are used quite frequently to study the targets of ligand-mediated receptor activation. TLR receptors activate an NF-κB driven promoter, the IL-12 p40 promoter, and the iNOS promoter in macrophage cells lines and non-myeloid TLR-transfected cell lines following activation by microbial ligands (Yang *et al.*, 1998; Kirschning *et al.*, 1998; Brightbill *et al.*, 1999). These constructs can serve as an important resource for studying TLR activation.

Reporter assays

Various reporter genes are available to study the regulation of promoter activity downstream of Toll-like receptors. The most common reporter genes used in this manner are luciferase/Renilla luciferase (Promega), the CAT (Promega), and green fluorescence protein (GFP, Clontech). Whole cell extracts can be made from cells stimulated with the microbial ligand or not, and extracts can used in enzymatic assays with their individual substrate in a luminometer to measure luciferase activity, or by acetylation of a ^{14}C-labeled chloramphenical substrate in the case of the CAT assay. The luciferase assay is measured as relative light units, while the CAT assay can be quanitified by PhosphorImager as the percentage conversion of the substrate that is acetylated. Assay protocols for each assay are available through Promega. GFP expression can be determined by FACs. Important controls for the assays should include the use of a reporter gene with no promoter, in addition to a constitutively active promoter, such as CMV or SV40 that would theoretically be unaffected by stimulation through the Toll-like receptor. This will ensure that the fold activation observed is specific to the signaling pathways downstream of the receptor being studied (Plevy *et al.*, 1997).

Transfection

Transfection of cell lines and primary macrophages can often be difficult and efficiencies vary widely. A number of protocols exist including Electrophoration (BioRad), Superfect (Qiagen), calcium phosphate, and lipid based reagents. Reporter gene constructs can be introduced as episomal DNA or stably integrated through drug selection (Ausubel *et al.*, 1987). In the transient assay, a second reporter gene, such as β-galactosidase or Renilla luciferase (Promega), under the control of a constitutively active promoter should be used to normalize for transfection efficiency. For stable reporter genes, multiple clones should be assayed in parallel.

Studying the promoter

Depending on the emphasis of the study, the promoter fragment can be used as a stably integrated reporter to show the connection between the receptor and the gene in question. However, to identify specific promoter elements involved in the regulation of the gene by TLR activation a series of deletion and substitution mutants can be constructed in the context of the cloned promoter fragment. Detailed description of this process and methods of studying regulated gene expression can be found in Carey and Smale (1999). Once a region of activity has been identified, the sequences can be input into the TRANSFAC database (www.embl-heidelberg.de/srs/srsc or transfac.gbf-braunschweig.de) to identify potential regulatory elements in the promoter. Once identified these sequences can be mutated or used as probes in the gel shift assay (outlined above) to determine the relevance of each element to TLR activation of the promoter.

◆◆◆◆◆◆ CONCLUSION

The assays described here review the approaches taken by researchers recently in the study of TLR function. There are, however, many more directions to be studied in TLR biology. Stimulation of Toll-like receptors by microbial ligands can lead to the activation of a variety of host defense mechanisms that reflect the arsenal used by the innate immune system to fight infection. With the rise of genomics and as our understanding of the molecular mechanisms behind gene expression improves, the essential role of Toll-like receptors in immune responses to infection will become increasingly more evident.

References

Adachi, O., Kawai, T., Takeda, K., Matsumoto, M., Tsutsui, H., Sakagami, M., Nakanishi, K. and Akira, S. (1998). Targeted disruption of the MyD88 gene results in loss of IL-1- and IL-18-mediated function. *Immunity* **9**, 143–150.

Ausubel, F. M., Brent, R., Kinston, R. E., Moore, D. D., Seidman, J. G., Smith, J. A. and Struhl, K. (1987). In *Current Protocols in Molecular Biology,* Introduction of DNA into mammalian cells, units 9.1–9.5. John Wiley and Sons, Inc., New York.

Aliprantis, A. O., Yang, R. B., Mark, M. R., Suggett, S., Devaux, B., Radolf, J. D., Klimpel, G. R., Godowski, P. and Zychlinsky, A. (1999). Cell activation and apoptosis by bacterial lipoproteins through Toll-like receptor-2. *Science* **285**, 736–739.

Baeuerle, P. A. and Baltimore, D. (1996). NF-kappa B: ten years after. *Cell* **87**, 13–20.

Barnes, P. F., Chatterjee, D., Abrams, J. S., Lu, S., Wang, E., Yamamura, M., Brennan, P. J. and Modlin, R. L. (1992). Cytokine production induced by Mycobacterium tuberculosis lipoarabinomannan. Relationship to chemical structure. *J. Immunol.* **149**, 541–7.

Belvin, M. P. and Anderson, K. V. (1996). A conserved signaling pathway: the *Drosophila* Toll-dorsal pathway. *Annu. Rev. Cell. Dev. Biol.* **12**, 3343–416.

Brightbill, H. D., Libraty, D. H., Krutzik, S. R. *et al.* (1999). Host defense mechanisms triggered by microbial lipoproteins through Toll-like receptors. *Science* **285**, 732–736.

Brightbill, H. D., Plevy, S. E., Modlin, R. L. and Smale, S. T. (2000). A prominent role for Sp1 during LPS-mediated induction of the IL-10 promoter in macrophages. *J. Immun.* **164**, 1940–1951.

Brightbill, H. D. and Modlin, R. L. (2000). Toll-like receptors: molecular mechanisms of the mammalian immune response. *Immunology* **101**, 1–10.

Cao, Z., Xiong, J., Takeuchi, M., Kurama, T. and Goeddel, D. V. (1996). TRAF6 is a signal transducer for interleukin-1. *Nature* **383**, 443–446.

Carey, M. and Smale, S. (1999). In *Transcriptional Regulation in Eukaryotes: Concepts, Strategies and Techniques.* Cold Spring Harbor Laboratory Press, New York, pp. 1–640.

Chan, J., Tanaka, K., Carroll, D., Flynn, J. and Bloom, B. R. (1995). Effects of nitric oxide synthase inhibitors on murine infection with Mycobacterium tuberculosis. *Infect. Immun.* **63**, 736–740.

Chan, V. W., Mecklenbrauker, I., Su, I., Texido, G., Leitges, M., Carsetti, R., Lowell, C. A., Rajewsky, K., Miyake, K. and Tarakhovsky, A. (1998). The molecular mechanism of B cell activation by toll-like receptor protein RP-105. *J. Exp. Med.* **188**, 93–101.

Coligan, J. E., Kruisbeek, A. M., Margulies, D. H., Shevach, E. M. and Strober, W. (1993). In *Current Protocols in Immunology,* In vitro assays for mouse lymphocyte function, unit 3.5, and Immunologic studies in humans, unit 7.1, John Wiley and Sons, New York.

DeOgny, B. C., Pramanik, B. C., Ardnt, L. L., Jones, J. D., Rush, J., Slaughter, C. A., Radolf, J. D. and Norgard, M. V. (1994). Solid-phase synthesis of biologically active lipopeptides as analogs for spirochetal lipoproteins. *Peptide Res.* **7**, 91–97.

Ezekowitz, R. A. B., Sastr, K., Bailly, P. and Warner, A. (1990). Molecular characterization of the human macrophage mannose receptor: demonstration of multiple carbohydrate recognition-like domains and phagocytosis of yeast in Cos-1 cells. *J. Exp. Med.* **172**, 1785–1794.

Fauve, R. M., Jusforgues, H. and Hevin, B. (1983). Maintenance of granuloma macrophages in serum-free medium. *J. Immunol. Methods.* **64**, 345–351.
Flo, T. H., Halaas, O., Lien, E., Ryan, L., Teti, G., Golenbock, D. T., Sundan, A. and Espevik, T. (2000). Human toll-like receptor 2 mediates monocyte activation by Listeria monocytogenes, but not by group B streptococci or lipopolysaccharide. *J. Immunol.* **164**, 2064–2069.
Haworth, R. and Gordon, S. (1998). In *Methods in Microbiology.* Isolation of and measuring the function of professional phagocytes: murine macrophages. Academic Press, London, pp. 25287–25311.
Haziot, A., Ferrero, E., Kontgen, F., Hijiya, N., Yamamoto, S., Silver, J., Stewart, C. L. and Goyert, S. M. (1996). Resistance to endotoxin shock and reduced dissemination of gram-negative bacteria in CD14-deficient mice. *Immunity* **4**, 407–414.
Heine, H., Kirschning, C. J., Lien, E., Monks, B. G., Rothe, M. and Golenbock, D. T. (1999). Cutting edge: cells that carry A null allele for toll-like receptor 2 are capable of responding to endotoxin. *J. Immunol.* **162**, 6971–6975.
Hemmi, H., Takeuchi, O., Kawai, T., Kaisho, T., Sato, S., Sanjo, H., Matsumoto, M., Hoshino, K., Wagnar, H., Takeda, K. and Akira, S. (2000). A Toll-like receptor recognizes bacterial DNA. *Nature* **408**, 740–745.
Hirschfeld, M., Kirschning, C. J., Schwandner, R., Wesche, H., Weis, J. H., Wooten, R. M. and Weis, J. J. (1999). Inflammatory signaling by Borrelia burgdorferi lipoproteins is mediated by toll-like receptor 2. *J. Immunol.* **163**, 2382–2386.
Hirschfeld, M., Ma, Y., Weis, J. H., Vogel, S. N. and Weis, J. J. (2000). Cutting edge: repurification of lipopolysaccharide eliminates signaling through both human and murine toll-like receptor 2. *J. Immunol.* **165**, 618–622.
Hoshino, K., Takeuchi, O., Kawai, T., Sanjo, H., Ogawa, T., Takeda, Y., Takeda, K. and Akira, S. (1999). Toll-like receptor 4 (TLR4)-deficient mice are hyporesponsive to lipopolysaccharide: evidence for TLR4 as the Lps gene product. *J. Immunol.* **162**, 3749–3752.
Johnson, R. B. J., Godzik, C. A. and Cohn, Z. A. (1978). Increased superoxide anion production by imunnologically activated and chemically elicited macrophages. *J. Exp. Med.* **148**, 115–127.
Kawai, T., Adachi, O., Ogawa, T., Takeda, K. and Akira, S. (1999). Unresponsiveness of MyD88-deficient mice to endotoxin. *Immunity* **11**, 115–122.
Kirschning, C. J., Wesche, H., Ayres, T. M. and Roth, M. (1998). Human Toll-like receptor 2 confers responsiveness to bacterial lipopolysaccharide. *J. Exp. Med.* **188**, 2091–2097.
Kojima, H., Takeuchi, M., Ohta, T., Nishida, Y., Arai, N., Ikeda, M., Ikegami, H. and Kurimoto, M. (1998). Interleukin-18 activates the IRAK-TRAF6 pathway in mouse EL-4 cells. *Biochem. Biophys. Res. Commun.* **244**, 183–186.
Koren, H. S., Handwerger, B. S. and Wunderlich, J. R. (1975). Identification of macrophage-like characteristics in a cultured murine tumor line. *J. Immunol.* **114**, 894–897.
Kurt-Jones, E. A., Popova, L., Kwinn, L., Haynes, L. M., Jones, L. P., Tripp, R. A., Walsh, E. E., Freeman, M. W., Golenbock, D. T., Anderson, L. J. and Finberg, R. W. (2000). Pattern recognition receptors TLR4 and CD14 mediate response to respiratory syncytial virus. *Nature Immunol.* **1**, 398–401.
Libraty, D. H., Airan, L. E., Uyemura, K., Jullien, D., Spellberg, B., Rea, T. H. and Modlin, R. L. (1997). Interferon differentially regulates interleukin-12 and interleukin-10 production in leprosy. *J. Clin. Invest.* **99**, 336–341.
Lien, E., Chow, J. C., Hawkins, L. D., McGuinness, P. D., Miyake, K., Espevik, T., Gusovsky, F. and Golenbock, D. T. (2000). A novel synthetic acyclic lipid A-like

agonist activates cells via the lipopolysaccharide/Toll-like receptor 4 signaling pathway. *J. Biol. Chem.* **276**, 1873–1880.

Ma, Y. and Weis, J. J. (1993). *Borrelia burgdorferi* outer surface lipoproteins OspA and OspB possess B-cell mitogenic and cytokine-stimulatory properties. *Infect. Immun.* **61**, 3843–3853.

Malinin, N. L., Boldin, M. P., Kovalenko, A. V. and Wallach, D. (1997). MAP3K-related kinase involved in NF-kappaB induction by TNF, CD95 and IL-1. *Nature* **385**, 540–544.

Means, T. K., Wang, S., Lien, E., Yoshimura, A., Golenbock, D. T. and Fenton, M. J. (1999). Human toll-like receptors mediate cellular activation by Mycobacterium tuberculosis. *J. Immunol.* **163**, 3920–3927.

Medzhitov, R., Preston-Hurlburt, P. and Janeway, Jr. C. A. (1997). A human homologue of the *Drosophila* Toll protein signals activation of adaptive immunity. *Nature* **388**, 394–397.

Medzhitov, R. and Janeway, Jr. C. A. (1997). Innate immunity: impact on the adaptive immune response. *Curr. Opin. Immunol.* **9**, 4–9.

Muzio, M., Natoli, G., Saccani, S., Levrero, M. and Mantovani, A. (1998). The human toll signaling pathway: divergence of nuclear factor kappaB and JNK/SAPK activation upstream of tumor necrosis factor receptor-associated factor 6 (TRAF6). *J. Exp. Med.* **187**, 2097–2101.

Muzio, M., Ni, J., Feng, P. and Dixit, V. M. (1997). IRAK (Pelle) Family member IRAK-2 and MgD88 approximal mediators of IL-1 signaling. *Science* **278**, 1612–1615.

Netea, M. G., Kullberg, B. J., Verschueren, I. and Van Der Meer, J. W. (2000). Interleukin-18 induces production of proinflammatory cytokines in mice: no intermediate role for the cytokines of the tumor necrosis factor family and interleukin-1beta. *Eur. J. Immunol.* **30**, 3057–3060.

Ogata, B. H., Su, I., Miyake, K., Nagai, Y., Akashi, S., Mecklenbrauker, I., Rajewsky, K., Kimoto, M. and Tarakhovsky, A. (2000). The Toll-like receptor protein RP105 regulates lipopolysaccharide signaling in B cells. *J. Exp. Med.* **192**, 23–29.

Ozinsky, A., Underhill, D. M., Fontenot, J. D., Hajjar, A. M., Smith, K. D., Wilson, C. B., Schroeder, L. and Aderem, A. (2000). The repertoire for pattern recognition of pathogens by the innate immune system is defined by cooperation between Toll-like receptors. *Proc. Natl Acad. Sci. USA* **97**, 13766–13771.

Plevy, S. E., Gemberling, J. H., Hsu, S., Dorner, A. J. and Smale, S. T. (1997). Multiple control elements mediate activation of the murine and human interleukin 12 p40 promoters: evidence of functional synergy between C/EBP and Rel proteins. *Mol. Cell. Biol.* **17**, 4572–4588.

Poltorak, A., He, X., Smirnova, I., Liu, M. Y., Huffel, C. V., Du, X., Birdwell, D., Alejos, E., Silva, M., Galanos, C., Freudenberg, M., Ricciardi-Castagnoli, P., Layton, B. and Beutler, B. (1998a). Defective LPS signaling in C3H/HeJ and C57BL/10ScCr mice: mutations in Tlr4 gene. *Science* **282**, 2085–2088.

Poltorak, A., Smirnova, I., He, X., Liu, M. Y., Van Huffel, C., McNally, O., Birdwell, D., Alejos, E., Silva, M., Du, X., Thompson, P., Chan, E. K., Ledesma, J., Roe, B., Clifton, S., Vogel, S. N. and Beutler, B. (1998b). Genetic and physical mapping of the Lps locus: identification of the toll-4 receptor as a candidate gene in the critical region. *Blood Cells Mol. Dis.* **24**, 340–355.

Qureshi, S. T., Lariviere, L., Leveque, G., Clermont, S., Moore, K. J., Gros, P. and Malo, D. (1999a). Endotoxin-tolerant mice have mutations in Toll-like receptor 4. *J. Exp. Med.* **189**, 615–625.

Qureshi, S. T., Gros, P. and Malo, D. (1999b). Host resistance to infection: genetic control of lipopolysaccharide responsiveness by Toll-like receptor genes. *Trends Genet.* **8**, 291–294.
Radolf, J. D., Arndt, L. L., Akins, D. R., Curetty, L. L., Levi, M. E., Shen, Y., Davis, L. S. and Norgard, M. V. (1995). Treponema pallidum and Borrelia burgdorferi lipoproteins and synthetic lipopeptides activate monocytes/macrophages. *J. Immunol.* **154**, 2866–2877.
Ralph, P., Prichard, J. and Cohn, M. (1975). Reticulum cell sarcoma: an effector cell in antibody-dependent cell-mediated immunity. *J. Immunol.* **114**, 898–905.
Raschke, W. C., Baird, S., Ralph, P. and Nakoinz, I. (1978). Functional macrophage cell lines transformed by ableson leukemia virus. *Cell* **15**, 261–267.
Rothe, M., Sarma, V., Dixit, V. M. and Goeddel, D. V. (1995). TRAF2-mediated activation of NF-kappa B by TNF receptor 2 and CD40. *Science* **269**, 1424–1427.
Sanjabi, S., Hoffmann, A., Liou, H. C., Baltimore, D. and Smale, S. T. (2000). Selective requirement for c-Rel during IL-12 p40 gene induction in macrophages. *Proc. Natl Acad. Sci. USA* **97**, 12705–12710.
Schwandner, R., Dziarski, R., Wesche, H., Rothe, M. and Kirschning, C. J. (1999). Peptidoglycan- and lipoteichoic acid-induced cell activation is mediated bytoll-like receptor 2. *J. Biol. Chem.* **274**, 17406–17409.
Shimazu, R., Akashi, S., Ogata, H., Nagai, Y., Fukudome, K., Miyake, K. and Kinoto, M. (1999). MD-2, a molecule that confers lipopoylsaccharide responsiveness on Toll-like receptor 4. *J. Exp. Med.* **189**, 1777–1782.
Takeuchi, O., Hoshino, K., Kawai, T., Sanjo, H., Takada, H., Ogawa, T., Takeda, K. and Akira, S. (1999). Differential roles of TLR2 and TLR4 in recognition of gram-negative and gram-positive bacterial cell wall components. *Immunity* **11**, 443–451.
Takeuchi, O., Kaufmann, A., Grote, K., Kawai, T., Hoshino, K., Morr, M., Muhlradt, P. F. and Akira, S. (2000). Preferentially the R-stereoisomer of the mycoplasmal lipopeptide macrophage-activating lipopeptide-2 activates immune cells through a toll-like receptor 2- and MyD88-dependent signaling pathway. *J. Immunol.* **164**, 554–557.
Ting, A. T., Pimentel-Muinos, F. X. and Seed, B. (1996). RIP mediates tumor necrosis factor receptor 1 activation of NF-kappaB but not Fas/APO-1-initiated apoptosis. *EMBO J.* **15**, 6189–6196.
Underhill, D. M., Ozinsky, A., Hajjar, A. M., Stevens, A., Wilson, C. B., Bassetti, M. and Aderem, A. (1999). The Toll-like receptor 2 is recruited to macrophage phagosomes and discriminates between pathogens. *Nature* **401**, 811–815.
Vogel, S. N. (1992). B. Beutler, In *Tumor Necrosis Factors: The Molecules and their Emerging Role in Medicine.* The *Lps* gene. Insights into the genetic and molecular basis of LPS responsiveness and macrophage differentiation. Raven Press, New York, pp. 485–513.
Wesche, H., Henzel, W. J., Shillinglaw, W., Li, S. and Cao, Z. (1997). MyD88: an adapter that recruits IRAK to the IL-1 receptor complex. *Immunity* **7**, 837–847.
Wright, S. D. (1995). CD14 and innate recognition of bacteria. *J. Immunol.* **155**, 6–8.
Wright, S. D. (1999). Toll, a new piece of the puzzle of innate immunity. *J. Exp. Med.* **189**, 605–609.
Xie, Q. W., Whisnant, R. and Nathan, C. (1993). Promoter of the mouse gene encoding calcium-independent nitric oxide synthase confers inducibility by interferon gamma and bacterial lipopolysaccharide. *J. Exp. Med.* **177**, 1779–84.
Yang, R. B., Mark, M. R., Gray, A., Huang, A., Xie, M. H., Zhang, M., Goddard, A., Wood, W. I., Gurney, A. L. and Godowski, P. J. (1998). Tall-like receptor-2 mediates lipopolysaccharide-induced cellular signaling. *Nature* **395**, 284–8.

Yang, R. B., Mark, M. R., Gurney, A. L. and Godowski, P. J. (1999). Signaling events induced by lipopolysaccharide-activated toll-like receptor 2. *J. Immunol.* **163**, 639–43.

Yoshimura, A., Lien, E., Ingalls, R. R., Tuomanen, E., Dziarski, R. and Golenbock, D. (1999). Recognition of Gram-positive bacterial cell wall components by the innate immune system occurs via Toll-like receptor 2. *J. Immunol.* **163**, 1–5.

Zhang, F. X., Kirschning, C. J., Mancinelli, R., Xu, X.-P., Jin, Y., Faure, E., Mantovani, A., Rothe, M., Muzio, M. and Moshe Arditi, M. (1999). Bacterial lipopolysaccharide activates nuclear factor-κB through interleukin-1 signaling mediators in cultured human dermal endothelial cells and mononuclear phagocytes. *J. Biol. Chem.* **274**, 7611–7614.

List of suppliers

American Red Cross
www.redcross.com/

American Type Culture
Collection (ATCC), 10801 Univeristy Blvd., Manassas, VA 20110-2209, USA
(703) 365-2700
www.ATCC.org/

Amersham-Pharmacia Biotech
800 Centennial Avenue, Piscataway, NJ 08855-1327, USA
(800) 526-3593
www.apbiotech.com/

BD Transduction labs
1 Becton Drive, Franklin Lakes NJ USA 07417, USA
(800) 227-4063
www.translab.com/

Bio-Whittaker Molecular Applications
191 Thomaston Street, Rockland, ME 04841, USA
(207) 594-3400
www.bioproducts.com/

Biorad
1414 Harbour Way South Richmond, CA 94804, USA
(415) 232-7000
www.bio-rad.com

Clontech Laboratories, Inc.
1020 East Meadow Circle Palo Alto, CA 94303-4230, USA
(800) 662-2566
www.clontech.com

Endogen
3747 N. Meridian Rd., P.O. Box 117, Rockford, IL 61105, USA
(800) 487-4885
www.endogen.com/

Pierce
3747 N. Meridian Road P.O. Box 117, Rockford, IL 61105, USA
(815) 968-0747
www.piercenet.com

Promega
2800 Woods Hollow Road Madison, WI 53711-5399, USA
(800) 356-9526
www.promega.com/

Qiagen
28159 Avenue Stanford,
Valencia, CA 91355, USA
(800) 426-8157
www.qiagen.com/

Santa Cruz Biotechnology
2161 Delaware Avenue, Santa Cruz
CA 95060, USA
(800) 457-3801
www.scbt.com/

Sigma Chemicals
P.O. Box 14508, St. Louis,
MO 63178, USA
(800) 325-3010
www.sigma-aldrich.com/

Upstate Biotechnology
1100 Winter Street, Suite 2300,
Waltham, MA 02451, USA
(800) 548-7853
www.upstatebiotech.com/

23 Methods for Studying the Mechanisms of Microbial Entry into the Central Nervous System

Sandrine Bourdoulous[1], Pierre Olivier Couraud[1] and Xavier Nassif[2]
[1]*CNRS UPR 415, Institut Cochin de Génétique Moléculaire, 22 rue méchain, 75014, Paris, France;*
[2]*INSERM U411, Laboratoire de Microbiologie, Faculté de Médecine Necker-Enfants Malades, 156 Rue de Vaugirard, 75730, Paris, France*

◆◆

CONTENTS

◆◆◆◆◆◆ INTRODUCTION

Homeostasy of the brain interstitial fluid is maintained by two physiological barriers between the blood and the central nervous system (CNS). The first and largest one is the blood–brain barrier (BBB), a structure composed of brain parenchyma, microvascular endothelial cells, astrocytes, foot processes and capillary pericytes, which exhibits a low and strictly controlled permeability to blood-borne compounds, immune cells and pathogens, due to the endothelial tight junctions and low transcellular flux. The second barrier is the blood–cerebrospinal fluid barrier (BCSFB) that is present at the choroid plexuses. This structure is composed of fenestrated capillaries in contact with a layer of highly differentiated ependymal cells that closely resemble peripheral epithelial cells and express impermeable tight junctions. A major function of these barriers is to regulate the passage of immune cells, proteins and other nutrients from the blood to the brain and to protect the CNS against circulating neurotoxic pathogens and toxins. However, it happens that bacteria may

METHODS IN MICROBIOLOGY, VOLUME 31
ISBN 0-12-521531-2

cross these barriers, invade the brain parenchyma or enter the cerebrospinal fluid compartment and cause encephalitis or meningitis through molecular mechanisms that remain poorly understood. The subject of this chapter is to provide a variety of physiological, cellular and molecular methods for studying the mechanisms of microbial entry into the CNS.

The blood–brain barrier: a challenge for bacterial pathogens

Among the invasive bacterial pathogens, few are capable of invading the brain to cause meningitis, which suggests that they have developed special attributes capable of circumventing the BBB. The majority of bacterial meningitis cases are caused by *Streptococcus pneumoniae*, *Neisseria meningitidis*, *Haemophilus influenzae* and, in newborn, by K1 *Escherichia coli* and Group B *Streptococcus* (Zhang and Tuomanen, 1999; Kim, 2000; Zhang *et al.*, 2000). All these pathogens multiply in the extracellular compartment and can possibly interact directly with the components of the BBB to invade the meninges. Alternatively, some intracellular pathogens, like *Mycobacterium tuberculosis* and *Listeria monocytogenes*, may infect leukocytes as Trojan horses to enter the CSF and can cause meningitis often associated with inflammatory process of the parenchyma.

Our current understanding of bacterial invasion of the brain remains sketchy. Extracellular pathogens which have invaded the bloodstream from their port-of-entry, usually the throat, multiply to high densities in the blood, the level of bacteremia correlating with the risk of meningeal invasion. Once the bacteria reach the luminal side of the cerebral capillaries, they can enter the CSF either at the choroid plexus level or through the endothelial cells of brain capillaries. For most bacteria, the precise entry site remains unknown. In the case of *Haemophilus influenzae*, experimental data obtained in monkeys demonstrate a higher density of bacteria in the ventricular CSF compared to the lumbar space, thus suggesting that bacteria enter the CSF at the choroid plexus (Smith, 1987). In the case of *N. meningitidis*, data obtained using postmortem samples of a patient who died of fulminant meningococcemia showed bacteria interacting with endothelial cells of the choroid plexuses (Figure 23.1) as well as the meninges (Figure 23.2) (Pron *et al.*, 1997).

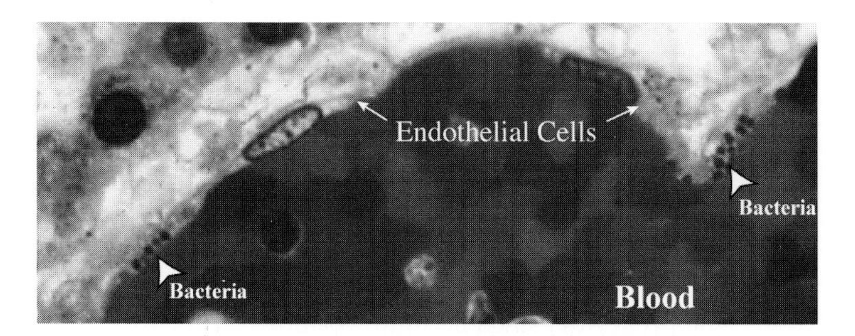

Figure 23.1. Postmortem section of a capillary in the choroid plexus of the brain showing *Neisseria meningitidis* adhering to the endothelial cells. (This figure is also reproduced in colour between pages 276 and 277.)

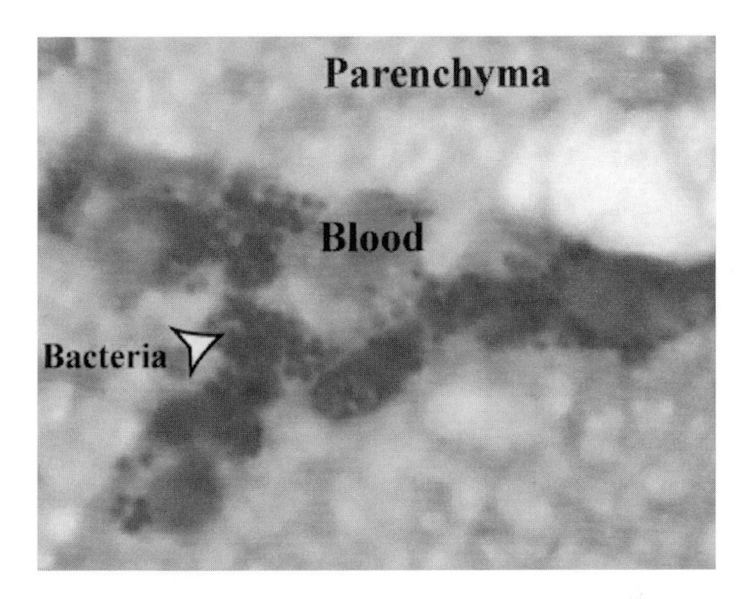

Figure 23.2. Postmortem section of the brain parenchyma showing *Neisseria meningitidis* adhering to brain capillaries. (This figure is also reproduced in colour between pages 276 and 277.)

The blood–brain barrier: molecular structure

The BBB is constituted of the endothelium of the brain capillaries, which differs from that of the other capillaries by its expression of tight junctions, the pericytes within the capillary basement membrane, and the astrocyte foot processes (Figure 23.3A) that invest the vast majority of the abluminal surface of the brain capillaries (Goldstein and Betz, 1986; Kniesel and Wolburg, 2000). This structure provides the dual feature of low rate fluid phase endocytosis and extremely high resistance tight junctions (trans-endothelial resistances higher than $1000\,\Omega\,cm^2$) that constitute the principal barrier to passive movement of fluid, electrolytes, macromolecules and cells through a paracellular pathway. The BBB permeability, in the physiological situation, is regulated in particular by astrocyte-derived factors that control brain capillary endothelial transport properties, by regulating the organization of their tight junction architecture.

The choroid plexuses, contain permeable ('fenestrated') capillaries close to the basal layer of epithelial cells which secrete the cerebrospinal fluid (CSF) and control the transport processes between the blood and the CSF (Figure 23.3B). These cells, which closely resemble a transporting epithelium, are covered with microvilli at their apical side, which greatly increases the contact surface between the cells and the CSF. An occluding band of tight junctions close to the CSF side of the cells restricts the paracellular pathway, although these junctions are slightly more permeable than those of the BBB. (For review see Segal, 1998 or Strazielle and Ghersi-Egea, 2000.)

The molecular composition of tight junctions has been the subject of intensive investigations over the past few years [for review see Tsukita *et al.*, 1999 or Zahraoui *et al.*, 2000]. To date, multiple integral membrane

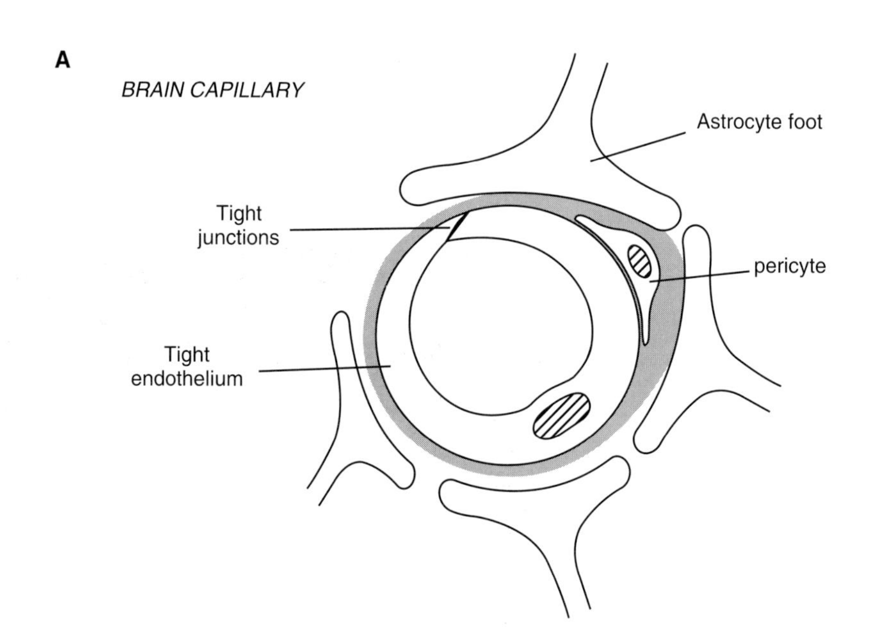

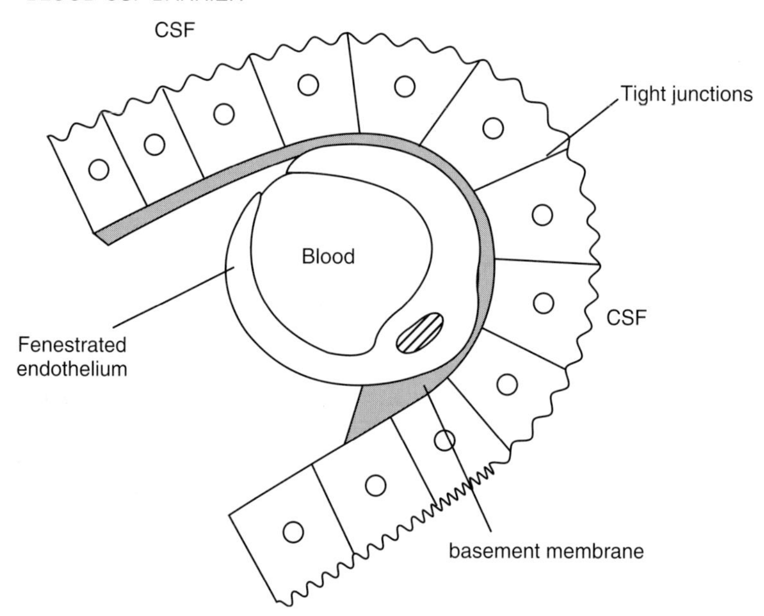

Figure 23.3. Schematic cross-section of a brain capillary (A) and a choroid plexus (B). The choroid plexus consists of numerous fronds projecting into the CSF, each frond composed of several villous processes. Each villus (detailed here) contains a large capillary of fenestrated endothelial cells, in close association with a layer of polarized epithelial cells, with their apical membrane facing the CSF. The blood brain barrier structure is a structure composed of brain parenchyma microvascular endothelial cells forming tight junctions, in close contact with the astrocyte foot processes and capillary pericytes.

proteins localized in both endothelial and epithelial tight junctions have been identified: occludin, claudins and junction-associated membrane (JAM) proteins (Tsukita and Furuse, 1999; Martin-Padura *et al.*, 1998). Occludin and claudins (an expanding family of 16 members) contain four transmembrane domains and a carboxy-terminal cytoplasmic domain that promotes its interaction with peripheral membrane proteins (see below). Claudin-1 and -5 are expressed at the BBB and play a crucial role in the tight junction organization and function, whereas occludin appears to be dispensable. JAM (known now as JAM-1 since the recent identification of the closely related JAM-2 protein), which is also expressed at the BBB, has a single transmembrane domain and belongs to the immunoglobulin superfamily (Paluver *et al.*, 2000). However, it remains unclear whether JAMs are components of the tight junction strand itself or are localized between the strands.

Several peripheral membrane proteins underlying tight junction strands have been identified, including three proteins, ZO-1, ZO-2 and ZO-3, of the MAGUK (membrane-associated guanylate kinase homologues) family, which have been localized to the cytoplasmic surface of tight junctions. These proteins, contain several protein–protein interaction domains: three PDZ domains, one SH3 domain, and one guanylate kinase-like (GUK) domain. Through their GUK domain, they can directly bind to the carboxy-terminal region of occludin, whereas their PDZ domains are responsible for their interaction with the carboxy-terminus of claudins. An interaction of JAM with ZO1 and occludin was recently described (Bazzoni *et al.*, 2000). In addition, ZO-1 and ZO-2 were shown to directly associate with actin filaments, thus forming molecular linker between the integral membrane proteins and the actin cytoskeleton. Moreover, ZO-1 can interact with both ZO-2 and ZO-3 and forms heterodimers. Other peripheral membrane proteins, cingulin, 7H6 antigen and symplekin, which do not belong to the MAGUK family, were shown also to be concentrated at tight junctions (Lapierre, 2000). However, it still remains unclear how these proteins are connected to tight junction strands at the molecular level. Moreover, several cytoplasmic signalling molecules have been reported to be concentrated at tight junctions, and are thought to be involved in inside-out as well as inside-in signalling at this level (Lapierre, 2000). Indeed, two types of heterotrimeric G-proteins, $G\alpha_0$ and $G\alpha_{i2}$ are localized around tight junctions and might function as negative regulators of tight junction assembly. Also, two atypical PKC isotypes, PKCζ and PKCλ, and their specific binding protein (ASIP: atypical PKC isotype-specific interacting protein) were shown to be concentrated at tight junctions and to play a role in the establishment of cell polarity. In addition, several proteins responsible for vesicular transport or fusion, such as the small G proteins Rab3B and Rab13 were also reported to be concentrated at tight junctions. Thus a multi-protein complex appears to be involved in the molecular architecture of tight junctions, and might be involved in their functional regulation.

The molecular basis of tight junction regulation is of key importance for understanding how BBB permeability is regulated under normal and pathological conditions. Both assembly and barrier properties are influenced by several signalling pathways, including the calcium/PKC, and the cAMP/PKA pathways (Clarke *et al.*, 2000; Weqeuer *et al.*, 2000). The formation and regulation of tight junctions are correlated with changes in the phosphorylation state of the proteins that localize at the tight junctions. As an example, occludin is heavily phosphorylated on serine/threonine in tight junction strands, whereas the less phosphorylated forms are distributed in the basolateral membrane and in the cytoplasm. Indeed, occludin switches from a non (or less) phosphorylated form to a hyperphosphorylated form as the junction assembles. Tyrosine phosphorylation at cellular junctions, in response to the tyrosine phosphatase inhibitor pervanadate, was also accompanied by an increase in brain endothelial monolayer permeability, as measured by a drop in electrical resistance (Staddon *et al.*, 1995). More specifically, tyrosine phosphorylation of occludin is necessary for tight junction reassembly, whereas that of ZO-1 and ZO-2 is correlated with altered permeability and redistribution of tight junction components. The actual kinase(s) and or phosphatase(s) that alter occludin or ZO-1 phosphorylation have not yet been identified. The barrier function of the intercellular tight junction is also dynamically regulated by the interaction between scaffolding proteins and the actin cytoskeleton. Indeed, regulation of the actin cytoskeleton organization has fundamental implication for the control of tight junction permeability (Hopkins *et al.*, 2000). This is documented by the regulation of epithelial or endothelial permeability by the Rho family of small GTPases (Rho, Rac, Cdc42) (Nusrat *et al.*, 1995; Adamson *et al.*, 1999). In the same line, disassembly of the cadherins/catenins complexes, which constitute the adherens junctions, by calcium deprivation or phosphorylation of catenins has been shown to affect tight junction organization and increase permeability (Staddon and Rubin, 1996).

In physiological conditions, the BBB and BCSFB barriers protect the brain efficiently against any pathogen infections. Pathogenic bacteria may bypass these barriers by using different mechanisms: (i) bacterial toxins like lipopolysaccharide or the release of the host inflammatory cytokines induced by bacterial components, can drastically alter paracellular permeability of brain barriers, presumably by disturbing the structural organization of tight junction proteins and the associated actin cytoskeleton (Walsh *et al.*, 2000); (ii) a direct interaction of the bacteria and the components of the BBB can trigger a cascade of intracellular events leading, through inside-in signalling, to the opening of intercellular junctions and/or transcytosis of the bacteria. To further investigate the molecular mechanisms by which invasive bacteria can enter the CNS, we describe in this chapter some *in vitro* and *in vivo* systems currently available, with special emphasis on brain microvessel cell culture systems and the study of the regulation of the molecular architecture of tight junctions and on the associated actin cytoskeleton.

◆◆◆◆◆◆ CELL CULTURE SYSTEMS

Choroid Plexus epithelial cell culture systems

Choroid Plexus (CP) epithelial cell culture systems have been reviewed recently (Strazielle and Ghersi-Egea, 2000) and will not be extensively described here. Briefly, epithelial cells from porcine (Engelbertz *et al.*, 2000) or rat (Strazielle and Ghersi-Egea, 1999) CP are obtained virtually pure by trypsin digestion and seeded on laminin-coated culture dishes or semi-permeable filter inserts (Transwell®). This culture system on filters allows the CP epithelial cells to fully differentiate, especially following serum deprivation, leading to the establishment of an *in vitro* model of the BCSFB. In this model, the apical compartment represents the ventricular side *in vivo*, and the basolateral side represents the side facing the fenestrated CP capillaries. These cultured CP epithelial cells are highly polarized, with a dense and homogeneous presence of microvilli at their apical surface, and continuous tight junctions, which confer to these monolayers a very high electrical resistance (up to $1500\,\Omega\,cm^2$). No such human *in vitro* model has been described yet. This is the reason why some human epithelial cell lines of other origins have been used to study the interaction of bacterial pathogens with tight junction forming monolayers of epithelial cells (Herz *et al.*, 1996; Pujol *et al.*, 1997).

Brain microvessel endothelial cell culture systems

The pioneering work by F. Joo (for review see Joo, 1993) and others on the isolation and *in vitro* pharmacological characterization of brain microvessels paved the way to the *in vitro* culture of brain microvessel endothelial cells (BMECs), which has now been reported with cells of various species. Primary cultures of BMECs express some features of the BBB *in situ*, such as receptors or transporters, but usually fail to develop tight junctions associated with high transendothelial electrical resistance, and in addition, rapidly lose their differentiated phenotype along passages. Several attempts have been made to optimize these tentative *in vitro* models of BBB, by using co-culture with astrocytes, and by producing stable, immortalized brain endothelial cell lines which retain in culture most of the characteristics of brain endothelial cells. The following sections will describe briefly the three major types of BMECs, which are currently used to evaluate and investigate the biochemical and molecular basis and regulation of the BBB: primary cultures, co-culture systems and cell lines.

Primary cultures of brain endothelial cells

Several groups have reported the detailed experimental approach for *in vitro* culture of BMECs of various origins: bovine, porcine, murine or human (Bowman *et al.*, 1983; Audus and Borchardt, 1986; Dorovini-Zis *et al.*, 1991; Abbott *et al.*, 1992; Franke *et al.*, 2000).

Method

Briefly, brain microvessels are isolated from cerebral grey matter by enzymatic digestion, using dispase and/or collagenase, after meninges and large surface vessels have been removed. Brain microvessels are then purified by percoll gradient centrifugation or by absorption onto glass beads and grown on substrate-coated tissue culture dishes or microporous filter membrane inserts (Transwell). In general the culture surface is coated with Type I (rat tail) collagen I and/or fibronectin. The cells are grown in the presence of serum and exogenous endothelial cell growth supplements or fibroblast growth factor-2 (FGF-2 or bFGF). The cultured BMECs usually reach confluence at 1 week and form a tight monolayer on approximately day 10, before undergoing senescence and de-differentiation after about 2 weeks in culture or after a few passages. Of particular interest, human BMECs can be grown out of small fragments of cerebral cortex obtained from surgical resections of children with seizure disorders or epileptic young adults (Stius *et al.*, 1997).

The confluent monolayers of primary BMECs retain an endothelial phenotype which can be checked by immunostaining (when species-cross reactive antibodies are available) of some endothelial markers, including PECAM-1 (CD31), wW-factor, the cytokine-inducible adhesion molecule VCAM-1, VE-cadherin (or cadherin-5), and by assessing the capacity of acetylated low density lipoprotein uptake using a fluorescent molecule, or of specific lectin binding (Bandereia simplicifolia or Ulex Europaeus Agglutinin). In addition, the purity of the endothelial culture can be assessed by identifying contaminating pericytes or astrocytes, using antibodies against smooth muscle actin or glial fibrillary acidic protein (GFAP), respectively. While bovine or porcine BMEC cultures can be found virtually pure, human BMECs may contain as much as 50% or more non-endothelial cells and may need to be purified by flow cytometry sorting before use (Stins *et al.*, 1997).

BMECs can retain many of the morphological and biochemical properties of the BBB *in vivo*: the expression of transferrin and insulin receptors can be detected by immunostaining, whereas the expression of the drug-transporting P-glycoprotein (Pgp) can be documented by functional assays of drug ([3H] colchicin or vinblastin) accumulation in and efflux out of BMECs. However, the limitation of primary BMEC cultures is their high paracellular permeability, reflected by the measurement of low (generally lower than $100\,\Omega\,\text{cm}^2$) electrical resistance across the BMEC monolayer. This apparently results from a reduced complexity of tight junctions in cell culture due to the absence of the brain environmental inputs. Several pharmacological treatments have been described, which appear to substitute at least partially for these brain inputs, inasmuch as they can improve the characteristics of BMECs. Addition of dexamethasone (100–500 nM) in the BMEC culture medium has been shown to up-regulate the expression and the organization of tight junction-associated proteins and to decrease the permeability of monolayers. Also, an elevation of the intracellular level of cyclic AMP, through the addition of a cell-permeable cAMP analogue or of a phosphodiesterase inhibitor, leads to a profound structural reorganization of the actin-based cytoskeleton and a striking increase in cell junction tightness (Rubin *et al.*, 1991). In addition, retinoic acid was shown to re-induce the activity of the BBB metabolic enzyme γ-glutamyl transpeptidase, which is readily lost in culture.

Astrocyte co-cultures

Brain capillaries are almost completely ensheathed by astrocyte processes which are believed to provide the cerebral endothelium with specific stimuli responsible for the development and the maintenance of the BBB phenotype *in vivo*. Considering that the absence of such environmental stimuli might prevent BMECs from retaining in culture the fully differentiated phenotype of BBB *in vivo*, some investigators proposed a co-culture system in which BMECs are grown in the presence of primary astrocytes or astrocyte-conditioned medium.

Methods

In these set-ups, primary cultures of rat or porcine astrocytes are co-cultured with BMECs on the two opposite sides of a semi-permeable filter of culture inserts (Transwell) or grown in the bottom chamber of the Transwell with the BMECs grown on the filter. In the former set-up, direct contact between BMECs and astrocyte process is made through the 0.4 μm pores of the filter, while, in the latter one, only astrocyte-secreted factors can freely diffuse in the culture medium and through the pores and reach the BMECs. It has been shown that astrocyte co-culture or conditioned medium largely improve the *in vitro* BBB model, with regard to structural, metabolic and permeability characteristics.

In these experimental conditions, *in vitro* models of BBB, based on bovine or porcine BMECs (Rubin *et al.*, 1991; Cecchelli *et al.*, 1999; Engelbertz *et al.*, 2000) have been proposed and validated: (i) they display high transendothelial electrical resistance (500–600 $\Omega\,cm^2$ even up to 1000 $\Omega\,cm^2$); (ii) BMECs are highly polarized regarding the expression of receptors and transporters; (iii) the pattern of permeability for a large number of standard molecules, over a wide range of hydrophilicity, is well correlated with the brain bio-availability of the same molecules measured *in vivo*.

Brain endothelial cell lines

Despite the high value of these models, a number of drawbacks still limit their use in basic research as well as in drug-screening processes: (1) their use is time consuming and needs considerable know-how which may hamper their routine use in non-expert laboratories; (2) such available models are based on bovine or porcine BMECs only, which may constitute a serious limitation for studies with species-specific bacterial pathogens (*Neisseria meningitidis*), viruses (HIV, or HTLV-I) or for immunological studies; (3) BMECs rapidly lose the characteristics of BBB endothelial cells with passages in culture, which limits their availability for biochemical or pharmacological studies. In order to bypass these drawbacks, numerous efforts have been made over many years to establish continuous, immortalized cell lines which would retain in culture a stable phenotype reminiscent of BBB endothelium *in vivo*. Following earlier reports describing the production of various transformed cell lines with anchorage-independent growth and/or noticeable morphological and functional changes, only a few immortalized cell lines have been reported, which display a non-transformed phenotype and retain most of the BBB endothelial markers. The RBE4 cells were isolated from rat brain cortex and immortalized with the adenovirus E1A-encoding sequence (Durieu-Trautmann *et al.*, 1993; Roux *et al.*, 1994), whereas the GP8 cells, also from rat brain cortex, were immortalized with SV40 large T-antigen (Greenwood *et al.*, 1996). RBE4 and GP8 cells have been extensively characterized and maintain in culture the differentiated phenotype of cerebral endothelium. These cells, together with the GP8-derived GPNT cell line (Regina *et al.*, 1999), are now widely used as validated models of brain endothelium for biochemical, pharmacological, toxicological or immunological purposes. However, these cells still exhibit a reduced complexity of tight junctions associated with a low electrical resistance, which limits their use in permeability regulation studies. In summary, it should be pointed out that the limiting step of all these models remains, that they do not reproduce the efficient limitation of the paracellular flux observed *in vivo*.

◆◆◆◆◆◆ PATHOPHYSIOLOGICAL CONSEQUENCES OF MICROBIAL INTERACTION WITH BRAIN ENDOTHELIAL CELLS

The effect of bacterial interaction on brain endothelial cell permeability can be assessed both *in vitro* and *in vivo* by a few technical approaches that we have detailed in this section.

In vitro physiological assays

Culture on filter inserts

Analysis of modifications in BMEC permeability or transcytosis requires culturing the cells onto permeable filters or filter inserts.

Methods

The most common type of filter insert used with BMECs is the Costar Tranwell® culture-insert system (Costar, Cambridge, MA). Transwell® inserts are available in many sizes, but the 24 or 12 mm inserts provide the easiest access for transport studies. BMECs are cultured onto Transwell(®) inserts with a 0.4 μm pore polycarbonate filter. The 3.0 μm pore filters are not recommended since BMECs can migrate through these pores and form a monolayer on both sides of the filter. Pre-coated filters are commercially available, with a variety of different substrates (Type I collagen, fibronectin, laminin) which may affect the morphology and phenotypic characteristics of the cells. Indeed, BMECs grown on type IV collagen or laminin were reported to exhibit a more differentiated phenotype than on

Type I collagen (Engelbertz *et al.*, 2000). Alternatively, the filters can be coated by the investigator with the various substrates, most often however with Type I (rat-tail) collagen I (3 mg ml^{-1} for 3 h) occasionally followed by coating with fibronectin (10 µg ml^{-1} for 1 h) and the excess washed away. The BMECs are plated at 50 000 to 250 000 cells cm^{-2} in a volume of 1.5 ml in the insert chamber, with 2.5 ml of medium in the bottom chamber. Confluent monolayers are obtained in about 10 days under standard culture conditions. The apical side of the monolayer is defined as the top of the insert and the basal side is defined as the side facing the collagen/fibronectin matrix.

Measure of the paracellular flux

Modification by bacterial pathogens in the tightness of the endothelial junctions can be assessed by the measure of the paracellular flux of diffusible markers through a BMEC monolayer grown on a semi permeable insert.

Methods

Radioactive tracers such as [^{3}H] Inuline or [^{14}C] sucrose are added to the apical surface (insert chamber) of non-infected and infected monolayers and samples (50 µl) are taken from the basal (bottom) chamber at various times and the permeability is calculated (Cecchelli *et al.*, 1999). To avoid the use of radioactive material, the same kind of experiments may be performed by using fluorescent Dextran polymers of increasing sizes (4 to 250 kDa) (Romero *et al.*, 2000).

Measure of the transcellular electrical resistance

Another way to evaluate the paracellular flux is to determine the consequence of the bacteria/cell interaction on the electrical resistance.

Methods

Transwell® inserts can be easily removed from the cluster plate and transcellular electrical resistance measurements performed using an Endohm apparatus (World Precision Instruments), which has been documented by some investigators to allow more reproducible results than the more commonly used chopstick electrodes (Engelbertz *et al.*, 2000; Cecchelli *et al.*, 1999). Resistance values are expressed in ohm-cm^2 (Ω cm^2), following subtraction of the resistance of a cell-free filter. It has been mentioned above that the electrical resistance of a primary culture of BMECs is lower than 200 (Ω cm^2) whereas it can be of 500–600 Ω cm^2, and even up to 1000 Ω cm^2, in improved cell culture systems, especially with astrocyte co-culture. The effect on the transcellular electrical resistance of bacterial interaction or invasion of a BMEC monolayer can thus be easily assessed by this method. Measurement of the resistance is performed before the infection of the monolayer, bacteria are then added in the insert chamber of the Transwell® on the apical surface of the BMECs and the resistance can be measured at different time points during the infection process.

It should be pointed out that the presence in the media of growing bacteria may be responsible for modifications which alter the intercellular junctions and that it is indispensable to perform the appropriate controls, ideally an isogenic mutant which grows in the media but does not affect the electrical resistance.

Measure of bacterial internalization and/or transcytosis

To measure the ability of bacteria to invade and transcytose through brain barriers, bacteria are put in contact with a tight monolayer of endothelial cells grown on a transwell filter for 2 to 4 h.

Methods

The number of invading bacteria can be determined by a classical gentamycin protection assay. Briefly, cells are washed three times and an antibiotic-containing medium is added in both upper and lower chambers for 1 h to kill the extracellular bacteria. Monolayers are lysed by addition of 0.025% ice-cold Triton X-100 or resuspend by scraping into GCB liquid medium, serially diluted and spread onto GCB plates. Bacteria are grown overnight and colony-forming units are counted. To determine the number of bacteria able to transcytose, the monolayers are washed three times after the infection process and the antibiotic-containing medium is added in both upper and lower chambers for 1 h to kill the extracellular bacteria, the monolayers are then washed again three times and antibiotic-containing medium is added in the upper chamber to avoid any interfering proliferation of bacteria in this chamber, and the lower chamber is filled with antibiotic-free medium. Samples are taken from the lower chamber at various times and the number of bacteria counted corresponds to the bacteria which have entered the endothelial cells.

In vivo physiological assays

Animal models for some bacteria inducing meningitis

Several animal models have been used to study the mechanisms that lead to the development of an inflammation into the meninges following invasion of the CSF (Nassif and So, 1995). In these models, bacteria are directly injected into the CSF through the cisterna magna. Alternatively, a model using new-born rats has been used with *E. coli* and *N. meningitidis* to study the meningeal invasion. The bacteria are injected intraperitoneally and the course of infection can be established by determining the number of bacteria in the blood and the cerebrospinal fluid (CSF) at various time points. The major limitation concerning this model, at least with *N. meningitidis*, is that this pathogen does not interact with rat cells. Therefore the crossing of the BBB by these bacteria in new-born rats do not result from a specific interaction of the bacteria with the cellular barrier, as in the human disease, but from the ability of the bacteria to induce a sustained bacteremia and to bypass the BBB. This model is therefore suitable to assess the ability of the bacteria to multiply in the bloodstream but not to evaluate the pathological mechanisms responsible for bacterial invasion of the CSF.

◆◆◆◆◆◆ STRUCTURAL EFFECT OF MICROBIAL INTERACTION ON BARRIER PROPERTIES

As mentioned above, alteration of tight junction function may eventually be an important pathogenic process of meningitis-causing bacteria. They can be affected in two ways: either indirectly via a variety of agents, such as bacterial toxins or cytokines released by host cells during the inflammatory response to microbial invasion (i.e. TNFα, IL1β), or directly following the specific interaction of the bacteria with the endothelial cells. The recent identification of several tight junction proteins now permits direct examination of the effects of microbial infection on the molecular structure of tight junctions. Such an analysis should provide important insights into the consequences of bacterial interaction with BBB components.

In vitro culture of endothelial cells on Transwell filter inserts as described above eventually may be used to study the structural changes in tight junctions induced by bacterial pathogens. Since an alteration of the paracellular permeability can occur by more than one mechanism, via direct effect on the structural tight junction proteins (localization, phosphorylation) and/or via changes in the perijunctional actin cytoskeleton, a few approaches are described here to assess the effect of microbial infection on the architecture of the tight junctions.

Immunocytochemistry

Studies have documented that cytokines may induce a delocalization of the tight junction-associated proteins occludin, ZO-1 and ZO-2 or of the adherens junction proteins cadherin or catenins. An analysis of the cellular localization of the junctional proteins can then be performed in the infected cells by immunofluorescence analysis.

Methods

Cells are fixed in 4% paraformaldehyde made up in PBS containing 0.5 mM $CaCl_2$ and 0.5 mM $MgSO_4$. After 15 min incubation at room temperature, the cells are washed and then permeabilized by incubation for 10 min with 0.5% Triton X-100 in PBS. After washing, the cells are blocked by incubation for 30 min in PBS containing 3% Bovine Serum Albumin (BSA) plus 0.1 M lysine, pH 7.4 and incubated with primary antibody in PBS containing 0.3% BSA for either 2 h at room temperature or overnight at 4°C. After washing, the cells are then incubated for 30–60 min at room temperature with appropriate FITC- or rhodamine-conjugated secondary antibody in PBS containing 0.3% BSA. After another washing step, the filters are mounted and analysed with a fluorescence or confocal microscope.

ZO-1 is known to play a key role in linking the tight junction complexes to the apical perijunctional actin-myosin ring. A displacement of ZO-1 from the junction may thus coincide with a disruption of apical actin that can also be assessed by immunofluorescence analysis of the actin cytoskeleton by using rhodamine-labelled phalloidin. Loss of actin from the submembrane cortical region results in the formation of intracytoplasmic aggregates of F-actin, depolymerization of F-actin filaments and an increase in G-actin. The role of apical actin disorganization in the redistribution of the junctional proteins, the opening of tight junctions and an increase in permeability has been documented in response to inflammatory cytokines, such as TNF-α (Descamps *et al.*, 1997). A similar analysis may be performed following infection of the endothelial monolayer, and the role of the actin cytoskeleton confirmed in the presence of cytochalasin-B, that should inhibit bacteria-induced F-actin reorganization and ZO-1 redistribution.

Biochemistry

Analysis of tyrosine phosphorylation: To examine the level of tyrosine phosphorylation of individual proteins, immunoprecipitation with an anti-phosphotyrosine antibody followed by protein-specific blotting can be performed. Reciprocally, protein-specific immunoprecipitation followed by anti-phosphotyrosine antibody blotting should provide similar information.

Methods

For such analysis, cells are lysed at 4°C in RIPA buffer (Tris-Hcl 5 mM, pH 7.5, NaCl 70 mM, EDTA 1 mM, Nonidet P40 1%, SDS 0.1%, Deoxycholate 0.75%) containing phosphatase and protease inhibitors (2 mM sodium orthovanadate, 5 mM natrium fluoride, 1 mM PMSF, 10 μg ml^{-1} each of aprotinin/leupeptin/pepstatin) for 15–30 min then scraped off the culture dishes or filter inserts with a cell scraper, transferred in an eppendorf tube, and the cell lysate is cleared by centrifugation at 10 000 g for 15 min. Immunoprecipitations are performed with appropriate antibodies and resolved by SDS PAGE as described earlier. In general, phosphotyrosine immunoprecipitation followed by protein-specific blotting is a fairly sensitive method to reveal changes in the amount of phosphorylation of junctional or cytoskeleton-associated proteins (Staddon *et al.*, 1995; Etienne *et al.*, 1998).

Analysis of serine/threonine phosphorylation: As described in the introduction to this chapter, modification in serine and/or threonine phosphorylation also takes an important part in the regulation of the intercellular junctions. Whereas anti-phosphotyrosine antibodies in general recognize phosphotyrosine in proteins regardless of sequence, the anti-phosphoserine/threonine antibodies are sequence specific. So far, no phosphopeptide antibody directed against junctional proteins is available. To assess modification in serine and/or threonine phosphorylation, cells are usually labelled with [^{32}P] phosphate and subject to protein analysis as described earlier. The phosphoaminoacid content of protein can be analysed by subsequent limited acid hydrolysis followed by high voltage electrophoretic separation.

Analysis of protein complexes: Adherens and tight junction proteins exist as complexes that can be disrupted by modification of the phosphorylation status of some of their components. When associated, the junctional proteins can be co-immunoprecipitated under mild, non-denaturating detergent condition. Regulation of junctional complex organization may thus be easily assessed by biochemistry analysis.

Methods

Cells are lysed in a NP-40-based buffer (0.1% Nonidet P40, 50 mM Tris-HCl pH 7.4, 250 mM NaCl, 5 mM EDTA) containing phosphatase and protease inhibitors (2 mM sodium orthovanadate, 5 mM natrium fluoride, 1 mM PMSF, 10 μg ml^{-1} each of aprotinin/leupeptin/pepstatin). The lysates, cleared by centrifugation, are used for immunoprecipitation with an antibody against the protein of interest (as example: ZO-1, ZO-2, β-catenin). After resolution by SDS PAGE, the proteins that associate with the protein of interest are found in the immune complex and are revealed by immunoblotting with the appropriate antibody (i.e. occludin, claudin, vascular endothelial cadherin).

Analysis of protein interaction with the cytoskeleton: An alteration in the association between the molecular linker ZO-1, ZO-2 or β-catenin with the actin cytoskeleton can also be determined by using Triton X-100 as detergent. Since cytoskeleton-associated proteins are non-soluble in Triton X-100, alteration in their interaction with the actin cytoskeleton thus results in decrease of the detergent-insoluble pools of these proteins.

Methods

Soluble fractions are collected after 15 min incubation on ice with a Triton X-100-based buffer (25 mM Mes pH 6.5, 150 mM NaCl, Triton X-100 0.5%) containing phosphatase and protease inhibitors (2 mM sodium orthovanadate, 5 mM Natrium Fluoride, 1 mM PMSF, 10 μg ml^{-1} each of aprotinin/leupeptin/pepstatin). Triton-insoluble material is extracted with a buffer containing 10 mM Tris-HCl pH 8, 150 mM NaCl, Triton X-100 1%, 60 mM octyl glucoside and phosphatase and protease inhibitors. Both Triton-soluble and -insoluble fractions are resolved by SDS PAGE and immunoblots with the appropriate antibodies.

◆◆◆◆◆◆ CONCLUSION

In summary, cerebro-vascular endothelium and choroid plexus epithelium are the two faces (the BBB and BCSFB, respectively) of the physiological barrier which protects the brain against environmental pathogens. The understanding of the molecular organization of the tight and adherens junctions expressed at cell–cell contacts in these barriers has dramatically increased over the last few years, raising hopes that *in vitro* models of the BBB might be designed which would make possible studies of the mechanisms of microbial entry into the CNS. Indeed, valuable models with low permeability are available, based on co-culture of bovine or porcine brain endothelial cells with astrocytes, or on porcine or murine choroid plexus epithelial cells. However, not only are they still time-consuming and require significant know-how, but moreover no such model has been proposed yet with human epithelial or endothelial cells, which constitutes a limitation for studying human pathogens. The cell culture procedures and the physiological, cellular and molecular methods presented above will be useful for studying the mechanisms of para-cellular infiltration of extracellular bacteria or leukocytes infected by intracellular bacteria. The study of the putative transcellular route of invasion will require alternative immunochemical and imaging methods which will need to be optimized with endothelial cells. It is anticipated that these *in vitro* experimental approaches will allow the unravelling of the mechanisms of cerebral infection by bacterial pathogens and pave the way to the development of therapeutic strategies.

References

Abbott, N. J., Hughes, C. C., Revest, P. A. and Greenwood, J. (1992). Development and characterisation of a rat brain capillary endothelial culture: towards an *in vitro* blood-brain barrier. *J. Cell Sci.* **103**, 23–37.

Adamson, P., Etienne, S., Couraud, P. O., Calder, V. and Greenwood, J. (1999). Lymphocyte migration through brain endothelial cell monolayers involves signaling through endothelial ICAM-1 via a rho-dependent pathway. *J. Immunol.* **162**, 2964–2973.

Audus, K. L. and Borchardt, R. T. (1986). Characteristics of the large neutral amino acid transport system of bovine brain microvessel endothelial cell monolayers. *J. Neurochem.* **47**, 484–488.

Bazzoni, G., Martinez-Estrada, O. M., Orsenigo, F., Cordenonsi, M., Citi, S. and Dejana, E. (2000). Interaction of junctional adhesion molecule with the tight junction components ZO-1, cingulin, and occludin. *J. Biol. Chem.* **275**, 20520–20526.

Bowman, P. D., Ennis, S. R., Rarey, K. E., Betz, A. L. and Goldstein, G. W. (1983). Brain microvessel endothelial cells in tissue culture: a model for study of blood–brain barrier permeability. *Ann. Neurol.* **14**, 396–402.

Cecchelli, R., Dehouck, B., Descamps, L., Fenart, L., Buee-Scherrer, V., Duhem, C., Lundquist, S., Rentfel, M., Torpier, G. and Dehouck, M. P. (1999). In vitro model for evaluating drug transport across the blood–brain barrier. *Adv. Drug. Deliv. Rev.* **36**, 165–178.

Clarke, H., Marano, C. W., Peralta Soler, A. and Mullin, J. M. (2000). Modification of tight junction function by protein kinase C isoforms. *Adv. Drug Deliv. Rev.* **41**, 283–301.

Descamps, L., Cecchelli, R. and Torpier, G. (1997). Effects of tumor necrosis factor on receptor-mediated endocytosis and barrier functions of bovine brain capillary endothelial cell monolayers. *J. Neuroimmunol.* **74**, 173–184.

Dorovini-Zis, K., Prameya, R. and Bowman, P. D. (1991). Culture and characterization of microvascular endothelial cells derived from human brain. *Lab. Invest.* **64**, 425–436.

Durieu-Trautmann, O., Federici, C., Creminon, C., Foignant-Chaverot, N., Roux, F., Claire, M., Strosberg, A. D. and Couraud, P. O. (1993). Nitric oxide and endothelin secretion by brain microvessel endothelial cells: regulation by cyclic nucleotides. *J. Cell Physiol.* **155**, 104–111.

Engelbertz, C., Korte, D., Nitz, T., Franke, H., Haselbach, M., Wegener, J. and Galla, H. J. (2000). In *The Blood-Brain Barrier and Drug Delivery to the CNS.* (eds D. Begley, M. W. Bradbury, J. Kreuter) The development of *in vitro* models for the blood–brain and blood–CSF barriers, pp. 33–63. Marcel Dekker, New York.

Etienne, S., Adamson, P., Greenwood, J., Strosberg, A. D., Cazaubon, S. and Couraud, P. O. (1998). ICAM-1 signaling pathways associated with Rho activation in microvascular brain endothelial cells. *J. Immunol.* **161**, 5755–5761.

Franke, H., Galla, H. and Beuckmann, C. T. (2000). Primary cultures of brain microvessel endothelial cells: a valid and flexible model to study drug transport through the blood–brain barrier *in vitro. Brain Res. Brain Res. Protoc.* **5**, 248–256.

Goldstein, G. W. and Betz, A. L. (1986). The blood–brain barrier. *Sci. Am.* **255**, 74–83.

Greenwood, J., Pryce, G., Devine, L., Male, D. K., dos Santos, W. L., Calder, V. L. and Adamson, P. (1996). SV40 large T immortalised cell lines of the rat blood–brain and blood-retinal barriers retain their phenotypic and immunological characteristics. *J. Neuroimmunol.* **71**, 51–63.

Hopkins, A. M., Li, D., Mrsny, R. J., Walsh, S. V. and Nusrat, A. (2000). Modulation of tight junction function by G protein-coupled events. *Adv. Drug Deliv. Rev.* **41**, 329–340.

Joo, F. (1993). The blood–brain barrier in vitro: the second decade. *Neurochem. Int.* **23**, 499–521.

Kim, K. S. (2000). E. Coli invasion of brain microvascular endothelial cells as a Pathogenetic Basis of meningitis. In (eds Oelschlaeger, Hacker) *Subcellular Biochemistry: Bacterial Invasion into Eukaryotic Cells.* Kluwer Academic/Plenum Publishers, New York, 47–59.

Kniesel, U. and Wolburg, H. (2000). Tight junctions of the blood–brain barrier. *Cell Mol. Neurobiol.* **20**, 57–76.

Lapierre, L. A. (2000). The molecular structure of the tight junction. *Adv. Drug Deliv. Rev.* **41**, 255–264.

Martin-Padura, I., Lostaglio, S., Schneemann, M., Williams, L., Romano, M., Fruscella, P., Panzeri, C., Stoppacciaro, A., Ruco, L., Villa, A., Simmons, D. and Dejana, E. (1998). Junctional adhesion molecule, a novel member of the immunoglobulin superfamily that distributes at intercellular junctions and modulates monocyte transmigration. *J. Cell Biol.* **142**, 117–127.

Merz, A. J., Rifenbery, D. B., Arvidson, C. G. and So, M. (1996). Traversal of a polarized epithelium by pathogenic Neisseriae: facilitation by type IV pili and maintenance of epithelial barrier function. *Mol. Med.* **2**, 745–754.

Nassif, X. and So, M. (1995). Interaction of pathogenic neisseriae with nonphagocytic cells. *Clin. Microbiol. Rev.* **8**, 376–388.

Nusrat, A., Giry, M., Turner, J. R., Colgan, S. P., Parkos, C. A., Carnes, D., Lemichez, E., Boquet, P. and Madara, J. L. (1995). Rho protein regulates tight junctions and perijunctional actin organization in polarized epithelia. *Proc. Natl Acad. Sci. USA* **92**, 10629–10633.

Palmeri, D., van Zante, A., Huang, C. C., Hemmerich, S. and Rosen, S. D. (2000). Vascular endothelial junction-associated molecule, a novel member of the immunoglobulin superfamily, is localized to intercellular boundaries of endothelial cells. *J. Biol. Chem.* **275**, 19139–19145.

Pron, B., Taha, M. K., Rambaud, C., Fournet, J. C., Pattey, N., Monnet, J. P., Musilek, M., Beretti, J. L. and Nassif, X. (1997). Interaction of Neisseria maningitidis with the components of the blood–brain barrier correlates with an increased expression of PilC. *J. Infect. Dis.* **176**, 1285–1292.

Pujol, C., Eugene, E., Marceau, M. and Nassif, X. (1999). The meningococcal PilT protein is required for induction of intimate attachment to epithelial cells following pilus-mediated adhesion. *Proc. Natl Acad. Sci. USA* **96**, 4017–4022.

Regina, A., Romero, I. A., Greenwood, J., Adamson, P., Bourre, J. M., Couraud, P. O. and Roux, F. (1999). Dexamethasone regulation of P-glycoprotein activity in an immortalized rat brain endothelial cell line, GPNT. *J. Neurochem.* **73**, 1954–1963.

Romero, I. A., Prevost, M. C., Perret, E., Adamson, P., Greenwood, J., Couraud, P. O. and Ozden, S. (2000). Interactions between brain endothelial cells and human T-cell leukemia virus type 1-infected lymphocytes: mechanisms of viral entry into the central nervous system. *J. Virol.* **74**, 6021–6030.

Roux, F., Durieu-Trautmann, O., Chaverot, N., Claire, M., Mailly, P., Bourre, J. M., Strosberg, A. D. and Couraud, P. O. (1994). Regulation of gamma-glutamyl transpeptidase and alkaline phosphatase activities in immortalized rat brain microvessel endothelial cells. *J. Cell Physiol.* **159**, 101–113.

Rubin, L. L., Hall, D. E., Porter, S., Barbu, K., Cannon, C., Horner, H. C., Janatpour, M., Liaw, C. W., Manning, K. and Morales, J. *et al.* (1991). A cell culture model of the blood-brain barrier. *J. Cell Biol.* **115**, 1725–1735.

Segal, M. (1998). The Blood–CSF barrier and the choroid plexus, In (ed. W. M. Pardridge) *Introduction to the Blood–Brain Barrier* pp.251–258. Cambridge University Press.

Smith, A. L. (1987). Pathogenesis of *Haemophilus influenzae* meningitis. *Pediatr. Infect. Dis. J.* **6**, 783–786.

Staddon, J. M., Herrenknecht, K., Smales, C. and Rubin, L. L. (1995). Evidence that tyrosine phosphorylation may increase tight junction permeability. *J. Cell Sci.* **108**, 609–619.

Staddon, J. M. and Rubin, L. L. (1996). Cell adhesion, cell junctions and the blood–brain barrier. *Curr. Opin. Neurobiol.* **6**, 622–627.

Stins, M. F., Gilles, F. and Kim, K. S. (1997). Selective expression of adhesion molecules on human brain microvascular endothelial cells. *J. Neuroimmunol.* **76**, 81–90.

Strazielle, N. and Ghersi-Egea, J. F. (1999). Demonstration of a coupled metabolism-efflux process at the choroid plexus as a mechanism of brain protection toward xenobiotics. *J. Neurosci.* **19**, 6275–6289.

Strazielle, N. and Ghersi-Egea, J. F. (2000). Choroid plexus in the central nervous system: biology and physiopathology. *J. Neuropathol. Exp. Neurol.* **59**, 561–574.

Tsukita, S., Furuse, M. and Itoh, M. (1999). Structural and signalling molecules come together at tight junctions. *Curr. Opin. Cell Biol.* **11**, 628–633.

Tsukita, S. and Furuse, M. (1999). Occludin and claudins in tight-junction strands: leading or supporting players? *Trends Cell Biol.* **9**, 268–273.

Tuomanen, E. (1999). Molecular and cellular biology of pneumococcal infection. *Curr. Opin. Microbiol.* **2**, 35–39.

Walsh, S. V., Hopkins, A. M. and Nusrat, A. (2000). Modulation of tight junction structure and function by cytokines. *Adv. Drug Deliv. Rev.* **41**, 303–313.

Wegener, J., Hakvoort, A. and Galla, H. J. (2000). Barrier function of porcine choroid plexus epithelial cells is modulated by cAMP-dependent pathways in vitro. *Brain Res.* **853**, 115–124.

Zahraoui, A., Louvard, D. and Galli, T. (2000). Tight junction, a platform for trafficking and signaling protein complexes. *J. Cell Biol.* **151**, 31–36.
Zhang, J. R. and Tuomanen, E. (1999). Molecular and cellular mechanisms for microbial entry into the CNS. *J. Neurovirol.* **5**, 591–603.
Zhang, J. R., Mostov, K. E., Lamm, M. E., Nanno, M., Shimida, S., Ohwaki, M. and Tuomanen, E. (2000). The polymeric immunoglobulin receptor translocates pneumococci across human nasopharyngeal epithelial cells. *Cell* **102**, 827–837.

24 Methods for Studying Bacteria-induced Host Cell Death

David S Weiss[1] and Arturo Zychlinsky[2]
[1]*Skirball Institute and Department of Microbiology, New York University School of Medicine, 540 First Avenue, New York, NY 10016, USA;* [2]*Max Planck Institute for Infection Biology, Campus Charité Mitte, Berlin D-10017, Germany*

◆◆

CONTENTS

◆◆◆◆◆◆ INTRODUCTION

Many pathogens modulate the death-pathways of host cells by actively inducing or inhibiting host cell death in order to manipulate host defenses (Roulston *et al.*, 1999; Weinrauch and Zychlinsky, 1999). The importance of the control of host cell death is underscored by findings that pathogens harboring mutations in components involved in cell death modulation are often avirulent. Clearly, regulation of cell death can have important consequences on the outcome of infection and disease.

Over the past 10 years there has been an explosion in research and knowledge in the field of cell death. This chapter is designed to provide the researcher with techniques to evaluate the role of cell death in bacterial infection. We describe methods to detect and characterize cell death. Elucidation of cell death modulation pathways may help to explain observed pathologies and identify new avenues for anti-bacterial therapy.

METHODS IN MICROBIOLOGY, VOLUME 31
ISBN 0–12–521531–2

◆◆◆◆◆◆ EUKARYOTIC CELL DEATH

Cell death is a vital process in eukaryotic organisms. It is required for morphological changes such as digit formation during development and for selection of immune cells (Jacobson *et al.*, 1997). Inappropriate cell death is involved in the pathology of many diseases. Two types of cell death are apoptosis and necrosis (reviewed in Allen *et al.*, 1997; Majno and Joris, 1995).

Apoptosis is observed in most cell types in animals ranging from the nematode *Caenorhabditis elegans* to humans (Jacobson *et al.*, 1997). Apoptosis is characterized by chromatin condensation and the accumulation of dense masses of chromatin adjacent to the nuclear membrane. Apoptotic cells shrink in size and lose contacts with adjacent cells. Adherent cells in culture often lift up and float during apoptosis. Plasma membrane convolution results in the formation of blebs which eventually pinch off the cell, enclosing cellular contents including nuclear fragments in vesicles called apoptotic bodies. *In vivo*, apoptotic cells are recognized by phagocytes and rapidly cleared (Savill *et al.*, 1993).

During necrosis, chromatin is scattered in an irregular pattern throughout the nucleus. Necrotic cells swell in size, as do their cytosolic organelles. The cytosol appears vacuolated and the plasma membrane lyses, releasing cellular contents. It is important to note that there are examples of cell death that do not fully adhere to the definitions of either apoptosis or necrosis (Majno and Joris, 1995). If such an event is observed, the researcher should describe the features of the cell death rather than forcing an inappropriate classification as either apoptosis or necrosis.

◆◆◆◆◆◆ BACTERIA-INDUCED HOST CELL DEATH

A wide array of pathogenic bacteria and bacterial products induce host cell death. The mechanisms include (reviewed in Weinrauch and Zychlinsky, 1999) the following: *Staphylococcus aureus* and *Listeria monocytogenes* secrete pore-forming toxins which disrupt the host cell membrane; *Corynebacterium diphtheriae* and *Pseudomonas aeruginosa* secrete protein synthesis inhibitors; many gram-negative pathogens, including *Salmonella*, *Shigella* and *Yersinia* spp. as well as several plant pathogens, encode type III secretion systems which secrete cell death inducers into host cells (Galan and Collmer, 1999).

◆◆◆◆◆◆ POTENTIAL PITFALLS IN CELL DEATH ASSAYS

Cell deaths can vary tremendously, both in kinetics and in observed morphological and biochemical features. The events and pathways involved differ depending on cell type, differentiation state and the death stimulus. The occurrence of cell death cannot be ruled out if particular features are not detected. Thus, it is important to use several methods and perform a time course when characterizing and demonstrating the presence of cell death. One method should always be based on morphological changes as these are among the most conserved cell death events.

Many substances are toxic to cells and induce cell death at high concentrations. To be sure that a death stimulus is specific, appropriate controls must be used. When using bacteria, it is ideal to use isogenic mutants that do not induce cell death, assuming they are comparable in all other functions. If such mutants have not been identified, use of a related non-pathogenic strain is recommended.

In vitro, apoptotic cells are not cleared as they are *in vivo*. These cells eventually lyse and gain a necrotic appearance, a process called secondary necrosis (Majno and Joris, 1995). Characterization of initially apoptotic cells as necrotic cells must be taken into account when analyzing cell death *in vitro*.

◆◆◆◆◆◆ QUICK AND EASY ASSAYS FOR CELL DEATH

Morphology

An easy way to detect cell death *in vitro* is to observe morphological changes using phase contrast or Nomarsky microscopy. Visible apoptotic changes include cell shrinkage and rounding, loss of adherence, blebbing and apoptotic body formation and chromatin condensation. In contrast, necrotic cells swell in size and vacuolate before lysing, and appear like cellular 'ghosts'. Alternatively, using light microscopy, counterstaining of the nucleus and cytoplasm allows visualization of morphological changes, especially chromatin condensation. Standard stains like eosin are used to stain the nucleus and hematoxylin or methylene blue/azure can be used to stain the cytoplasm (Allen *et al.*, 1997).

More advanced microscopic techniques are also useful for detecting morphological changes of dying cells. Chromatin structure can be observed after staining cells with the cell-permeable fluorescent probes Hoechst 33342 or acridine orange, which stain DNA. Cells can then be analyzed by fluorescence microscopy or flow cytometry. Propidium iodide (PI), which is not cell-permeable, can also be used to stain DNA in fixed and permeabilized cells.

Transmission electron microscopy detects ultrastructural changes (condensed chromatin, membrane blebbing) and scanning electron microscopy is used to observe membrane blebbing and apoptotic body formation (Allen *et al.*, 1997). Video time-lapse microscopy follows the progression of cell death in individual cells and is a useful method to prove the occurrence of cell death (Willingham, 1999).

Vital dye exclusion

Intact plasma membranes are impermeable to large molecules. A simple technique to assay for cell death is to test for the exclusion of the large dye trypan blue (960 MW). Both necrotic and secondary necrotic cells will stain with trypan blue, which can be seen using a light microscope. An advantage of this method is that few cells are needed and the assay is quick, although quantification is operator-dependent. PI and other fluorescent DNA binding dyes are also excluded from cells with intact plasma membranes and can be visualized by fluorescence microscopy or quantified by flow cytometry.

Vital dye exclusion

Trypan blue staining

- Mix a solution of cells in a 1.5 ml eppendorf tube 1:1 by volume with Trypan Blue solution from Sigma (cat. #T-8154). Pipette thoroughly to break apart any cell clumps.
- After 5 min, count 10 μl aliquots of the mixture using a hemocytometer. Cells must be counted soon after staining since viable cells will begin to take up trypan blue after long periods of time. Using a light microscope, count cells within a defined area which are trypan blue positive (dark blue) and those that are negative (not stained or light blue). Count multiple 10 μl aliquots for better accuracy. Cell viability is determined using the following formula: % viable cells = (viable cells/total cells)*100%.

Propidium iodide staining

- Add propidium iodide (PI) solution (Roche—cat. #1348639) to cells to a final concentration of 50 $\mu g\, ml^{-1}$. Allow cells to incubate for 30 min. For flow cytometry, 10^6 cells are required. Few cells are required if fluorescence is assayed using a fluorescence microscope.
- PI staining is detected as red fluorescence (above 630 nm). PI positive cells are considered to be non-viable (necrotic or secondary necrotic).

Lactate dehydrogenase (LDH) assay

In the past, some cytotoxicity assays measured the release of radioactive isotopes from pre-labeled cells. These assays can still be used, but since easier non-radioactive techniques are now available, we recommend use of the LDH or MTT assays.

Cells undergoing necrosis or secondary necrosis lyse and release their cellular contents. A relatively simple assay for cytotoxicity measures the release of one intracellular protein, LDH (Nachlas, 1960). LDH catalyzes the oxidation of lactate to pyruvate, resulting in the reduction of NAD+ to NADH. The NADH is oxidized in a subsequent reaction carried out by an exogenous catalyst, causing the reduction of a colorless tetrazolium salt (INT) to a red formazan salt. The color change can be quantified using an ELISA plate reader.

The LDH assay can be performed in a 96-well plate. After the assay, a sample of the supernatant is removed, added to a new 96-well plate, and mixed with the LDH substrate mix containing INT, diaphorase (catalyst), sodium lactate and NAD+. After a 30-min incubation in the dark, the reaction is stopped by addition of 6% acetic acid and the samples are read at 490 nm. The percentage of cytotoxicity is determined by comparison of LDH released from experimental samples with total LDH contained in cells and LDH released from untreated samples. LDH is a stable enzyme, so the activity measured reflects the accumulation of this enzyme in the culture media at a given time point. The stability, however, depends on the levels of proteases in the culture media, and therefore may vary depending on the cell type and media used.

It is important to note that LDH is only released upon membrane lysis and therefore apoptotic cells which have not undergone secondary necrosis will not be counted as dying cells. Although many bacteria have LDH, it is contained intracellularly and should not interfere with the assay. Control samples containing bacteria alone should always be included to assay for background readings. If the bacteria are lysed during the course of the assay, bacterial LDH would contribute to total LDH activity and make the interpretation of results impossible.

LDH assay

This is a modified version of the protocol provided with the Promega CytoTox 96 Non-Radioactive Cytotoxicity Assay (cat.# G1780).

- Perform the cell death assay in a 96-well plate.
- Different cell types contain varying amounts of LDH. Therefore, the number of cells plated per well must be optimized for each cell type. The absorbance value of the total LDH contained in cells (assayed by lysing cells with Triton X-100, see below) should be at least threefold higher than the absorbance value of untreated cells.
- Cells are plated in triplicate. Set up wells for control samples: untreated cells for background readings, and untreated cells to be lysed for 'total' samples. Since serum contains LDH, the assay must be carried out in serum-free media or if necessary, media containing no more than 1% serum.
- Add the death stimulus to the cells and incubate until cell death occurs.
- To assay for cell death, carefully remove 50 μl of supernatant from the samples without disturbing the cells and transfer them to a new 96-well plate. To measure the 'total' LDH in the 'total' control cells, lyse them with Triton X-100 to a final concentration of 1% and pipette up and down prior to removal of a 50 μl aliquot.
- Add 50 μl of reconstituted Substrate Mix (INT, diaphorase, sodium lactate and NAD+) to each well containing supernatants.
- Incubate in the dark for 30 minutes.
- Add 100 μl of 6% acetic acid to each well to stop the reaction.
- Pop any large bubbles in the wells before measuring the OD490 using an ELISA plate reader. % cytotoxicity is calculated using the following formula: % cytotoxicity = (Experimental − Untreated)/('Total' − Untreated)*100%.

MTT assay

The MTT assay measures metabolically active cells. MTT (3,(4,5-dimethylthiazol-2-yl)2,5-diphenyl-tetrazolium bromide), a yellow tetrazolium salt, is converted into insoluble purple formazan crystals by living cells (Mosmann, 1983). Similar to the LDH assay, this assay can be performed in a 96-well plate. At the end of the assay, samples are incubated with MTT for 2 h. The cells are lysed with detergents and the crystals are solubilized with organic solvents, allowing the measurement of the color change, which is proportional to the number of living cells.

Bacteria can catalyze the conversion of MTT into formazan crystals. Therefore, the MTT assay cannot be used if live bacteria are present either

MTT assay

The protocol outlined below is an adaptation of the MTT protocol described by Hansen *et al.* (1989).

Reagents:

MTT solution—MTT (3,(4,5-dimethylthiazol-2-yl)2,5-diphenyl-tetrazolium bromide) can be purchased from Sigma (cat. # M-2128). MTT is dissolved in sterile PBS to a concentration of $5\,mg\,ml^{-1}$. The MTT stock solution must be kept in the dark at 4°C.
Extraction buffer—For 100 ml of buffer—20 g SDS, 50 ml N,N-dimethyl formamide (DMF) from Sigma (cat.# D-8654), 45 ml dH_2O, 2 ml 17N acetic acid, 2.5 ml 1N HCl.

Protocol:

- The cell death assay is performed on cells in a 96-well plate. The final volume in each well before addition of MTT should be 80 µl. The optimal number of cells per well must be determined for each cell type such that the absorbance of untreated cells is above 1.0 and the absorbance of 'MTT late' cells (see below) is under 0.1.
- All samples are set up in triplicate. Set up wells for control samples: untreated cells to which MTT is added at the same time as to experimental samples ('Untreated'), and untreated cells to which MTT is added after the 2 h MTT incubation ('MTT late' cells).
- Add the death stimulus to cells and incubate.
- Once cell death is to be assayed, add 20 µl of the MTT solution (final concentration of MTT is $1\,mg\,ml^{-1}$). Do not add MTT to 'MTT late' samples. Allow cells to incubate for 2 h at 37°C, 5% CO_2. Dark purple crystals should be visible by light microscopy inside viable cells.
- Add MTT to 'MTT late' samples. Add 100 µl of extraction buffer to all samples. Incubate for 2 h to overnight. Check that crystals have dissolved before reading samples.
- Measure the OD570 using an ELISA plate reader. Calculate % cytotoxicity using the following formula: % cytotoxicity = 100% − (Experimental − 'MTT late')/('Untreated' − 'MTT late')*100

intra- or extracellularly. Providing bacteria are not located intracellularly, bacteria-induced cell death can be assayed if bacteria are washed away prior to the addition of MTT. This assay can also be used to measure the cell death induced by bacterial products.

◆◆◆◆◆◆ BIOCHEMICAL DETECTION OF CELL DEATH

DNA fragmentation

DNA laddering

Fragmentation of nuclear DNA is a hallmark of apoptosis. During apoptosis, linker DNA between nucleosomes is cleaved, generating 180–200 base pair fragments of DNA and multiples thereof (Wyllie *et al.*,

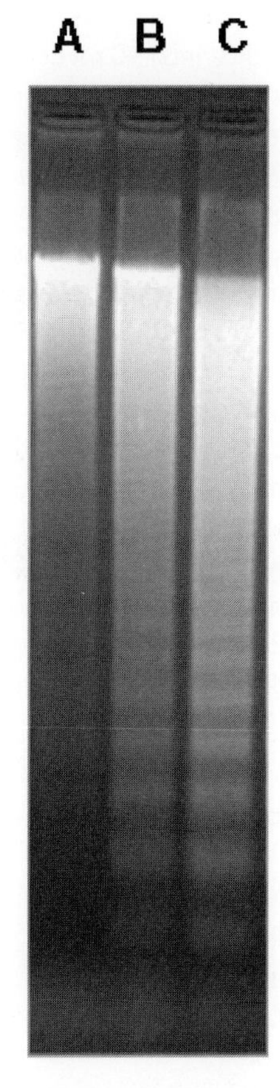

Figure 24.1. DNA fragmentation assay: Apoptosis was induced in a system where THP-1 cells were permeabilized in 100 μl of buffer containing 100 μg/ml digitonin. DNA was extracted from 50 μl aliquots 3 hours after the addition of mock buffer (lane A), dATP/cytochrome c (lane B), or recombinant caspase-3 (lane C). DNA was purified using silica resin in the presence of guanine hyrdochloride.

DNA fragmentation

The procedure requires: (a) silica resin; (b) a chaotropic lysis buffer; (c) wash buffer; and (d) a lightly buffered, low ionic strength elution buffer.

Preparation of silica resin

The silica resin can be any silica slurry such as 'glass milk' found with kits used to extract DNA from agarose gels. One source of inexpensive silica, not described further here, is a 50% slurry of silicic acid in water, which does not need to be prepared further. The following recipe uses silica from Sigma (cat. # S-5631) as another inexpensive alternative. As supplied, however, the silica grains are heterogeneous and contain particles of many different sizes. Centrifugation is employed to remove the very small silica 'fines' and retain the larger silica particles. These larger particles have ideal DNA binding and elution properties. The silica can be cleaned with nitric acid during this preparative step. Here we describe our adaptation of a recipe found on Paul Hengen's website at *http://www-lmmb.ncifcrf.gov/~pnh/FAQlist.html.*

- Put 10 g of silica into 40 ml ddH_2O in a 50 ml blue-capped conical tube. Put on a shaker platform for at least 10–15 min to make sure the slurry of powder is well hydrated before starting the centrifugation steps.
- Spin out the larger silica particles in a clinical centrifuge at 150 g at room temperature for 15 min.
- Pour off the supernatant. Approximately 5 to 8 ml of resin will be packed at the bottom of the tube. Resuspend the resin with 40 ml ddH_2O and centrifuge again. This time the pellet may be slightly smaller. Decant the supernatant again, resuspend and spin one more time as before. Approximately 5 ml of resin should remain.
- Resuspend the pellet in 10 ml ddH_2O (or 2× the volume of the pellet), add 15 ml of nitric acid and heat to 80°C for 1 h in a water bath (keep the cap slightly loose).
- Spin out the cleaned resin at 1000 g for 5 minutes. Wash the pellet in 30 ml of ddH_2O four times to remove traces of nitric acid.
- After the last wash, resuspend the pellet in an equal volume of water (e.g. add 5 ml ddH_2O to each 5 ml pellet). Aliquot into eppendorf tubes and store at −20°C.

Chaotropic lysis buffer:

- 30% Isopropanol
- 4.5 M guanidine HCl
- 15% Triton X-100
- 15 mM Tris pH 5.0

Wash buffer:

- 25 mM NaCl
- 75% Ethanol

Elution buffer (TE plus RNase):

- 10 mM Tris pH 8.0
- 1 mM EDTA
- 50 – 100 μg ml^{-1} RNase A (Qiagen – cat.# 19101)

Procedure:

It is important to include a positive control (i.e. treat your cells with a known inducer of apoptosis such as etoposide or staurosporine).

- Resuspend the apoptotic cells into a small volume (not to exceed 500 μl) of buffer, such as PBS or HBSS, in a microcentrifuge tube. If the cells are in a culture dish or multi-well plate, remove media and replace with a small volume of buffer to cover the cells. The optimal number of cells will have to be determined by experimentation. Typically, 1 million cells per lane are sufficient to visualize DNA fragmentation if a significant percentage of the cells are undergoing apoptosis. Too many cells can lead to an excessively viscous lysate that is difficult to manipulate at later steps.
- Add 1.5 volumes of chaotropic lysis buffer to the cell suspension. This step both lyses the cells and enables the DNA to bind to the silica resin. If the cells were on a dish or other plastic surface, add the lysis buffer and gently scrape the cells to lyse them thoroughly and transfer to microcentrifuge tubes for further processing. Take care not to be too vigorous as this may lead to unnecessary shearing of chromosomal DNA which will mask the DNA ladder.
- Add silica resin. Typically we add 100 μl per 1 ml total volume of lysed cell suspension. Put the mix on a rotating platform at room temperature for 15 min to allow the DNA to bind to the resin.
- Pellet the DNA-silica complexes in a microcentrifuge for 5 min at 14 000 rpm. Aspirate the supernatant. At this point the resin may appear stringy due to its binding to chromosomal DNA.
- Resuspend the pellet into 1 ml of the cell lysis solution by pipeting (not vortexing, which shears chromosomal DNA). Dissociate the pellet back into a loose slurry. Recentrifuge the sample and remove and discard the supernatant. This step removes residual protein and other cellular contaminants from the DNA-silica.
- Wash the pellet by resuspending into 1 ml wash buffer and pipeting as described in the previous step. Centrifuge the sample and discard the supernatant. Repeat at least once. This step removes both protein as well as chaotropic salts and detergents from the silica resin. After the last wash, remove all traces of wash buffer.
- Allow the wash buffer to evaporate by leaving the tubes open. The resin will typically change from an off-white to a bone-white color when dry. It is important that the wash buffer has completely evaporated before eluting the DNA. Ethanol will interfere with the elution of the DNA from the resin and subsequent loading on the agarose gel.
- Elute the DNA by adding 60 μl of elution buffer to the dried resin. Flick the tube to resuspend the resin and incubate at 50°C for 15 min. Spin out the resin by centrifugation in a microfuge, remove the supernatant which contains the DNA and transfer to a new tube (you will be able to recover only 30 to 40 μl).
- Load samples on a 1% agarose gel and run at 80 V for at least an hour to achieve good separation of the fragments. Stain with ethidium bromide and visualize the DNA under UV illumination.

1984). When resolved electrophoretically on an agarose gel and stained with ethidium bromide, the DNA fragments show a characteristic ladder-like appearance (Figure 24.1). This appearance is indicative of apoptosis, although the inability to detect DNA laddering does not rule out apoptosis since DNA fragmentation is not required (Allen *et al.*, 1997). DNA degradation during necrosis generates much larger DNA fragments and is easily distinguished. Drawbacks in using this method include the fact that a large number of cells ($>10^6$) are required per sample and that only a small number of samples can be run in each assay. Furthermore, the assay is not quantitative.

There are several methods of extracting cellular DNA which can be employed to isolate apoptotic DNA fragments. These include phenol extractions as described in a review by Allen *et al.* (1997), and binding of DNA to glass fibers such as silica resin in the presence of chaotropic salts. The latter is an easier procedure, and is the basis of a kit available from Roche (cat. #1835246). Here we describe a shorter and inexpensive protocol that has been developed in our lab by Dr William Navarre. In short, cells are treated with an apoptotic stimulus, collected, lysed, and after intact nuclei are removed, the lysates are passed over a silica resin which binds the DNA. The DNA is subsequently eluted, resolved on an agarose gel and stained with ethidium bromide.

Detection of mono- and oligonucleosomes by ELISA

As mentioned earlier, DNA between nucleosomes (linker DNA) is cleaved during apoptosis. The mono- and oligonucleosomes that are generated are released into the cytosol prior to disruption of the plasma membrane. Cytosolic nucleosomes can be detected by a sandwich ELISA, using anti-histone and anti-double-strand DNA antibodies.

After incubation of cells with the pro-apoptotic stimulus, cells are pelleted and the supernatants removed. This removes nucleosomes and DNA which have been released from necrotic cells. The pelleted cells are lysed in a low concentration of detergent that leaves the nuclei intact. After pelleting the nuclei, the supernatants containing the cytosolic nucleosomes are collected and added to a 96-well plate coated with anti-histone antibody. Nucleosomes are bound by the antibody. Next, an anti-double strand DNA antibody conjugated to horseradish peroxidase (HRP) is added. Since DNA is still complexed with nucleosomes during apoptosis, only nucleosome/DNA complexes are detected, while free DNA bound to antibody is washed away. A substrate for HRP is added and the plate is read in an ELISA plate reader.

This assay is quantitative and very sensitive, so very few cells are needed. Unfortunately, the commercially available kit from Roche (cat.# 1774425) is quite expensive. It is important to collect samples before secondary necrosis occurs, as their nucleosomes will be washed away, and will result in underestimation of apoptosis. One group suggests pooling supernatants and lysates, to account for secondary necrotic cells (Geng *et al.*, 1996). However, this modification allows the detection of nucleosomes from true necrotic cells, thus overestimating levels of apoptosis.

DNA content

Loss of nuclear DNA can be measured as an assay for apoptosis. Low molecular weight DNA that leaks into the cytosol during apoptosis can be extracted from cells. The high-molecular weight DNA remaining in the nucleus is stained with a fluorescent dye such as PI and analyzed by flow cytometry. Cells with a DNA content lower than that of cells in the G1 phase of the cell cycle, or 'hypodiploid' cells, are considered apoptotic.

This assay is quick and cheap if a flow cytometer is available. It does not, however, discriminate between apoptotic and necrotic cells, unless combined with other techniques (see below). Another drawback of the assay is that cells in G2 which are undergoing apoptosis may appear to have DNA contents similar to G1 cells. This would lead to overestimation of cells in G1 and underestimation of the level of apoptosis. Importantly, if detergent lysis is used, particles containing sub-G1 levels of DNA cannot be counted to determine the number of cells undergoing apoptosis, since one apoptotic cell may break into several sub-G1 particles (apoptotic bodies and cell fragments).

A light fixative, like ethanol, can be used to prevent the separation of nuclear fragments and the assay can be used quantitatively (Darzynkiewicz and Juan, 1999). High-molarity phosphate-citrate buffer is used to extract DNA. Importantly, cells must not be fixed with a strong fixative such as formaldehyde, because this does not allow extraction of DNA. Using the ethanol fixation method, extracted DNA can be run on an agarose gel to assay for DNA laddering, as described above. This allows detection of both DNA fragmentation and loss of DNA content from the same sample. Immunocytochemistry can also be performed on ethanol fixed cells to allow the detection of additional markers.

TUNEL

The TUNEL (Terminal dUTP Nick End Labeling) assay detects cell death by labeling the free 3′-OH ends of degraded DNA. The enzyme terminal deoxynucleotidyl transferase (TdT), which is added exogenously to cells, catalyzes the addition of labeled dUTP to 3′-OH ends of nicked DNA and double strand breaks. In order to prevent loss of DNA during the procedure, cells are fixed with formaldehyde. They are then permeabilized to allow the addition of TdT. Cleavage of DNA and generation of free 3′-OH ends occurs during necrosis as well as apoptosis. However, DNA is cleaved into much smaller and therefore more numerous fragments during apoptosis, generating a more intense staining with TUNEL.

dUTP can be tagged with a variety of labels. Fluorescein allows direct detection of labeled DNA by fluorescence microscopy or flow cytometry and is the most common method used (Figure 24.2). Fluorescein-labeled DNA can also be detected using the light microscope and an HRP-labeled anti-fluorescein antibody. Alternatively, biotin-labeled dUTP can be used

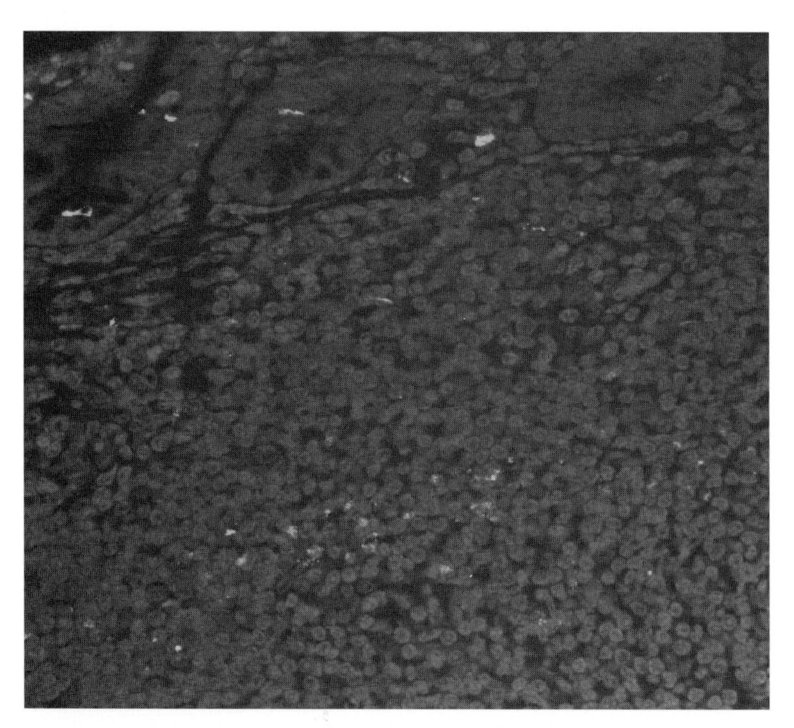

Figure 24.2. Detection of apoptotic cells using TUNEL. Rabbit ileal loops were inoculated with wild-type *Shigella flexneri*. Sections of Peyer's patches were stained with TUNEL (apoptotic cells appear green) and counterstained with propidium iodide. (This figure is also reproduced in colour between pages 276 and 277.)

TUNEL

This protocol uses the Promega Apoptosis Detection System, Fluorescein (cat.# G3250). It is a modified version of the protocol for analysis by flow cytometry provided with the kit from Promega. Protocols for TUNEL staining of cells on slides and tissue sections are provided with the kit.

Preparation of samples for flow cytometry

- For cells in suspension: Collect each sample into an eppendorf tube on ice, centrifuge samples at 1500 rpm for 10 min at 4°C. Aspirate supernatants and resuspend cells in 200 µl pre-chilled PBS. Transfer cells to a *round-bottom* 96-well plate.
- For adherent cells: Collect supernatants into eppendorf tubes on ice. Collect cells using HBSS-/0.05% Trypsin/0.5mM EDTA to make the cells lift off the plate. Pipette up and down to break apart clumps of cells. For each sample, add cells to the eppendorf already containing the corresponding supernatant. Spin at 1500 rpm for 10 min at 4°C. Aspirate the supernatant and resuspend cells in 200 µl pre-chilled PBS. Transfer cells to a *round-bottom* 96-well plate.
- A multi-channel pipetman can be used for all subsequent pipeting steps.
- Centrifuge at 1500 rpm for 5 min at 4°C.

- Wash once with cold PBS.
- Resuspend cells in 200 μl 1% formaldehyde/PBS (cold) and leave on ice for 5 min.
- Centrifuge at 1500 rpm for 5 min at 4°C.
- Wash once with cold PBS.
- Resuspend cells in 0.2% Triton X-100/PBS (cold) and leave on ice for 5 min.
- Centrifuge at 1500 rpm for 5 min at 4°C.
- Wash twice with cold PBS. During this step the TUNEL mix should be made up (see below).
- Resuspend cells in 80 μl of equilibration buffer (which should have been thawed already) from the Promega kit and leave at room temperature for 5 min.
- Centrifuge at 1500 rpm for 5 min at 4°C.
- ***Nucleotide mix and TdT enzyme *MUST be kept in the DARK* and on ice.***

  TUNEL mix (for one sample)
  45 μl equilibration buffer
  5 μl nucleotide mix
  1 μl TdT enzyme
- First make the TUNEL mix without adding the TdT enzyme. Add aliquots to the '−TdT' samples. Then add TdT to the TUNEL mix and add aliquots to the other samples.
- Resuspend the cells in the TUNEL mix. Cover the 96-well plate with a clean plastic cover and use parafilm to seal the cover to the plate (this prevents water from entering the plate). Float the plate in a 37°C water bath. It is very important that the temperature of the water bath is *exactly* 37°C. Keep the water bath covered so the reaction can proceed in the dark. Leave the plate at 37°C for 1 h. Every 15 min, take the plate out and shake it gently for 15–30 s.
- Add 200 μl of 20 mM EDTA to each well to stop the reaction after 1 h.
- Centrifuge at 1500 rpm for 5 min at 4°C (without the cover).
- Wash twice with 200 μl PBS/ 0.1% Triton X-100/ 5 mg ml^{-1} BSA.
- Resuspend cells in 200 μl PBS/ 5 μg ml^{-1} propidium iodide/ 500 μg ml^{-1} RNase A (Qiagen—cat.# 19101) and leave in the dark for 30 min.
- Transfer the samples into FACS tubes (USA Scientific—cat.# 1412–0400). FACS samples, reading red fluorescence (for PI) on the x-axis and green fluorescence (for TUNEL) on the y-axis.
- '−TdT' samples should be run first. Green fluorescence in these samples should be regarded as background. Green fluorescence above this background level in experimental samples should be counted as TUNEL positive cells.

and detected using labeled avidin. One drawback of the indirect detection methods is increased background.

In situ nick translation (ISNT) is another method which can be used to detect nicked DNA. The enzyme DNA polymerase I catalyzes the addition of labeled nucleotides to nicks, but does not label double strand breaks. Therefore, this method is not as sensitive as TUNEL and is not recommended.

Plasma membrane asymmetry

Annexin V is a protein that binds to negatively charged phospholipids like phosphatidlyserine (PS) in a calcium-dependent manner. Anionic phospholipids including PS are normally located on the inner leaflet of the plasma membrane. During apoptosis, PS is 'flipped' to the outer leaflet of the plasma membrane, resulting in membrane asymmetry (van Engeland *et al.*, 1998). Detection of translocated PS by annexin V can be used as a quantitative assay for apoptosis.

Fluorescein-4-isothiocyanate (FITC)-labeled annexin V can be detected by flow cytometry or fluorescence microscopy, while biotin-labeled annexin V can be detected using avidin-HRP and light microscopy (kits available from R&D Systems—cat.# TA4638, TA4619 and Roche—cat.# 1828690). PS on the inner leaflet of the plasma membrane stains with annexin V in necrotic cells which have lysed. Therefore, annexin V staining should be combined with PI staining to allow the differentiation between apoptotic and necrotic cells.

When using adherent cells, the method by which cells are collected can affect PS externalization. It is possible to add annexin V to cells prior to collection to circumvent this problem (van Engeland *et al.*, 1996). An advantage of this method is that since the plasma membrane is intact during annexin V staining, simultaneous detection of other surface markers using antibodies conjugated to different labels is possible. One drawback of this assay is that a large number of cells are required per sample.

Mitochondrial permeability assays

Mitochondria store energy in the form of an electrochemical gradient across the inner mitochondrial membrane, generating a transmembrane potential. During many types of apoptoses, pores open in the mitochondrial membrane, disrupting the gradient, causing a loss of transmembrane potential and allowing increased permeability (Petit *et al.*, 1996). The increase in permeability can be measured by loading cells with membrane-permeable fluorochromes that accumulate in the mitochondrial matrix and are released when the mitochondria become 'leaky'. Importantly, mitochondrial potential is pH-sensitive, so it is essential to keep the pH between 7.0 and 8.0 during any of the following assays.

Two membrane-permeable fluorochromes are $DiOC_6$ and Rhodamine 123, both of which emit green fluorescence. By double labeling cells with Rh123 and PI, apoptotic and necrotic cells can be distinguished. Living cells only stain green because PI is excluded. Early apoptotic cells have diminished green fluorescence due to mitochondrial damage and do not stain with PI because their membranes are intact. Necrotic and secondary necrotic cells do not exhibit green fluorescence but stain with PI.

JC-1 (Molecular Probes cat.# T-3168 or kit from R&D Systems cat.# TA700) is a fluorochrome which is green when monomeric and turns orange when it aggregates in intact mitochondria. Induction of apoptosis in cells pre-loaded with JC-1 can be analyzed by fluorescence microscopy or flow cytometry.

Mitotracker Red dye (Molecular Probes cat.# M-7512,M-7513) is a mitochondrial-selective dye that is retained in the mitochondria after fixation of cells. It is added to cells after cell death has been induced. Living cells stain red while dead cells do not. Once inside the mitochondria, Mitotracker reacts with thiol-groups to form fixation-stable complexes. This is useful because cells can be fixed and then labeled with TUNEL or for immunocytochemistry.

◆◆◆◆◆◆ MOLECULAR MECHANISMS OF APOPTOSIS

This section describes methods to detect activation of proteins often involved in apoptosis.

Caspases

Introduction

Caspases are a family of conserved intracellular cysteine proteases involved in cell death. Synthesized as inactive zymogens, caspases contain an N-terminal prodomain followed by a large and small subunit (Cohen, 1997; Thornberry and Lazebnik, 1998). Cleavage between domains liberates the large and small subunits, both of which are required to form the active enzyme. Cleavage of caspase substrates occurs C-terminal to a required aspartate residue contained within a divergent tetrapeptide motif. Different caspases recognize distinct cleavage sites, accounting for their substrate specificity. Some active caspases can cleave and activate other caspases, initiating a protease cascade.

Caspases have a wide spectrum of cellular substrates ranging from structural proteins to signal transducers (Thornberry and Lazebnik, 1998; Widmann *et al.*, 1998). Cleavage of these substrates is often linked to progression of cell death. An array of tools are available to study caspases.

Many of these tools are modifications of peptides representing substrate cleavage sites. Most caspase inhibitors are substrate peptides modified by the addition of a C-terminal fluoromethyl ketone (fmk) group. The ketone group forms an irreversible complex with and blocks the catalytic activity of caspases that recognize the peptide. Alternatively, the peptides can be linked to a fluorogenic group which is released by caspase cleavage, and can be used to measure caspase activity. In addition, FITC-labeled peptides can be used to localize active caspases.

Caspases have overlapping cleavage site specificity. Therefore, peptides designed to represent the cleavage site recognized by one caspase will often be recognized by other caspases as well, albeit with different affinities. For this reason, despite the many claims of both commercial vendors and several laboratories, none of the peptides are truly 'specific'

(Garcia-Calvo *et al.*, 1998). These limitations must be taken into account when utilizing these peptides to study caspases.

Pharmacological caspase inhibitors

Inhibition of caspase activity has been demonstrated to inhibit cell death in many instances, and is one method by which caspase involvement can be implicated. Briefly, the caspase inhibitor is pre-incubated with cells for about an hour, prior to addition of the death stimulus. zVAD-fmk (z-valine-alanine-asparate-fmk) is a peptide substrate inhibitor that is advertised as 'broad-range', but its affinity for specific caspases can vary up to 1000 fold (Garcia-Calvo *et al.*, 1998). Inhibition of cell death with zVAD-fmk is a good indication that caspases are involved in cell death, however, lack of inhibition does not prove that caspases are not involved. Furthermore, the permeability of caspase inhibitors can vary and should be taken into consideration when interpreting data.

Negative control peptides should be included in all assays. It is not recommended to rely solely on pharmacological inhibition of caspases to demonstrate caspase involvement. Instead, this method should be combined with some of the methods described below to demonstrate caspase activity or detect active caspase fragments.

Caspase activity assays

One way to determine if caspases have been activated is to assay for their proteolytic activity. Peptides representing caspase cleavage sites have been coupled to fluorochromes and are commercially available (see R&D Systems and Bachem catalogs). Active caspases recognizing the peptide release the fluorochrome through a cleavage event. Once released, the fluorochrome emits fluorescence if excited with light. This is detected using a fluorometer. Alternatively, if a fluorometer is not available, a similar assay can be performed using peptides coupled to chromophores. Release of chromophores is measured using a spectrophotometer.

Since the peptides are not completely specific, one method to make the assay more specific is to immunoprecipitate a particular caspase from a lysate, and then test its activity. This is the basis of a caspase-3-specific fluorometric assay from Roche (cat.# 2012952). This principle should be applicable to assays for any caspase, provided a good antibody is available for the immunoprecipitation step.

Detection of active caspases

Direct detection of active caspase fragments is another way to show caspase involvement in cell death. FITC-labeled zVAD-fmk, which is commercially available (Promega—cat.# G7461), irreversibly binds to and inhibits active caspases. Labeled caspases can be detected using fluorescence microscopy or flow cytometry. Alternatively, biotin-labeled

VKD-fmk (R&D Systems—cat.# FMK011) can be used, allowing detection by avidin-coupled labels. Avidin-HRP allows detection of active caspases in cell extracts by Western blotting.

Alternatively, active caspase fragments can be detected by Western blotting of cell lysates with anti-caspase antibodies. Antibodies to almost all known caspases are commercially available (see Pharmingen and R&D Systems catalogs), although their quality can vary. In some instances, it may be very difficult to detect active fragments by Western blotting. This may depend on expression levels of the particular caspase, or may result from the fact that some caspases induce apoptosis when only very small amounts are active.

Detection of caspase substrate cleavage

During apoptosis, caspases cleave a variety of cellular substrates. Provided these cleavage events are specific to apoptosis, detection of any of the cleaved substrates can be used as a measure of apoptosis. Two examples of caspase substrates, detectable using commercially available antibodies, are described below.

Poly (ADP-ribose) polymerase (PARP) is a 113 kD enzyme that binds to DNA strand breaks and is involved in base-excision repair (Trucco *et al.*, 1998). Full-length PARP is cleaved and inactivated by caspase-3 during apoptosis, generating p85 and p26 fragments (Tewari *et al.*, 1995). Anti-PARP antibodies are available commercially from several companies. Promega sells a p85-specific anti-PARP antibody which does not react with full-length PARP (cat.# G7341). This antibody can be used in Western blots or to stain cells in culture as well as tissue sections. The fact that only cells in which PARP has been cleaved react with the antibody makes this a powerful tool to detect apoptosis. In addition, cells can be double-labeled with anti-PARP and TUNEL to provide a more detailed analysis of apoptosis.

Epithelial cells express intermediate filament proteins called cytokeratins (Southgate *et al.*, 1999). Cytokeratin 18 (CK18) is cleaved by caspases during cell death (Caulin *et al.*, 1997). One CK18 antibody, M30, has been shown to detect an epitope in cleaved CK18 which is not detected in full-length CK18 (Leers *et al.*, 1999). M30 is commercially available from Roche as a FITC-labeled antibody for direct detection by fluorescence microscopy or flow cytometry (cat.# 2156857). Alternatively, an unlabeled antibody is also available (cat.# 2140322).

Cytochrome c release

During apoptosis, cytochrome c is released from mitochondria into the cytosol where it binds and activates Apaf-1. Apaf-1 subsequently binds and activates caspase-9, initiating a caspase cascade leading to apoptosis (Green and Reed, 1998). Release of cytochrome c from mitochondria is indicative of apoptosis, but not required, and can be detected by several techniques.

Anti-cytochrome c antibodies can be used to localize cytochrome c within cells. Briefly, cells are fixed and permeabilized before being incubated with an anti-cytochrome c antibody (Bossy-Wetzel and Green, 2000). Depending on the secondary antibody used, cytochrome c localization can be detected by either light microscopy or fluorescence microscopy. Anti-cytochrome c antibodies are available from Pharmingen (cat.# 65971A—immunoprecipitation and immunocytochemistry) and Promega (cat.# G7421—immunocyto- and immunohistochemistry and Western blotting). Since cells are permeabilized, PI or other DNA staining dyes can be used to visualize condensed chromatin and fragmented nuclei, distinguishing between apoptotic and necrotic cells.

Cytochrome c localization can also be detected by separating cells into cytosolic and mitochondrial fractions and Western blotting for cytochrome c. Cells are initially broken in the presence of sucrose which prevents disruption of mitochondria (Bossy-Wetzel and Green, 2000). Cytosolic and mitochondrial lysates are then resolved by SDS-PAGE, followed by Western blotting for cytochrome c. One drawback of this method is that 1×10^7 cells are required.

Endogenous inhibitors of cell death

Multiple proteins involved in inhibition of cell death have been identified. These proteins exert their effects on different cell death pathways. One way to characterize cell death is to determine if expression of these inhibitors blocks cell death.

The Bcl-2 family of proteins are the best known inhibitors of cell death involving mitochondrial pathways. This family of proteins includes both anti- and pro-apoptotic members (Adams and Cory, 1998; Gross *et al.*, 1999). Expression of anti-apoptotic members, such as Bcl-2 and Bcl-XL, prevents apoptosis induced by cytotoxic insults including γ-irradiation, staurosporine, dexamethasone and cytokine removal (Adams and Cory, 1998). It does not protect cells from apoptosis induced by the death receptor Fas.

IAPs (Inhibitors of Apoptosis) have been shown to directly inhibit caspases 3, 7 and 9 and block apoptosis induced by staurosporine, taxol and the death receptors TNF receptor (TNFR) and Fas (Deveraux and Reed, 1999). IAPs can therefore block apoptosis involving mitochondrial and death receptor pathways.

In order to study the effects of these inhibitors on cell death, the inhibitors are first expressed in cells by transient or stable transfection. The death stimulus is then added and cell death is assayed. The same strategy is used for expression of dominant-negative constructs, discussed in the next section. Importantly, overexpression of proteins by this method does not represent physiological levels. Therefore, results implicating the involvement of a particular protein using overexpression systems must be verified by other methods such as using specific pharmacological inhibitors or using cells from knockout mice.

◆◆◆◆◆◆ ADDITIONAL TOOLS FOR STUDYING PROTEINS INVOLVED IN CELL DEATH

Dominant-negatives

Dominant-negative constructs of proteins involved in cell death, such as caspases, can be used to inhibit cell death. Point mutation of the catalytic site cysteine results in a dominant-negative caspase (Cohen, 1997; Friedlander *et al.*, 1997). Dominant-negative caspases can complement results obtained with chemical caspase inhibitors.

The death receptors Fas, TNFR, and TRAIL-R have been shown to signal for apoptosis through the adapter molecule Fas-associated death domain protein (FADD) (Ashkenazi and Dixit, 1998; Bodmer *et al.*, 2000). FADD constructs lacking the N-terminal death effector domain (DED) function as dominant-negatives and inhibit death-receptor induced apoptosis (Chinnaiyan *et al.*, 1996; Schneider *et al.*, 1997). Dominant-negative FADD can be used to investigate death-receptor involvement in cell death induced by bacteria and bacterial components.

Toll-like receptors (TLRs) define a newly discovered family of receptors which signal in response to bacterial components such as lipopolysaccharide and bacterial lipoproteins (BLP) (Bowie and O'Neill, 2000; Kopp and Medzhitov, 1999). TLR2 has been shown to signal for cell death through a pathway involving FADD and caspase-8, similar to the pathway used by death receptors (Aliprantis *et al.*, 2000). Dominant-negative constructs of both FADD and caspase-8 block BLP-induced, TLR2-dependent cell death (Aliprantis *et al.*, 2000). Signaling through TLRs may contribute to bacteria-induced cell death.

Cells from knockout mice

Cells deficient in a specific gene can be used to determine if the gene product is involved in specific processes. Resistance of such cells to bacteria-induced host cell death implicates the missing host protein in the cell death process. Results obtained with cells from knockout mice are often more specific than results obtained with pharmacological or dominant-negative inhibitors. However, in the absence of a specific protein, other proteins may provide compensatory functions not normally observed. Such a mechanism has been shown to occur in cells deficient in specific caspases, and therefore this phenomenon must be taken into account (Zheng *et al.*, 2000).

◆◆◆◆◆◆ *IN VIVO* CELL DEATH

Detection of cell death in *in vivo* infection models is one of the most relevant ways to demonstrate bacteria-induced host cell death. Sections from tissues of infected animals are often used for histology. Observation of changes in histology and cellular morphology is possible, as is staining of sections for specific cell death markers. In addition, some organs (liver and spleen) can be easily broken apart into single cell suspensions and analyzed by many of the *in vitro* techniques described earlier. Blood cells can be analyzed similarly.

Detection of apoptotic cells *in vivo* is difficult since they are rapidly cleared by phagocytic cells. Thus, when measuring apoptosis *in vivo*, it is important to select an assay that measures an event occurring prior to clearance. TUNEL is the most widely used method for detecting apoptosis on tissue sections. When using TUNEL, it is important to counterstain sections with a nuclear-staining dye, allowing visualization of both TUNEL positive and negative nuclei.

Staining with antibodies specific for cleavage products produced during cell death is another method to detect cell death. Antibodies against cleaved PARP and CK18 can be used. CK18 is an epithelial cell specific protein whose cleavage appears to be an early event in cell death, preceding annexin V labeling (Leers *et al.*, 1999). Dying epithelial cells should therefore be detectable before they are cleared.

Finally, some tissues can be homogenized, allowing experiments such as caspase activation assays or Western blots to be performed on tissue lysates. This approach works in tissues in which cell death is widespread and dying cells make up a large proportion of the total cells.

◆◆◆◆◆◆ CONCLUSION

Host cell death during bacterial infection is an important component of pathogenesis. This chapter describes many techniques for the study of bacteria-induced host cell death. Elucidation of the mechanisms by which bacteria induce host cell death may contribute to our knowledge of basic cell and molecular biological processes. In addition, better understanding of this process may lead to the generation of more effective anti-bacterial agents.

Acknowledgments

We would like to thank Yvette Weinrauch, William Navarre and Molly Ingersoll for critical reading of this chapter. AZ is supported by NIH grants AI 42780-01 and AI 37720 and by the Irma T. Hirschl Trust.

References

Adams, J. M. and Cory, S. (1998). The Bcl-2 protein family: arbiters of cell survival. *Science* **281**, 1322–6.

Aliprantis, A. O., Yang, R. B., Weiss, D. S., Godowski, P. and Zychlinsky, A. (2000). The apoptotic signaling pathway activated by Toll-like receptor-2. *Embo J.* **19**, 3325–36.

Allen, R. T., Hunter, W. J., 3rd and Agrawal, D. K. (1997). Morphological and biochemical characterization and analysis of apoptosis. *J. Pharmacol. Toxicol. Methods* **37**, 215–28.

Ashkenazi, A. and Dixit, V. M. (1998). Death receptors: signaling and modulation. *Science* **281**, 1305–8.

Bodmer, J. L., Holler, N., Reynard, S., Vinciguerra, P., Schneider, P., Juo, P., Blenis, J. and Tschopp, J. (2000). TRAIL receptor-2 signals apoptosis through FADD and caspase-8. *Nat. Cell. Biol.* **2**, 241–3.

Bossy-Wetzel, E. and Green, D. R. (2000). Assays for cytochrome c release from mitochondria during apoptosis [In Process Citation]. *Methods Enzymol.* **322**, 235–42.

Bowie, A. and O'Neill, L. A. (2000). The interleukin-1 receptor/toll-like receptor superfamily: signal generators for pro-inflammatory interleukins and microbial products. *J. Leukoc. Biol.* **67**, 508–14.

Caulin, C., Salvesen, G. S. and Oshima, R. G. (1997). Caspase cleavage of keratin 18 and reorganization of intermediate filaments during epithelial cell apoptosis. *J. Cell. Biol.* **138**, 1379–94.

Chinnaiyan, A. M., Tepper, C. G., Seldin, M. F., O'Rourke, K., Kischkel, F. C., Hellbardt, S., Krammer, P. H., Peter, M. E. and Dixit, V. M. (1996). FADD/MORT1 is a common mediator of CD95 (Fas/APO-1) and tumor necrosis factor receptor induced apoptosis. *J. Biol. Chem.* **271**, 4961–5.

Cohen, G. M. (1997). Caspases: the executioners of apoptosis. *Biochem. J.* **326**, 1–16.

Darzynkiewicz, Z. and Juan, G. (1999). Selective extraction of fragmented DNA from apoptotic cells for analysis by gel electrophoresis and identification of apoptotic cells by flow cytometry. *Methods Mol. Biol.* **113**, 599–605.

Deveraux, Q. L. and Reed, J. C. (1999). IAP family proteins–suppressors of apoptosis. *Genes Dev.* **13**, 239–52.

Friedlander, R. M., Gagliardini, V., Hara, H., Fink, K. B., Li, W., MacDonald, G., Fishman, M. C., Greenberg, A. H., Moskowitz, M. A. and Yuan, J. (1997). Expression of a dominant negative mutant of interleukin-1β converting enzyme in transgenic mice prevents neuronal cell death induced by trophic factor withdrawal and ischemic brain injury. *J. Exp. Med.* **185**, 933–940.

Galan, J. E. and Collmer, A. (1999). Type III secretion machines: bacterial devices for protein delivery into host cells. *Science* **284**, 1322–8.

Garcia-Calvo, M., Peterson, E. P., Leiting, B., Ruel, R., Nicholson, D. W. and Thornberry, N. A. (1998). Inhibition of human caspases by peptide-based and macromolecular inhibitors. *J. Biol. Chem.* **273**, 32608–13.

Geng, Y. J., Wu, Q., Muszynski, M., Hansson, G. K. and Libby, P. (1996). Apoptosis of vascular smooth muscle cells induced by *in vitro* stimulation with interferon-gamma, tumor necrosis factor-alpha, and interleukin-1 beta. *Arterioscler. Thromb. Vasc. Biol.* **16**, 19–27.

Green, D. R. and Reed, J. C. (1998). Mitochondria and apoptosis. *Science* **281**, 1309–12.

Gross, A., McDonnell, J. M. and Korsmeyer, S. J. (1999). BCL-2 family members and the mitochondria in apoptosis. *Genes Dev.* **13**, 1899–911.

Hansen, M. B., Nielsen, S. E. and Berg, K. (1989). Re-examination and further development of a precise and rapid dye method for measuring cell growth/cell kill. *J. Immunol. Methods* **119**, 203–10.

Jacobson, M. D., Weil, M. and Raff, M. C. (1997). Programmed cell death in animal development. *Cell* **88**, 347–54.

Kopp, E. B. and Medzhitov, R. (1999). The Toll-receptor familly and control of innate immunity. *Curr. Opinion Immunol.* **11**, 13–18.

Leers, M. P., Kolgen, W., Bjorklund, V., Bergman, T., Tribbick, G., Persson, B., Bjorklund, P., Ramaekers, F. C., Bjorklund, B., Nap, M., Jornvall, H. and Schutte, B. (1999). Immunocytochemical detection and mapping of a cytokeratin 18 neo- epitope exposed during early apoptosis. *J. Pathol.* **187**, 567–72.

Majno, G. and Joris, I. (1995). Apoptosis, oncosis, and necrosis. An overview of cell death [see comments]. *Am. J. Pathol.* **146**, 3–15.

and Mosmann, T. (1983). Rapid colorimetric assay for cellular growth and survival: application to proliferation and cytotoxicity assays. *J. Immunol. Methods.* **65**, 55–63.

and Nachlas, M. M. (1960). The determination of lactic dehydrogenase with a tetrazolium salt. *Anal. Biochem.* **1**, 317.

Petit, P. X., Susin, S. A., Zamzami, N., Mignotte, B. and Kroemer, G. (1996). Mitochondria and programmed cell death: back to the future. *FEBS Lett.* **396**, 7–13.

Roulston, A., Marcellus, R. C. and Branton, P. E. (1999). Viruses and apoptosis. *Annu. Rev. Microbiol.* **53**, 577–628.

Savill, J., Fadok, V., Henson, P. and Haslett, C. (1993). Phagocyte recognition of cells undergoing apoptosis. *Immunol. Today* **14**, 131–136.

Schneider, P., Thome, M., Burns, K., Bodmer, J. L., Hofmann, K., Kataoka, T., Holler, N. and Tschopp, J. (1997). TRAIL receptors 1 (DR4) and 2 (DR5) signal FADD-dependent apoptosis and activate NF-kappaB. *Immunity* **7**, 831–6.

Southgate, J., Harnden, P. and Trejdosiewicz, L. K. (1999). Cytokeratin expression patterns in normal and malignant urothelium: a review of the biological and diagnostic implications. *Histol. Histopathol.* **14**, 657–64.

Tewari, M., Quan, L. T., O'Rourke, K., Desnoyers, S., Zeng, Z., Beidler, D. R., Poirier, G. G., Salvesen, G. S. and Dixit, V. M. (1995). Yama/CPP32β, a mammalian homolog of CED-3, is a CrmA-inhibitable protease that cleaves the death substrate poly(ADP-ribose) polumerase. *Cell* **81**, 801–9.

Thornberry, N. A. and Lazebnik, Y. (1998). Caspases: enemies within. *Science* **281**, 1312–16.

Trucco, C., Oliver, F. J., de Murcia, G. and Menissier-de Murcia, J. (1998). DNA repair defect in poly(ADP-ribose) polymerase-deficient cell lines. *Nucleic Acids Res.* **26**, 2644–9.

van Engeland, M., Nieland, L. J., Ramaekers, F. C., Schutte, B. and Reutelingsperger, C. P. (1998). Annexin V-affinity assay: a review on an apoptosis detection system based on phosphatidylserine exposure. *Cytometry* **31**, 1–9.

van Engeland, M., Ramaekers, F. C., Schutte, B. and Reutelingsperger, C. P. (1996). A novel assay to measure loss of plasma membrane asymmetry during apoptosis of adherent cells in culture. *Cytometry* **24**, 131–9.

Weinrauch, Y. and Zychlinsky, A. (1999). The induction of apoptosis by bacterial pathogens. *Annu. Rev. Microbiol.* **53**, 155–187.

Widmann, C., Gibson, S. and Johnson, G. L. (1998). Caspase-dependent cleavage of signaling proteins during apoptosis. A turn-off mechanism for anti-apoptotic signals. *J. Biol. Chem.* **273**, 7141–7.

Willingham, M. C. (1999). Cytochemical methods for the detection of apoptosis. *J. Histochem. Cytochem.* **47**, 1101–10.

Wyllie, A. H., Morris, R. G., Smith, A. L. and Dunlop, D. (1984). Chromatin cleavage in apoptosis: association with condensed chromatin morphology and dependence on macromolecular synthesis. *J. Pathol.* **142**, 67–77.

Zheng, T. S., Hunot, S., Kuida, K., Momoi, T., Srinivasan, A., Nicholson, D. W., Lazebnik, Y. and Flavell, R. A. (2000). Deficiency in caspase-9 or caspase-3 induces compensatory caspase activation [In Process Citation]. *Nat. Med.* **6**, 1241–7.

25 Alternative Models in Microbial Pathogens

Man-Wah Tan[1] **and Frederick M Ausubel**[2]

[1]*Department of Genetics, and Department of Microbiology and Immunology, Stanford University School of Medicine, Stanford, CA 94305, USA;* [2]*Department of Genetics, Harvard Medical School and Department of Molecular Biology, Massachusetts General Hospital, Boston, MA 02114, USA*

◆◆

CONTENTS

◆◆◆◆◆◆ INTRODUCTION

The struggle for survival is one of the most potent forces that drives evolution. To survive, an organism needs to overcome a variety of biotic and abiotic insults. The most common biotic insult comes in the form of antagonistic interactions between organisms, such as interactions between bacterial pathogens and their hosts. Bacterial pathogens have developed a variety of offensive and defensive weapons to defeat their hosts. To defend themselves against pathogens, plant and animal hosts have evolved the ability to recognize the aggressors and have developed a relay system to communicate this information within the cell so that appropriate defenses can be mounted. Ideally, to make full use of powerful genomic methods to uncover the pathogen-derived virulence determinants and the host-derived defense factors, both the pathogen and its host should be amenable to genetic analysis and their genomes completely sequenced. This would permit both the host and pathogen to be genetically altered and the effects of these alterations on pathogenesis to be readily tested. The availability of complete genome sequences of both the host and pathogen also provides a unique opportunity to utilize DNA chip technology to perform coordinate genome-wide gene expression analyses on the host and pathogen as they interact.

METHODS IN MICROBIOLOGY, VOLUME 31
ISBN 0-12-521531-2

Recent work has shown that there are common mechanisms underlying host–pathogen interactions: pathogens use similar virulence factors to infect divergent hosts (reviewed in Finlay, 1999; Finlay and Falkow, 1997). Similarly, the innate immune response to pathogen appears to be ancient and conserved in plants, vertebrates, and invertebrates (reviewed in Tan and Ausubel, 2000; Wilson *et al.*, 1997). Our laboratories took advantage of the conservation of these mechanisms in host–pathogen interactions by developing a non-vertebrate multi-host pathogenesis system. This has enabled us to use genome-wide strategies to dissect the molecular basis of host–pathogen interactions. The first multi-host pathogenesis system developed was based on the ability of *Pseudomonas aeruginosa* to infect and/or kill divergent hosts. *P. aeruginosa* is a ubiquitous Gram-negative bacterium and an important pathogen in cystic fibrosis patients, or in patients whose immune system is compromised by medical intervention, infection or burn. The strain used in our studies is PA14, which was originally isolated from the blood of a burn wound patient. PA14 was subsequently shown to infect a variety of plant species, including *Arabidopsis thaliana* (Rahme *et al.*, 1995, 1997, 2000), and to kill invertebrate hosts such as the soil nematode *Caenorhabditis elegans* (Mahajan-Miklos *et al.*, 1999; Tan *et al.*, 1999a,b; Tan and Ausubel, 2000) and insects, such as *Galleria mellonella* (Jander *et al.*, 2000) and *Drosophila melanogaster* (G. Lau, S. Mahajan-Miklos, E. Perkins and L. Rahme, personal communication). Because pathogenesis in these hosts involves a shared set of *P. aeruginosa* virulence determinants (see below and reviewed in Finlay, 1999; Mahajan-Miklos *et al.*, 2000; Rahme *et al.*, 2000; Tan and Ausubel, 2000), this multi-host system allows the entire *P. aeruginosa* genome to be scanned systematically, efficiently and economically for any gene that affects pathogenesis *in vivo*. This is accomplished by screening a mutagenized library of *P. aeruginosa* for less pathogenic mutants following infection of individual mutant clones on individual plants or insects or feeding mutant clones of bacteria growing on separate petri plates to *C. elegans*. In addition, in the case of *A. thaliana, C. elegans* and *D. melanogaster*, host immunity mutants can be isolated by screening for host mutants that are either more resistant or more susceptible to pathogen attack.

The concept of using non-vertebrate hosts to dissect pathogen- and host-derived factors has recently been extended to include other pathogens and hosts. For example, the gram-negative pathogens *Salmonella enterica, Serratia marscesens, Burkholderia cepacia* and *B. pseudomallei* and the gram-positive pathogens *Enterococcus faecalis* and *E. faecium* have also been shown to kill *C. elegans* (Aballay *et al.*, 2000; Garsin *et al.*, 2000; Tan and Ausubel, 2000; J. Jedelloh, personal communication). The cellular slime mold *Dictyostelium discoideum* has also been shown to be an effective host for *P. aeruginosa* (S. Putatzki, H. Ennis and R. Kessin, personal communication) and *Legionella pneumophila* (Solomon *et al.*, 2000) and certain strains of the plant pathogens *Erwinia caratovora* and *E. chrysanthemi* have been shown to infect *D. melanogaster* (Basset *et al.*, 2000). In this chapter, we describe protocols for assessing the virulence of various pathogens using simple non-vertebrate hosts.

◆◆◆◆◆◆ *P. AERUGINOSA*—MULTI-HOST SYSTEM

A. thaliana

There are many species of plants that are susceptible to *P. aeruginosa* (Table 25.1, Cho *et al.*, 1975; Schroth *et al.*, 1977). In *A. thaliana*, *P. aeruginosa* PA14 causes severe soft-rot symptoms that correspond to bacterial proliferation in the leaves (Plotkinova *et al.*, 2000; Rahme *et al.*, 2000). *P. aeruginosa* PA14 can invade *A. thaliana* leaves directly through the stomata without the requirement for wounding or mechanical infiltration. PA14 primarily colonizes the intercellular spaces, causing disruption of plant cell walls and membrane structures. Ultimately, PA14 causes a systemic infection that is characterized by basipetal movement along the vascular parenchyma of the leaf, resulting in rotting of the petiole and central bud and death of the plant. Distinctive features of *P. aeruginosa* pathogenesis are that the surface of mesophyll cell walls adopt an unusual convoluted or undulated appearance, that PA14 cells orient themselves perpendicularly to the outer surface of mesophyll cell walls, and that PA14 cells make circular perforations, approximately equal to the diameter of *P. aeruginosa*, in mesophyll cell walls. Like other plant pathogens, the ability of *P. aeruginosa* PA14 to cause disease symptoms and proliferate in *A. thaliana* leaves is ecotype (wild-type variety) specific. For example, 5 days after inoculation into ecotype Llagostera (Ll) with 10^3 cfu cm^{-2} leaf surface area, 3×10^7 cfu cm^{-2} leaf surface area could be recovered from these leaves. However, similar inoculation in ecotype Argentat (Ag) leaves did not produce any disease symptom, and the bacteria were only able to grow to a density of 3×10^5 cfu cm^{-2} leaf surface area, 5 days post-inoculation (Rahme *et al.*, 1995). Moreover, virulence is also determined by the genotype of the pathogen. Proliferation in *Arabidopsis* Ll leaves of isogenic strains of PA14 that have lesions in the *gacA*, *plcS*, or *toxA* genes was significantly less than their parental wild-type, attaining a density of 6×10^3, 1×10^5 and 2×10^6 cfu cm^{-2} leaf surface area, respectively (Rahme *et al.*, 1995). Similarly, *P. aeruginosa* strains PAK and PA01 were also less virulent compared to PA14, with the maximal levels of growth reached in Ll *Arabidopsis* at 6×10^4 and 8×10^5, respectively (Rahme *et al.*, 2000).

P. aeruginosa PA14 also causes soft-rot symptoms to develop when inoculated into the midrib of lettuce leaf stems, and unlike the infection of *A. thaliana*, both PAK and PA01 strains were also infectious in lettuce (Rahme *et al.*, 1997). The severity of symptom development in lettuce directly correlated with the extent of growth in *A. thaliana*. Because of the ease of testing several strains on a single lettuce leaf stem, a lettuce screen (described in Rahme *et al.*, 1997) was used to identify *P. aeruginosa* PA14 transposon insertion mutants that failed to elicit disease symptoms. In principle, one can use any of the plant species listed in Table 25.1 for the above screen. The lettuce screen led to the identification of nine mutants out of 2500 prototrophic mutants tested. Importantly, all nine mutants identified from the plant screen also exhibited reduced pathogenicity when tested in a burned mouse pathogenicity model at a dose of 5×10^5 cfu (Rahme *et al.*, 1997).

Table 25.1 Plant species that can be infected by *P. aeruginosa*

Cruciferae
Arabidopsis thaliana
Lettuce (*Lactuca sativa 'Great Lakes' and 'Romaine'*)
Rutabaga (*Brassica campestris*)
Brussels sprouts (*Brassica oleracea gemmifera*)
Umbelliferae
Celery (*Apium graveolens var. Dulce*)
Carrot (*Daucus carrota* var. *sativa*)
Solanaceae
Tomato (*Lycopersicon eculentum*)
Potato tuber (*Solanum tuberosum* 'Whiterose')
Cucurbitaceae
Cucumber (*Cucumis sativus*)

The protocol described below (Protocol 1) to quantify *P. aeruginosa* PA14 pathogenicity in *A. thaliana* is based on the method described by Rahme *et al.* (1997). It is a variant of a leaf-infiltration assay developed for *P. syringae* (Davis *et al.*, 1991; Dong *et al.*, 1991). Conditions used for growing *A. thaliana* are as follows: Germinate *A. thaliana* ecotype Llagostera (Ll) seeds (available from the *Arabidopsis* Biological Resource Center, Columbus, OH) and grow in Metromix 200 (W.R. Grace, Inc.) in a climate-controlled greenhouse (20 ± 2°C, relative humidity 70 ± 5%) with supplemental fluorescent lighting (16 h photo period, 150–300 $\mu Em^{-2} s^{-1}$). After 10 days to 2 weeks, transplant seedlings into fresh soil and incubate in a growth chamber at 22°C with a 12 h photo period and a light intensity of 100–150 $\mu Em^{-2} s^{-1}$, but at the same temperature and relative humidity as the greenhouse.

Protocol 1

1. Grow individual bacterial strains or mutants aerobically overnight at 37°C to saturation in 5 ml King's B (Appendix 1, King *et al.*, 1954) or Luria broth (LB) (Miller, 1972).
2. Pellet 1 ml of each overnight culture in a microcentrifuge for 1 minute. Remove the supernatant and resuspend the cells in 1 ml of 10 mM $MgSO_4$. Pellet again and resuspend cells in 1 ml of 10 mM $MgSO_4$. Dilute the cells 1 : 10 into 10 mM $MgSO_4$ and measure absorbance at 600 nm (OD_{600}). Dilute this suspension to $OD_{600} = 0.002$ with 10 mM $MgSO_4$ (corresponding to a bacterial density of approximately $4–5 \times 10^3$ per cm^2 of leaf area).
3. To inoculated leaves with the bacterial suspension, force the bacterial suspension through the stomatal openings on the abaxial surface

(underside) of mature leaves of 6-week-old *A. thaliana* ecotype Llagostera (Ll) using a 1 ml syringe without a needle. Infiltrate control plants with sterile 10 mM $MgSO_4$.
4. Label the inoculated leaves by attaching tape to toothpicks, writing the inoculum used on the tape, and inserting the toothpicks into the soil on the clockwise side of the inoculated leaves.
5. Incubate inoculated plants in a growth chamber at 28–30°C and 90 to 100% relative humidity.
6. Bacterial growth at each time point is measured by determining the average of the logarithm of the number of colony forming units in five leaf discs. Each leaf disc is obtained by punching with a 0.28 cm^2 cork borer outside the initial infiltration site. Grind the leaf discs in 10 mM $MgSO_4$ in Eppendorf tubes using a plastic pestle. The number of viable bacteria is determined by plating appropriate dilutions on King's B medium supplemented with appropriate antibiotics.

Insect

P. aeruginosa has also been reported to be a pathogen of insects (Bulla *et al.*, 1975). For example, the 50% lethal dose (LD_{50}) of *P. aeruginosa* when injected into the hemolymph of the greater wax moth, *Galleria mellonella* larvae is fewer than 10 bacteria (Jander *et al.*, 2000; Jarosz, 1995; Lysenko, 1963). *P. aeruginosa* PA14 is also highly virulent when fed to the diamondback moth, *Plutella xylostella* (Jander *et al.*, 2000). Additionally, *P. aeruginosa* strains PA14 and PA01 kill adult *D. melanogaster* following an abdominal prick using a pin dipped in a suspension of *P. aeruginosa* (G. Lau, S. Mahajan-Miklos, E. Perkins and L. Rahme, personal communication).

Recently, our laboratory showed that there is a positive correlation between virulence of *P. aeruginosa* PA14 mutants in the wax moth caterpillars and mice, suggesting that these insects would serve as a good model system to identify and characterize bacterial genes involved in mammalian pathogenesis (Jander *et al.*, 2000). This result further suggests that the use of insects as hosts can be extended to identify virulence determinants of other human pathogens that infect insects, such as the bacterial pathogens *Proteus vulgaris, P. mirabilis, Serratia marcescens*, and the fungal pathogens, *Fusarium oxysporum* and *Aspergillus fumigatus* (Chadwick *et al.*, 1990; Jander *et al.*, 2000; Tanada and Kaya, 1993).

The following is a protocol (Protocol 2) for infecting *G. mellonella* larvae with *P. aeruginosa* and for determining the 50% lethal dose (LD_{50}). Protocols for infecting *Drosophila* are given later.

Protocol 2

1. Grow from a single colony *P. aeruginosa* PA14 wild-type or mutant in LB containing 100 $\mu g\, ml^{-1}$ rifampicin at 37°C overnight. Dilute this overnight culture 1 : 100 in LB and grow to an optical density at 600 nm (OD_{600}) of 0.3 to 0.4.

2. Pellet the culture and resuspend in 10 mM $MgSO_4$. After dilution to an OD_{600} of 0.1 in 10 mM $MgSO_4$, make 10-fold serial dilutions of the bacterial suspension in 10 mM $MgSO_4$, 2 mg ml^{-1} rifampicin.
3. Use a 10 µl Hamilton syringe to inject 5 µl aliquots of the serial dilutions (containing from 10^6 to 0 bacteria) into separate 5th *instar G. mellonella* larvae via the hindmost left proleg (Figure 25.1). Inject 10 larvae at each dilution, three replicates per dilution. As a negative control, inject in triplicate, 10 larvae per replicate, 5 µl each of 10 mM $MgSO_4$, 2 mg ml^{-1} rifampicin.

   G. mellonella larvae can be obtained from Van der Horst Wholesale, St. Marys, OH. The addition of rifampicin prevents infection by bacteria naturally present on the surface of the larvae.
4. Score the larvae as live or dead after 60 h at 25°C. The normally white larvae became mellanotic (black) about 5 h before they die and are readily distinguished from live ones.
5. To determine the LD_{50}, use a computer program such as Systat to fit a curve to the infection data of the following form:

$$Y = A + (1 - A)/(1 + \exp[B - G \cdot \ln(X)]),$$

   where X is the number of bacteria injected, Y is the fraction of larvae killed by infections, and B and G are parameters which are varied for optimal fit of the curve to the data points.

C. elegans

P. aeruginosa kills *C. elegans* by at least three largely distinct mechanisms, which are dependent on growth conditions and the genotype of the bacteria. Strain PA14, when grown in low salt medium (NG), kills worms over a period of 2–3 days ('slow killing') by an infection-like process that

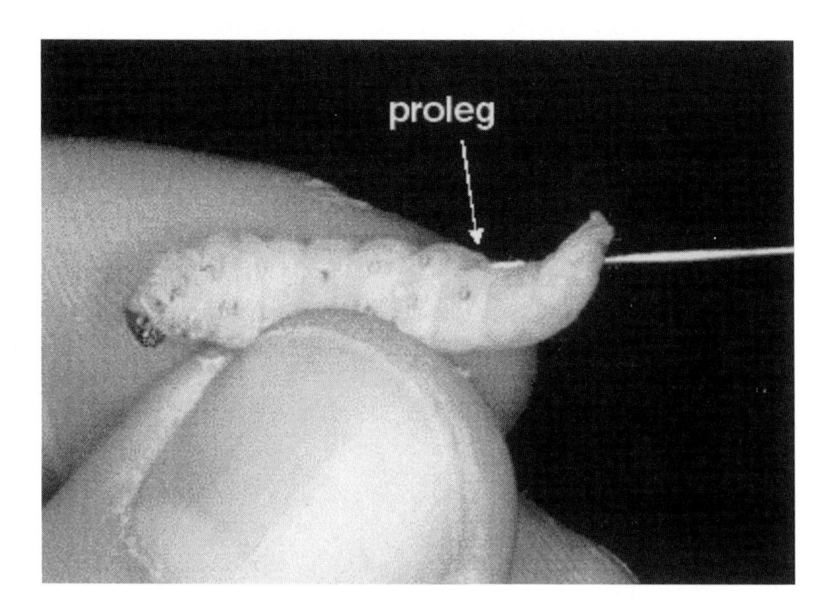

Figure 25.1. Injection of a *C. mellonella* 5th instart larva with a Hamilton syringe via the hindmost left proleg (arrow).

correlates with the accumulation of bacteria in the worm gut (Tan *et al.*, 1999a). When PA14 is grown in a high salt medium (PGS), it kills worms within 4–24 h ('fast-killing') by the production of low molecular weight toxin(s) (Mahajan-Miklos *et al.*, 1999). Another strain of *P. aeruginosa*, PA01, kills rapidly by yet another mechanism. When PA01 is grown on brain-heart infusion agar, worms become paralyzed within 4 h upon contact with the bacterial lawn (Darby *et al.*, 1999).

There are several host-associated factors that affect the susceptibility of nematodes to bacteria-mediated killing. Gravid adult hermaphrodites are more susceptible than adult males due to embryos hatching from within the gravid adults (Tan *et al.*, 1999a). This non-specific effect can be eliminated by using either *C. elegans* males or temperature sensitive sterile hermaphrodite mutants, such as *fer-1* or *glp-4* (available from *Caenorhabditis* Genetics Center at the University of Minnesota). *C. elegans* mutants with defects in feeding and defecation are also more susceptible to *P. aeruginosa* slow killing. *C. elegans* is a filter-feeder; taking in liquid with bacteria and then spitting out the liquid while retaining and grinding up the bacteria when they reach the terminal bulb (Avery, 1993). *C. elegans* mutants defective in grinding, such as *eat-13* and *phm-2*, receive a higher bacteria inoculum than wild-type because they allow the entry of more intact pathogens into the intestines. Consequently, these mutants are more sensitive to pathogen-mediated killing (M.-W. Tan, G. Alloing and F.M. Ausubel, unpublished data). *C. elegans* defecates at regular intervals by a series of sequential muscle contractions. A defect in any of these steps leads to failure to remove intestinal contents at regular intervals and causes a 'constipated' phenotype. Constipated mutants, such as *aex-2* and *unc-25*, which retain pathogens in the intestines longer are also more sensitive than wild-type worms to pathogen-mediated killing (M.-W. Tan and F.M. Ausubel, unpublished data).

In addition to *P. aeruginosa*, several other bacterial species have been reported to be pathogens of *C. elegans* (Aballay *et al.*, 2000; Garsin *et al.*, 2000; Tan and Ausubel, 2000). These include several human pathogens such as *B. cepacia*, *S. marcescens*, *S. typhimurium*, *E. faecalis* and *E. faecium*. Protocol 3 describes conditions used to assay *C. elegans* slow killing by *P. aeruginosa* strain PA14 (Tan *et al.*, 1999a). This protocol is also suitable for testing *B. cepacia* and *S. marcescens*. Other variations or pathogen-specific conditions are summarized in Table 25.2. Protocols for general growth and maintenance of *C. elegans* are described in Sulston and Hodgkin (1987).

Protocol 3

1. Seed 3.5 mm diameter NG agar (a modification of the NGM agar described in Sulston and Hodgkin, 1987, Appendix 1) plates with 10 µl of an overnight bacterial culture.
2. Incubate the plates at 37°C for 24 h. Equilibrate plates to ambient temperature (23–25°C) for several hours prior to adding 30–40 worms at a specific developmental stage. Usually 1-day old hermaphrodites or the final stage larvae (L4) are used. For statistical analysis, have

three-four replicates per trial. Use *E. coli* OP50 as a negative control.

3. Incubate plates at 25°C and score for the number of dead worms every 4–6 h (after the initial 24 h) until all the worms exposed to the wild-type pathogen are dead. A worm is considered dead if it does not respond to touch. Exclude from analysis worms that die as a result of getting stuck to the walls of petri plates. This can be minimized by ensuring that there is no condensate on the walls prior to the addition of worms.
4. The LT_{50} (time required to kill 50% of the nematodes) can be determined as follows: For each replicate, a curve can be fitted to the data with the aid of a computer program such as SYSTAT, using the equation: proportion of worms killed at time $Ti = A + (1 - A)/\{1 + \exp[B - G \times \ln(\text{hours after exposure}, Ti)]\}$, where A is the fraction of worms that died in an OP50 control experiment, and B and G are parameters which are varied for optimal fit of the curve to the data points. Once B and G have been determined, LT_{50} can be calculated by the formula,

$$LT_{50} = \exp(B/G) \times (1 - 2 \times A)^{(1/G)}$$

Table 25.2 *C. elegans* killing assays for various pathogens

P. aeruginosa PA14 fast killing assay (Mahajan-Miklos *et al.*, 1999)

1. Inoculate 5 ml of LB with single colony of *P. aeruginosa* PA14 and grow at 37°C overnight.
2. Spread the center of a 3.5 cm diameter plate containing 5 ml PGS agar (see Appendix 1) with 5 μl of the overnight culture. Incubate plates for 24 h at 37°C.
3. After equilibrating the plates to room temperature, add 30–40 L4-stage hermaphrodite *C. elegans* to each plate and incubate at 25°C.
4. Assay for worm mortality at 4-h intervals.

Note: For this assay, it is important to use fresh PGS plates (1–7 days after pouring). Worms used for this assay must be cultivated on OP50 that has not turned 'slimy' or mucoidy. Slimy plates give inconsistent results.

P. aeruginosa PA01 paralysis assay (Darby *et al.*, 1999)

1. Grow *P. aeruginosa* PA01 overnight at 37 °C in Brain Heart Infusion (BHI) broth.
2. Dilute the overnight culture 100-fold in fresh BHI broth, then spread 400 μl of the dilution over the entire surface of BHI agar plates (10 cm diameter). Incubate the plates for 24 h at 37°C.
3. Wash nematodes off culture plates with M9 buffer (pH 6.5). Spin down the nematodes and resuspend in a minimal volume of M9 buffer. Seed the *P. aeruginos* a lawn with nematodes by placing droplets of the suspension onto the bacterial lawn and incubate at 21–23°C. The droplets will dry within 30 minutes.

4. Examine the worms after 4 h; worms that do not move spontaneously or do not respond to touch are scored as paralyzed.

E. faecalis OG1RF and V583 and *E. faecium* E007 (Garsin *et al.*, 2000)

1. Inoculate 2 ml of BHI medium with a single colony of the test strain and grow at 37 °C for 4 h or overnight.
2. Spread 10 μl of this culture on BHI agar (with appropriate antibiotics that permit the growth of test *Enterococcus* strains but prevent the growth of *E. coli* OP50) in 3.5 cm tissue culture plate.
3. After incubating the plates overnight at 37°C, bring to room temperature for 2 to 5 h before the addition of *C. elegans*. Pick 20–30 *C. elegans* L4 hermaphrodites from *E. coli* OP50 lawns to the *Enterococcus* lawn and incubate at 25°C.
4. Score worm mortality over time by examining the plates at approximately 24 h intervals for 150–200 h. Both *E. faecalis* and *E. faecium* either prevent the hatching of eggs or kill the hatchlings. Therefore, the production of progeny does not obscure the killing assay.

S. typhimurium (Aballay *et al.*, 2000)

1. Inoculate 5 ml of LB with a single colony of test strain and grow at 37°C overnight.
2. Spread 10 μl of the overnight culture on NG agar media (Appendix 1) in 3.5 cm diameter plates. Incubate the seeded plates at 37°C for 2–12 h. Longer incubation times can negatively affect the killing assay.
3. After equilibrating the plates to room temperature, add 10–20 L4-stage hermaphrodite *C. elegans* to each plate and incubate at 20–25°C.
4. Transfer the originally seeded worms daily to fresh plates, which are prepared in the same manner as described above, until the end of their reproductive stage. (This step is necessary because the killing is relatively slow, thus permitting the seeded worms (Po) to produce F1 generation that would develop into adults making it impossible to distinguish them from the Po parents. The transferring step can be eliminated when *C. elegans* males or temperature sensitive sterile mutants, such as *fer-1* or *glp-4*, are used.)
5. Examine the plates daily for 8–10 days for dead worms.

◆◆◆◆◆◆ OTHER MODELS OF HOST–PATHOGEN INTERACTIONS

Recently, the concept of using genetically amenable hosts to study human pathogen has been extended to a system that uses the free-living unicellular organism, *Dictyostelium discoideum*, and its interaction with an intracellular pathogen, *Legionella pneumophila* (Solomon and Isberg,

2000; Solomon *et al.*, 2000). Similarly, a system to study the interactions between plant pathogens from the genus *Erwinia* and their host/vector, *D. melanogaster* have been developed (Basset *et al.*, 2000).

Legionella pneumophila and *Dictyostelium discoideum*

L. pneumophila, the causative agent of Legionaires disease, is found naturally as parasites of freshwater amoebae of the genera *Acanthamoeba* and *Naegleria* (Rowbotham, 1980). *L. pneumophila* enjoys a wide host range: more than 13 species of amoebae and two species of ciliated protozoa can support its growth (reviewed in Fields, 1996). In humans, it replicates in macrophages. *L. pneumophila* is able to grow intracellularly in *D. discoideum*, requiring the same genes that it requires for growth in macrophages and amoeba, such as the *dot* genes (Gao *et al.*, 1997; Segal and Shuman, 1999; Solomon *et al.*, 2000).

The following protocol (Protocol 4) for quantifying growth of *L. pneumophila* in *D. discoideum* in liquid culture is based on Solomon *et al.* (2000). Techniques for maintaining *D. discoideum* axenically in liquid medium or as plaques on a lawn of *Klebsiella aerogenes* are detailed by Sussman (1987).

Protocol 4

1. Culture *D. discoideum* axenically in a shaken flask in HL-5 liquid medium (Appendix 1). When cells are at the exponential growth phase, pellet cells by spinning at 600*g* for 5 minutes. Resuspend in an equal volume of PBS buffer. Pellet cells again by a 5 min spin at 600*g*, then remove PBS by aspiration. Resuspend cells in MB medium (Appendix 1) and plate *D. discoideum* as an adherent monolayer in 24-well tissue culture dishes at a density of 10^6 cells ml^{-1}. Incubate cells at 25.5°C. Allow plates to equilibrate for 4 h before the addition of bacteria.
2. Grow *L. pneumophila* at 37°C for 48 h on CYE plates (Appendix 1). Alternatively, *L. pneumophila* can also be grown at 37°C in AYE liquid medium (same as CYE without agar and charcoal) to an OD_{600} of 3 to 3.5, to allow maximal infectivity. Pellet bacterial cells by spinning for 5 min at 16 000*g*. Resuspend bacterial cells in MB medium; the concentration of bacteria is determined by assuming that $OD_{600} = 1.0$ is equivalent to 10^9 bacteria ml^{-1}.
3. Infect *D. discodeum* at a multiplicity of infection (MOI) of approximately 1 : 1.
4. Quantify the number of viable *L. pneumophila* and *D. discoideum* each day by measuring colony-forming units (CFU) or plaque-forming units (PFU), respectively.
   a. *Quantitation on L. pneumophila.* Add 0.02% saponin (Sigma S-4521) to infected wells to lyse *D. discoideum* and to release the intracellular bacteria. Prepare a dilution series of the lysed *D. discoideum* and plate on CYE plates. Incubate CYE plates at 37°C for 3 to 4 days before counting the colonies (CFU).
   b. *Quantitation of D. discoideum.* Make a dilution series of harvested *D. discoideum* in PBS and plate on lawns of *K. aerogenes*. *K. aerogenes* can

be cultured in LB and plated on SM/5 agar medium (Appendix 1). Incubate plates at 21°C and count the number of plaques (PFU) 3 to 4 days after plating.

Drosophila melanogaster and *Erwinia*

Studies on the *Drosophila* immune responses have yielded great insights into how a host responds to microbial infections. Recently, it was shown that the signaling pathways in *Drosophila* defense response are strikingly similar to innate immune responses in mammals (reviewed in Hoffmann *et al.*, 1999). Studies on bacterial infection of *Drosophila* have relied on direct introduction of bacteria into the body cavity by injection or pricking, or by ingestion of bacteria by the fly larvae. Interestingly, it appears that different pathways may be involved in activating antibacterial responses, depending on the mode of infection. For example, in the absence of physical injury, hemocytes play a significant role in activating a systemic antibacterial response, whereas introduction of bacteria by pricking triggers other pathways that bypass the requirement of hemocytes (Basset *et al.*, 2000). Here we describe both the pricking and feeding protocols for infecting *Drosophila*.

Grow from a single colony the bacterial strains in LB overnight at 37°C. Pellet the overnight bacterial culture. Anesthetize 2–4-day-old adult flies with CO_2. Prick the thorax of the flies with a thin needle that has been dipped into a concentrated bacterial pellet (approximately 4×10^{11} cells ml^{-1}) from an overnight culture in LB (Basset *et al.*, 2000). Alternatively, inject 0.15 µl stationary phase culture of bacteria, diluted 1:10 in *Drosophila* Ringer's solution, into the thorax using a thin pulled glass pipette (Hedengren *et al.*, 1999). Transfer injected fly to fresh vials and assess for survivals at regular intervals. Transfer surviving flies to fresh vials every third day. Do not include flies that die within 3 h of injection in the analysis (Lemaitre *et al.*, 1996). A similar procedure can be used for third instar larvae as well.

The following feeding protocol was first used to demonstrate the infection of *Drosophila* larvae by two phytopathogens, *Erwinia carotovora carotovora* and *E. chrysanthemi paradisiaca* (Basset *et al.*, 2000). Mix the following in a 2 ml microfuge tube: (1) 200 µl of a concentrated bacterial pellet from overnight culture in LB; (2) 400 µl of crushed banana; and (3) approximately 200 third instar larvae. Cover the tube with a foam plug and incubate at room temperature. After 30 min, transfer the mixture to a standard corn-meal fly medium and incubate at 29°C. Larvae infected by the two species of *Erwinia* do not die, however, 90% of the treated larvae show induction of genes encoding antimicrobial peptides (Basset *et al.*, 2000).

◆◆◆◆◆◆ CONCLUSION

The underlying mechanisms of interactions between bacterial pathogens and their animal and plant hosts are complex. The recent development of alternate non-vertebrate hosts that are amenable to genetic analysis are providing important tools to dissect complex host–pathogen interactions. An important consideration when interpreting results obtained from all the systems described in this chapter is that the temperature used to assess pathogenesis in all the non-vertebrate hosts ranges from 25°C–30°C, whereas many mammalian pathogens appear to express virulence determinants only at relatively high temperatures such as 37°C. None the less, the common offensive strategies employed by a wide range of prokaryotic pathogens and the conservation of multiple aspects of the innate immune response of their eukaryotic hosts suggest that the analysis of non-vertebrate/bacterial–pathogen interactions will have global relevance.

References

Aballay, A., Yorgey, P. and Ausubel, F. M. (2000). *Salmonella typhimurium* proferates and establishes a persistent infection in the intestine of *Caenorhabditis elegans*. *Curr. Biol.* **10**, 1539–1542.

Avery, L. (1993). The genetics of feeding in *Caenorhabditis elegans*. *Genetics* **133**, 897–917.

Basset, A., Khush, R. S., Braun, A., Gardan, L., Boccard, F., Hoffmann, J. A. and Lemaitre, B. (2000). The phytopathogenic bacteria *Erwinia carotovora* infects *Drosophila* and activates an immune response. *Proc. Natl Acad. Sci. USA* **97**, 3376–3381.

Bulla, L. A., Rhodes, R. A. and St. Julian, G. (1975). Bacteria as insect pathogens. *Ann. Rev. Microbiol.* **29**, 163–190.

Chadwick, J. S., Caldwell, S. S. and Chadwick, P. (1990). Adherence patterns and virulence for *Galleria mellonella* larvae of isolates of *Serratia marcescens*. *J. Invertebr. Pathol.* **55**, 133–134.

Cho, J. J., Schroth, M. N., Kominos, S. D. and Green, S. K. (1975). Ornamental plants as carriers of *Pseudomonas aeruginosa*. *Phytopathology* **65**, 425–431.

Darby, C., Cosma, C. L., Thomas, J. H. and Manoil, C. (1999). Lethal paralysis of *Caenorhabditis elegans* by *Pseudomonas aeruginosa*. *Proc. Natl Acad. Sci. USA* **96**, 15202–15207.

Davis, K. R., Schott, E. and Ausubel, F. M. (1991). Virulence of selected phytopathogenic pseudomonads in *Arabidopsis thaliana*. *Mol. Plant–Microbe. Interact.* **4**, 477–488.

Dong, X., Mindrinos, M., Davis, K. R. and Ausubel, F. M. (1991). Induction of *Arabidopsis* defense genes by virulent and avirulent *Pseudomonas syringae* strains and by a cloned avirulence gene. *Plant Cell* **3**, 61–72.

Feeley, J. C., Gibson, R. J., Gorman, G. W., Langford, N. C., Rasheed, J. K., Mackel, D. C. and Baine, W. B. (1979). Charcoal-yeast extract agar: primary isolation medium for *Legionella pneumophila*. *J. Clin. Microbiol.* **10**, 437–441.

Fields, B. S. (1996). The molecular ecology of legionellae. *Trends Microbiol.* **4**, 286–290.

Finlay, B. B. (1999). Bacterial disease in diverse hosts. *Cell* **96**, 315–318.

Finlay, B. B. and Falkow, S. (1997). Common themes in microbial pathogenicity revisited. *Microbiol. Mol. Biol. Rev.* **61**, 136–169.

Gao, L. Y., Harb, O. S. and Abu Kwaik, Y. (1997). Utilization of similar mechanisms by *Legionella pneumophila* to parasitize two evolutionarily distant host cells, mammalian macrophages and protozoa. *Infect. Immun.* **65**, 4738–4746.

Garsin, D. A., Sifri, C. D., Mylonakis, E., Qin, X., Singh, K. V., Murray, B. E., Calderwood, S. B. and Ausubel, F. M. (2001). A simple model host for identifying Gram-positive virulence factor. *Proc. Natl Acad. Sci. USA* in press.
Hedengren, M., Asling, B., Dushay, M. S., Ando, I., Ekengren, S., Wihlborg, M. and Hultmark, D. (1999). *Relish*, a central factor in the control of humoral but not cellular immunity in *Drosophila*. *Mol. Cell.* **4**, 827–837.
Hoffmann, J. A., Kafatos, F. C., Janeway, C. A. and Ezekowitz, R. A. (1999). Phylogenetic perspectives in innate immunity. *Science* **284**, 1313–1318.
Jander, G., Rahme, L. G. and Ausubel, F. M. (2000). Positive correlation between virulence of *Pseudomonas aeruginosa* mutants in mice and insects. *J. Bacteriol.* **182**, 3843–3845.
Jarosz, J. (1995). Interaction of *P. aeruginosa* proteinase with the non-self response of insects. *Cytobios.* **83**, 71–84.
King, E. O., Ward, M. K. and Raney, D. E. (1954). Two simple media for the demonstration of phycocyanin and fluorescin. *J. Lab. Clin. Med.* **44**, 301–307.
Lemaitre, B., Nicolas, E., Michaut, L., Reichhart, J. M. and Hoffmann, J. A. (1996). The dorsoventral regulatory gene cassette spatzle/Toll/cactus controls the potent antifungal response in *Drosophila* adults. *Cell* **86**, 973–983.
Lysenko, O. (1963). The mechanisms of pathogenicity of *Pseudomonas aeruginosa* (Schroeter) Migula 1. The pathogenicity of strain N-06 for the larvae of the Greater Wax Moth, *Galleria mellonella* (Linnaeus). *J. Insect Pathol.* **5**, 78–82.
Mahajan-Miklos, S., Rahme, L. G. and Ausubel, F. M. (2000). MicroReview: elucidating the molecular mechanisms of bacterial virulence using non-mammalian hosts. *Mol. Microbiol.* **37**, 981–988.
Mahajan-Miklos, S., Tan, M.-W., Rahme, L. G. and Ausubel, F. M. (1999). Molecular mechanisms of bacterial virulence elucidated using a *Pseudomonas aeruginosa-Caenorhabditis elagans* pathogenesis model. *Cell* **96**, 47–56.
Miller, J. H. (1972). *Experiments in Molecular Genetics*. 2nd edition., Cold Spring Harbor, New York, Cold Spring Harbor Laboratory Press.
Plotkinova, J. M., Rahme, L. G. and Ausubel, F. M. (2000). Pathogenesis of the human opportunistic pathogen *Pseudomonas aeruginosa* PA14 in *Arabidopsis*. *Plant Physiol.* **124**, 1766–1744.
Rahme, L. G., Ausubel, F. M., Cao, H., Drenkard, E., Goumnerov, B. C., Lau, G. W., Mahajan-Miklos, S., Plotnikova, J., Tan, M. W., Tsongalis, J., Walendziewicz, C. L. and Tompkins, R. G. (2000). Plants and animals share functionally common bacterial virulence factors. *Proc. Natl Acad. Sci. USA* **97**, 8815–8821.
Rahme, L. G., Stevens, E. J., Wolfort, S. F., Shao, J., Tompkins, R. G. and Ausubel, F. M. (1995). Common virulence factors for bacterial pathogenicity in plants and animals. *Science* **268**, 1899–1902.
Rahme, L. G., Tan, M. W., Le, L., Wong, S. M., Tompkins, R. G., Calderwood, S. B., and Ausubel, F. M. (1997). Use of model plant hosts to identify *Pseudomonas aeruginosa* virulence factors. *Proc. Natl Acad. Sci. USA* **94**, 13245–13250.
Rowbotham, T. J. (1980). Preliminary report on the pathogenicity of *Legionella pneumophila* for freshwater and soil amoebae. *J. Clin. Pathol.* **33**, 1179–1183.
Schroth, M. N., Cho, J. J., Green, S. K. and Kominos, S. D. (1977). Epidemiology of *Pseudomonas aeruginosa* in agricultural areas. In: *Pseudomonas aeruginosa: Ecological Aspects and Patient Colonization*, (ed. V. M. Young), Raven Press, pp.1–29. New York.
Segal, G. and Shuman, H. A. (1999). *Legionella pneumophila* utilizes the same genes to multiply within *Acanthamoeba castellanii* and human macrophages. *Infect. Immun.* **67**, 2117–2124.

Solomon, J. M. and Isberg, R. R. (2000). Growth of *Legionella pneomophila* in *Dictyostelium discoideum*: a novel system for genetic analysis of host–pathogen interactions. *Trends Microbiol.* **8**, 478–480.

Solomon, J. M., Rupper, A., Cardelli, J. A. and Isberg, R. R. (2000). Intracellular growth of *Legionella pneumophila* in *Dictyostelium discoideum*, a system for genetic analysis of host–pathogen interactions. *Infect. Immun.* **68**, 2939–2947.

Sulston, J. and Hodgkin, J. (1987). Methods. In *The Nematode Caenorhabditis elegans*. (W. B. Wood, ed.), pp. 587–606. Cold Spring Harbor, New York.

Sussman, M. (1987). Cultivation and synchronous morphogenesis of *Dictyostelium* under controlled experimental conditions. *Methods Cell Biol.* **28**, 9–29.

Tan, M.-W. and Ausubel, F. M. (2000). *Caenorhabditis elegans*: a model genetic host to study *Pseudomonas aeruginosa* pathogenesis. *Curr. Opin. Microbiol.* **3**, 29–34.

Tan, M.-W., Mahajan-Miklos, S. and Ausubel, F. M. (1999a). Killing of *Caenorhabditis elegans* by *Pseudomonas aeruginosa* used to model mammalian bacterial pathogenesis. *Proc. Natl Acad. Sci. USA* **96**, 715–720.

Tan, M.-W., Rahme, L. G., Sternberg, J. A., Tompkins, R. G. and Ausubel, F. M. (1999b). *Pseudomonas aeruginosa* killing of *Caenorhabditis elegans* used to identify *P. aeruginosa* virulence factors. *Proc. Natl Acad. Sci. USA* **96**, 2408–2499.

Tanada, Y. and Kaya, H. (1993). *Insect Pathology*. Academic Press, San Diego, Calif.

Wilson, I., Vogel, J. and Somerville, S. (1997). Signalling pathways: A common theme in plant and animals? *Curr. Biol.* **7**, R175–R178.

◆◆◆◆◆◆ APPENDIX I

King's B medium (King *et al.*, 1954)
1.0% Difco Proteose Peptone #3
0.15% K_2HPO_4, anhydrous
1.5% glycerol
adjust pH to 7.0 with HCl.
After autoclaving, add sterile $MgSO_4$ to 5 mM

NG agar
0.3% NaCl
0.35% Bacto Peptone (Difco)
1 ml l^{-1} cholesterol (5 mg ml^{-1} in 95% ethanol)
2% Bacto Agar (Difco)
After autoclaving, add the following sterile solutions:
1 ml l^{-1} $CaCl_2$ 1 M
1 ml l^{-1} $MgSO_4$ 1 M
25 ml l^{-1} KH_2PO_4 1 M, adjusted to pH 6.0 with solid KOH.

PGS agar
1.0% Bacto Peptone (Difco)
1.0% glucose
1.0% NaCl
2.74% sorbitol
1.7% Bacto Agar (Difco)

Dispense 5 ml agar per 3.5 cm diameter plate. Use after 1 day. Store the remainder at 4°C and use within 1 week.

CYE agar (Feeley *et al.*, 1979)
0.2% activated charcoal (Norit A or Norit SG)
1.0% yeast extract
1.7% agar (Difco)
After autoclaving, add filter-sterilized 0.04% L-cysteine $HCl \cdot H_2O$ and 0.025% ferric pyrophosphate. Adjust the medium to pH 6.9 by adding 4 to 4.5 ml of 1 N KOH.

SM/5 agar
0.2% glucose
0.2% Bacto Peptone (Difco)
0.2% yeast extract
0.02% $MgSO_4 \cdot 7H_2O$
0.19% KH_2PO_4
0.1% K_2HPO_4
2% agar
pH = 6.4

HL5 Medium
1.4% glucose
0.7% yeast extract
1.4% peptone*
0.095% $Na_2HPO_4 \cdot 7H_2O$
0.05% KH_2PO_4
pH = 6.5
Autoclave (a) glucose, (b) yeast extract and peptone and (c) Na and K phosphates in three separate solutions.
* Thiotone (BBL) and Bacteriological peptone (Oxoid) are preferred. Difco proteose peptone tend to give variable results that is batch dependent (Sussman, 1987).

MB medium
20 mM ME [2(*N*-morpholinoethanesulfonic acid)], pH 6.9
0.7% yeast extract
1.4% Thiotone E peptone

26 Antimicrobial Activity of Host Cells

Jerrold Weiss[1], Frank DeLeo[2] and William M Nauseef[1]
[1]*Department of Internal Medicine and The Inflammation Program, University of Iowa, Iowa City, IA 52242 and Iowa City Veterans' Administration Medical Center, Iowa City, IA 52246;* [2]*Laboratory of Human Bacterial Pathogenesis, Rocky Mountain Laboratories, NIAID, NIH, Hamilton, MT 59840*

◆◆

CONTENTS

List of abbreviations

AOAH	Acyloxyacylhydrolase
BPI	Bactericidal/permeability-increasing protein
BSA	Bovine serum albumin
CFU	Colony-forming unit
DCF	2′,7′-dihydrochlorofluorescein
DE	Dihydroethidine
DHR	Dihydrorhodamine-123
DTNB	5,5′-dithiobis(2-nitrobenzoic acid)
DPBS/G	Dulbecco's phosphate-buffered saline/0.1% glucose
FFA	Free fatty acid(s)
HBSS	Hanks' balanced salts solution
LPS	Lipopolysaccharide(s)
MPO	Myeloperoxidase
PBSA	Phosphate-buffered saline containing 0.2% albumin
PE	Phosphatidylethanolamine
PLA2	Phospholipase A2
PL	Phospholipid(s)
PMN	Polymorphonuclear leukocyte(s)
ROS	Reactive oxidant species
TLC	Thin-layer chromatography
TNB	5-thiobis-2-nitrobenzoic acid

METHODS IN MICROBIOLOGY, VOLUME 31
ISBN 0–12–521531–2

◆◆◆◆◆◆ INTRODUCTION

Essential host defense responses to invading microbes include sequestration of viable organisms to prevent further dissemination and creation of a highly noxious microenvironment to render the targeted microbe non viable. Professional phagocytes play major roles in these responses and, hence, will be our focus. There has been great progress in the identification of a wide array of protein and metabolic products with microbicidal capacity and, in some cases, in delineation of the structural and mechanistic determinants of the action of individual products studied in isolation (Elsbach *et al.*, 1999; Klebanoff, 1999; Lehrer and Ganz, 1999; Zanetti *et al.*, 2000). However, characterization of the role and cellular determinants of the intracellular action of specific antimicrobial agents remains a difficult challenge (Elsbach *et al.*, 1999). In fact, in some circumstances, even the ability to discern live and dead bacteria that structurally are grossly intact can be problematic. The antimicrobial role of professional phagocytes, especially of macrophages, includes microbial digestion to facilitate disassembly and removal of microbial remnants, down-regulation of the inflammatory response and return of the host to a resting state. In this chapter, we will focus on methods to assess intracellular bacterial killing and digestion and the mobilization and action of individual anti-microbial agents within professional phagocytes.

◆◆◆◆◆◆ DISTINGUISHING SUBLETHAL AND LETHAL BACTERIAL INJURY

Much like the actions of exogenous antibiotics, the cytotoxic effects of phagocytes may be manifest as potentially reversible microbial growth inhibition, death and/or lysis. Death is not necessarily accompanied, at least initially, by gross structural disorganization. Therefore, direct microscopic observation of ingested bacteria is insufficient to discern bacterial viability. Differing degrees of bacterial injury during phagocytosis may also be missed by conventional plating assays. A striking example is illustrated in Figure 26.1. As shown, ingestion of *Escherichia coli* by isolated polymorphonuclear leukocytes (PMN) is accompanied by a rapid loss of bacterial colony-forming ability in nutrient agar. Depending on the strain of *E. coli* used and the source of PMN, these effects may occur readily in buffered salts solutions (e.g. Hanks' buffered salts or RPMI) or require the addition of opsonic factors to promote bacterium-phagocyte interaction (Mannion *et al.*, 1990b). Serum is a convenient source of opsonins, including fragments of C3 generated *de novo* during complement activation by the bacteria. Downstream products of complement activation generate a membrane-active complex (MAC; C5b-9_n) that, while not needed for phagocytosis, can confer extracellular cytotoxicity and greatly enhance bacterial injury after ingestion by PMN (Taylor, 1983; Elsbach *et al.*, 1994). The latter is achieved by insertion of ≤ 100 complexes per

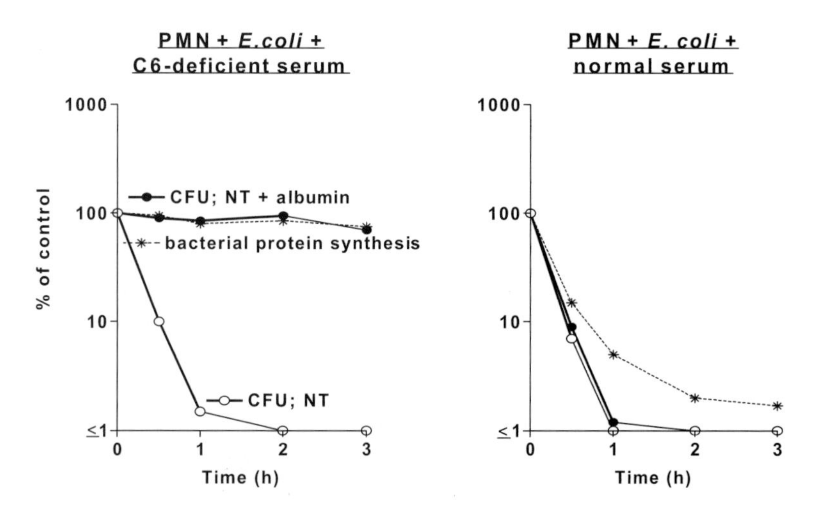

Figure 26.1. Colony-forming ability and protein-biosynthetic activity of *E. coli* J5 exposed to rabbit peritoneal exudate PMN in the presence of 2% C6-deficient (left panel) or normal (right panel) serum. Incubations contained 5×10^6 PMN and 5×10^7 bacteria in 1 ml of HEPES-buffered HBSS (with divalent cations) and serum. Data are expressed as a percentage of values obtained with initial bacterium inoculum.

bacterium, well below that needed for direct antibacterial cytotoxicity by complement (Madsen *et al.*, 1996). After deposition of sub-cytotoxic amounts of MAC on its surface, ingested *E. coli* lose the ability to incorporate radioactive amino acids into bacterial protein concomitant with a decline in bacterial colony-forming units (CFU) in nutrient agar (Figure 26.1). In contrast, *E. coli* ingested from medium without MAC, suffer the same loss of CFU without loss of overall protein biosynthetic activity. The latter bacteria form colonies if nutrient agar is supplemented with serum albumin (≥ 1 mg ml^{-1}) indicating that the bacteria are not yet dead but simply unable to form colonies in nutrient agar. These findings emphasize the importance of having an 'on-line' assay of bacterial health that can be linked kinetically as well as physiologically to bacterial viability.

Assays with several purified PMN antibacterial proteins (e.g. the bactericidal/permeability-increasing protein; BPI) as well as intact PMN strongly suggest that pulse labeling of bacteria with radioactive amino acids provides such an assay (Mannion *et al.*, 1990; Elsbach *et al.*, 1999). Whereas the inability to form colonies in nutrient agar (or several other bacteriological media) may correspond to potentially reparable bacterial injury, our experience suggests that the loss of ability to incorporate fed radioactive precursors into bacterial protein is closely linked to lethal damage.

Protein biosynthesis by ingested bacteria

1. Host cells and bacteria are mixed at desired cell : bacteria ratio (usually ranging from 1:1 to 1:10) in a buffered-balanced salts' solution

(e.g. 10 mM HEPES-buffered Hanks' balanced salts solution (HBSS), pH 7.4, containing divalent cations). If necessary bacteria are preopsonized in the same medium supplemented with serum (e.g. 10% pooled human serum). To prepare serum, peripheral blood is collected without anticoagulant from several healthy human adult volunteers and allowed to clot at 37°C. The clot is retracted by chilling on ice for ca. 30 min. Serum is recovered as the supernatant obtained by sedimentation of the clot at 10 000*g* for 5 min at 4°C. Serum is stored at −80°C and thawed just once prior to use to retain maximal complement activity.

2. Incubate the cell suspension at 37°C with rotation (30–60 rpm) to facilitate mixing.
3. At timed intervals, cell suspensions are treated with cycloheximide (ca. 0.5 mM) to inhibit host cell protein synthesis.
4. After 15 min treatment with cycloheximide, add radiolabeled amino acids (e.g. 0.1–1 $\mu Ci\,ml^{-1}$ of 14C-labeled L-amino acid mixture; ca. 60 $mCi\,mmol^{-1}$; New England Nuclear).
5. Continue incubation for another 15 min at 37°C.
6. Stop incubation by addition of an equal volume of 20% ice-cold trichloroacetic acid (TCA); place on ice for $\geq$ 15 min.
7. Collect TCA precipitates by either vacuum filtration through a 0.45 µm membrane filter (e.g. HAWP membrane, Millipore) or by sedimentation at 10 000*g* for 10 min at 4°C.
8. Wash the TCA precipitate twice with cold 5% TCA.
9. If TCA precipitate was collected on a membrane filter, dry the filter under a heat lamp and count in a liquid scintillation counter. If the TCA precipitate was collected by sedimentation, wash twice with ethanol/diethyl ether (1:1, v/v) to remove traces of TCA and resuspend as desired. For example, for analysis by SDS-PAGE, resuspend in SDS-PAGE buffer: apply a measured volume to the gel and use another aliquot to measure total cpm by liquid scintillation counting. The pattern of bacterial protein synthesis can be quantitatively assessed by phosphor-image analysis/densitometry (Molecular Dynamics). Using a tritium screen and conventional mini-gels, labeled protein species containing as little as 3000 cpm can be detected within 24 h of exposure.

Comments

1. As few as 10^5 bacteria per sample can be analyzed but may require use of higher levels of radiolabeled amino acid precursors and/or removal of serum used for pre-opsonization to reduce dilution of labeled precursors with unlabeled amino acids present in serum. If so desired, bacteria are washed after pre-opsonization by centrifugation at 10 000*g* for 5 min before mixing bacteria with host cells.
2. 35S-methionine/cysteine of high specific radioactivity can be used to further increase detection of small amounts of bacterial protein synthesis but may somewhat skew data by over-representing biosynthesis of methionine- and cysteine-rich proteins.

3. The ability of cycloheximide to inhibit host cell protein synthesis should be pre-tested to determine the minimum inhibitory concentration. Two-dimensional gels can be used to monitor bacterial protein synthesis $\pm$ cycloheximide, relying on prominent and well-resolved bacterial proteins to determine whether short-term cycloheximide treatment alters host–bacterium interactions. In experiments similar to those shown in Figure 26.1, effects of PMN on *E. coli* were unaffected by cycloheximide treatment as described here.
4. Under conditions in which the ingested bacteria appear to remain fully viable (Figure 26.1), bacterial protein synthesis/CFU of intracellular bacteria is roughly similar to that of a corresponding number of extracellular bacteria suggesting that added radiolabeled amino acids are efficiently delivered to the phagocytic vacuoles.
5. With the rapid development of proteomics, it should be possible to map bacterial biosynthetic patterns by juxtaposing the radioautographic image obtained from the labeled bacterial proteins to stained gels containing unlabeled bulk bacterial proteins in sufficient quantity to permit downstream mass spectroscopic analysis of eluted and fragmented species. Using bacterial species and strains in which the genome has been fully defined should make it possible to infer protein identity directly from the size and charge properties of the labeled species deduced from 2D gel electrophoresis.
6. To minimize contribution of non-ingested (extracellular and surface-associated) bacteria to the overall bacterial protein synthesis measured, conditions should be first established that best support efficient bacterial uptake. Methods are described in later sections to monitor phagocytosis by confocal microscopy and improve the synchrony of the process by temperature and centrifugal manipulations. In addition, cell suspensions can be pre-washed at low speed (80–100 *g* for 7 min) to remove extracellular bacteria and/or pretreated with gentamicin (100 $\mu g\, ml^{-1}$) before addition of radioactive amino acid precursors.

◆◆◆◆◆◆ IDENTIFICATION OF INTRACELLULAR PROTEIN–BACTERIUM INTERACTIONS

Relatively few host antibacterial proteins have actions sufficiently specific to permit product accumulation to serve faithfully as evidence of their interaction with ingested microbes. Many of those are enzymes; assays of the actions of some will be described in sections below. For many other host antimicrobial proteins, direct evidence of interactions with microbial targets is limited to assay of protein–microbe contacts. Because of the very small volumes within most phagocytic vacuoles, truly attached vs merely adjacent proteins can often not be adequately distinguished by direct histo- or immunocytochemistry, even by electron microscopy. This limitation, however, can be readily circumvented by lysing the host cell (and enclosed phagocytic compartments) to release the ingested organisms. Although the released organisms can be further purified by differential or density gradient centrifugation, direct immunostaining of the cell lysates can be sufficient. Based on the binding properties of the protein and organism of interest, previously established in cell-free assays with purified protein or cell extracts/lysates, lysis is performed under conditions that preclude *de novo* associations after cell disruption and dissociation of previously bound protein. Lysis conditions are carefully chosen to maximize host cell disruption without inducing additional damage to the microbe that might cause shedding of outer envelope layers containing bound host proteins. Prelabeling of bacterial lipids (described below) provides a very sensitive marker of microbial envelope integrity. To illustrate application of this method, conditions we have used to monitor intracellular associations of BPI with *E. coli* are outlined below.

Immunofluorescent detection of intracellular host–protein–bacterium interactions

1. As described above, host cells and (pre-opsonized) bacteria are mixed and incubated at 37°C.
2. At desired time(s), cell suspensions are diluted in medium containing heparin (final concentration, 10 international units ml^{-1}) or $MgCl_2$ (final concentration, 4 mM), chilled on ice and sonicated to lyse the host cells.
3. An aliquot (ca. 10 μl) of the sonicated cell suspension is placed on a pre-cleaned microscope slide, air-dried and heat fixed gently.
4. Smears are incubated in a moist chamber by overlaying first with serum diluted in phosphate-buffered saline containing 0.2% (w/v) albumin (PBSA), then with specific primary antibodies diluted in PBSA and last with a fluorescent conjugate of a secondary antibody diluted in PBSA and directed against the primary antibodies. Between each step, smears are washed several times with PBSA.
5. After drying, smear is viewed by epifluorescence microscopy under oil.

Comments

1. This procedure takes advantage of the much greater native resistance of bacterial cell envelopes to physical trauma in comparison to eukaryotic cells and membranes. However, this difference may gradually disappear as bacteria are subjected to cytotoxic onslaught by the host cell. Thus, this procedure works best when applied soon after ingestion of the bacteria.
2. Probe sonicators provide a more reproducible delivery of energy to cells within cell suspensions and therefore are preferred over water-bath sonciators. Sonicated samples are immersed in wet ice to control temperature during sonication. At a power setting of ca. 40 W in samples containing 5×10^6 PMN in 0.5 ml in tubes of 12 mm diameter, sonication is of 10 s duration. Efficacy of lysis of host cells can be monitored by phase contrast microscopy.

3. Essential controls include: (A) mixture on ice of host cells and bacteria to preclude ingestion and intracellular protein–bacterium interaction. There should be no fluorescent staining of bacteria under these conditions. (B) Pre-incubation of bacteria with purified host anti bacterial protein before mixing with host cells, incubation, sonication, etc. Smears should be made after pre-incubation and after mixing of bacterial suspensions with host cells (further incubation). These samples should show positive fluorescence but may also reveal loss of fluorescence due to biological and/or physical alterations secondary to exposure to host cells and subsequent sonication.
4. Serum for pretreatment should originate from a different animal species than the source of primary antibody. The secondary antibody should be reactive with conserved epitopes present in the primary antibody but not in antibodies of serum used for pretreatment to reduce non-specific binding of primary and secondary antibodies to bacteria.

◆◆◆◆◆◆ DIGESTION OF BACTERIAL LIPIDS

It is often assumed that successful elimination of invading bacteria by the host includes digestion of bacterial macromolecular constituents. In general, however, direct experimental evidence is remarkably sparse. The breakdown of major envelope components is the means by which certain host antibacterial proteins exert their cytotoxic effects as shown for the Group IIA phospholipase A2 (Weinrauch *et al.*, 1996) and perhaps elastase as well (Belaaouaj *et al.*, 2000). In addition, degradation within the microbial envelope may be important for penetration of cytotoxic species to target sites that reside within the microbial cell as well as overall microbial disassembly. Conversely, the persistence of structurally intact envelope molecules can have major inflammatory and immunological consequences (Schwab, 1993; Heumann *et al.*, 1994; Rietschel *et al.*, 1996). This is well illustrated by the lipopolysaccharides (LPS) of Gram-negative bacteria which trigger profound pro-inflammatory host cell responses during acute exposure and may induce long-term immunological sequelae during chronic presentation. Both professional phagocytes and antigen-presenting cells (e.g. dendritic cells) possess a specialized LPS deacylase (acyloxyacylhydrolase; AOAH) that partially deacylates and thereby detoxifies LPS (Munford and Hall, 1986; Munford, 1991; Munford and Erwin, 1992).

Studies of bacterial phospholipid (PL) metabolism during bacterium–host cell interactions have been more common, in large part because of the ease of achieving uniform and high specific radioactive labeling of bacterial PL by growth of bacteria in medium supplemented with radioactive free fatty acids (FFA) (Elsbach and Weiss, 1988, 1991). In both Gram-negative and Gram-positive bacteria, uptake and incorporation of fed FFA into PL is highly efficient (routinely $>50\%$). Because labeling coincides with the production of new membrane lipids produced over several generations, the labeled lipids mirror in composition the overall envelope PL content. Long-chain FA ($\geq$ 14 carbon chain length) are incorporated, without further modification, selectively into either the sn_1 or sn_2 position depending on the structure of the FA and the properties of bacterial acyltransferases (Cronan and Rock, 1996; Fischer, 1994). Together with the relatively simple lipid profile of prokaryotic cells, this yields a well-defined and reproducible labeled lipid profile. Typically, one-dimensional thin-layer chromatography (TLC) is sufficient to monitor bacterial lipid metabolism although, within host cells, bacterial lipid breakdown products (especially FFA) are utilized by the host cell to make a different and more complex array of lipids (Patriarca *et al.*, 1972; Weinrauch *et al.*, 1999).

Multiple pathways exist by which specific lipid metabolites can be generated complicating identification of participating enzymes. To infer the action of the antibacterial Group IIA PLA2, a labeled FA should be used that is selectively incorporated into the 1-acyl position of PL: palmitate (C16:0) in *E. coli* and many other Gram-negative bacteria; palmitate or oleate (C18:1) in *S. aureus* and many Gram-positive bacteria. The primary labeled product of PLA2 attack under such circumstances is a labeled 1-acyl lyso-PL. Kinetic analyses of labeled lipid formation will reveal whether this labeled lipid product is a primary product of PLA2 action or secondarily generated by a different and more circuitous route. All cells contain complex lipid metabolizing machinery for diverse physiological needs including stress responses that may be activated in bacteria under attack. However, major bacterial deacylases are not PLA2 and are dispensable at least under laboratory conditions permitting construction of PLA-deficient strains (Cronan and Rock, 1996). Moreover, the Group IIA PLA2 can bind at low doses to extracellular bacteria without degrading bacterial PL until bacteria and PLA2 are co-ingested by phagocytes (Wright *et al.*, 1990; Weiss *et al.*, 1994). This manipulation can be used to show intracellular PLA2 action, define the products generated and test its role in overall intracellular bacterial destruction.

Assay of digestion of bacterial PL

1. Pre-label bacterial PL during growth by supplementing growth medium with radiolabeled long-chain FFA (e.g. 14C-16:0 or 18:1). These FFA are typically stored in organic solvent at −20°C. Just before use, the FFA are warmed to room temperature. The desired aliquot is taken, transferred to a sterile vessel to be used for bacterial growth, dried under nitrogen, resuspended in 0.1 volume of growth medium supplemented with 0.1% albumin and swirled for 5 min at 37°C to resolubilize the FFA. Growth medium is added to desired final volume followed by an aliquot of a freshly harvested overnight culture of the bacteria → $A_{600} \approx 0.03$.
2. The bacteria are grown to mid-logarithmic phase ($A_{600} \approx 0.5$) at 37°C with vigorous shaking (~200 rpm) and aeration.
3. The bacteria are washed once, resuspended in 0.5 volume of fresh medium and incubated at 37°C with shaking for 15–20 min to

maximize incorporation of cell-associated FFA → PL without significant further bacterial growth and dilution of specific radioactivity of PL.

4. Albumin is added (final concentration = 0.5% (w/v)) and bacteria are washed to remove cell-associated unesterified FFA.
5. Washed bacteria are resuspended in the desired medium to the desired concentration for subsequent assays.
6. After incubation with host cells, cell suspensions are extracted with 6 vol of chloroform : methanol (1 : 2, v/v) (Bligh and Dyer, 1959). Extractions are carried out overnight at room temperature or at 37°C for $\geq$30 min.
7. After extraction, 3 vol of 50 mM KCl and 2 vol of chloroform are added, samples are vortexed and spun at 5000*g* for 5 min to optimize the separation of the upper (aqueous-methanol) and lower (chloroform) phases (Figure 26.2).
8. These separated phases are carefully collected by pasteur pipette. Each recovered phase is washed with fresh solvent corresponding to other phase (i.e. wash recovered aqueous-methanol phase with fresh chloroform and vice versa) and the recovered aqueous-methanol (or chloroform) phases are separately combined.
9. The combined recovered chloroform phase is dried under nitrogen and the recovered aqueous-methanol phase, if desired, is dried by rotary evaporation. The dried material is resuspended in a small volume (< 50 μl) of chloroform : methanol (2 : 1; v/v) and subjected to TLC for resolution of individual lipid species.
10. Labeled lipid species resolved by TLC are detected and quantified by phosphor-image analysis and densitometry. Labeled lipid standards are run in parallel lanes to aid identification of labeled lipid species in experimental samples.

Comments

1. Many bacterial species possess a high capacity for uptake and incorporation of long-chain FFA → mature lipids, especially PL. In *E. coli* and *S. aureus*, up to 10 μCi ml^{-1} of 14C-FFA (~60 mCi mmol^{-1}) can be incorporated during growth.
2. To identify positional specificity (i.e. sn_1 or sn_2) of FA incorporation into PL, PL are purified by extraction and TLC as described, eluted from silica gel with chloroform/methanol/acetic acid/water (55/33/9/4; v/v/v/v), dried under nitrogen and resuspended by sonication in aqueous buffer. The resuspended PL are incubated with purified PLA2 (e.g. pig pancreatic enzyme commercially available from Sigma) for increasing time and the primary labeled products (14C-FFA and/or –lyso-PL) identified by extraction and TLC as described above. FFA and lyso-PL are resolved during TLC in silica gel G in many different solvent systems including petroleum ether/diethyl ether/glacial acetic acid (70/30/1; v/v/v) and chloroform/methanol/water/acetic acid (65/25/4/1; v/v/v/v). In the former, the R_f of FFA, lyso-PL and PL is ca. 0.4, 0.0 and 0.0, respectively,

irrespective of the PL head group whereas in the latter solvent system the R_f is ~0.9, 0.3 and 0.5 (using lyso-phosphatidylethanolamine (PE) and PE). The migration of other lyso-PL and PL species depends on the head group but is always PL > lyso-PL.

3. Incubations with intact bacteria and host cells are normally carried out in the presence of high concentrations of albumin (e.g. 1.5%; w/v) to increase the efficiency of recovery of the primary lipid degradation products. This is dependent on the ability of bacterial metabolites such as FFA and lyso-PL to diffuse across membranes and complex with extracellular albumin. However, depending on the chemical nature of the lipid metabolites, concentration of albumin and metabolic machinery of the host cell, the primary lipid breakdown products may be subject to further metabolism including recycling into host lipids (Weinrauch *et al.*, 1999). Because of the difference in composition of the bulk, metabolically active lipid pools of bacteria and host cells, metabolism of bacterial lipids can generally still be recognized but the precise metabolic events may be difficult to deduce. During Bligh and Dyer extraction, lipid metabolites complexed to albumin are released and albumin denatured. It is often advantageous to remove denatured albumin by sedimentation (e.g. 5000*g* for 5 min) before addition of KCl and chloroform to facilitate phase separation.
4. Nearly all lipids will be recovered in the chloroform phase but more polar species of lyso-PL may partially partition into the aqueous methanol phase.
5. Labeled lipid species in experimental samples are presumptively identified by co-migration with labeled lipid standards during TLC in at least two different separation systems.

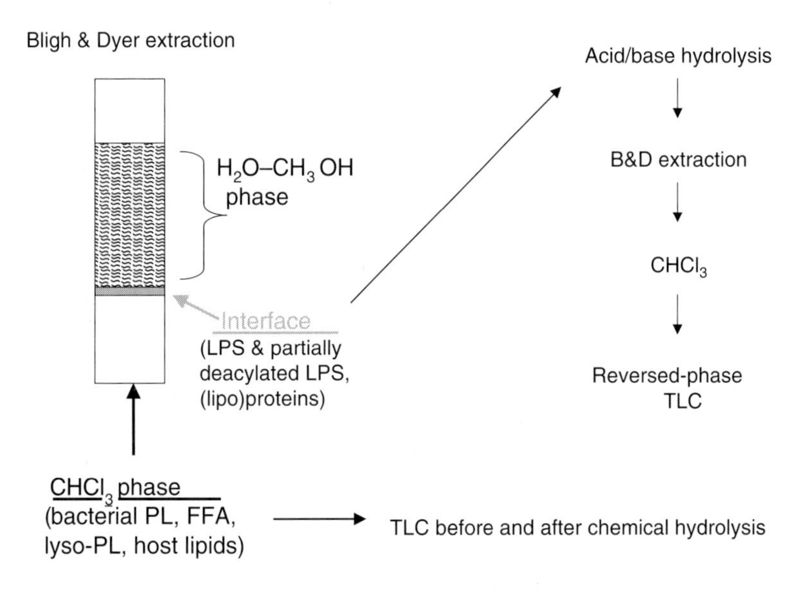

Figure 26.2. Schematic of analytical procedures used for assay of bacterial lipid alterations produced during incubation of bacteria with host cells.

In contrast to the acyl chains of bacterial PL, the acyl chains of lipopolysaccharides (LPS) derive solely from FA made endogenously within the bacterium (Cronan and Rock, 1996). Therefore, LPS cannot be metabolically labeled by feeding growing bacteria labeled FFA, even with those FA present in LPS. However, efficient labeling of the acyl chains of LPS has been achieved by genetically disrupting the enzyme machinery (e.g. pyruvate dehydrogenase complex) needed to make acetate, the 2-C precursor of all FA synthesis by the cell (Munford *et al.*, 1992; Giardina *et al.*, 2001). These strains require acetate for growth. Hence, growth in defined minimal medium with radiolabeled acetate yields the full spectrum of labeled lipid species, including LPS, in proportion to the chemical abundance of their fatty acyl constituents.

The FA composition of LPS is distinct from that of PL as well as other acylated bacterial species (e.g. lipoproteins) making it possible to distinguish and quantitatively monitor deacylation of these different bacterial lipid families (Katz *et al.*, 1999). For example, in *E. coli* grown to mid-late log phase, the major LPS-derived FA are 3-OH-14 : 0, 12 : 0 and 14 : 0 whereas PL contain mainly 16 : 0, 16 : 1 and 18 : 1 with only small amounts of 14 : 0 (Nikaido *et al.*, 1996). Lipoproteins are enriched in 16 : 0. Moreover, intact and partially deacylated LPS and lipoproteins can be separated from PL, FFA and lyso-PL as well as di- and tri-glycerides by organic solvent extraction and phase separation as described above (Figure 26.2). Loss of 3-OH-FA from the interface where intact and partially deacylated LPS partition and reciprocal recovery of these FA either as FFA or re-esterified as part of host lipids reflects partial deacylation of LPS by AOAH—to date the only host enzyme known to produce this deacylation pattern. Analogous bacterial deacylases have not been described. In all recovered lipid samples, the composition of the 14C-FA can be determined by chemical hydrolysis to release FFA and, after extraction, reversed-phase TLC to resolve major FA species (Figure 26.2).

Assay of digestion of bacterial LPS

1. Pre-label acylated bacterial species of bacterial acetate auxotrophs by growth in defined medium supplemented with 14C-acetate.
2. Grow bacteria at 37°C with vigorous shaking to mid-late log phase ($A_{600} \sim 0.5–1.0$).
3. Harvest bacteria and wash with medium supplemented with 0.5% albumin (w/v).
4. After experimental incubations, extract bacterial lipids as described above and schematized in Figure 26.2.
5. Interface is recovered as a thin meniscus after careful removal of chloroform and aqueous-methanol phases and washing with these solvents.
6. Recovered labeled lipids in chloroform phase (and in aqueous-methanol phase if significant cpm is present in this phase) are resolved by TLC, as described above, before and after sequential treatment with 4N HCl and 4N NaOH at 100°C to release FA from ester- and amide-linked lipid pools.

7. Labeled lipids recovered at interface chemically hydrolyzed, as above, and released 14C-FA recovered after Bligh and Dyer extraction (Figure 26.2) are resolved by reversed-phase TLC on KC_{18} silica gel G plates (Whatman, Clifton, NJ) with glacial acetic acid/acetonitrile (1 : 1, v/v) as the solvent.
8. Resolved 14C-lipids are detected and quantified by phosphor-image analysis (Molecular Dynamics) and identified by co-migration with labeled lipid standards.

Comments

1. LPS deacylation is defined as the percentage of LPS-specific FA recovered in the chloroform phase/total recovery of these FA in combined chloroform phase and interface. Because of the exclusive presence of 3-OH-FA and 12 : 0 in LPS, their release from LPS can be unambiguously monitored. The presence of small amounts of 14 : 0 in large bacterial PL pools makes analysis of this FA more difficult but may still be possible by analysis of FA of lipids recovered at the interface.
2. The purity and efficiency of recovery of 14C-LPS at the interface can be assessed by: (a) comparing the FA composition of the interface with that of LPS purified from the same labeled bacterial population; and (b) measuring recovery of 14C-FA unique to LPS in samples extracted with host cells without prior incubation.
3. Acetate auxotrophs require from 0.4–2 mM acetate for optimal growth. If the acetate is provided entirely as 14C-acetate, specific radioactivity of 14C-LPS produced is nearly 1000 cpm ng^{-1}. This roughly corresponds to the amount of LPS present in 10^5 bacteria. 14C-LPS represents around 15% of the total labeled population; 14C-PL ~ 50–60%. Using a tritium screen (Molecular Dynamics), as little as 200 cpm per resolved labled lipid species can be detected within 24 h.
4. PL deacylation is defined as the percentage of PL-specific FA recovered as either FFA, lyso-PL or as part of host lipids or as loss from intact bacterial PL (mainly PE in *E. coli*). The overlap in the lipid compositions of the host cell and bacteria $< 20\%$.
5. The procedures described can be used to generate and purify labeled lipid standards. Recovered species can be further analyzed by gas chromatography or mass spectroscopy to confirm compositional identity with correction for 14C/12C content of FA.

◆◆◆◆◆◆ OXYGEN-DEPENDENT ANTIMICROBIAL ACTION

Prominent among the array of defensive responses mounted by phagocytes is the agonist-dependent generation of reactive oxygen species (ROS) by neutrophils, monocytes, and macrophages (reviewed by

Klebanoff, 1999). The enzyme responsible for the conversion of molecular oxygen to superoxide anion is a multicomponent complex composed of a heterodimeric membrane protein, flavocytochrome b_{558}, and at least three cytosolic proteins (reviewed by Babior, 1999; Figure 26.3). In unstimulated phagocytes, flavocytochrome b_{558}, composed of gp91*phox* and p22*phox*, resides in the plasma membrane and in the membranes of granules, whereas p47*phox*, p67*phox*, and rac2 exist in the cytosol. During phagocytosis p47*phox* undergoes post-translational modification, including phosphorylation, with secondary conformational reorganization (Shiose and Sumimoto, 2000) and the cytosolic proteins subsequently translocate to the phagosomal membrane to associate with flavocytochrome b_{558} and form an assembled oxidase (reviewed by Nauseef, 1999; Figure 26.3). The assembled oxidase converts molecular oxygen to superoxide anion using electrons derived from oxidation of NADPH, supplied by the hexose monophosphate shunt. Although superoxide anion represents the most proximal product of the functioning oxidase, it spontaneously dismutates to hydrogen peroxide (H_2O_2) and spawns a cascade of reactions that generate a wide variety of reactive oxygen species (ROS). Among these products are those whose activity or half-life is potentiated by the action of the granule protein myeloperoxidase (MPO), also released into the phagosome during stimulation. For example, MPO oxidizes Cl^- in the presence of H_2O_2 to the Cl^+ state, generating HOCl and chlorinating target proteins (Hazen *et al.*, 1999). The highly reactive HOCl is toxic but relatively short-lived. In contrast, chloramines produced in the phagosome by the MPO–H_2O_2-chloride system provide a stable and durable means to effect damage to target organisms. Taken together, the capability of PMNs to generate ROS and release granule proteins into the phagolysosome equip phagocytes with a rapidly deployable system to deliver a host of toxic products to an organism trapped within a confined intracellular compartment (Hampton *et al.*, 1998; Elsbach *et al.*, 1999; Figure 26.3).

Within the context of this phagocyte–microbe interaction, survival of microbes within phagocytes can reflect employment of several different strategies. Successful pathogens can elude this orchestrated system, sabotage the generation of reactive oxygen species and their derivatives, and/or simply endure their cytotoxic effects. To that end, micro-organisms can inhibit their phagocytosis or, once ingested, block the normal maturation that eventually culminates in phagosome–lysosome fusion and thus subvert delivery of granule proteins into phagosomes. Likewise, oxidant generation in response to an invading microbe can be undermined, by inhibiting the assembly and/or the activity of the NADPH oxidase at the phagosomal membrane. Given that the determinants of oxidase regulation, including factors contributing to the termination of its activity, are incompletely characterized, one can imagine that both blocking oxidase assembly as well as accelerating its disassembly and inactivation might be mechanisms exploited by successful pathogens. Alternatively, some micro-organisms may simply endure the onslaught and prolong survival by releasing products that act as competing substrates for ROS, thereby scavenging the toxic products. Identification

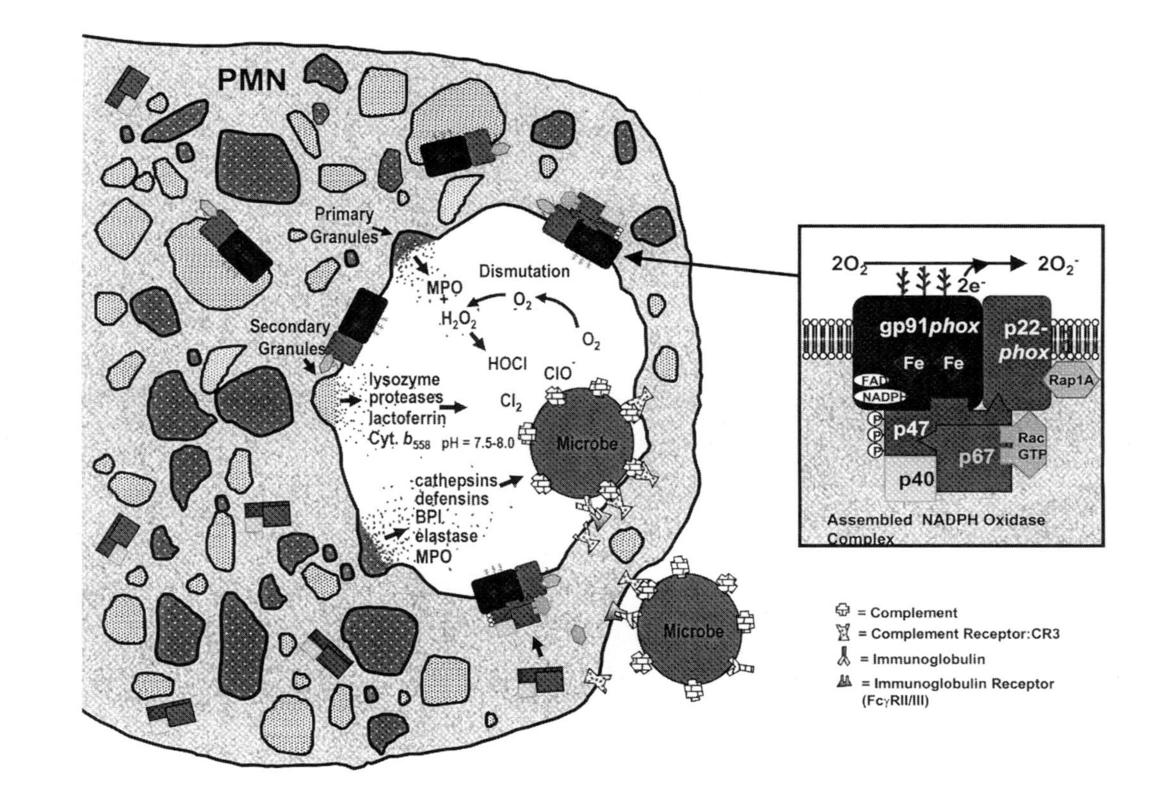

Figure 26.3. Schematic of subcellular organization of key components of antimicrobial arsenal of PMN and molecular and subcellular changes that occur during phagocytosis.

of the specific ploy(s) adopted by a given microbe requires a careful evaluation of respiratory burst oxidase activity, including an assessment of its activation per se and, in many cases, determination of the topology of oxidase assembly.

♦♦♦♦♦♦ MEASUREMENT OF NADPH OXIDASE ACTIVITY

Since the ROS generated by the NADPH oxidase are derived from molecular oxygen, measurement of oxygen consumption from the medium by stimulated cells should yield the same results as assessment of the production of specific oxygen-derived radicals. However, the lack of specific substrates that exclusively interact with given elements in the mix of ROS and the spatial constraints limiting easy access to the intraphagosomal space leave oxygen consumption as the most reliable, albeit less sensitive, assay of oxidase activity.

Oxygen consumption

Polarigraphic measurement of oxygen pressure in solution requires careful calibration of the Clark electrode used. A 0.8 V polarizing voltage is applied to a 10 mM solution of sodium dithionite stirred in the sample chamber. The dithionite reduces oxygen in solution, eliminating any current, and thus providing the 'zero point' for the recorder. To set the high end of the recorder, thoroughly clean the sample chamber and replace with air-saturated water, generated by bubbling air continuously through water at 37°C for 1 h. Apply 0.8 V to the electrode while continuously stirring the saturated water and adjust the recorder to 100%. Given that oxygen-saturated water at 37°C and 760 mm Hg contains 217 nmoles of O_2 per ml H_2O (Handbook of Chemistry and Physics) the full scale for measurements will be now be set.

The standard Clark electrode sample chamber accommodates volumes of 3 ml, although smaller chambers are available.

1. Equilibrate in the sample chamber a 2.9 ml suspension of phagocytes (e.g. final concentration of 2.5×10^6 PMNs ml^{-1}) in a calcium- and magnesium-containing buffer (e.g. supplemented Hanks balanced salt solution) for 2 min at 37°C without stirring.
2. Insert the magnetic stirrer and probe, without introducing bubbles, set recorder to 100%, mix, and measure the oxygen concentration continuously for 5 min.
3. Stop the stirrer, added the warmed suspension of particles in 100 μl, and carefully add to the sample cuvette. Restart mixing and recording for an additional predetermined time. From the tracings one can calculate the latency between addition of agonist and a detectable change in oxygen concentration, the maximal rate of oxygen consumption, and the total amount of oxygen consumed over the

period of stimulation. Previous studies have reported that human neutrophils consume 141 $\pm$ 15.7 nmoles of O_2 per 2.5×10^6 PMNs in 20 min in response to opsonized *S. aureus* (500 : 1, particle to cell ratio) (Root *et al.*, 1975).

Comments

For accurate and reproducible measurement of oxygen consumption it is essential to avoid generating bubbles in the sample chamber, a common complication when the stimulus is added to the cell suspension. Because the amount of oxygen present in a bubble far exceeds that consumed by stimulated PMNs, the presence of bubbles will negate detection of cellular oxygen consumption. The integrity of the probe membrane and cleanliness of the sample chamber are equally important factors for obtaining reliable measurements.

In relation to microbicidal action against ingested organisms, the amount and identity of ROS generated in the phagosome is what is relevant. Although the complexity of the chemistry of competing interactions among molecules within the phagosome makes predictions difficult, modeled reactions predict that 5–10 $nmols^{-1}$ ml^{-1} sec^{-1} of O_2^- are generated within phagosomes in PMN during ingestion of 15–20 bacteria (Hampton *et al.*, 1998). A wide variety of very sensitive assays are available to detect and/or quantitate ROS. During phagocytosis, neutrophils chemiluminesce as a result of activation of the NADPH oxidase (Cheson *et al.*, 1976). Levels of induced chemiluminescence can be greatly amplified by adding lucigenin, luminol, or isoluminol providing amongst the most sensitive assays of oxidase activity (Allen, 1986). Lucigenin and (iso)luminol are equivalent in their quantum yields after activation but have very different reactivity. Whereas the luminol-based system is principally MPO-dependent, lucigenin-enhanced chemiluminescence detects ROS that are more proximally derived from oxidase activation. Although the identities of the species detected are not firmly established, superoxide dismutase (SOD) decreases lucigenin-enhanced chemiluminescence, indicating that O_2^- contributes to this response.

The concentration of phagocytes, agonist, and substrate (luminol or lucigenin) as well as the temperature at which phagocytosis is assayed all influence the kinetics and magnitude of the chemiluminescence. These issues have been reviewed (Allen, 1986) and will not be reiterated here. A protocol for measuring luminol/lucigenin-enhanced chemiluminescence by purified PMNs ingesting opsonized zymosan (OPZ) is provided as an example. It is worth noting that the activity of PMNs can be measured in whole blood, if desired, thereby bypassing any untoward effects of cell isolation. The sample should be diluted $\geq 1:1000$ to reduce the number of erythrocytes, because the absorption spectrum of hemoglobin overlaps those of luminol and lucigenin. Fortunately the sensitivity of the assay is such that even after dilution there will be sufficient PMNs to produce a detectable signal.

Intact phagocyte lucigenin/luminol-enhanced chemiluminescence

Stock solutions of the chemilumigenic probe should be stored at 4°C in the dark. Prepare lucigenin as a 5 mM stock solution in DMSO. The concentration of probe diluted in an aqueous buffer can be verified spectrophotometrically using millimolar extinction coefficients of 37.3 at 369 nm and 9.65 at 430 nm (Allen, 1986). Stock solutions of luminol are prepared at 20 mM in DMSO and the concentration of probe diluted in aqueous buffer confirmed spectrophotometrically using the millimolar extinction coefficient of 7.63 at 347 nm (Allen, 1986). Chemiluminescence by stimulated phagocytes is measured using a thermostated microplate luminometer (Lucy, Anthos Labtec Instruments; Salzburg, Austria). Cells are suspended in Optiplate 96-well plates (Packard). The incubation chamber is pre-warmed to 37°C and programmed to take readings on each well of pre-set duration and frequency (e.g. 2 s per well every 4 min). Applications of using chemiluminescence to assay the activity of the broken cell system have been reported (Inanami *et al.*, 1998) but are not presented here.

1. The total assay volume is 200 µl per well and the reaction buffer, calcium and magnesium-containing Hanks balanced salt solution containing 0.1% D-glucose and 1.0% human serum albumin, are endotoxin-free.
2. Cells are suspended at desired concentration (e.g. 10^7 PMN ml^{-1}) and 100 µl added to each well. Add Lucigenin or luminol to each well for a final concentration of 100 µM or 1 µM, respectively.
3. At the selected time, add agonist (1 µM fMLP, 10–100 ng ml^{-1} PMA, or 5×10^6 OPZ) and return the plate to the reader for measurement. Raw data are generated as relative light units emitted per unit time.

Reduction of ferricytochrome c

The measurement of superoxide dismutase-inhibitable reduction of ferricytochrome c represents a useful method to quantitate respiratory burst activity (Babior *et al.*, 1973). Ferricytochrome c reduction by O_2^- is equimolar, so the amount of O_2^- produced can be calculated by measuring the concentration of reduced ferricytochrome c (the molar extinction coefficient is 21 100 M^{-1} cm^{-1}). Although used to measure responses both to soluble and particulate stimuli, the assay accurately measures extracellular O_2^- only and therefore cannot detect intracellular respiratory burst activity or O_2^- produced within phagosomes.

Assays in 96-well microplates have replaced many of the early applications of this method, which included endpoint assays performed in polystyrene tubes and/or kinetic assays done individually in spectrophotometer cuvettes. The microplate format conserves reagents and allows the investigator to test several experimental conditions simultaneously, reducing trial-to-trial variation. Contemporary microplate readers have excellent optical performance and kinetic capabilities simplifying the assay further, as detailed in the protocol described below.

Microplate-based kinetic assay using ferricytochrome c reduction

1. The plate reader should be warmed to 37°C and set to read at an absorbance of 550 nm every 15 s for 10–30 min, with plate agitation prior to each measurement or twice per minute.
2. Preparation of a 'reagent' 96-well plate, containing rows of (1) cells (e.g. PMN at 5–10 × $10^6/\text{ml}^{-1}$ in 1 mM in Dulbecco's phosphate-buffered saline with 10 mM D-glucose (DPBS/G)); (2) ferricytochrome c (1 mM in DPBS/G); and (3) buffer (DPBS/G) greatly facilitates the execution of this assay. Reagents can be added simultaneously to 8–12 wells in the 'reaction plate', allowing for rapid and reproducible initiation of cell activation.
3. Add agonist (1 µl of 0.2 mg ml^{-1} PMA from a 2 mg ml^{-1} stock in DMSO; add particulate stimuli suspended in a volume of 50 µl) to wells in the reaction plate. Add 20 µl 0.4 mg ml^{-1} superoxide dismutase (from a 4 mg ml^{-1} stock solution in DPBS/G and held on ice) to a least one well for each experimental condition. Inclusion of this well provides assessment of the amount of ferricytochrome c reduction that is not inhibited by SOD and, by definition, is not mediated by superoxide anion. Add sufficient buffer to each well such that the total reaction volume will be 200 µl once the cells and ferricytochrome c are added.
4. Using a multi-channel pipettor, transfer 20 µl ferricytochrome c from the reagent plate to each well.
5. Transfer 100 µl cells from the reagent plate to each well using a multi-channel pipettor.
6. Quickly transfer the reaction plate to the reader and begin recording.
7. Most plate reader software can perform velocity calculations using kinetic data. The most common way to express these data is as a maximum rate (V_{max}). For ferricytochrome c reduction, the V_{max} should be calculated from a 1–2 min linear portion of the absorbance plot using the molar extinction coefficient for ferricytochrome c.

Comments

1. Depending on the turbidity of the assays containing particulate stimuli, it may be necessary to perform an endpoint reading on clarified assay wells. To clarify assays, centrifuge the plate for 8 min at 500*g* following the kinetic read and then transfer 100 µl to a new plate and read absorbance at 550 nm.
2. For experiments where a large number of variables are tested or when particulate stimuli are used, it may be desirable to synchronize cell activation and start of the assay. Chill the reagent plate, all reagents, and cells on ice and then combine in the reaction plate on ice. The plate is warmed in the plate reader at the start of the measurement. If particulate stimuli are used, chilled plates can be centrifuged at 300 *g* for 5 min at 4–10°C prior to collecting measurements in a warm plate reader. When the centrifugation step is incorporated into the assay, the

plates should be agitated during measurement to give accurate absorbance readings.

Use of fluorometric substrates for assessing NADPH oxidase activity

Ferricytochrome c does not enter cells to an appreciable extent and thus does not effectively monitor intracellular O_2^- generation. In contrast, several commercially available probes readily cross cell membranes and are oxidized to fluorescent derivatives by ROS produced by phagocytic leukocytes. This capacity represents a significant advantage over the ferricytochrome c, especially when it is desirable to measure oxidants produced intraphagosomally, although the probes are neither specific for O_2^- nor are quantifiable. Commonly used fluorometric probes for ROS include 2′,7′-dihydrodichlorofluorescein (DCF) homovanillic acid and dihydroethidine (DE) (Ruch *et al.*, 1983; Rothe and Valet, 1990; Emmendorffer *et al.*, 1990). Detection of hydrogen peroxide using homovanillic acid is the least sensitive method of those described here and, given the availability of sensitive alternatives, is infrequently used. DCF and DHR are most effectively oxidized by H_2O_2 in the presence of peroxidase although their ability to interact with other ROS is incompletely characterized. O_2^- or H_2O_2 in combination with peroxidase oxidize DE to ethidium bromide, which then intercalates into cellular DNA. The advantages and limitations of each have been described in detail (Vowells *et al.*, 1995; van Pelt *et al.*, 1996). Although DHR is more sensitive in its ability to detect ROS, DCF is more widely used. DE use is limited somewhat by its spontaneous oxidation in quiescent cells (Rothe and Valet, 1990).

Flow cytometric analysis of stimulated PMNs using DCF or DHR has proven especially useful for quantifying heterogeneous populations of PMNs in individuals who are carriers of X-linked chronic granulomatous disease (Rothe and Valet, 1990; Vowells *et al.*, 1995). However, flow cytometric assays do not provide kinetic data points such as the time of onset to respiratory burst, V_{max}, time to V_{max}, and the time to respiratory burst inactivation.

We have modified a 96-well microplate assay to measure the kinetics of ROS production during phagocytosis (DeLeo *et al.*, 1999). The protocol below can be modified for use with DHR or DE.

Detecting respiratory burst activity using DCF

1. As with the 96-well microtiter plate format for the ferricytochrome c assay, preparation of a 'reagent plate' greatly facilitates speed and accuracy in this assay. Coat each well of the reaction plate with 50–100 μL of 10% normal human serum and incubate at 37°C for ≥1 hour. Remove serum, wash two times with 0.9% NaCl or DBPS/G, and chill the plate on ice.
2. Prepare a 10 mM solution of DCF (Molecular Probes, Inc., Eugene, OR) from a 100 mM stock-solution prepared in DMSO. Prepare 25 μM DCF

in DPBS/G by adding 25 μl of the 10 mM solution to 10 ml DPBS/G: vortex slowly while adding. Wrap in foil and keep at room temperature. Equilibrate cells (e.g. 10^7 PMN ml^{-1} in DBPS/G containing 25 μM DCF at room temperature for 30–45 min with very gentle agitation). Chill cells on ice afterward. Cover all tubes containing DCF or DCF-loaded cells with foil to protect from light.

3. Set microplate fluorometer to read at excitation and emission wavelengths of 485 nm and 538 nm, respectively. The reader should be programmed to read at least once per minute for 60–90 min at 37°C without agitation if particulate stimuli are used.
4. Add cold buffer (DPBS/G) to each well as necessary so that once all components of the assay are added the final volume will be 200 μl.
5. Add agonist (1 μl of 0.2 mg ml^{-1} PMA from a 2 mg ml^{-1} stock in DMSO; add particulate stimuli suspended in a volume of 50 μl) to wells in the reaction plate. OPZ works well at 5 particles per PMN; use 10–50 bacteria per PMN.
6. Using a multi-channel pipettor, transfer 100 μl DCF-loaded PMNs from the reagent plate to appropriate wells of the reaction plate.
7. Transfer 50 μl opsonized bacteria (or other stimuli) from the reagent plate to appropriate wells of the read plate. Centrifuge the read plate at 300–500 g for 5 min at 4–10°C and then very carefully transfer to the plate reader and start the assay.

Data derived from fluorescent-based measurements of reactive oxygen species are commonly expressed as maximum rates (V_{max}). Given the significantly increased length of the assay in comparison to that of ferricytochrome c reduction, rates can be calculated from 5–10 min portions of the fluorescence plot. Data can then be expressed as change in fluorescence per min and rates plotted for each time interval or as a single data point representing the maximum rate of fluorescence increase for a 5–10 min period during the entire assay (V_{max}).

◆◆◆◆◆◆ LOCALIZATION OF THE ASSEMBLED OXIDASE

Assessment of oxidase activity in a population of stimulated phagocytes may overlook localized host–microbe interactions that may be important for the successful invasion of a specific target cell. Recognizing that assembly of the oxidase is a prerequisite for its activity, one can assess the successful translocation of the cytosolic oxidase components to the phagosomal membrane either by analyzing oxidase assembly on isolated phagosomes or on phagosomes, *in situ*, viewed by immunofluorescence microscopy. Assays with isolated phagosomes require relatively large numbers of phagosomes, thus making the synchronization of phagocytosis more challenging. However, both p47*phox* and p67*phox* associate with the phagosome early (~1 min) and remain bound for a relatively prolonged period (~30 min) before dissociating (DeLeo *et al.*, 1999), providing an ample window for the synchronization of their assembly during the experiment.

Phagosomes can be isolated on sucrose density gradients, using a modification of a previously described technique (Kaufman *et al.*, 1996). Alternatively, phagosomes can be separated from other intracellular compartments using free-flow electrophoresis, as recently applied to the study of uptake of mycobacteria by murine macrophages (Ferrari *et al.*, 1999).

Sucrose density gradient isolation of phagosomes

1. Gently tumble isolated cells (e.g. PMN) suspended in HBSS (10^7 cells ml^{-1}) with zymosan that had been opsonized with heat-inactivated serum (OPZ) (1 mg ml^{-1}; 5 particles to 1 PMN) with tumbling at 37°C for the desired time.
2. Pellet and wash cells with cold HBSS and suspended in 1 ml of cold homogenization buffer (250 mM sucrose, 3 mM imidazole, pH 7.4 containing 1 mM PMSF, 0.3 μM aprotinin, 2 μM leupeptin and 3 μM pepstatin A). Disrupt cells in a 2 ml Dounce homogenizer with a tight-fitting pestle at 4°C. Generally 20 to 30 passes are optimal but one should monitor cell disruption by phase microscopy, adjusting the number of passes to obtain maximal cellular disruption with minimal nuclear lysis.
3. Remove nuclei and unbroken cells by centrifugation (500*g*, 5 min) and adjust the recovered supernatant to 40% sucrose (w/w), using 62% sucrose in 3 mM imidazole, pH 7.4. Carefully layer the phagosome-containing fraction over a 1 ml 62% sucrose cushion and then successively add 2 ml of 35%, 25% and 10% sucrose solutions, respectively, to complete the step gradient.
4. Spin the gradient at 100 000 *g* for 1 h at 4°C (SW41 rotor, Beckman Instruments, Palo Alto, CA). OPZ-containing phagosomes, sedimented at the interface of the 25% and 35% sucrose, are collected using a capillary tube and peristaltic pump and suspended in 15 ml of cold HBSS. Phagosomes can be studied immediately or frozen for later analysis. For certain applications, phagosome suspensions can be concentrated using a Centriprep 30 (Amicon, Beverly, MA) (spinning twice at 3000 rpm for 20 min at 4°C) and/or pelleted at 50 000× *g* for 15 min at 4°C. Associated p47*phox* or p67*phox* can be determined by boiling phagosome pellets in SDS-sample buffer and then analyzing by SDS-PAGE and immunoblotting.

Monitoring phagocytosis and NADPH oxidase assembly by immunofluorescence microscopy

Biochemical methods used to isolate phagosomes are technically challenging and require large-scale experiments to detect phagosome-associated proteins by immunoblotting. Synchronization of phagocytosis in large-scale samples is also more difficult. Methods employing flow cytometry can accurately monitor phagocytosis, respiratory burst activity, bacterial integrity simultaneously (Hampton and Winterbourn, 1999; van Eeden *et al.*, 1999; Lehmann *et al.*, 2000) but do not permit assessment of enrichment of host or bacteria-derived proteins at or in phagosomes.

In contrast, immunofluorescence techniques can be used to monitor phagocytosis by individual cells under conditions in which phagocytosis can be synchronized precisely (Wright and Silverstein, 1984; Drevets and Campbell, 1991; Allen and Aderem, 1996). Use of low bacteria-to-PMN ratios allows better resolution of the kinetics and we have applied this approach to monitor NADPH oxidase assembly during phagocytosis (Allen *et al.*, 1999; DeLeo *et al.*, 1999). Phagocytosis and NADPH oxidase assembly can be examined using p47*phox* and/or p67*phox*-specific antibodies in combination with an antibody specific for the ingested bacteria or with bacteria endogenously expressing a fluorescent protein such as GFP. Antibodies to the flavocytochrome b_{558} subunits, gp91*phox* or p22*phox*, can be used to demonstrate co-localization of p47/67*phox* with flavocytochrome b_{558} at phagosomes and to assess degranulation. Enrichment of these proteins at phagosomes-ingested bacteria can be demonstrated by determining co-localization with a confocal laser-scanning microscope and appropriate software. By analyzing multiple confocal planes in a given cell and reconstructing a three-dimensional composite (z-series), it is possible to distinguish bound versus ingested bacteria.

Microscopic assessment of phagocytosis and NADPH oxidase assembly

1. Cells (e.g. PMNs) are resuspended in HEPES-buffered RPMI 1640 or DBPS/G to 10^7 cells ml^{-1} and kept on ice.
2. Opsonize zymosan or bacteria with 10–100% human serum for 30 min at 37°C. Prepare enough so that you will have 2–4 OPZ or ≤ 10 bacteria per PMN in each assay. Add OPZ to HEPES-buffered RPMI 1640 so there are $2–4 \times 10^6$ particles ml^{-1}. Alternatively, use 10^7 bacteria in 1 ml. Each time point will require 1 ml of cells and 1 ml of bacteria.
3. Coat glass microscope coverslips with human serum for PMN adherence. Glass coverslips should be pre-soaked in an acid solution overnight and then washed and stored in ethanol. Prior to use, flame each one twice after immersing in ethanol (70–95%). After flaming, add 2–3 per 35 mm petri dish. Coat each 12 mm round glass coverslip with 30 µl of 10–100% human serum for at least 1 h at 37°C. Wash by adding DPBS to the dish, swirling and then aspirating into a biohazardous waste container. Repeat wash and then add 1 ml PMNs (10^6 PMNs) and transfer dish to 37°C for at least 10–15 minutes to allow cells to adhere. It is best to use a metal tray for chilling and heating the dishes, as styrofoam will slow the heat transfer significantly.
4. After the incubation at 37°C, transfer the dishes/tray to a refrigerator at 4°C and incubate for 10–15 min to chill cells.

5. To synchronize phagocytosis, add 1 ml chilled bacteria to the dishes and then transfer to a centrifuge set at 10–16°C. Phagocytosis will occur at 16–18°C. The cooler the centrifuge, the longer the lag time before the onset of phagocytosis and PMN activation. The advantage of slowing down phagocytosis with the colder temperature (10°C) and styrofoam racks is that PMN proteins transiently recruited to phagosomes can be detected more readily. Centrifuge samples at 500 g for 2 min with no brake.
6. At the end of the spin, very carefully transfer the dishes to a 37°C incubator and start timing the assay.
7. Process one of the samples immediately and without incubating at 37°C (0 min incubation).
8. Aspirate media with unbound cells and then fill the dish with DPBS and swirl to wash away unbound opsonized zymosan or bacteria and cells.
9. Partially fill the dish with 10% formalin and fix cells at room temperature for 10–15 min.
10. Using fine tweezers, transfer each coverslip to a glass petri dish containing 100% acetone pre-chilled at −20°C and incubate for 5 min at −20°C to permeabilize cells.
11. During the acetone incubation, aspirate the formalin from the plastic 35 mm petri dish and then wash with DPBS and partially re-fill with DSBS. Return coverslips to the plastic petri dishes and swirl to remove any residual acetone.
12. Aspirate DPBS and add at least 2 ml blocking solution (10% normal goat serum, 0.5% BSA, and 0.05% sodium azide in PBS) per dish and incubate overnight at 4°C or 1 h at room temperature.
13. Wash each coverslip three times in wash buffer (0.5% BSA, and 0.05% sodium azide in PBS). Use fine tweezers to hold coverslips and each wash should entail immersing and swirling sequentially in three separate small beakers. Wick away excess liquid from the edge and dry the back with a tissue. Lay the coverslip on the lid of 24-well cell culture dish, using the circular ridges embossed on the lid to isolate each coverslip.
14. Add 30 µl of the appropriate primary antibody dilution (usually 1:200–500 for polyclonal antibodies) to each glass coverslip. Put the lid with the coverslips into a sealed container with dampened paper towels and incubate for 1 h at room temperature.
15. Wash each coverslip six times as described above being very careful to maintain top–bottom orientation. Remove residual primary antibody from the lid with a paper towel or tissue. Dry the back of the coverslip as above and replace onto the lid. Add 30 µl of the appropriate secondary antibody conjugated to a fluorochrome and incubate again for 1 h. Dilutions vary according to source of antibody, but 1:200 to 1:1000 is usually appropriate. Using labeled primary antibodies or rhodamine conjugated to phalloidin eliminates the need for a secondary antibody.
16. Wash the coverslips again six times in wash buffer and then once in water. Dry the back of each coverslip and apply cells facedown into

mounting medium (Either Permount (Fisher Scientific) or a solution prepared by mixing 2.4 g polyvinyl alcohol (Sigma), 6.0 g glycerol, 6.0 ml dH_2O and 12.0 ml 0.2 M Tris-Cl, pH 6.5 for 2 h. Heat for 10 min at 50°C and then centrifuge at 1000 g to remove unsolubilized material. Transfer supernatant to a new tube and add 0.625 g 1,4-diazabicyclo-[2.2.2]-octane (Sigma) and then aliquot and store at −20°C) on a glass microscope slide. The slide should contain only a small volume of mounting medium, approximately 75 μl in small drops for six coverslips, and six fit well on one standard microscope slide.

17. Incubate in the dark at 4°C overnight. View the following day on a fluorescent microscope or a confocal laser-scanning scope.

Expected results

The number of cell-associated bacteria can be identified directly, using microbe-specific antibodies, and those that are enriched for p40/47/67*phox*, Rac2 and/or flavocytochrome b_{558} will reflect the extent of NADPH oxidase assembly at the phagosome. With time and continued cell activation, more p40/47/67*phox* and Rac2 will translocate to phagosomes under normal conditions and phagosomes will also become enriched with flavocytochrome b_{558} as secretory vesicles and specific granules fuse with maturing phagosomes. For inert particles such as OPZ, ~95% of phagosomes are enriched for p47*phox* or p67*phox* during peak assembly. The percentage of phagosomes with associated cytosolic components of the oxidase and the time to reach peak assembly vary significantly among various bacterial species. For *Neisseria meningitidis* we found that ~60–70% of phagosomes were p47*phox* or p67*phox*-enriched.

◆◆◆◆◆◆ DETECTION OF MYELOPEROXIDASE-MEDIATED MODIFICATIONS

Fusion of the mature phagosome with azurophilic granules delivers high concentrations of MPO to the ingested microbe, estimated to be ~1–2 mM during the uptake of 15–20 organisms (Hampton *et al.*, 1998). MPO is unique among members of the family of animal peroxidases in its capacity to catalyze the two-electron oxidation of Cl^- to Cl^+ and thereby generate HOCl and chlorinate a wide range of protein and lipid substrates (Hazen *et al.*, 1999; Kettle, 1999). The participation of the MPO-H_2O_2-halide system can be assessed by quantitating its chlorinating activity or by detecting MPO-mediated modifications of proteins or lipids.

Several assays can be used to quantitate the chlorinating activity of MPO. The optimal method for on-line measurement and kinetic analysis is use of the H_2O_2 electrode, directly measuring H_2O_2 consumption due to formation of HOCl. More commonly used assays, such as those using monochlorodimedon, ascorbate, or trimethylbenzidine as substrates, have serious shortcomings, detailed in Kettle and Winterbourn (1994).

We use the taurine chloramine assay to measure PMN mediated chlorination activity. The principle of this asssay derives from the ability of HOCl to react with taurine to produce taurine monochloride, which is measured by its reaction with 5-thio-2-nitrobenzoic acid (TNB). TNB is a yellow compound (ε_{412nm} 14 100 $M^{-1}cm^{-1}$) that is oxidized to a colorless compound, 5,5′-dithiobis(2-nitrobenzoic acid) (DTNB), by reaction with taurine monochloride.

Taurine chloramine assay for chlorination

1. Prepare the substrate, TNB, as a 2 mM solution by dissolving 1 mM DTNB in 50 mM phosphate buffer (pH 7.4) that is titrated first to pH 12 with NaOH for 5 min, to promote hydrolysis, and subsequently adjusted to pH 7.4 with HCl. Because TNB is sensitive to light and is oxidized in air, the stock solution should be prepared fresh weekly and stored at 4°C under N_2 in a brown bottle containing 1 mM EDTA.
2. Incubate isolated PMNs suspended in phosphate-buffered saline supplemented with calcium and magnesium (10^6 PMNs in 500 μl) with 12 mM taurine at 37°C for 10 min.
3. No stimulus or OPZ is added and the cell suspension is gently tumbled at 37°C for the period of interest, typically 30 min.
4. Terminate reactions by adding 20 μg/ml^{-1} catalase and plunging the tubes into an ice bath for 10 min.
5. Collect supernatants by centrifugation (5 min at 14 400*g* at 4°C) and add 400 μl to 600 μl of PBS containing TNB. Because normal PMNs generate ~ 50 nmoles of HOCl per 10^6 cells in 30 min and TNB should be in excess in the final reaction buffer, we adjust the addition of TNB stock solution so that its final concentration in the reaction buffer is 100 μm. The supernatant and reaction buffer are mixed on a vortex mixer and allowed to sit in the dark for 5 min for the reaction to be completed.
6. Measure the absorbance of the solution at 412 nm, using for reference supernatants generated under the same conditions but with 2 mm methionine added to the cell suspension. Because methionine will scavenge HOCl, this provides a more appropriate negative control than does using the reaction buffer alone in the reference cuvette.
7. The amount of HOCl produced is determined by first measuring the difference in A_{412} in the presence and absence of methionine. This value is then multiplied by the dilution factor of supernatant in the reaction buffer (e.g. 2.5 in this experiment), and divided by 28 200 (the molar extinction coefficient for TNB multiplied by 2, the number of TNB molecules reacting with each taurine chloramine).

In addition to quantitation of the chlorinating activity of stimulated PMNs, several sensitive assays can be employed to detect evidence of MPO-dependent modification of biological molecules, including proteins and lipids. The relative merits and shortcomings of these various analytical techniques and their relevance to biological processes are thoroughly discussed elsewhere (Hazen *et al.*, 1999; Kettle, 1999;

Winterbourn and Kettle, 2000). However, what merits development is the application of these approaches to the identification and quantitation of MPO-dependent modification of bacterial biomolecules within the phagolysosome. Identification of MPO-specific alterations in bacterial components would not only illuminate the targets critical for effective cytostatic, bactericidal, and degradative events but also identify the contribution of other granule components that synergize with MPO to create an environment inhospitable to the ingested micro-organism.

References

Allen, R. C. (1986). Phagocytic leukocyte oxygenation activities and chemiluminescence: a kinetic approach to analysis. *Methods Enzymol.* **133**, 449–493.

Allen, L.-A. and Aderem, H. A. (1996). Molecular definition of distinct cytoskeletal structures involved in complement- and Fc receptor-mediated phagocytosis in macrophages. *J. Exp. Med.* **184**, 627–637.

Allen, L.-A.H., DeLeo, F. R., Gallois, A., Toyoshima, S., Suzuki, K. and Nauseef, W. M. (1999). Transient association of the nicotinamide adenine dinucleotide phosphate oxidase subunits p47*phox* and p67*phox* with phagosomes in neutrophils from patients with X-linked chronic granulomatous disease. *Blood* **93**, 3521–3530.

Babior, B. M. (1999). NADPH oxidase: an update. *Blood* **93**, 1464–1476.

Babior, B. M., Kipnes, R. S. and Curnutte, J. T. (1973). Biological defense mechanisms: The production by leukocytes of superoxide, a potential bactericidal agent. *J. Clin. Invest.* **52**, 741–744.

Belaaouaj, A., Kim, K. S. and Shapiro, S. D. (2000). Degradation of outer membrane protein A in Escherichia coli killing by neutrophil elastase. *Science* **289**, 1185–1188.

Bligh, E. G. and Dyer, W. J. (1959). A rapid method of total lipid extraction and purification. *Can. J. Biochem. Physiol.* **37**, 911–917.

Cheson, B. D., Christensen, R. L., Sperling, R., Kohler, B. E. and Babior, B. M. (1976). The origin of the chemiluminescence of phagocytosing granulocytes. *J. Clin. Invest.* **58**, 789–796.

Cronan, J. E., Jr. and Rock, C. O. (1996). Biosynthesis of membrane lipids. In (eds J. L. Ingraham, K. B. Low, B. Magasanik, M. Schaechter and H. E. Umbarger) *Escherichia coli and Salmonella typhimurium: Cellular and Molecular Biology*. vol. 1., pp. 612–636. American Society for Microbiology, Washington DC, 2nd edition.

DeLeo, F. R., Allen, L.-A.H., Apicella, M. A. and Nauseef, W. M. (1999). NADPH oxidase activation and assembly during phagocytosis. *J. Immunol.* **163**, 6732–6740.

Drevets, D. A. and Campbell, P. A. (1991). Macrophage phagocytosis: use of fluorescence microscopy to distinguish between extracellular and intracellular bacteria. *J. Immunol. Meth.* **142**, 31–38.

Elsbach, P. and Weiss, J. (1988). Phagocytosis of bacteria and phospholipid degradation. *Biochim. Biophys. Acta (Reviews on Biomembranes).* **947**, 29–52.

Elsbach, P. and Weiss, J. (1991). Utilization of labeled Escherichia coli as phospholipid substrate. *Meth. Enzymol.* **197**, 24–31.

Elsbach, P., Weiss, J. and Levy, O. (1994). Integration of antimicrobial host defenses: Role of the bactericidal/permeability-increasing protein. *Trends Microbiol.* **2**, 324–328.

Elsbach, P., Weiss, J. and Levy, O. (1999). Oxygen-independent antimicrobial systems of phagocytes. *Inflammation. Basic Principles and Clinical Correlates.* third edition (eds Gallin, J. I., Snyderman, R. and Nathan, C.) Lippincott-Raven, pp. 801–817.

Emmendorffer, A., Hecht, M., Lohmann-Matthes, M. L. and Roesler, J. (1990). A fast and easy method to determine the production of reactive oxygen intermediates by human and murine phagocytes using dihydrorhodamine. 123 *J. Immunol. Meth.* **131**, 269–275.

Ferrari, G., Langen, H., Naito, M. and Pieters, J. (1999). A coat protein on phagosomes involved in the intracellular survival of mycobacteria. *Cell* **97**, 435–447.

Fischer, W. (1994). Lipoteichoic acid and lipids in the membrane of Staphylococcus aureus. *Methods Microbiol. Immunol.* **183**, 61–76.

Giardina, P. C., Gioaninni, T., Buscher, B. A., Zaleski, A., Zhang, D. S., Stoll, L., Teghanemt, A., Apicella, M. A. and Weiss, J. (2001). Construction of acetate auxotrophs of Neisseria meningitidis to study host-meningococcal endotoxin interactions. *J. Biol. Chem.* **276**, 5883–5891.

Hampton, M. B. and Winterbourn, C. C. (1999). Methods for quantifying phagocytosis and bacterial killing by human neutrophils *J. Immunol. Meth.* **232**, 15–22.

Hampton, M. B., Kettle, A. J. and Winterbourn, C. C. (1998). Inside the neutrophil phagosome: oxidants, myeloperoxidase, and bacterial killing. *Blood* **92**, 3007–3017.

Hazen, S. L., Hsu, F. F., Gaut, J. P., Crowley, J. R. and Heinecke, J. W. (1999). Modification of proteins and lipids by myeloperoxidase. *Methods Enzymol.* **300**, 88–105.

Heumann, D., Barras, C., Severin, A., Glauser, M. P. and Tomasz, A. (1994). Gram-positive cell walls stimulate synthesis of tumor necrosis factor alpha and interleukin-6 by human monocytes. *Infect. Immun.* **62**, 2715–2721.

Inanami, O., Johnson, J. L., McAdara, J. K., El Benna, J., Faust, L. R. P., Newburger, P. E. and Babior, B. M. (1998). Activation of the leukocyte NADPH oxidase by phorbol ester requires the phosphorylation of $p47^{PHOX}$ on serine 303 or 304. *J. Biol. Chem.* **273**, 9539–9543.

Katz, S. S., Weinrauch, Y., Munford, R. S., Elsbach , P. and Weiss, J. (1999). Deacylation of lipopolysaccharide in whole Escherichia coli during destruction by cellular and extracellular components of a rabbit peritoneal inflammatory exudate. *J. Biol. Chem.* **274**, 36579–36584.

Kaufman, M., Leto, T. and Levy, R. (1996). Translocation of annexin I to plasma membranes and phagosomes in human neutrophils upon stimulation with opsonized zymosan: Possible role in phagosome function. *Biochem. J.* **316**, 35–42.

Kettle, A. J. (1999). Detection of 3-chlorotyrosine in proteins exposed to neutrophil oxidants. *Methods Enzymol.* **300**, 111–120.

Kettle, A. J. and Winterbourn, C. C. (1994). Assays for the chlorination activity of myeloperoxidase. *Methods Enzymol.* **233**, 502–512.

Klebanoff, S. J. (1999). Oxygen metabolites from phagocytes. In *Inflammation: Basic Principles and Clinical Correlates.* (eds J. I. Gallin and R. Snyderman) pp. 721–768. Lippincott Williams & Wilkens Philadelphia.

Lehmann, A. K., Sornes, S. and Halstensen, A. (2000). Phagocytosis: measurement by flow cytometry. *J. Immunol. Meth.* **243**, 229–242.

Lehrer, R. I. and Ganz, T. (1999). Antimicrobial peptides in mammalian and insect host defence. *Curr. Opin. Immunol.* **11**, 23–27.

Madsen, L. M., Inada, M. and Weiss, J. (1996). Determinants of activation by complement of Group II phospholipase A2 acting against *Escherichia coli. Infect. Immun.* **64**, 2425–2430.

Mannion, B. A., Weiss, J. and Elsbach, P. (1990). Separation of sublethal and lethal effects of polymorphonuclear leukocytes on *Escherichia coli. J. Clin. Invest.* **86**, 631–641.

Munford, R. S. and Hall, C. L. (1986). Detoxification of bacterial lipopolysaccharides (endotoxins) by a human neutrophil enzyme. *Science* **234**, 203–205.

Munford, R. S. (1991). How do animal phagocytes process bacterial lipopolysaccharides? *APMIS* **99**, 487–491.

Munford, R. S. and Erwin, A. L. (1992). Eukaryotic lipopolysaccharide deacylating enzyme. *Meth. Enzymol.* **209**, 485–492.

Munford, R. S., DeVeaux, L. C., Cronan, Jr. J. E. and Rick, P. D. (1992). Biosynthetic radiolabeling of bacterial lipopolysaccharide to high specific activity. *J. Immunol. Meth.* **148**, 115–120.

Nauseef, W. M. (1999). The NADPH-dependent oxidase of phagocytes. *Proc. Assn. Am. Phys.* **111**, 373–382.

Nikaido, H. (1996). Outer membrane. In J. L., Ingraham, K. B., Low, B., Magasanik, M., Schaechter, H. E., Umbarger (eds) *Escherichia coli and Salmonella typhimurium: Cellular and Molecular Biology.* vol. 1, American Society for Microbiology, Washington DC, 2nd edition, pp. 29–47.

Patriarca, P., Beckerdite, S., Pettis, P. and Elsbach, P. (1972). Phospholipid metabolism of phagocytic cells. VII. The degradation and utilizaion of phospholipids of various microbial species by rabbit granulocytes. *Biochim. Biophys. Acta.* **280**, 45–56.

Rietschel, E. T., Brade, H., Holst, O., Brade, L., Muller-Loennies, M., Amet, U., Zahringer, U., Beckmann, F., Seydel, U., Brandenburg, K., Ulmer, A. J., Mattern, T., Heine, H., Schletter, J., Loppnow, H., Schonbeck, U., Flad, H.-D., Haushildt, S., Schade, U. F., Di Padova, F., Kusumoto, S. and Schumann, R. R. (1996). Bacterial endotoxin: chemical constitution, biological recognition, host response, and immunological detoxification. *Curr. Top. Microbiol. Immunol.* **216**, 39–80.

Root, R. K., Metcalf, J. A., Oshino, N. and Chance, B. (1975). H2O2 release from human granulocytes during phagocytosis. *J. Clin. Invest.* **55**, 945–955.

Rothe, G. and Valet, G. (1990). Flow cytometric analysis of respiratory burst activity in phagocytes with hydroethidine and 2′,7′-dichlorofluorescin. *J. Leukocyte Biol.* **47**, 440–448.

Ruch, W., Cooper, P. H. and Baggiolini, M. (1983). Assay of H_2O_2 production by macrophages and neutrophils with homovanillic acid and horse-radish peroxidase *J. Immunol. Methods* **63**, 347–357.

Schwab, J. H. (1993). Phlogistic properties of peptidoglycan-polysaccharide polymers from cell walls of pathogenic and normal flora bacteria which colonize humans. *Infect. Immun.* **61**, 4535–4539.

Shiose, A. and Sumimoto, H. (2000). Arachidonic acid and phosphorylation synergistically induce a conformational change of $p47^{phox}$ to activate the phagocyte NADPH oxidase. *J. Biol. Chem.* **275**, 13793–13801.

Taylor, P. W. (1983). Bactericidal and bacteriolytic activity of serum against Gram-negative bacteria. *Microbiol. Rev.* **47**, 46–83.

Van Eeden, S. F., Klut, M. E., Walker, B. A. M. and Hogg, J. C. (1999). The use of flow cytometry to measure neutrophil function. *J. Immunol. Methods* **232**, 23–43.

Van Pelt, L. J., van Zwieten, R., Weening, R. S., Roos, D., Verhoeven, A. J. and Bolscher, B. G. J. M. (1996). Limitations on the use of dihydrorhodamine 123 for flow cytometric analysis of the neutrophil respiratory burst. *J. Immunol. Methods* **191**, 187–196.

Vowells, S. J., Sekhsaria, S., Malech, H. L., Shalit, M. and Fleisher, T. A. (1995). Flow cytometric analysis of the granulocyte respiratory burst: a comparison study of fluorescent probes. *J. Immunol. Methods* **178**, 89–97.

Weinrauch, Y., Elsbach, P., Madsen, L. M., Foreman, A. and Weiss, J. (1996). The potent anti-staphylococcus aureus activity of a sterile rabbit inflammatory fluid is due to a 14-kD phospholipase A2. *J. Clin. Invest.* **97**, 250–257.

Weinrauch, Y., Katz, S. S., Munford, R. S., Elsbach, P. and Weiss, J. (1999). Deacylation of purified lipopolysaccharides by cellular and extracellular components of a sterile rabbit peritoneal inflammatory exudate. *Infect. Immun.* **67**, 3376–3382.

Weiss, J., Inada, M., Elsbach, P. and Crowl, R. M. (1994). Structural determinants of the action against *Escherichia coli* of a human inflammatory fluid phospholipase A_2 in concert with polymorphonuclear leukocytes. *J. Biol. Chem.* **269**, 26331–26337.

Winterbourn, C. C. and Kettle, A. J. (2000). Biomarkers of myeloperoxidase-derived hypochlorous acid. *Free Radic. Biol. Med.* **29**, 403–409.

Wright, G., Weiss, J., Kim, K. S. and Elsbach, P. (1990). Bacterial phospholipid hydrolysis enhances the destruction of *Escherichia coli* ingest by rabbit granulocytes. Role of cellular and extracellular phospholipases. *J. Clin. Invest.* **85**, 1925–1935.

Wright, S. D. and Silverstein, S. C. (1984). Phagocytosing macrophages exclude proteins from the zones of contact with opsonized targets. *Nature* **309**, 359–361.

Zanetti, M., Gennaro, R., Scocchi, M. and Skerlavaj, B. (2000). Structure and biology of cathelicidins. *Adv. Exp. Med. Biol.* **479**, 203–218.

27 Methods for Studying Infection and Immunity in *Drosophila*

Phoebe Tzou[1], Marie Meister[2] and Bruno Lemaitre[1]

[1]*Centre de Génétique Moléculaire du CNRS, 91198 Gif-sur-Yvette, France;*
[2]*UPR9022 du CNRS, Institut de Biologie Moléculaire et Cellulaire, 15 rue Descartes; 67084 Strasbourg Cedex, France*

◆◆

CONTENTS

List of abbreviations

AMP: Antimicrobial peptide
Ecc15: *Erwinia carotovora carotovora* 15

◆◆◆◆◆◆ INTRODUCTION

Research on pathogens such as *Listeria, Yersinia, Salmonella, Shigella, Escherichia*, which are particularly suitable to genetic manipulation, have increased our understanding of the molecular interactions between bacterial factors and host cellular components (Finlay and Cossart, 1997). Despite present advances, however, the overall spectrum of interactions between infectious microbes and their hosts remain poorly understood. To date, the model host organism systems used to analyze host/bacteria interaction (cell culture or mouse) have not allowed for a systematic identification by genetic screening of the host factors involved in the infection process and corresponding host immune responses.

METHODS IN MICROBIOLOGY, VOLUME 31
ISBN 0-12-521531-2

Nevertheless, many of the mechanisms underlying host innate immune responses, as well as invasive strategies used by pathogenic microbes, appear to be conserved across phylogeny, pointing to their ancient origin (Hoffmann *et al.*, 1999; Tan and Ausubel, 2000). These results highlight the potential of non-mammalian model organisms that are amenable to genetic analysis for studying host–pathogen interactions.

During the last century *Drosophila melanogaster*, the fruit fly, has been a widely-used model organism for genetic studies. A facile genetic system, reliable husbandry techniques, and fully sequenced genome all contribute to the usefulness of this organism. In addition, *Drosophila*, like other insects, shows efficient constitutive and inducible host defense responses that display striking parallels with mammalian innate immune responses (phagocytosis by macrophage-like hemocytes, antimicrobial peptides, proteolytic cascades). Consequently, *Drosophila* is especially suitable for the analysis of the interplay between microbes and the innate immune defense. Flies, like mammals, possess respiratory and digestive tract tissues that are also the target for invading pathogens. Although the physiology of these organs are significantly different from their mammalian counterparts they share some basic properties as barriers to microbial infection. Finally, our good understanding of the evolution as well as the ecology of *Drosophila* in relation to natural pathogens in the wild can be relevant for host–pathogen analysis. In keeping with this idea, it should be pointed out that flies function as vectors in the spread of many human and plant pathogens.

In this review, we describe the basic techniques currently used to both infect *Drosophila* and to monitor corresponding immune responses.

◆◆◆◆◆◆ MICROBIAL INFECTION OF *DROSOPHILA* LARVAE AND ADULTS

In our laboratory, we use two methods for infecting *Drosophila*: (1) Microbial injection—a direct introduction of bacteria and fungi into the body cavity of the fly; and (2) Natural infection without injury.

Introduction of microbes into larvae and adults

Microbial injection is used to assay *Drosophila*'s immune responses and survival to different bacteria and fungal species. Basically, microbes are introduced inside the body cavity using either a needle (septic injury) or a micro-injector. In addition to the introduction of microbes inside the body cavity, this stimulus results in an injury and the activation of melanization reactions at the wound site. Clean injury experiments differ from a microbial injection only by the fact that the needle used to puncture the fly has not been previously dipped in a bacterial or fungal solution.

Septic injury

Third instar larvae

Wandering third instar larvae are washed in water and placed in a small drop of water on a black rubber block (Figure 27.1A). From an overnight bacterial culture, a bacterial 'pellet' is generated with an OD close to 200, a thin needle (entomology needle used for dissection) is dipped into it, and larvae are punctured on their posterior lateral side (Figure 27.1B). Injected larvae are transferred to a filter paper moistened with water (or 0.2% glucose solution) in a vial containing fly medium. The site of the puncture heals and remains visible as a pronounced dark spot (due to the melanization reaction). Nearly all third instar larvae that are injured with a needle will die a few hours after injury (before or during the pupal stage). To obtain a better survival rate, we challenge larvae with a more appropriate needle, such as a 0.2 mm diameter tungsten wire which has been sharpened in a 0.1 M NaOH solution by electrolysis. To avoid damaging the needle, a drop (50 to 100 µl) of bacterial solution can be placed directly onto the same rubber block recipient where the larvae will be pricked.

Under these conditions, nearly 50% of injured larvae give rise to adults. Overall, larval injury induces a rather high lethality rate, which is possibly due to the strong internal pressure, or to a critical developmental stage just prior to metamorphosis.

Adult flies

To inject adult flies, a thin metal needle is used, mounted on a small handle (Figure 27.1C). The bacterial and fungal pellet (see above) is deposited inside a cut-off microfuge tube lid. We dip the thin needle into the bacterial pellet and puncture the dorsal or lateral side of the thorax of a CO_2-anesthetized fly (Figure 27.1D), then gently separate the fly from the needle with a paint brush. The use of cold light is recommended to avoid dessication of the challenged animals. They are then transferred to a clean vial of standard corn-meal fly medium where they normally recover within 5 to 20 minutes. Less than 5% of them immediately die when injected with a non-pathogenic strain probably due to gross inner lesions. Flies that recover exhibit normal viability. The site of injury heals and remains visible as a small dark spot. This method is straightforward and allows large samples to be tested in a short period of time, i.e. 300 flies or more can be infected per hour.

Injection

Septic injury with a small needle is probably the simplest way to infect *Drosophila*. This technique, however, does not allow for accurate quantification of injected microbes or microbial components. When necessary, we use the Drummond's Nanoject (automatic) injector to deliver a defined volume of microbial solution (Figure 27.1E). Glass capillary tips can be pulled under high heat, backfilled with mineral oil and then mounted onto

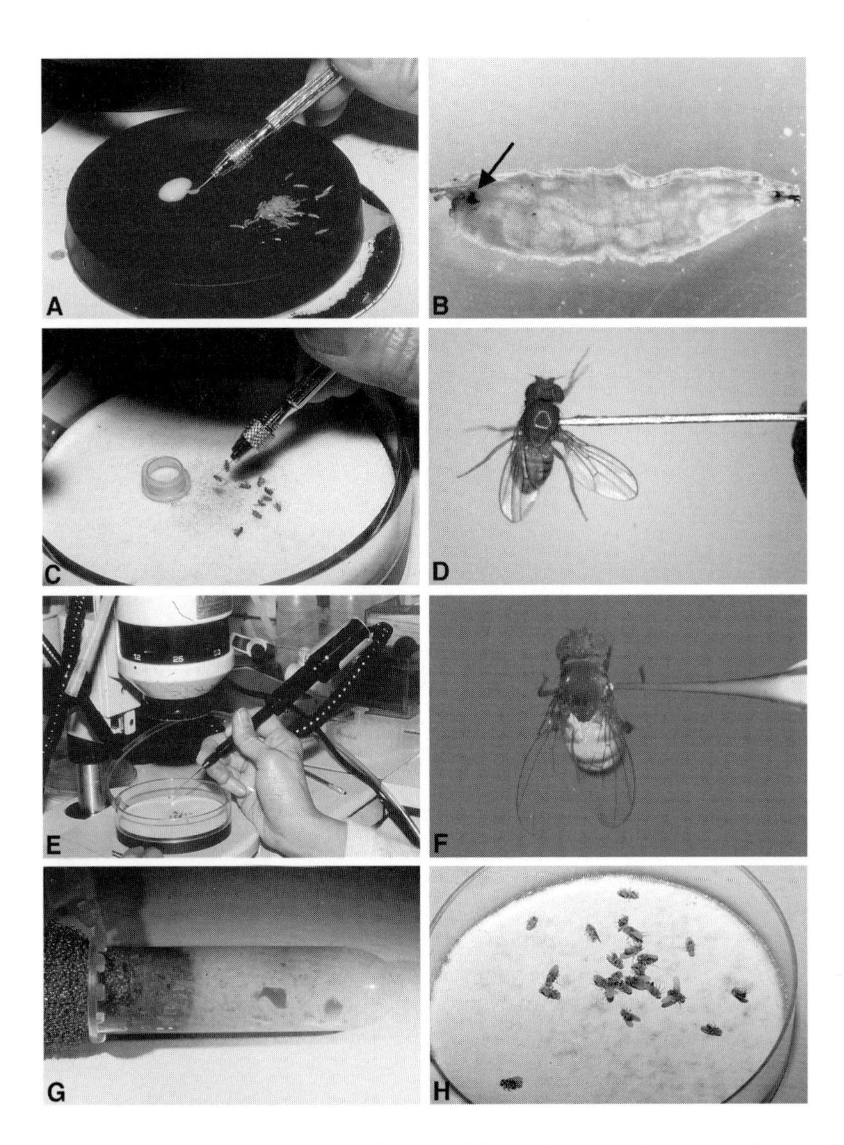

Figure 27.1. Microbial infection of *Drosophila*. (A) Septic injury of larvae is performed with a tungsten needle. A rubber block is employed to protect the needle from damage. A drop of microbial solution is placed on the rubber block, and larvae are placed inside a drop of water. (B) Larva are punctured on their posterior lateral side, triggering a melanization reaction at the injury site (arrow). (C) The needle is mounted on a handle to prick adult flies. The CO_2 pad (Inject + Matic^{TM}, Geneva) provides a convenient way to anesthetize flies. The microbial solutions are concentrated and placed into the cap of a microfuge tube. (D) An adult fly is pricked on the dorsal side of the thorax. (E) Drummond Nanoject injector: a fine capillary tip is backfilled with mineral oil before mounting onto the injector handle, and dipping into microbial solution to load. After specifying the quantity, at each pulse, the injector will release the exact amount (varies from 4 to 73 nl) into the body cavity of the fly. (F) An adult fly being injected with 9.2 nl of GFP expressing bacteria (OD = 100). (G) Natural infection of larvae is performed by mixing the following in a centrifuge vial: crushed banana, bacterial pellet, and third instar larva. After 30 min of incubation at room temperature, the larvae and the bacterial mixture are directly transferred to a standard fly vial. (**H**) Natural fungal infection is done by covering the flies thoroughly with fungal spores. Flies are anesthetized and shaken on a Petri dish containing a sporulating fungal species. (This figure is also reproduced in colour between pages 276 and 277.)

the Nanoject device calibrated for specified injection amount. Injections of 4 to 73 nl into an adult fly are possible (Figure 27.1F). This apparatus can be used for the injection of micro-organisms, chemicals, and purified bacterial compounds, where highly accurate conditions are required. These compounds may be diluted in Ringer solution (Ashburner, 1989). Injection, however, tends to provoke more trauma than septic injury due to the larger needle and the introduction of high amounts of solution.

Natural infection of larvae and adults

Bacterial natural infection

Larvae

Natural infections are performed with bacterial strains such as *Erwinia carotovora carotovora 15 (Ecc15)* that can induce a strong immune response in *Drosophila* in the absence of physical injury (Basset *et al.*, 2000). Approximately 200 third instar larvae are placed into a 2 ml microfuge tube containing a mixture of 200 µl of an overnight bacterial culture concentrated to OD = 200 and 400 µl of crushed banana. For multiple samples, banana may be crushed in a hand-held spice grinder (Krupps). The larvae, bacteria and banana are thoroughly mixed by strongly shaking the capped microfuge tube; afterwards a piece of foam is inserted into the reopened tube to prevent larvae from wandering away from the bacterial mixture (Figure 27.1G). The infection process takes place at room temperature for 30 min. The mixture is then transferred to a standard corn-meal fly medium and incubated at 29°C. When *Ecc15* is used in this infection procedure, 80% of the treated larvae express genes encoding antimicrobial peptides in fat body cells; lower concentrations of *Ecc15* reduce the percentage of larvae that express these genes.

Adults

Adults are dehydrated for a few hours in a dry vial in the absence of food then transferred into a vial containing filter paper hydrated with 5% sucrose/concentrated bacterial solution, and incubated at 25°C. Each day, the paper is rehydrated with the same solution. To date, we have not identified a bacterial species that naturally infects adults and stimulates a systemic immune response. Infection of *Drosophila* adults by *Ecc15* induces the expression of antimicrobial peptide genes in several epithelial tissues (Tzou *et al.*, 2000). The feeding of *Drosophila* with *Serratia marcescens Db11* causes high lethality, likely due to toxin secretion (Flyg *et al.*, 1980). This bacterial strain, however, does not reproducibly induce the expression of antibacterial peptides.

Natural infection by entomopathogenic fungi

The genes encoding the antifungal peptides *drosomycin* and *metchnikowin* can be selectively induced after massively covering the adults with spores of the entomopathogenic fungi *Beauvaria bassiana* or *Metharizium anosiplae*. The level of antifungal peptide gene expression increases over several days and is similar to the level obtained after microbial injection challenges (Lemaitre *et al.*, 1997). These two fungi are pathogenic for many insect species and have the ability to cross the cuticle of insects through the secretion of proteases and lipases (Clarkson and Charnley, 1996). Natural infection or injection of these fungi causes a significant mortality in wild-type adult flies.

Production of fungal spores

The seeding of fungal spores should be carried out under sterile conditions. *Beauvaria bassiana* spores are spread onto 5.5 cm *Petri* dishes (or 9 cm and larger for collecting spores) of malt-agar (1 g peptone, 20 g glucose, 20 g malt extract, 15 g agar, in 1 l water, autoclaved). The use of glass beads facilitates homogeneous plating. Incubate the spores at 25 to 29°C. The fungal hyphae will germinate after 3 to 5 days at 25°C. After 10 to 30 days, check for the presence of dust-like spores. Well-sporulated plates are stored at 4°C for infection experiments.

To collect spores, wash the 9 cm dishes with 10 ml of sterile water. Separate spores from hyphal bodies through a funnel lined with glassfibre. Spores are collected into a 50 ml vial and centrifuged at 5000 rpm for 15 min at 4°C. Discard supernatant. Quantify the number of spores per ml using a hemacytometer (generally around 10^9 to 10^{12} sp ml^{-1}) under high magnification. Aliquots of fungal spores in 20% glycerol can be stored for several months at 4°C or several years at −20°C.

Infection of flies

One way to naturally infect adults with fungi is to use a pencil or dropper to place a droplet of concentrated liquid solution of spores onto CO_2-anesthetized adults. This is tedious and requires a high quantity of spores, but it is useful for some fungal species that do not grow easily on Petri dishes. Alternatively, the most efficient and natural way to infect flies with fungi is to place the CO_2-anesthesized flies on a 5.5 cm or 9 cm dish with a well-sporulated carpet of fungi. Hand-shake the Petri dish until the flies are totally covered with spores (Figure 27.1H). Infected flies are transferred into clean vials of normal medium and are incubated at 29°C. Vials should be changed every 2 days.

Larvae can well be rolled on sporulated plates. This treatment induces the formation of melanotic tumors and the induction of the *drosomycin* gene (Braun *et al.*, 1998).

Parameters that influence infection

Several parameters can influence the infection process.

- *The infection procedure*. As mentioned earlier, natural infection provides a cleaner picture of the infection process, without the interference of host reaction induced by wounding. In the case of fungal infection, for instance, challenge via septic injury with *B. bassiana* triggers an immune response with characteristics and kinetics different from that elicited by a natural infection with the same fungus, suggesting that a different set of recognition signals is switched on by different infection methods. In the case of injection, the size of the needle and the site of injection may influence the infection process. At the larval stage, differences in needle thickness can mean life or death. Adults, however, can better cope with thick needles.

- *The nature of the microbes and their concentration.* The use of various types of bacteria is recommended to compare the pattern of antimicrobial peptide gene expression, since the latter differs according to the microorganisms used (Lemaitre *et al.*, 1997). In our laboratory, we currently use a mixture of Gram-negative (*E. coli*) and Gram-positive (*Microccocus luteus*) bacteria to induce a high level of all AMP gene expression.
- *The temperature of infection.* Flies live well between 16 and 32°C. Temperature can influence both the growth of the microbes and the physiology of the insect. Many fungal species and some *Bacillus*, as well as *Erwinia* species, favor growth at 30°C.
- *The rearing conditions.* Crowded conditions may induce more trauma in flies. Reduced food amount and contaminated medium can also lead to immune-compromised larvae or adult flies. Thus, all lines to be tested should be taken from healthy stocks to minimize pre-existing disadvantages.

Unfortunately, there is no rigorous analysis of these parameters and their influence on the experimental outcome. Differences in infection procedures can lead to divergent conclusions—and may explain a number of contradictory interpretations of results obtained in different laboratories. However, the possibility to infect a high number of flies over a short time course allows the comparison of many samples in one experiment.

Axenic larvae and adult

For different purposes, it can be informative to rear flies under axenic conditions, namely to assess the role of an infection, in the absence of other living microbial contamination. None the less, even if axenic conditions can eliminate other living microorganisms besides the one introduced experimentally, it does not limit the effect of dead fungi as well as tissue contamination that can elicit an immune response. Rearing axenic flies begins by collecting embryos in a small basket (without contamination by larvae), sterilizing them by dechorionation in 50% chlorox bleach for 5 min, rapid washes in sterile water, then in 60% ethanol. They are eventually transferred in a drop of ethanol solution into recently autoclaved glass vials that contain standard fly medium. Note that the development of axenically raised *Drosophila* is severely delayed and asynchronous.

♦♦♦♦♦♦ MEASURING SURVIVAL TO MICROBIAL INFECTION

It is clear that *Drosophila* can be used as a model to study the pathogenicity of microbes, including the mechanisms by which they kill and the mechanisms by which they escape the host immune responses. Alternatively, weakly pathogenic strains can be useful in analyzing the host defense in immuno-deficient mutants. Microbial infection and stress also induce various physiological and behavioral modifications in a fly (e.g. delay or acceleration of metamorphosis, change of behavior) and this is a field that has not yet been investigated.

Survival results are highly dependent on the parameters described above. Previous studies have identified several classes of bacteria that exhibited different interactions with *Drosophila*. Highly pathogenic bacteria kill flies after injection of low doses (such as *Pseudomonas aeruginosa, Serratia marcescens, Staphylococcus aureus*) weakly pathogenic bacteria induce low lethality as in the case of *Erwinia carotovora*; and non-pathogenic strains *such as E. coli, Salmonella typhimurium, Bacillus megaterium* produce little lethality (Boman *et al.*, 1972; Flyg *et al.*, 1980; PT and BL unpublished data).

In our laboratory, survival experiments are carried out in the same conditions for each fly sample tested. Groups of 20 or more 2–4-day-old adults are infected in the same conditions (methods, needle, experimenter, time) and transferred to a fresh vial every 2 to 3 days to ensure fresh growth conditions. Flies that die within 3 h after infection are excluded from the analysis (less than 5% as a norm, see above). Previous observations showed that survival rates may depend on the genetic background. For example, we noted that some homozygous *ebony* (*e*) fly stocks exhibit a low viability after infection as reported by Flyg *et al.* (1980). In order to examine the survival due exclusively to the mutation under analysis, we chose mutated chromosomes carrying a minimal number of markers. The fly strain must also exhibit a good viability in the absence of immune challenge.

In the case of infected larvae, a common practice is to transfer them onto a Petri dish containing apple juice agar or fly medium (Ashburner, 1989) to facilitate the sorting of dead vs live animals. A basic survival count includes the number of pupae and adults that emerge.

♦♦♦♦♦♦ OBSERVING AND COUNTING BACTERIA INSIDE *DROSOPHILA*

We have little information about the fate of bacteria in *Drosophila* larvae and adults. The use of bacteria carrying genomic mutation that confers antibiotic resistance (e.g. rifampicine) is a way to monitor the amount of bacteria introduced into the host. Larvae and adults are collected and washed in water, briefly sterilized in 70% ethanol to eliminate bacteria sticking to the external cuticle, transferred to microfuge tubes containing

LB medium and homogenized. Dilutions are then plated on LB plates with antibiotic selection for bacterial count.

Due to the transparent cuticle of larva, as well as to the ease of dissecting both larvae and adults, GFP reporter gene can be used to monitor bacteria within the host. With a GFP-expressing strain of *Ecc15*, we can observe the localization of live bacteria in larvae of different mutant backgrounds under epifluorescent illumination at high magnification (Basset *et al.*, 2000). We can also study the exact distribution of the bacteria in dissected tissues. Finally, microbes can be observed using classical histology techniques.

◆◆◆◆◆◆ MEASURING *DROSOPHILA* ANTIMICROBIAL PEPTIDE GENE EXPRESSION

In response to infection, *Drosophila* produces a battery of peptides that exhibit distinct activity spectra. This response is regulated at the transcriptional level and AMP gene expression is regulated by distinct signaling cascades that are evolutionary conserved (TOLL, IMD) (reviewed in Khush and Lemaitre, 2000).

The patterns of expression can be classified in three groups (Ferrandon *et al.*, 1998; Tzou *et al.*, 2000): (i) Systemic response injection of microbes into the body cavity induces a strong expression of AMP genes in the fat body and a low expression in a fraction of hemocytes. (ii) Local response; recent studies have shown that many epithelia can express a subset of AMP genes, and that the expression can be enhanced upon natural microbial infection. (iii) Constitutive expression; a number of tissues constitutively express AMP genes (e.g. *drosomycin* is constitutively expressed in the spermathecae of females).

Analysis of transcripts by Northern blots together with the use of reporter genes are the most common ways to monitor the pattern of AMP gene expression. Direct analysis of peptide expression is more troublesome due to the difficulty of obtaining good antisera against these small cationic peptides. Fortunately, HPLC chromatography and Maldi-Tof mass spectrometry have been useful for purification and monitoring of these peptides (for more details on these methods see Hetru and Bulet, 1997; Uttenweiler-Joseph *et al.*, 1998).

Northern blot

This technique has been extensively used to analyze the kinetics of infection-induced AMP gene expression. Total RNA is extracted from samples of 20 flies treated by standard procedures. The possibility to quantify radioactive hybridization signals using a Phosphorimager and to successively re-probe the same nylon or nitrocellulose membrane with the various AMP cDNAs probes makes this approach suitable to compare the expression of AMP genes. cDNA encoding Actin or rp49 are generally used as internal controls since their expression is not modulated by the immune response. Northern blot of Poly(A^+) RNA, or RT-PCR is used to monitor the expression of genes that are weakly expressed.

Reporter genes

The use of reporter genes is an informative method to analyze the expression pattern of *Drosophila* immune genes. However, it is important to ascertain that the reporter gene faithfully reproduces the pattern of endogenous gene expression.

LacZ reporter genes

Many *Drosophila* lines carrying a *P*-transgene wherein various AMP gene promoter sequences are fused upstream of *lacZ* have been described (*diptericin-lacZ, cecropin-lacZ, drosocin-lacZ, drosomycin-lacZ* (Charlet *et al.*, 1996; Engstrom *et al.*, 1993; Manfruelli *et al.*, 1999; Reichhart *et al.*, 1992)). In addition, *P-lacZ* enhancer trap insertions in *cactus* and *thor* loci that are inducible upon bacterial challenge have also been reported (Bernal and Kimbrell, 2000; Nicolas *et al.*, 1998). These lines allow the analysis of the expression pattern of the corresponding genes by X-gal staining, and the quantification of the expression level by titration of *lacZ* activity.

X-gal staining

This method provides an easy way to study reporter gene expression in larval or adult tissues. Dissect larvae (or adults) in PBS and quickly place the dissected tissues in a small basket immersed in 1 × PBS on ice. Fix 5 to 10 min in 1 × PBS with 0.5 or 1% glutaraldehyde on ice. Wash three times in 1 × PBS on ice. Incubate at 37°C or room temperature in staining buffer from 10 min to overnight. Staining buffer is obtained by adding 30 µl of X-gal (5-bromo-4-chloro-3-indolyl β-D-galactoside, 5% in dimethylformamide) per ml of staining solution (10 mM NaH_2PO_4/Na_2HPO_4 pH 7.2, 150 mM NaCl, 1 mM $MgCl_2$, 3.5 mM K_3FeCN_6, and 3.5 mM K_4FeCN_6). Preincubation at 37°C and centrifugation of this solution can prevent the formation of undesirable crystals. Staining solution can be stored for several months at 4°C. *Drosophila* expresses an endogenous galactosidase in the midgut and few other tissues, the optimal pH of which is 6.5.

β-galactosidase titration

Five larvae, pupae or adults carrying a *lacZ* reporter gene are collected at different time intervals after infection and homogenized in 500 µl of Z Buffer pH 8.0 (stored at −20°C; 60 mM Na_2HPO_4, 60 mM NaH_2PO_4, 10 mM KCl, 1 mM $MgSO_4$, 50 mM β-Mercaptaethanol, adjust pH to 8). After a 10 min centrifugation, the supernatant is collected and vortexed. Protein concentration is estimated by classical methods such as the Bradford assay using BSA as a protein standard. *lacZ* activity is determined by measuring the OD at 420 nm in a cuvette incubated at 37°C containing an appropriate volume of the supernatant mixed with Buffer Z + o-nitrophenol-β-D-galactoside (ONPG) ($[ONPG]_{final} = 0.35\,mg\,ml^{-1}$). Three or more independent measurements for each test are necessary. According to Miller (1972), β-galactosidase activity $= ((\Delta OD)/\Delta T_{min})_V/(\text{protein concentration in v}) \times 1/0.0045$. The use of microtiter dishes allows the measurement of 100 samples at once.

GFP-reporter genes

Lines carrying AMP gene promoters fused to GFP have recently been described (Ferrandon *et al.*, 1998; Tzou *et al.*, 2000). Thus the expression of AMP genes can be monitored in living larvae and adults. The use of GFP reporter genes has revealed the complex expression patterns of the AMP genes. This is due to the ability to analyze the expression of many AMP genes in large samples of individuals and in tissues that are less accessible to classical staining methods, such as the trachea. GFP-expressing *Drosophila* are analyzed directly under a stereomicroscope (e.g. Leica MZFLIII) equipped with epifluorescent illumination (excitation filter 480/40 nm; dichroic filter 505 nm; emission filter 510 nm). Analysis with GFP reporter gene has two major drawbacks. In order to fluoresce, the protein requires cyclization, which results in a lag time and thus GFP detection occurs long after that of *lacZ* β-galactosidase. Second, GFP activity is difficult to quantify although quantification of a *drosomycin-GFP* reporter expression has been used to screen for regulators of the immune response using a spectrophotometer (Jung *et al.*, 2001).

It is thus clear that GFP and *lacZ* reporter genes are complementary tools. Lines carrying both a *drosomycin-GFP* and a *diptericin-lacZ* reporter gene on the *X* chromosome are currently used to monitor the pattern of expression of both AMP genes in the same animal (Manfruelli *et al.*, 1999).

◆◆◆◆◆◆ PROTEIN ANALYSIS

One advantage of *Drosophila* is the possibility to collect sufficient material from larvae or adults to perform basic biochemical experiments. Larval fat body can easily be isolated (for large-scale collection, see Ashburner, 1989), although the fat body of adults is a loose tissue difficult to excise. However, careful preparation of the adult abdominal dorsal carcass provides predominantly fat body. RNA from such preparations contains minor contamination of epidermal and muscle RNA. The collection of either adult carcass or larval fat body has allowed the analysis of the degradation of Cactus and of the processing of Relish upon infection (Nicolas *et al.*, 1998; Stöven *et al.*, 2000).

Several protocols have been described for immunolocalization of proteins in the fat body (Ip *et al.*, 1993; Lemaitre *et al.*, 1995b; Rutschmann *et al.*, 2000b; Stöven *et al.*, 2000). In our lab, we proceed as follows. Fat bodies are dissected on ice in 1 × PBS and transferred to a small basket immersed in 1 × PBS in 24-well titer plates used in cell culture. Do not let the tissue dry. All subsequent steps are performed under moderate agitation. Fat bodies are fixed in 4% paraformaldehyde, 2 mM $MgSO_4$, 1 mM EGTA and 0.1 M PIPES buffer for 15 min at room temperature and rinsed in 1 × PBS at 4°C. A brief subsequent fixation at 4°C in 0.5% glutaraldehyde/1 × PBS for 20 s or less can prevent the fat bodies from degradation by Triton. Three 5 min washes in 1 × PBS are followed by permeabilization for 2 h in PBT-A (1% BSA , 0.1% Triton X-100 in PBS). The primary antibody is applied at an appropriate dilution in PBT-A and incubated overnight at 4°C. The preparation is then washed three times for 30 min in PBT-B (0.1% BSA , 0.1% Triton X-100 in PBS). The secondary antibody, usually linked to alkaline phosphatase, is first pre-adsorbed on fixed fat body, then diluted and applied for 4 h to the sample in PBT-B at room temperature. The preparation is fixed for 10 min in 0.5% glutaraldehyde/1 × PBS, washed three times in AP-Sol (100 mM Tris-HCl, pH 9.5, 100 mM NaCl, 50 mM $MgCl_2$, 0.1% Triton X-100) and incubated for 2 h in the staining solution (0.34 mg ml^{-1} NBT, 0.17 mg ml^{-1} X-Phosphate in AP-Sol). The fat bodies are subsequently mounted in glycerol/ethanol (1/1) on glass slides.

◆◆◆◆◆◆ ANALYSIS OF BLOOD CELL FUNCTIONS

In *Drosophila*, hematopoiesis and blood cell types exhibit specific features according to the developmental stage. During embryogenesis, a macrophage population differentiates in the anterior mesoderm and rapidly migrates to colonize the whole embryo (Tepass *et al.*, 1994). A second blood cell population appears simultaneously in the foregut region that corresponds to crystal cells (Lebestky *et al.*, 2000). At the end of embryogenesis, the larval hematopoietic organ differentiates anteriorly on the dorsal vessel.

During the three larval stages, most of the circulating blood cells are produced by the hematopoietic organ, the lymph glands, that are composed of a variable number of paired lobes along the dorsal vessel (Lanot *et al.*, 2001; Rizki and Rizki, 1984; Shresta and Gateff, 1982). The vast majority of larval circulating hemocytes consist of small round cells, called plasmatocytes, that are characterized by strong phagocytic capacity. Less than 5% of the hemocytes are crystal cells, with typical crystalline inclusion, that are proposed to contain the enzymes and substrate necessary for melanization reactions. A third cell type that only differentiates under given immune conditions is the lamellocyte, a large flattened cell devoted to encapsulation. Such an immune reaction occurs when an invader is too large to be phagocytosed, as is the case for wasp parasitization. At the onset of metamorphosis, the circulating plasmatocytes become highly active macrophages that ingest histolyzing larval tissues. In the lymph glands, large numbers of such phagocytes differentiate in all lobes and are released from the glands. The latter are empty 15 h after pupariation, and subsequently, a typical hematopoietic organ cannot be identified in pupae or adults.

In adults, the only circulating blood cell type is the plasmatocyte: crystal cells and lamellocytes do not differentiate at this stage.

Observation of *Drosophila* hemocytes

In embryos

Embryonic hemocytes can be identified with specific enhancer trap lines or antibodies. Two *lacZ* enhancer trap lines are commonly used: line 197 (Abrams *et al.*, 1992) and line E8-2-18 (Hartenstein and Jan, 1982): the expression of lacZ is detected by the classical reaction using X-gal substrate (Ashburner, 1989), or with anti-β-galactosidase antibodies. Three antibodies directed against hemocyte-specific antigens are available to date. These antigens are peroxidasin, a protein combining both peroxidase and extracellular matrix motifs, which was proposed to participate in extracellular matrix consolidation and in defense mechanisms (Nelson *et al.*, 1994), croquemort, a CD36 homolog that is required for phagocytosis of apoptotic cells by embryonic macrophages (Franc *et al.*, 1996; Figure 27.2A), and lozenge, a Runt-domain transcription factor that in embryos is specifically expressed in crystal cells (Lebestky *et al.*, 2000).

In larvae

It is easy to obtain hemocytes from larvae, especially at the third instar. Individuals are washed in distilled water and dried on a filter paper, then punctured posteriorly and gently squeezed to deposit a droplet of hemolymph (<1 μl) on a poly-lysine coated glass-slide. After 5 min drying, the preparations are fixed for 5 min in a 1% formaldehyde in 0.1 M sodium phosphate buffer (pH 7.3) solution and mounted in glycerol gelatin then observed by interference phase contrast microscopy (Figure 27.1B). They can also be stained for 1 min in an aqueous 1% toluidin blue/0.1% eosin solution, washed in 95% ethanol, successively transferred to 100% ethanol and to xylene, and finally mounted in Eukitt.

To better visualize hemocytes, or to detect specific cell types, it is possible to use *lacZ* transgenic fly lines. A transgenic strain in which all hemocytes express *lacZ* was produced by Govind (1995). Lineage specific enhancer trap lines were reported by Braun *et al.* (1997), at least for plasmatocytes and lamellocytes (Figure 27.1 C,D).

Method: larvae are washed and dried, then punctured to deposit hemolymph on a glass coverslip. After 5 min drying, the preparations are fixed for 30 s in a 0.5% glutaraldehyde/PBS solution. Staining is in 0.2% 5-bromo-4-chloro-3-indolyl-β-D-galactopyranoside (X-gal), 3.5 mM $K_4Fe(CN)_6$, 3.5 mM $K_3Fe(CN)_6$, 1 mM $MgCl_2$, 150 mM NaCl, 10 mM Na_2HPO_4/NaH_2PO_4 buffer pH 7.2, overnight at 37°C. Preparations are mounted in glycerol gelatin for observation.

Crystal cells can be visualized in whole animals by heating them at 70°C for 10 min in a water bath (Rizki *et al.*, 1980). This induces specific

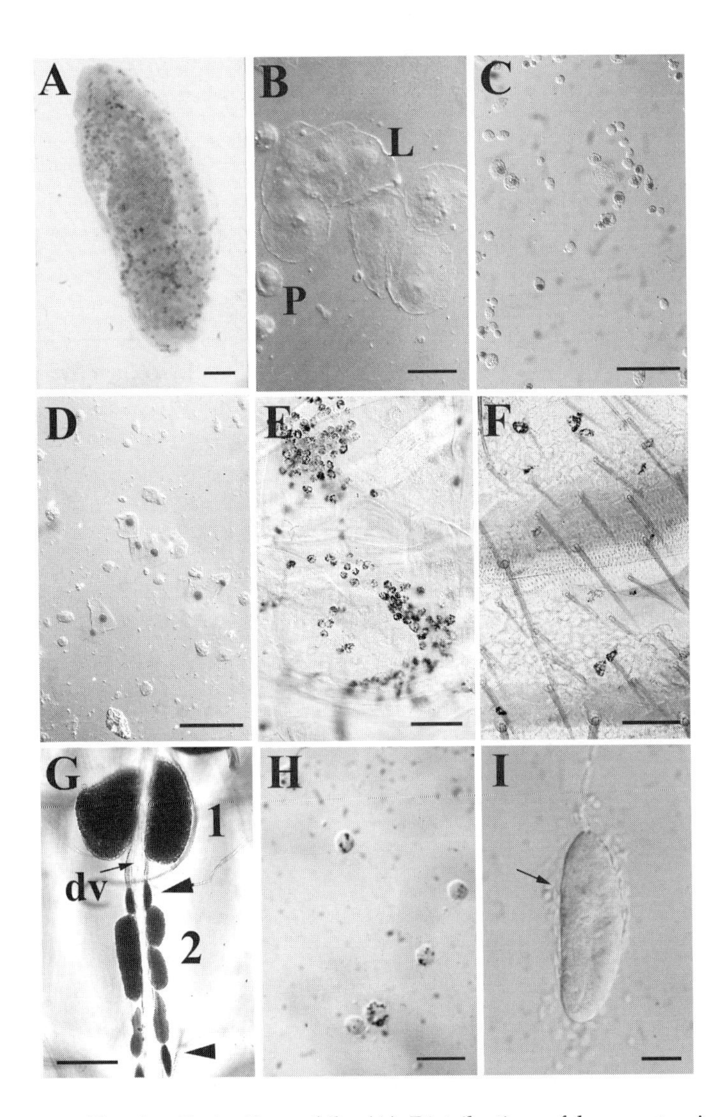

Figure 27.2. Blood cells in *Drosophila*. (A) Distribution of hemocytes in a late embryo, as evidenced by anti-croquemort antibody. Bar: 50 μm. (B) Larval blood cells observed by interference phase contrast microscopy. P: plasmatocyte; L: lamellocyte. Bar: 20 μm. (C, D) *lacZ* expression in larval hemocytes: staining in plasmatocytes in line *l(3)05309* (C) and in lamellocytes in line *l(3)06946* (D) (Braun *et al.*, 1997). Bar: 50 μm. (E, F) Observation of sessile hemocytes through the cuticle after Indian ink injection into a third instar larva (E) or into a *yellow* adult (F). Bar: 50 μm. (G) Dissection of a larval lymph gland attached to the dorsal vessel and stained with osmium tetroxide. Arrow heads: pericardial cells; 1 and 2 designate the first and second lobes of the lymph glands: the second lobes are well developed when larvae are raised at 18°C; dv: dorsal vessel. Bar: 100 μm. (H) Indian ink phagocytosis by larval plasmatocytes observed 2 h after injection. Bar: 20 μm. (I) Encapsulated *L. boulardi* egg in larval hemocoel 24 h after parasitization. The wasp egg is surrounded by lamellocytes (arrow); blackening has not yet occurred. Bar: 50 μm. (This figure is also reproduced in colour between pages 276 and 277.)

blackening of crystal cells that become easily visible in circulation and within the hematopoietic organ, although they disappear during the early hours of metamorphosis.

Immunohistochemistry can be performed on an air-dried droplet of hemolymph following the protocol below.

Method: fixation is carried out for 5 min in 4% paraformaldehyde, or 4% glutaraldehyde, or in a mix (4% paraformaldehyde; 0.5% glutaraldehyde for instance) in 0.1 M phosphate buffer or in PBS on ice. The fixation conditions depend on the antigen. After three 10 min rinses in PBS-0.1% Tween-20 (PBT), saturation is achieved with 3% pre-immune serum in PBT, for 1h at room temperature. Primary antibody at the appropriate dilution is applied in 3% serum/PBT overnight at 4°C, then after three 10 min rinses in PBT, the preparation is treated with secondary antibody (commercial, usually diluted 1/100 to 1/500) in 3% serum/PBT, 1 h at room temperature. Depending on the secondary antibody (fluorescent, alkaline phosphatase or peroxidase), the samples are processed according to the selected system.

In situ hybridization technique for blood cells is adapted from Tautz and Pfeiffle (1989), with modifications from J.M. Reichhart and G. Grossnibach.

Method: Glass slides are boiled for 5 min in distilled water with soap, rinsed successively under tap water, distilled water and in 95% ethanol before sterilization at 180°C. Hemocytes are deposited on the glass slide, then fixed in Carnoy's medium for 10 min. The preparation is rinsed for 2 min in 95% ethanol and stored (up to several months) at −20°C in absolute ethanol. Before further treatment, the slides must be rehydrated for 2 min in 70% ethanol and 2 min in PBS. For proteinase K treatment, 50 μl of freshly thawed proteinase K are added to 50 ml prewarmed PBS (25 $\mu g\, ml^{-1}$ final) and digestion on the slides is allowed to proceed at 37°C for 5 min. Proteinase K digestion is crucial for the procedure and every new batch should be tested first for incubation time. Digestion is arrested by 2 min in 2 $mg\, ml^{-1}$ glycine in PBT followed by 5 min in PBT. A post-fixation step of 20 min in PBT/5% formaldehyde is followed by a 5 min rinse in PBT, and acetylation. Acetylation solution is obtained by mixing 20 ml of 1 M triethanolamine with 180 ml water, to which 500 μl acetic anhydride are rapidly added. The slides are immediately incubated in the solution for 10 min, then dehydrated in graded ethanol and air dried. For hybridization, the probe is denatured by 4 min boiling then chilling on ice-water: 20 μl of hybridization solution (hybridization solution: 50% deionized formamide; 5 × SSC; 100 $\mu g\, ml^{-1}$ salmon sperm; 100 $\mu g\, ml^{-1}$ tRNA *E. coli*; 50 $\mu g\, ml^{-1}$ heparine; 0.1% Tween-20, in water) containing 2 μl of probe (10 ng of dig-labeled DNA) are deposited on the glass slide. The reaction is covered with an 18 × 18 coverslip and sealed with rubber cement. Hybridization is allowed to proceed at 48°C overnight in a humid chamber. Next day, the coverslip is removed in hybridization solution, washes are successively for 20 min in hybridization solution at 48°C, for

20 min in 50% hybridization solution/50% PBT at 48°C, for 20 min in PBT at 48°C and three times for 5 min in PBT at room temperature. 300 μl of pre-adsorbed anti-dig antibody at appropriate dilution in PBT are deposited on the preparation, covered with a large coverslip and incubation is performed at room temperature for 2 h in a humid chamber. After four 5 min rinses in PBT, it is possible to introduce an amplification step (Vectastain kits) if a low signal is expected. Detection of the signal follows two 2 min rinses in levamisol-AP-9.5 buffer (100 mM NaCl; 50 mM $MgCl_2$; 100 mM Tris pH 9.5; 1 mM levamisol ; 0.1% Tween-20). The staining mixture contains 1 ml levamisol-AP-9.5 buffer, 4.5 μl NBT (Nitro-blue tetrazolium 75 mg ml^{-1} in 70 DMF/water) and 3.5 μl BCIP (5-bromo-4-chloro-3-indolyl phosphate 50 mg ml^{-1} in DMF). For staining, 300 μl of this solution are added to the slide and incubated in a humid chamber at room temperature. Incubation time varies between 15 min and 16 h depending on the probe, with replacement of the staining solution every 2 h, which should greatly reduce the background. The reaction is stopped by 2 min in PBS and 10 min in water, the preparation counter-stained with 0.5% acridine orange in water and mounted in glycerol gelatin.

In adults

Adult hemocytes are more difficult to analyze as it is difficult to obtain hemolymph at this stage; moreover the hemocytes are mostly attached to internal tissues. An efficient way to visualize them is to inject Indian ink into the hemocoel (see below): the nested plasmatocytes take up the Indian ink particles and become visible through the epidermis, in adults as well as in larvae (Figure 27.2 E, F).

Observation of *Drosophila* larval hematopoietic tissue

Lymph glands are very often lost in crude dissections of larvae, as they are tiny and loose. To preserve them for observation, we dissect larvae in PBS, ventral part facing up, after pinning them anteriorly and posteriorly with thin needles. The abdominal integument is cut medio-longitudinally in order to open the larva like a book, exposing the different organs in place. The brain complex is toppled over anteriorly. The trachaea, fat body and gut are successively removed, without pulling, but by delicately detaching them from the nervous and tracheal networks. Thus the dorsalmost organs are displayed: the two dorsal tracheal trunks framing the dorsal vessel, and, anteriorly, the paired lymph gland lobes (see Figure 27.1 G).

Immunohistochemistry can be performed on such preparations, which are fixed for 5 min in 4% formaldehyde/0.5% glutaraldehyde in PBS, washed several times in PBS, then blocked for 2 h in 3% serum/PBT, and incubated overnight in primary antibody. The following steps are as previously described above.

Measure of phagocytosis

Plasmatocytes are the most abundant hemocytes, and they are very efficient phagocytes. Several techniques have been developed to visualize phagocytosis.

Method: The injected material is either Indian ink (Pébéo, Gemenos, France) diluted 1/30 to 1/60 in PBS ((Lanot *et al.*, 2000), Figure 27.2 H), GFP-labeled bacteria (Basset *et al.*, 2000), FITC-labeled bacteria (Elrod-Erickson *et al.*, 2000) or fluorescent polystyrene beads (Elrod-Erickson *et al.*, 2000) at the appropriate dilution. Injection is carried out with a Nanoject injector (Drummond); the volume is usually 40–50 nl per individual. Whereas injection into adults can be done under the routine fly room dissecting microscope into the thorax of anesthesized flies, injection in larvae requests more precaution. The injector is fixed under the dissecting microscope. The larva is impaled onto the needle while keeping it under tension with a pair of tweezers. Ideally, the needle should penetrate the larva in the last intersegmental space at the rear. Before removing the needle after the injection, it is recommended to wait a couple of seconds to allow the injected volume to diffuse throughout the larva. It is often necessary to humidify the contact zone between the needle and the cuticle to prevent damage to the cuticle. Once the larva has been removed from the needle with a paint brush, it is deposited on a humid filter paper. Injected adults or larvae can be examined as soon as 2 h later for phagocytosis, directly under the dissecting microscope. In the case of adults, ink-labeled plasmatocytes are best observed in a *yellow* background.

Measure of encapsulation

Differentiation of lamellocytes is induced when parasitoid wasps lay eggs in *Drosophila* larvae. Some 50 drosophilid parasitoids have been reported (Carton *et al.*, 1986), among which *Leptopilina boulardi* has been well investigated. Wasp eggs are deposited in second instar larvae and, in resistant flies, the parasitoid egg is rapidly neutralized by encapsulation/melanization (Figure 27.2 I).

Method: We use 2–10-day-old *L. boulardi* adults which have not been in contact with *Drosophila* over the last 48 h. *Drosophila* 24 h egg-laying from ca. 50 females are allowed to develop until the second larval instar. 20 *L. boulardi* females and five males are added to the vial for 4 h, then withdrawn. This treatment results in an efficient parasitization rate of the larvae, and black capsules can be seen on live individuals 48 h later. Differentiation of lamellocytes is already detected 10 h following wasp parasitization. It is better ascertained with the use of a lamellocyte-specific enhancer trap line such as *l(3)06946* (Braun *et al.*, 1997).

Measure of melanization

Humoral melanization is a function of blood cells in *Drosophila*. Melanin formation involves the activity of phenol and polyphenol oxidases that catalyze the conversion of tyrosine to DOPA, dopamine, N-acetyldopamine, quinone and subsequently melanin. Tests for phenoloxidase activity in cell-free hemolymph and blood cells have been described in detail (Shresta and Gateff, 1982). Two very simple tests are given below.

Assay of phenoloxidase activity

Filter paper is soaked with 10 mM phosphate buffer (pH 6.5) containing 2 mg ml^{-1} L-DOPA (L-3,4-dihydroxyphenylalanine, Sigma) and phenoloxidase activity is detected in blood samples by dropping the hemolymph of a single larva on the paper. For quantification of phenoloxidase activity, 3 μl hemolymph samples are added to 50 μl of 10 mM phosphate buffer (pH 5.9) containing 10 mM L-DOPA, and the OD is recorded for 30 min at 470 nm.

◆◆◆◆◆◆ MUTATIONS AFFECTING *DROSOPHILA* IMMUNE RESPONSE

Several mutations affecting different components of *Drosophila* immune response have been described (listed in Table 27.1). More information on these genes can be found on Flybase (http://flybase.bio.indiana.edu:82/). A more extensive list of genes involved in *Drosophila* immunity is available on the web (http://www.cnrs-gif.fr/cgm/immunity/Drosophila_immunity_genes.html).

◆◆◆◆◆◆ *DROSOPHILA* BLOOD CELL LINES

Several cell lines are available from *Drosophila* that either exhibit hemocyte features (capability of phagocytosis, inducibility of defense genes, expression of hemocyte markers), or are derived from tumorous hemocytes. For the analysis of immune gene regulation, the most commonly used cell lines are SL2 (Schneider, 1972) and mbn-2 cells (Gateff *et al.*, 1980), in which endogenous antimicrobial genes can be induced by bacteria or LPS treatment (Kappler *et al.*, 1993; Samakovlis *et al.*, 1992). The well-known Kc cell line was tested independently in several laboratories and, as no immune-inducible gene expression was evidenced, it has not been exploited so far as a model system for immunity. Kc cells, however, produce many proteins constitutively expressed by blood cells (Fessler *et al.*, 1994).

Transfected SL2 and mbn-2 cells have been successfully used in a number of studies to dissect antimicrobial gene promoters, to analyze interactions between immune transactivators and promoters, or to investigate protein/protein interactions between various immune effectors (Engström *et al.*, 1993; Han and Ip, 1999; Kappler *et al.*, 1993; Kim *et al.*, 2000; Silverman *et al.*, 2000; Stöven *et al.*, 2000; Tauszig *et al.*, 2000).

Table 27.1 List of mutations known to affect *Drosophila* immune response

L(3)*hem*, *dom*	Severe reduction of the hemocyte number and melanization	(Braun *et al*., 1998; Gateff, 1994)
nec, *spz*, Tl, *tub*, *pll*, *cact*, *dif*, *dl*	These mutations affect the Toll pathway that regulates a subset of AMP genes and many other effector genes. This pathway is required for survival after fungal infection. Some of these mutations also alter cellular response (hemocyte count, lamellocyte differentiation). Mutations that constitutively activate or block this pathway have been described.	(Lemaitre *et al*., 1996; Levashina *et al*., 1999; Manfruelli *et al*., 1999; Meng *et al*., 1999; Qiu *et al*., 1998; Rutschmann *et al*., 2000a)
imd, dredd, rel, IKKγ	These mutations affect the Imd pathway which regulates a subset of AMP genes and is required for survival after Gram-negative infection. These mutations are homozygous viable and seem to have no apparent effect on the cellular response.	(Hedengren *et al*., 1999; Lemaitre *et al*., 1995a; Leulier *et al*., 2000; Rutschmann *et al*., 2000b)
Bc, *lz*	These mutations affect crystal cell function and block proPhenolOxidase activation.	(Lebestky *et al*., 2000; Rizki and Rizki, 1984)
hop, stat92	These mutations affect components of the JAK-STAT pathway that controls the expression of complement-like protein as well as hemocyte differentiation and division. Mutations that constitutively activate or block this pathway have been described.	(Hanratty and Dearolf, 1993; Harrison *et al*., 1995; Yan *et al*., 1996)

◆◆◆◆◆◆ CONCLUSIONS

Most of the techniques described here are simple and do not require extensive skills or complex devices. Therefore, it requires very little to test the effect of each experimenter's favorite pathogen on *Drosophila*; if the pathogen shows an effect, powerful genetic and molecular tools can be applied to identify host factors that are the target of these microorganisms. In this context, we should keep in mind that some of the medically important bacteria are also pathogenic for fly. In this case, *Drosophila* also provides an easy assay to screen mutated microbes for the loss of virulence factors.

Acknowledgments

We thank Dr René Lanot and Dr Ranjiv Khush for stimulating discussions and Nathalie Mansion for help with the figures. This work was supported by ATIPE CNRS, the Fondation pour la Recherche Médicale (FRM) and Programme Microbiologie (PRMMIP98).

References

Abrams, J. M., Lux, A., Steller, H. and Krieger, M. (1992). Macrophages in Drosophila embryos and L2 cells exhibit scavenger receptor-mediated endocytosis. *Proc. Natl Acad. Sci. USA* **89**, 10375–10379.

Ashburner, M. (1989). *Drosophila, a Laboratory Manual* (Cold Spring Harbor, Cold Spring Harbor Laboratory Press).

Basset, A., Khush, R., Braun, A., Gardan, L., Boccard, F., Hoffmann, J. and Lemaitre, B. (2000). The phytopathogenic bacteria, *Erwinia carotovora*, infects *Drosophila* and activates an immune response. *Proc. Natl Acad. Sci. USA* **97**, 3376–3381.

Bernal, A. and Kimbrell, D. A. (2000). Drosophila Thor participates in host immune defense and connects a translational regulator with innate immunity. *Proc. Natl Acad. Sci. USA* **97**, 6019–24.

Boman, H. G., Nilsson, I. and Rasmuson, B. (1972). Inducible antibacterial defence system in Drosophila. *Nature* **237**, 232–235.

Braun, A., Hoffmann, J. A. and Meister, M. (1998). Analysis of the Drosophila host defense in domino mutant larvae, which are devoid of hemocytes. *Proc. Natl Acad. Sci. USA* **95**, 14337–14342.

Braun, A., Lemaitre, B., Lanot, R., Zachary, D. and Meister, M. (1997). Drosophila immunity: analysis of larval hemocytes by P-element-mediated enhancer trap. *Genetics* **147**, 623–634.

Carton, Y., Bouletreau, M., van Alphen, J. J. M. and van Lenteren, J. C. (1986). The Drosophila parasitic wasps. In *The Genetics and Biology of Drosophila*, M. Ashburner, H. L. Carson and J. N. Thompson (eds), Academic Press.

Charlet, M., Lagueux, M., Reichhart, J., Hoffmann, D., Braun, A. and Meister, M. (1996). Cloning of the gene encoding the antibacterial peptide drosocin involved in Drosophila immunity. Expression studies during the immune response. *Eur. J. Biochem.* **241**, 699–706.

Clarkson, J. M. and Charnley, A. K. (1996). New insights into the mechanisms of fungal pathogenesis in insects. *Trends Microbiol.* **4**, 197–203.

Elrod-Erickson, M., Mishra, S. and Schneider, D. (2000). Interactions between the cellular and humoral immune responses in Drosophila. *Curr. Biol.* **10**, 781–4.

Engström, Y., Kadalayil, L., Sun, S. C., Samakovlis, C., Hultmark, D. and Faye, I. (1993). Kappa B-like motifs regulate the induction of immune genes in Drosophila. *J. Mol. Biol.* **232**, 327–33.

Ferrandon, D., Jung, A. C., Criqui, M., Lemaitre, B., Uttenweiler-Joseph, S., Michaut, L., Reichhart, J. and Hoffmann, J. A. (1998). A drosomycin-GFP reporter transgene reveals a local immune response in Drosophila that is not dependent on the Toll pathway. *EMBO J.* **17**, 1217–27.

Fessler, L., Nelsson, R. and Fessler, J. (1994). Drosophila extracellular matrix. *Methods Enzymol.* **245**, 271–294.

Finlay, B. B. and Cossart, P. (1997). Exploitation of mammalian host cell functions by bacterial pathogens. *Science* **276**, 718–725.

Flyg, C., Kenne, K. and Boman, H. G. (1980). Insect pathogenic properties of Serratia marcescens: phage-resistant mutants with a decreased resistance to Cecropia immunity and a decreased virulence to Drosophila. *J. Gen. Microbiol.* **120**, 173–181.

Franc, N., Dimarcq, J., Lagueux, M., Hoffmann, J. and Ezekowitz, R. (1996). Croquemort, a novel Drosophila hemocyte/macrophage receptor that recognizes apoptotic cells. *Immunity* **4**, 431–443.

Gateff, E. (1994). Tumor-suppressor genes, hematopoietic malignancies and other hematopoietic disorders of Drosophila melanogaster. *Ann. N Y Acad. Sci.* **712**, 260–279.

Gateff, E., Gissmann, L., Shresta, R., Plus, N., Pfister, H., Schröder, J. and Zur Hausen, H. (1980). Characterization of two tumorous blood cell lines of Drosophila melanogaster and the viruses they contain. In *Invertebrate Systems in vitro,* E. Kurstak, K. Maramorosch and A. Dübendorfer (eds), Amsterdam, Elsevier/North Holland Biomedical Press, pp. 517–533.

Govind, S. (1995). Rel signalling pathway and the melanotic tumor phenotype of Drosophila. *Biochem. Soc. Trans.* **24**, 39–44.

Han, Z. S. and Ip, Y. T. (1999). Interaction and specificity of Rel-related proteins in regulating Drosophila immunity gene expression. *J. Biol. Chem.* **274**, 21355–21361.

Hanratty, W. P. and Dearolf, C. R. (1993). The Drosophila Tumorous-lethal hematopoietic oncogene is a dominant mutation in the hopscotch locus. *Mol. Gen. Genet.* **238**, 33–37.

Harrison, D., Binari, R., Stines Nahreini, T., Gilman, M. and Perrimon, N. (1995). Activation of a Drosophila Janus kinase (JAK) causes hematopoietic neoplasia and developmental defects. *EMBO J.* **14**, 2857–2865.

Hartenstein, V. and Jan, Y. N. (1982). Studying Drosophila embryogenesis with P-lacZ enhancer trap lines. *Roux's Arch. Dev. Biol.* **201**, 194–220.

Hedengren, M., Asling, B., Dushay, M. S., Ando, I., Ekengren, S., Wihlborg, M. and Hultmark, D. (1999). Relish, a central factor in the control of humoral but not cellular immunity in Drosophila. *Mol. Cell.* **4**, 827–37.

Hetru, C. and Bulet, P. (1997). Strategies for isolation and characterization of antimicrobial peptides of invertebrates. In *Antibacterial Peptide Protocols,* W. Shafer (ed.) Totowa, NJ, Humana Press Inc., pp. 35–49.

Hoffmann, J. A., Kafatos, F. C., Janeway, C. A. and Ezekowitz, R. A. (1999). Phylogenetic perspectives in innate immunity. *Science* **284**, 1313–1318.

Ip, Y., Reach, M., Enstrom, Y., Kadalayil, L., Cai, H., Gonzalez-Crespo, S., Tatei, K., and Levine, M. (1993). Dif, a dorsal-related gene that mediates an immune response in Drosophila. *Cell* **75**, 753–763.

Jung, A., Criqui, M. C., Rutschmann, S., Hoffmann, J. A. and Ferrandon, D. (2001). A microfluorometer assay to measure the expression of ß-galactosidase and GFP reporter genes in single Drosophila flies. *Biotechniques* **30**, 594–8, 600–1.

Kappler, C., Meister, M., Lagueux, M., Gateff, E., Hoffmann, J. and Reichhart, J. (1993). Insect immunity. Two 17bp repeats nesting a κB-related sequence confer inducibility to the diptericin gene and bind a polypeptide in bacteria-challenged Drosophila. *EMBO J.* **12**, 1561–1568.

Khush, R. S. and Lemaitre, B. (2000). Genes that fight infection: what the Drosophila genome says about animal immunity. *Trends Genet.* **16**, 442–449.

Kim, Y. S., Ryu, J. H., Han, S. J., Choi, K. H., Nam, K. B., Jang, I. H., Lemaitre, B., Brey, P. T. and Lee, W. J. (2000). Gram-negative bacteria-binding protein, a pattern recognition receptor for lipopolysaccharide and beta-1,3-glucan that mediates the signaling for the induction of innate immune genes in Drosophila melanogaster cells. [In Process Citation] *J. Biol. Chem.* **275**, 32721–32727.

Lanot, R., Zachary, D., Holder, F. and Meister, M. (2001). Post-embryonic hematopoiesis in Drosophila. *Devel. Biol.* **230**, 243–57.

Lebestky, T., Chang, T., Hartenstein, V. and Banerjee, U. (2000). Specification of Drosophila hematopoietic lineage by conserved transcription factors. *Science* **288**, 146–149.

Lemaitre, B., Kromer-Metzger, E., Michaut, L., Nicolas, E., Meister, M., Georgel, P., Reichhart, J. and Hoffmann, J. (1995a). A recessive mutation, immune deficiency (imd), defines two distinct control pathways in the Drosophila host defense. *Proc. Natl Acad. Sci. USA* **92**, 9365–9469.

Lemaitre, B., Meister, M., Govind, S., Georgel, P., Steward, R., Reichhart, J. M. and Hoffmann, J. A. (1995b). Functional analysis and regulation of nuclear import of dorsal during the immune response in Drosophila. *EMBO J.* **14**, 536–545.

Lemaitre, B., Nicolas, E., Michaut, L., Reichhart, J. and Hoffmann, J. (1996). The dorsoventral regulatory gene cassette spätzle/Toll/cactus controls the potent antifungal response in Drosophila adults. *Cell* **86**, 973–983.

Lemaitre, B., Reichhart, J. and Hoffmann, J. (1997). Drosophila host defense: differential induction of antimicrobial peptide genes after infection by various classes of microorganisms. *Proc. Natl Acad. Sci. USA* **94**, 14614–14619.

Leulier, F., Rodriguez, A., Khush, R. S., Chen, P., Abrams, J. M. and Lemaitre, B. (2000). The Drosophila caspase Dredd is required to resist Gram-negative bacterial infection. *EMBO R.* **1**, 353–358.

Levashina, E. A., Langley, E., Green, C., Gubb, D., Ashburner, M., Hoffmann, J. A., and Reichhart, J. M. (1999). Constitutive activation of Toll-mediated antifungal defense in serpin-deficient Drosophila. *Science* **285**, 1917–1919.

Manfruelli, P., Reichhart, J. M., Steward, R., Hoffmann, J. A. and Lemaitre, B. (1999). A mosaic analysis in Drosophila fat body cells of the control of antimicrobial peptide genes by the Rel proteins Dorsal and DIF. *EMBO J.* **18**, 3380–3391.

Meng, X., Khanuja, B. S. and Ip, Y. T. (1999). Toll receptor-mediated Drosophila immune response requires Dif, an NF- kappaB factor. *Genes Dev.* **13**, 792–797.

Miller, J. H. (1972). *Experiments in Molecular Genetics.* (Cold Spring Harbor, Cold Spring Harbor Publishers).

Nelson, R., Fessler, L., Takagi, Y., Blumberg, B., Keene, D., Olson, P., Parker, C. and Fessler, J. (1994). Peroxidasin, a novel enzyme-matrix protein of Drosophila development. *EMBO J.* **13**, 3438–3447.

Nicolas, E., Reichhart, J. M., Hoffmann, J. A. and Lemaitre, B. (1998). In vivo regulation of the IkappaB homologue cactus during the immune response of Drosophila. *J. Biol. Chem.* **273**, 10463–9.

Qiu, P., Pan, P. C. and Govind, S. (1998). A role for the Drosophila Toll/Cactus pathway in larval hematopoiesis. *Development* **125**, 1909–1920.

Reichhart, J., Meister, M., Dimarcq, J., Zachary, D., Hoffmann, D., Ruiz, C., Richards, G. and Hoffmann, J. (1992). Insect immunity: developmental and inducible activity of the Drosophila diptericin promoter. *EMBO J.* **11**, 1469–1477.

Rizki, T. and Rizki, R. (1984). The cellular defense system of Drosophila melanogaster. In *Insect Ultrastructure.* R. King, H. and Akai (eds), Plenum Publishing Corporation, pp. 579–604.

Rizki, T., Rizki, R. and Grell, E. (1980). A mutant affecting the crystal cells in Drosophila melanogaster. *Roux's Arch. Dev. Biol.* **188**, 91–99.

Rutschmann, S., Jung, A. C., Hetru, C., Reichhart, J. M., Hoffmann, J. A. and Ferrandon, D. (2000a). The Rel protein DIF mediates the antifungal but not the antibacterial host defense in Drosophila. *Immunity* **12**, 569–580.

Rutschmann, S., Jung, A. C., Zhou, R., Silverman, N., Hoffmann, J. A., and Ferrandon, D. (2000b). Role of Drosophila IKKγ in a Toll-independent antibacterial immune response. *Nature Immunol.* **1**, 342–347.

Samakovlis, C., Asling, B., Boman, H., Gateff, E. and Hultmark, D. (1992). In vitro induction of cecropin genes: an immune response in a Drosophila blood cell line *BBRC* **188**, 1169–1175. [Biochemistry Biophysics Research Communication]

Schneider, I. (1972). Cell lines derived from late embryonic stages of Drosophila melanogaster. *J. Embryol. Exp. Morphol.* **27**, 353–365.

Shrestha, R. and Gateff, E. (1982). Ultrastructure and cytochemistry of the cell types in the larval hematopoietic organs and hemolymph of Drosophila melanogaster *Dev. Growth Differ.* **24**, 65–82.

Silverman, N., Zhou, J., Stöven, S., Pandey, N., Hultmark, D. and Maniatis, T., (2000). A Drosophila $I_{\kappa}B$ kinase complex required for Relish cleavage and antibacterial immunity. *Genes & Dev.* **14**, 2461–2471.

Stöven, S., Ando, I., Kadalayil, L., Engström, Y. and Hultmark, D. (2000). Activation of the Drosophila NF-κB factor Relish by rapid endoproteolytic cleavage. *EMBO R.* **1**, 347–352.

Tan, M. and Ausubel, F. (2000). *Caenorhabditis elegans*: a model genetic host to study *Pseudomonas aeruginosa* pathogenesis. *Curr. Opin. Immunol.* **3**, 29–34.

Tauszig, S., Jouanguy, E., Hoffmann, J. A. and Imler, J. L. (2000). Toll-related receptors and the control of antimicrobial peptide expression in Drosophila. *Proc. Natl Acad. Sci. USA* **97**, 10520–10525.

Tautz, D. and Pfeiffle, C. (1989). A non-radioactive in situ hybridization method for the localisation of specific RNAs in Drosophila embryos reveals translational control of the segmentation gene hunchback. *Chromosoma* **98**, 80–85.

Tepass, U., Fessler, L. I., Aziz, A. and Hartenstein, V. (1994). Embryonic origin of hemocytes and their relationship to cell death in Drosophila. *Development* **120**, 1829–1837.

Tzou, P., Ohresser, S., Ferrandon, D., Capovilla, M., Reichhart, J. M., Lemaitre, B., Hoffmann, J. A. and Imler, J. L. (2000). Tissue-specific inducible expression of antimicrobial peptide genes in Drosophila surface epithelia. *Immunity* **13**, 737–748.

Uttenweiler-Joseph, S., Moniatte, M., Lagueux, M., Van Dorsselaer, A., Hoffmann, J. A. and Bulet, P. (1998). Differential display of peptides induced during the immune response of Drosophila: a matrix-assisted laser desorption ionization time-of-flight mass spectrometry study. *Proc. Natl Acad. Sci. USA* **95**, 11342–11347.

Yan, R., Small, S., Desplan, C., Dearolf, C. R. and Darnell, J. E., Jr. (1996). Identification of a Stat gene that functions in Drosophila development. *Cell* **84**, 421–430.

28 Electron Microscopy in Cellular Microbiology: Ultrastructural Methods for the Study of Cell Surface Antigens and Pathogenic Host–Cell Interactions

P Gounon

Station Centrale de Microscopie Electronique, Institute Pasteur, 28 rue de Dr Roux 75724 Paris, France

◆◆

CONTENTS

List of abbreviations

PBS	Phosphate buffered saline
TBS	Tris buffered saline
BSA	Bovine serum albumin
NGS	Normal goat serum
PB	Phosphate buffer 0.1 M pH 7.5
EC	Eppendorf centrifuge
PVP	Polyvinyl pirrolidone
SEM	Scanning electron microscopy
PLT	Progressive lowering of temperature

METHODS IN MICROBIOLOGY, VOLUME 31
ISBN 0–12–521531–2

◆◆◆◆◆◆ INTRODUCTION

Microbial and viral pathogens have evolved a wide range of sophisticated strategies to interact with cellular hosts. The analysis of the cascade of signals involved in complex interactions between cells and pathogens has lead us to analyse at the molecular level both microbes and cells. A new field of science has emerged called 'Cellular Microbiology' (Cossart *et al.*, 2000).

Among many techniques used in cellular microbiology, electron microscopy remains an essential tool due to its 'incredible ability (...) to provide information at the molecular organellar and cellular levels (...). The question of whether a signaling molecule localizes to the cytoplasmic surface of the membrane, in the actin cortex or on the membrane of an early endosome compartment just beneath the plasma membrane is obviously crucial to evaluating its function. But even the best confocals cannot obtain this resolution, which is trivial to achieve at the EM level' (Griffiths, 2001).

In ultrastructural research, the ideal method for the localization of extra or intra-cellular antigens is not readily available. However, a network of different and complementary methods or of strategic approaches can lead to firm conclusions to the structure of the bacterial and cell surfaces (Figure 28.1).

The cell surface antigens, despite their apparent availability or accessibility, present many preparative pitfalls mainly due to their sensitivity to the most common fixatives and moreover their tendency to collapse during conventional dehydration procedures. Therefore it is advisable to localize such antigens by combining the results obtained by different techniques before a firm conclusion. Among available methods, the neglected whole mount procedure, a very simple and rapid technique, seems likely to be an important tool for examining immunologically accessible surface antigens especially when used in combination with the conventional thin section method. The main advantage is that both procedures offer a clear insight of the amount, labelling efficiency and exact localization of the antigen under study. Samples screened with the whole mount procedure can be fixed in glutaraldehyde and OsO_4, stained *en bloc* before being embedded conventionally, thus combining strong labelling and optimal preservation of ultrastructure.

Observation of interaction at high resolution between cells and pathogens needs many other preparative techniques. As we cannot cover all available procedures for electron microscopy in detail, we indicate the most useful protocols for immunostaining including a short discussion about troubleshooting and precision of colloidal gold immunolabelling, for preparation of thin sections either after embedding in acrylic resins or for cryosectioning. We propose, in addition, less detailed although valuable techniques for precise immunostaining of cell membranes of gram-negative bacteria, enhancement of cell membrane visualization and high resolution scanning electron microscopy.

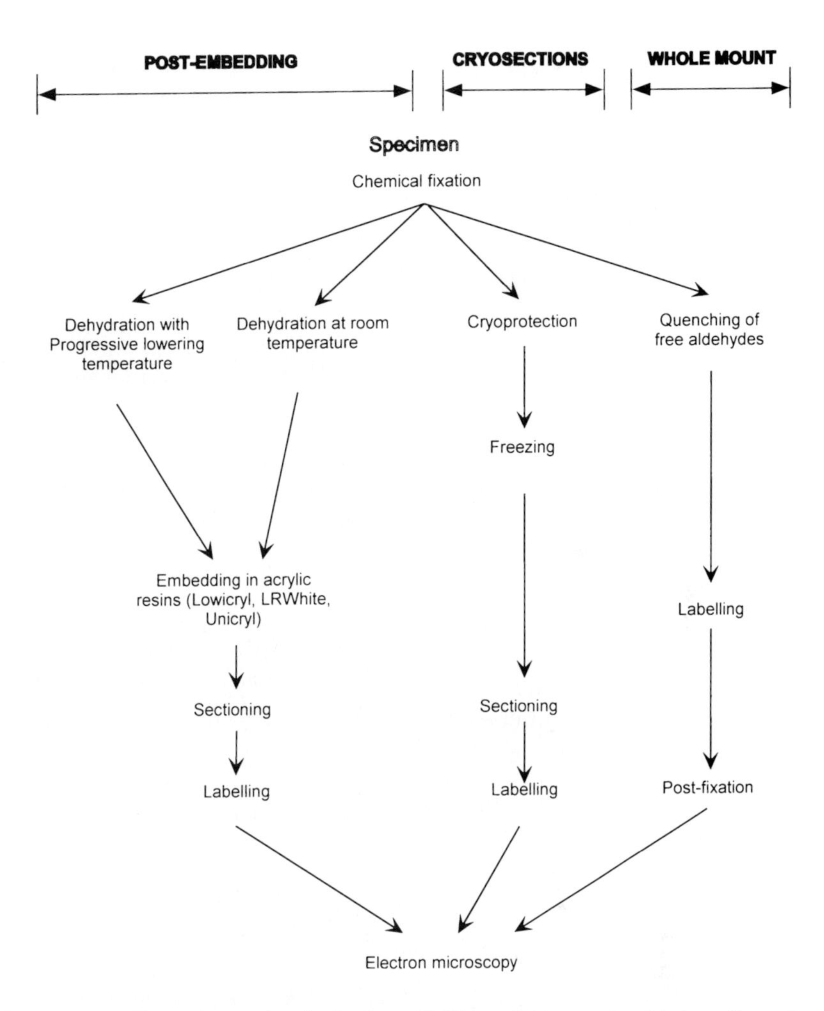

Figure 28.1. Flow chart of principal available techniques to obtain cell surface antigen localization by immunocytochemical techniques. Whole mount procedure gives fast and reliable information, which is nevertheless partial, as only the surface of cells is readily accessible to probes. This method should be completed with cryosectioning or post-embedding procedures.

◆◆◆◆◆◆ PREPARATION OF SUPPORT FILM

Support films are most commonly prepared with Formvar, Butvar, carbon or collodion. Collodion films are easy to prepare, but they must be strengthened by a carbon layer as they have a tendency to be very fragile, to drift and to lose their mass under electron beam illumination. Formvar films are strong and elastic, they don't need to be reinforced by an evaporated carbon layer. The surface of the film is rather hydrophobic and thus must previously be treated by glow-discharge. Butvar films are equivalent to Formvar films. We routinely prepare Formvar films on nickel grids (instead of copper grids) as copper ions could reduce the

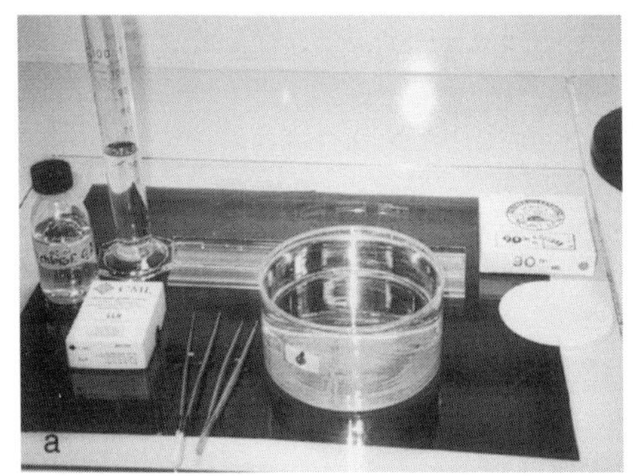

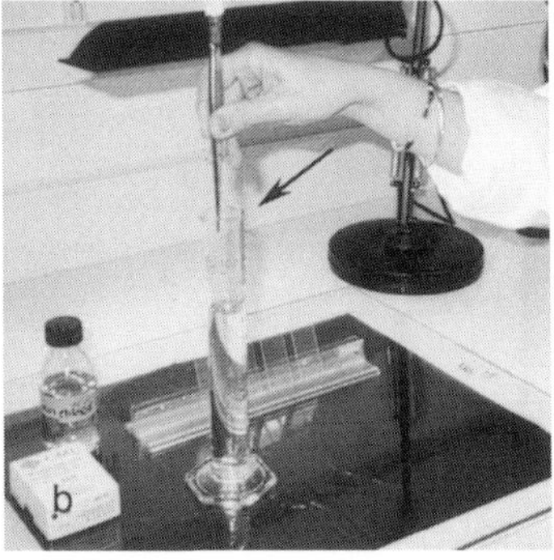

Figure 28.2. Preparation of support films for whole mount procedure and cryosections. (a) Material for preparation of film. From left to right, bottle with fresh solution of Formvar, cylinder partly filled with Formvar dissolved in chloroform, glass slides, after cleaning they are placed on a tray (in the back), forceps, dish with a clean water surface, filter papers. (b) A clean glass slide is immersed in the solution in the cylinder, then lifted slowly and kept for 15–20 seconds in the cylinder (arrow). In the back of the picture a neon tube ramp is used to check the quality of the film detached from the slide.

antigen–antibody reaction. A Formvar film is a replica of the glass surface of a microscopic slide (76 × 26 mm).

Procedure

- Dissolve 0.35% Formvar in ultrapure chloroform without agitation taken from a fresh or new bottle. This is done the day before as Formvar dissolves slowly.
- Pour the Formvar solution into a graduated cylinder (previously rinsed with a small quantity of chloroform) of size slightly larger than the microscopic slide, leaving 8 – 10 cm free (Figure 28.2a).
- Clean a microscope glass slide with soap or liquid detergent (dish-washing liquid), rinse in de-ionised water and dry with a clean dust-free towel. Do not polish the surface too energetically as the glass becomes electrostatically charged and attracts dust particles. Also the Formvar film becomes difficult to separate from the slide.
- Immerse $\frac{2}{3}$ of the glass slide in the Formvar solution for a few seconds, then lift the slide from the solution but keep it inside the cylinder just above the fluid level for 15 – 30 s. Then lift the slide out, drain the excess Formvar at the bottom edge on a filter paper and leave the slide to dry on a tray (Figure 28.2b).
- Score the Formvar film along the edges of the slide with a razor blade.
- Float the Formvar film off the glass slide onto a clean water surface by lowering the slide slowly into the water at an angle less than 30°.
- Inspect the film under reflected light (use a neon tube ramp). A thin film should be of a faint gray, almost invisible showing no wrinkles, irregularities or dust. Put washed grids on the film.
- Pick up the film with grids on it using a Parafilm® sheet. It is important to keep the water clean, fingers must never enter the water.

Before using, check a few grids in the electron microscope. They must be free of contamination, dust or wrinkles.

- It is advisable to make grids hydrophilic by glow discharge just before use.

◆◆◆◆◆◆ WHOLE MOUNT PROCEDURE

Bacteria in suspension in water or PBS are adsorbed onto Formvar film using the standard procedure for negative staining. It is advisable to check the number of cells stuck on the surface after washing and staining with 4–5 drops of 1% ammonium molybdate or 2% sodium phosphotungstate. Too few bacteria will make the final observation extremely time consuming, too many will increase background or false-positive immunolabelling.

- Grids are incubated on drops (place on Parafilm sheet in a moist chamber). Cells can be fixed appropriately (usually 0.1% glutaraldehyde in phosphate buffer for 2–3 min) and subjected to immunolabelling following the general protocol described below.

◆◆◆◆◆◆ STANDARD IMMUNOCYTOCHEMICAL PROTOCOL

The protocol given below is routinely used in our laboratory for all applications. It can be used on thin sections (acrylic embedding or cryosections) and on whole mount cells as well. The reader can find in the literature many variations of this procedure (Bendayan, 1995; Griffiths, 1993; Newman and Hobot, 1993). They all give identical results and can be used without difficulty. Sometimes we introduce another step in the case of high background and very often on acrylic thin sections. Steps 2 and 3 could be modified by introducing 0.1% Tween-20 (final solution PBS, 1% BSA, 1% NGS, 0.1% Tween-20).

- PBS (or TBS), 50 mM NH_4Cl for 5–10 min.
- PBS, 1% BSA + 1% NGS for 5–10 min.
- Specific antibody diluted in PBS, BSA 1%, NGS 1% for 30 min to many hours. This step could be realized at 4–8°C in a moister chamber.
- PBS BSA 0.1% for 3 × 5 min.
- PBS gelatine 0.01% (from skin fish gelatine) for 5 min.
- Colloidal gold antibodies diluted in PBS 0.01% gelatine for 30 min.
- PBS for 3 × 5 min.
- 1% Glutaraldehyde in PBS for 5 min.
- PBS for 5 min.
- Distilled water for 3 × 5 min.
- Dry sections.
- Counterstaining (optional).

After drying, grids are generally observed directly or after a light counterstaining with 1% ammonium molybdate or 1–2% phosphotungstic acid. Preparations could also be stabilized with a thin layer (2–3 nm) of evaporated carbon (Figures 28.3, 28.4 and 28.5).

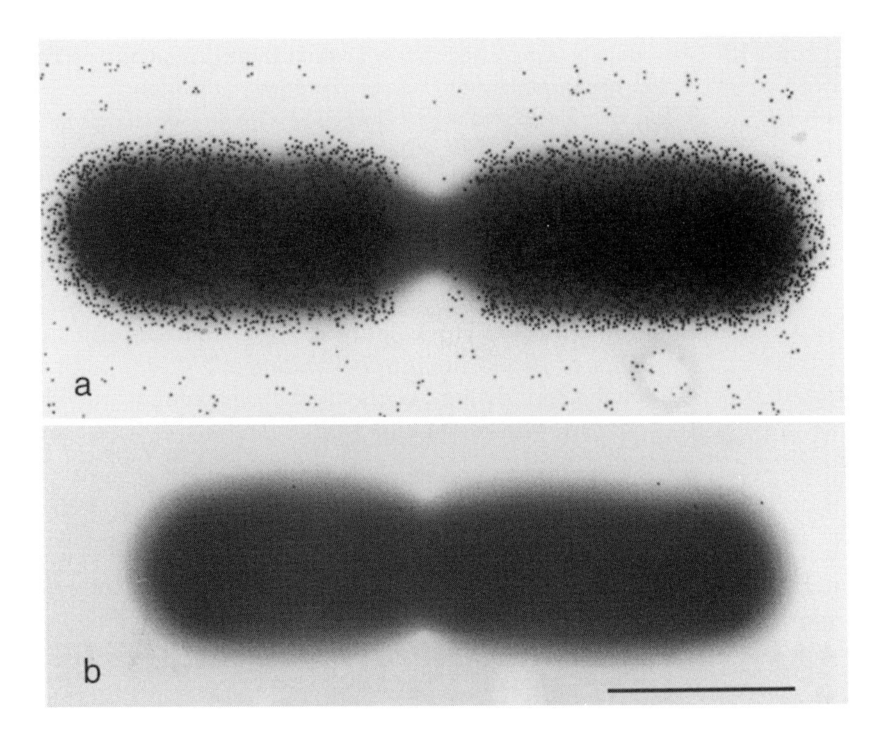

Figure 28.3. (a) Example of whole mount immunostaining of bacteria. The bacterium *Listeria monocytogenes* was subjected to anti ActA antibodies then to goat anti-rabbit antibodies labelled with 10 nm colloidal gold. Cells were observed unstained at 100 kV. (b) Control experiment where cells were only treated with colloidal gold secondary antibodies. For more details see Kocks *et al.* (1993). Bar = 0.5 μm.

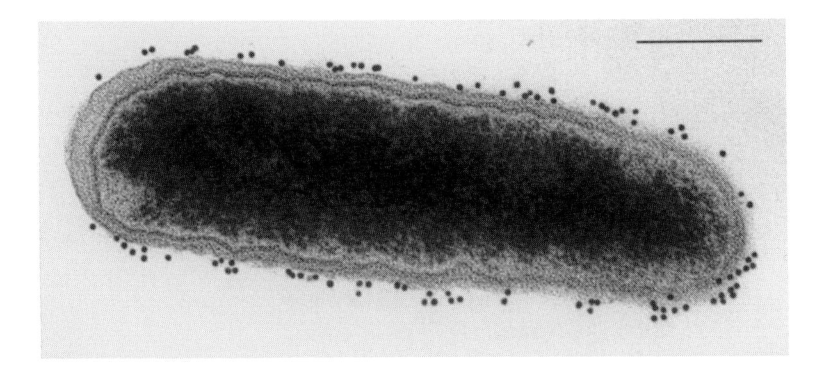

Figure 28.4. Bacteria *Listeria monocytogenes* were treated using the whole mount procedure as described (see Figure 28.3) then fixed with glutaraldehyde and osmium tetroxyde, dehydrated and embedded in Epoxy resins. Blocks were thin sectioned, stained and conventionally observed. One pole of the bacterium is not labelled which reflects the cell polarity involved in unidirectional actin-based movement of *L. monocytogenes*. Bar = 0.5 μm. (From Kocks *et al.*, 1993, with permission from the Company of Biologists Limited.)

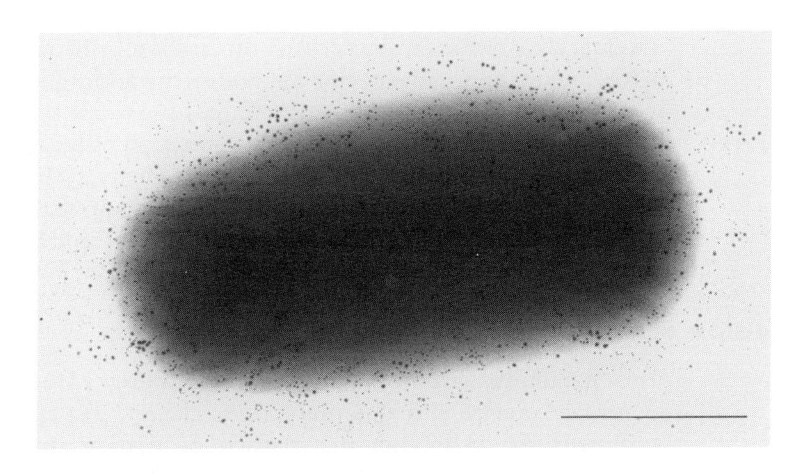

Figure 28.5. Double immuno staining of *E. coli* A30 expressing two surfaces proteins (AfaE-III and AfaD). Primary antisera were raised in rabbit and in mouse. Anti-rabbit IgG was conjugated to 5 nm gold particles (invasin AfaE-III) and anti-mouse IgG was conjugated with 10 nm gold particles. (From Garcia *et al.*, 1996, Blackwell Science Ltd, with permission.) Bar = 0.5 μm.

Trouble shooting in immunolabelling

Many pitfalls arise during immunogold labelling. In addition to the problems of weak antibodies, a low amount of available antigens or too strong and inappropriate fixation may occur and possible sources of background are very high. However, many solutions are possible. The quenching of free aldehyde can be achieved in various ways. PBS-50 mM NH_4Cl appears to be the most efficient method. This step reduces immunocytochemical background. In the case of non-specific attraction of antibodies to cellular components through fixative residues and free aldehyde remaining in the tissue, which is fairly common, it is advisable to block the surface of the sections with BSA before and during incubations. The concentration range is usually from 1 to 5%. However, high concentrations of proteins can dramatically decrease the binding of specific antibodies.

Another possible source of background is the presence in cells or tissues of highly positively charged components (collagen, elastin, lysine residues in histones) attracting negatively charged gold particles. To overcome this it is possible: (i) to block the surface of the sections with BSA; (ii) to incubate with buffer at a higher pH (i.e., 7.5–8.0); and (iii) to increase the salt content of the incubation medium (usually 0.15 M NaCl, can be increased to 0.2 M NaCl or higher).

Hydrophobic tissue components (tryptophan residues) can attract hydrophobic gold particles. In this case it will be necessary to add surfactant to the incubation buffer. The best is Tween-20 (0.1–1%). Wash the section thoroughly after this incubation step as Tween-20 is generally not recommended for use with colloidal gold.

Although not very often encountered, the presence of receptors in tissue can attract non-specifically, secondary labelled antibodies. It will be necessary to block the surface of the section with normal goat serum before incubation with specific antibodies, by adding normal serum (1%) in all incubation buffers. It could be interesting to use Protein A, Protein G, Protein AG or Fab fragment conjugates instead of IgG gold conjugates.

The presence of sulfur-containing components in the tissue, cystein or epoxy resin strongly attract gold particles. This can be overcome, at least partially, by increasing BSA in the incubation medium. Do not add gelatine since it will increase the attraction of gold to the surface of the section.

Precision of immunogold labelling

It is generally assumed that immunostaining with colloidal gold is extremely precise and this precision is close to the size of the colloidal gold particles. Although this particulate marker provides very good precision, one must consider such problems as (i) the fall down phenomenon arising after final drying of the section, and (ii) the size of the two antibodies used in the so-called 'sandwich method'.

Fall down after drying is well illustrated in Figure 28.6 where the colloidal gold particles lie a short distance around the true localization of the probed epitope with the sandwich method. Using a bridge of two immunoglobulins, the distance of the particle to the epitope can reach 20–30 nm which represents two to three diameters of the gold particle (assuming labelling with 10 nm gold particles). These facts must be kept in mind when looking for very precise localization of membrane antigens; however, many specialized methods exist to enhance the final precision of particulate immunostaining (see plasmolysis and gelatin embedding, p. 549).

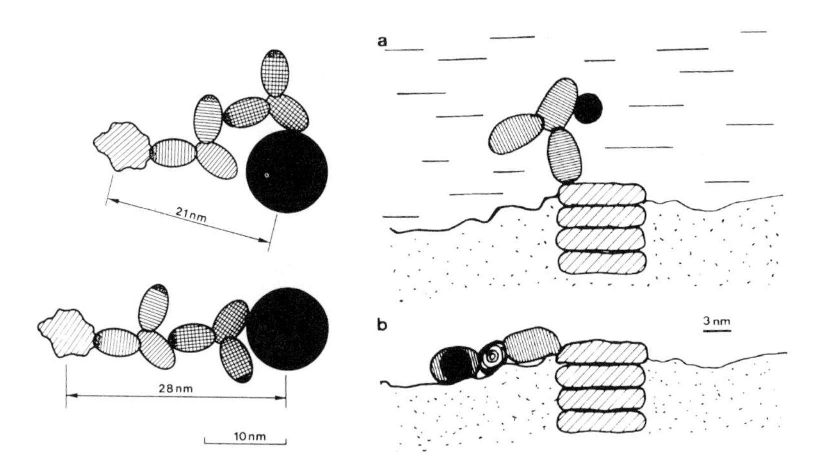

Figure 28.6. (a) Schematic representation of the binding of IgG gold complex to an epitope probed with a specific IgG (indirect method). After drying (b), the gold particles 'fall' at a short distance, however the particle is not exactly at the epitope position. With the sandwich method, the use of two antibodies (one from rabbit or mouse, the second one from goat linked to colloidal gold) can lead to a distance of 20–30 nm between the gold marker and the epitope (from E. Kellenberger and Hayat, M. A. 1991, in *Colloidal Gold, Principles, Methods and Applications*, Vol. 3, M. A. Hayat, ed. Academic Press, with permission).

◆◆◆◆◆◆ PREPARATION, IMMUNOSTAINING AND OBSERVATION OF THIN SECTIONS

Many different protocols can be used for immunocytochemical probing of antigens in bacteria, cells and tissue. Two main pathways, although non-exclusive, are equally acceptable (Figure 28.1). The first one is to realize permanent embedding in acrylic resins. Acrylic resins are available from different suppliers (Lowicryl K4M, HM20, K11M, HM23, LR White, Unicryl), all quite similar in their application.

The second pathway is to make ultrathin sections at low temperature from cells or tissues after mild fixation after infusion and cryoprotection in sucrose (Tokayasu protocol). This technique can be improved to allow better preservation of antigenicity and ultrastructure (Liou *et al.*, 1996).

Low temperature embedding in acrylic resins

It has been demonstrated that enzymes can maintain their structure and activity at very low temperatures in concentrated organic solvent. Therefore, in order to minimize molecular thermal vibration, which can have adverse effects on specimens weakly fixed with paraformaldehyde, one can dehydrate samples partially or totally at low temperature. Carlemalm *et al.* (1982) introduced the PLT technique (progressive lowering of temperature) that combines increasing organic solvent concentration with decreasing temperature, after which infiltration and polymerization are carried out. The results obtained with Lowicryl clearly show the advantages of this approach to obtain good structural preservation of cellular content and ultrastructure. Furthermore, the PLT method uses low temperature to reduce protein denaturation and to maintain the degree of hydration, which may be important in preserving protein structural conformation and antigenicity. Specimens often suffer during dehydration by organic solvents, mainly ethanol, whereas final infiltration by resin monomers and polymerization seems to be less critical. For low-temperature embedding, it is best to have specialized equipment (Gounon, 1999). However, one can achieve the required low temperatures using home made equipment or freezers.

Hundreds of protocols are available throughout the literature for low-temperature embedding and the spectrum of resins is so large that selection is not easy. An extensive study on the effect of tissue processing has been reported by Bendayan (1995) giving interesting comparisons. The following simple rules will give good guidance to the beginner.

Lowicryl K4M seems very often more popular than Lowicryl HM20. This may be due to the old classification of HM20 as 'hydrophobic' or 'apolar', similar to epoxide or polyester families. Certainly the miscibility with water is less than Lowicryl K4M, but HM20 cannot be considered a true apolar resin. Lowicryl HM20 polymerizes easily within two days at −50°C under UV light. Thin sectioning is easier since the block surface does not have a tendency to wet, as with Lowicryl K4M, thus the water level in the knife trough does not have to be very low. Nevertheless the rough surface obtained after thin sectioning allows good efficiency of immunolabelling (Figure 28.7). The immunocytochemical background observed with Lowicryl HM20 is equivalent to that observed with Lowicryl K4M.

Unicryl is easy to manipulate as it is very fluid at −20°C and even at −35°C. Although Unicryl polymerizes slowly, this does not affect either staining properties, or immunocytochemical sensitivity. Generally, Lowicryl K4M, Lowicryl HM20, and Unicryl give similar results as far

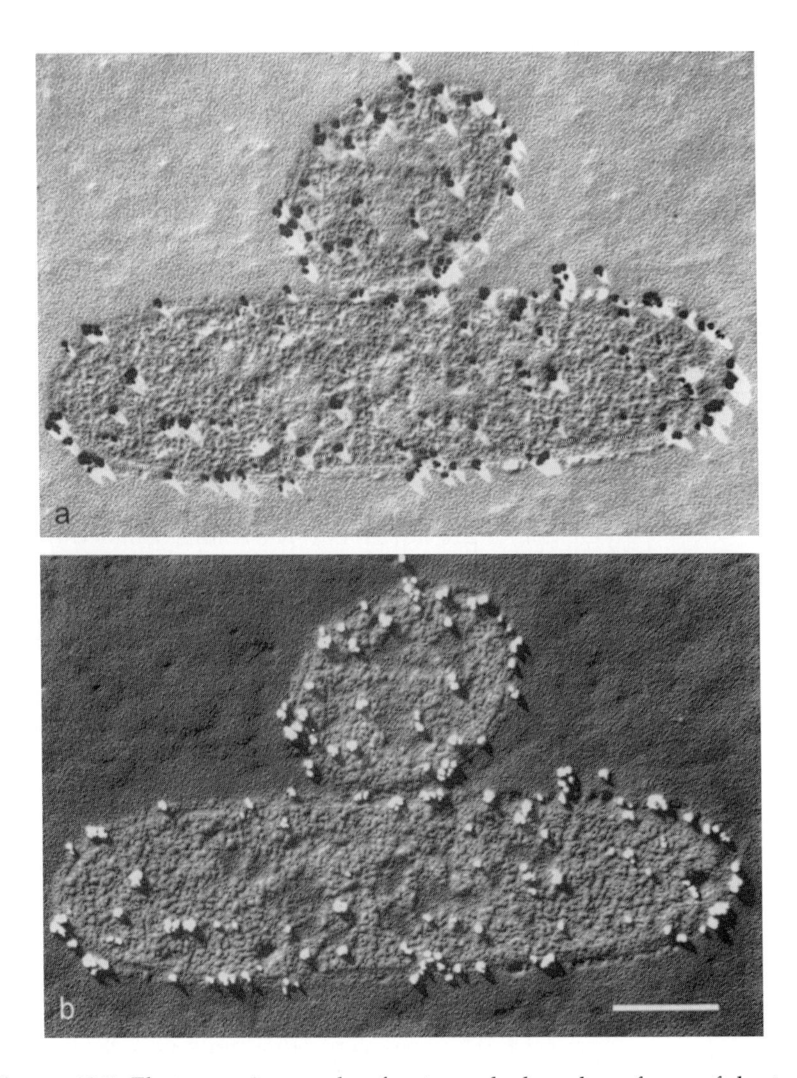

Figure 28.7. Electron micrograph of rotary shadowed surfaces of bacteria embedded in acrylic resin (Unicryl) immunostained with colloidal gold. Thin sections were stained with UMP kinase antibodies then with anti-IgG gold conjugates. The surface of the section is not flat, thus allowing binding of emerging cellular epitopes and antibodies. However, the extreme surface is only accessible to antibodies. In reverse contrast (b) the perception of colloidal gold placed on the surface of the section is better, as our brain is more used to seeing shadows as dark. (Unpublished data, for more details, see Landais *et al.*, 1999.)

as ultrastructural details and immunocytochemical sensitivity are concerned. However, the author prefers Lowicryl HM20 for its high performance at very low temperature, and Unicryl for its staining properties and fast infiltration of specimens at −20°C.

LR White (hard grade) can be used with simple protocols (see below) generally giving very good results (Newman and Hobot, 1993). Various dehydrating agents can be used for PLT protocols with acrylic resins, ethanol, methanol and acetone are miscible with Lowicryl. Ethylene glycol and dimethyl formamide, which are proposed in the literature, give excellent results in terms of structural preservation but are less employed owing to their limitations. Dimethyl formamide (DMF) and ethylene glycol are miscible with Lowicryl K4M, but do not mix with Lowicryl HM20.

Acetone can be used but care should be exercised during PLT because 70% acetone freezes around −25°C. Moreover, acetone prevents acrylic resin polymerization. Thus, if any amount of acetone is left in the specimen, polymerization can be partial and blocks are difficult to cut into thin sections. Normally, excess solvent must be eliminated by infiltration with pure resin.

Ethanol is most commonly used with Lowicryl at any temperature, although methanol offers excellent preservation of nuclear ultrastructure and organization of DNP and RNP fibrils.

Fixation

The choice of fixative is crucial in maintaining good ultrastructure and antigenicity. It is important to test different fixatives to achieve optimal results. As reported by Raposo *et al.* (1997) the following fixatives can be recommended for cells and tissues.

- 2% formaldehyde (made freshly from paraformaldehyde) in 0.1 M phosphate buffer pH 7.4. Higher concentrations can be used, but not exceeding 8%.
- Mixture of 2% formaldehyde and 0.1 – 0.3% glutaraldehyde in phosphate buffer, pH 7.4.
- Mixture of 2% formaldehyde and 1% acrolein in phosphate buffer, pH 7.4.
- Mixture of 0.1% glutaraldehyde and 1% acrolein in phosphate buffer pH 7.4.

Fixation is carried out at 4°C on crushed ice for a period not exceeding few hours. Nevertheless cells and tissues can be stored in 1% formaldehyde in phosphate buffer for up to 1 month at 4°C without noticeable loss of ultrastructural quality and immuno reactivity.

Preparation of formaldehyde containing fixatives

The required amount of paraformaldehyde is directly placed in the appropriate buffer (generally 0.1 M phosphate buffer) in a conic flask loosely closed by a small funnel. The flask is warmed-up to 60°C in a bain-marie under rotary agitation. Within a few minutes the paraformaldehyde dissolves slowly. As soon as the powder is dissolved, the flask is cooled down on ice. The solution is filtered under the chemical hood.

Glutaraldehyde or other additives can be added at this step. Do not keep this fixative solution for long; best results are obtained from freshly prepared solutions.

Embedding in Lowicryl resins

Lowicryl resins are manufactured by Polysciences Inc (USA). Lowicryl K4M (polar) and HM20 (non-polar) are now available as MonoStep resins which do not need the mixing of three components by the user.

PLT dehydration protocol

The sample (volume less than 0.5 mm^3) is transferred into 5 ml of the dehydrating agent and processed according to the protocol for PLT-dehydration of biological specimens with ethanol or methanol for embedding.

Standard PLT dehydration for Lowicryl K4M and HM20

Resin	Concentration of ethanol (%)	Time (min)	*T* (°C)
MonoStep polar K4M	30	30	0
	50	60	−20
	70	60	−30
	100	2 × 60	−30
MonoStep nonpolar HM20	30	30	0
	50	60	−20
	70	60	−35
	100	2 × 60	−50

During dehydration, the sample should slowly and carefully be agitated from time to time either by hand (gentle swirling) or using a mechanical stirrer (which should not come into contact with the sample).
Important: The samples should be kept in the same vial during the whole procedure. Any exposure of the sample to air which leads to condensation of water or drying) has to be avoided (i.e. while changing the dehydrating agent, the sample is kept submerged). The dehydrating agent for the next step of the dehydration protocol has to be added in pre-chilled small aliquots. During the dehydration steps, the vials are kept closed by polyethylene lids.

Infiltration with resin

The samples must be kept very small (0.5 mm^3).

Standard infiltration procedure for Lowicryl

Concentration of the resin (%)	Temperature (°C)		Time (min)
	polar	non-polar	
50	−30	−50	60
66	−30	−50	60
100	−30	−50	60
100	−30	−50	overnight

During resin infiltration, the samples are kept in closed vials as used for dehydration. Agitation from time to time with a toothpick or using a slow rotator is recommended.

Embedding and UV polymerization

The sample is rapidly transferred to 0.4 or 0.5 ml gelatine capsules prefilled with chilled resin. After complete filling and closing of the capsules, they are kept for about 1 hour in the cold before UV-polymerization is started. Optimal polymerization should be induced by indirect (i.e. reflected) UV-irradiation. UV sources from various manufacturers may be used provided their emission maximum is close to 360 nm wavelength.

Minimal temperatures and UV-irradiation time required for the polymerization of MonoStep resins

Resin	Minimal temperature for polymerization (°C)	Time of UV-irradiation
Polar	−30	1 day
Non-polar	−50	1 day

After low-temperature polymerization the resin blocks are slowly warmed to room temperature and irradiated for another day in direct UV-light, to improve the cutting properties of the blocks.

Embedding in Unicryl resin

Unicryl is a fairly new resin introduced by Scala *et al.* (1992) and initially known as Bioacryl. This resin is used in histology where it gives good staining on semithin sections and in immunocytochemistry (see Figures 28.7 and 28.8) where it offers high sensitivity and good structural preservation. Bogers *et al.* (1996) compared Unicryl and Lowicryl K4M, and concluded that embedding in Unicryl gives more sensitive results in tissues where the concentration of antigen is low. A comparison between LR White and Unicryl also demonstrated the superiority of Unicryl. The author proposes this protocol, which is suitable for many types of cells and tissues.

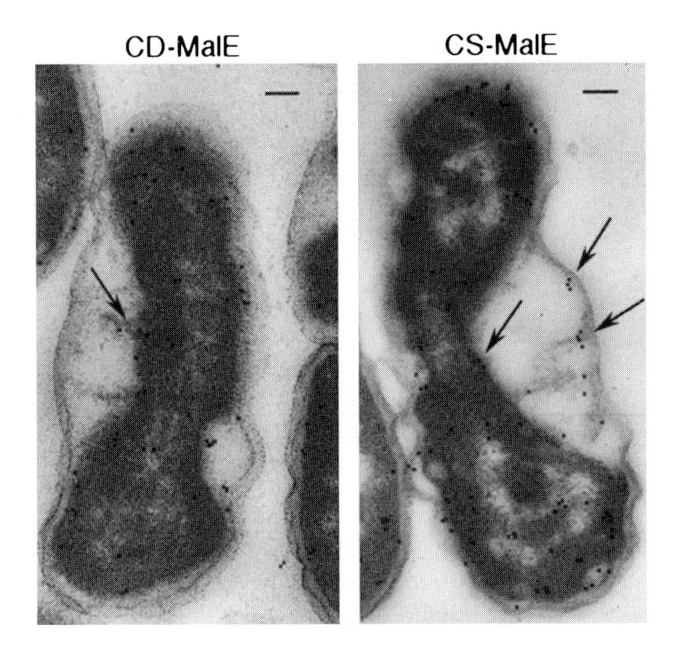

Figure 28.8. Immunoelectron microscopic analysis of localization of MalE on plasmolysed *E. coli*. Cells were plasmolysed as described and embedded in Unicryl with progressive lowering of temperature. In these two strains MalE is localized either in the cytoplasmic membrane (CD-MalE) or in the outer membrane (CS-MalE). For more details see Seydel *et al.* (1999). Reprinted with permission of Blackwell Science Ltd.

Dehydration and Infiltration

Standard schedule for embedding in Unicryl by PLT method

Resin	Minimal temperature (°C)	Time (min)
30% ethanol	4	10
50% ethanol	4	10
75% ethanol	−20	15
90% ethanol	−20	15
100% ethanol	−20	15
Unicryl ethanol 1:2	−20	30
Unicryl ethanol 2:1	−20	30
Unicryl (pure)	−20	60
Unicryl (pure)	infiltration	overnight

Optionally, add a step with fresh resin (60 min −20°C) before final embedding of the samples in capsules.

Embedding and polymerization for Unicryl

Embedding and polymerization are performed at low temperature (−20°C). The actual polymerization is carried out for 4 days in a polymerization chamber fitted with UV light using two 6W tubes (type Sylvania blacklite 350. F6W/BL 350), and then followed by final curing for 7 days at 4°C. It is possible to shorten the time needed for curing either by reducing the distance between the sample and the UV strip lights or by increasing the curing temperature (see following Table).

Another modification for fast curing has been proposed by Gounon and Rolland (1998). Benzoin ethyl ether at 0.1% or 0.05% is added to the neat resin during infiltration and for final embedding. In this case, polymerization can occur in less than 2 days at temperatures between −30°C and −35°C.

Minimal temperature and UV-irradiation time for Unicryl resin polymerization

Temperature (°C)	Specimen–lamp distance (cm) for 2 × 6 W lamps[a]			
	5 cm	10 cm	15 cm	20 cm
−20	Not advisable[b]	Not advisable	1–2 days	2–3 days
+4	Not advisable	1–2 days	2–3 days	3–4 days
−10	2 days	2–3 days	3–4 days	4–5 days

[a] Times given are approximate and depend on the quantity of resin, opacity of the vials, and efficiency of the lamps at different temperatures.

[b] These combinations of distance–temperature are not advisable as they give brittle blocks induced by overheating during fast polymerization.

Embedding protocol in LR White

There are numerous options and alternatives for embedding in LR White. The literature contains many different protocols which are precisely adapted to each specific case (Newman and Hobot, 1993).

LR White is supplied in three grades, soft, medium and hard. Only the hard grade is acceptable for electron microscopy. LR White can be supplied in uncatalyzed form. The catalyst is supplied in a small plastic vial as 'LR White catalyst benzoyl peroxide'. All the powder is mixed with the resin (following the manufacturer's information sheet) taking care not to introduce air and moisture in the resin (use bubbling with dry nitrogen).

We propose below a standard protocol for LR White embedding not requiring special or expensive equipment and giving very often excellent results compared with more sophisticated protocols.

Dehydration and embedding in LR White

Solution	Min temperature	Time
Ethanol 50%	RT or 4°C on ice	15 min
Ethanol 70%	RT or 4°C on ice	2 baths 30 min in all
Ethanol 90–95%	RT or 4°C on ice	2 baths 30 min in all
2 : 1 neat resin : Ethanol 90–95%	RT or 4°C on ice	1 × 30 min
Neat resin	RT or 4°C on ice	4 × 30 min

After fixation and quenching free aldehydes, blocks of tissues or cells previously embedded in gelatine are dehydrated.

Samples are transferred to small gelatine capsule (0.3 ml or '0 Gauge') fitted in an aluminum holder. Capsule must be completely filled with resin. Then they are tightly closed with the caps leaving a minimal amount of air at the top. Polymerization is realized in an oven accurately adjusted to 50°C, for precisely 24 h.

After polymerization a small amount of unpolymerized resin could remain at the top of the gelatine capsule, and this should be removed.

Ultrathin cryosections

This method and its variants are now well documented (Griffiths, 1993). A general protocol will be given below which gives excellent results (see Figure 28.9). However cryoultramicrotomy, which is not a complex technique, requires skill and training.

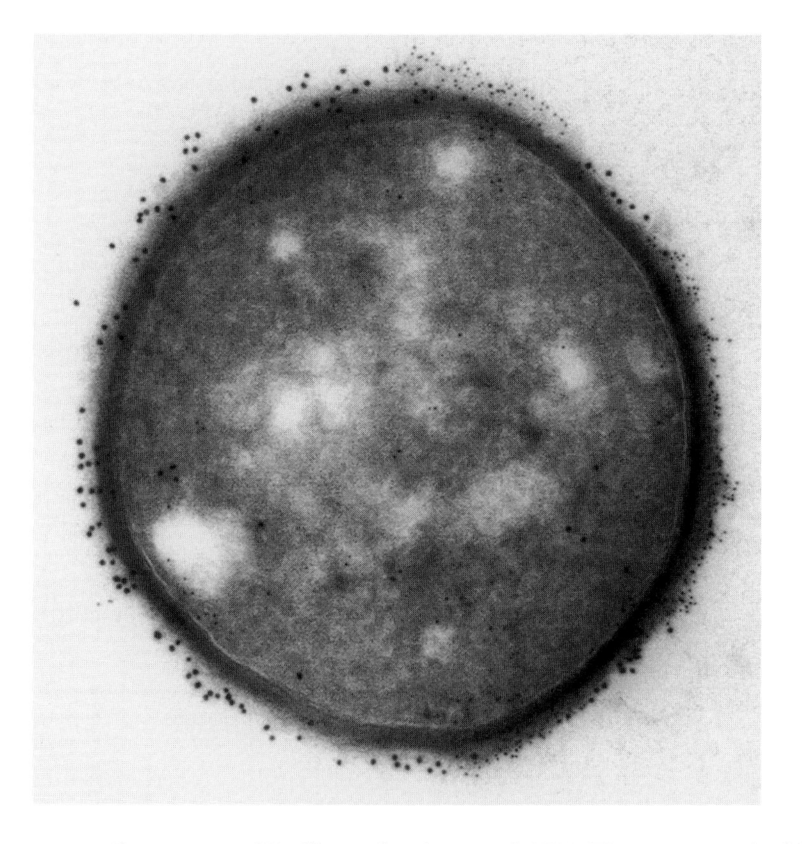

Figure 28.9. Cryosection of *Bacillus anthracis* strain CAF20. This section is double immunolabelled for two surface co-expressed S-layer proteins, EA1 and Sap. EA1 was probed by a specific rabbit serum and 10 nm anti-rabbit IgG gold conjugate. Sap was labelled with mouse antibodies and 5 nm gold conjugate. Unpublished results, for details see Mesnage *et al.* (1997, 1998).

The processing of samples requires the following steps: fixation, support, cryoprotection and freezing, sectioning, section retrieval and storage, immunolabelling, contrast enhancement and post-embedding.

Fixation

Fixation is carried out as recommended for low temperature embedding (see above).

Support

Tissues can be processed further as small coherent pieces. However, cultured cells or bacteria need support to handle them prior to sectioning. Support can be achieved by embedding the specimen in 5 or 10% gelatine in PB. This concentration of gelatine must be adjusted to obtain a medium gel viscosity which depends on the batch of gelatine available.

Embedding in gelatine can be performed in different ways (see Raposo *et al.*, 1997). We use capillary vessels (microvette CB 300 from Sarstedt, RFA) (Figure 28.10) which can be centrifuged in EC. The capillary vessel is filled with host (37°C) gelatine solution (maximum 250 μl) and placed in the centrifuge vessel filled with water. This vessel acts as a 'bain-marie'. Cells are transferred to the top of melt gelatine and the whole setting is centrifuged at low speed in EC. After solidification of gelatine on ice, the pellet is pushed away from the bottom of the inner vessel with a rod acting as a syringe.

Cryoprotection and freezing

Usually a solution of 2.3 M sucrose is used for infiltration of small gelatine blocks. Other mixtures have been proposed (sucrose and PVP, PVP added in the range 5–15%) producing thinner and well stretched sections.

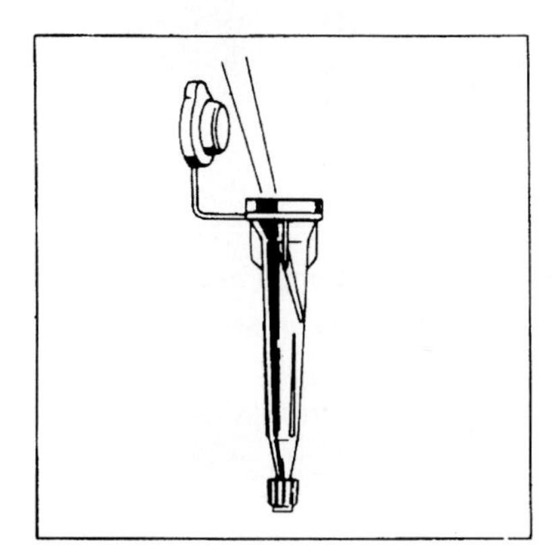

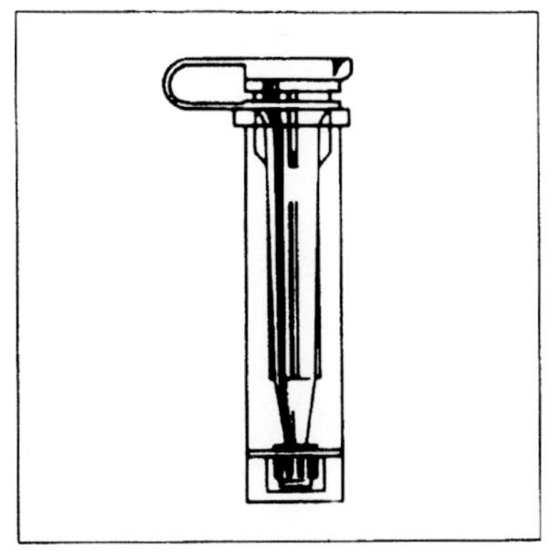

Figure 28.10. Drawing of capillary vessel Microvette CB 300 from Sarstedt, Germany. Cells or bacteria are centrifuged at the bottom of the capillary vessel (b) through gelatine solution kept warm by the centrifuge vessel (a) filled with water. A small plug of packed cells is pushed away from the bottom of the tube with a rod acting as a syringe.

Specimens are infused by sucrose or sucrose–PVP for at least two hours at 4°C in small glass vials. The specimens are then glued to the metallic holder of the microtome and immediately frozen in liquid nitrogen.

Sectioning

Cryosectioning is performed either on glass knives or on diamond knives. For ultrathin sectioning only the best knives should be used. Thus it is recommended to use diamond knives and trimming tools now commercially available (Diatome, Bienne, Switzerland; note: Drukker has stopped producing diamond knives).

The temperature of the specimen in the cryochamber is critical to obtain thin sections. It is difficult to give a precise temperature, which must be adjusted depending on the specimen and the characteristics of the knife and the microtome. We can indicate an operating temperature of −80°C for trimming and semi-thin sectioning and a temperature ranging from −100°C to −120°C for ultrathin sectioning. Semi-thin sections (0.5 μm) are easily checked by phase contrast microscopy. Ultrathin sections are handled with an eyelash mounted on a wooden stick. For continuous ultrathin sectioning it is advisable to attach an adjustable ionizer (Diatome, Bienne, Switzerland) to decrease static electricity in the chamber therefore preventing loss of sections. The ionizer should be switched off while collecting the sections, otherwise the sucrose drop will stick to the metal holder of the knife. The thickness of the sections is determined by their interference colour. White to yellow corresponds to the thinnest sections and is best seen using a source of cold light incorporated into the chamber of the cryoultramicrotome. If the cells or tissue are fixed with a mild fixative (2% paraformaldehyde, for example), thicker sections (gold/purple) can be collected, otherwise the ultrastructure will be lost due to the over-stretching in the sucrose solution used to collect the sections. However, this is largely overcome when using the newly developed method to retrieve the sections from the knife as detailed below.

Section retrieval

The ultrathin sections are collected from the knife with a loop made of nickel wire (0.1 mm) (loop diameter 2–3 mm) mounted on a wooden stick. A special loop has been designed by Diatome (Perfect loop) to collect sections and represents a breakthrough in section handling. Picking up the ultrathin sections from the knife is one of the most critical steps in cryosectioning. The sucrose solution allows the sections to stretch onto the drop thereby creating flat sections. The sucrose droplet should not be allowed to freeze before collecting the sections. On the other hand, if the sucrose droplet is still too warm the sections will be over-stretched resulting in damage to the ultrastructure. After retrieval of the sections from the cryochamber the sucrose droplet is thawed at room temperature and the sections are then rapidly placed on Formvar-coated nickel grids. Recently, Liou *et al.* (1996) examined the effects of different ultrastructure and antigenicity. They showed that structural damage due to over-stretching could be reduced by mixing the 2.3 M sucrose pick-up solution with a 2% methylcellulose solution. For mildly fixed cells and tissues this method has great advantage because it gives support to the ultrathin section, which considerably improves the ultrastructure. A freshly prepared 3:2 mixture of 2.3 M sucrose and 2% methyl cellulose is routinely used (Liou *et al.*, 1996, 1997).

◆◆◆◆◆◆ PLASMOLYSIS AND GELATINE EMBEDDING

Due to the relative lack of precision of immunogold labelling, it could sometimes be difficult to determine the exact location of an antigen in the cell wall of bacteria. The antigen could be present on the external membrane (which is assumed by the whole mount procedure, see above) in the periplasm or in the internal membrane. To overcome this discrepancy it would be interesting to undertake plasmolysis of the bacteria, which separates, in appropriate conditions, the two membranes (Figure 28.7).

Protocol

2 ml cells from culture grown overnight is more than sufficient.

- Centrifuge cells for 2 min in Eppendorf centrifuge (EC; 14 000 rpm).
- Resuspend cells in 500 μl PBS + 20% (w/v) sucrose.
- Incubate for 5 min at RT.
- Add 500 μl 2% paraformaldehyde, 0.2% glutaraldehyde, 20% sucrose in PBS.
- Incubate for 30 min at RT.
- Centrifuge for 1 min in EC.
- Re-suspend cells in 1 ml PBS + 50 mM glycin, lysin or NH_4Cl.
- Incubate for 5 min at RT.
- Centrifuge for 1 min in EC.
- Re-suspend cells in 1 ml 5% (w/v) gelatine in PBS (37°C) in microvette CB300 (Figure 28.8).
- Centrifuge for 5 min in EC.
- Leave at least 15 min on ice.
- Extract the pellet at the bottom of the microvessel as indicated above and put in fixation solution (2% PFA, 0.2% GA in PBS).
- Leave for 30 min on ice.
- Remove fixation solution and add 1 ml PBS + 50 mM glycin for 5 min at 0°C
- Eventually cut the gelatin block in smaller pieces.
- Incubate overnight at 4°C in 2.3 M sucrose in PBS. If equilibrated, the gelatine pieces will float in the middle of the flask.
- Fix the pieces onto a small holder and freeze with liquid N_2. Store in N_2.
- Samples are ready for cryoultramicrotomy. Ultrathin cryosections of 70 – 80 nm are cut on a microtome.

◆◆◆◆◆◆ METHODS TO ENHANCE MEMBRANE VISUALIZATION

With conventional fixation methods, it is often difficult to see precisely cytoplasmic membranes, notably in culture cells.

Potassium ferrocyanide added to the osmium tetroxyde probably forms a stable complex (De Bruijn, 1973). This method renders certain cellular components, such as membranes, microfilaments and microtubules, more electron dense than post-fixation in osmium tetroxyde alone. The triple layered appearance of the cytoplasmic membranes becomes accentuated. The distinct but balanced contrast of membranes and other cellular components facilitates evaluation of the overall architecture (Figure 28.11).

The chemical interactions are only poorly understood and it has not been possible to give a detailed explanation. The interactions of osmium

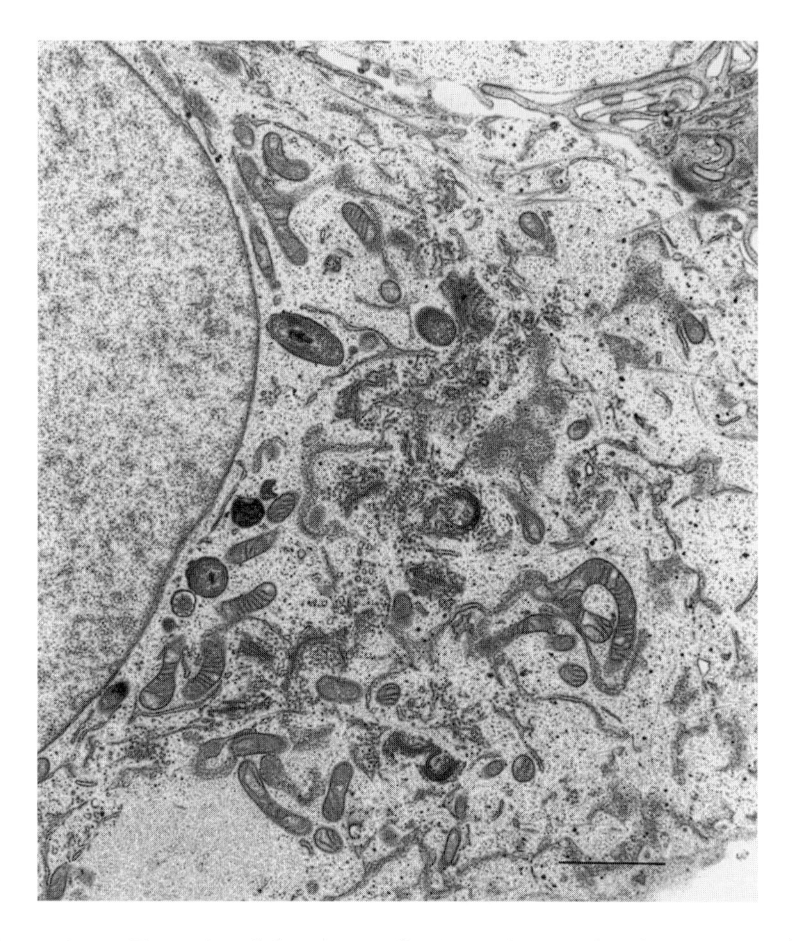

Figure 28.11. Example of low magnification micrograph of experimentally infected cell with *Bordella pertussis* fixed with osmium tetroxyde and potassium ferricyanide to enhance membrane visualization. Golgi apparatus and internal cell membranes (mitochondria, endoplasmic reticulum) are clearly seen over the cytoplasmic background (bar = 2 μm).

compounds and complex cyanides with fixed material are obviously extremely complicated. Of the many formulations which have been proposed we currently use a mixture of 1% osmium tetroxyde and 1% potassium ferricyanide (formula $K_3Fe(CN)_6$). This fixative is slow to penetrate tissues and only very small specimens can be used. However, this method can be recommended for cells in culture. The samples have to be well washed in distilled water after fixation before dehydration.

◆◆◆◆◆◆ ELECTRON MICROSCOPY OF ACTIN MICROFILAMENTS

Fixation of actin microfilaments

Actin is the single most abundant protein in animal cells. As part of the cytoskeleton, actin microfilaments are associated with many actin-binding proteins and are deeply involved in cytoplasmic transport, endocytosis and secretion.

At the light microscopy level, it is rather simple to observe the meshwork of actin filaments using fluorescently labelled molecules, the most common are phalloidin or phallacidin. However, it is not easy to clearly observe at the electron microscopy level actin microfilaments, specially in culture cells. The exact reason is relatively obscure but this could be due to the high sensitivity of actin filaments to metabolic arrest, to their thinness (7 nm) and to their sensitivity to the chemical fixatives used in electron microscopy.

It is important to mention that actin microfilaments are notoriously sensitive to osmium tetroxyde. It was demonstrated in the late 1970s that if a solution of actin microfilaments is fixed with osmium tetroxyde, the filaments are rapidly shortened within seconds into small pieces, as noted after examination of the treated solution by negative staining as well as by viscosimetry measurements. To overcome this discrepancy, the authors tried various conditions to stabilize microfilaments including fixation with glutaraldehyde, different buffers, pH conditions and reduced temperature for fixation with aldehydes and osmium tetroxyde. They found that the best condition (which means the minimal damage) was a brief osmication in a phosphate buffer at pH 6.0 and 0°C. Tilney *et al.* (1990), Tilney and Portnoy (1989) and Tilney and Tilney (1994) use these findings and other empirical approaches to define an optimal method which is to fix cells by immersion in a solution containing 1% OsO_4, 1% glutaraldehyde in 50 mM phosphate buffer pH 6.3, for no more than 45 min at 0°C. This fixative is of low-osmolarity thus allowing soluble components to diffuse freely though pores created in the membrane by osmium. Moreover this fixative is poorly rational as one mixes a highly oxidizing substance (osmium tetroxyde) with strongly reducing aldehydes (glutaraldehyde). This fixative is fairly unstable, thus it must be prepared extemporaneously keeping all the solutions in melting crushed ice. The fixation time must also be as brief as possible. After fixation, tissues or cells are washed with distilled water to remove phosphate ions as well as any cellular components poorly fixed or cross-linked. Samples are then treated overnight in 1% aqueous solution of uranyl acetate to enhance the final contrast of the specimen. They are dehydrated with ethanol or acetone and embedded in epoxy resin (Figure 28.12).

Decoration of actin filaments: Visualization of actin filament polarity in thin sections

The polarity of actin filaments can be determined from the specific binding of heavy mero-myosin to form an arrowhead configuration. The arrowhead points in the direction in which the filament generates forces by interacting with myosin. Although the polarity of actin–myosin complexes can be clearly visualized in negatively stained preparations of isolated filaments, negative staining has only limited application to the study of the distribution of actin filaments *in situ* (Figure 28.13).

Tilney and Tilney (1994) developed an improved method to clearly visualize actin filament polarity in thin sections. In the study of

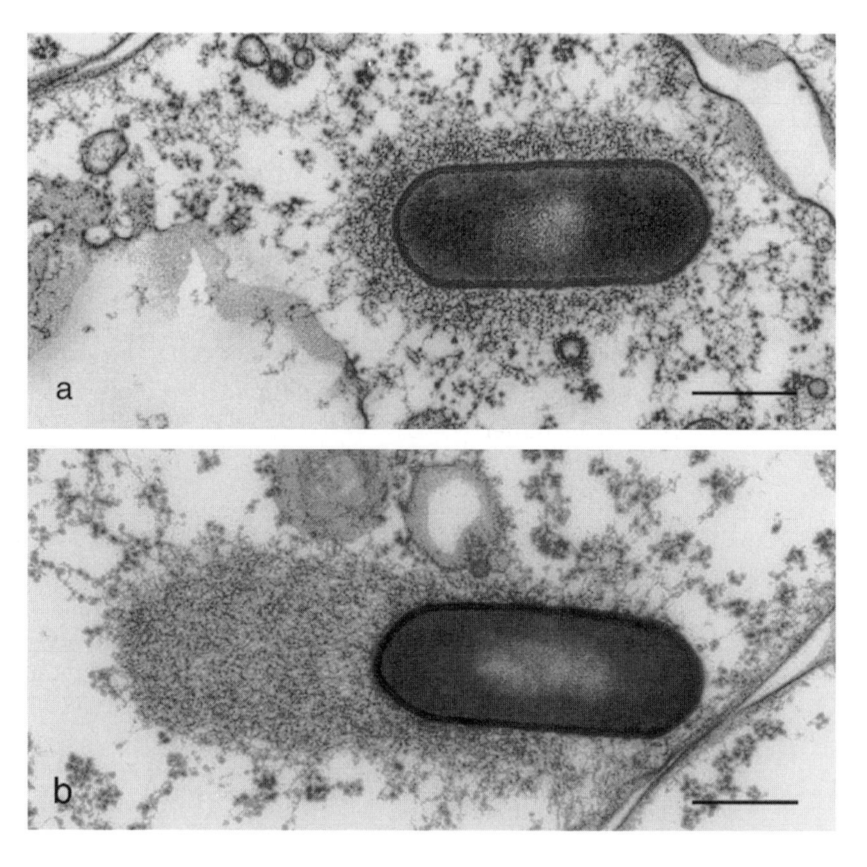

Figure 28.12. Ultrathin sections of longitudinally cut *Listeria monocytogenes* in the cytoplasm of macrophages that had been infected for 35 h. Bacteria display actin microfilaments (electron dense material) on their surface. Bar = 0.1 μm. (From Kocks *et al.*, 1993 with permission of The Company of Biologists Limited.)

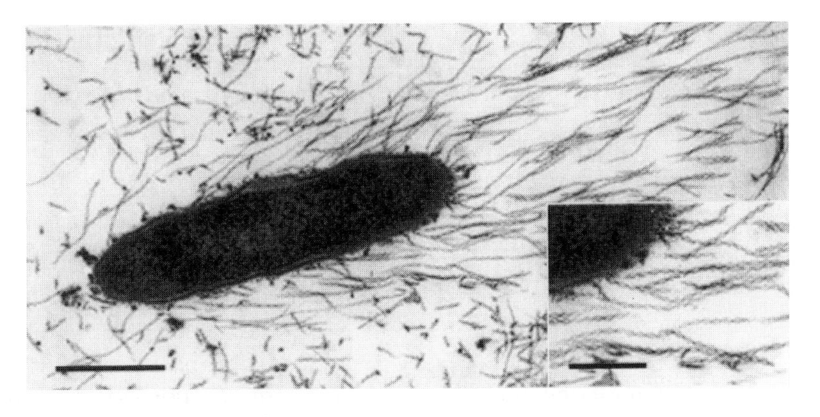

Figure 28.13. Ultrastructural aspect of myosin S_1 decorated actin tails of *Rickettsia conorii*. Individual microfilaments are easily distinguishable. With higher magnification (inset) one can see the polarity of actin microfilaments by their characteristic arrowhead decoration (bar = 0.5 μm). (From Gouin *et al.*, 1999; with permission of The Company of Biologists Limited.)

actin-based motility of many intracellular bacterial pathogens this latter method has proved highly efficient (Gouin *et al.*, 1999).

S_1 Myosin subfragments are prepared from extraction of rabbit muscle. To decorate actin filaments, cells were first detergent extracted with Triton X-100 in 50 mM phosphate buffer, pH 6.8 and 3 mM $MgCl_2$ for 10 min at 4°C. The detergent solution is carefully decanted and a solution of 5 mg ml^{-1} S_1 myosin subfragment in 0.1 M phosphate buffer pH 6.8 is added for 30 min. The first 10 min takes place on ice, the last 20 min at room temperature on a slow moving oscillating table. The S_1 solution is then discarded and cells are rinsed in 0.1 M phosphate buffer for 20 min to remove unbound S_1.

Samples are fixed in 1% glutaraldehyde, 2% tannic acid, 50 mM phosphate buffer pH 6.8 for 30 min. In the original recipe specimens were fixed in 1% glutaraldehyde, 0.2% tannic acid in 0.1 M phosphate buffer pH 7.0. The two protocols work equally well; the final contrast could be rather low with the second one. Tannic acid must be added as a powder immediately before use. For high concentration, tannic acid must be dissolved previously in phosphate buffer shortly before use and glutaraldehyde is added to it immediately before use.

Fixation is carried out at room temperature because if the tannic acid/glutaraldehyde mixture is cooled to 4°C, it will precipitate. As pointed out by Tilney and Tilney, no detergent should be present at the beginning of fixation as it also precipitates the tannic acid. The preparation is then washed in 0.1 M phosphate buffer and post fixed in 1% OsO_4 in 0.1 M phosphate buffer, pH 6.2 for 30 min at 4°C. Following osmication, the preparation is washed carefully with water, then stained *en bloc* with uranyl acetate before dehydration in ethanol and embedding in epoxy resin. Before observation, thin sections are conventionally stained with uranyl acetate and lead citrate. The use of high quality S_1 myosin fragment is the key to good decoration. In our experience, commercially available subfragments of myosin are generally unsatisfactory and we have obtained the best results with our own preparations.

◆◆◆◆◆◆ SCANNING ELECTRON MICROSCOPY

As a nice alternative to the whole mount procedure described previously, it would be interesting to probe the expression of surface protein directly on cells in culture. Many specialized SEM procedures have been described (Dykstra, 1993), however, a very simple and reliable protocol has been defined by Gounon *et al.* (2000). This approach allows one to analyse immunocytochemically the early steps of interactions between cell hosts and pathogens (Figure 28.13).

Preparation of infected cells

HeLa cells were grown on 12 mm round coverslips. At semi-confluence they were infected with *E. coli* HB101 (pILL 1101) containing the afa-3 gene cluster from uropathogenic A30 isolate. This plasmid codes for an afimbrial adhesive sheath composed of AfaD and AfaE-III (Garcia *et al.*, 1996). After 3 h incubation, non-adherent bacteria were washed with culture medium. The cells were then fixed with 1% glutaraldehyde in 0.1 M phosphate buffer pH 7.4 for 30 min at 4°C. After fixation they were carefully washed with PBS and processed for immunocytochemistry.

Immunocytochemistry

Coverslips with cells attached were deposited face down on top of a small drop (10 μl) of the incubation solutions as described above. After final fixation and washing, coverslips were processed for observation by SEM.

Preparation for scanning electron microscopy

Coverslips with cells attached to their surface were dehydrated in an increasing ethanol series then treated with Hexamethyldisilazane (Polysciences, EppelHeim, Germany) (Bray *et al.*, 1993) for 10 min and excess HMDS was evaporated under airflow in a hood. The coverslips were then coated with a 3-nm-thick carbon layer. Preparations were observed with a JEOL 6340F SEM (Jeol, Tokyo, Japan) operating at 5.0 kV and equipped with 'semi-in lens' detector for secondary electron imaging and YAG detector (Autrata, USA) for back-scattered electron imaging.

The resolution of the microscope is currently 1.5 nm in the range 5–7 kV. Images were stored as electronic files with 1280 × 1204 × 8 bits per pixel in bitmap format. Secondary electron images and corresponding backscattered images were sometimes electronically merged with the 'lightest' pixel function (Figures 28.14 and 28.15).

Figure 28.14. Immunogold staining of the AfaE adhesin of *E. coli* strain HB101 (PILLI 101). AfaE protein was probed with 15 nm gold conjugate. Cells were then prepared for SEM. The image presented here is observed by back-scattered electron imaging where colloidal gold staining, limited to the surface of the bacteria, is shown as bright dots. The intensity of the immunostaining is closely correlated with the extent to which the bacteria are wrapped in extensions of the cell surface (From Gounon *et al.*, 2000, with permission of Elsevier.)

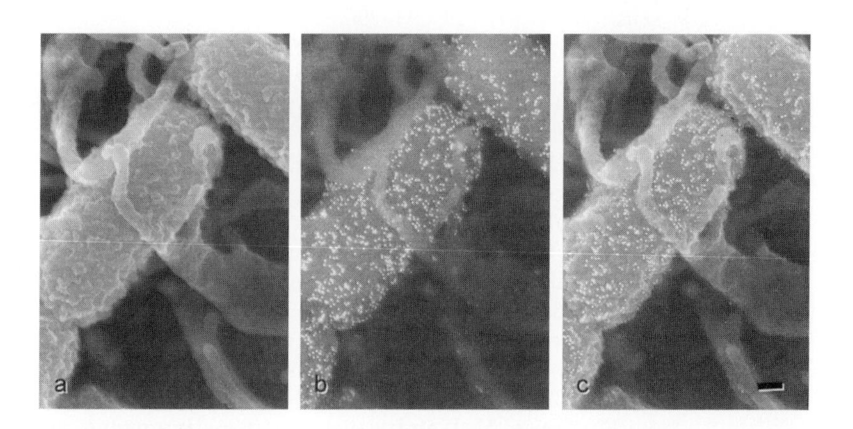

Figure 28.15. Scanning electron microscopy of immunogold staining of AfaE adhesin produced by *E. coli* strain. (a) Secondary electron imaging. Colloidal gold cannot be distinguished. (b) Back-scattered electron imaging. Colloidal gold appears as bright dots. (c) The two images can be merged either on-line using the software of the microscope or off-line with Classical Pictures software. Bar = 100 nm. (From Gounon *et al.*, 2000, with permission of Elsevier.)

References

Bendayan, M. (1995). Colloidal gold post-embedding immunocytochemistry. *Progr. Histochem. Cytochem.* **29**, 1–163.

Bogers, J. J., Nibbeling, H. A., Deelder, A. M. and VanMarck, E. A. (1996). Quantitative and morphological aspects of Unicryl versus Lowicryl K4M embedding in immunoelectron microscopic studies. *J. Histochem. Cytochem.* **44**, 43–48.

Bray, D. F., Bagu, J. and Koegler, P. (1993). Comparison of hexamethyldisilazane (HMDS), Peldri II, and critical-point drying methods for scanning electron microscopy of biological specimens. *Microsc. Res. Techn.* **26**, 489–495.

Carlemalm, E., Garavito, R. M. and Villiger, W. (1982). Resin development for electron microscopy and an analysis of embedding at low temperature. *J. Microsc. (Oxford)* **126**, 123–143.

Cossart, P., Boquet, P., Normark, S. and Rappuoli, R. (2000). *Cellular Microbiology.* Washington DC: ASM Press.

De Bruijn, W. C. (1973). Glycogen, its chemistry and morphological appearance in the electron microscope. I. A modified OsO4 fixative which selectively contrasts glycogen. *J. Ultrastruct. Res.* **42**, 29–50.

Dykstra, M. J. (1993). *A Manual of Applied Techniques for Biological Electron Microscopy.* New York: Plenum Press.

Garcia, M. I., Gounon, P., Courcoux, P., Labigne, A. and Le Bouguenec, C. (1996). The afimbrial adhesive sheath encoded by the afa-3 gene cluster of pathogenic Escherichia coli is composed of two adhesins. *Mol. Microbiol.* **19**, 683–693.

Gouin, E., Gantelet, H., Egile, C., Lasa, I., Ohayon, H., Villiers, V., Gounon, P., Sansonetti, P. J. and Cossart, P. (1999). A comparative study of the actin-based motilities of the pathogenic bacteria Listeria monocytogenes, Shigella flexneri and Rickettsia conorii. *J. Cell Sci.* **112**, 1697–708.

Gounon, P. (1999). Low-temperature embedding in acrylic resins. *Methods in Molecular Biology* **117**, *Electron Microscopy Methods and Protocols* **117**, 359–365.

Gounon, P. and Rolland, J. P. (1998). Modification of Unicryl composition for rapid polymerization at low temperature without alteration of immunocytochemical sensitivity. *Micron* **29**, 293–296.

Griffiths, G. (1993). *Fine Structure Immunocytochemistry.* New York, Berlin, Heidelberg: Springer-Verlag.

Griffiths, G. (2001). Bringing electron microscopy back into focus for cell biology. *Trends Cell Biol.* **11**, 153–154.

Kocks, C., Hellio, R., Gounon, P., Ohayon, H. and Cossart, P. (1993). Polarized distribution of Listeria monocytogenes surface protein ActA at the site of directional actin assembly. *J. Cell Sci.* **105**, 699–710.

Landais, S., Gounon, P., Laurent-Winter, C., Mazie, J. C., Danchin, A., Barzu, O. and Sakamoto, H. (1999). Immunochemical analysis of UMP kinase from Escherichia coli. *J. Bacteriol.* **181**, 833–840.

Liou, W., Geuze, H. J., Geelen, M. J. and Slot, J. W. (1997). The autophagic and endocytic pathways converge at the nascent autophagic vacuoles. *J. Cell Biol.* **136**, 61–70.

Liou, W., Geuze, H. J. and Slot, J. W. (1996). Improving structural integrity of cryosections for immunogold labelling. *Histochem. Cell Biol.* **106**, 41–58.

Mesnage, S., Tosi-Couture, E., Gounon, P., Mock, M. and Fouet, A. (1998). The capsule and S-layer: two independent and yet compatible macromolecular structures in Bacillus anthracis. *J. Bacteriol.* **180**, 52–58.

Mesnage, S., Tosi-Couture, E., Mock, M., Gounon, P. and Fouet, A. (1997). Molecular characterization of the Bacillus anthracis main S-layer component: evidence that it is the major cell-associated antigen. *Mol. Microbiol.* **23**, 1147–1155.

Newman, G. R. and Hobot, J. A. (1993). *Resin Microscopy and On-section Immunocytochemistry*. Berlin: Springer-Verlag.

Raposo, G., Kleijmer, M. J., Posthuma, G., Slot, J. W. and Geuze, H. G. (1997). Immunogold labeling of ultrathin cryosections: application in immunology, in *Weir's Handbook of Experimental Immunology*, pp. 1–208, edited by L. A. Herzenberg and D. M. Weir, Oxford: Blackwell Science Inc.

Scala, C., Cenacchi, G., Ferrari, C., Pasqualini, G., Preda, P. and Manara, G. C. (1992). A new acrylic resin formulation: A useful tool for histological, ultrastructural and immunocytochemical investigations. *J. Histochem. Cytochem.* **40**, 1799–1804.

Seydel, A., Gounon, P. and Pugsley, A. P. (1999). Testing the '+2 rule' for lipoprotein sorting in the Escherichia coli cell envelope with a new genetic selection. *Mol. Microbiol.* **34**, 810–21.

Tilney, L. G., Connelly, P. S. and Portnoy, D. A. (1990). Actin filament nucleation by the bacterial pathogen, Listeria monocytogenes. *J. Cell Biol.* **111**, 2979–88.

Tilney, L. G. and Portnoy, D. A. (1989). Actin filaments and the growth, movement, and spread of the intracellular bacterial parasite, Listeria monocytogenes. *J. Cell Biol.* **109**, 1597–608.

Tilney, L. G. and Tilney, M. S. (1994). Methods to visualize actin polymerization associated with bacterial invasion. *Methods Enzymol.* **236**, 476–481.

29 Combining Gnotobiotic Mouse Models with Functional Genomics to Define the Impact of the Microflora on Host Physiology

Lora V Hooper[1], Jason C Mills[1,2], Kevin A Roth[1,2], Thaddeus S Stappenbeck[1], Melissa H Wong and Jeffrey I Gordon[1]

Departments of Molecular Biology and Pharmacology[1] and Pathology and Immunology[2], Washington University School of Medicine, St Louis, MO 63110, USA

◆◆

CONTENTS

◆◆◆◆◆◆ INTRODUCTION

Interactions between humans and microbes are often viewed from the perspective of pathogenic relationships. However, we are colonized from birth to the end of our lives by a vast and complex consortium of non-pathogenic microbes that are much more likely to function as allies than enemies. Our relationships with these micro-organisms have been forged during a long history of co-evolution. Because of this shared history, and the vastness and variety of our affiliated microbial populations, commensal bacteria are likely to have a pronounced impact on our development and physiologic functions. Unfortunately, at present we have a very limited understanding of the extent to which our biology is influenced by our commensals, or of the molecular mechanisms that establish and maintain such host–microbial relationships. Knowledge in this area should provide important insights about the contributions of commensals to human health. It should also provide new perspectives about what constitutes a pathologic relationship, and how pathogens are able to emerge from, or to invade our commensal populations.

ISBN 0-12-521531-2

Determining the effects of commensals on host biology has been challenging. Because host–microbial relationships are generally characterized by dynamic and reciprocal interactions among bacteria, host cells, and components of mucosal immune systems, cultured cells may not accurately portray *in vivo* responses to commensals. However, the fusion of several recently developed technologies with other more established methods has made it possible to use *in vivo* models to define the effects of specific components of the microflora on host gene expression, in quantitative terms, with a high degree of cellular resolution. First, gnotobiotic techniques and involving the use of germ-free mice, allow the impact of colonization by a single bacterial species to be monitored in a genetically defined host. Second, high-density DNA microarrays provide a means for conducting comprehensive and relatively unbiased assessments of host responses to microbes (see Lockhart and Winzeler, 2000 for review). Third, laser capture microdissection (LCM; Bonner *et al.*, 1997) can be used to recover specified cell populations from complex tissues prior to and following colonization of gnotobiotic mice. LCM thus preserves the influence of surrounding cells and the environment on cellular gene expression. Fourth, quantitative real-time RT-PCR (Heid *et al.*, 1996) allows quantitative measurements of colonization-associated changes in the levels of specific mRNAs within microdissected cells.

In this chapter, we describe each of these methods in detail, and where appropriate, illustrate them with examples from our own investigations of the impact of commensal microbes on intestinal gene expression.

◆◆◆◆◆◆ USE OF GNOTOBIOTICS TO CREATE SIMPLIFIED EXPERIMENTAL MODELS OF HOST–MICROBIAL INTERACTIONS

A key experimental strategy for defining the impact of micro-organisms on host physiology is to first examine cellular function in the absence of bacteria (i.e. under germ-free conditions) and then to evaluate the effects of adding a single species, or defined number of species of bacteria. Rearing mammals under germ-free conditions has developed into a scientific field of its own, termed gnotobiology from the Greek 'gnosis', meaning knowledge, and 'bios', meaning life.

Maintenance of germ-free mice in flexible film isolators

Germ-free mice must be housed in a sterile environment that is protected from contamination, but which allows for easy manipulation of the animals. Currently, there are two types of housing. Lightweight stainless steel isolators (Gustafsson, 1959) are extremely durable and can be steam sterilized in a large autoclave. Although they form very reliable barriers, they are expensive. Flexible film isolators (Trexler and Reynolds, 1957) are more commonly used. They are more readily punctured and cannot be steam sterilized, but their cost is low.

Setting up a flexible film isolator

In flexible film isolators, a clear vinyl 'bag' forms an envelope that completely encloses a sterile space (Figure 29.1A). These isolators come in several different sizes, ranging from those that can hold a single cage to those that can accommodate numerous breeding cages. We routinely use a 24″ × 24″ × 48″ chamber that accommodates ~eight mouse cages and can be placed on a cart or a table. Manipulations are carried out within the sterile workspace using neoprene gloves directly attached to the bag by metal compression rings, thus forming an integral part of the barrier (Figure 29.1A). A double-door port is located in the wall of the isolator opposite the gloves. This port is used to bring supplies into the isolator. Flexible film isolators are available from Standard Safety Equipment Company (McHenry, Illinois), along with detailed instructions for their set-up and maintenance. Flexible film isolators can also be obtained from Class Biologically Clean (www.cbclean.com).

Air is supplied to each isolator by a dedicated blower (i.e. one per isolator) attached to a filter by a flexible vinyl hose. This set-up allows air to be sterilized before it enters the isolator (Figure 29.1A). The sterilizing filter is constructed by wrapping three or four layers of 4 μm fiberglass filter media mats (12 mm thick) around a stainless steel cylindrical support and securing the device with Mylar tape. A vinyl-coated glass screen is then used to protect the filter media from damage. Air exits the isolator through an identical filter assembly (Figure 29.1A). All filter unit components are available from Standard Safety along with detailed assembly instructions.

Sterilization of the interior of flexible film isolators is done chemically, using a dilute solution of Clidox (chlorine dioxide liquid; Pharmacal, Naugatuck, CT). Items made of glass, plastic, or stainless steel can be placed inside the isolator prior to sterilization. A stainless steel atomizer with a plastic reservoir is used to spray Clidox into the isolator. The isolator is then left sealed for ~2 h. The air blower is started subsequently, inflating the isolator. Air pressure within the isolator is kept slightly above atmospheric, so that the isolator remains inflated and immediate entrance of airborne contaminants is prevented in the case of accidental puncture.

Sterile supplies, such as autoclaved food, water, and bedding, are brought into the isolator through the side port. These supplies are autoclaved in a stainless steel drum whose diameter is exactly that of the port. Its open end is covered prior to autoclaving with a sheet of Mylar film (Figure 29.1B). After autoclaving, the drum is attached to the isolator port with a plastic transfer sleeve. The interior of the sleeve is sterilized for ~30 min by fogging with Clidox. The inner port cap is then removed from inside the isolator, the Mylar seal on the mouth of the cylinder is punctured, and the supplies are imported into the isolator. The plastic transfer sleeve can also be used to transfer cages or supplies between isolators. Supplies that cannot be autoclaved are brought into the isolator by placing them in the entry port, sealing the outer cap, fogging the port with Clidox for 30 min, and removing the inner port cap.

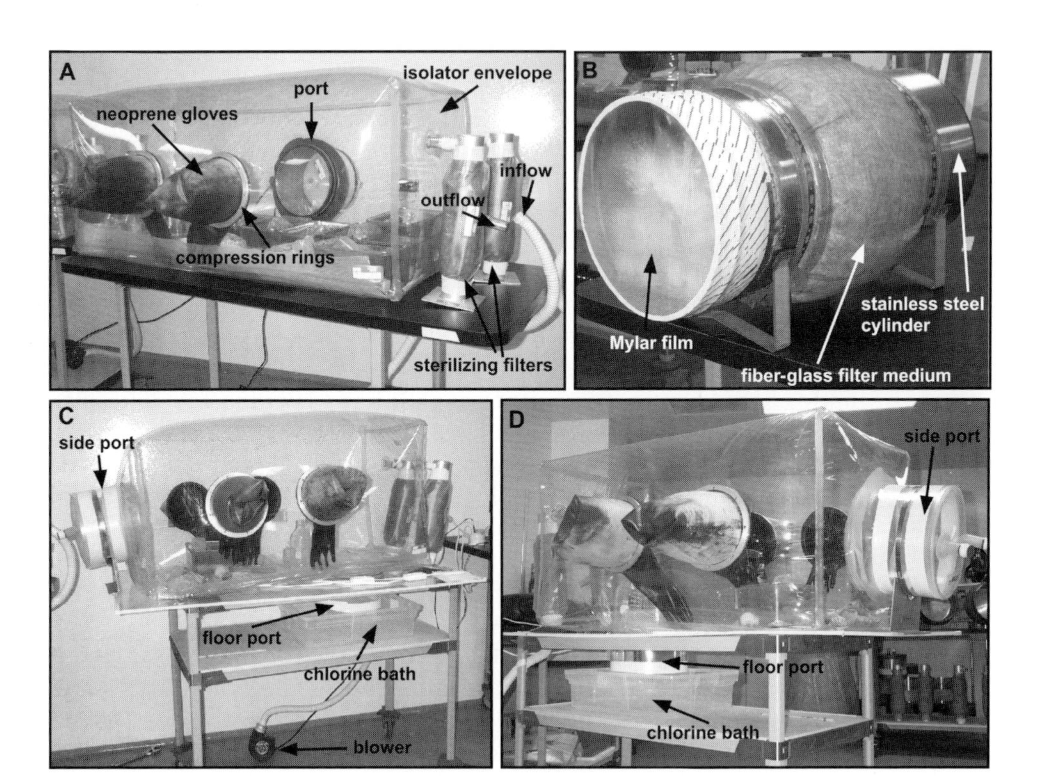

Figure 29.1. Use of flexible film isolators for maintaining and re-deriving germ-free mice. (A) *Flexible film isolator.* A clear vinyl envelope, placed on a cart, completely encloses a sterile space containing cages and supplies. Neoprene gloves attached to the envelope by metal compression rings allow manipulations to be carried out within the sterile workspace. A double-doored port located in the wall of the isolator opposite the gloves is used to import supplies into the isolator. Air is supplied to the isolator by a blower attached to a sterilizing filter by a flexible vinyl hose. Air exits the isolator through an identical filter assembly. (B) *Sterilizing drum.* Sterile supplies such as food, water, and bedding, are autoclaved in a stainless steel cylinder whose diameter is exactly that of the isolator port. Its open end is covered prior to autoclaving with a sheet of Mylar film. After autoclaving, the drum is attached to the isolator port with a plastic transfer sleeve. The sleeve interior is sterilized by fogging with Clidox. The inner port cap is then removed from inside the isolator, the Mylar seal on the mouth of the cylinder is punctured, and the supplies are imported into the isolator. (C and D) *Two views of a surgical isolator for Cesarean re-derivation.* This isolator differs from the normal flexible film isolator by having two sets of gloves and an additional port in the floor. Prior to re-derivation, the outer floor port cap is removed and the mouth of the port is submerged in a chlorine bath. The uterus containing live pups is removed aseptically from the female, and each horn of the uterus is clamped using sterile hemostats. The uterus is passed into the chlorine bath, and brought up into the germ-free isolator. Pups are immediately removed and fostered to germ-free lactating females. (This figure is also reproduced in colour between pages 276 and 277.)

Special attention must be paid to the diets of germ-free mice. Since germ-free mice lack a microflora, required nutrients synthesized by microbes (e.g. vitamin K) are lacking. In addition, autoclaving causes some nutrient destruction. An enriched formula is therefore critical (e.g. BeeKay autoclavable rodent diet; B&K Universal).

Microbiological surveillance

Routine monitoring of isolators for the presence of bacteria and fungi is imperative when maintaining a germ-free colony. In our laboratory, animal feces are monitored on a weekly basis for microbial contamination. In addition, the inside surfaces of the isolator are swabbed and cultured. All samples are cultured aerobically and anaerobically in three different media: nutrient broth (a general purpose medium for cultivating micro-organisms with non-exacting nutritional requirements), brain/heart infusion broth (allows cultivation of a wide variety of fastidious micro-organisms), and Sabouraud Dextrose Broth (supports growth of yeasts, molds and aciduric micro-organisms). All three media are available from BD Difco. Anaerobic incubations are done at 37°C in GasPak jars (Fisher) with activated GasPak Hydrogen + CO_2 envelopes (Fisher). Aerobic incubations are done at 37°C and 42°C.

Sources of germ-free mice

Currently, there are only a few commercial suppliers of germ-free mice. Taconic, Inc. offers two strains: NIH(S) and Black Swiss (www.taconic.com). The University of Wisconsin Gnotobiotic Laboratory (www.medsch.wisc.edu/gnotolab/) offers germ-free BALB/c, NIH(S) and Black Swiss strains, as well as several transgenic and knockout lines.

Re-derivation of mice as germ-free

Any conventionally-raised mouse line can, theoretically, be made germ-free. This allows assessment of the effects of overexpression, misexpression, or knockout of specific genes in animals lacking a microflora, or harboring only one or a few bacterial species. Germ-free re-derivations can be performed using one of two techniques: embryo transfer or Cesarian section. Both methods take advantage of the fact that embryos/fetuses developing in the uterus of a healthy mother are free of microbes. Embryo transfer into the uterus of a pseudopregnant germ-free female is useful for recovering embryos stored in frozen embryo banks, and aids in eliminating certain viral pathogens (Carthew *et al.*, 1983). However, because of its relative ease and simplicity, Cesarian re-derivation is the more commonly used method.

To re-derive a mouse line by Cesarian section, timed matings are performed, and the pregnant females are sacrificed just before delivery (i.e. at 19 days gestation, although the duration of pregnancy varies somewhat depending on genetic background). Cesarean re-derivations require a special surgical isolator that differs from the normal flexible film isolator by having two sets of gloves and an additional port in the floor (Figure 29.1C,D). Prior to the re-derivation, the outer cap of this port is removed and the port mouth is submerged in a 0.9–1.2% chlorine bath held at 27–30°C. (Chlorine can be purchased as a 10–12% solution at local pool supply stores; the exact concentration can be determined using the DPD total chlorine test from Hach Co., www.hach.com). The uterus, containing live pups, is removed aseptically from the gravid conventionally raised female, and each horn of the uterus is clamped using sterile hemostats. The uterus is then passed into the temperature-controlled chlorine bath for exactly 90 s, and is brought up into the germ-free isolator. Pups are immediately removed from the uterus, revived by rubbing with sterile cotton swabs, and fostered to germ-free lactating females.

Colonization of germ-free mice with defined components of the microflora

How a particular bacterial species is introduced into mice varies depending on the microbe and its host niche. For example, *Bacteroides thetaiotaomicron*, a prominent component of the distal small intestinal and colonic microflora of humans and mice, readily colonizes the mouse intestine after stationary phase cultures are spread on the fur (Bry *et al.*, 1996; Hooper *et al.*, 1999). In contrast, *Helicobacter pylori*, which normally colonizes the human stomach, is inoculated by gavage of 10^7 colony forming units (Guruge *et al.*, 1998).

Once animals have been colonized for a period of time, the density of colonization should be determined so that it can be correlated with host responses. For example, in the case of *B. thetaiotaomicron*, serial dilution platings of 1 μl of duodenal, jejunal, ileal, cecal, and colonic contents are done at the time of sacrifice (Bry *et al.*, 1996; Hooper *et al.*, 1999). For *H. pylori*, complete stomach contents are homogenized and subjected to serial dilution plating (Guruge *et al.*, 1998).

Options for small-scale gnotobiotic research

Investigators who have no previous experience with gnotobiotics, or who are conducting one-time investigations or pilot studies, have several options. The University of Wisconsin Gnotobiotic Laboratory offers a number of services, ranging from breeding and housing germ-free animals to carrying out specific experiments on germ-free or gnotobiotic animals. Taconic provides re-derivation services, and can also perform certain types of experiments on germ-free animals. In addition, Taconic offers an isolator rental service, in which a fully equipped gnotobiotic isolator is shipped to the investigator. Charles River Laboratories (www.criver.com) will also perform re-derivations.

◆◆◆◆◆◆ DNA MICROARRAY ANALYSIS OF COLONIZATION-ASSOCIATED CHANGES IN TISSUE GENE EXPRESSION

Any of a number of analytical techniques can be applied to gnotobiotic animal models to assess the impact of bacterial colonization on host physiology. However, there currently is a limited pre-existing conceptual and experimental framework for anticipating the spectrum of host responses elicited by commensals. The development of high-density oligonucleotide- or cDNA-based microarrays makes it possible to monitor expression of tens of thousands of genes simultaneously. These microarrays offer an unprecedented opportunity to comprehensively assess the impact of microbial colonization on host gene expression. For example, we have used such an approach to conduct a global analysis of the impact of *B. thetaiotaomicron* colonization on intestinal gene expression in a gnotobiotic mouse model (Hooper *et al.*, 2001).

Currently, there are two basic microarray technologies: cDNA-based and oligonucleotide-based. cDNA microarrays are usually constructed by affixing PCR products, representing portions of genes, to ordinary glass slides. To compare two RNA populations, cDNA incorporating a red-channel fluorophore (e.g. Cy5) is synthesized from one RNA, and cDNA incorporating a green-channel fluorophore (e.g. Cy3) is generated from the other RNA. The ratio of red to green signal produced from each DNA on the microarray allows quantitation of the relative amounts of the corresponding mRNA species in the two RNA preparations (Ferea and Brown, 1999). A number of cDNA-based microarrays are commercially available (e.g. Incyte Genomics (www.incyte.com) and NEN Life Sciences (www.nen.com)). In addition methods for designing and constructing your own custom cDNA arrays can be found at cmgm.stanford.edu/pbrown/mguide/index.html.

High density oligonucleotide-based microarrays

Our discussion focuses on the widely used commercial oligo-based arrays (known as GeneChips) from Affymetrix, Inc.(www.affymetrix.com). Genes are represented on GeneChips by sets of 25 base oligonucleotides (termed 'probes'; the labeled cRNAs prepared from RNA samples and hybridized to GeneChips are known as 'targets'). The oligonucleotide probes are generated *in situ* by light-direct chemical synthesis at high density on a chip (details about this photolithographic process can be found in Lipshutz *et al.*, 1999). Each probe sequence is located within a single 24 μm by 24 μm spot or 'feature' on the chip (Figure 29.2A).

The entire cohort of probes used to recognize a specific cRNA target is known as a 'probe set'. Probe sets comprise multiple (usually ~20) oligonucleotide 'probe pairs'. Each of the 20 probe pairs represents a different region of the transcript (Figure 29.2A). Each probe pair, in turn, consists of a 'perfect match' (PM) oligo, whose 25 nucleotide sequence is designed to exactly complement the target cRNA sequence, and a 'mismatch' (MM) oligo that differs from the target cRNA by a one base mismatch at the middle (13th) position. Hybridization of a cRNA target species with the perfect match oligo represents signal, while hybridization with the corresponding mismatch oligo represents noise. This probe-pair-based system acts as a sophisticated internal control for possible non-specific hybridization to relatively short nucleotide sequences. Each 1.28 × 1.28 cm GeneChip contains ~6500 to 12000 probe sets. Details regarding the genes represented on these arrays are available from www.affymetrix.com.

Isolation of total RNA and preparation of cRNA targets

Microarray-based expression analysis involves a comparison of transcript levels between at least two tissues. Tissue harvested immediately after sacrifice is snap-frozen in liquid nitrogen. Total cellular RNA can be extracted using the Qiagen RNeasy Midi Kit. (The Mini Kit can be employed for tissue samples that weigh less than 100 mg.) The minimum amount of total RNA needed to generate a cRNA target for hybridization to Affymetrix GeneChips is 5 μg. However, procedures requiring much smaller amounts of starting material have been reported (see Luo *et al.*, 1999; Wang *et al.*, 2000 and below).

Detailed protocols for target preparation are available from the WhiteheadMIT Center for Genome Research website (www-genome.wi.mit.edu/MPR/methodsPUBg99.html) and from Affymetrix. Therefore, we will provide only a brief overview of the process (see Figure 29.2B). The first step in target preparation for oligonucleotide arrays is synthesis of double-stranded cDNA using total cellular RNA as the template. First-strand cDNA synthesis is primed using oligod$(T)_{24}$ with an attached antisense T7 promoter sequence (obtained from GenSet; www.gensetoligos.-com). After second strand synthesis is completed, an *in vitro* RNA transcript is generated from the second strand DNA template using T7 RNA polymerase and biotinylated NTPs. The double stranded cDNA synthesis step is stoichiometric: i.e. one template mRNA molecule is converted to one double stranded cDNA molecule. However, the generation of cRNA results in a linear amplification of the template, since multiple transcripts can be primed from a single second strand DNA template. The biotinylated cRNA target is then incubated at 94°C in the presence of magnesium, yielding 30–200 base fragments to minimize the influence of RNA secondary structures on hybridization.

Hybridization, washing, staining, and scanning of microarrays

Hybridization and detection of bound cRNA targets require specialized equipment available from Affymetrix, including a rotisserie hybridization oven, a GeneChip Fluidics Station, and a confocal laser scanner. In brief, targets are hybridized to the arrays overnight at 42°C with rapid rotation. After washing away unhybridized target, the bound biotinylated target is detected by staining the array with streptavidin-phycoerythrin. In many cases it is advantageous to further amplify the signal by adding biotinylated anti-streptavidin, followed by another round of staining with streptavidin-phycoerythrin. After staining, patterns of hybridization are detected by scanning the array with the confocal laser. Hybridization data are collected as light emitted from the fluorescent reporter group (phycoerythrin) bound to the target which, in turn, is bound to the oligonucleotide probe. Detailed protocols for hybridization, washing,

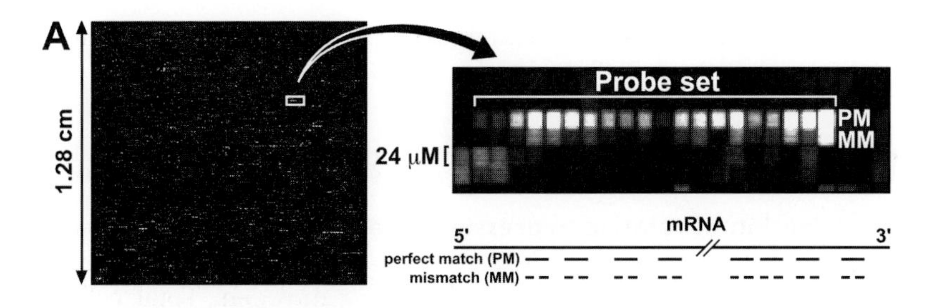

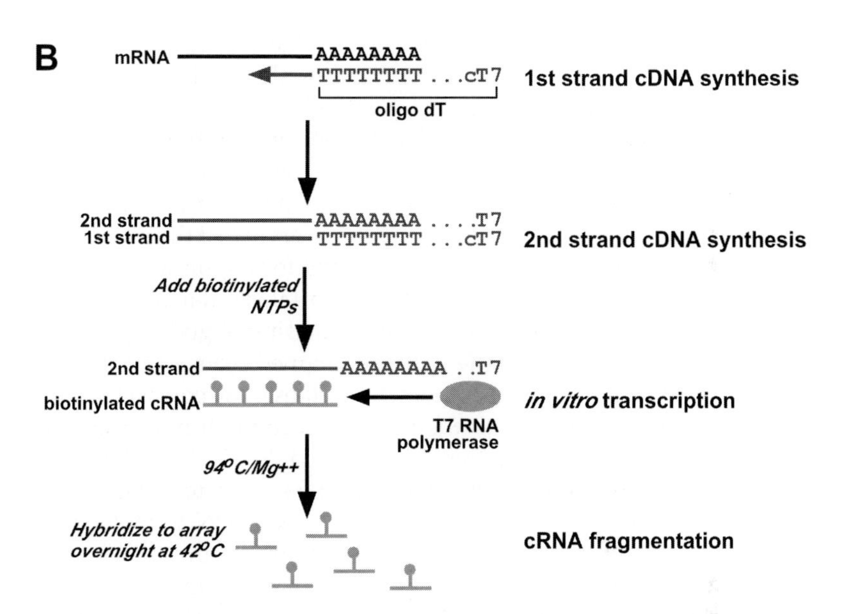

Figure 29.2. Probe set and probe pair representation in high density oligo-based microarrays and preparation of cRNA targets. (A) *Array design*. Genes are represented on $1.28\,cm^2$ arrays by sets of 25 base oligonucleotides known as 'probes'. 10^7 copies of each probe sequence is present within a single $24\,\mu m \times 24\,\mu m$ area. A gene is represented by multiple (≤ 20) oligonucleotide 'probe pairs' comprising a 'probe set', with each probe pair representing a different region of the transcript. A probe pair consists of a 'perfect match' (PM) oligo, whose 25 nucleotide sequence is designed to exactly complement the target RNA sequence, and a 'mismatch' (MM) oligo differing from the target transcript by a one base mismatch at the middle (13th) position. Hybridization of a cRNA target species with the perfect match oligo represents signal, while hybridization with the corresponding mismatch oligo represents noise. Currently available chips feature 6500 to 12000 probe sets arrayed on a single $1.28\,cm \times 1.28\,cm$ chip. (B) *Target preparation*. First strand cDNA synthesis is primed from total cellular RNA using oligod$(T)_{24}$ with an attached anti-sense T7 promoter (cT7). A sense T7 promoter site is generated in the second cDNA strand. *In vitro* transcription is then performed using T7 RNA polymerase and biotinylated NTPs, resulting in a population of biotinylated cRNAs. The cRNAs are incubated at 94°C in the presence of magnesium, yielding 30–200 base fragments to minimize the influence of RNA secondary structures on hybridization.

staining, and scanning are available at www-genome.wi.mit.edu/MPR/methodsPUBg99.html.

Analysis of GeneChip microarray datasets

Determination of relative expression levels

The proprietary GeneChip software provided by Affymetrix determines the level of a given target RNA species by first comparing the signal intensity of hybridization at each PM oligonucleotide (signal) relative to that at each corresponding MM oligonucleotide (noise) and then integrating PM to MM relationships across all 20 oligonucleotide probe pairs comprising a probe set. To compare the relative levels of expression of a gene in two RNA samples, one sample is always designated 'baseline' and the other 'partner'. The most important parameters derived from the PM to MM comparison are:

(a) Average Difference—the average of the PM-MM differences across all 20 probe pairs, a rough measure of signal intensity or transcript abundance
(b) Absolute Call—whether the PM-MM signal is strong enough such that the target transcript is deemed Present.
(c) Difference Call—whether a given transcript is Increased or Decreased on the partner GeneChip relative to the baseline GeneChip.
(d) Fold Change—the relative increase or decrease of a given transcript.

Strategies for eliminating noise (false-positives)

When looking for changes in gene expression, most investigators consider only those genes showing a Difference Call of Increased or Decreased, a Fold Change of $\geq$2, and an Absolute Call of Present in either the partner or the baseline sample. Using these criteria, Affymetrix GeneChips typically have a false positive rate of 1–2% (Lipshutz *et al.*, 1999; Lee *et al.*, 2000). Given the scale of a typical chip experiment, even such a low error rate can produce a large number of genes falsely called Increased or Decreased between two conditions. For example, since the Affymetrix Mu11K GeneChip set represents ~11 000 different mouse genes, a 1% error rate means 110 false positive calls (Claverie, 1999; Lockhart and Winzeler, 2000). However, our experience has shown that comparisons of biologically different RNA populations often yield no more than 100 to 200 changes in gene expression. Thus, without any way to filter false positives, $\geq$50% of the changes in expression may be noise.

One way to reduce noise is to perform replicate experiments. This can be done in one of two ways. In a *biological replicate*, two or more samples are collected independently, under similar experimental conditions. cRNA targets are then prepared in parallel and each preparation is hybridized to a separate microarray. In an *analytical replicate*, several RNA samples are prepared under the same conditions, pooled, and then split into two or more aliquots prior to target cRNA preparation and hybridization to separate chips. In both cases, only those genes whose expression level changes in the same direction in all replicate comparisons are scored as exhibiting true changes. Analytic replicates provide a way of filtering noise due to variations in the microarraying procedure or target preparation. Biological replicates also facilitate reduction of biological noise. However, replication arithmetically increases the cost of experiments. Replication also does not help the chip user to assess the overall quality of comparisons. For example, if comparison A results in 100 replicated changes and comparison B results in 50 replicated changes, are the tissues being compared in A really more divergent than those compared in B, or is A just a 'noisier' chip study? In addition, replication does not help the chip user stratify the quality of individual, replicated gene changes. Is a Fold-Change of ~5 with lower baseline and partner chip Average Differences more likely to be real than a Fold-Change of 2.5 with high baseline and partner Average Differences? For that matter, what are 'high' and 'low' Average Differences?

To address some of these issues, we have developed a method for noise filtration that does not rely on the use of replicates (Mills and Gordon, 2001). Although the approach was developed using the mouse Mu11K GeneChip set, the theoretical framework is applicable to other GeneChip microarrays. Briefly, we generated multiple separate comparisons of single RNA samples, treating them as if each were actually a different RNA (a 'same–same' comparison). Each gene with a Difference Call of Increased or Decreased was considered a false positive, and, for each such probe set, we plotted the signal intensity of the probe set on the baseline chip versus its signal intensity on the chip to which it was compared (the partner chip). Using this approach, we then prepared a grid plotting the distribution of paired partner versus baseline signal intensities for every false positive (Figure 29.3A). This grid represents combinations of signal intensities in any given comparison that were likely to yield false positives. After compiling multiple same–same comparisons and multiple resulting false positive distributions, we prepared look-up tables (LUTs) ranking the ranges of signal intensities most prone to error. The LUTs could then be used to stratify changes in gene expression in comparisons between 'truly different' RNA populations in order of their likelihood of being noise. The LUTs rank genes from 0 to 6, with a score of 0 for a given pair of partner and baseline chip signal intensities bearing about a 15% chance of being reproduced in subsequent replicate experiments, and an LUT score of 6 carrying about a 70% chance for replication. In practice, we usually set a threshold LUT value of 4. Our studies show that LUT scores <4 account for 90% of false positives in same–same comparisons.

If readers are interested in establishing LUTs for their own microarray system, they need only make a few same–same comparisons (i.e. for each same–same comparison, duplicate cRNAs are generated from the same starting RNA; each cRNA preparation is hybridized to a separate microarray; the two datasets are then compared to one another using the standard GeneChip comparison software). The user can then analyze comparisons between biologically different RNA populations based on how they overlap with the same-same comparisons. Genes whose paired signal intensities fall within the most common paired signal intensities of the false positive distribution are the least likely to be true positives.

Not surprisingly, in our experience most false positives have either low Fold Changes or low Average Differences on both chips. However, we found that another subset of false positives also occurs when hybridized target transcript intensity is so strong that it is near maximal on both chips. Both these systematic patterns of error are a by-product of the global scaling protocols that the GeneChip software uses to compare two chips (see Figure 29.3B). The software scales all probe sets on both chips in a comparison so that the average chip-wide intensities are equal. After scaling, GeneChip software can compare individual genes on equal footing. In theory, scaling corrects for chip-wide, arbitrary differences in intensity due, for instance, to variation in hybridization conditions or labeling. In practice, global scaling works for genes whose expression levels are in the midrange of chip intensity signals. However, it creates artefacts at the extremes of chip signal detection.

Gene annotation and categorization

Once genes have been identified whose expression has changed during the experimental treatment, they must be categorized and interpreted.

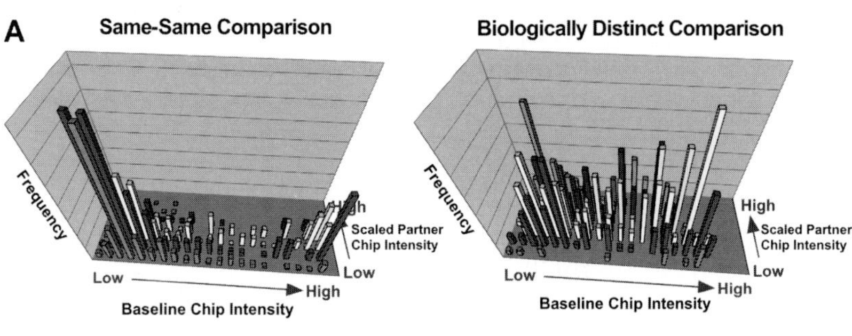

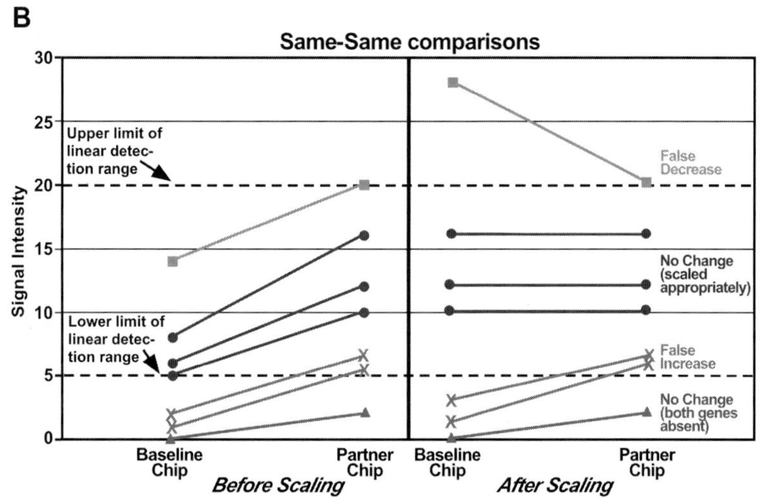

Lists of these genes can be exported from the GeneChip proprietary software into an Excel spreadsheet, where much of the annotation process can be automated. The issue of annotation is of greatest concern for arrays such as the Mu11K, where a substantial fraction of probe sequences on the array recognize uncharacterized genes from the public UniGene EST database. This is an advantage to users interested in identifying novel sequences in their experiments. However, it means extra research.

The first step in identification is to link the probe set metrics (e.g. Fold Change, Average Difference, etc.) with Affymetrix's gene identification information, as well as with the GenBank identification number of the sequence Affymetrix used to design the probe set. Once information from the internal database (supplied by Affymetrix) has been appended to each gene, users must then consult relevant external, web-based databases to definitively identify genes. In our experience, the most

---

Figure 29.3. Strategy for eliminating noise from GeneChip datasets. (A) *Distributions of Partner vs. Baseline chip intensities in Same–Same GeneChip comparisons differ from comparisons of biologically distinct RNA.* Examples of two types of comparisons are depicted: (1) Same–Same, where chips are hybridized to cRNAs prepared from the same starting RNA and (2) Biologically Distinct, where the two chips in the comparison are hybridized to cRNAs prepared from different starting populations of RNA. The entire range of baseline chip signal intensities ('Average Differences') is represented by the horizontal axis. The corresponding entire range of partner chip intensities is represented by the axis going into the plane of the page. The frequency of probe sets having specified combinations of Baseline and Partner chip intensities is represented by the vertical axis. Only the probe sets called Increased or Decreased are plotted. Note how Same–Same comparisons result in probe sets at the extremes of Baseline chip intensity and at the low end of Partner chip intensity, whereas Biologically Distinct comparisons result in probe sets in the intermediate Baseline chip intensity values and/or high Partner chip intensity values. (B) *Microarray scaling as a source of false-positive changes in gene expression.* Affymetrix GeneChips typically have a false positive rate of 1 to 2% (Lipshutz *et al.*, 1999; Lee *et al.*, 2000). Given the scale of a typical chip experiment, even such a low error rate can produce a large proportion of genes falsely called Increased or Decreased. A significant source of false positive errors is global scaling, in which the signal intensities of all probe sets on both chips are multiplied by a scaling factor so that the average chip-wide signal intensities are equal. Scaling corrects for chip-wide differences in signal intensity due to variations in hybridization conditions or target labeling. Following scaling, the GeneChip software can compare individual genes across multiple chips and determine whether there are changes in their expression levels. In this example, signal intensities for specific probe sets from a comparison of two microarrays probed with duplicate cRNA targets (a 'Same–Same' comparison) are represented, both before and after scaling. Although in practice both partner and baseline chips are scaled to an arbitrary chip-wide target intensity, to simplify the example we have assumed here that the partner chip needs no scaling. Because the targets are derived from the same tissue RNA sample, the signal outputs from each of the probe sets theoretically should be equal between the two chips following scaling. However, in practice, global scaling works well only for genes whose expression levels are in the linear range of signal detection (represented by blue lines). Artefacts may be created at the extremes of chip signal detection. This can happen for genes having low Fold Changes or low Average Differences (gray lines), or for genes whose signal intensity is so strong that it is near maximal on both chips (green lines). Adapted from Mills and Gorden, 2001. (This figure is also reproduced in colour between pages 276 and 277.)

useful databases for gene identification are: (i) BLAST (at NCBI, www.ncbi.nlm.nih.gov; provides the most up-to-date information about cloned and sequenced cDNAs); (ii) UniGene (www.ncbi.nlm.nih.gov/UniGene; clusters ESTs and performs homology searches for all six possible reading frames, providing putative homologs and orthologs); and (iii) TIGR (www.tigr.org/; has its own clusters as well as mirroring the UniGene clusters, providing cluster sequence and orientations of all ESTs that fall within the cluster). Identification of genes uncharacterized at the time of GeneChip design can be performed by integrating BLAST and UniGene information. We have developed a software tool that does this, as well as automating many other parts of the annotation procedure, as long as the microarray datasets are stored in Excel spreadsheets (one row per gene). This software (GenQuery Engine; Mills and Gordon, 2001) calculates LUT scores, adds GenBank and SWISS-PROT hyperlinks, returns gene-specific information from the Affymetrix-supplied database, queries TIGR and UniGene, and returns UniGene gene identification and homology information. The software is freely available online at gordonlab.wustl.edu/mills.

Once genes are identified, they are categorized after extensive searches of the literature. NCBI's PubMed, at www.ncbi.nlm.nih.gov/entrez/query.fcgi, provides an excellent gateway to the biomedical literature. Categorization can occur at multiple hierarchical levels and may result in identification of gene response 'modules' (Hartwell *et al.*, 1999; Tamayo *et al.*, 1999; Eisen *et al.*, 1998). Of course the annotation process is subjective, and will vary as a function of both user and experiment.

Independent verification of microarray results

Changes in gene expression detected by microarrays often need to be validated by an independent method. At one level, it may be necessary to verify changes in expression using the same RNAs used to generate microarray targets. It may also be necessary to determine biological reproducibility by repeating the experiment and auditing the results with some other method for detecting changes in gene expression.

Independent validation can be done using one of several techniques for quantitating levels of specific mRNAs. The approach is often dictated by the amount of RNA available and the sensitivity required to detect the transcript of interest. Traditional methods include Northern blotting (Zhu *et al.*, 1998; Harkin *et al.*, 1999) and RNase-protection. A newer high-throughput method, known as real-time quantitative reverse-transcriptase polymerase chain reaction (qRT-PCR) requires only small amounts of total RNA (< 20 ng per reaction), and is sensitive enough to detect even low abundance transcripts (Heid *et al.*, 1996). Because of its reproducibility, sensitivity and specificity, qRT-PCR is ideal for quantitatively validating microarray results (Lee *et al.*, 2000; Hooper *et al.*, 2001).

Principles of SYBR-Green based real-time quantitative RT-PCR

In qRT-PCR the amount of amplicon (PCR product) generated after each PCR cycle is monitored in real time. Increasing amounts of double-stranded DNA product are reflected by increasing fluorescence emitted from a dye that binds to the DNA, or from a fluorogenic probe that anneals to the middle of the amplicon (see below). A threshold cycle (C_T) for each amplicon is defined as the PCR cycle at which the fluorescence intensity crosses a user-established threshold (Figure 29.4A). C_T is inversely proportional to the log of the starting mRNA transcript copy number (Heid *et al.*, 1996).

There are two basic approaches for performing qRT-PCR. One employs a labeled fluorogenic probe (TaqMan probe) that anneals to sequences within the amplicon. The TaqMan probe has a 5′-conjugated fluorescent tag that serves as the reporter and a 3′-conjugated tag that quenches the reporter fluorescence. The intact probe does not fluoresce. Amplicon is detected when the 5′ nuclease activity of Taq polymerase cleaves the TaqMan probe during the extension phase of PCR, releasing the reporter dye and resulting in a fluorescence emission (monitored in real time). A second approach is based on the use of SYBR-Green I (Molecular Probes, Eugene, OR) which fluoresces when bound to double stranded DNA. Applied Biosystems (www.appliedbiosystems.com) has developed conditions compatible with their 5700 and 7700 Sequence Detection Systems that allow qRT-PCR to be performed with SYBR-Green I.

The fluorogenic TaqMan probe allows detection of multiple transcripts within a single sample and is highly specific for a given sequence. None the less, SYBR-Green offers a number of advantages. First, the cost is lower. Second, the method is more versatile, since the dye can be used to detect amplicon from any transcript. Third, binding of multiple SYBR-Green molecules to each amplicon increases the sensitivity of detection. The principal drawback to the SYBR-Green method is its lack of specificity, since any contaminating double-stranded DNA species (such as primer-dimer) will contribute to the fluorescence signal. However, this can be overcome by establishing a melting temperature for the amplicon that exceeds that of any primer-dimer complexes (see below). Fluorescence emission is then measured at this melting temperature after the extension phase of each PCR cycle.

cDNA synthesis

As with any RT-PCR application, the first step in qRT-PCR is cDNA synthesis. 10 μg of total RNA is DNase-treated for 15 min at 37°C in a reaction mixture that includes 10 mM $MgCl_2$, 1 mM DTT, 4 U of human placental RNase inhibitor, and 0.5 μg of RNase-free DNase (final volume of 100 μl; all reagents from Roche Biochemicals; biochem.roche.com). Residual DNase is then removed using the Qiagen RNeasy Mini Kit, and the purified RNA quantitated by measuring absorption at 260 nm. Next, 0.25 μg of oligod$(T)_{15}$ (Roche) or 0.1 μg of random hexamers (Life Technologies; www.lifetech.com) is added to 2 μg of the DNase-treated

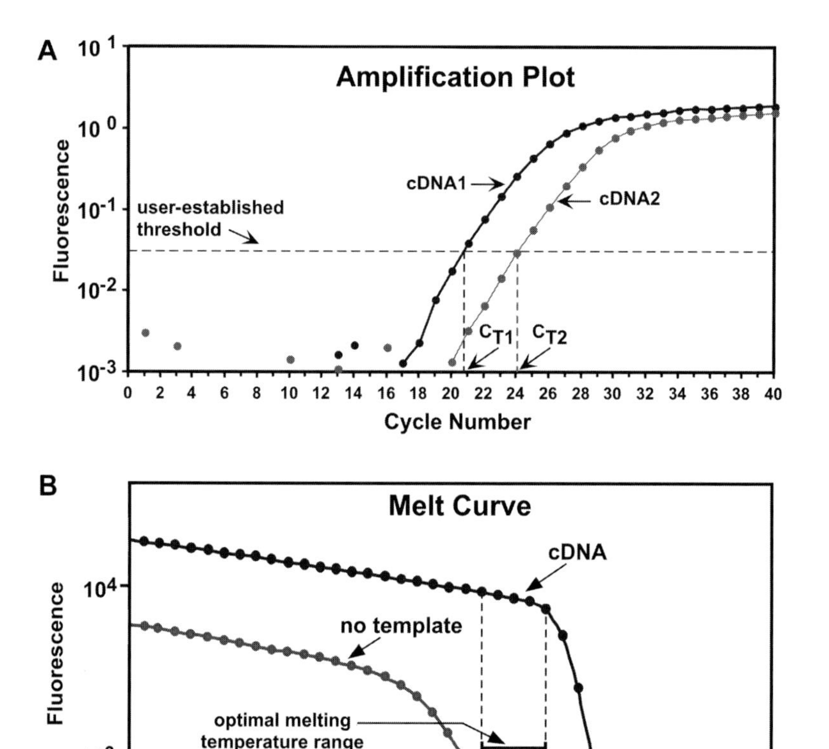

Figure 29.4. SYBR-Green based real-time quantitative RT-PCR (qRT-PCR). (A) *qRT-PCR amplification plot.* The amount of amplicon present after each PCR cycle is monitored in real time. Increasing amounts of amplicon are reflected by increasing fluorescence from SYBR-Green, a dye that binds to double-stranded DNA. The threshold cycle (C_T) is defined as the PCR cycle at which the fluorescence intensity crosses a user-established threshold. C_T is inversely proportional to the log of the starting mRNA copy number. In this example, the amount of transcript represented in cDNA1 is higher than in cDNA2. (B) *Melt curve.* In SYBR-Green based qRT-PCR, contaminating primer-dimer will contribute to the fluorescence signal. Therefore, an amplicon melting temperature exceeding that of any primer-dimer complexes is established. See the text for details on how to establish a melt curve. Because the stability of double-stranded DNA decreases with increasing temperature, comparison of the amplicon melt curve (generated with cDNA as template) versus the primer-dimer melt curve (generated with no template) allows determination of an optimal melting temperature range in which the majority of the fluorescence signal is derived from the amplicon. In a normal qRT-PCR assay, fluorescence emission is measured at this temperature following the extension step of each PCR cycle.

RNA in a final volume of 12 µl. The reactions are incubated for 5 min at 70°C, cooled to 42°C in the same water bath over the course of 1 h, quickly spun in a microfuge (to collect condensate), and returned to 42°C. Two µl of the mixture are transferred to another tube for the minus reverse transcriptase (−RT) reaction. Ten µl of a +RT reaction mixture [200U MMLV reverse transcriptase (Roche), 1 × RT buffer (50 mM Tris-HCl, pH 8.3, 7.5 mM KCl, 3 mM $MgCl_2$), 2 mM dNTPs, 20 mM DTT, 24U human placental RNase Inhibitor (Roche)] is added to the tube containing 10 μl of RNA/primer mixture. 2 μl of the −RT mixture is added to the tube containing 2 μl of the RNA/primer mixture. The reactions are then incubated for an additional hour at 42°C. cDNAs (+RT reactions) are diluted to 200 μl with deionized H_2O and controls (−RT reactions) with 20 μl of dH_2O. Finally, a PCR assay using primers specific to actin is performed on 2 μl aliquots of both the +RT and −RT reactions to verify the efficiency of cDNA synthesis and to confirm the absence of genomic DNA contamination. This actin assay uses primers that span an intron–exon junction, so that different sized PCR products are produced depending upon whether the cDNA template or a contaminating genomic DNA template is amplified (forward actin primer = 5′-CACCACACCTTCTACAATGCTG, reverse actin primer = 5′-TCATCAGGTAGTCAGTGAGGTCGC; cDNA product = 300 bp, genomic DNA product = 450 bp).

Primer design

Design and selection of a primer pair that yields a single, specific amplicon is perhaps the most critical step in performing SYBR-Green based qRT-PCR. Primers that comply with the following requirements yield the best results: (i) Tm = 58–60°C; (ii) GC content = 20–80%; (iii) length ~20 nucleotides; and (iv) no more than two G or C residues among the five bases at the 3′ end. Additionally, the amplicon should be 50–150 bp to minimize PCR cycle time and ideally should span an intron–exon junction to allow differentiation between the genomic DNA and cDNA product. Prior to qRT-PCR, all primer pairs are tested by conventional PCR using cDNA and genomic DNA templates, as well as a water blank, in order to establish that a single amplicon is produced in the cDNA sample and that minimal primer-dimer bands are produced in the absence of template.

To ensure that the fluorescence measurement after each cycle is due only to the specific amplicon, a melting temperature must be identified that results in a maximal difference between the amount of amplicon and primer-dimer. Normal qRT-PCR amplifications are performed (95°C for 15 s; anneal at 55°C for 45 s; extend at 72°C for 30 s for a total of 40 cycles). The product obtained after 40 cycles of PCR is then subjected to a melt curve analysis (Figure 29.4B). To generate this melt curve, fluorescence is measured at 0.2°C intervals between 60 and 92°C. We exploit the PE Biosystems 7700 Sequence Detection System software to collect and analyze the 160 separate incremental readings as follows. The machine is programmed to collect increments of four 0.2°C fluorescence readings. The first four readings span the temperature range of 60.0 to 60.6°C. This collection of four readings is treated as one 'round' of data collection. The second round collects four fluorescence measurements at 0.2°C increments between 60.8 and 61.4°C. This iterative process is continued for a total of 40 rounds (160 data points) in order to span 60–92°C. We then plot the fluorescence measurement from the first or third data point from each round to generate a melt curve such as that shown in Figure 29.4B.

Comparison of the amplicon melt curve, generated with cDNA as template, versus the primer-dimer melt curve, generated with no template, provides a temperature range at which the majority of the fluorescence signal is derived from the amplicon (see Figure 29.4B). Note that when plotting this data, round number is converted to temperature using the following equation:

$$(\text{round number} - 1)(0.8^\circ\text{C}) + 60^\circ\text{C}$$

Once this temperature has been determined from a melt curve analysis, subsequent experimental qRT-PCR assays are performed by simply programming the Sequence Detection System to ramp to the melt temperature following the extension phase of each cycle so that fluorescence data can be collected.

Assembling the reaction mixture

Assays are performed in triplicate. Each 25 μl reaction mixture contains:

1. 2 μl of cDNA.
2. 2× SYBR Green Mix [8% glycerol, 2.5 mM $MgCl_2$, 2× PCR buffer (40 mM Tris-HCl, pH8.4, 100 mM KCl), 200 μM dATP, dGTP, dCTP, 400 μM dUDP, 0.32× SYBR Green I dye in DMSO (Molecular Probes, Eugene, OR), 225 nM ROX standard I (6 mer with 5′ ROX dye conjugated and 3′ phosphorylated; catalog number 4105, Synthegen, Houston, TX; www.synthegen.com) and 0.25 U Platinum Taq Polymerase (Life Technologies). Note that the 2× SYBR Green stock can be made in large quantities, aliquoted and stored at 4°C for at least 2 weeks. Applied Biosystems also sells 2× SYBR Green Mix.
3. 900 nM of each primer.
4. 0.25 U Uracil DNA glycosylase (UDG; Life Technologies). The glycosylase eliminates problems with amplicon contamination from previous runs, since it recognizes and cleaves double stranded nucleic acid sequences containing UDP. Although UDG is not required, we highly recommend that it be used. If UDG is included, the PCR reaction mixture must be incubated for 2 min at 50°C prior to initiating thermocycling.

Relative quantitation of gene expression

Comparison of mRNA levels between samples can be done using either the Standard Curve Method or the Comparative C_T Method. Each of these methods is detailed in the Applied Biosystems ABI Prism 7700 Sequence Detection System User Bulletin #2 (available in PDF format from the Applied Biosystems website listed in Table 29.1). In both methods, an endogenous reference control transcript is used to correct for variations in the efficiency of cDNA synthesis between samples. We routinely use mRNA encoding the glycolytic pathway enzyme glyceraldehyde-3 phosphate dehydrogenase (Gapdh; forward primer = 5′ TGGCAAAG TGGAGATTGTTGCC, reverse primer = 5′ AAGATGGTGATGGGCTTC

CCG), or 18S rRNA (forward primer = 5′ CATTCGAACGTCTGCCCT ATC, reverse primer = 5′ CCTGCTGCCTTCCTTGGA). Using 18S rRNA requires cDNA synthesis to be primed with random hexamers.

In the Standard Curve Method, amplicon quantity is determined by comparison to a standard curve generated using serial dilutions of a stock solution of any cDNA preparation that contains a high concentration of the sequence of interest. (Note that a stock solution of recombinant plasmid DNA containing the sequence can also be used, but there are caveats to this approach if you want to calculate the absolute amount of the sequence; see below.)

A separate standard curve is prepared for both the target sequence and reference control. Within a sample, the target value is divided by the reference value to obtain a normalized value. To compare the relative levels of target mRNA in two or more samples, one sample is designated the 'calibrator' and the normalized target values in the 'experimental' samples are divided by the normalized calibrator value. Thus, quantities in the experimental samples are expressed as n-fold differences relative to the calibrator.

It is also possible to use the standard curve method to quantify absolute amounts of a transcript in a sample, but this requires that the absolute quantities of the standard be established by an independent method (such as A_{260}). Plasmid DNA cannot be used as a standard for quantitating absolute transcript levels from cDNA since there is no control for the efficiency of *in vitro* transcription.

In the Comparative CT Method, a mathematical formula is used to determine a fold change in normalized target levels between an experimental and calibrator sample. A standard curve is not necessary, provided that the target and reference control sequences are amplified with equal efficiency (this is established in a separate validation experiment). We favor this approach since it permits direct comparison of multiple samples and allows more samples to be assayed in a single plate due to elimination of the standard curve. First, the difference in CT between target and reference sequences in each sample is calculated (_CT). Second, the difference between the _CT of the experimental versus the calibrator sample is determined ($\Delta\Delta C_T$). Since the fluorescence data collected from the qRT-PCR assay is plotted in a log-linear fashion, $\Delta\Delta C_T$ can be converted to a fold-difference with the formula $2^{-\Delta\Delta C_T}$. (To read more about the mathematics behind this equation see the Applied Biosystems ABI Prism 7700 Sequence Detection System User Bulletin #2.) In order to use the $\Delta\Delta C_T$ method, it is critical that the target sequence be present in the calibrator sample, since division by an exceedingly small denominator will give an inflated fold change between two samples. One way to circumvent this problem is to run a standard curve with each assay, allowing absolute quantitation of mRNA levels in each sample.

Table 29.1 Sources of protocols and other information relevant to gnotobiotics and functional genomics

Web Site	URL/contact info	Description
Gnotobiotics		
Standard Safety Equipment Co.	Website pending; Phone: (815)363–8565	Sells flexible film isolators and gnotobiotic supplies; provides excellent protocols for setting up gnotobiotic equipment
Class Biologically Clean	www.cbclean.com	Sells flexible film isolators and other gnotobiotic supplies
Taconic, Inc.	www.taconic.com	Supplier of germ-free mice and re-derivation services
University of Wisconsin Gnotobiotic Laboratory	www.medsch.wisc.edu/gnotolab/	Offers germ-free mouse strains and breeding and housing services for germ-free animals
Charles River Laboratories	www.criver.com	Offers germ-free re-derivation services
Microarrays		
Brown Lab Microarraying Guide	cmgm.stanford.edu/pbrown/mguide/index.html	Detailed methods for design, construction, use, and analysis of cDNA-based microarrays
Affymetrix, Inc.	www.affymetrix.com	Manufacturer of oligonucleotide-based microarrays
Whitehead/MIT Center for Genome Research	www-genome.wi.mit.edu/MPR/methodsPUBg99.html	Provides detailed protocols for microarray target preparation, hybridization, staining, and scanning
BLASTn	www.ncbi.nlm.nih.gov/blast	Allows sequences to be compared against all cloned genes and ESTs in public databases.
UniGene	www.ncbi.nlm.nih.gov/UniGene	Database of clustered ESTs; provides putative homologs and orthologs

The Institute for Genomic Research	www.tigr.org/	Database of clustered ESTs; provides putative homologs and orthologs
GenQuery Engine	gordonlab.wustl.edu/mills	Freely-available software for calculating LUT scores and annotating Affymentrix Microarray data
NCBI's PubMed	www.ncbi.nlm.nih.gov/entrez/query.fcgi	Gateway to the biomedical literature
Real-time quantitative RT-PCR		
Applied Biosystems	www.appliedbiosystems.com	Sells equipment and reagents for real-time quantitative RT-PCR; literature on qRT-PCR is downloadable in PDF format
Laser Capture Microdissection		
NIH Laser Capture Microdissection Website	dir.nichd.nih.gov/lcm/lcm.htm	Protocols for laser capture microdissection and for RNA isolation from microdissected material
Arcturus Engineering, Inc.	www.arctur.com/technology	Sells LCM equipment and consumable reagents
Tyramide signal amplification		
NEN Life Sciences	www.nen.com	Provider of tyramide signal amplification kits and protocols; web site provides good descriptions of tyramide signal amplification

◆◆◆◆◆◆ DEFINING THE CELLULAR BASIS FOR HOST RESPONSES TO COLONIZATION

Although DNA microarray analysis of a tissue allows initial identification of host responses to microbes, a detailed understanding of such responses requires pinpointing their cellular basis. Three methods for delineating the cellular localization of *in vivo* host responses are described below. *Laser capture microdissection* (LCM) allows precise collection of a specified population of cells from a frozen histologic section of a complex tissue (Emmert-Buck *et al.*, 1996). LCM offers a way to isolate cells while minimizing artefactual perturbations in the population of mRNAs expressed *in vivo*. Colonization-associated transcriptional responses in the dissected cell population can then be quantitated using qRT-PCR (Wong *et al.*, 2000; Hooper *et al.*, 2001). New tyramide signal amplification-based methods for *in situ hybridization* permit sensitive and semi-quantitative assessment of changes in transcript levels in specific cell types. Finally, sensitive single and *multilabel immunohistochemical* methods allow definition of cellular responses at a protein level.

Laser capture microdissection

This method of cell procurement involves the transfer of visually identified cells to an overlying transparent polymer film by activated laser pulses. The transfer film is made of ethylene vinyl acetate and is attached to an optical quality 'cap' which is placed over the section (Figure 29.5A,B). LCM allows specified subsets of cells to be isolated from a complex tissue. For example, in the intestine, we routinely recover three populations: epithelium from crypts (containing proliferating undifferentiated cells plus differentiated members of the Paneth cell lineage), epithelium overlying villi (containing post-mitotic, differentiated members of the other three intestinal epithelial lineages), and mesenchyme underlying crypt-villus units (Wong *et al.*, 2000; Hooper *et al.*, 2001; see Figure 29.5C).

The methods described in this section are largely modifications of protocols published on the websites of the National Institutes of Health, where the technique was first developed (dir.nichd.nih.gov/lcm/lcm.htm) and Arcturus Engineering, Inc., which sells LCM equipment and consumable reagents (www.arctur.com/technology).

Preparation of tissue for laser capture microdissection using mouse intestine as an example

Maintenance of RNA integrity during tissue procurement and histologic manipulation is crucial for analysis of gene expression in microdissected cell populations. Goldsworthy *et al.* (1999) compared a variety of methods. Frozen sections fixed briefly in 70% ethanol had the best preserved morphology and RNA integrity. In our hands, this method works very well for mouse intestine.

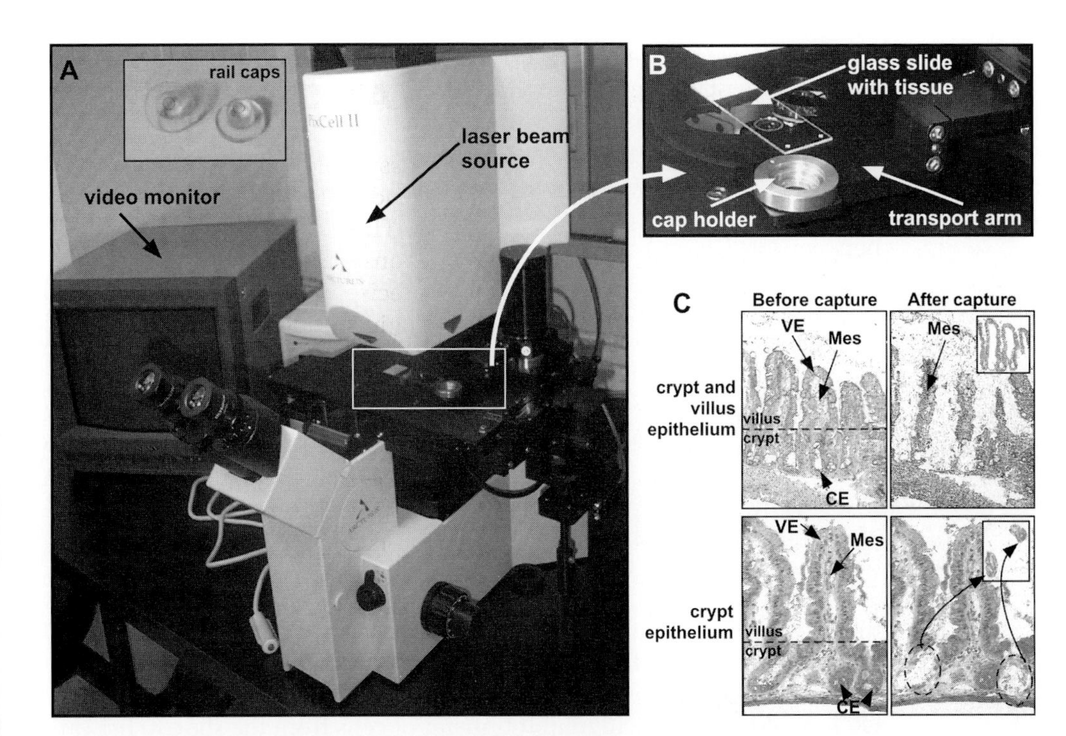

Figure 29.5. Laser capture microdissection. (A and B) *Laser capture microscope.* Frozen tissue sections are placed on the microscope stage and can be viewed on the monitor screen. A 'rail cap' is placed over the tissue and activated laser pulses result in the transfer of selected cells to the cap. (C) *Example of laser capture microdissection of small intestinal epithelial cells.* Sections are stained with nuclear fast red. Insets show captured villus and crypt epithelial populations as they appear on the cap. VE = villus epithelium; CE = crypt epithelium; Mes = mesenchyme. (This figure is also reproduced in colour between pages 276 and 277.)

To prepare mouse intestine for LCM, it must be frozen as quickly as possible to obtain high quality sections and prevent RNA degradation. We routinely complete the process of harvesting and freezing at −80°C within 2–3 min. A specified portion of the gut (typically 8–10 cm long) is recovered and perfused with cold PBS using a 12 ml syringe and $1\frac{1}{2}''$ 18 gauge blunt needle. To preserve tissue morphology, the segment is then gently flushed with TissueTek OCT (Miles Scientific, Elkhart, IN). The end of the segment (~0.5 cm), held by forceps during the flushing process, is then removed, discarded and the remaining segment cut into three 2-cm-long sub-segments. The sub-segments are placed in the base of a standard TissueTek cryomold (Miles Scientific), and then carefully overlaid with additional OCT. Note that the sub-segments should lie flat at the base of the mold to maximize the amount of tissue represented when the contents of the mold are sectioned. After OCT is added, the cryomold is immediately submerged in liquefied Cryo-cool (Stephens Scientific), and stored, wrapped in foil, at −80°C until sectioning.

Preparation of sections for LCM is similar to cutting routine frozen sections with some important differences. Since care must be taken to prevent exogenous RNase contamination, we use disposable blades.

The choice of glass slides is critical since the tissue must stick to the slide during staining and retrieval of targeted cell populations must be allowed. We recommend testing a variety of slide types to determine which one is optimal for a particular tissue. Superfrost/Plus coated slides from Fisher Scientific (Pittsburgh, PA; www1.fishersci.com/) work well for capturing cells from the mouse intestine, although there is some variation between lots.

Five to seven μm thick sections of intestine are cut parallel to the cephalocaudal axis in a cryostat chamber equilibrated to −15°C. To obtain sections containing well-oriented small intestinal crypt-villus units, the tissue block must first be cut until the lumen of the gut is exposed. For good laser capture, sections must also be relatively free of folds, and have a consistent thickness.

Sections can either be stained immediately after cutting or placed into a slide holder on dry ice. (Unstained and unprocessed sections can be stored at −80°C in a dessicator for as long as a week without significant RNA degradation.) Frozen tissue sections should be stained one slide at a time, at room temperature, with RNase-free solutions. To do so, the slide is thawed for 1 min at room temperature (unless fixed directly after cutting), fixed in 70% ethanol for 15 s, and rinsed by dipping 10 times in nuclease-free water (Ambion). This step is critical in order to completely remove OCT. Residual OCT will inhibit transfer of cells from the slide to the cap. However, too much time in water will activate endogenous RNases.

The section is subsequently stained for 10–15 s in a general purpose, water-soluble dye. We routinely use Nuclear Fast Red. Not all stains are compatible with RNA isolation (see the Arcturus website listed in Table 29.1 for more information). After staining, the slide is dehydrated with two successive washes in 95% ethanol, two washes in 100% ethanol (1 min each), followed by three washes (1 min each) in histology grade xylene (Fisher). After air drying for 1 min, slides with sections can be safely stored for 2–3 days at room temperature in an air-tight jar containing dessicant without significant RNA degradation.

For some tissues, immunohistochemical (IHC) identification of specific cell subpopulations prior to LCM is compatible with subsequent qRT-PCR detection of specific transcripts (Fend *et al.*, 1999; Jin *et al.*, 1999). However, applying immunohistochemical techniques can result in rapid mRNA degradation by endogenous RNases (see Wong *et al.*, 2000; Murakami *et al.*, 2000). One solution is to perform navigated LCM. With navigated LCM, one tissue section is prepared for LCM, quickly stained (e.g. with Nuclear Fast Red), and stored until use. The adjacent section is marked using more time consuming histochemical or immunohistochemical methods. The marked, stained section is then used as a 'roadmap' to guide (navigate) microdissection of the adjacent section (see Wong *et al.*, 2000 for further details about how navigated LCM is performed).

RNA isolation and real-time quantitative RT-PCR from LCM cell populations

Microdissected epithelium from a 5–7 μm section of mouse intestine typically yields ~2 pg of RNA per cell. Thus, 5000 cells (requiring

~20–30 min of LCM) will provide ~10 ng of total RNA. A variety of protocols are now available for purifying such small quantities of RNA. The 'gold standard' is a scaled-down guanidinium extraction method involving denaturation of captured cellular material in guanidinium buffer, extraction in phenol chloroform, and ethanol precipitation. Details of this method can be found at dir.nichd.nih.gov/lcm/protocol.htm. Stratagene (La Jolla, CA; www.stratagene.com) currently produces a useful kit for micro-isolation of RNA.

Synthesizing cDNA from LCM cellular RNA is similar to synthesizing cDNA from whole tissue RNA (described above). However, there are three important differences. First, we do not quantitate the RNA prior to cDNA synthesis, owing to the extremely small quantities obtained from LCM. Oligo $d(T)_{15}$ or random-primed cDNA synthesis is simply done using 10 μl of the LCM cellular RNA preparation (5 μl is used for a control reaction containing no reverse transcriptase). Second, we do not DNase treat the RNA prior to cDNA synthesis. However, all cDNA samples are screened for genomic DNA contamination by the actin PCR assay (see above). Third, cDNA (+RT) reactions are diluted with 40 μl of deionized H_2O and control (−RT) reactions with 20 μl of deionized H_2O. Subsequent qRT-PCR analysis of specific transcripts is carried out exactly as for whole tissue analyses.

Another goal is to use these RNA preparations and microarrays for global gene expression profiling. The maximum number of cells that can be realistically dissected for such an experiment is of the order of tens of thousands (yield = 10–100 ng of total RNA). Most microarray platforms require microgram quantities of labeled cRNA probe. One solution is to perform one round of the amplification procedure outlined in Figure 29.2B followed by a second round that uses the cRNA produced from the first round as the template (Luo *et al.*, 1999; Wang *et al.*, 2000). An alternative solution is to use a hybridization detection method that has enhanced sensitivity. The cDNA-based MICROMAX gene chip system from NEN Life Science Products (www.nen.com) employs tyramide signal amplification to enhance detection (TSA; see below). A MICROMAX array, containing ~2400 unique DNAs, can be probed with as little as 500 ng of labeled target cDNA. Currently, these chips are only available for human genes.

In situ hybridization and immunohistochemistry based on tyramide signal amplification

LCM with follow-up real-time quantitative RT-PCR is an attractive way to define gene expression at the cellular level. However, there may be no prior information about the cell type that normally supports expression of the gene of interest. Two additional approaches allow determination of the cellular origins of gene expression: *in situ* hybridization (ISH) detection of selected mRNAs, and immunohistochemical (IHC) localization of their protein products. Each method has advantages and disadvantages. Both are ultimately limited by the detection system used to visualize a probe:target or an antibody:antigen interaction. The application of tyramide signal amplification (TSA; www.nen.com) methods to cell and tissue-based gene product localization protocols has provided a relatively simple, very sensitive, non-radioactive detection tool.

TSA is based on the horseradish peroxidase (HRP) catalyzed conversion of tyramine into a highly reactive oxidized intermediate that binds rapidly and covalently to cell-associated proteins at, or near, the HRP-linked probe (Bobrow *et al.*, 1989; Speel *et al.*, 1999). The short half-life of the oxidized intermediate ensures rapid binding and excellent signal resolution compared to typical HRP substrates. Tyramine can be used either as a fluorescently labeled conjugate ('TSA direct'), or attached to a non-fluorescent intermediate molecule such as biotin ('TSA indirect') which is subsequently detected with additional reagent (e.g. fluorescently labeled or enzyme-linked streptavidin). TSA can improve ISH and IHC sensitivity 10–1000 fold over conventional techniques (Komminoth and Werner, 1997; Shindler and Roth, 1996).

A number of factors must be considered when deciding to use protein and/or mRNA localization procedures. Proteins are typically present in far greater abundance than their mRNAs, and unlike mRNAs that are susceptible to degradation, fixed cellular proteins are very stable. IHC protocols are relatively easy to perform and can be used by non-specialists. In addition, numerous IHC staining kits are commercially available. However, there are three major problems with using IHC to survey the cellular patterns of expression of numerous gene products. First, primary antibodies against the desired proteins may not be available. Antibody generation is time-consuming, relatively costly, and does not guarantee that useful preparations will be obtained. Second, cellular localization, abundance, and post-translational processing will vary between proteins. Therefore, there is no fixation and processing protocol for IHC detection that can be universally applied to all proteins. Experiments have to be performed to define conditions that optimize the immunoreactivity for each protein of interest. Third, establishing the specificity of IHC results can be troublesome. With sensitive detection methods, virtually all monoclonal and polyclonal antibodies will label something in tissue sections. Performing control experiments with Western blots, pre-immune control serum, pre-absorbing the primary antibody with its antigen, or omitting the primary antibody may not be adequate to verify the specificity of an observed IHC signal (Willingham, 1999; Burry, 2000). If these three problems can be overcome, IHC provides a very powerful way to define the cellular origins of gene expression. With multi-label applications, IHC also permits detailed assessment of interactions within and between cells.

ISH detection of mRNA has been a difficult technique to master, requiring specialized equipment, labor-intensive protocols, and slow turn-around time. These problems stem from the low abundance of most mRNAs, and mRNA lability. The development of sensitive, non-radioactive ISH techniques, particularly TSA, has alleviated some of these problems. There are at least two potential advantages in using ISH to define the cellular origins of expression of multiple genes. First, once tissue fixation and processing protocols are optimized for one mRNA species, modifications are not required for detection of different mRNAs. Second, unlike antibody generation, probe generation is relatively rapid and straightforward. Typically, cRNAs are generated to the mRNA of interest, although labeled oligos can be effective in detecting more abundant transcripts.

TSA-*in situ* hybridization

We recently developed a relatively simple protocol for ISH that utilizes digoxigenin labeled cRNA probes and TSA-Plus reagents (www.nen.com), a second generation of TSA reagents with increased sensitivity compared to the original commercially available conjugates. This protocol can be performed either on frozen or paraffin-embedded tissue sections (Zaidi *et al.*, 2000). We typically use 4–8 μm thick Bouin's fixed, paraffin-embedded tissue sections for TSA-ISH, although formalin or 4% paraformaldehyde fixed and/or frozen sections can also be employed. Sections are de-paraffinized in xylene and graded alcohols followed by rehydration in diethylpyrocarbonate (DEPC)-treated phosphate buffered saline (PBS). Sections are digested for 15 min in Proteinase K (15 $\mu g\, ml^{-1}$ in PBS) at room temperature, washed in PBS for 5 min, and fixed for 15 min in 4% paraformaldehyde in PBS. Following a 5 min PBS wash, sections are treated with 0.3% DEPC in PBS for 15 min, washed in PBS for 5 min, incubated for 15 min in 0.5% Triton X-100 (in PBS), and washed for 15 min each in PBS, 2× SSC, and 5× SSC. The sections are then hybridized overnight at 65°C with digoxigenin-labeled probe (100–500 $ng\, ml^{-1}$ of probe in 5× SSC, 50% formamide, and 50 $\mu g\, ml^{-1}$ salmon sperm DNA, pH 7.5; note that probes are labeled using the DIG RNA Labeling Kit from Roche Molecular Biochemicals; biochem.roche.com). Post hybridization washes are performed in the following order: (i) 2× SSC for 15 min at 65°C; (ii) 1× SSC for 15 min at room temperature; and (iii) PBS for 5 min. To decrease background from endogenous peroxidases, sections are incubated for 15 min in 0.3% H_2O_2 in PBS, washed in PBS for 5 min, and incubated for 30 min in PBS-blocking buffer (PBS with 1% BSA, 0.2% powdered skim milk, and 0.3% Triton X-100). Hybridized probe is detected with HRP-conjugated mouse anti-digoxigenin antibodies (Jackson Immunoresearch Laboratories, West Grove, PA; www.jackson-immuno.com) diluted in PBS-blocking buffer, applied either for 1 h at room temperature or overnight at 4°C. The sections are then washed three times in PBS for 5 min each and fluorescently labeled (either fluorescein or cyanine 3) by applying TSA-Plus detection reagent for 15 to 60 min. Sections are washed three times for 5 min in PBS, counter-stained with Hoechst 33258 (0.1 $mg\, ml^{-1}$ in PBS), coverslipped in PBS : glycerol (1 : 1), and viewed with an epifluorescence microscope.

Several simple modifications to the protocol can be made to enhance the signal-to-noise ratio for any given probe. For example, for abundant mRNAs all the steps between the Proteinase K and hybridization steps can be eliminated since these are included largely to decrease background signals. For low abundance mRNAs and/or weak signals, adjustments in the Proteinase K digestion step, hybridization temperature, and/or

TSA-Plus detection step can improve the sensitivity of detection significantly.

Dual TSA *in situ* hybridization and immunohistochemical detection studies

In situ localization of mRNA and/or immunohistochemical detection of protein for a specific gene is oftentimes inadequate for defining its cell-specific expression pattern in complex, heterogeneous tissue. To accomplish this task, multi-label detection of protein and/or mRNA of the gene of interest and specific phenotypic markers (e.g. mRNA, protein, or lectin-binding sites) is ideal. We have developed protocols for performing multi-label ISH, dual ISH and IHC, and multi-label IHC detection (Shindler and Roth, 1996; Zaidi *et al.*, 2000). Multi-label ISH detection can be readily accomplished using cRNA probes with different hapten labels. For example, one probe can be digoxigenin labeled and the second biotinylated. Both probes can be incubated simultaneously on the tissue section using the ISH protocol described above. The two probes can then be detected using sequential secondary reagents and TSA-Plus fluorophores with an intervening incubation in 3% H_2O_2 in PBS for 5 minutes to destroy any residual HRP activity from the first probe detection prior to beginning the second probe detection. Multi-label detection can be extended to a third probe, labeled with a unique hapten (e.g. fluorescein), and a third TSA-Plus fluorophore (e.g. cyanine 3, cyanine 5, and coumarin).

To perform dual ISH and IHC detection, the standard ISH protocol must be modified to avoid Proteinase K digestion, which typically interferes with IHC antigen preservation. To accomplish this task, we have replaced the Proteinase K digestion step in the standard ISH protocol with an antigen 'retri-review step' utilizing heat-induced epitope retri-review in either 10 mM citrate buffer, pH 6.0 or Trilogy solution (Cell Marque; Austin, TX; www.cellmarque.com). The remainder of the ISH protocol is unchanged. Following TSA-Plus ISH detection, residual HRP activity is destroyed with 3% H_2O_2 in PBS and the slides are processed for IHC detection. Since Proteinase K is not used and most IHC detection is improved by heat-induced epitope retri-review, the subsequent IHC detection of cell-specific antigens is readily accomplished. This protocol permits detailed analysis of mRNA and protein co-localization. Moreover, it is easily modified for detection of lectin binding and/or multi-epitope comparisons.

Simultaneous detection of two proteins using antibodies raised in the same species

Numerous protocols for multi-label IHC detection are available and will not be discussed in detail here. However, a method that we developed for simultaneous detection of immunoreactivity from two unconjugated primary antisera raised in the same species has proven very effective and will be briefly described (Shindler and Roth, 1996).

It is common to want to compare the immunostaining patterns of two different antibodies. If the antibodies are raised in separate species, dual labeling can be simply performed using two different secondary antibodies conjugated with distinct fluorophores, or detected with sequentially performed TSA reactions. However, if both antibodies are raised in the same species, dual detection becomes more complicated. A simple solution is to initially detect one of the primary antibodies using TSA methods followed by detection of the second primary antibody using conventional techniques. Because of the incredible sensitivity of TSA detection over that of conventional fluorophore-conjugated secondary antibodies, the first primary antibody preparation can be used at a concentration that is undetectable by conventional means. Thus, the TSA detectable first primary antibody is 'invisible' to the reagent used to detect the second primary antibody (we refer to this method as 'dilutional neglect'). Dilutional neglect requires that you first determine the concentration of primary antibody that is detectable by TSA but not by conventional methods. Once this is established, one simply performs the initial primary antibody detection with TSA, followed by application of the second primary antibody and fluorophore-conjugated secondary antibodies.

Acknowledgments

Work from our lab cited in this review was supported in part by grants from the NIH (DK30292, DK58529, and DK51929). L.V.H. is the recipient of a Career Development Award from the Burroughs Wellcome Fund. J.C.M. is a post-doctoral fellow of the Howard Hughes Medical Institute.

References

Bobrow, M. N., Harris, T. D., Shaughnessy, K. J. and Litt, G. J. (1989). Catalyzed reporter deposition, a novel method of signal amplification. Application to immunoassays. *J. Immunol. Methods* **125**, 279–285.

Bonner, R. F., Emmert-Buck, M., Cole, K., Pohida, T., Chuaqui, R., Goldstein, S. and Liotta, L. A. (1997). Laser capture microdissection: molecular analysis of tissue. *Science* **278**, 1481–1483.

Bry, L., Falk, P. G., Midtvedt, T. and Gordon, J. I. (1996). A model of host-microbial interactions in an open mammalian ecosystem. *Science* **273**, 1380–1383.

Burry, R. W. (2000). Specificity controls for immunocytochemical methods. *J. Histochem. Cytochem.* **48**, 163–166.

Carthew, P., Wood, M. J. and Kirby, C. (1983). Elimination of Sendai (parainfluenza type I) virus infection in mice by embryo transfer. *J. Reprod. Fertil.* **69**, 253–257.

Claverie, J. M. (1999). Computational methods for the identification of differential and coordinated gene expression. *Hum. Mol. Genet.* **8**, 1821–1832.

Eisen, M. B., Spellman, P. T., Brown, P. O. and Botstein, D. (1998). Cluster analysis and display of genome-wide expression patterns. *Proc. Natl Acad. Sci. USA* **95**, 14863–14868.

Emmert-Buck, M. R., Bonner, R. F., Smith, P. D., Chuaqui, R. F., Zhuang, Z., Goldstein, S. R., Weiss, R. A. and Liotta, L. A. (1996). Laser capture microdissection. *Science* **274**, 998–1001.

Fend, F., Emmert-Buck, M. R., Chuaqui, R., Cole, K., Lee, J., Liotta, L.A. and Raffeld, M. (1999). Immuno-LCM: laser capture microdissection of immunostained frozen sections for mRNA analysis. *Am. J. Pathol.* **154**, 61–66.

Ferea, T. L. and Brown, P. O. (1999). Observing the living genome. *Curr. Opin. Genet. Dev.* **9**, 715–722.

Goldsworthy, S. M., Stockton, P. S., Trempus, C. S., Foley, J. F. and Maronpot, R. R. (1999). Effects of fixation on RNA extraction and amplification from laser capture microdissected tissue. *Mol. Carcinog.* **25**, 86–91.

Guruge, J. L., Falk, P. G., Lorenz, R. G., Dans, M., Wirth, H. P., Blaser, M. J., Berg, D. E. and Gordon, J. I. (1998). Epithelial attachment alters the outcome of *Helicobacter pylori* infection. *Proc. Natl Acad. Sci. USA* **95**, 3925–3930.

Gustafsson, B. E. (1959). Lightweight stainless steel systems for rearing germfree animals. *Ann. NY Acad. Sci.* **78**, 17–28.

Harkin, D. P., Bean, J. M., Miklos, D., Song, Y. H., Truong, V. B., Englert, C., Christians, F. C., Ellisen, L. W., Maheswaran, S., Oliner, J. D. and Haber, D. A. (1999). Induction of GADD45 and JNK/SAPK-dependent apoptosis following inducible expression of BRCA1. *Cell* **97**, 575–586.

Hartwell, L. H., Hopfield, J. J., Leibler, S. and Murray, A. W. (1999). From molecular to modular cell biology. *Nature* **402**, C47–C52.

Heid, C. A., Stevens, J., Livak, K. J. and Williams, P. M. (1996). Real time quantitative PCR. *Genome Res.* **6**, 986–994.

Hooper, L. V., Xu, J., Falk, P. G., Midtvedt, T. and Gordon, J. I. (1999). A molecular sensor that allows a gut commensal to control its nutrient foundation in a competitive ecosystem. *Proc. Natl Acad. Sci. USA* **96**, 9833–9838.

Hooper, L. V., Wong, M. H., Thelin, A., Hansson, L., Falk, P. G. and Gordon, J. I. (2001). Molecular analysis of commensal host–microbial relationships in the intestine. *Science* **291**, 881–884.

Jin, L., Thompson, C. A., Qiun, X., Kuecker, S. J., Kulig, E. and Lloyd, R. V. (1999). Analysis of anterior pituitary hormone mRNA expression in immunophenotypically characterized single cells after laser capture microdissection. *Lab. Invest.* **79**, 511–512.

Komminoth, P. and Werner, M. (1997). Target and signal amplification: approaches to increase the sensitivity of in situ hybridization. *Histochem. Cell Biol.* **108**, 325–333.

Lee, C. K., Weindruch, R. and Prolla, T. A. (2000). Gene-expression profile of the ageing brain in mice. *Nat. Genet.* **25**, 294–297.

Lipshutz, R. J., Fodor, S. P., Gingeras, T. R. and Lockhart, D. J. (1999). High density synthetic oligonucleotide arrays. *Nat. Genet.* **21**, 20–24.

Lockhart, D. J. and Winzeler, E. A. (2000). Genomics, gene expression and DNA arrays. *Nature* **405**, 827–836.

Luo, L., Salunga, R. C., Guo, H., Bittner, A., Joy, K. C., Galindo, J. E., Xiao, H., Rogers, K. E., Wan, J. S., Jackson, M. R. and Erlandert, M. G. (1999). Gene expresion profiles of laser-captured adjacent neuronal subtypes. *Nat. Med.* **5**, 117–122.

Mills, J. C. and Gordon, J. I. (2001). A new approach for filtering noise from high-density oligonucleotide microarray datasets. *Nucleic Acids Res.* **29**, e72.

Murakami, H., Liotta, L. A. and Star, R. A. (2000). IF-LCM: laser capture microdissection of immunofluorescently defined cells for mRNA analysis. *Kidney Int.* **58**, 1346–1353.

Shindler, K. and Roth, K. A. (1996). Double immunofluorescent staining using two unconjugated primary antisera raised in the same species. *J. Histochem. Cytochem.* **44**, 1331–1335.

Speel, E. J. M., Hopman, A. A. and Komminoth, P. (1999). Amplification methods to increase the sensitivity of in situ hybridization: play CARD(S). *J. Histochem. Cytochem.* **47**, 281–288.

Tamayo, P., Slonim, D., Mesirov, J., Zhu, Q., Kitareewan, S., Dmitrovsky, E., Lander, E. S. and Golub, T. R. (1999). Interpreting patterns of gene expression with self-organizing maps: methods and application to hematopoietic differentiation. *Proc. Natl Acad. Sci. USA* **96**, 2907–2912.

Trexler, P. C. and Reynolds, L. I. (1957). Flexible film apparatus for the rearing and use of germfree animals. *Appl. Microbiol.* **5**, 406–412.

Wang, E., Miller, L.D., Ohnmacht, G. A., Liu, E. T. and Marincola, F. M. (2000). High-fidelity mRNA amplification for gene profiling. *Nat. Biotechnol.* **18**, 457–459.

Willingham, M. C. (1999). Conditional epitopes: is your antibody always specific? *J. Histochem. Cytochem.* **47**, 1233–1235.

Wong, M. H., Saam, J. R., Stappenbeck, T. S., Rexer, C. H. and Gordon, J. I. (2000). Genetic mosaic analysis based on *cre* recombinase and navigated laser capture microdissection. *Proc. Natl Acad. Sci. USA* **97**, 12601–12606.

Zaidi, A. U., Enomoto, H., Milbrandt, J. and Roth, K. A. (2000). Dual fluorescent in situ hybridization and immunohistochemical detection with tyramide signal amplification. *J. Histochem. Cytochem.* **48**, 1369–1376.

Zhu, H., Cong, J. P., Mamtora, G., Gingeras, T. and Shenk, T. (1998). Cellular gene expression altered by human cytomegalovirus: global monitoring with oligonucleotide arrays. *Proc. Natl Acad. Sci. USA* **95**, 14470–14475.

Index

Note: Page references in *italics* refer to Figures

Index

Index